D1269159

Praise for *Energy Systems Engineering: Evaluation and Implementation*

"A carefully written book, providing good breadth as well as depth on major conventional and sustainable energy systems. The students enrolling in our course come from a variety of engineering disciplines and have a range of backgrounds, but the text is suited to the mix."

PROF. DAVID DILLARD
Dept. of Engineering Science and Mechanics
Virginia Tech

"What I find best about the text is the presentation of the material from a systems analysis perspective. Understanding the goals of design is as, if not more, important than how to do the design. The text really is superb with regard to the modern sustainability concept of 'triple bottom line.' Each section delivers not only the technology but also a means to financially or economically evaluate the costs and benefits of such a technology."

PROF. JEFF CASELLO
Dept. of Civil and Environmental Engineering
University of Waterloo, Ontario

"Energy systems and their engineering are the foundation for our society and nation's future. Broad in scope, with focused instructional detail, this text offers a uniquely excellent, student-accessible educational resource for integrating thermodynamic, alternative and renewable energy conversion processes."

PROF. RANDY L. VANDER WAL
Dept. of Materials Science and Engineering
Penn State University

"I found the instructor materials, including solutions, PowerPoints, and spreadsheets, to be very helpful for our junior-level energy course."

PROF. A. J. BOTH, PhD
Dept. of Environmental Sciences, Bioenvironmental Engineering
Rutgers University

"Well-organized and neither too specific nor too general on important topics, this book has proven both useful and thought-provoking for our students studying energy issues."

PROF. JAN DEWATERS
Institute for a Sustainable Environment
Clarkson University

"One of the best researched and most up-to-date books available on the subject for our renewable energy course."

PROF. CHAD JOSHI
College of Professional Studies
Northeastern University

About the Authors

Francis M. Vanek, Ph.D., is a Senior Lecturer and Research Associate in the School of Civil and Environmental Engineering, and previously the Systems Engineering Program, at Cornell University, where he specializes in the areas of energy efficiency, alternative energy, and energy for transportation. He was previously a consultant with Taitem Engineering of Ithaca, NY. He is also lead author of *Sustainable Transportation Systems Engineering* from McGraw-Hill Education.

Louis D. Albright, Ph.D., is a Professor Emeritus of Biological and Environmental Engineering at Cornell University. He is also a Fellow of the American Society of Agricultural and Biological Engineers (ASABE).

Largus T. Angenent, Ph.D., is a Professor in the Department of Biological and Environmental Engineering at Cornell University. He specializes in converting organic biomass and waste materials into bioenergy, and also works in the areas of biosensors and bio-aerosols.

Energy Systems Engineering
Evaluation and Implementation

Francis M. Vanek, Ph.D.

Louis D. Albright, Ph.D.

Largus T. Angenent, Ph.D.

Third Edition

New York   Chicago   San Francisco
Athens   London   Madrid
Mexico City   Milan   New Delhi
Singapore   Sydney   Toronto

Library of Congress Control Number: 2016931149

Energy Systems Engineering: Evaluation and Implementation, Third Edition

1 2 3 4 5 6 7 8 9 0   DOC/DOC   1 2 1 0 9 8 7 6

ISBN 978-1-25-958509-8
MHID 1-25-958509-3

This book is printed on acid-free paper.

Sponsoring Editor
Lauren Poplawski

Copy Editor
Ratika Gambhir, Cenveo Publisher Services

Editing Supervisor
Stephen M. Smith

Proofreader
Sudhir Babu, Cenveo Publisher Services

Production Supervisor
Pamela A. Pelton

Art Director, Cover
Jeff Weeks

Acquisitions Coordinator
Lauren Rogers

Composition
Cenveo Publisher Services

Project Manager
Tanya Punj, Cenveo® Publisher Services

To my wife, Catherine Johnson, to my parents, Jaroslav and Wilda Vanek,
and to my children, Raymond and Mira Vanek-Johnson
—Francis M. Vanek

To my wife and partner in life, Marilyn Albright
—Louis D. Albright

To my wife and son, Ruth Ley and Miles Ley Angenent
—Largus T. Angenent

Contents at a Glance

Contents

Preface to the Third Edition

The goal of the third edition of *Energy Systems Engineering* remains the same as that of the first two: to provide both professional engineers and engineering students interested in energy systems with essential knowledge of major energy technologies, including how they work, how they are quantitatively evaluated, what they cost, and what is their benefit or impact on the natural environment. A second goal is to provide the reader with an overview of the context within which these systems are being implemented and updated today and into the future. Perhaps at no time in recent history has society faced such challenges in the energy field: the yearning to provide a better quality of life to all people, especially those in the more impoverished countries, coupled with the twin challenges of a changing energy resource base and the effects of climate change due to increased concentration of CO_2 in the atmosphere. Energy systems engineers from many disciplines, as well as nonengineers in related fields, will serve at the forefront of meeting these challenges.

It has now been 12 years since 2004 when we first began drafting chapters that would eventually become part of the first edition of the book, which was published in 2008. During that time, some energy trends have hardly changed at all. For example, the routine process of updating figures on CO_2 emissions and on the installation of renewable energy capacity highlights the challenge for the energy field around the world. Figure 4-3 gives world emissions of carbon from fossil fuel use, which in 2004 stood at 7.6 gigatonnes of carbon equivalent. That figure gives emissions for 2010 (the most recent year available from the U.S. Carbon Dioxide Information and Analysis Center) of 9.2 gigatonnes—an increase of 1.6 gigatonnes in the annual emissions rate in just six years. Not only does this increase represent the growing challenge of meeting energy needs without increasing atmospheric CO_2, but the increased emissions also symbolize newly added energy-consuming devices, ranging from large power plants to small individual private vehicles, for which populations will expect continued access to energy supplies in the years going forward.

On a more positive note, we have documented the many-fold increase in the installed capacity of solar and wind photovoltaic capacity around the world in Chaps. 10 and 13—by a factor of 27 for solar and a factor of 6 for wind from 2005 to 2013, measured in total installed gigawatts. Even the global CO_2 emissions data may be changing. Although they do not appear in Fig. 4-3, preliminary data from the International Energy

Agency show that 2014 was the first year in which the world economy grew compared to the previous year but with global CO_2 emissions remaining constant. These hopeful indicators suggest that individuals, businesses, and governments are beginning to embrace this challenge.

Turning from motivation to content, chapter topics have been chosen in the first part of the book to provide key background for the analysis of energy systems, and in the second part to give a representative view of the energy field across a broad spectrum of possible approaches to meeting energy needs. In Chaps. 1 to 3, we present tools for understanding energy systems, including a discussion of sustainable development, a systems approach to energy and energy policy, and economic tools for evaluating energy systems as investments. In Chaps. 4 and 5, we consider climate change and fossil fuel availability, two key factors that will shape the direction of energy systems in the twenty-first century. Chapters 6 through 14 present a range of technologies for generating energy for stationary applications, including fossil fuel combustion, carbon sequestration, nuclear energy, solar energy, wind energy, and biological energy. Chapters 15 and 16 turn to energy conversion for use in transportation systems, and Chap. 17 provides a brief overview of some emerging technologies not previously covered, as well as the conclusions for the book.

The contents of the book assume a standard undergraduate engineering background, or equivalent, in physics, chemistry, mathematics, and thermodynamics, as well as a basic introduction to statistics, fluid mechanics, and heat transfer. Each technology area is introduced from first principles, and no previous knowledge of the specific technologies is assumed.

This book originated in two courses taught at Cornell University, one in the School of Mechanical and Aerospace Engineering entitled "Future Energy Systems," and the other in the Department of Biological and Environmental Engineering entitled "Renewable Energy Systems." In addition, a third course, "Civil Infrastructure Systems," taught in the School of Civil and Environmental Engineering, influenced the writing of passages on sustainable development and systems engineering. Energy system concepts, example problems, and end-of-chapter exercises have been developed through introduction in the classroom. In both courses, we have focused on solar and wind energy systems, so we have also placed a special emphasis on these two fields in this book. Interest in solar and wind energy is growing rapidly at the present time, but information about these fields may not be as accessible to some engineers, so we aim to provide a useful service by giving them extensive treatment in Chaps. 9 through 13.

Presentation of technical content in the book adheres to several premises for energy systems engineering that are independent of the technologies themselves. The first is that *energy systems choices should be technology-neutral*. No energy system is perfect, and every system has a range of advantages and disadvantages. Therefore, to the extent possible, the choice of any system should be based on choosing criteria first and then finding a system, or mixture of systems, that best meets those criteria, rather than preordaining that one type of system or another be chosen. Fossil fuels have the major drawback of greenhouse gas emissions, and nuclear energy must contend with the management of high-level nuclear waste, so these factors may point to renewables as the preferred option. However, renewables in turn suffer from intermittency, diffuseness, and high initial capital cost.

Members of the general public see that our current energy system is imperfect, and sometimes respond by insisting that they be provided a perfect energy system. It is the task of the energy professional to inform the public that no such system exists.

A second premise is that *there is value to a portfolio approach to energy options*. The value of the portfolio approach still holds, but the interpretation has changed since 2004, and this may represent one of the largest shifts as the book goes to a third edition. Several factors at this time point to renewable energy emerging as the eventual dominant energy source, rather than an energy system that is evenly shared among different resources. First, the rate of greenhouse gas emissions, and the extent to which the climate is changing rapidly, is greater than was anticipated a decade ago. This development puts great urgency on accelerated action to address climate change. Second, the other two primary energy sources—fossil with sequestration and nuclear—are not growing as rapidly as might have been anticipated, given the severity of the climate change problem. For sequestration, there has been only modest investment in capacity, although several projects continue to demonstrate proof of concept. For nuclear, the Fukushima accident and its aftermath dampen enthusiasm for the high capital cost of new plants in many countries. The third and final development is the rapid growth of renewable energy sources such as solar and wind, led by the growing manufacturing capacity of China and the low cost and improving reliability of the technology.

Nevertheless, there are educational reasons, separate from the desirability of one technology over another, to continue to include all three sources in this book, both in detail and keeping up-to-date with the latest developments. First, investment in nuclear energy and sequestration continue. Nuclear plants are under construction in China, South Korea, and elsewhere, and are approaching completion in the states of Georgia, Tennessee, and South Carolina in the United States. If the commissioning of these new plants goes smoothly, the rate of plant starts could grow. Also, SaskPower in Saskatchewan, Canada, recently commissioned an entire unit within the Boundary Dam coal-fired power plant to divert by-product CO_2 to sequestration. Again, if this system is successful, it could be reproduced elsewhere.

Another factor is that all three primary sources continue to be significant contributors to overall world demand, and this situation cannot change instantly. Fossil fuels stand at 82% of the market by energy content, down from 85% in 1990 but still the dominant source by far. Nuclear is still larger than any one renewable source, including hydropower, wind, or solar. As long as all sources continue to contribute, it is beneficial to compare them side by side in the context of a book on energy systems.

A third premise is that *where long-term technologies will take time to develop fully, there is value to developing "bridge" technologies for the short-to-medium term*. Some of the technologies presented in this book eliminate harmful emissions or use only renewable resources, but will take time to deploy, due to the slow nature of infrastructure transitions. In this situation, there is value to bridge technologies that are cost-effective now and also reduce nonrenewable resource consumption or CO_2 emissions, even if they do not eliminate them entirely. Typically, these technologies consume fossil fuels but are more efficient or have higher utilization, so that they deliver more energy service per unit of resource consumed or CO_2 emitted.

Although the book is written by American authors in the context of the U.S. energy system, we maintain an international focus. This is important because of the increasingly global nature of the energy industry, in terms of both the resource base and the transfer of technology between countries. We hope that non-U.S. readers of the book will find the material accessible, and that U.S. readers can apply the content to energy projects in other countries, and also to understanding the energy situation around the world.

For simplicity, all costs are given in dollars; however, other world currencies can of course be substituted into equations dealing with financial management.

Both a systems approach and an engineering economics approach to designing and costing projects are emphasized. The use of good systems engineering techniques, such as the systematic consideration of the project scope, evaluation of tradeoffs between competing criteria, and consideration of project life-cycle cost and energy consumption, can deliver more successful projects. Consideration of the cost of and revenues from a project, as well as technical efficiency, helps us to better understand the profitability of a project.

For the purposes of cost analysis, approximate prices for the cost of energy resources and the purchase of energy conversion equipment have been introduced in places at their appropriate values. These values are intended to give the reader a general sense of the financial dimensions of a technology, for example, approximately what proportion of an energy technology's life-cycle cost is spent on fuel versus capital or nonfuel operating costs. Note, however, that theses values should not be used as a basis for detailed decision making about the viability of a project, since up-to-date costs for specific regions are the required source. It is especially important to find up-to-date numbers for one's particular project of interest because of the volatility in both energy and raw material prices that has developed since 2004 or 2005. With rapid economic growth in the two largest countries by population in the world, namely China and India, there is burgeoning demand not only for energy commodities but also for materials such as steel or copper that are essential for fabricating energy conversion technologies. This affects not only operating costs of fossil-fuel-driven energy systems but also capital cost of both renewable and nonrenewable energy systems. To give a concrete example from Chap. 5, oil prices were at or below $20/barrel up to 2005, then fluctuated between $80 and $140 per barrel for most of the period 2008–2014, and finally fell to the range of $30–$50/barrel in the period from 2014 on. It is thus difficult to predict what will happen next.

Earlier books on energy systems have placed an emphasis on equipment to prevent release of air pollutants such as scrubbers or catalytic converters. As presented in the body of this book, emerging technologies that use new energy resources and eliminate CO_2 emissions also tend to eliminate emissions of harmful air pollutants, so air quality as a separate objective is deemphasized here. In some cases, it appears instead as a constraint on energy systems development: where air quality problems may be aggravated by emerging technologies that are beneficial for other objectives but increase emissions of air pollutants, regulations or targets related to air quality may restrict our ability to use that technology.

In conclusion, we offer a word of "truth in advertising" about the contents of the book: it provides some answers, and also many unanswered questions. It is humbling to write a book about energy systems, just as it is to teach a course or give a presentation about them: one ends up realizing that it is a very challenging area, and that many aspects of future solutions remain hidden from us at present. We hope that in publishing this book, we have helped to answer some of the questions about energy systems where possible, and, where not, posed them in such a way that the act of exploring them will move the field forward. The extent and complexity of the challenge may seem daunting at times, yet there are and will be great opportunities for energy professionals, both now and in the future. We wish each of you success in your part of this great endeavor.

Acknowledgments

There were many, many people whose support made such a difference in the writing of this book. I would first like to thank my family, including my wife Catherine Johnson, my parents Jaroslav and Wilda Vanek, and my children Ray and Mira, for their patience as I spent so many hours on the book, for their encouragement, and for their curiosity about the topic, which affects both young and old alike.

Within the Sibley School of Mechanical and Aerospace Engineering at Cornell, I would like to thank Prof. Zellman Warhaft, who taught MAE 501 "Future Energy Systems" before I did, and who helped me to establish my teaching the course the first year and more recently with various questions about the textbook. I would also like to thank Profs. Juan Abello, Michel Louge, Betta Fisher, Sid Leibovich, and Lance Collins and other members of the Sibley School faculty for their support for my teaching the course between 2004 and 2006, and June Meyermann and Bev Gedvillas for clerical support. Thanks also to several students who worked with me on the course as graders and proofreaders, including Tim Fu, Adam Fox, Julia Sohnen, Rob Young, and Cory Emil, and an additional thank you to all the rest of the students whose input and discussion made teaching the course so interesting and rewarding. We had the good fortune of hosting a number of excellent guest speakers during the course whose input benefited the book as well, including Dr. Jim Lyons, PhD, from the General Electric wind energy program in Niskayuna, NY; Dr. Gay Cannough, PhD, from ETM Solar Works of Endicott, NY; Sandy Butterfield, PE, from the National Renewable Energy Laboratories in Golden, CO; and Tom Ruscitti from General Motors' Honeoye Falls, NY fuel cell R&D facility. My thanks to all of these individuals as well.

Outside of the Sibley School, I would like to thank Profs. Mark Turnquist and Linda Nozick of the School of Civil and Environmental Engineering at Cornell for input on various issues related to transportation, nuclear energy, and systems engineering. From the Department of Biological and Environmental Engineering, I would also like to thank Prof. Norm Scott and my coauthor, Prof. Louis Albright, for their input into the "Future Energy Systems" course that helped to guide the development of the curriculum.

Many thanks are due to the various colleagues who helped with the production of the book, especially Ellie Weyer, who worked closely with me on proofreading many of the chapters and providing technical input and feedback. Additional thanks to Greg Pitts, Carol Henderson, Rob Young, Phil Glaser, my mother Wilda Vanek (PhD in history from Harvard University, 1961), and my wife Catherine Johnson (MFA in film and media studies from Temple University, 1999) for proofreading help and production assistance. Thanks also to everyone at McGraw-Hill who collaborated with me on this book, especially Larry Hager, David Zielonka, and Rick Ruzycka for the first edition, and Mike Penn and David Fogarty for the second edition, all of whom worked tirelessly to answer questions, provide information, and, in general, help the writing process to go smoothly. Thanks as well to Tanya Punj, Ratika Gambhir, and everyone at Cenveo Publisher Services for their efforts on the third edition. A general thank you, too, to all the individuals who provided images and figures for reprinting, as credited throughout the book.

The contents of this book benefited in a number of places from research collaboration with a number of colleagues. I would like to especially thank the late Prof. Emeritus Edward Morlok for his collaboration with me on the study of freight transportation energy use presented in Chap. 14, and moreover for giving me my start in the area of sustainable energy systems during the time that he was my dissertation supervisor in Systems Engineering at the University of Pennsylvania from 1993 to 1998. I would also like to thank my colleagues at Taitem Engineering (Technology as if the Earth Mattered) of Ithaca, NY, especially Ian Shapiro, PE; Lou Vogel, PE; and Susan Galbraith, for their work with me on combined heat and power systems, and on alternative fuel vehicles. Thanks are due to James Campbell and Prof. Alan McKinnon of the Logistics Research Centre at Heriot-Watt University, Edinburgh, Scotland, for their collaboration on the study of energy use in freight and logistics in the United Kingdom. Thanks to the students who worked on the hydrogen feasibility M.Eng project from 2003 to 2005, especially Julien Pestiaux, Audun Ingvarsson, and Jon Leisner. Thanks also to the students who worked on the wind energy feasibility M.Eng project in 2009 and 2010, especially Christine Acker, Kim Campbell, Stephen Clark, Jesse Negherbon, Alex Hernandez, Christina Hoerig, Happiness Munedzimwe, Reginald Preston, Karl Smolenski, and Jun Wan. Lastly, thanks to the students who worked on the alternative fuel vehicle M.Eng projects in 2009 and 2010, including Auret Basson, Mert Berberoglu, Ryan Cummiskey, Gurkan Gunay, Sara Lachapelle, Torsten Steinbach, Jackson Wang, Alice Yu, and Naji Zogaib.

For more recent work on alternative energy systems, thanks are due to Itotoh Akhigbe, Corey Belaief, Omer Yigit Gursoy, Pouyan Khajavi, Alejandro Martinez, Graham Peck, Ptah Plummer, Brad Sandahl, Taylor Schulz, Manuel Garcia Vilches, Michelle Chau, Hwan Choi, Benting Hu, Josh Lazoff, Molly McDonough, Sandra Quah, Justin Steimle, Kartik Shastri, Wan Hua Xie, Marrisa Yang, Martin Yu, Dan Zhao, Robert Ainslie, Nitesh Donti, Jacqueline Maloney, Ruju Mehta, Yeswanth Subramanian, and Yilin Wang.

Special thanks to reviewers of specific chapters in the book are as follows: Prof. Emeritus Jaroslav Vanek, Department of Economics in the College of Arts and Sciences, Cornell University, for reviewing Chap. 3 on economic tools (thanks, Dad!); Dr. Rajesh Bhaskaran of the Sibley School for reviewing Chap. 4 on climate change; Prof. Dave Hammer, School of Electrical and Computer Engineering at Cornell, for reviewing Chap. 8 on nuclear energy systems; Prof. David Dillard, Virginia Tech, for reviewing Chap. 9 on the solar energy resource; Paul Lyons, PE, of Zapotec Energy, Cambridge, MA, for reviewing Chap. 10 on photovoltaics; Howard Harrison, President of GTR Energy Ltd. of Oakville, Ontario, Canada, and also Sandy Butterfield (introduced above) for reviewing Chap. 13 on wind energy systems; and Dr. John DeCicco of the University of Michigan for reviewing Chap. 15 on transportation energy technologies. Thanks to Prof. Mike Timmons of Cornell University for helpful corrections and suggestions throughout on engineering economics, solar energy, and wind energy.

Thanks also to Rob Garrity, Principal of Finlo Energy Systems, for input on current conditions in the solar PV marketplace; Prof. Beth Clark, Department of Physics, Ithaca College, for providing information about the Ithaca College wind energy feasibility study referenced in Chap. 13; and John Manning, PE, Principal of Earth Sensitive Solutions, LLC, for background on ground source heat pumps used in Chap. 17. Additional thanks to Mike Timmons of Cornell University and Mike Elmore of Binghamton University for helpful comments on several chapters.

While revising the book, we had the benefit of input from a number of faculty members from other universities and colleges who offered feedback from their experience with using the book in the classroom or who reviewed the book in general. This list includes Prof. Anwar Abu-Zarifa, Islamic University of Gaza; Prof. Matthew Barth, U.C. Riverside; Prof. A. J. Both, Rutgers University; Prof. C. S. Chen, Miami University, Ohio; Profs. David Dillard and Michael Ellis, Virginia Tech; Prof. Kerry Patterson, Lipscomb University; Prof. Mark Schumack, University of Detroit Mercy; Jeff Casello, University of Waterloo; Erik Dahlquist, Malardalen University; Jan DeWaters, Clarkson University; Gary Dorris, Ascend Analytics; Katrina Hay, Pacific Lutheran University; Chad Joshi, Northeastern University; Rahim Khoie, University of the Pacific; Sam Landsberger, California State University at Los Angeles; Shinya Nagasaki, McMaster University; Donald Newlin, Oakland University; Shawn Reeves, EnergyTeachers.org; Randy Vander Wal, Penn State University; Grayson Walker, Old Dominion University; and Mary Lou Zeeman, Bowdoin College. I would also like to remember Abeeku Brew-Hammond of Kwame Nkrumah University of Science and Technology, who passed away in 2013 and was an enthusiastic supporter of the book. While contributions to the book in terms of chapter reviews, technical input to chapter content, and feedback from instructors are gratefully acknowledged, responsibility for any and all errors rests with the authors.

Lastly, thanks to all faculty, students, and friends who expressed support and encouragement during the project—it has always been appreciated—and thanks to anyone else who deserves mention but whom I may have overlooked.

—Francis M. Vanek

No book such as this should be attempted without the benefit of first teaching the material to interested and interactive students. The students over the years who questioned, contributed, and succeeded (yes, especially those in the earlier days of my renewable energy systems course when it met, for a decade, three times a week at 8 a.m.) have helped me immeasurably in understanding the material of this book and how it can be presented. Without you, my part of this book would never have been possible and life would have been much less enjoyable.

—Louis D. Albright

Note to Instructors

If you are an instructor of an energy systems course using this book as a textbook, please be aware that the end-of-chapter exercises included in the book are a subset of a larger collection of homework and exam problems, instructor's manual, spreadsheet solutions, PowerPoint slides, and computer scripts. Please visit www .mhprofessional.com/energysystemsengineering for information about how you can download these materials to assist your teaching.

CHAPTER 1

Introduction

1-1 Overview

The purpose of this chapter is to introduce the engineer to the worldwide importance of energy systems, and to the historic evolution of these systems up to the present time. We discuss how energy use in various countries is linked to population and level of economic activity. Next, we consider current trends in total world energy use and energy use by different countries, and discuss how the pressure from growing energy demand and growing CO_2 emissions poses a substantial challenge for the world in the coming years and decades. Thereafter, we present an overview of how a *sustainable* world energy situation might appear at some point in the mid to latter part of the twenty-first century, including contributions from different energy resources, and allocation to major end uses. We also provide a *road map* for how the chapters in the book contribute to the understanding needed for the success of this sustainable energy solution. The chapter concludes with a review of basic units used to measure energy in the metric and U.S. customary systems.

1-2 Introduction

When an energy engineer[1] thinks of access to energy or energy systems, whether as a professional responsible for the function of some aspect of the system, or as an individual consumer of energy, a wide range of applications come to mind. These applications include electricity for lighting or electronics, natural gas for space heating and industrial uses, and petroleum products such as gasoline or diesel for transportation. Access to energy in the industrialized countries of Asia, Europe, and North America is so pervasive that consumption of energy in some form is almost constant in all aspects of modern life—at home, at work, or while moving from place to place. In the urban areas of industrializing and less-developed countries, not all citizens have access to energy, but all live in close proximity to the local power lines and motorized vehicles that are part of this system. Even in the rural areas of these countries, people may not be aware of modern energy systems on an everyday basis, but may come into occasional contact with them through access to food shipments or the use of rural bus services. Indeed, there are very few human beings who live in complete isolation from this system.

If it is true that use of energy is omnipresent in modern life, then it is also true that both individuals and institutions (governments, private corporations, universities,

---

[1]The term "energy engineer" is used here to refer both to readers with formal engineering training and those without formal training who are interested in understanding and implementing energy systems.

schools, and the like) *depend* on reliable access to energy in all its forms. Without this access, the technologies that deliver modern amenities including comfortable indoor temperature, safe food products, high-speed transportation, and so on, would quickly cease to function. Even the poorest persons in the less-developed countries may rely on occasion for their survival on shipments of food aid that could not be moved effectively without mechanized transportation.

In order for an energy resource to be reliable, it must, first of all, deliver the service that the consumer expects. Second, it must be available in the quantity desired, when the consumer wishes to consume it (whether this is electricity from a wall outlet or gasoline dispensed from a filling station). Lastly, the resource must be available at a price that is economically affordable.

The above qualities define the reliability of the energy resource, for which the consumer will experience adverse consequences if not met—immediately, if the energy system or device stops functioning correctly, or within a short period of time, if the price of the resource is unaffordable. Longer term, there is another criterion for the energy resource that must be met, and one for which society as a whole suffers negative consequences, even if the individual user does not experience consequences directly from her or his actions and choices: environmental sustainability. At the beginning of the twenty-first century, this dimension of energy use and energy systems is increasingly important. Practices that may have placed negligible stress on the planet 100 or 150 years ago, simply because so few people had access to them, are no longer acceptable today, when billions of human beings already have access to these practices or are on the verge of achieving sufficient wealth to have access to them. Thus the need to deliver energy in a way that is both *reliable* and *sustainable* places before humanity a challenge that is both substantial and complex, and one that will require the talent and efforts of engineers as well as research scientists, business managers, government administrators, policy analysts, and so on, for many years, if not decades, to come.

1-2-1   Historic Growth in Energy Supply

Looking back at how humanity has used energy over the millennia since antiquity, it is clear that the beginning of the industrial revolution marks a profound change from gradual refinement of low-power systems to rapid advances in power-intensive systems of all sorts. Along with this acceleration of evolution came a rapid expansion of the ability of human beings to multiply their maximum power output through the application of technology.

Of course, ever since the dawn of recorded history, it has been human nature to improve one's quality of life by finding alternatives to the use of human force for manipulating and moving objects, transforming raw materials into the necessities of life, and conveying oneself between points A and B. The earliest *technologies* used to these ends include the use of horses and other draft animals for mechanical force or transportation, and the use of water currents and sails for the propulsion of boats, rafts, and other watercraft. Over time, humans came to use wood, charcoal, and other *biofuels* for space heating and "process heat," that is, heat used for some creative purpose such as cooking or metallurgy, for various activities. The sailboat of the ancient Mediterranean cultures evolved into the sophisticated sail-rigging systems of the European merchant ships of the 1700s; in Asia, Chinese navigators also developed advanced sail systems. By the year 1800, coal had been in use for many centuries as an alternative to the combustion of plants grown on the Earth's surface for providing heat, and efforts to mine coal were expanding in European countries, notably in Britain.

The evolution of wind power for mechanical work on land prior to 1800 illustrates the gradual refinement of a technology prior to the industrial revolution. The earliest *windmills* in the Middle East used the force of the wind against a vertical shaft to grind grain. Later, the rotation around a horizontal axis was adopted in the *jib mill* of Crete and other Mediterranean locations. Jib mills also used removable "sails" on a hollow wooden frame of the mill "wing" so that the operator could adjust the amount of surface area to the availability of wind, and especially protect the mill from excessive force on a windy day by operating the mill with only part of the wing cover. Several more centuries brought the advent of the windmill that could be rotated in response to the direction of the wind. At first, it was necessary to rotate the entire mill structure around its base on the ground in order to face into the wind. A later refinement was the rotating mill in which only the top "turret" of the mill turned (specimens of which are still seen today in the Netherlands, Belgium, and Northern France), while the rest of the structure was affixed permanently to its foundation. This entire evolution took place over approximately 1000 years.

Compared to this preindustrial evolution, changes have taken place much more rapidly since 1800. In short order, water and wind power gave way to steam power, which in turn has given way to the ubiquitous use of the electric motor and electric pump in all manner of applications. On the transportation side, the two centuries since 1800 have seen the rise of the steam-powered water vessel, the railroad, the internal combustion engine, the automobile, the airplane, and ultimately the spacecraft. Along the way, the use of electricity not only for transmitting energy but also for storing and transmitting information has become possible.

A comparison of two events from antiquity with events of the present age illustrates how much has changed in the human use of energy. In the first event, the Carthaginian army under General Hannibal met and defeated the Roman army under General Paulus at Cannae, to the east of modern day Naples on the Italian peninsula, in the year 216 BC.[2] On the order of 100,000 men and 10,000 horses took part in this battle on both sides. At around the same time, Emperor Qin Shihuang ordered the construction of the Great Wall of China, which required approximately 10 years around the year 200 BC. This project is thought to have involved up to 300,000 men at a time using manual labor to assemble materials and erect the wall.

An "energy analysis" of either of these endeavors, the battle or the building project, reveals that in both cases, the maximum output of even such a large gathering of men and horses (in the case of Cannae) is modest compared to that of a modern-day energy system. For purposes of this analysis, we use a metric unit of power, the *watt* (abbreviated W), that is defined in Sec. 1-6-1. Using a typical figure for long-term average output of 70 W or 200 W for either a human body or a horse, respectively, one can estimate the output at Cannae at approximately 9×10^6 W, and at the Great Wall at approximately 2.1×10^7 W.[3] By comparison, a modern fossil- or nuclear-fired power

---

[2]This example is due to Lorenzo (1994), p. 30, who states that the average output for a horse was much less than a modern "horsepower" due to the lack of horseshoes.

[3]Note that these figures are significantly less than the maximum instantaneous output of either human or horse, for example in the case of a laborer exerting all his strength to lift a stone into position in the Great Wall. Inability to sustain a high instantaneous output rate is in fact one of the main limitations of using human or animal power to do work.

plant in China, Italy, or any other industrial country can deliver 5×10^8 W or more uninterrupted for long periods of time, with a crew of between 10 and 50 persons only to maintain ongoing operations. Similarly, a string of four or five modern railroad locomotives pulling a fully loaded train up a grade can deliver between 4×10^7 W of power with a crew of just three persons.

Two further observations on this comparison between the ancient and the modern are in order. First, if we look at the growth in maximum power output per human from 70 W in 200 BC to a maximum of approximately 5×10^7 W in 2000 for a power plant—six orders of magnitude—we find that most of the advancement has come since the year 1800. Some part of the scientific understanding necessary for the subsequent inventions that would launch the modern energy system came prior to this year, during the advances of Chinese scientists in the preceding centuries, or the Renaissance and Age of Enlightenment in Europe. However, most of the actual engineering of these energy technologies, that is, the conversion of scientific understanding into productive devices, came after 1800. Second, having created a worldwide, resource-intensive system to convert, distribute, and consume energy, human society has become *dependent* on a high rate of energy consumption and cannot suddenly cut off all or even part of this energy consumption without a significant measure of human stress and suffering. For example, a modern office high-rise in any major metropolis around the globe—Sao Paulo, New York, London, Johannesburg, Dubai, Hong Kong, or Sydney, to name a few—is both a symbol of industrial strength and technological fragility. These buildings deliver remarkable comfort and functionality to large numbers of occupants, but take away just three key inputs derived from energy resources—electricity, natural gas, and running water—and they become uninhabitable.

1-3   Relationship between Energy, Population, and Wealth

Changes in levels of energy consumption, both for individual countries and for the world as a whole, are in a symbiotic relationship with levels of both population and wealth. That is, increasing access to energy makes it possible for human society to support larger populations and increasing levels of wealth, while at the same time, a growing population and increasing wealth will spur the purchase of energy for all aspects of daily life.

A comparison of estimates of world population and energy production [measured in either joules (J) or British thermal units (BTUs)—see Secs. 1-6-1 and 1-6-2, respectively] from 1850 to 2010 is shown in Fig. 1-1. The growth in world energy production intensity per capita, or amount of energy produced per person, is also shown, measured in either billion joules per capita, or million BTU per capita. The values shown are the total energy production figures divided by the population for each year. While population growth was unprecedented in human history over this period, growing nearly fivefold to approximately 6 billion, growth in energy consumption was much greater, growing more than twentyfold over the same period.

From analysis of Fig. 1-1, the energy production growth trend can be broken into five periods, each period reflecting events in worldwide technological evolution and social change. From 1850 to 1900, industrialization and the construction of railroad networks was underway in several parts of the world, but much of the human population did not yet have the financial means to access manufactured goods or travel by rail, so the effect of industrialization on energy use per capita was modest,

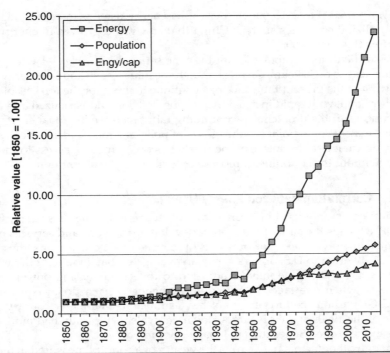

FIGURE 1-1   Relative growth in world population, world energy production, and average energy production per capita 1850–2015, indexed to 1850 = 1.00.

Notes: In 1850, population was 1.26 billion, energy was 25 exajoules, abbreviated EJ, or 23.7 quadrillion BTU, or quads, and per capita energy use was 19.8 gigajoules, abbreviated GJ, per person or 18.8 million BTU per person.[4] For all data points, energy per person is the value of the energy curve divided by the value of the population curve; see text. Total energy and energy per capita figure for 2015 based on 2013 observed figure of 568 EJ and extrapolation of 2010–2013 energy consumption trend out to 2015, resulting in 2015 value of 589 EJ. 2015 population figure is 7.214 billion people according to United Nations. (*Sources:* Own calculations based on energy data from Energy Information Administration of the U.S. Department of Energy and population data from U.S. Bureau of the Census.)

and energy production grew roughly in line with population. From 1900 to 1950, both the part of the population using modern energy supplies and the diversity of supplies (including oil and gas as well as coal) grew, so that energy production began to outpace population and energy intensity per capita doubled by 1950, compared to 1850. From 1950 to 1975, energy production and energy intensity grew rapidly in the post–World War II period of economic expansion. From 1975 to 2000, both energy and population continued to grow, but limitations on output of some resources for energy, notably crude oil, as well as higher prices, encouraged more efficient use of energy around the world, so that energy production per capita remained roughly constant.

---

[4]In other words, 1 EJ = 10^{18} joules, 1 GJ = 10^9 joules. See Sec. 1-6 on energy units at the end of this chapter.

Finally, from 2000 to 2015, global per capita energy use has increased slightly, led by industrializing countries such as China that saw rapid growth in energy consumption after the year 2000.

It is difficult to extrapolate from the time series data in Fig. 1-1 what will happen over the next 25 or 50 years. On one hand, the absolute magnitude of energy production is growing at the present time and may continue to grow for the foreseeable future. On the other hand, with rapid increase in awareness in both industrialized and industrializing countries of the problems of continuing rapid increase in global energy consumption, we may at some point see the effects of policies aimed at dramatically improving energy efficiency so that more of the world's people can have access to the benefits of energy without the continued rapid rise of total energy consumption.

1-3-1   Correlation between Energy Use and Wealth

The most commonly used measure of wealth at present is the gross domestic product, or GDP, which is the sum of the monetary value of all goods and services produced in a country in a given year. For purposes of international comparisons, these values are usually converted to U.S. dollars, using an average exchange rate for the year. In some instances, the GDP value may be adjusted to reflect purchasing power parity (PPP), since even when one takes into account exchange rates, a dollar equivalent earned toward GDP in one country may not buy as much as in another country (e.g., in the decade of the 1990s and early 2000s, dollar equivalents typically had more purchasing power in the United States than in Japan or Scandinavia).

Though other factors play a role, the wealth of a country measured in terms of GDP per capita is to a fair extent correlated with the energy use per capita of that country. In recent data, both the GDP per capita and per capita energy consumption vary by one to two orders of magnitude between the countries with the lowest and highest values. Consider three countries, namely, Bahrain, the United States, and Zimbabwe, using 2011 data. For GDP per capita, the values are $1131 per person for Zimbabwe versus ~$55,900 per person for the United States, not adjusting for PPP. For energy, they are 14.3 GJ/person (13.5 million BTU/person) for Zimbabwe, versus 476 GJ/person (451 million BTU/person) for Bahrain. Table 1-1 presents a subset of the world's countries including Bahrain, the United States, and Zimbabwe, selected to represent varying degrees of wealth as well as different continents of the world, to illustrate the connection between wealth and energy.

Plotting these countries' GDP per capita as a function of energy use per capita (see Fig. 1-2) shows that GDP per capita rises with energy per capita, especially if one excludes countries such as Russia and Bahrain, which may fall outside the curve due to their status as major oil producers or due to extreme climates. Also, among the most prosperous countries in the figure, namely, those with a per capita GDP above $30,000 (Germany, Australia, Japan, Canada, and the United States), there is a wide variation in energy use per capita, with U.S. and Canadian citizens using on average about twice as much energy as those of Japan or Germany.

1-3-2   Human Development Index: An Alternative Means of Evaluating Prosperity

In order to create a measure of prosperity that better reflects broad national goals beyond the performance of the economy, the United Nations has, since the early 1990s, tracked the value of the human development index (HDI). The HDI is measured on a

Country	Population (millions)		Energy (EJ)		GDP (billion US$)	
	2004	2011	2004	2011	2004	2011
Australia	20.2	21.8	6.0	6.5	699.7	1444.2
Bahrain	0.7	1.2	0.5	0.6	11.7	33.9
Brazil	186.4	197.6	10.6	12.5	804.5	2353.0
Canada	32.3	34.0	15.1	14.1	1131.5	1788.7
China	1315.8	1336.7	47.7	109.4	2240.9	10,380.4
Gabon	1.4	1.6	0.1	0.1	9.0	17.2
Germany	82.7	81.5	17.3	14.2	2805.0	3859.5
India	1103.4	1189.2	15.4	24.8	787.8	2049.5
Israel	6.7	7.5	0.9	1.1	122.8	303.8
Japan	128.1	127.5	26.4	22.1	4583.8	4616.3
Poland	38.5	38.4	4.3	4.3	303.4	546.6
Portugal	10.5	10.8	1.3	1.1	183.0	230.0
Russia	143.2	142.5	33.9	32.1	768.8	1857.5
Thailand	64.2	66.7	3.5	5.2	165.5	373.8
United States	298.2	311.6	116.8	102.8	12,555.1	17,418.9
Venezuela	26.7	27.6	3.5	3.4	134.4	205.8
Zimbabwe	13.0	12.1	0.3	0.2	3.9	13.7

Note: GDP values in current dollars.

Sources: UN Department of Economics and Social Affairs (2006) and Population Reference Bureau (2015), for population; U.S. Energy Information Administration (2015), for energy use; International Monetary Fund (2015), for economic data.

TABLE 1-1   Population, Energy Use, and GDP of Selected Countries, 2004 and 2011

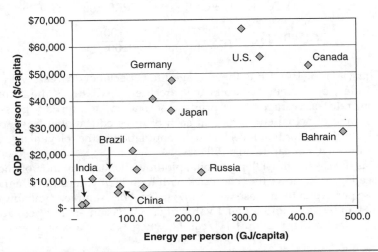

FIGURE 1-2   Per capita GDP as a function of per capita energy consumption in gigajoules (GJ) per person for selected countries, 2011.

Note: Conversion is 1 million BTU = 1.055 GJ. *Sources:* UN Department of Economics and Social Affairs (2015), for population; U.S. Energy Information Administration (2015), for energy consumption; International Monetary Fund (2015), for GDP.

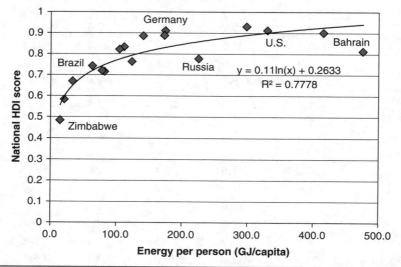

FIGURE 1-3   Human development index as a function of per capita energy consumption in GJ per person for selected countries, 2011.

scale from 0 (worst) to 1 (best),[5] and is an average of the following three general indices for life expectancy, education, and GDP per capita:

$$\text{Life expectancy index} = (LE-25)/(85-25) \qquad (1\text{-}1)$$

$$\text{Education index} = 2/3(ALI) + 1/3(CGER) \qquad (1\text{-}2)$$

$$CGER = 1/3(\text{GER Prim} + \text{GERSecond} + \text{GERTerti}) \qquad (1\text{-}3)$$

$$\text{GDP index} = \frac{[\log(GDPpc) - \log(100)]}{[\log(40000) - \log(100)]} \qquad (1\text{-}4)$$

In Eqs. (1-1) to (1-4), LE is the average life expectancy in years; ALI is the adult literacy index, or percent of adults that are literate; CGER is the combined gross enrollment rate, or average of the primary, secondary, and tertiary gross enrollment rates (i.e., ratio of actual enrollment at each of three educational levels compared to expected enrollment for that level based on population in the relevant age group);[6] and GDPpc is the GDP per capita on a PPP basis.

As was the case with GDP per capita, plotting HDI as a function of energy use per capita (see Fig. 1-3) shows that countries with high HDI values have higher values of energy use per capita than those with a low value. For example, Zimbabwe, with a life expectancy of 59 years and an HDI value of 0.484, has an energy per capita value of 14.3 GJ/capita or 13.5 million BTU/capita, whereas for Canada, the corresponding

---

[5]In the future, it may be necessary to adjust benchmark life expectancy and GDP per capita values in order to prevent the HDI from exceeding a value of 1 for the highest ranked countries.

[6]For example, suppose a country has a population of age to attend primary school (approximately ages 6–11) of 1,000,000 children. If the actual enrollment for this age group is 950,000, then GERPrim = 0.95.

values are 0.950, 4.7×10^{11}, and 442, respectively. Also, among countries with a high value of HDI (> 0.80), there is a wide range of energy intensity values, with Bahrain consuming 476 GJ/capita (451 million BTU/capita) but having an HDI value of 0.813, which is somewhat lower than that of Canada.

1-4   Pressures Facing World due to Energy Consumption

As shown in Fig. 1-1, world energy consumption increased dramatically between 1950 and 2015, from approximately 100 EJ to over 500 EJ. This trend continues at present due to ongoing expansion in consumption of energy. Many wealthy countries have slowed their growth in energy consumption compared to earlier decades, notably countries such as Japan or the members of the European Union. Some countries have even stabilized their energy consumption, such as Denmark, which consumed 0.92 EJ (0.86 quad) in 1980 and 0.78 EJ (0.74 quad) in 2012. Such countries are in a minority, especially since it is the emerging economies of the developing world where demand for energy is growing most rapidly. In these countries, a historical transformation is under way: as their economies grow wealthier, they adopt many of the features of the wealthy countries such as the United States, Japan, or Canada, and their per capita energy consumption therefore approaches that of the wealthy countries.

1-4-1   Industrial versus Emerging Countries

The data in Fig. 1-4, which include all gross energy consumption such as energy for transportation, conversion to electricity, space heating and industrial process heat, and as a raw material, illustrate this point. In the figure the countries of the world are grouped into either the "industrial" category, which includes the European Union and Eastern Europe, North America, Japan, Australia, and New Zealand, or an "emerging"

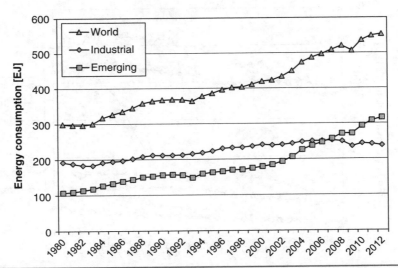

FIGURE 1-4   Comparison of total energy consumption of industrial and emerging countries, 1980–2012.

[Source: U.S. Energy Information Administration (2015).]

category, which includes all other countries. This division is made for historical reasons, because at the beginning of the expansion in energy use in 1950, the countries in the industrial category were generally considered the *rich* countries of the world. Today, some of the emerging countries have surpassed a number of the industrial countries in GDP per capita or energy use per capita, such as South Korea, which has passed all the countries of the former Eastern European bloc (Czech Republic, Poland, and others) in both measures. In any case, thanks to their faster growth rate since approximately 2000, the emerging countries overtook the industrial countries in 2007, even as both groups of countries followed the ups and downs of global economic trends (e.g., slower growth in periods of relative economic weakness such as the early 1980s, early 1990s, early 2000s, or recession of 2009-2010). In 2012, the emerging countries consumed 316 EJ (299 quads) versus 237 EJ (225 quads) for the industrial countries.

Furthermore, a comparison between total energy use and total population for the two groups of countries suggests that the total energy use for the emerging countries might grow much more in the future. The industrial countries comprise approximately 1.4 billion people, with the emerging countries representing the remaining 5.3 billion people on the planet (2008 values). On this basis, the industrial countries consume more than three times the energy per capita than emerging countries. If, in the future, the economies of the emerging countries were to grow to resemble those of the industrial economies, such that the per capita energy consumption gap between the two were to be largely closed, total energy for the emerging countries might exceed that of the industrial group by a factor of 2 or 3.

To further illustrate the influence of the industrial and emerging countries on world energy consumption, we can study the individual energy consumption trends for a select number of countries that influence global trends, as shown in Fig. 1-5. The "countries" include an agglomeration of 38 European countries called "Europe38," which consists of

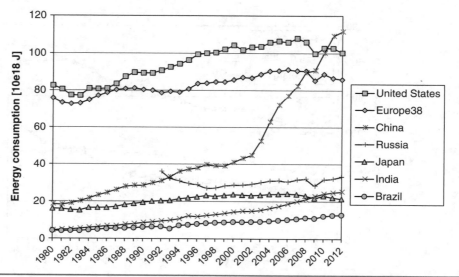

Figure 1-5 Comparison of select countries total energy consumption, 1980–2012.

Note: Russia's trend line appears in year 1992 because in prior years its energy consumption was part of the former Soviet Union. [*Source:* U.S. Energy Information Administration (2015).]

38 countries from Western and Eastern Europe, but not the former Soviet Union. The other countries included are the United States, Japan, China, Russia, India, and Brazil.[7] Together, the countries represented in the figure account for 71% of the world's energy consumption in 2012.

Two important trends emerge from Fig. 1-5. First, among the rich countries represented, energy consumption growth in the United States outpaces that of Europe38 and Japan, so that the United States is becoming an increasingly dominant user of energy among this group. In the late 1980s, the United States and Europe38 were almost equal in terms of energy use, but energy use in Europe has grown more slowly since that time, measured either in percentage or absolute magnitude. Second, the percent growth rate, and absolute magnitude in the case of China, is much higher for China, Brazil, and India from 2002 onward than for the other countries in the figure. These three countries grew by 101%, 24%, and 43%, respectively, during this time period. Particularly, striking was the growth of energy consumption in China by some 45 EJ (43 quads) in the period from 2002 to 2008 alone; this spike in energy consumption was enough to visibly affect the trend in world energy consumption in Fig. 1-4. This trend reflects the transformation of the Chinese economy from that of a low-cost industrial producer to that of a modern consumer economy, similar to that of the rich countries, as symbolized by the two photographs in Fig. 1-6.

The energy growth path of China can be viewed from a different perspective by comparing it to that of the United States in terms of the annual growth since reaching a threshold of 40 quads of gross energy consumption, as shown in Fig. 1-7. The United States reached this threshold in 1955 and China in 2001. The figure shows the subsequent growth pathway, where the horizontal axis gives the number of years elapsed since 1955 or 2001, depending on the country, and the vertical axis gives the total energy consumption in that year. For the United States, growth from 40 to 90 quads required 40 years (1955–1995). In the case of China, this growth took place over a period of just 8 years (2001–2009).

While growth in energy use in China may be rapid, especially recently, the per capita value remains lower than that of industrialized countries. In 2011, the average person in China consumed 82 GJ (78 million BTU), compared to 330 GJ (313 million BTU) for the average person in the United States. Figure 1-8 illustrates the trend for China, the United States, and five other countries for the period 1980–2011: while per capita energy consumption for China has risen markedly since 2000, it remains well below that of Japan, the United Kingdom, Russia, and especially the United States.

The preceding data suggest that, for rapidly growing countries like China, Brazil, and India, there is much room for both total energy consumption and energy consumption per capita to grow. Recall also the strong connection between energy use and GDP. Whereas the industrial countries typically grow their GDP 2–3% per year, China's economy has been growing at 7–10% per year in recent years. The values in 2011 of $7.8 trillion GDP and $7766 per capita for China, compared to $17.4 trillion and $55,900, respectively, for the United States, indicate that there is much room for GDP and GDP per capita in China to grow, barring some major shift in the function of the world's economy. This growth is likely to continue to put upward pressure on energy consumption in China.

---

[7]Brazil, Russia, India, and China are included in the figure because they comprise the four "BRIC" countries, which represent a large fraction of the world's energy consumption, population, and economic activity.

(a)

(b)

FIGURE 1-6   An example of rapid technological transformation in China.

Figure 1-6(a) shows a Chinese steam locomotive in regular service, Guangdong Province, 1989.
Figure 1-6(b) shows a Maglev train in Shanghai, 2006. Just 17 years separate these two photographs.
Although the adoption of modern technology has generally increased energy consumption in China, it
has also increased efficiency in some cases, as happened with the phasing out of the use of inefficient
steam locomotives by the mid-1990s. [*Source:* Jian Shuo Wang, wangjianshuo.com (Fig. 1-6b). Reprinted
with permission.]

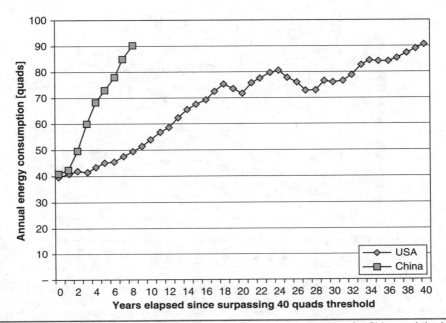

FIGURE 1-7   40 to 90 quads in 8 years? Figure of annual energy consumption for China and the United States for the number of years required to grow from 40 to 90 quads of total energy. Year 0 in the figure is equivalent to the years 2001 and 1955 for China and the United States, respectively.

(*Source:* U.S. Energy Information Administration.)

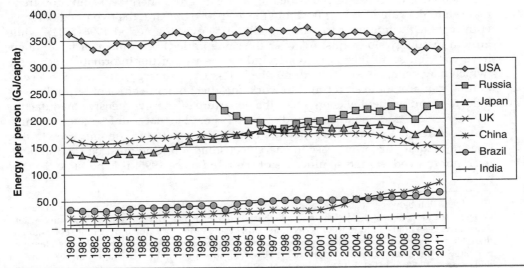

FIGURE 1-8   Per capita energy consumption for select industrialized and BRIC countries, 1980–2011.

Note: Russia trend line appears in year 1992 because in prior years its energy consumption was part of former Soviet Union. Per capita energy use in the United Kingdom is used to represent typical values for the European Union as a whole. [*Source:* U.S. Energy Information Administration.]

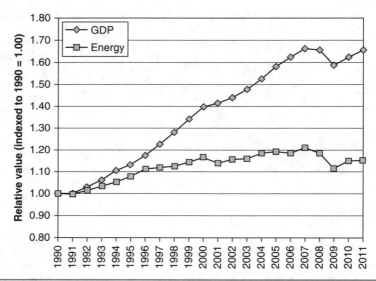

FIGURE 1-9   Relative values of energy use and real GDP in the United States indexed to 1990 = 1.00, 1990–2011. In 1990, the value of GDP in 2005 dollars was $8.0 trillion, and the value of energy consumption was 89.3 EJ (84.6 quads). [*Source:* U.S. Energy Information Administration.]

At the same time, there is another trend emerging in the industrial countries such as the United States that is more promising in terms of energy supply and demand, namely the potential breaking of the link between GDP and energy. Figure 1-9 shows U.S. real GDP (in constant 2005 dollars)[8] and U.S. gross energy consumption trend indexed to the value 1993 = 1.00. From the year 1996 onward, the two trends diverge with real GDP continuing to grow rapidly (43% over the period 1996–2011), while growth in energy slows (just 3% over the same period). A number of factors may have contributed to this situation, including the rise of the information economy (which may generate wealth with relatively less use of energy), the use of telecommuting in place of physical travel to work, the effect of more efficient technology on industrial and residential energy use, the departure of energy-intensive manufacturing to overseas trading partners[9] and so on. It is possible that countries such as China may soon reach a point where they too can shift to a path of economic growth that does not require major increase in energy consumption. One must be careful, however, not to overstate the significance of the trend since 1996. It is possible that the

[8]The distinction between "constant" and "current" dollars is as follows: An amount given in current dollars is the face value of the dollar amount in the year in which it is given; an amount given in constant dollars has been adjusted to reflect the declining purchasing power of a dollar over time. For example, the 1993 and 2000 U.S. GDP in current dollars was $6.66 billion and $9.82 billion, respectively. However, the value of the dollar in 2000 was 88% of a 1993 dollar, so adjusting the GDP values to 2000 constant dollars gives $7.53 and $9.82 billion, respectively. See Chap. 3.
[9]Caveat: If, in fact, the shifting of manufacturing overseas was shown to be the main contributor to the slowdown in the growth of energy consumption in the USA, it could be argued that in fact there is no breaking of the link between GDP and energy growth, since the energy consumption would still be taking place, but accounted for under the energy balances of other countries.

break between GDP and energy is temporary and that future values will show a return to a closer correlation between energy and GDP; furthermore, even in the period 1996–2011, the value of energy use per capita for the United States is very high and could not be sustained if it were applied to the world's population of nearly 7 billion, given current technology.

1-4-2   Pressure on CO_2 Emissions

Negative effects of energy use do not come directly from the amount of gross energy consumption itself, but rather from the amount of nonrenewable resources consumed in order to deliver the energy, the side effects of extracting energy resources, the total emissions of pollution and greenhouse gases emitted from the use of that energy, and so on. We can examine the case of CO_2 emissions by looking at the trend for these emissions for industrial and emerging countries, as shown in Fig. 1-10. In the figure, the industrial countries are divided into "North America" (i.e., Canada and the United States, but not Mexico), Europe38, and "Other Industrial" (Japan, Australia, and New Zealand), while the emerging countries are divided between the four BRIC countries and "Rest of World"(ROW), or all remaining countries in the world not previously included. The pattern is similar to the one for energy consumption shown in Fig. 1-4, with emissions from the ROW, and especially BRIC countries, growing faster than those for the industrial countries from approximately 2000 onward. The emerging countries also surpassed the industrial countries sooner in terms of CO_2 emissions compared to energy consumption, with these two events taking place in the years 2003 and 2007,

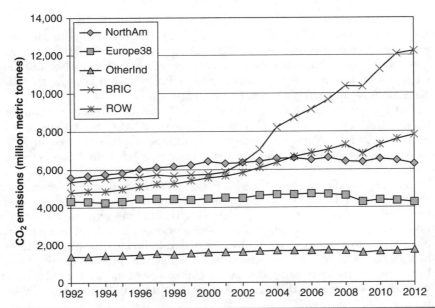

FIGURE 1-10   Carbon dioxide emissions for North American, European, other industrial, BRIC, and ROW countries, 1992–2008.

Note: Other industrial countries include Japan and Oceania. Rest of world includes all countries not in previous four groups. (*Source:* U.S. Energy Information Administration.)

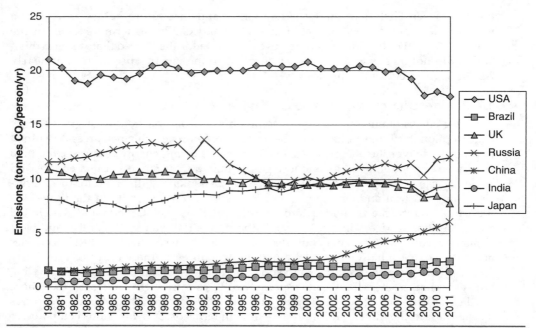

Figure 1-11   Per capita CO_2 emissions for select countries, 1980–2011. [*Source:* U.S. Energy Information Administration.]

respectively. Another way of saying this is that the industrial countries emit less CO_2 per unit of energy consumed (in 2012, 52 kg/GJ, versus 64 kg/GJ for the emerging countries). The industrial countries have an advantage both because they convert energy resources to energy products more efficiently, and because they have greater access to low- and zero-CO_2 energy resources.

At the worldwide level, emissions of CO_2 per GJ of energy consumed declined slightly from 1980 to 2012, by about 6% (62 versus 58 kgCO_2 per GJ), using the data from Figs. 1-4 and 1-10. Worldwide emissions of CO_2 per capita increased slightly from 1980 to 2011, from 4.0 to 4.6 tonnes per person per year, after doubling between 1950 and 1980.[10]

Per capita CO_2 emissions trends for select industrial and emerging countries are shown in Fig. 1-11. The BRIC countries show growing per capita emissions since 2000, with China in particular showing a large rise due to both growing industrial output for domestic and global markets and a growing middle class. Although it is not as easy to discern due to the scale of the figure, Brazil and India also grew in per capita emissions by 25% and 50%, respectively, and Russian per capita emissions grew as well. U.S. and European emissions (represented by the United Kingdom as a typical European energy user) declined on the other hand. Among the industrial countries only emissions in

---

[10]See Marland et al. (2003) for comparison of 1950 and 1980: ~2.4 metric tonnes of CO_2 equivalent in 1950, 4.5 tonnes in 1980.

Japan actually increased from 1998 to 2011, influenced by the closure of nuclear power plants after the Fukushima accident in 2011.

In conclusion, trends in Figs. 1-10 and 1-11 suggest that there is reason for great concern regarding CO_2 emissions in the short to medium term. While the value of CO_2 per unit of energy or per capita may be more or less stable—possibly increasing slightly, but not substantially—it is still high for a planet that needs to dramatically cut greenhouse gas emissions. Furthermore, since both total energy use and total world population are climbing, total CO_2 emissions will climb with them if steps are not taken to decrease CO_2 per unit of energy or per capita.

1-4-3   Observations about Energy Use and CO_2 Emissions Trends

Two observations round out the discussion of energy and CO_2. The first observation is in regard to the distinct roles of the industrial and emerging countries in addressing the energy challenges of the present time. At first glance, it might appear from Fig. 1-4 that the emerging countries are not doing as much as the industrial countries in using energy as efficiently as possible, and from Fig. 1-5 that China in particular is not doing its part to address the sustainable use of energy. This type of thinking is incomplete, however. Countries such as China, and the emerging countries in general, consume much less energy per capita than do the industrial countries (Fig. 1-8). Therefore, in creating a global solution for energy, the countries of the world must recognize that while one factor is a sustainable overall level of energy use, another factor is the right of emerging countries to expand their access to energy in order to improve the everyday well-being of their citizens. A possible solution is one in which the industrial countries greatly increase their efficiency of energy use so as to decrease both energy use and CO_2 emissions per capita, while the emerging countries work toward achieving a quality of life equal to that of the industrial countries, but requiring significantly less energy and CO_2 emissions per capita than the current values for the industrial countries.

The second observation is in regard to the overall trend in energy and CO_2 emissions and how best it should be addressed. Looking at Figs. 1-4, 1-5, 1-10, and 1-11, it is clear that the growth in energy consumption and CO_2 emissions has been continuous over many years. Even though the industrial countries appear to be curbing their growth since approximately 2008, emerging countries are in a position to continue to push the trend upward. A key factor is therefore to enable increased energy use without increasing negative impact on the environment through technology. It will be shown in this book that a great deal of physical potential exists to deliver energy in the substantial amounts needed, and without emitting correspondingly large amounts of CO_2 to the atmosphere, using sources that we already understand reasonably well. The challenge lies in developing the technology to deliver the energy without degrading the environment. Renewable resources are available worldwide in quantities that dwarf current worldwide energy consumption by humans. Renewables also have the strong advantage that they do not create a waste stream of either CO_2 from fossil fuels or high-level radioactive waste, which are disadvantages of fossil and nuclear energy. However, resources for certain types of nuclear energy are available in such large quantities as to be almost limitless. Fossil fuels are available in amounts that could last for two or three centuries as well, if they can be extracted safely and combusted without emitting CO_2 to the atmosphere.

1-4-4   Discussion: Contrasting Mainstream and Deep Ecologic Perspectives on Energy Requirements

Underlying the previous discussion of addressing world poverty and global CO_2 emissions is a fundamental premise, namely, that increased access to energy per capita is a requirement for wealth, and that increased wealth will lead to improvements in human well-being. Based on earlier figures (Fig. 1-2 for GDP or Fig. 1-3 for HDI), the correlation between energy consumption and economic indicators support the idea that, as poor countries increase their access to energy, their position will rise along the wealth or HDI curve. Success in the area of wealth or HDI, it is reasonable to assume, will in turn lead to success in providing quality of life. By this logic, the wealthy countries of the world should not reduce their per capita energy consumption, since this would imply a loss of quality of life. Instead, these countries should make the inputs and outputs of energy systems environmentally benign. As long as this is done, the environment is protected at whatever level of energy consumption emerges, and the mix of activities in which their citizens engage (manufacturing, commerce, retailing, tourism, mass media entertainment, and the like) is no longer a concern. The technologies and systems thus developed can be made available to the emerging countries so that they too can have greater access to energy consumption and achieve a high quality of life in a sustainable way.

This "mainstream" perspective has been widely adopted by economists, engineers, and political leaders, but it is not the only one. Another approach is to fundamentally rethink the activities in which humans engage. Instead of allowing the mix of activities to continue in its present form, activities and lifestyles are chosen to reduce energy and raw material requirements as well as waste outputs, so that the effect on the environment is reduced to a sustainable level. This approach is one of the fundamental tools of *deep ecology*, an alternative philosophy for society's relationship with the natural world that has emerged in recent decades

By implication, deep ecology criticizes the mainstream approach of being one of "shallow ecology," which only addresses the surface of the ecological crisis by making technical changes to resource extraction and by-product disposal, and does not address the core problem of excessive interference in the natural systems of the world. For example, in the case of energy systems, instead of extracting energy resources or generating energy at the same rate but in a more environmentally friendly way, the deep ecologist seeks to change people's choices of end uses of energy so that much less energy is required. A primary target of this effort is *consumer culture*, that is, the purchase, especially by middle- and upper-class citizens of goods and services that are beyond what is necessary for basic existence, but that are thought to improve quality of life. Deep ecologists seek to reorient purchases and consumption away from consumer culture toward focusing on *essential* goods and services. In the long term, many among this group also advocate gradually reducing the total human population so as to reduce resource consumption and pollution.

The difference between the mainstream and deep ecologic approach can be illustrated using different countries' energy consumption per capita values, and considering how they might change in the future in each case. For instance, the country of Bangladesh is on the low end of the energy per capita spectrum, with a value of 6.8 GJ/person (6.4 million BTU/person). Zimbabwe was presented earlier as another country with low per-capita energy use, at 14.3 GJ/person. From Table 1-1, the equivalent value for the United States is 330 GJ/person (313 million BTU/person). Both approaches agree that the low value of energy per capita limits the human potential of Bangladesh, since it implies difficulty in delivering adequate food, medical care, or other basic necessities. However, they diverge on what should be the *correct* target for energy intensity. In the mainstream approach, there is no restriction on Bangladesh eventually rising to the level of energy intensity of the United States, so long as the requirement for protecting the environment is met. By contrast, the deep ecologic approach envisions both countries aspiring to some intermediate value, perhaps on the order of 50 to 100 GJ/person (50 to 100 million BTU/person), due to improved access to environmentally friendly energy sources in Bangladesh and changes in lifestyle choices in the United States. In practical terms, both mainstream and deep ecologic approaches would approve of the development in Bangladesh of basic health care, adequate access to food to avoid hunger and starvation, and universal primary education—all of which are unmet needs at present. However, the mainstream approach would in addition encourage the development of universal access to mechanized transportation and widespread automobile ownership, a comprehensive network of power plants and electric grid, shopping malls and other commercial amenities, and so on, whereas deep ecology would oppose them.

The deep ecologic alternative not only incorporates some attractive advantages but also some significant challenges. The reduction of energy and resource consumption obtained by following the deep ecologic path can yield real ecological benefits. With less consumption of energy services, the rate of coal or oil being mined, renewable energy systems being produced and installed, CO_2 being emitted to the atmosphere, and so on, is reduced. In turn, reducing the throughput of energy, from resource extraction to conversion to the management of the resulting by-products, can simplify the challenge of reconciling energy systems with the environment. This problem is made more difficult in the mainstream case, where the throughput rate is higher.

However, in the short to medium term, the deep ecologic approach would challenge society because many of the activities in a modern service economy that appear not to be essential from a deep ecologic perspective are nevertheless vital to the economic well-being of individuals who earn their living by providing them. If demand for these services disappeared suddenly, it might reduce energy consumption and emissions but also create economic hardship for many people. Therefore, the adoption of a deep ecologic approach to solving the energy problem would require a transformation of the way in which economies in wealthy countries provide work for their citizens, which could only be carried out over a period measured in years or decades.

The number of persons fully committed to a deep ecologic path is relatively small at the present time, especially compared to the numbers adhering to the more mainstream idea of continued high levels of economic activity reconciled with adequate environmental safeguards. However, a related approach to personal behavior of giving environmental considerations priority in lifestyle and consumption decisions has started to influence the choices of some individual consumers. Understanding the environmental impact of choices concerning size of home, consumer purchases, travel decisions, and other available options, these consumers choose to forgo purchases that are economically available to them, so as to reduce their personal environmental footprint. This shift has in turn led to some curbing of growth in energy consumption in the industrialized countries among certain segments of society, for example, those who identify themselves as "environmentalists." It is likely that its influence will grow in the future, as will its impact on energy demand.

1-5 Energy Issues and the Contents of This Book

1-5-1 Motivations, Techniques, and Applications

The contents of this book can be divided into three parts: *motivations and drivers* (a part of which is provided by the preceding discussion in this chapter), *tools*, and *applications*. The flowchart in Fig. 1-12 graphically represents the logical connection between different parts of the book.

Three major challenges make up the *motivations and drivers* for the contents of the book, and are covered in the first five chapters. The first is "sustainable development" (covered in Chaps. 1 and 2), meaning continued development that meets human needs, especially in the emerging countries so that they may rise out of poverty, without irreversibly damaging the natural environment. Chief among the environmental concerns is climate change (Chap. 4), which requires that society stabilize (and possibly some day reduce) the amount of CO_2 in the atmosphere, through decisions about energy systems, among other things. Lastly, as the world continues to consume its supply of nonrenewable fossil fuels, oil and gas production will peak sometime in the next several decades, requiring the development of alternative energy sources, such as nonconventional fossil fuels with no CO_2 emissions to the atmosphere, or nonfossil fuels (Chap. 5).

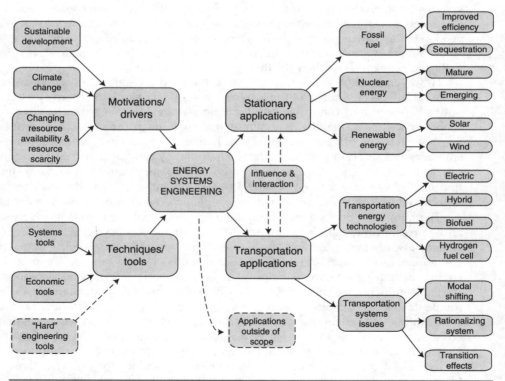

FIGURE 1-12 Flowchart of contents of book, including motivations/drivers, techniques/tools, and applications.

At the lower left in the figure are "techniques and tools." First among these are the "hard" engineering tools, shown with a dashed line to indicate that they are the assumed background that the energy engineer brings to the study of energy systems— thermodynamics, heat transfer, fluid mechanics, electricity and magnetism, chemistry, and statistics. Techniques included in the book appear in Chaps. 2 and 3, including systems tools (Chap. 2), which are presented to help with the modeling of energy use as taking place in a system with interacting parts. These systems have a "life cycle" with distinct stages (planning, implementation, end use or operational lifetime, and dismantling/disposal), where each stage contributes to the overall success or failure of the system. Economic tools (Chap. 3) are also included, and these make the connection between energy systems as technologies and as financial investments, which include both upfront capital costs and ongoing maintenance and operating costs that must be repaid over time through revenues (energy sales, financial benefit of energy savings, and so on).

The motivations/drivers and tools/techniques feed directly in to the engineering of energy systems, that is, the applications covered in the remainder of the book. These are divided between stationary applications, represented by the generation of electricity and space or process heat; and transportation applications, covering all types of move-ment of vehicles by land, sea, or air. Linkages of "interaction and influence" connect the two application areas in several ways; for instance, the primary energy resources which they consume overlap in some areas, and there are synergies whereby demand in both areas can sometimes be met simultaneously. These two areas do not cover 100% of all energy end uses, so applications outside the scope of the book—representing no more than 10 to 15% of all end-use energy consumption—are shown with a dashed line.[11]

The stationary applications side is represented by the major conversion technolo-gies covered in the book, with the assumption that the electricity and/or heat generated is then used for whatever purpose the end-user desires: indoor climate control, lighting, operations of appliances, and so on (the end uses themselves are largely outside the scope of the book). The stationary applications are divided among fossil fuel systems (Chaps. 6 and 7), nuclear energy systems (Chap. 8), and renewable energy systems (Chaps. 9–14). Transportation applications are divided between the individual vehicle propulsion technologies (Chap. 15), and the systems issues that arise when the propul-sion systems are manifested, in the form of a great number of vehicles, vessels, and aircraft, in the world's multimodal transportation network (roads, railroads' water-ways, air networks, and pipelines, as discussed in Chap. 16). Chapter 17 briefly dis-cusses some technologies not treated in greater depth in the body of the book, along with providing conclusions regarding future prospects for energy.

1-5-2   Initial Comparison of Three Underlying Primary Energy Sources

While the number of possible applications in Fig. 1-12 is large, the primary energy sources that underlie them consist of just three: fossil, nuclear, and renewable. Since the fundamental characteristics of each of these primary sources has a strong impact on how each is subsequently converted, transmitted, and used, it is useful in this introduc-tory chapter to (1) make some basic comparisons before delving into greater depth in

---

[11]For example, some fraction of the total fossil fuels extracted annually is used as feedstocks for the petrochemical industries or agricultural fertilizers, a use that is not considered in this book.

Chaps. 6 to 16, and (2) to present a possible pathway for how the mix of primary energy might evolve in the twenty-first century in response to the motivations and drivers discussed above. The remainder of this section focuses on these two objectives.

Table 1-2 presents some basic advantages and disadvantages of the three types of primary energy. *Fossil energy* includes any carbon- or hydrocarbon-based resource extracted from subsurface resources and provides approximately 82% of world energy at present (using 2012 figures), down from 85% in the year 2000. Although convenient due to their universal familiarity and relatively low direct cost (i.e., not including the cost of pollution or accidents), fossil fuels are relatively finite in supply compared to their current rate of global consumption, and are the leading contributor to the increase in concentration of CO_2 in the atmosphere. *Nuclear energy* includes all types of primary energy derived from nuclear bonds as opposed to chemical bonds in the case of fossil fuels. Nuclear energy emits no greenhouse gases or other air pollution during production at a nuclear power plant, but carries risks from radiation in case of nuclear plant accidents and from the accumulation of long-lived, highly radioactive wastes that result from the use of nuclear fuel. *Renewable energy* comprises any systems that convert reoccurring energy fluxes in nature, many of them originating from the incidence of solar energy on the Earth, into a form of energy that can be used in human-built systems. Renewable energy is vast in its total global availability in nature and causes no greenhouse gas emissions or pollution in its conversion and end use.

Turning to the second objective, it is useful not only to evaluate the three primary sources of energy today, but also to consider how they might be used in the future to achieve a sustainable global energy solution for the twenty-first century. At present

	Advantages	Disadvantages
Fossil	The dominant primary energy source at present (85% of world demand). Relatively low capital cost. Often has relatively low direct total cost including fuel cost. "Dispatchable," i.e., user controls when fuel is converted (to electricity, mechanical motion, etc.).	Fuel cost can be volatile. Emits CO_2 to atmosphere, unless otherwise controlled. Extraction can be destructive. Risk of accidents during extraction. Limits on ultimately recoverable supply.
Nuclear	Relatively low fuel cost, and fuel cost not volatile. Emits no CO_2 or other air pollutants during generation. Dispatchable. Ultimately recoverable fuel supply potentially much larger than fossil fuel supply.	High capital cost of construction. Risk of radiation in case of accidents. Need for long-term storage of highly radioactive, long-lived waste products.
Renewable	"Fuel" does not cost anything in many cases (wind, sun). Energy supply is inexhaustible. Total energy supply based on total sunlight intercepted by Earth is very large. Emits no CO_2 or other air pollutants during generation.	High capital cost of construction. Intermittent rather than dispatchable nature for many renewable sources. Energy supply in nature is often diffuse, requiring large area as well as large amount of capital equipment.

TABLE 1-2   Basic Advantages and Disadvantages of Three Types of Primary Energy: Fossil, Nuclear, and Renewable

nuclear energy contributes about 7% of global primary energy, and renewable energy about 11%, with fossil energy supplying the rest. We begin with two assumptions related to population and per capita energy consumption:

1. Global population growth will slow and reach a plateau of around 9 billion people around the year 2060, and will remain approximately constant at that level from 2060 to 2100. This scenario is based on a United Nations midrange projection of the path of population growth in the twenty-first century.

2. Average global per capita energy consumption will rise for the rest of the century due to gradually rising values in industrializing countries such as China, India, or Brazil, and also falling values in wealthy countries, so that the average across all countries stabilizes at 100 million BTU/person by the end of the century.

In addition, it is assumed that to create a representative outcome, each of the three primary sources will contribute one-third of the energy supply at the end of the century. Since conventional fossil (i.e., fossil combustion with uncontrolled release of CO_2 to the atmosphere) is incompatible with protecting the climate, carbon capture with sequestration (CCS), will displace conventional fossil so that eventually, all remaining fossil energy use will emit no CO_2.

The results, which displace the 85% conventional fossil used in 2000 with a mix of noncarbon emitting sources by 2100, are shown in Fig. 1-13. The mix, comprising a contribution of a third each, from fossil, nuclear, and renewable was chosen arbitrarily, so no inferences regarding the relative strength or weakness of different sources should be derived from it. Nevertheless, if this situation were achieved, since both per capita energy use and population would stabilize by 2100, the world might continue to meet the global need for energy for some time thereafter, using the mix of resources shown. Eventually, the one-third portion contributed by CCS would need to be replaced by some mix of the other two sources, as fossil resources decline in availability and are

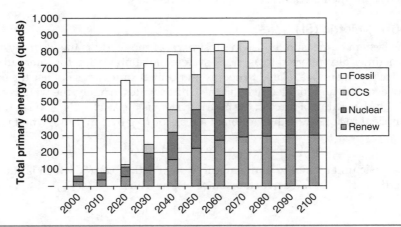

FIGURE 1-13   A possible pathway for primary energy production in the twenty-first century arriving at 100 million BTU/person, on average, available for 9 billion people in 2100.

Note: "Fossil" = fossil energy without sequestration, "CCS" = fossil energy with carbon capture and sequestration, "Renew" = all types of renewables. See text.

ultimately exhausted. Indeed, if the actually recoverable amount of fossil resource cannot sustain the production for the twenty-first century shown for Fossil/CCS in Fig. 1-13, then the substitution of some combination of nuclear and renewable for CCS would need to happen sooner than the year 2100.

Two other comments are in order regarding the rate of transition from fossil to CCS. First, rate of substitution of CCS for fossil is chosen to approximately achieve the commonly stated goal of reducing CO_2 emissions by 80% in the industrialized countries and by some lesser amount in the emerging countries by the year 2050. Thereafter fossil drops to less than 5% in 2060 and disappears by 2070. Nevertheless, even though the goal of 100% CO_2-emission-free energy is achieved by 2070, there is still a large increase in total fossil (i.e., non-CCS) energy use between 2000 and 2020, and only thereafter does fossil energy use gradually decline. Second, this path is in line with currently observed rapid growth in global fossil energy use, but it may not arrive quickly enough for the community of nations to effectively counteract climate change, the impacts of which are growing increasingly severe. If stronger measures were taken than are currently in place, the phase-out of conventional fossil might take place more rapidly than what is shown in the figure.

A more detailed consideration of the strengths and weaknesses of the three types of energy and possible pathways to energy sustainability are included in the conclusions in Chap. 17. The intervening Chaps. 2 to 16 provide the content to support a more sophisticated treatment at the end of the book.

1-6   Units of Measure Used in Energy Systems

Many units of measure used in energy systems are already familiar to readers, and some, namely the watt, joule, and BTU, and multiples thereof such as the GJ or EJ, have been used in the preceding sections. Units in common use throughout this book are defined here, while some units unique to specific technologies are defined later in the book where they arise. It is assumed that the reader already understands basic metric units such as the meter, gram, seconds, and degrees Celsius.

1-6-1   Metric (SI) Units

The system of measure in use in most parts of the world is the *metric* system, which is sometimes also referred to as the SI system (an abbreviation for International System). For many quantities related to energy, the United States does not use the metric system; however, the unit for electricity in the United States, namely, the watt, is an SI unit.

The basic unit of force in the metric system is the *newton* (N). One newton of force is sufficient to accelerate the motion of a mass of 1 kg by 1 m/s in 1 s, that is

$$1\,N = 1\,kg \cdot m/s^2 \tag{1-5}$$

The basic unit of energy in the metric system is the *joule* (J), which is equivalent to the exertion of 1 N of force over a distance of 1 m. Thus

$$1\,J = 1\,N \cdot m \tag{1-6}$$

The basic unit of power, or flow of energy per unit of time, is the *watt*, which is equivalent to the flow of 1 J of energy per second. Therefore

$$1\,W = 1\,J/s \tag{1-7}$$

A convenient alternative measure of energy is the flow of 1 W for an hour, which is denoted 1 *watthour* (Wh). A commonly used measure of electrical energy is the *kilowatthour*, which is abbreviated kWh and consists of 1000 Wh. Since both joules and watthours are units of energy, it is helpful to be able to convert easily between the two. The conversion from watthours to joules is calculated as follows:

$$(1 \text{ Wh})\left(3600 \frac{s}{h}\right)\left(\frac{1 \text{ J/s}}{W}\right) = 3600 \text{ J} \tag{1-8}$$

In other words, to convert watthours to joules, multiply by 3600, and to convert joules to watthours, divide by 3600.

The quantity of energy in an electrical current is a function of the current flowing, measured in *amperes* (A), and the change in potential, measured in *volts* (V). The transmission of 1 W of electricity is equivalent to 1 A of current flowing over a change in potential of 1 V, so

$$1 \text{ W} = 1 \text{ VA} \tag{1-9}$$

The unit *voltampere* (VA) may be used in place of watts to measure electrical power.

Metric units are adapted to specific applications by adding a prefix that denotes a multiple of 10 to be applied to the base unit in question. Table 1-3 gives the names of the prefixes from micro- (10^{-6}) to exa- (10^{18}), along with abbreviations, numerical factors, and representative uses of each prefix in energy applications. A familiarity with the practical meaning of each order of magnitude can help the practitioner to avoid errors in calculation stemming from incorrect manipulation of scientific notation. For instance, the annual output of a coal- or gas-fired power plant rated at 500 MW should be on the order of hundreds or thousands of GWh, and not hundreds or thousands of MWh.

Prefix	Symbol	Factor	Example
Micro-	μ	10^{-6}	Microns (used in place of "micrometers") ~ wavelength of visible light
Milli-	m	10^{-3}	Milliampere ~ current flow from a single photovoltaic cell
Kilo-	k	10^{3}	Kilowatthour ~ unit of sale of electricity to a residential customer
Mega-	M	10^{6}	Megawatt ~ maximum power output of the largest commercial wind turbines
Giga-	G	10^{9}	Gigawatthour ~ measure of the annual output from a typical fossil-fuel-powered electric power plant
Tera-	T	10^{12}	Terawatt ~ measure of the total rated capacity of all power plants in the world
Peta-	P	10^{15}	Petajoule ~ measure of all the energy used by the railroads in the United States in 1 year
Exa-	E	10^{18}	Exajoule ~ measure of all the energy used by an entire country in 1 year

TABLE 1-3 Prefixes Used with Metric Units

Large amounts of mass are measured in the metric system using the *metric ton*, which equals 1000 kg; hereafter it is referred to as a *tonne* in order to distinguish it from U.S. customary units. The tonne is usually used for orders of magnitude above 10^6 g and units based on the tonne such as the kilotonne (1000 tonnes, or 10^9 g), megatonne (10^6 tonnes), and gigatonne (10^9 tonnes) are in use. The abbreviation "MMT" is used to represent units of million metric tonnes.

As an example of conversion between units involving metric units that have prefixes, consider the common conversion of kWh of energy to MJ, and vice versa. The conversion in Eq. (1-6) can be adapted as follows:

$$(1 \text{ kWh})\left(3600 \,\frac{\text{s}}{\text{h}}\right)\left(1000 \,\frac{\text{W}}{\text{kW}}\right)\left(1 \,\frac{\text{J/s}}{\text{W}}\right)\left(\frac{1 \text{ MJ}}{1\times10^6 \text{ J}}\right) = 3.6 \text{ MJ} \qquad (1\text{-}10)$$

In other words, to convert kWh to MJ, multiply by 3.6, and to convert MJ to kWh, divide by 3.6.

Example 1-1   A portable electric generator that is powered by diesel fuel produces 7 kWh of electricity during a single period of operation. (a) What is the equivalent amount of energy measured in MJ? (b) Suppose the fuel consumed had an energy content of 110 MJ. If the device were 100% efficient, how much electricity would it produce?

Solution

(a)   7 kWh × 3.6 MJ/kWh = 25.2 MJ

(b)   110 MJ/3.6 MJ/kWh = 30.6 kWh

1-6-2   U.S. Standard Customary Units

The basic units of the U.S. customary units (sometimes referred to as *standard* units) are the pound mass (lbm) or ton for mass; the pound force (lbf) for force; the degree Fahrenheit (F) for temperature; and inch (in), foot (ft), or mile (mi) for distance. One pound is equivalent to 0.454 kg. One standard ton, also known as a *short ton*, is equal to 2000 lb and is equivalent to 907.2 kg or 0.9072 tonne. One pound force is equivalent to 4.448 N. A change in temperature of 1°F is equivalent to a change of 0.556°C.

The most common unit of energy is the *British thermal unit*, or BTU. One BTU is defined as sufficient energy to raise the temperature of 1 lbm of water by 1°F at a starting temperature of 39.1°F, and is equivalent to 1055 J. At other starting temperatures, the amount of heat required varies slightly from 1 BTU, but this discrepancy is small and is ignored in this book. A unit of 1 quadrillion BTU (10^{15} BTU) is called a *quad*. The units BTU/second (equivalent to 1.055 kW) and BTU/hour (equivalent to 3.798 MW) can be used to measure power. Typical quantities associated with increasing orders of magnitude of BTU measurements are given in Table 1-4.

Number of BTUs	Typical Measurement
Thousand BTUs	Output from portable space heater in 1 h
Million BTUs	Annual per capita energy consumption of various countries
Billion BTUs	Annual energy consumption of an office park in the United States
Trillion BTUs	Total annual energy consumption of all railroads in the United States or European Union from train and locomotive movements
Quadrillion BTUs (quads)	Annual energy consumption of an entire country

TABLE 1-4   Orders of Magnitude of BTU Measurements and Associated Quantities

An alternative unit for power to the BTU/hour or BTU/second is the horsepower, which was developed in the late 1700s by James Watt and others as an approximate measure of the power of a horse. The inventors of the unit recognized that no two horses would have exactly the same output, but that a typical horse might be able to raise the equivalent of 33,000 lb by 1 ft of height in the time interval of 1 min, which is equivalent to 746 W.

In the United States, the unit used to measure the average efficiency of electric power production is a hybrid standard-metric unit known as the *heat rate*, which is the average amount of heat supplied in thermal power plants, in BTUs, needed to provide 1 kWh of electricity. For example, a heat rate of 3412 BTU/kWh is equivalent to 100% efficiency, and a more realistic efficiency of 32% would result in a heat rate of 10,663 BTU/kWh.

1-6-3   Units Related to Oil Production and Consumption

In some cases, the energy content of crude oil is used as a basis for making energy comparisons. A common measure of oil is a *barrel*, which is defined as 42 U.S. gallons of oil, or approximately the volume of a barrel container used to transport oil. (Note that the U.S. gallon is different from the *imperial gallon* used in Britain, one imperial gallon containing approximately 1.2 U.S. gallons.) A unit in common use for measuring the energy content of large quantities of energy resources is the *tonne of oil equivalent*, abbreviated "toe," which is the typical energy content of a tonne of oil. One barrel of oil has 0.136 toe of energy content, and one toe is equivalent to 42.6 GJ of energy. National and international energy statistics are sometimes reported in units of toe, ktoe (1000 toe), or mtoe (million toe), in lieu of joules; the International Energy Agency, for example, reports the energy balances of member countries in ktoe on its Website.

1-7   Summary

Modern energy supplies, whether in the form of electricity from the grid or petroleum-based fuels for road vehicles, have a profound influence on human society, not only in the industrialized countries but also in the industrializing and less developed countries. While humanity began to develop these systems at the dawn of recorded history, their evolution has greatly accelerated since the advent of the industrial revolution around the year 1800. Today, energy use in the various countries of the world is highly correlated with both the GDP and HDI value for that country, and as countries grow wealthier, they tend to consume more total energy and energy per capita. This situation creates a twin challenge for the community of nations, first of all to provide sufficient energy to meet growing demand and second, to reduce the emission of CO_2 to the atmosphere. There are two systems for measuring quantities of energy in use in the world, namely, the metric system, used by most countries, and the U.S. customary system used in the United States; some common units are joules or BTUs for energy or watts for power.

References

Energy Information Agency (2014). *International Energy Outlook 2014*. Web-based resource, available at www.eia.gov. Accessed June 17, 2015.

International Monetary Fund (2015). *World Economic Outlook Database*. IMF, Washington, DC.

Lorenzo, E. (1994). *Solar Electricity: Engineering of Photovoltaic Systems*. PROGENSA, Sevilla, Spain.

Marland, G.,T. Boden, and R. Andres (2003). "Global, Regional, and National Fossil Fuel CO_2 Emissions." Report, Carbon Dioxide Information Analysis Center, Oak Ridge National Laboratory, U.S. Department of Energy, available on line at http://cdiac .esd.ornl.gov/trends/emis/meth_reg.htm.

Population Reference Bureau (2014). *Population world data sheet*. Web-based resource, available at www.prb.org. Accessed June 17, 2015.

United Nations Dept. of Economics and Social Affairs (2006). *World Population Prospects*. United Nations, New York.

Further Reading

Devall, B., and G. Sessions (1985). *Deep Ecology: Living as if Nature Mattered*. Peregrine Smith Books, Salt Lake City, UT.

Energy Information Administration (2015). *Annual Energy Outlook 2015*. Web-based resource, available at www.eia.gov. Accessed June 16, 2015.

Hinrichs, R., and M. Kleinbach (2002). *Energy: Its Use and the Environment*. Brooks-Cole, Toronto.

Hoffert, M., K. Caldeira, G. Benford, et al. (2002). "Advanced technology paths to global climate stability: Energy for a greenhouse planet." *Science*, Vol. 298, pp. 981–987.

Merchant, C. (1992) *Radical Ecology: The Search for a Livable World.* Routledge, New York.

Priest, J. (1979). *Energy for a Technological Society*, 2nd ed. Addison-Wesley, Reading, MA.

Smith, H. (1936). "The Origin of the Horsepower Unit." *American Journal of Physics*, Vol. 4, No. 3, pp. 120–122.

Temple, R. (1986). *The Genius of China: 3000 Years of Science, Discovery, and Invention*. Touchstone Books, New York.

Tester, J., E. Drake, M. Driscoll, et al. (2012). *Sustainable Energy: Choosing Among Options*, 2nd ed. MIT Press, Cambridge, MA.

UN Development Program (2006). *Human Development Report 2006*. United Nations, New York.

Exercises

1-1. Use the Internet or other resources to chart the development of an energy technology, from its earliest beginnings to the present day. Did the roots of this technology first emerge prior to the start of the industrial revolution? If so, how? If not, when did the technology first emerge? In what ways did the industrial revolution accelerate the growth of the technology? More recently, what has been the impact of the information age (e.g., computers, software, electronically controlled operation, and the Internet, and the like) on the technology?

1-2. For a country of your choice, obtain time series data for total energy consumption, CO_2 emissions, population, and GDP and/or HDI. If possible, obtain data from 1980 to the most recent year available, or if not, obtain some subset of these yearly values. These data can be obtained from the U.S. Energy Information Agency (www.eia.gov), the International Energy Agency (www.iea.org), or other source. Compare the trend for this country for measures such as energy use per capita and GDP earned per unit of energy to that of the United States or China. In what ways is your chosen country similar to the United States and/or China? In what ways is it different?

1-3. The country of Fictionland has a population of 31 million and consumes on average 12.09 exajoule (EJ), or 11.46 quads, of energy per year. The life expectancy of Fictionland is 63 years, and the GDP per capita, on a PPP basis, is $13,800. The adult literacy rate is 75%. The eligible and actual student enrollments for primary, secondary, and college/university levels of education are given in the table below:

	Eligible	Enrolled
Primary	2,500,000	2,375,000
Secondary	2,100,000	1,953,000
University	1,470,000	558,600

   a. Calculate the HDI for Fictionland.
   b. How does Fictionland's HDI to energy intensity ratio compare to that of the countries in the scatter charts in the chapter? Is it above, below, or on a par with these other countries?

1-4. Regression analysis of population, economic, and environmental data for countries of the world. For this exercise, download from the Internet or other data source values for the population, GDP in either unadjusted or PPP form, energy consumption, and land surface area of as many countries as you can find. Then answer the following questions:
   a. From the raw data you have gathered, create a table of the countries along with their GDP per capita, energy use per capita, and population density in persons per square kilometer or square mile.
   b. In part (a), did your data source allow you to include figures for all three measures for all the major countries of all the continents of the world? If not, what types of countries was it not possible to include, and why do you suppose this might be the case?
   c. Using a spreadsheet or some other appropriate software, carry out a linear regression analysis of energy consumption per capita as a function of GDP per capita. Produce a scatter chart of the values and report the R^2 value for the analysis.
   d. One could also speculate that population density will influence energy consumption, since a densely populated country will require less energy to move people and goods to where they are needed. Carry out a second regression analysis of energy consumption per capita as a function of population density. Produce a scatter chart of the values and report the R^2 value for the analysis.
   e. Discussion: Based on the R^2 value from parts (c) and (d), how well do GDPpc and population density predict energy consumption? What other independent variables might improve the model? Briefly explain.
   f. Given the global nature of the world economy, what are some possible flaws in using energy consumption figures broken down by country to make statements about the relative energy consumption per capita of different countries?

1-5. According to the U.S. Department of Energy, in 2005 the United States' industrial, transportation, commercial, and residential sectors consumed 32.1, 28.0, 17.9, and 21.8 quads of energy, respectively. What are the equivalent amounts in EJ?

1-6. From Fig. 1-8, the energy consumption values in 2000 for the United States, Japan, China, and India are 104, 23.7, 40.9, and 14.3 EJ, respectively. What are these same values converted to quads?

1-7. Also for 2000, the estimated total CO_2 emissions for these four countries were 5970 MMT, 1190 MMT, 2986 MMT, and 1048 MMT, respectively. Create a list of the four countries, ranked in order of decreasing carbon intensity per unit of energy consumed. Give the units in either tonnes CO_2 per terajoule TJ or tons per billion BTU consumed.

1-8. Convert the energy consumption values for the countries of Australia, Brazil, Israel, Portugal, and Thailand from Table 1-1 into units of million toe.

1-9. Use the description of the derivation of the horsepower unit by James Watt and others to show that 1 hp = 746 W, approximately.

CHAPTER 2

Systems and Policy Tools

2-1 Overview

The goal of this chapter is to present and explore a number of systems tools that are useful for understanding and improving energy systems. We first discuss the difference between conserving existing energy sources and developing new ones, as well as the contemporary paradigm of sustainable development. Next, we introduce the terminology for understanding systems and the systems approach to implementing an energy technology or project. The remainder of the chapter reviews specific systems tools. Some are qualitative or "soft" tools, such as conceptual models for describing systems or the interactions in systems. Other tools are quantitative, including exponential and asymptotic growth curves, multi-criteria analysis, optimization, Divisia analysis, and Monte Carlo simulation.

2-2 Introduction

Energy systems are complex, often involving combinations of thermal, mechanical, and electrical energy, which are used to achieve one or more of several possible goals, such as electricity generation, climate control of enclosed spaces, propulsion of transportation vehicles, and so on. Each energy system must interact with input from human operators and with other systems that are connected to it; these inputs may come from sources that are distributed over a wide geographic expanse. Energy systems therefore lend themselves to the use of a systems approach to problem solving.

When engineers make decisions about energy systems, in many instances they must take into account a number of goals or criteria that are local, regional, or global in nature. These goals can be grouped into three categories:

1. *Physical goals:* Meeting physical requirements that make it possible for the system to operate. For example, all systems require primary energy resources that power all end uses. For renewable energy systems, this means the wind, sun, tide, water, and the like, from which mechanical or electrical energy is extracted. For nonrenewable energy systems, this means the fossil fuels, fissile materials for nuclear fission, and the like. Any type of energy system must function efficiently, reliably, and safely.

2. *Financial goals:* Monetary objectives related to the energy system. For a privately held system (small-scale renewable energy system, electric generating system in a factory or mill, and so on), the goal may be to reduce expenditures on the purchase of energy products and services, for example, from an electric or gas supplier. For a commercial energy system, the goal may be to return profits to shareholders by generating energy products that can be sold in the marketplace.

3. *Environmental goals:* Objectives related to the way in which the energy system impacts the natural environment. Regional or global impacts include the emissions of greenhouse gases that contribute to climate change, air pollutants that degrade air quality, and physical effects from extracting resources used either for materials (e.g., metals used in structural support of a solar energy system) or energy (e.g., coal, uranium, and wood). Goals may also address impacts in the physical vicinity of the energy system, such as noise and vibrations, thermal pollution of surrounding bodies of water, the land footprint of the system, or disruption of natural habitats.

Depending on the size of the energy system, not all objectives will be considered in all cases. For example, a private individual or small private firm might only consider the financial goals in detail when changing from one energy system to another. On the other hand, a large-scale centralized electric power plant may be required by law to consider a wide range of environmental impacts, and in addition be concerned with the choice of primary energy resource and the projections of costs and revenues over a plant lifetime that lasts for decades.

2-2-1   Conserving Existing Energy Resources versus Shifting to Alternative Resources

Physical, financial, and environmental objectives can be achieved by choosing from a wide range of energy options. In general, these options can be grouped into one of four categories:

1. *Conserving through personal choice:* An individual or group (business, government agency, or other institution) chooses to use *existing* energy technology options in a different way in order to achieve objectives. Turning down the thermostat in the winter or choosing to forego an excursion by car in order to reduce energy consumption are both examples of this option.

2. *Conserving by replacing end-use technology:* The primary energy source remains the same, but new technology is introduced to use the energy resource more efficiently. For example, instead of turning down the thermostat, keeping it at the same temperature setting but improving the insulation of a building so that it loses heat more slowly, whereby less energy is required to keep the building at the same temperature. Switching to a more efficient automobile (conserves gasoline or diesel fuel) is another example.

3. *Conserving by replacing energy conversion technology:* Continue to rely on the same primary source of energy (e.g., fossil fuels), but conserve energy by replacing the energy conversion device with a more efficient model. For example, replace a furnace or boiler with a more efficient current model (conserves natural gas or heating oil).

4. *Replacing existing energy sources with alternatives:* Total amount of end-use energy consumption is not changed, but a change is made at the energy source. For example, an organic food processing plant might switch to a "green" electricity provider (one that generates electricity from renewable sources and sells it over the grid) in order to reduce the environmental impact of its product and promote an environmentally friendly image. In principle, this change can be made without addressing the amount of electricity consumed by the plant: whatever amount of kWh/month was previously purchased from a conventional provider is now provided by the green provider.

The definition of "energy conservation" used here is a broad interpretation covering both technological and behavioral changes, as in options 1 and 2 above. In other contexts, the term "conservation" may be used to refer strictly to changing habits to reduce energy (option 1 above), or strictly to upgrading technologies that affect consumption of conventional energy resources (options 2 and 3 above). Also, the three options are not mutually exclusive, and can be used in combination to achieve energy goals. For example, a homeowner might upgrade a boiler and insulation, turn down thermostat settings, and purchase some fraction of their electricity from a green provider, in order to reduce negative consequences from energy use.

A useful concept in considering steps to reduce the effects of energy use or other activities that affect the environment is the distinction between *eco-efficiency* and *eco-sufficiency*. Eco-efficiency is the pursuit of technologies or practices that deliver more of the service that is desired per unit of impact, for example, a vehicle that drives further per kilogram of CO_2 emitted. Eco-sufficiency is the pursuit of the combination of technologies and their amount of use that achieves some overall target for maximum environmental impact. For example, a region might determine a maximum sustainable amount of CO_2 emissions from driving, and then find the combination of more efficient vehicles and programs to reduce kilometers driven that achieves this overall goal. In general, eco-sufficient solutions are also eco-efficient, but eco-efficient solutions may or may not be eco-sufficient.

2-2-2   The Concept of Sustainable Development

One of the best-known concepts at the present time that considers multiple goals for the various economic activities that underpin human society, among them the management of energy systems, is "sustainable development." The sustainable development movement is the result of efforts by the World Commission on Environment and Development (WCED) in the 1980s to protect the environment and at the same time eradicate poverty. The report that resulted from this work is often referred to as the Brundtland report, after its chair, the Norwegian prime minister and diplomat Gro Harlem Brundtland. According to the report,

> "Sustainable development is development that meets the needs of the present without compromising the ability of future generations to meet their own needs."
>
> (WCED, 1987, p. 43)

The Brundtland report has influenced the approach of the community of nations to solving environmental problems in the subsequent years. Within its pages, it considers many of the same goals that are laid out for energy systems at the beginning of this chapter. Three goals stand out as being particularly important:

1. *The protection of the environment:* The Brundtland report recognized that the growing world population and increased resource and energy consumption were adversely affecting the natural environment in a number of ways, which needed to be addressed by the community of nations.

2. *The rights of poor countries to improve the well-being of their citizens:* This goal is implied by the words "to meet present needs" in the quote above. The report recognized that all humans have certain basic needs, and that it is the poor for whom these needs are most often not being met. The report then outlined a number of concrete steps for addressing poverty and the quality of life in poor countries.

3. *The rights of future generations:* According to the report, not only did all humans currently living have the right to a basic quality of life, but the future generations of humanity also had the right to meet their basic needs. Therefore, a truly sustainable solution to the challenge of environmentally responsible economic development could not come at the expense of our descendents.

These three components are sometimes given the names *environment, equity,* and *intertemporality.* The effect of the sustainable development movement was to raise environmentalism to a new level of awareness. Advocates of environmental protection, especially in the rich countries, recognized that the environment could not be protected at the expense of human suffering in poor countries. These countries were guaranteed some level of access to resources and technology, including energy technology, which could help them to improve economically, even if these technologies had some undesirable side effects. It was agreed that the rich countries would not be allowed to "kick the ladder down behind themselves," that is, having used polluting technologies to develop their own quality of life, they could not deny other countries access to them. On the other hand, the Brundtland report sees both poor and rich countries as having a responsibility to protect the environment, so the poor countries do not have unfettered access to polluting technologies. The report does not resolve the question of what is the optimal balance between protecting the environment and lifting the poor countries out of poverty, but it does assert that a balance must be struck between the two, and this assertion in itself was an advance.

The rise of the sustainable development movement has boosted efforts to protect and enhance the natural environment by invoking a specific objective, namely, that the environment should be maintained in a good enough condition to allow future generations to thrive on planet Earth. National governments were encouraged to continue to monitor present-time measures of environmental quality, such as the concentration of pollutants in air or water, but also to look beyond these efforts in order to consider long-term processes whose consequences might not be apparent within the time frame of a few months or years, but might lead to a drastic outcome over the long term. Indeed, many of the possible consequences of increased concentration of CO_2 in the atmosphere are precisely this type of long-term, profound effects.

Sustainable Development in the Twenty-First Century

Since its inception in the 1980s, many people have come to divide the term sustainable development into three "dimensions" of sustainability:

1. *Environmental sustainability:* Managing the effects of human activity so that they do not permanently harm the natural environment.

2. *Economic sustainability:* Managing the financial transactions associated with human activities so that they can be sustained over the long term without incurring unacceptable human hardship.

3. *Social/cultural sustainability:* Allowing human activity to proceed in such a way that social relationships between people and the many different cultures around the world are not adversely affected or irreversibly degraded.

Some people use variations on this system of dimensions, for example, by modifying the names of the dimensions, or splitting social and cultural sustainability into two separate dimensions. In all cases, the intent of the multidimensional approach to sustainable development is that in order for a technology, business, community, or nation to be truly sustainable, they must succeed in all three dimensions. For example, applied to energy systems, this requirement means that a renewable energy system that eliminates all types of environmental impacts but requires large subsidies from the government to maintain its cash flow is not sustainable, because there is no guarantee that the payments can be maintained indefinitely into the future. Similarly, a program that converts a country to completely pollution-free energy supply by confiscating all private assets, imposing martial law, and banning all types of political expression might succeed regarding environmental sustainability, but clearly fails regarding social sustainability!

The business world has responded to the sustainable development movement by creating the "triple bottom line" for corporate strategy. In the traditional corporation, the single objective has been the financial "bottom line" of generating profits that can be returned to investors. The triple bottom line concept keeps this economic objective, and adds an environmental objective of contributing to ecological recovery and enhancement, as well as a social objective of good corporate citizenship with regard to employees and the community. Thus the triple bottom line of economics, environment, and society is analogous to the three dimensions of sustainable development. A growing number of businesses have adopted the triple bottom line as part of their core practice.

2-3   Fundamentals of the Systems Approach

2-3-1   Initial Definitions

In Sec. 2-2, we have introduced a number of concepts that support a systems-wide perspective on energy. We have discussed the existence of multiple goals for energy systems. We have also considered how the presence of different stakeholders that interact with each other (such as the rich countries and the poor countries) or influence one another (such as the influence of the current generation on the future generations), favors thinking about energy from a systems perspective. The following definitions are useful as a starting point for thinking about systems.

First, what is a *system*? A system is a group of interacting *components* that work together to achieve some common *purpose*. Each system has a *boundary*, either physical or conceptual, and entities or objects that are not part of the system lie outside the boundary (see example of a system in Fig. 2-1). The area outside the boundary is called the *environment*, and *inputs* and *outputs* flow across the boundary between the system and its environment. The inputs and outputs may be physical, as in raw materials and waste products, or they may be virtual, as in information or data. Also, each component within a system can (usually) itself be treated as a system (called a *subsystem*), unless it

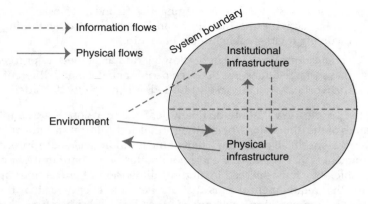

Figure 2-1 Conceptual model of built infrastructure system, surrounded by natural environment.
The built infrastructure system consists of all physical infrastructure built and controlled by humans. It has two components, institutional infrastructure and physical infrastructure. The institutional infrastructure represents the functioning of human knowledge in a conceptual sense; it receives information flows from the environment and the physical infrastructure about the current condition of each, and then transmits information to the physical infrastructure, usually in the form of commands meant to control the function of the physical infrastructure. The boundary shown is conceptual in nature; in physical space, the human physical infrastructure and natural environment are interwoven with each other. [*Source:* Vanek (2002).]

is a fundamental element that cannot be further divided into components. Furthermore, each system is (usually) a component in a larger system. Take the case of the electric grid. Each power plant is a subsystem in the grid system, and the power plant can be further subdivided into supporting subsystems, such as the boiler, turbine, generator, and pollution control apparatus, and the like. Going in the other direction, the broadest possible view of the grid system is the complete set of all national grids around the globe. Even if they are not physically connected, the various regional grids of the world are related to one another in the world because they draw from the same world fossil fuel resource for the majority of their electricity generation. One could conceivably argue that the world grid is a subsystem in the system of all electric grids in the universe, including those on other planets in other galaxies that support life forms that have also developed electrical grids, but such a definition of system and subsystem would serve no practical purpose (!). Failing that, the world grid system is the broadest possible system view of the grid.

In the field of systems engineering, a distinction is made between the term "system" as a concept and the *systems approach*, which is the process of conceptualizing, modeling, and analyzing objects from a systems perspective in order to achieve some desired outcome. When implementing the systems approach, the purpose, components, and boundary of a system are usually not preordained, but rather defined by the person undertaking the analysis. This definition process can be applied to physical systems, both in a single physical location (e.g., power plant) or distributed over a large distance (the electrical grid). Systems can also be purely conceptual, or mixtures of physical and conceptual, for example, the energy marketplace system with its physical assets, energy companies, government regulators, and consumers.

A system can be "abstracted" to whatever level is appropriate for analysis. For example, a large central power plant may, as a system, consist of mechanical, electrical/ electronic, safety, and pollution control subsystems. If an engineer needs to solve a problem that concerns primarily just one of these subsystems, it may be a legitimate and useful simplification to create an abstract model of the power plant system as being a single system in isolation. In other words, for electrical and electronics systems, the plant could be treated as consisting only of an electrical network and computer network. This abstracting of the system can be enacted provided that by doing so no critical details are lost that might distort the results of the analysis.

It can be seen from the power plant example that two different meanings for the term "subsystem" are possible. On the one hand, each major component of the power plant is a subsystem, such as the turbine, which has within its boundaries mechanical components (axis, blades, and so on), electrical components (sensors, wires, and so on), and components from other systems. Other subsystems such as the boiler, generator, and so on could be viewed in the same way. This definition of subsystem is different from isolating the entire electrical subsystem in all parts of the plant, including its presence in the turbine, boiler, generator, and all other areas of the plant, as a subsystem of the plant. Neither definition of subsystems is "correct" or "incorrect"; in the right context, either definition can be used to understand and improve the function of the plant as a whole.

In regard to identifying a boundary, components, and purpose of a system, there are no absolute rules that can be followed in order to implement the systems approach the *right* way. This approach requires the engineer to develop a sense of judgment based on experience and trial-and-error. In many situations, it is, in fact, beneficial to apply the systems model in more than one way (e.g., first setting a narrow boundary around the system, then setting a broad boundary) in order to generate insight and knowledge from a comparison of the different outcomes. It may also be useful to vary the number and layers of components recognized as being part of the system, experimenting with both simpler and more complex definitions of the system.

2-3-2   Steps in the Application of the Systems Approach

The engineer can apply the systems approach to a wide range of energy topics, ranging from specific energy projects (e.g., a gas-fired power plant or wind turbine farm) to broader energy programs (e.g., implementing a policy to subsidize energy efficiency and renewable energy investments). The following four steps are typical:

1. *Determine stakeholders:* Who are the individuals or groups of people who have a stake in how a project or program unfolds? Some possible answers are customers, employees, shareholders, community members, or government officials. What does each of the stakeholders want? Possible answers include a clean environment, a population with adequate access to energy, a return on financial investment, and so on.

2. *Determine goals:* Based on the stakeholders and their objectives, determine a list of goals, such as expected performance, cost, or environmental protection. Note that not all of the objectives of all the stakeholders need to be addressed by the goals chosen. The goals may be synergistic with each other, as is often the case with technical performance and environmental protection—technologies that perform the best often use fuel most efficiently and create the smallest

waste stream. The goals may also be in conflict with each other, such as level of technical performance and cost, where a high-performing technology often costs more.

3. *Determine scope and boundaries:* In the planning stage, decide how deeply you will analyze the situation in planning a project or program. For example, for the economic benefits of a project, will you consider only direct savings from reduced fuel use, or will you also consider secondary economic benefits to the community of cleaner air stemming from your plant upgrade? It may be difficult to decide the scope and boundaries with certainty at the beginning of the project, and the engineer should be prepared to make revisions during the course of the project, if needed (see step 4).

4. *Iterate and revise plans as project unfolds:* Any approach to project or program management will usually incorporate basic steps such as initial high-level planning, detail planning, construction or deployment, and launch. For a large fixed asset such as a power plant, "launch" implies full-scale operation of the asset over its revenue lifetime; for a discrete product such as a microturbine or hybrid car, "launch" implies mass production. Specific to the systems approach is the deliberate effort to regularly evaluate progress at each stage and, *if it is judged necessary,* return to previous stages in order to make corrections that will help the project at the end. For example, difficulties encountered in the detail planning stage may reflect incorrect assumptions at the initial planning stage, so according to the systems approach one should revisit the work of the initial stage and make corrections before continuing. This type of *feedback* and *iteration* is key in the systems approach: in many cases, the extra commitment of time and financial resources in the early stages of the project is justified by the increased success in the operational or production stages of the life cycle.

The exact details of these four steps are flexible, depending in part on the application in question, and the previous experience of the practitioner responsible for implementing the systems approach. Other steps may be added as well.

The Systems Approach Contrasted with the Conventional Engineering Approach

The opposite of the systems approach is sometimes called the *unit approach*. Its central tendency is to identify one component and one criterion at the core of a project or product, generate a design solution in which the component satisfies the minimum requirement for the criterion, and then allow the component to dictate all other physical and economic characteristics of the project or product. For a power plant, the engineer would choose the key component (e.g., the turbine), design the turbine so that it meets the criterion, which in this case is likely to be technical performance (total output and thermal efficiency), and then design all other components around the turbine.

In practice, for larger projects it may be impossible to apply the unit approach in its pure form, because other criteria may interfere at the end of the design stage. Continuing the example of the power plant, the full design may be presented to the customer only to be rejected on the grounds of high cost. At that point the designer iterates by going back to the high-level or detail design stages and making revisions. From a systems perspective, this type of iteration is undesirable because the unproductive work embedded in the detail design from round 1 could have been avoided with sufficient thinking about high-level design versus cost at an earlier stage. In practice, this type of pitfall is well recognized by major engineering organizations, which are certain to carry out some type of systems approach at the early stages in order to avoid such a financial risk.

While the systems approach may have the advantages discussed, it is not appropriate in all situations involving energy. The systems approach requires a substantial commitment of "administrative overhead" beyond the unit approach in the early stages in gathering additional information and considering trade-offs. Especially in smaller projects, this commitment may not be merited. Take the case of an electrician, who is asked to wire a house once the customer has specified where they would like lights and electrical outlets. The wiring requirements may be specified by the electrical code in many countries, so the electrician has limited latitude to optimize the layout of wires under the walls of the house. She or he might reduce the amount of wiring needed by using a systems approach and discussing the arrangement of lights, outlets, switches, and wires with the customer so as to reduce overall cost. However, the homeowner or architect has by this time already chosen the lighting and electrical outlet plan that they desire for the house, and they are likely not interested in spending much extra time revisiting this plan for saving a relatively small amount of money on wiring. In short, there may be little to learn from a systems approach to this problem. A "linear" process may be perfectly adequate, in which the electrician looks at the wiring plan, acquires the necessary materials, installs all necessary wire, and bills the customer or the general contractor at the end of her or his work, based either on a fixed contract price or on the cost of time and materials.

Examples of the Systems Approach in Action

The following examples from past and current energy problem solving reflect a systems approach. In some cases, the players involved may have explicitly applied systems engineering, while in others they may have used systems skills without even being aware of the concept (!).

Example 1   *Demand-side management.* Demand-side management (DSM) is an approach to the management of producing and selling electricity in which the electric utility actively manages demand for electricity, rather than assuming that the demand is beyond its control. Demand-side management was pioneered in the 1970s by Amory Lovins of the Rocky Mountain Institute, among others, and has played a role in energy policy since that time.

The distinction in energy decision making with and without DSM is illustrated in Fig. 2-2. The shortcoming for the utility in the original case is that system boundary is drawn in such a way that the customer demand for electricity is outside the system boundary. Not only must the utility respond to the customer by building expensive new assets in cases where consumer demand increases, but there is a chance that consumer demand will be highly peaked, meaning that some of the utility's assets only run at peak demand times and therefore are not very cost-effective to operate (they have a high fixed cost, but generate little revenue). This outcome is especially likely in a regulated electricity market where the consumer is charged the same price per kWh no matter what the time of day is or the total amount of demand on the system at the time.

In the DSM system, the boundary has been redrawn to incorporate consumer demand for electricity. The new input acting from outside the boundary is labeled "consumer lifestyle"; in the previous system this input did not appear, since the only aspect of the consumer seen by the system was her/his demand for electricity. The effect of

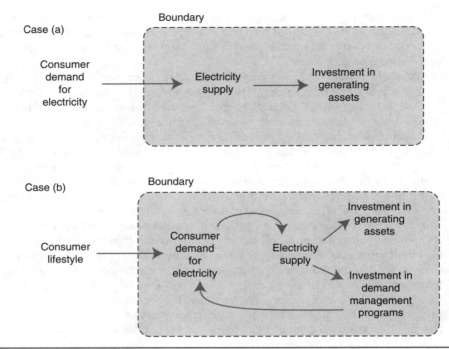

FIGURE 2-2   System boundaries and function: (a) conventional case, (b) case with DSM.

consumer lifestyle on the system is important, because it means that the consumer has the right to expect a certain quality of life, and the system cannot meet its needs by arbitrarily cutting off power to the customer without warning. Within the system, a new arrow has been added between demand and electric supply, representing the supplier's ability to influence demand by varying prices charged with time of day, encouraging the use of more efficient light bulbs and appliances, and so on. This tool is given the name "invest in DSM," and has been added to the preexisting tool of "invest in generating assets."

Example 2   *Combined energy production for electricity and transportation.* This example involves a comparison of the generation of electricity for distribution through the electric grid and liquid fuels (gasoline, diesel, and jet fuel) for transportation. Currently, power plants that generate electricity do not also provide fuels for transportation, and refineries used to produce liquid fuels do not also generate electricity for the grid in a significant way.

With the current demand for new sources of both electricity and energy for transportation, engineers are taking an interest in the possibilities of developing systems that are capable of supporting both systems. This transition is represented in Fig. 2-3. The "original" boundaries represent the two systems in isolation from each other. When a new boundary is drawn that encompasses both systems, many new solutions become available that were not previously possible when the two systems were treated in isolation. For example, an intermittent energy source such as a wind farm might sell electricity to the grid whenever possible, and at times that there is excess production, store energy for use in the transportation energy system that can be dispensed to

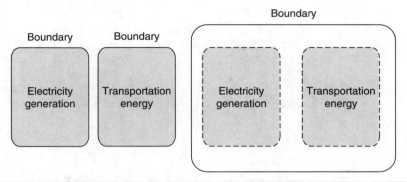

FIGURE 2-3   Electricity and transportation energy as two systems in isolation, and combined into a single larger system.

vehicles later. Plug-in hybrid vehicles might take electricity from the grid at night during off-peak generating times, thereby increasing utilization of generating assets and at the same time reducing total demand for petroleum-based liquid fuels. Also, new vehicle-based power sources such as fuel cells might increase their utilization if they could plug in to the grid and produce electricity during the day when they are stationary and not in use.

Example 3   *Modal shifting among freight modes.* In the transportation arena, an attractive approach to reducing energy consumption is shifting passengers or goods to more energy efficient types of transportation, also known as "modes" of transportation (see Chap. 14). This policy goal is known as *modal shifting.* In the case of land-based freight transportation, rail systems have a number of technical advantages that allow them to move a given mass of goods over a given distance using less energy than movement by large trucks.

How much energy might be saved by shifting a given volume of freight from truck to rail? We can make a simple projection by calculating the average energy consumption per unit of freight for a given country, or the total amount of freight moved divided by the total amount of energy consumed, based on government statistics. Subtracting energy per unit freight for rail from that of truck gives the savings per unit of modal shifting achieved. The upper diagram in Fig. 2-4 represents this view of modal shifting.

This projection, however, overstates the value of modal shifting because it does not take into account differences between the movement of high-value, finished goods and bulk commodities in the rail system. The latter can be packed more densely and moved more slowly than high-value products, which customers typically expect to be delivered rapidly and in good condition. The movement of high-value goods also often requires the use of trucking to make connections to and from the rail network, further increasing energy use. In the lower diagram in Fig. 2-4, the rail system is divided into two subsystems, one for moving high-value goods and one for moving bulk commodities. Connecting the modal shifting arrow directly to the high-value goods subsystem then makes the estimation of energy savings from modal shifting more accurate. A new value for average energy consumption of high-value rail can be calculated based on total movements and energy consumption of this subsystem by itself, and the difference between truck and rail recalculated. In most cases, there will still be energy savings to be had from modal shifting of freight, but the savings are not as great as in the initial calculation—and also more realistic.

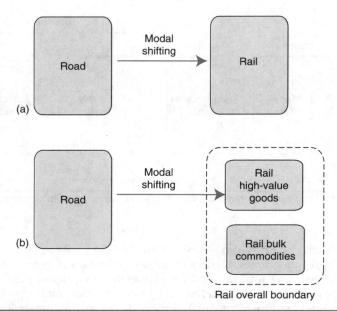

FIGURE 2-4   Two views of modal shifting of freight: (a) rail treated as uniform system, (b) rail divided into two subsystems.

2-3-3   Stories, Scenarios, and Models

The terms stories, scenarios, and models describe perspectives on problem solving that entail varying degrees of data requirements and predictive ability. Although these terms are widely used in many different research contexts as well as in everyday conversation, their definitions in a systems context are as follows. A *story* is a descriptive device used to qualitatively capture a cause-and-effect relationship that is observed in the real world in certain situations, without attempting to establish quantitative connections. A story may also convey a message regarding the solution of energy systems problems outside of cause-and-effect relationships. A *scenario* is a projection about the relationship between inputs and outcomes that incorporates quantitative data, without proving that the values used are the best fit for the current or future situation. It is common to use multiple scenarios to bracket a range of possible outcomes in cases where the most likely pathway to the future is too difficult to accurately predict. Lastly, a *model* is a quantitative device that can be calibrated to predict future outcomes with some measure of accuracy, based on quantitative inputs. Note that the term model used in this context is a *quantitative* model, as opposed to a *conceptual* model such as the conceptual model of the relationship between institutional infrastructure, physical infrastructure, and the natural environment in Fig. 2-1.

The relationship between the three terms is shown in Fig. 2-5. As the analysis proceeds from stories to scenarios to models, the predictive power increases, but the underlying data requirements also increase. In some situations, it may not be possible to provide more than a story due to data limitations. In others, the full range of options may be available, but it may nevertheless be valuable to convey a concept in a story in order to provide an image for an audience that can then help them to understand a

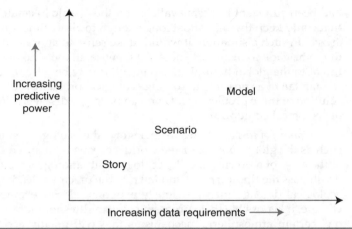

FIGURE 2-5   Predictive ability versus data requirements for stories, scenarios, and models.

scenario or model. In fact, in some situations, a modeling exercise fails because the underlying story was not clear or was never considered in the first place.

Examples of Stories in an Energy Context
The following examples illustrate the use of stories to give insight into our current situation with energy. When using stories like these, it is important to be clear that stories are devices that help us to conceptualize situations and problems, but that they are not necessarily the solutions to the problems themselves.

1. *A rising tide lifts all boats:* Improvements in the national or global economy will improve the situation for all people, and also make possible investments in environmental protection including cleaner energy systems. This story does not address the question of whether the improvements in well-being will be spread evenly or unevenly, or whether some "boats" will be left behind entirely. However, it describes with some accuracy the way in which economic resources of the wealthy countries (North America, Europe, and Japan) lifted newly industrialized countries such as Taiwan or Korea to the point where they have recently been spending more on environmental protection.

2. *Deck chairs on the Titanic* (also sometimes referred to as *string quartet on the Titanic*): If one system-wide goal of the world economy is to lift all countries through the interactions of a global marketplace, another goal is to be certain to address the most important problems so as to avoid any major setbacks. As in the case of the Titanic, there is no point in determining the best ways to arrange the deck chairs, or to perfect the performance of the string quartet on board the ship, if the threat of icebergs is not addressed. Similarly, for a given energy system or technology, there may be little value in finding a solution to some narrowly defined problem related to its performance, if it has fundamental problems related to long-term availability of fuel or greenhouse gas emissions.

3. *Bucket brigade on the Titanic:* The metaphor of the sinking of the Titanic is useful for another story. In the actual disaster in 1912, another ship in the vicinity arrived on the scene too late to rescue many of the passengers. Suppose there

had been sufficient buckets available on the Titanic to create a bucket brigade that could keep the ship afloat long enough to allow time for the rescue ship to arrive. In such a situation, it would make sense to allocate all available people to the brigade to insure its success. Likewise, in the situation with the current threat to the global natural environment, it may be wise to reallocate as much human talent as possible to solving ecological problems, so as to accelerate the solution of these problems and avert some larger disaster prior to the regaining of ecological equilibrium.

4. *The tipping point:* According to this story, the ecological situation of systems such as the global climate or the world's oceans is balanced on a tipping point at the edge of a precipice. If the ecological situation gets too far out of balance, it will pass the tipping point and fall into the precipice, leading to much greater ecological damage than previously experienced. For example, in the case of climate, it is possible that surpassing certain threshold levels of CO_2 will "turn on" certain atmospheric mechanisms that will greatly accelerate the rate of climate change, with severe negative consequences.

5. *The infinite rugby/American football game:* Many readers are familiar with the rules of the games rugby or American football, in which opposing teams attempt to move a ball up or down a field toward an "endline" in order to score points, within a fixed length of time agreed for the game. In this story, the ecological situation can be likened to a game in which the goal is to make forward progress against an opponent, but with no endline and no time keeping. The opponent in this game is "environmental degradation," and the "home team" is ourselves, the human race. If we are not successful against the opponent, then we are pushed further and further back into a state of environmental deterioration. Since there is no endline at our end of the field, there is the possibility that the environmental deterioration can continue to worsen indefinitely. However, if our team is successful, we can make forward progress to an area of the unbounded field where the ecological situation is well maintained. Since there is no time clock that runs out, the presence of environmental degradation as an opponent is permanent, and the commitment to the game must similarly be permanent. However, a successful strategy is one that gets us to the desirable part of the field and then keeps us there.

6. *The airplane making an emergency landing:* The current challenge regarding energy is likened to an airplane that is running out of fuel and must make an emergency landing. If the pilot is impatient and attempts to land the plane too quickly, there is a danger of crashing; however, if the pilot stays aloft too long, the plane will run completely out of fuel and also be in danger of crashing because it will be difficult to control. Similarly, draconian action taken too quickly in regard to changing the energy system may lead to social upheaval that will make it difficult for society to function, but if action is taken too slowly, the repercussions to the environment may also have a critical effect on the functioning of society.

It should be noted that some of these stories contradict each other. In story 4, the emphasis is on a point of no return, whereas in story 5, there is no tipping point on which to focus planning efforts. However, it is not necessary to subscribe to some stories and not to others. From a systems perspective, stories are tools that can be used in specific situations to gain insights. Thus it may be perfectly acceptable to use the tipping point story in one situation, and the rugby game story in another, as circumstances dictate.

An Example of Stories, Scenarios, and Models Applied to the "Boats in the Rising Tide"

It is possible to build scenarios and models based on many of the stories in the previous section. Take story 1, the rising tide lifting all boats. At the level of a story, we describe a connection between the growing wealth of the rich countries, the growing market for exports from the industrializing "Asian Tigers" (Taiwan, Korea, and so on), the growing wealth of these industrializing countries, and their ability to begin to invest in environmental protection. All of this can be done verbally or with the use of a flowchart, but it need not involve any quantitative data.

At the other extreme of gathering and using data, we might build an economic model of the relationship between the elements in the story. In this model, as wealth grows in the rich countries, imports also grow, leading to increased wealth in the industrializing countries and the upgrading of the energy supply to be cleaner and more energy efficient. The model makes connections using mathematical functions, and these are calibrated by adjusting numerical parameters in the model so that they correctly predict performance from the past up to the present. It is then possible to extrapolate forward by continuing to run the model into the future. If the rate of adoption of clean energy technology is unsatisfactory, the modeler may introduce the effect of "policy variables" that use taxation or economic incentives to move both the rich and industrializing countries more quickly in the direction of clean energy. Again, these policy variables can be calibrated by looking retrospectively at their performance in past situations.

While mathematical models such as these are powerful tools for guiding decisions related to energy systems, in some situations the connections between elements in the model may be too complex to capture in the form of a definitive mathematical function that relates the one to the other. As a fallback, scenarios can be used to create a range of plausible outcomes that might transpire. Each scenario starts from the present time and then projects forward the variables of interest, such as total economic growth, demand for energy, type of energy, technology used, and total emissions. A mathematical relationship may still exist between variables, but the modeler is free to vary the relationship from scenario to scenario, using her or his subjective judgment. Some scenarios will be more optimistic or pessimistic in their outcome, but each should be plausible, so that, when the analysis is complete, the full range of scenarios brackets the complete range of likely outcomes for the variables of interest. In the rising tide example, a range of scenarios might be developed that incorporate varying rates of economic growth among the emerging economies, the economic partners that import goods from these economies, and rates of improvement in clean energy technology, in order to predict in each scenario a possible outcome for total emissions in some future year. Also, a distinction might also be made between scenarios with and without the "policy intervention" to encourage clean energy sources.

2-3-4   Systems Approach Applied to the Scope of This Book: Energy/Climate Challenges Compared to Other Challenges

As the broad goals of sustainable development from the beginning of the chapter suggest, the challenge of sustainability is broader than the issues of energy and climate.

Globally, at least three other major issues face both the natural world and human society at the present time, as follows:

1. *Water issues:* Changes to the amount of water available (for instance due to droughts and floods) as well as the quality of water (pollution, fresh versus salt water) have an impact on agriculture and water use in the human built environment, and also the health of nonhuman species.

2. *Biodiversity:* Decline in population or extinction of some species may disrupt the function of the "web of life" in some ecosystems (e.g., the balance between predator and prey species), while negative consequences may come from the arrival of invasive pest species that have no predators in a region to which they are newly arrived.

3. *Land and food security:* Changes to the quality of land and soil (desiccation, loss of topsoil, eventual rise in sea level due to thermal expansion of water, and melting of glaciers) lead to threats to the adequacy of the food supply and again to nonhuman species health.

For the purposes of this book, it is reasonable to limit the scope of the book to energy and climate, while at the same time briefly exploring the impact of energy and climate on water/biodiversity/food, as shown in Fig. 2-6. In fact, this exercise exemplifies the "systems approach" in action: before moving on from the boundaries we have created,

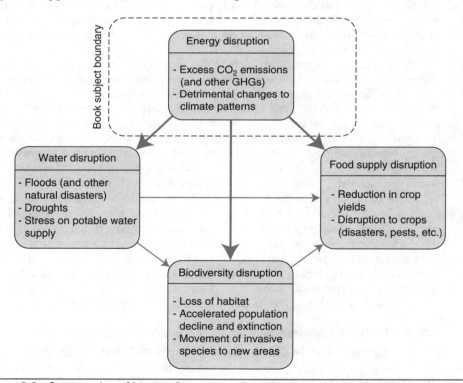

FIGURE 2-6 Systems view of impacts from energy disruptions on water, biodiversity, and food security.

presumably to work within them for the remainder of the book, we consider what impacts are generated when the effects of energy and climate trends impact on the other issues, and vice versa. It should be emphasized that issues besides energy and climate are excluded only to keep the scope of the book and list of prerequisite skills for the reader manageable. Each of the other three problems is critical in its own right and must be given due attention if society is to succeed in the goals of sustainable development.

In Fig. 2-6, "energy disruptions," shown at the top of the figure, have direct impact on the other three areas; thick arrows flow across the "book subject boundary" to the other three areas, signifying cross-boundary impacts. In the case of water, climate change leads to change in rainfall and temperature patterns that may result in the amount of water in a region being above or below historically observed ranges, resulting in flooding or drought. Total available water may also fall below what is needed for the sum of human needs in a given region, due to drought or possibly to weather-related disasters that affect water infrastructure, leading to stress on the water supply. In the case of biodiversity, climate-related changes in average temperature may threaten some species or allow others to move into regions that they could not previously inhabit, causing impact on both the natural world and human-controlled agriculture. In the case of food security, a warmer climate may reduce crop yields. Figure 2-6 shows secondary impacts emanating from water and biodiversity (thin as opposed to thick arrows); for example, as climate change stresses the water supply, lack of water in turn stresses food security.

We can now extend the scoping exercise by exploring what would happen if we redrew the boundaries of the book to include water in addition to energy/climate, as shown in Fig. 2-7. The title "water disruption" at the left side is now capitalized, and the energy/climate title is not, to signify that the starting point for disruptions is water.

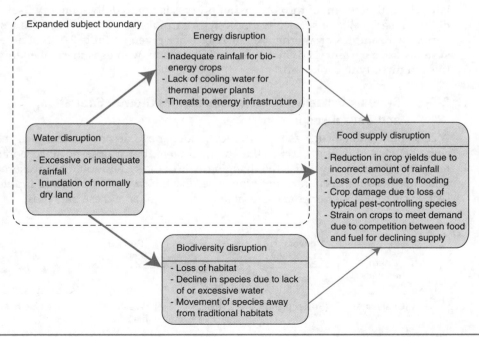

FIGURE 2-7   Systems view of impacts from water disruptions on energy, biodiversity, and food security.

Also, the other three titles (energy/biodiversity/food) remain the same but the descriptions populating the lists change to reflect the new source of challenges. For instance, other developments besides climate change, such as human population growth or the expectation of a higher standard of living around the world, can stress water supplies directly. When this situation arises, the energy system may be disrupted due to lack of water for bio-energy crops or for cooling in thermal power systems (coal, nuclear, concentrating solar power). Other observations, for example those under food security, may be unchanged regardless of whether the starting point for disruption is energy or water. While acknowledging that the scope of this book does not allow us to develop these relationships here, we can recognize through this second step in the scoping exercise that we should be aware of the impact of outside developments on energy as well as the other way around (water on energy as well as energy on water, as in this case).

2-4   Other Systems Tools Applied to Energy

One prominent feature of human use of energy is that energy resources must often be substantially converted from one form to another before they can be put to use. This conversion can be captured in a conceptual model by considering various energy "currencies." These are intermediate energy products that must be produced by human effort for eventual consumption in energy end uses.

The steps in the model are shown in Fig. 2-8. The initial input is an energy resource, such as a fossil fuel. At the conversion stage, a conversion technology is used to create a "currency" from the energy resource. Although the conversion to a currency inevitably incurs losses, these losses are sustained in order to gain the benefits of transmitting or storing a currency, which is easier to handle and manipulate. The ultimate aim of the conversion process is not to deliver energy itself but to deliver a service, so a service technology is employed to create a service from the currency. Examples of alternative currencies are given in Table 2-1. Not all energy systems use a currency. For instance, a solar cooking oven uses thermal energy from the sun directly to cook food.

2-4-1   Systems Dynamics Models: Exponential Growth, Saturation, and Causal Loops

A conceptual model such as the currency model above describes functional relationships between different parts of the energy system, but it does not incorporate any temporal dimension by which the changing state of the system could be measured over time, nor does it incorporate any type of feedback between outputs and inputs. This section presents time-series and causal loop models to serve these purposes.

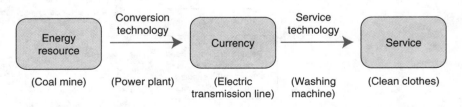

FIGURE 2-8   Currency model of energy conversion.

Energy Resource	Conversion Technology	Currency	Service Technology	Service (Typical Example)
Coal	Power plant	Electricity	Washing machine	Washing clothes
Crude oil	Refinery	Gasoline	Car with internal combustion engine	Transportation
Natural gas	Steam reforming plant	Hydrogen	Car with hydrogen fuel cell stack	Transportation

Note: The pathways to meeting the service needs are not unique; for example, the need for clothes washing could also be met by hand washing, which does not require the use of a currency for the mechanical work of washing and wringing the clothes.

TABLE 2-1   Examples of Energy Currencies

Exponential Growth

In the case of energy systems, exponential growth occurs during the early stages of periods of influx of a new technology, or due to a human population that is growing at an accelerating rate. Exponential growth is well known from compound interest on investments or bank accounts, in which an increasing amount of principal accelerates the growth from one period to the next. Exponential growth in energy consumption over time can be expressed as follows:

$$E(t) = a \cdot \exp(bt) \tag{2-1}$$

where t is the time in years, $E(t)$ is the annual energy consumption in year t, and a and b are parameters that fit the growth to a historic data trend. Note that a may be constrained to be the value of energy consumption in year 0, or it may be allowed other values to achieve the best possible fit to the observed data. Example 2-1 illustrates the application of the exponential curve to historical data, where a and b are allowed to take best-fit values.

Example 2-1   The table below provides annual energy consumption values in 10-year increments for the United States from 1900 to 2000. Use the exponential function to fit a minimum-squares error curve to the data, using data points from the period 1900 to 1970 only. Then use the curve to calculate a projected value for the year 2000, and calculate the difference in energy between the projected and actual value.

Year	Actual
1900	9.6
1910	16.6
1920	21.3
1930	23.7
1940	25.2
1950	34.6
1960	45.1
1970	67.8
1980	78.1
1990	84.7
2000	99.0

Solution   For each year, the estimated value of the energy consumption $E_{est}(t)$ is calculated based on the exponential growth function and the parameters a and b. Let $t = y - 1900$, where y is the year of the data. Then E_{est}(year) is calculated as E_{est} (year) $= a \cdot \exp[b(\text{year} - 1900)]$.

The value of the squared error term $(\text{ERR})^2$ in each year is

$$(\text{ERR})^2 = [E_{actual} \, (\text{year}) - E_{est} \, (\text{year})]^2$$

Finding the values of a and b that minimize the sum of the errors for the years 1900 through 1970 gives $a = 10.07$ and $b = 0.02636$, as shown in the next table. For example, for the year 1970, we have

$$E_{est}(1970) = 10.07 \cdot \exp[0.02636(1970 - 1900)] = 63.7$$

$$(\text{ERR})^2 = (67.8 - 63.7)^2 = 17.0$$

Creating a table for actual energy, estimated energy, error, and square of error gives the following:

Year	Actual	Estimated	Error	(Error)²
1900	9.6	10.1	−0.5	0.2
1910	16.6	13.1	3.5	12.0
1920	21.3	17.1	4.3	18.4
1930	23.7	22.2	1.5	2.2
1940	25.2	28.9	−3.7	13.6
1950	34.6	37.6	−3.0	9.0
1960	45.1	49.0	−3.9	15.0
1970	67.8	63.7	4.1	17.0

Finally, solving for $year = 2000$ using Eq. (2-1) gives an estimated energy value of 140.5 quads. Thus the projection overestimates the 2000 energy consumption by $140.5 - 99.0 = 41.5$ quads, or 42%. The actual and estimated curves can be plotted as shown in Fig. 2-9.

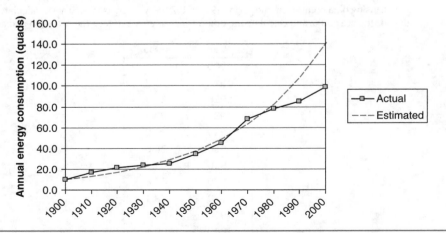

FIGURE 2-9   Comparison of actual and estimated energy curves.

Asymptotic Growth, Saturation, Logistics Function, and Triangle Function

The result from Example 2-1 shows that the predicted energy consumption value of 140.5 quads for the year 2000 is substantially more than the actual observed value. Compared to the prediction, growth in U.S. energy use appears to be slowing down. Such a trend suggests *asymptotic* growth, in which the growth rate slows as it approaches some absolute ceiling in terms of magnitude. Since the planet Earth has some finite carrying capacity for the number of human beings, and since there is a finite capacity for each human being to consume energy, it makes sense that there is also a limit on total energy consumption that is possible.

Just as the exponential function considers growth without impact of long-term limits, two other mathematical functions, the *logistics curve* and the *triangle function*, explicitly include both accelerating growth near the beginning of the model lifetime and slowing growth as the limits are approached near the end. Taking the first of these, the logistics function is an empirically derived formula that has been found to fit many observed real-world phenomena ranging from the penetration of a disease into a population to the conversion of a market over to a new technology (e.g., the penetration of personal computers into homes in the industrialized countries). Although it can be written in different ways, one possible form is the following:

$$f(t) = \frac{F \cdot e^{(c_1 + c_2 t)}}{1 + e^{(c_1 + c_2 t)}} \tag{2-2}$$

Here t is the value of time in appropriate units (e.g., years, months), $f(t)$ is the fraction of the population that has converted to the new technology at time t, F is the maximum fraction or penetration that is expected (i.e., $F = 100\%$ if the entire market is eventually expected to convert, or less if some fraction is expected to remain unconverted), and c_1 and c_2 are parameters used to shape the function so that it fits the historically observed rate of penetration.

In practical applications, the fraction of penetration in each time period t may be given along with the value of F, and the goal may be to predict when F is approximately reached. (Since the logistics curve approaches F asymptotically, it cannot be reached exactly; however, reaching a threshold within a few percentage points is often adequate.) The values of c_1 and c_2 should be chosen to minimize the error between the observed values and the fit curve, as was illustrated in Example 2-1.

Unlike the logistics function, where the value of $f(t)$ approaches but never actually achieves the limiting value F, the triangle function depends on the user to set in advance the time period at which 100% penetration or the cessation of all further growth will occur. In the triangle function, the value of f is the percentage relative to the completion of growth, which can be multiplied by an absolute quantity (e.g., maximum amount of energy at end of growth phase measured in EJ or quads) to compute the absolute value of penetration. Let a and b be the start and end times of the transition to be modeled. The triangle function then has the following form:

$$f(t) = 2\left(\frac{t-a}{b-a}\right)^2 \quad \text{if } a \le t \le \frac{a+b}{2}$$

$$f(t) = 1 - 2\left(\frac{b-t}{b-a}\right)^2 \quad \text{if } \frac{a+b}{2} \le t \le b \tag{2-3}$$

For times less than a, $f(t) = 0$, and for times greater than b, $f(t) = 1$.

In terms of the difference between the two functions, the logistics curve can be fit to the initial values to project when the population will get within some increment of full saturation, in forward-looking cases where the penetration is still relatively low (e.g., less than 10 or 15%) and the goal is to predict penetration in the future. It is also possible to use the logistics curve retrospectively to fit a curve to a technology penetration by adjusting the parameters c_1 and c_2 to minimize error. With the triangle function, the values of a and b are set at the endpoints (for instance, 30 years apart) in order to estimate the likely population at intermediate stages. Example 2-2 illustrates the application of both the logistics and triangle functions.

Example 2-2 Automobile production in a fictitious country amounts to 1 million vehicles per year, which does not change over time. Starting in the year 1980, automobiles begin to incorporate fuel injection instead of carburetors in order to reduce energy consumption. Compute future penetration of fuel injection as follows:

1. *Using the logistics function:* Suppose we treat 1980 as year 0, and that 5000 vehicles are produced with fuel injection in this year. In 1981, the number rises to 20,000 vehicles. The expected outcome is 100% saturation of the production of vehicles with fuel injection. How many vehicles are produced with fuel injection in 1983?

2. *Using the triangle function:* Suppose now that only the total time required for a full transition to fuel injection is known, and that this transition will be fully achieved in the year 1990. How many vehicles are produced with fuel injection in 1983, according to the triangle model?

Solution

1. *Logistics function:* Since the new technology is expected to fully penetrate the production of vehicles, we can set $F = 1 = 100\%$. In general, with several years in a historical curve we would use a numerical solver to find values of c_1 and c_2 that minimize error, but in this illustrative example there are two equations (for years 1980 and 1981) along with two unknown values of c_1 and c_2, so we can solve exactly. We first use an equation for 1980 to solve for c_1, and then solve for c_2 using an equation for 1981. In 1980, or year $t = 0$, since the penetration is 5000 out of 1 million, we have $f(0) = 0.5\%$. We can therefore solve the following equation for c_1 (note that since $t = 0$, c_2 cancels in the equation):

$$f(0) = \frac{1 \cdot \exp(c_1 + c_2 \cdot 0)}{1 + \exp(c_1 + c_2 \cdot 0)} = \frac{1 \cdot \exp(c_1)}{1 + \exp(c_1)} = 0.5\%$$

Solving gives $c_1 = -5.29$.[1] In 1981 at time $t = 1$, we have $f(1) = 2 \times 10^4 / 1 \times 10^6 = 2.0\%$, so we can now solve for c_2 in a similar manner:

$$f(1) = \frac{1 \cdot \exp(c_1 + c_2 \cdot t)}{1 + \exp(c_1 + c_2 \cdot t)} = \frac{1 \cdot \exp(-5.29 + c_2 \cdot 1)}{1 + \exp(-5.29 + c_2 \cdot 1)} = 2.0\%$$

[1]It is assumed here and in example problems throughout this book (in the interest of conserving space) that the reader is familiar with techniques either in spreadsheets or in computational software products for setting up and solving equations or systems of equations with one or more unknowns, without needing to present the algebraic manipulation necessary to solve for, e.g., in this case c_1, in a symbolic form.

Solving gives $c_2 = 1.40$. Lastly, we solve for the case of 1983, or $t = 3$:

$$f(3) = \frac{1 \cdot \exp(c_1 + c_2 \cdot t)}{1 + \exp(c_1 + c_2 \cdot t)} = \frac{1 \cdot \exp(-5.29 + 1.4 \cdot 3)}{1 + \exp(-5.29 + 1.4 \cdot 3)} = 25.2\%$$

Since $f(3) = 0.252$, the prediction is that $0.252(1 \times 10^6) = 252{,}000$ cars will be made with fuel injection in 1983.

2. *Triangle function:* We adopt the same convention as in part 1, namely, $t = 0$ in 1980, $t = 1$ in 1981, and so forth. Therefore, $a = 0$, $b = 10$, and $(a + b)/2 = 5$. Since $t = 3 < (a + b)/2$, we use $p(t) = p(3) = 2[(t - a)/(b - a)]^2$, which gives $p(3) = 0.18$ and a prediction that 180,000 cars are built with fuel injection in 1983.

The two curves from Example 2-2 are plotted in Fig. 2-10. Notice that the logistics function in this case has reached 85% penetration in the year 1985, when the triangle function is at 50%, despite having very similar curves for 1980–1982. A triangle curve starting in 1980 and ending in 1987 ($a = 0$, $b = 7$) would have a more similar shape to the logistics function.

Causal Loop Diagrams
The cause-and-effect relationships underlying the trend in the fuel injection influx example can be described as follows. After the initial start-up period (years 1980 and 1981), the industry gains experience and confidence with the technology, and resistance to its adoption lessens. Therefore, in the middle part of either of the curves used, the rate of adoption accelerates (1982–1985). As the end of the influx approaches, however, there may be certain models which are produced in limited numbers or which have other complications, such that they take slightly longer to complete the transition. Thus the rate of change slows down toward the end. Note that the transition from carburetors to fuel injection is not required to take the transition path

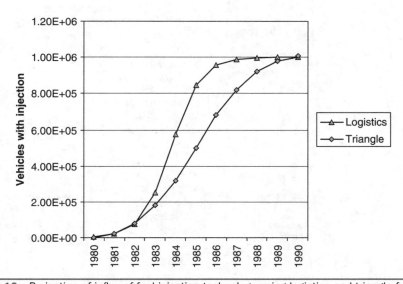

FIGURE 2-10 Projection of influx of fuel injection technology using logistics and triangle functions.

shown, it is also possible that transition could end with a massive transition of all the remaining vehicles in the last year, for example, in the year 1986 or 1987 in this example. However, experience shows that the "S-shaped" curve with tapering off at the end is a very plausible alternative.

These types of relationships between elements and trends in a system can be captured graphically in *causal loop diagrams*. These diagrams map the connections between components that interact in a system, where each component is characterized by some quantifiable value (e.g., temperature of the component, amount of money possessed by an entity, and the like). As shown in Fig. 2-11, linkages are shown with arrows leading from sending to receiving component and, in most cases, either a positive or negative sign. When an increase in the value of the sending component leads to an increase in the value of the receiving component, the sign is positive, and vice versa. Often the sending component receives either positive or negative feedback from a loop leading back to it from the receiving component, so that the systems interactions can be modeled. A component can influence, and be influenced by, more than one other component in the model.

In Fig. 2-11, the symbol in the middle of the loop ("B" for balancing or "R" for reinforcing) indicates the nature of the feedback relationship around the loop. On the left-hand side, increasing average temperature leads to reduced emissions of CO_2 from the combustion of fossil fuels in furnaces, boilers, and the like, so the loop is balancing. On the right, increased CO_2 increases the need for the use of air conditioning, which further increases CO_2 concentration. The diagram does not tell us which loop is the more dominant, and hence whether the net carbon concentration will go up or down. Also, it only represents two of the many factors affecting average CO_2 concentration in the atmosphere; more factors would be required to give a complete assessment of the expected

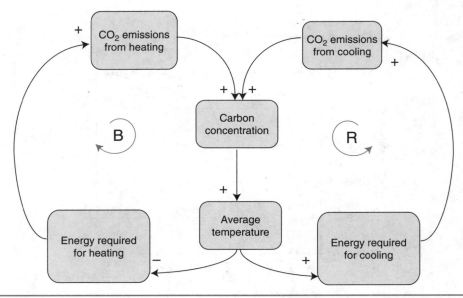

FIGURE 2-11   Causal loop diagram relating outdoor climate and indoor use of climate control devices.

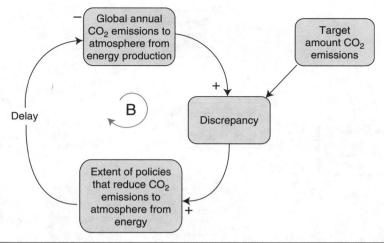

FIGURE 2-12   Causal loop diagram incorporating a goal, a discrepancy, and a delay: the relationship between CO_2 emissions and energy policy.

direction of CO_2 concentration. Thus it is not possible to make direct numerical calculations from a causal loop diagram. Instead, its value lies in identifying all relevant connections to make sure none are overlooked, and also in developing a qualitative understanding of the connections as a first step toward quantitative modeling.

The causal loop diagram in Fig. 2-12 incorporates two new elements, namely, a discrepancy relative to a goal and a delay function. In this case, a goal for the target annual CO_2 emissions is set outside the interactions of the model; there are no components in the model that influence the value of this goal, and its value does not change. The "discrepancy" component is then the difference between the goal and the actual CO_2 emissions. Here we are assuming that these emissions initially exceed the goal, so as they increase, the discrepancy also increases. Since the target does not change, it does not have any ability to positively or negatively influence the discrepancy, so there is no sign attached to the arrow linking the two. The discrepancy then leads to the introduction of policies that reduce CO_2 emissions toward the goal. As the policies take hold, emissions decline in the direction of the target, and the discrepancy decreases, so that eventually the extent of emissions reduction policies can be reduced as well. The link between policy and actual emissions is indicated with a "delay," in that reducing emissions often takes a long lead time in order to implement new policies, develop new technologies, or build new infrastructures. It is useful to explicitly include such delays in the causal loop diagram since they can often lead to situations of overshoot and oscillation.

The simple examples of causal loop diagrams presented here can be incorporated into larger, more complete diagrams that are useful for understanding interactions among many components. The individual components and causal links themselves are not complex, but the structures they make possible allow the modeler to understand interactions that are quite complex. As an example, Fig. 2-13 shows a causal loop diagram relating the level of economic activity to the choice of renewable or nonrenewable energy sources, the availability of nonrenewable fuels, and the impact on the climate. The general outcome in this causal loop diagram is that loss of resources and climate change will eventually cause a shift toward renewable energy sources.

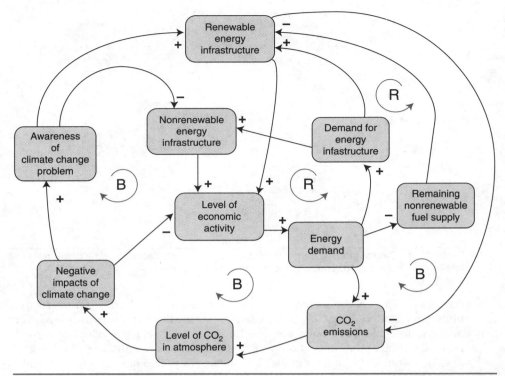

FIGURE 2-13   Causal loop diagram relating economic activity, choice of energy source, availability of fuel, and impact on climate.

2-5   Other Tools for Energy Systems

The remainder of this chapter discusses a range of tools used in analyzing energy systems and making decisions that best meet the goals of developing energy resources that are economical and sustainable. While these tools are diverse, they have a common theme in that they all emphasize taking a systems view of energy, in order to not overlook factors that can turn an attractive choice into an unattractive one.

2-5-1   Kaya Equation: Factors That Contribute to Overall CO_2 Emissions

The Kaya equation was popularized by the economist Yoichi Kaya as a way of disaggregating the various factors that contribute to the overall emissions of CO_2 of an individual country or of the whole world. It incorporates population, level of economic activity, level of energy consumption, and carbon intensity, as follows:

$$CO_{2\,emit} = (P) \times \left(\frac{GDP}{P}\right) \times \left(\frac{E}{GDP}\right) \times \left(\frac{CO_2}{E}\right) \qquad (2\text{-}4)$$

If this equation is applied to an individual country, then P is the population of the country, GDP/P is the population per capita, E/GDP is the energy intensity per unit of GDP, and CO_2/E is the carbon emissions per unit of energy consumed. Thus the latter three terms in the right-hand side of the equation are performance measures, and in particular the last two terms are metrics that should be lowered in order to benefit the environment. The measure of CO_2 in the equation can be interpreted as CO_2 emitted to the atmosphere, so as to exclude CO_2 released by energy consumption but prevented from entering the atmosphere, a process known as "sequestration." Because the terms in the right-hand side of the equation multiplied together equal CO_2 emitted, it is sometimes also called the "Kaya Identity."

The key point of the Kaya equation is that a country must carefully control all the measures in the equation in order to reduce total CO_2 emissions or keep them at current levels. As shown in Table 2-2, values for five of the major players in determining global CO_2 emissions, namely, China, Germany, India, Japan, and the United States, vary widely between countries. Although it would be unreasonable to expect countries with the largest populations currently, namely, China and India, to reduce their populations to be in line with the United States or Japan, all countries can contribute to reducing CO_2 emissions by slowing population growth or stabilizing population. China and India have low GDP per capita values compared to the other three countries in the table, so if their GDP per capita values are to grow without greatly increasing overall CO_2 emissions, changes must be made in either E/GDP, or CO_2/E, or both. As for Germany, Japan, and the United States, their values for E/GDP and CO_2/E are low relative to China and India, but because their GDP/P values are high, they have high overall CO_2 emissions relative to what is targeted for sustainability. This is true in particular of the United States, which has the highest GDP/P in the table and also an E/GDP value that is somewhat higher than that of Japan or Germany. Using these comparisons as an example, individual countries can benchmark themselves against other countries or also track changes in each measure over time in order to make progress toward sustainable levels of CO_2 emissions.

From a systems perspective, one issue with the use of these measures is that they may mask a historic transfer of energy consumption and CO_2 emissions from rich to industrializing countries. To the extent that some of the products previously produced

Country	P (million)	GDP/P ($/person)	E/GDP (MJ/$)	CO$_2$/E (gram/MJ)	CO$_2$ (mil. tonnes)
Year 2004					
China	1298.8	1725	28.05	74.9	4707.0
Germany	82.4	34,031	5.53	55.8	865.3
India	1065.1	740	20.65	68.3	1111.6
Japan	127.3	35,998	5.21	52.9	1263.0
USA	293.0	42,846	8.44	55.7	5901.7
Year 2011					
China	1336.7	7766	10.53	74.3	8121.7
Germany	81.5	47,373	3.68	55.2	784.3
India	1189.2	1723	12.09	70.8	1753.9
Japan	127.5	36,215	4.78	54.4	1200.7
USA	311.6	55,903	5.90	53.3	5477.9

TABLE 2-2   Kaya Equation Values for Selection of Countries, 2004 and 2011

in the United States or Europe but now produced in China and India are energy intensive, their transfer may have helped to make the industrial countries "look good" and the industrializing countries "look bad" in terms of E/GDP. Arguably, however, the rich countries bear some responsibility for the energy consumption and CO_2 emissions from these products, since it is ultimately their citizens who consume them.

2-5-2   Life-Cycle Analysis and Energy Return on Investment

The preceding example about CO_2 emissions from products manufactured in one country and consumed in another illustrates an important point about taking into account all of the phases of the "life cycle" of either an energy conversion system or product. In this example, boundaries were drawn within the overall "system" of countries such that one stage of the product's life cycle, namely, the manufacturing stage, appeared in the energy consumption total of one country but the final purchase and consumption of another. *Life-cycle analysis*, or LCA, is a technique that attempts to avoid such pitfalls by including all parts of the life cycle. Although LCA is applied to energy consumption in this book, it can also be applied to other types of environmental impacts or to cash flow. In this section, we focus on LCA applied to energy consumption as a single dimension. In the next section, we incorporate multiple criteria, so as to compare the effects on energy with effects on other objectives.

LCA can help the practitioner overcome at least two important problems in measuring and comparing energy consumption:

- *Incomplete information about the true impact of a technology on energy efficiency:* If only one stage of the life cycle, or some subset of the total number of stages, is used to make claims about energy savings, the analysis may overlook losses or quantities of energy consumption at other stages that wash out projected savings. For example, in the case of comparing vehicle propulsion systems, one propulsion system may appear very efficient in terms of converting stored energy on board the vehicle into mechanical propulsion. However, if the process that delivers the energy to the vehicle is very energy intensive, the overall energy efficiency of the entire process from original source may be poor.

- *Savings made at one stage may be offset by losses at another stage:* This effect can occur in the case of a manufactured product that has one or more layers of both manufacturing and shipping before it is delivered to a consumer as a final product. It is often the case that concentrating manufacturing operations in a single large facility may improve manufacturing efficiency, since large versions of the machinery may operate more efficiently, or the facility may require less heating and cooling per square foot of floor space. However, bringing raw materials to and finished goods from the facility will, on average, incur a larger amount of transportation than if the materials were converted to products, distributed, and sold in several smaller regional systems. In this case, gains in the manufacturing stage may be offset, at least in part, by losses in the transportation stage.

The eight steps in the life cycle shown in Table 2-3 can be applied to both energy products, such as electricity or motor fuels, as well as manufactured products such as appliances and automobiles. The list of examples of each stage shown in the right column is not necessarily exhaustive, but does provide a representative list of the type of activities that consume energy in each stage. Some energy resources, such as coal, may be converted to currencies (electricity) after the raw material stage and therefore

Activity Category	Typical Activities
Resource extraction	Machinery used for mining, quarrying, and so on, of coal, metallic ores, stone resources Energy consumed in pumping crude oil and gas Energy consumption in forestry and agriculture to maintain and extract crops, including machinery, chemical additives, and so on
Transportation of raw materials	Movement of bulk resources (coal, grain, bulk wood products) by rail, barge, truck Movement of oil and gas by pipeline Intermediate storage of raw materials
Conversion of raw materials into semi-finished materials and energy products	Refining of crude oil into gasoline, diesel, jet fuel, other petroleum products Creation of chemical products from oil, gas, and other feedstocks Extraction of vegetable oils from grain Production of bulk metals (steel, aluminum, and so on) from ores Creation of bulk materials (fabric, paper, and so on) from crop and forest feedstocks
Manufacturing and final production	Mass production of energy-consumption-related devices (steam and gas turbines, wind turbines, motorized vehicles of all types, and so on) Mass production of consumer goods Mass production of information technology used to control energy systems
Transportation of finished products	Transportation by truck, aircraft, rail, or marine Intermediate storage and handling of finished products in warehouses, regional distribution centers, airports and marine ports, and so on
Commercial "overhead" for system management	Energy consumed in management of production/distribution system (office space) Energy consumed in retail space for retail sale of individual products
Infrastructure construction and maintenance	Construction of key components in energy system (power plants, grid, and so on) Construction of resource extraction, conversion, and manufacturing plants Construction of commercial infrastructure (office space, shopping malls, car dealerships, and so on) Construction of transportation infrastructure (roads, railroads, airports, marine ports) Energy consumed in maintaining infrastructure
Demolition/ disposal/recycling	Disposal of consumer products and vehicles at end of lifetime Dismantling and renovation of infrastructure Recycling of raw materials from disposal/demolition back into inputs to resource extraction stage

Note: Activities are given in approximately the order in which they occur in the life cycle of an energy product or manufactured good.

TABLE 2-3   Examples of Energy Use within Each Category of Energy Consumption Life Cycle

cease to exist in a physical form. Other energy resources, such as motor vehicle fuels, require extensive refining, distribution, and retailing before they are consumed. Consumer products generally depend on several "chains" of conversion of raw materials to components before final assembly, such as a motor vehicle, which incorporates metal resources and wires, plastics, fabrics, and other inputs, all arising from distinct forms of resource extraction. Energy is consumed to greater or lesser degrees in all of these stages, as well as in the building of the infrastructure required to support the system, and the energy consumed in commercial enterprises needed to control the system.

In theory, one could estimate energy consumption values for each stage of the life cycle, allocate these quantities among different products where there are multiple products relying on a single activity (e.g., energy consumed in office used to administer product), and divide by the number or volume of products produced to estimate an energy consumption value per unit. In practice, it may prove extremely laborious to accurately estimate the energy consumed in each activity of each stage. Therefore, as a practical simplification, the engineer can "inspect" stages for which energy consumption values are not easily obtained. If they can make a defensible argument that these stages cannot, for any range of plausible values, significantly affect the life cycle of total energy consumption, then they can either include a representative fixed value resulting from an educated guess of the actual value, or explicitly exclude the component from the LCA.

For energy resources, it is important not only to understand the life-cycle energy consumption, but also to verify that end-use energy derived from the resource justifies energy expended upstream in its extraction, conversion, and delivery. On this basis, we can calculate the *energy return on investment*, or EROI, which is the energy delivered at the end of the life cycle divided by the energy used in earlier stages. For example, it would not make sense to expend more energy on extracting some very inaccessible fossil fuel from under the Earth than the amount of energy available in the resource, on a per tonne basis. Similarly, the manufacture and deployment of some renewable energy device that requires more energy in resource extraction, manufacturing, and installation than it can deliver in its usable lifetime would fail the EROI test. Practitioners may also use the term "energy payback," meaning the amount of energy returned from the use of an energy system after expending energy on manufacture, installation, and so on.

2-5-3   Multi-Criteria Analysis of Energy Systems Decisions

The previous section presented the use of LCA using a single measure, namely, that of energy consumption at the various stages of the life cycle. The focus on energy consumption as a sole determinant of the life cycle value of a product or project may be necessary in order to complete an LCA with available resources, but care must be taken that other factors that are even more critical than energy use are not overlooked. For example, a project may appear attractive on an energy LCA basis, but if some other aspect that causes extensive environmental damage is overlooked, for example, the destruction of vital habitat, the natural environment may end up worse off rather than being improved.

Multi-criteria analysis of decisions provides a more complete way to incorporate competing objectives and evaluate them in a single framework. In this approach, the decision maker identifies the most important criteria for choosing the best alternative among competing projects or products. These criteria may include air pollution, water pollution, greenhouse gas emissions, or release of toxic materials, as well as direct economic value for money. The criteria are then evaluated for each alternative on some common basis, such as the dollar value of economic benefit or economic cost of different types of environmental damage, or a qualitative score based on the value for a given alternative relative to the best or worst competitor among the complete set of alternatives. An overall multi-criteria "score" can then be calculated for alternative i as follows:

$$\text{Score}_i = \sum_c w_c x_{ic}, \ \forall i \qquad (2\text{-}5)$$

Here score_i is the total score for the alternative in dimensionless units, w_c is the weight of criterion c relative to other criteria, and x_{ic} is the score for alternative i in terms of criterion c. At the end of the multi-criteria analysis, the alternative with the highest or lowest score is selected, depending on whether a high or low score is desirable.

Naturally, how one weights the various criteria will have an important impact on the outcome of the analysis, and it is not acceptable to arbitrarily or simplistically assign weights, since the outcome of the decision will then appear itself arbitrary, or worse, designed to reinforce a preselected outcome. One approach is therefore to use as weight sets of values that have been developed using previous research, and which those in the field already widely accept. Table 2-4 gives relative weights from two widely used studies from the U.S. Environmental Protection Agency and from Harvard University. In the tables, the category most closely related to energy consumption is global warming, since energy consumption is closely correlated with CO_2 emissions. This category has the highest weight in both studies (tied with indoor air quality in the case of EPA), but other types of impact figure prominently as well. In both studies, acidification (the buildup of acid in bodies of water), nutrification (the buildup of agricultural nutrients in bodies of water), natural resource depletion, and indoor air quality all have weights of at least 0.10. Only solid waste appears to be a less pressing concern, according to the values from the two studies. In order to use either set of weights from the table correctly, the scoring system for the alternatives should assign a higher score to an inferior alternative, so that the value of score_i will correctly select the alternative with the lowest overall score as the most attractive.

In one possible approach, the decision maker can adopt either EPA or Harvard values as a starting point for weights and then modify values using some internal

Category	EPA Weight	Harvard Weight
Global warming	0.27	0.28
Acidification	0.13	0.17
Nutrification	0.13	0.18
Natural resource depletion	0.13	0.15
Indoor air quality	0.27	0.12
Solid waste	0.07	0.10

Sources: USEPA (1990) and Norberg-Bohm (1992), as quoted in Lippiatt (1999).

TABLE 2-4   Comparison of USEPA and Harvard Weights for Important Environmental Criteria

process that can justify quantitative changes to weights. In Table 2-4, the weights shown add up to 1.00 in both cases. As long as value of weights relative to one another correctly reflects the relative importance of different criteria, it is not a requirement for multi-criteria analysis that they add up to 1. However, it may be simplest when adjusting weights to add to some while taking away from others, so that the sum of all the weights remains 1.

2-5-4   Choosing among Alternative Solutions Using Optimization

In many calculations related to energy systems, we evaluate a single measure of performance, either by solving a single equation (e.g., the multi-criteria score evaluated in the previous example), or by solving a system of equations such that the number of equations equals the number of unknown variables, and the value of each unknown is therefore uniquely identified. In other situations, there may be a *solution space* in which an infinite number of possible combinations of *decision variable* values, or *solutions*, can satisfy the system requirements, but some solutions are better than others in terms of achieving the system objective, such as maximizing financial earnings or minimizing pollution. *Optimization* is a technique that can be used to solve such problems.

We first look at optimization in an abstract form before considering examples that incorporate quantitative values for components of energy systems. In the most basic optimization problem relevant to energy systems, the decision variables regarding output from facilities can be written as a vector of values $x_1, x_2, \ldots, x_n$ from 1 to n. These facilities are constrained to meet demand for energy, that is, the sum of the output from the facilities must be greater than or equal to some demand value a. There are then constraints on the values of x_i, for example, if the ith power plant cannot produce more than a certain amount of power b_i in a given time period, this might be written $x_i \le b_i$. Then the entire optimization problem can be written

$$\text{Minimize } Z = \sum_{i=1}^{n} c_i x_i$$

subject to

$$\sum_i x_i \ge a \tag{2-6}$$

$$x_i \le b_i, \ \forall i$$

$$x_i \ge 0, \ \forall i$$

In Eq. (2-6), the top line is called the *objective function*, and the term "minimize" indicates that the goal is to find the value of Z that is as low as possible. The vector of parameters c_i are the cost coefficients associated with each decision variable x_i. For example, if the objective is to minimize expenditures, then cost parameters with units of "cost per unit of x_i" are necessary so that Z will be evaluated in terms of monetary cost and not physical units. The equations below the line "subject to" are the constraints for this optimization problem. Note that in the objective function, we could also stipulate that the value of Z be maximized, again subject to the equations in the constraint space. For example, the goal might be to maximize profits, and instead of costs per unit of production, the values of c_i might represent the profits per unit of production from each plant. Also, the *nonnegativity constraint* $x_i \geq 0$ does not appear in all optimization problems; however, in many problems, negative values of decision variables do not have any useful meaning (e.g., negative values of output from a power plant), so this constraint is required in order to avoid spurious solutions.

Example 2-3 illustrates the application of optimization to an energy problem that is simple enough to be solved by inspection, and also to be represented in two-dimensional space, so that the function of the optimization problem can be illustrated graphically.

Example 2-3 An energy company has two power plants available to meet energy demand, plant 1 that produces power for a cost of $0.015/kWh and has a capacity of 6 GW, and plant 2 that produces power for a cost of $0.03/kWh and has a capacity of 4 GW. What is the minimum cost at which the company can meet a demand of 8 GW for 1 h?

Solution From inspection, the company will use all 6 GW from plant 1 and then produce the remaining 2 GW with plant 2. It is convenient to convert costs to costs per GWh, that is, $c_1 = \$15,000/$GWh and $c_2 = \$30,000/$GWh. The minimum total cost is then $150,000.

To arrive at the same solution using a formal optimization approach, Eq. (2-6) can now be rewritten for the specific parameters of the problem as follows:

$$\text{Minimize } Z = 15,000x_1 + 30,000x_2$$

subject to

$$x_1 \leq 6$$
$$x_2 \leq 4$$
$$x_1 + x_2 \geq 8$$
$$x_1, x_2 \geq 0$$

Note that the third constraint equation says that the output from the two plants combined must be at least 8 GW.

Next, it is useful to present the problem graphically by plotting the feasible region for the solution space on a two-dimensional graph with x_1 and x_2 on the respective axes, as shown in Fig. 2-14. The constraint $x_1 + x_2 \geq 8$ is superimposed on the graph as well, and only combinations of x_1 and x_2 above and to the right of this line are feasible, that is, able to meet the power requirements. Two *isocost lines* are visible in the figure as well; and combination of x_1 and x_2 on each line will have the same total cost, for example, $150,000 or $180,000 in this case. The intersection of the constraint line with the isocost line $c_1x_1 + c_2x_2 = \$150,000$ gives the solution to the problem, namely, that since one cannot go to a lower isocost line without leaving the feasible space, the solution is at the point $(x_1 = 6, x_2 = 2)$.

Although it is possible to solve for a larger optimization problem by hand, it is usually more convenient to use a computer software algorithm to find the solution. The general goal of any such algorithm is to first find a feasible solution (i.e., a combination of decision variables that is in the decision space) and then move efficiently to consecutively improved solutions until reaching an optimum, where the value of Z cannot be improved. At each feasible solution, a "search direction" is chosen such that movement in that direction is likely to yield the largest improvement in Z compared to any other direction. The energy system analyst can accelerate the solution process by providing an initial feasible solution. For instance, suppose the initial feasible solutions given were $(x_1 = 4, x_2 = 4)$ in this example. The algorithm might detect that there were two search directions from this point, either along the curve $x_1 + x_2 = 8$ or along the curve $x_2 = 4$. Of the two directions, the former decreases the value of Z while the latter increases it, so the algorithm chooses the former direction and moves along it until reaching the constraint $x_1 = 6$, which is the optimum point. If no initial solution is provided, the algorithm may start at the origin (i.e., all decision variables equal to zero) and first test for feasibility

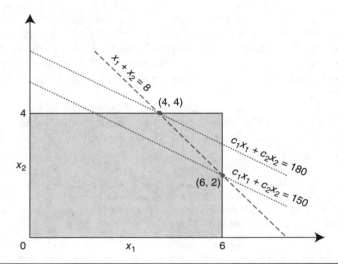

$x_1 + x_2 = 8$

$(4, 4)$

$c_1x_1 + c_2x_2 = 180$

$c_1x_1 + c_2x_2 = 150$

x_2

$(6, 2)$

x_1

FIGURE 2-14   Graphical representation of allocation of power production to plant 1 and plant 2.

of this solution. If it is feasible, the algorithm moves toward consecutively improved solutions until finding the optimal. If not, it moves in consecutive search directions until arriving at the boundary of the feasible space, and then continuing the optimization.

The problem presented in Example 2-3 is small enough that it can be presented in a visual form. Typical optimization problems are usually much larger than can be presented in 2-D or 3-D space. For example, suppose the problem consisted of n energy generating units, and output from each needed to be determined for m time periods. The total number of dimensions in this problem is $n \times m$, which is a much larger number than three, the largest number that can be represented graphically.

Optimization can be applied to a wide array of energy systems problems. As presented in Example 2-3, one possibility is the optimal allocation of energy production from different plants, where the cost per unit of production may or may not be constant. Another is the allocation of different energy resources to different finished products so as to maximize the expected profit from sales of the product. Optimization may also be applied to the efficient use of energy in networks, such as the optimal routing of a fleet of vehicles so as to minimize the projected amount of fuel that they consume.

Lastly, one of the biggest challenges with successful use of optimization is the need to create an objective function and set of constraints that reproduce the real-world system accurately enough for the recommendation output from the problem to be meaningful. An optimization problem formulation that is overly simplistic may easily converge to a set of decision variables that constitute a recommendation, but the analyst can see in a single glance that the result cannot in practice be applied to the real-world system. In some cases, adding constraints may solve this problem. In other cases, it may not be possible to capture in mathematical constraints all of the factors that impact on the problem. If so, the analyst may use the optimization tool for *decision support*, that is, generating several *promising* solutions to the problem, and then using one or more of them as a basis for creating by hand a *best* solution that takes into account subtle factors that are missing from the mathematical optimization.

2-5-5 Understanding Contributing Factors to Time-Series Energy Trends Using Divisia Analysis

A common problem in the analysis of the performance of energy systems is the need to understand trends over time up to the most recent year available, in order to evaluate the effectiveness of previous policies or discern what is likely to happen in the future. Specifically, if energy intensity of alternative technologies and the share of each technology in the overall market stay constant, one will most likely arrive at a different overall level of energy consumption than in the real world, in which both factors of energy intensity and market share are changing over time. In such situations, the energy systems analyst would like to know the contribution of the factors to the quantitative difference between the actual energy consumption value and the *trended* value, which is the amount that would have been consumed if factors had remained constant. *Divisia analysis*, also known as *Divisia decomposition*, is a tool that serves this purpose. (*LaSpeyres decomposition* is a related tool; for reasons of brevity, it is not presented here, but the reader may be interested in learning about it from other sources.[2])

In a typical application of Divisia analysis, there are multiple sources or sectors that contribute to overall energy production or consumption, each with a quantitative value of conversion efficiency or some other measure of performance that changes with time. Hereafter production or consumption is referred to as *activity*, denoted A, and the various contributors are given the subscript i. The share of activity allocated to the contributors also changes with time. Thus a contributor may be desirable from a sustainability point of view because of its high energy efficiency, but if its share of A declines over time, its contribution to improving overall system performance may be reduced.

The contribution of both efficiency and share can be mathematically "decomposed" in the following way. Let E_t be the total energy consumption in a given time period t. The relationship of E_t to A_t is then the following:

$$E_t = \left(\frac{E}{A}\right)_t \cdot A_t = e_t \cdot A_t \tag{2-7}$$

Here e_t is the intensity, measured in units of energy consumption per unit of activity. We can now divide the aggregate energy intensity and energy use into the different contributors i from 1 to n:

$$e_t = \left(\frac{E}{A}\right)_t = \sum_{i=1}^{n}\left(\frac{E_{it}}{A_{it}}\right)\left(\frac{A_{it}}{A_t}\right) = \sum_{i=1}^{n} e_{it}s_{it} \tag{2-8}$$

$$E_t = \left(\frac{E}{A}\right)_t A_t = \sum_{i=1}^{n}\left(\frac{E_{it}}{A_{it}}\right)\left(\frac{A_{it}}{A_t}\right)A_t = \sum_{i=1}^{n} e_{it}s_{it}A_t \tag{2-9}$$

---

[2]For example, see Appendix at the end of Schipper et al. (1997).

We have now divided overall intensity into the intensity of contributors e_{it}, and also defined the share of contributor i in period t as s_{it}, which is measured in units of (activity allocated to i in period t)/(total activity in t).

Next, we consider the effect of changes in intensity and share between periods $t-1$ and t. By the product rule of differential calculus, we can differentiate the right-hand side of Eq. (2-8):

$$\frac{de_t}{dt} = \sum_{i=1}^{n}\left[\left(\frac{de_{it}}{dt}\right)s_{it} + \left(\frac{ds_{it}}{dt}\right)e_{it}\right] \tag{2-10}$$

The differential in Eq. (2-10) can be replaced with the approximation of the change occurring at time t, written $\Delta e_t/\Delta t$, which gives the following:

$$\Delta e_t = \Delta t\frac{de_t}{dt}$$

$$= \Delta t\sum_{i=1}^{n}\left[\left(\frac{\Delta e_{it}}{\Delta t}\right)s_{it} + \left(\frac{\Delta s_{it}}{\Delta t}\right)e_{it}\right]$$

$$= \sum_{i=1}^{n}\left[(\Delta e_{it})s_{it} + (\Delta s_{it})e_{it}\right] \tag{2-11}$$

The change in contributors' intensity and share is now rewritten as the difference in value between periods $t-1$ and t, and the current intensity and share values are rewritten as the average between the two periods, giving:

$$\Delta e_t = \sum_{i=1}^{n}\left[(e_t - e_{t-1})_i\frac{(s_t + s_{t-1})_i}{2} + (s_t - s_{t-1})_i\frac{(e_t + e_{t-1})_i}{2}\right]$$

$$= \sum_{i=1}^{n}\left[(e_t - e_{t-1})_i\frac{(s_t + s_{t-1})_i}{2}\right] + \sum_{i=1}^{n}\left[(s_t - s_{t-1})_i\frac{(e_t + e_{t-1})_i}{2}\right] \tag{2-12}$$

It is the value of the two separate summations in Eq. (2-12) that is of particular interest. The first summation is called the "intensity term," since it considers the change in intensity for each i between $t-1$ and t, and the second term is called the "structure term," since it has to do with the relative share of the various contributors.

The steps in the Divisia analysis process are (1) for each time period t (except the first), calculate the values of the summations for the intensity and structure terms and (2) for each time period (except the first and second), to calculate the cumulative value of the structure term by adding together values of all previous intensity and structure terms. The cumulative values in the final time period then are used to evaluate the contribution of each term to the difference between the actual and trended energy consumption value in the final period. Example 2-4 demonstrates the application of Divisia analysis to a representative data set with intensity and structure terms and three time periods.

Example 2-4 The activity levels and energy consumption values in the years 1995, 2000, and 2005 for two alternative sources of a hypothetical activity are given in the table below. (The amount of the activity is given in arbitrary units; actual activities that could fit in this mode include units of a product produced by two different processes, kilometers driven by two different models of vehicles, and so on.) Construct a graph of the actual and trended energy consumption, showing the contribution of changes in structure and intensity to the difference between the two.

Source	Year	Activity (million units)	Energy (GWh)
1	1995	15	300
1	2000	14.5	305
1	2005	14.3	298
2	1995	22	350
2	2000	25.5	320
2	2005	26.5	310

Solution   The first step in the analysis process is to calculate total energy consumption and activity, since these values form a basis for later calculations. The following table results:

Year	Activity (million units)	Energy (GWh)
1995	37	650
2000	40	625
2005	40.8	608

From this table, the average energy intensity in 1995 is $(650\ \text{GWh})/(3.7 \times 10^7) = 17.6\ \text{kWh/unit}$, which is used to calculate the trended energy consumption values of 703 GWh and 717 GWh for the years 2000 and 2005, respectively.

The energy intensity e_{it} is calculated next using the energy consumption and activity data in the first table, and the share s_{it} is calculated from the first and second tables by comparing activity for i to overall activity. Both factors are presented for the two sources as follows:

	Source 1	Source 1	Source 2	Source 2
Year	Intensity	Share	Intensity	Share
1995	20.0	40.5%	15.9	59.5%
2000	21.0	36.3%	12.5	63.8%
2005	20.8	35.0%	11.7	65.0%

The intensity and structure terms for the years 2000 and 2005 are now calculated for each source using Eq. (2-12). It is convenient to create a table for each source, as follows:

For source 1:

	e_{it}	s_{it}	$(e_t - e_{t-1})_i$	$(s_t - s_{t-1})_i$	$(e_t + e_{t-1})_i/2$	$(s_t + s_{t-1})_i/2$	Intensity	Structure
1995	20.0	40.5%	—	—	—	—	—	—
2000	21.0	36.3%	1.0	−4.3%	20.5	38.4%	0.397	(0.880)
2005	20.8	35.0%	(0.2)	−1.3%	20.9	35.6%	(0.070)	(0.251)

For source 2:

	e_{it}	s_{it}	$(e_t - e_{t-1})_i$	$(s_t - s_{t-1})_i$	$(e_t + e_{t-1})_i/2$	$(s_t + s_{t-1})_i/2$	Intensity	Structure
1995	15.9	59.5%	—	—	—	—	—	—
2000	12.5	63.8%	(3.4)	4.3%	14.2	61.6%	(2.070)	0.611
2005	11.7	65.0%	(0.9)	1.2%	12.1	64.4%	(0.548)	0.146

Next, the overall intensity and structure term for each year is calculated by adding the values for sources 1 and 2. For example, for 2000, the intensity term is 0.397 − 2.070 = −1.673. The cumulative value for 2005 is the sum of the values for 2000 and 2005 for each term. A table of the incremental and cumulative values gives the following:

Year	Incremental Values		Cumulative Values	
	Intensity	Structure	Intensity	Structure
2000	(1.673)	(0.270)	(1.673)	(0.270)
2005	(0.617)	(0.106)	(2.290)	(0.376)

The effect on energy consumption in the intensity and structure of the sources can now be measured by multiplying activity in 2000 or 2005 by the value of the respective terms. Units for both intensity and structure terms are in kWh per unit. Therefore, the effect of intensity in 2000 and 2005 is calculated as

$$\text{Year 2000: } (4 \times 10^7 \text{ unit})(-1.673 \text{ kWh/unit})(10^{-6}\text{GWh/kWh}) = -66.9 \text{ GWh}$$

$$\text{Year 2005: } (4.08 \times 10^7 \text{ unit})(-2.29 \text{ kWh/unit})(10^{-6}\text{GWh/kWh}) = -93.4 \text{ GWh}$$

Using a similar calculation for the effect of structural changes, we can now show that the cumulative effect of both changes accounts for the difference in actual and trended energy consumption, as shown in the table below:

Factor	2000	2005
E(Trended), GWh	702.70	716.76
Intensity, GWh	(66.9)	(93.4)
Structure, GWh	(10.8)	(15.3)
E(Actual), GWh	625.00	608.00

This information can also be presented in graphical form, as shown in Fig. 2-15.

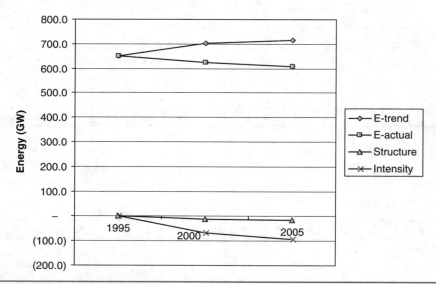

FIGURE 2-15 Difference between actual and trended energy consumption based on Divisia analysis.

In conclusion, from the preceding table and figure, both intensity and structure factors contribute to the reduction in energy consumption relative to the trended value, with the intensity factor accounting for most of the change, and the structure factor accounting for a smaller amount.

One practical example of the use of Divisia analysis is the decomposition of the contribution of different types of power generation to total CO_2 emissions, with activity measured in kWh generated, and CO_2 emissions by source used in place of energy consumption in Example 2-4. Another example is the contribution of different types of vehicles to total transportation energy consumption. A wide variety of other applications are also possible.

2-5-6   Incorporating Uncertainty into Analysis Using Probabilistic Approaches and Monte Carlo Simulation

When we solve forward-looking problems such as that of Example 2-3, where the goal is to meet a certain level of energy demand for the lowest possible cost, it is often assumed for simplicity that the exogenous values (such as the cost per kWh of production) are known with certainty, and the problem solved based on knowing the values completely accurately, leading to a *deterministic* solution. In reality, there may be *uncertainty* surrounding these values such that if values in the future prove to be different from expectation, the actual cost will be different from what was predicted. Furthermore, in the case of optimization, in some situations the optimal answer might even change if the actual exogenous values change sufficiently.

One way to compensate for this potential weakness in the deterministic approach is to simply describe qualitatively (e.g., in the form of a list of caveats) the level of uncertainty surrounding input values. Another is to create multiple scenarios, as described in Sec. 2-3-3, to calculate different possible outcomes based on a range of possible input values. Both of these approaches have drawbacks. Listing caveats may describe qualitatively the possible changes to the outcome, but does not quantify alternative outcomes. Use of scenarios is limited by the total number of scenarios that can be created, and the ability to judge what values are appropriate in place of the expected value.

Where data about the *statistical distribution* of specific inputs are available (e.g., the expected value and standard deviation of the cost of some energy resource), we can treat uncertainty in a more sophisticated way by incorporating the distribution data directly into the solution to the problem. Instead of solving the problem once as in the deterministic case, we repeat the calculation many times, each time generating new random values for independent input values (such as the cost per unit of energy), and in the end studying the distribution of answers returned. Each answer returned is called an *iterate*. This technique is called *Monte Carlo simulation* or alternatively *static simulation*, the latter name distinguishes it from, for example, *discrete-event simulation*, since the former does not capture the passage of time within the calculation of each iterate, whereas the latter focuses on how a system changes over time in response to random inputs. This section covers Monte Carlo (static) simulation only.

The steps in Monte Carlo simulation are the following:

1. *Preprocess:* Calculate the deterministic solution to the stated problem using expected values for each input. For instance, if a value is normally distributed, the expected value is the mean of the distribution, while if it is uniformly distributed, the expected value is the average of the two endpoint values. This first step helps to organize the calculations that occur in the simulation, and also provides an expected value around which the iterate values will fall.

Probabilistic values in Step 2 falling in a range significantly away from the deterministic solution usually mean an error in the calculation, so this step has troubleshooting value as well.

2. *Simulation process:* Calculate the desired number of probabilistic iterates. Typically 1000 or more iterates are calculated. Also, for each iterate, random numbers should be generated for each input value. For example, if a simulation involves fuel economy and the price of gasoline to calculate expected fuel expenses, the value for fuel economy should be generated separately from that of gasoline.

3. *Postprocess evaluation:* Tabulate the results of the simulation in a table or histogram and analyze the results. Output values can be put into "bins" of a given range and then the number of iterate responses in each bin can be tabulated in either a table or graphically in a histogram. The analyst can then evaluate the amount of variability in the histogram around the expected value from Step 1, and also assess whether the likelihood of a value falling above a critical threshold is acceptable (e.g., energy expenditure exceeding the amount in the budget plus contingency, which may have serious repercussions on the function of a business or government agency).

Monte Carlo simulation at its core depends heavily on the ability to generate random variates from any statistical distributions that might be necessary to model uncertainty in an input value, including not only the uniform and normal distributions mentioned above but also the many others available from the field of probability and statistics. As an example, the steps for generating from a uniform distribution a normally distributed value with known mean μ and standard deviation σ are the following:

1. Generate a random variate u from uniformly distributed $U \sim (0,1)$ (i.e., between 0 and 1). These functions are commonly included in spreadsheet or engineering calculation software packages.

2. Find a value z such that for the standard normally distributed random variable $Z, P(Z \leq z) = u$. Again, z can be evaluated using built-in functions, or the standard normal table that appears in any statistics textbook.

3. The random variate x that is used in the simulation is then evaluated:

$$x = \mu + z \cdot \sigma \tag{2-13}$$

Alternatively, for a uniformly distributed x between a and b, the calculation given u from $U \sim (0,1)$ is simpler:

$$x = a + u \cdot (b - a) \tag{2-14}$$

The entire simulation can be built in a spreadsheet using built-in functions, or the process can be streamlined using specialist Monte Carlo simulation packages such as @Risk® or Crystal Ball®. Example 2-5 illustrates the deterministic solution preprocess, the evaluation of random variates, and the presentation of results from a model Monte Carlo simulation.

Example 2-5 A common problem in energy systems is the effect of budgetary uncertainty on decision making for the future. Consider a municipality that must prepare a budget for fuel expenditures for the operation of snow removal equipment, or SRE, on streets and highways in the upcoming winter season. The simulation will consider three normally distributed random inputs: severity of winter storms (as represented by the total number of miles driven by the SRE fleet), fuel economy delivered by the SRE in terms of average miles per gallon, and the cost of diesel fuel per gallon. Mean and standard deviation values are given in the table below, as is the coefficient of variation (CV), that is, ratio of SD to mean, which indicates the amount of variability in the input value. For this simulation, (a) calculate the deterministic solution, (b) calculate the value of a single iterate if three samples of u are generated from the $U(0,1)$ distribution with values 0.823, 0.204, and 0.020 for the three required inputs, and (c) produce a histogram from a simulation of 100 iterates and discuss its implications if the municipality has a hard budget constraint of $140,000 for the season for SRE fuel expenditures. Note that the relatively small simulation size of 100 iterates is chosen here to make the example more transparent, although in general at least 1000 iterates are desirable to make the results more robust.

Variables	Units	Mean	SD	CV
Demand	Miles driven	80,000	24,000	30%
Fuel economy	MPG	3.5	0.525	15%
Fuel cost	$/gallon	$3.10	$0.775	25%

Solution

(a) The deterministic value is calculated from the mean values of demand, fuel economy, and fuel cost:

$$\text{Expense} = \frac{\text{Demand}}{\text{Fuel economy}}(\text{Fuel cost}) = \frac{80,000}{3.5}(\$3.10) = \$70,857$$

(b) For this particular iterate, we will calculate the value for demand in detail, with the assumption that the steps can be repeated for fuel economy and fuel cost. Given $u = 0.823$, first solve for z using the standard normal probability table or a built-in spreadsheet or math software function. From the standard normal table, $P(Z \leq 0.93) = 0.8238$, so for the purposes of the calculation, it is sufficiently accurate to set $z = 0.93$. Using Eq. (2-12) then gives:

$$x = \mu + z \cdot \sigma = 80,000 + 0.93 \cdot 24,000 = 102,320$$

The reader can verify that starting with random values 0.204 and 0.020 instead of 0.823 and repeating these steps for fuel economy and cost gives, to the nearest hundredth, 3.07 MPG and $1.51/gallon, respectively. Thus the value returned for this iterate is

$$\text{Expense} = \frac{\text{Demand}}{\text{Fuel economy}}(\text{Fuel cost}) = \frac{102,320}{3.07}(\$1.51) = \$50,327$$

Note that the municipality has "gotten lucky" in this iterate because they have experienced an unusually low fuel cost of $1.51/gal, which is more than two standard deviations below the mean cost of $3.10/gal. The result is an annual cost of ~$50K, well below the expected value of ~$71K from part (a). Returning such a low-valued variate in a Monte Carlo simulation is possible, but unusual.

(c) In principal, the simulation is carried out by repeating the steps in part (b) in an automated way. No two runs of the Monte Carlo simulation will be identical, but the results for a particular run of 100 iterates are tabulated in Table 2-5 and Fig. 2-16. No iterate returned a value lower than $20,000, and none higher than $200,000; however, within this range there was wide variability in returned values, reflecting relatively high CVs for both miles driven and the cost of fuel. This high level of variability may pose a concern for the municipality. In particular, 91% of the values fall in the bins $120–$140K or below, suggesting that there is a 9% probability that

Bin Values ($1000s)		Number
Minimum	Maximum	
<$40	$40	2
$40	$60	12
$60	$80	24
$80	$100	29
$100	$120	12
$120	$140	12
$140	$160	6
$160	$180	1
$180	>$180	2
	Total	100

Note: For example, 2 out of 100 iterates returned total expenditure values less than $40,000, 12 returned values between $40,000 and $60,000, and so on.

TABLE 2-5   Distribution of 100 Iterates into Bins for Representative Monte Carlo Simulation Run (Mean = $72,687, CV = 45%)

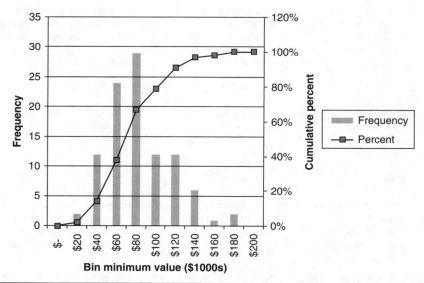

FIGURE 2-16   Graphical representation of distribution of 100 iterates into bins (left *y*-axis) and cumulative percentage of iterate values below bin upper value (right *y*-axis).

Note: Bin values shown are the minimum for a given bin, the maximum for that bin is the next value to the right (e.g., the bin marked '$20' is actually the $20K–$40K bin).

the budget constraint of $140K will be exceeded. This result may motivate the municipality to consider various possible responses, such as investments in improving SRE fuel economy, policy changes that might reduce the level of commitment to snow removal (to lower the expected value of demand), or increasing the size of the SRE budget by reducing commitments elsewhere. In any case, the example shows the power of the Monte Carlo simulation: if we only calculate the deterministic solution, then we know the expected value of total expenses for the winter, but we do not obtain any quantitative information about the risk that the upper bound will be exceeded.

2-6   Energy Policy as a Catalyst for the Pursuit of Sustainability

At times market forces align with the goals of sustainability, so that price signals encourage consumers to make decisions that benefit the environment, such as when high gasoline prices shift automobile choices toward smaller, more efficient vehicles. At other times, government leaders, in particular, may intervene in the economic market or in the function of society to achieve some desired outcome beyond what market forces generate. This phenomenon is an example of *energy policy* at work. Policies may seek to encourage the increasing use of "desirable" technologies and practices, or they may aim to discourage those that are undesirable. Also, generally it is a government body that must enact a policy because government carries the weight of legal enforcement behind its actions. However, government is not the only stakeholder: private business, public institutions other than the government itself, environmental groups (e.g., Sierra Club), and individual citizens also have a stake in policy and play a role in its evolution. It is the last stakeholder in particular, the individual citizen with their right to vote, that plays a special role because they act as a control on the behavior of government officials.

Degrees of Policy Intervention

One metric for energy policy is its degree of stringency, or the extent to which it either mildly or severely impacts the behavior of individuals and enterprises. The following list describes degrees of policy intervention, from least to most obtrusive:

- *"Libertarian" approach:* In a pure libertarian approach, there is no direct government policy intervention in decisions affecting energy. Instead, players are free to make what decisions they will. The role of government is limited to protection of the boundaries of the market (e.g., if we take the United States as an example, this would be U.S. national borders) so that players are not fearful of outside encroachment, and enforcement of legal agreements between individuals, so that one cannot cheat another with impunity. In this scenario, one player could still offer environmentally friendly technologies and others could purchase from them, improving sustainability. Also, although there might not appear to be any policy in place in this scenario, the decision not to regulate is itself a policy choice.

- *Encouraging voluntary action:* Under voluntary policies, government begins moving the system in sustainable direction, by encouraging (though not requiring) sustainable choices. Although less effective than legally binding policies, voluntary programs still have the potential to reduce negative impacts and may be the only feasible policy if governments are unable to agree on binding measures. Also, enterprises and individuals may implement voluntary programs to preempt government from passing binding requirements, which they may wish to avoid.

- *Use of financial instruments:* Financial instruments represent the first level of binding government policy, in the form of taxes or credits. They do not directly restrict behavior, but they do place an enforceable financial charge on certain choices. Financial instruments may either reward desired behavior (e.g., tax credit for choosing a technology with green benefits) or penalize undesirable behavior (e.g., surcharge on private automobile with low fuel economy).

- *Binding targets:* This level of policy entails enforceable targets, such as for the maximum allowable volume of pollution, or the minimum required level of use of a desirable technology such as renewable energy. Policies with binding targets may include quantitative targets that increase (for desirable practices) or decrease (undesirable practices) over time. They may also incorporate a trading system whereby a player that exceeds their target can sell "permits" to other players that are falling short.

- *Command and control:* In this most stringent of all policies, not only are targets binding, but the policy specifically regulates what type of technology is to be used to meet the target. Enforcement then consists of observation of players over the duration of the policy to make sure they remain in compliance.

Geographic Scope of Policy

Policies are enforceable at different degrees of stringency, and they also fall under the jurisdiction of different levels of government, from smallest to largest. At a more regional level, local governments are limited in their ability to enforce energy policy because of the complexity and administrative overhead involved. Packing recycling policies may indirectly help with energy efficiency since recycling of packaging materials can save energy compared to creating packaging from virgin raw materials. State government is usually a more appropriate policy level, especially in the case of the United States in large states such as California or New York, where states can issue binding regulations, levy surcharges on energy users to pay for programs or R&D, and so on.

The national government level is often the strongest location for issuing and enforcing policy, since national governments generally have the strongest mandates to oversee the evolution of technology, or protection of communities and the environment. Also, negative consequences from energy choices (e.g., pollution from burning of fuels) often spill across state boundaries, so the geographic scope is appropriate. However, in some cases opinions about policy are divergent between states to the point that it is not possible to reach a consensus on a national policy. In these cases, states may be the best level to proceed with more ambitious policies beyond those of the government. States are sometimes seen as "laboratories" for testing policies that if successful can be implemented at the national level.

Lastly, international bodies such as the United Nations or the European Union can have a role in setting policy. Groups of nations may come together in temporary unions to jointly implement policies as well. Issues such as regulation of greenhouse gas emissions are logically addressed at the global level since it is global emissions and atmospheric concentration that determine the level of harm. The complexity of reaching consensus can, however, be even more challenging than reaching policy agreement at the national level.

Advantages and Disadvantages of Policy Approaches

Advantages and disadvantages of policy approaches can be distilled to a trade-off between potential degree of impact on behavior and complexity of implementation. From a sustainability perspective, the critique of the libertarian perspective is that progress is possible but cannot be achieved at a speed commensurate with the urgency of the challenge. In the free market, some sellers will provide green options to buyers, leading to some overall environmental gain. Any seller who truly tries to fully meet their commitment to sustainability will be at such a cost disadvantage, however, that their product or service will not be economically viable.

At the other end of the spectrum, binding measures such as taxes, binding targets, or command and control have the potential for more rapid rate of change but also several potential flaws. Creation and enforcement of policies creates an economic burden that must be paid for by all the players in the system. Certain policies such as command and control may achieve the desired level of pollution reduction but at excessive cost, in cases where a regulated industry player may know of a more cost-effective way of meeting a pollution target but not be allowed to implement it because the policy requires a specific fix. Policy intervention may lead to unintended consequences, such as when more efficient technologies allow players to consume more units of some good or service because the cost per unit has been reduced. Lastly, policies may provide perverse incentives for some player to successfully introduce technologies that meet the letter of the law in terms of how the policy is written, but do not actually achieve any measurable improvement. For example, a policy may pay a player to introduce a technology or practice that is economically viable for the player, but since the technology or practice does not actually work, the government ends up spending the taxpayer's funds on a program that has no benefit.

2-7 Summary

During the course of the twenty-first century, energy systems will be required to meet several important goals, including conformance with the environmental, economic, and social goals of sustainable development. The existence of multiple goals, multiple stakeholders, and numerous available technologies lends itself to the use of a "systems approach" to solving energy problems, which emphasizes additional effort in the early conceptual and design stages in order to achieve greater success during the operational lifetime of a product or project. Qualitative systems tools available to the decision maker include the use of "stories" or causal loop diagrams to describe systems interactions. Quantitative tools include exponential, logistics, and triangle functions to project rates of growth or market penetration of energy technologies. They also include scenario analysis and quantitative modeling to estimate the future situation of energy-related measures such as consumption or pollution, and life-cycle analysis, energy return on investment, multi-criteria analysis, and optimization to determine whether a particular technology is acceptable or to choose among several competing technologies or solutions. Lastly, Divisia analysis provides a means of understanding the contribution of underlying factors to overall energy consumption or CO_2 emissions trends.

References

Lippiatt, B. (1999). "Selecting Cost-Effective Green Building Products: BEES Approach." *J Construction Engineering and Management*, Nov–Dec 1999, pp. 448–455.

Norberg-Bohm, V. (1992). "International Comparisons of Environmental Hazards: Development and Evaluation of a Method for Linking Environmental Data with the Strategic Debate Management Priorities for Risk Management." *Report from Center for Scientific and International Affairs*. John F. Kennedy School of Government, Harvard University, Cambridge, MA.

Schipper, L., L. Scholl, and L. Price (1997). "Energy Use and Carbon Emissions from Freight in 10 Industrialized Countries: An Analysis of Trends from 1973 to 1992." *Transportation Research Part D: Transport and Environment*, Vol. 2, No. 1, pp. 57–75.

U.S. Environmental Protection Agency (1990). "*Reducing Risk: Setting Priorities and Strategies for Environmental Protection.*" Report SAB-EC-90-021. Scientific Advisory Board, Washington, DC.

Vanek, F. (2002). "The Sector-Stream Matrix: Introducing a New Framework for the Analysis of Environmental Performance." *Sustainable Development*, Vol. 10, No. 1, pp. 12–24.

World Commission on Environment and Development (1987). *Our Common Future.* Oxford University Press, Oxford.

Further Reading

Ackoff, R. (1978). *The Art of Problem Solving*. Wiley & Sons, New York.

Greene, D., and Y. Fan (1995). "Transportation Energy Intensity Trends: 1972–1992." *Transportation Research Record*, No. 1475, pp. 10–19.

Hawken, P., A. Lovins, and L. Lovins (1999). *Natural Capitalism: Creating the Next Industrial Revolution*. Little, Brown and Co, Boston.

Jackson, P. (2010). *Getting Design Right: A Systems Approach*. CRC Press, Boca Raton, FL.

Koen, P., G. Ajamian, and S. Boyce. (2002). Fuzzy front end: Effective methods, tools, and techniques. In P. Belliveau, Ed., *Product Development Management Association Toolbook*, PDMA, Seattle.

Kossiakoff, A. (2003). *Systems Engineering: Principles and Practice*. Wiley & Sons, New York.

Motwani, J., A. Kumar, and J. Anthony. (2004) A business process change framework for examining the implementation of six sigma: A case study of Dow Chemical. *Total Quality Management Magazine* 16:4, pp. 273–283.

Sterman, J. (2002). "System Dynamics Modeling: Tools for Learning in a Complex World." *IEEE Engineering Management Review*, Vol. 30, No. 1, pp. 42–52.

Vanek, F., P. Jackson, and R. Grzybowski (2008). "Systems Engineering Metrics and Applications in Product Development: A Critical Literature Review and Agenda for Further Research." *Systems Engineering*, Vol. 11, No. 2, pp. 107–124.

Exercises

2-1. Consider a technical project with which you have been involved, such as an engineering design project, a research project, or a term project for an engineering course. The project can be related to energy systems, or to some other field. Consider whether or not the project used the systems approach. If it did use the approach, describe the ways in which it was applied, and state whether the use of the approach was justified or not. If it did not use the approach, do you think it would have made a difference if it had been used? Explain.

2-2.   Causal loop diagrams: In recent years, one phenomenon in the private automobile market of many countries of the world, especially the wealthier ones, has been the growth in the number of very large sport utility vehicles and light trucks. Create two causal loop diagrams, one "reinforcing" and one "balancing," starting with "number of large private vehicles" as the initial component.

2-3.   It is reasonable to expect that given growing wealth, the emergence of a middle class, access to modern conveniences, and so forth, the "emerging" countries of the world may experience exponential growth in energy consumption over the short to medium term. From data introduced earlier, these countries consumed 136 EJ of energy in 1990 and 198 EJ in 2004. Use these data points and the exponential curve to predict energy consumption in the year 2020. Discuss the implications of this prediction.

2-4.   From Fig. 1-1 in Chap. 1, the values of world energy consumption are given below. Calculate a best-fit curve to the six data points 1850–1975, and then use this curve to predict the value of energy consumption in 2000. If the actual value in 2000 is 419 EJ, by how much does the projected answer differ from the actual value?

1850	25 EJ
1875	27 EJ
1900	37 EJ
1925	60 EJ
1950	100 EJ
1975	295 EJ

2-5.   Current global energy consumption trends combined with per capita energy consumption of some of the most affluent countries, such as the United States, can be used to project the future course of global energy demand over the long term.
   a. Using the data in Table 2-2, calculate per capita energy consumption in the United States in 2004, in units of GJ per person. (Alternatively, solve using U.S. 2011 data.)
   b. If the world population ultimately stabilizes at 10 billion people, each with the energy intensity of the average U.S. citizen in 2004, at what level would the ultimate total annual energy consumption stabilize, in EJ/year?
   c. From data introduced earlier, it can be seen that the world reached the 37 EJ/year mark in 1900, and the 100 EJ/year mark in 1950. Using these data points and the logistics formula, calculate the number of years from 2004 until the year in which the world will reach 97% of the value calculated in part (b).
   d. Using the triangle formula, pick appropriate values of the parameters a and b such that the triangle formula has approximately the same shape as the logistics formula.
   e. Plot the logistics and triangle functions from parts (c) and (d) on the same axes, with year on the x-axis and total energy consumption on the y-axis.
   f. Discussion: The use of the United States as a benchmark for the target for world energy consumption has been chosen arbitrarily for this problem. If you were to choose a different country to project future energy consumption, would you choose one with a higher or lower per capita energy intensity than the United States? What are the effects of choosing a different country?

2-6. Multi-criteria decision making: A construction firm is evaluating two building materials in order to choose the one that has the least effect on the environment. The alternatives are a traditional "natural" material and a more recently developed "synthetic" alternative. Energy consumption in manufacturing is used as one criterion, as are the effect of the material on natural resource depletion and on indoor air quality, in the form of off-gassing of the material into the indoor space. Cost per unit is included in the analysis as well. The scoring for each criterion is calculated by assigning the inferior material a score of 100, and then assigning the superior material a score that is the ratio of the superior to inferior raw score multiplied by 100. For example, if material A emits half as much of a pollutant as material B, then material B scores 100 and material A scores 50. Using the data given below, choose the material that has the overall lower weighted score. (Note that numbers are provided only to illustrate the technique and do not indicate the true relative worth of synthetic or natural materials.)

Data for Problem 6:

	Units	Natural	Synthetic	Weight
Energy	MJ/unit	0.22	0.19	0.24
Natural resource	kg/unit	1000	890	0.15
Indoor AQ	g/unit	0.1	0.9	0.21
Cost	$/unit	$1.40	$1.10	0.40

2-7. Based on values in Table 2-2, for each of the five countries shown, calculate total annual GDP and energy consumption in EJ for the years 2004 and 2011. Then discuss the countries relative to one another in two years, and the relative rate of change for the period 2004–2011.

2-8. The table below gives the values for the year 2008 for China, Germany, India, Japan, and the United States. Calculate the three Kaya Identity factors (GDP/P, E/GDP, and CO_2/E) for each country and then discuss differences between them.

Country:	P (million)	GDP (billion $)	E (EJ)	CO_2(mil. tonnes)
China	1322.0	$7007	89.74	6166.6
Germany	82.4	$2835	15.15	812.6
India	1129.9	$3051	21.05	1449.0
Japan	127.4	$4306	23.07	1216.3
USA	301.1	$13,851	106.11	5840.5

2-9. A thrifty engineering student is concerned about not exceeding her budget for gasoline, so she conducts a Monte Carlo simulation to see what range of costs her driving might incur. The cost can be determined from miles driven, fuel economy in miles per gallon (MPG), and cost per gallon for fuel. Her budget is $700 for the year. The inputs to the cost are all normally distributed mean and standard deviation given in the form (mean, SD): mileage (6000, 1000); miles per gallon (30, 4); cost of fuel ($3.25, $0.25). (a) What is the expected value for her spending for the year? Does she expect to stay within budget? (b) Now suppose she runs the simulation, and for a single replicate or iterate, generates the following random numbers between 0 and 1, as the basis for calculating a value: for miles driven, 0.139; for fuel economy, 0.355; for cost of fuel, 0.437. What is the annual cost returned for this replicate? Is the value within budget? (c) Carry out a Monte Carlo simulation of the problem generating 1000 replicates and provide the average value of the sample, a histogram of probability of being in appropriate cost bins, and the percent of iterates that are above the $600 threshold.

CHAPTER 3

Engineering Economic Tools

3-1 Overview

This chapter gives an overview of economic tools related to energy systems. The differences between current and constant monetary amounts are first introduced, followed by two types of economic evaluation of energy projects, both with and without discounting of cash flows to take into account the time value of money. The latter part of the chapter considers direct versus external costs, and ways in which governments at times intervene in energy-related decision making in order to further social aims such as diversity of energy supply or a clean environment.

3-2 Introduction

As already mentioned in the previous chapter, if the engineer wishes to take a systems approach to energy, then the financial dimension of energy systems cannot be omitted. The economics of energy systems includes the initial cost of delivering components that function in the system (turbines, high-voltage transmission lines, and so forth); ongoing costs associated with fuel, maintenance, wages, and other costs; and the price that can be obtained in the market for a kWh of electricity or a gallon of gasoline equivalent of vehicle fuel. Energy economics is itself a highly specialized and elaborate field with a very large body of knowledge in support, and a full treatment of this field is beyond the scope of this chapter. However, the engineer can use a basic knowledge of the interface between energy technology and energy economics presented here to good advantage.

As a starting point before delving into the details of quantitative evaluation of the costs, revenues, and benefits of energy systems, we should consider two important functions of economics related to those systems. The first is project specific, namely, that most energy projects are scrutinized for cost-effectiveness, unless they are designated for some other demonstrative or experimental purpose. Any such economically driven energy projects must either convince the owner/operator that they are a good use of internally held resources, or convince an external funding source that they are worthy of financial backing. The second function is the relationship between society's collective choices about energy systems and the cost of energy products and services. Excessive investment in overly expensive energy projects will lead to higher energy costs for the public, but so will underinvestment in and neglect of the existing stock of energy infrastructure, so society must strike a careful balance. Whether it is to a large industrial corporation in one of the rich countries or to a villager in a poor country considering the cost of a bundle of firewood, the cost of energy is in fact one of the most important determinants of our well-being among the many types of economic prices that we pay.

3-2-1 The Time Value of Money

As consumers, we hardly need reminding that the purchasing power of our money in most industrialized economies is declining all the time. For most products and services, the amount that can be purchased by 1 dollar, 1 euro, or 100 yen is slightly less than it was a few years ago, due to inflation. In addition, investors who have money available for investment expect a return on their funds beyond the rate of inflation whenever possible, so that money invested in any opportunity, including an energy project, should meet these expectations. On this basis, money returned in the future is not worth as much as money invested today. The change in value of money over time due to these two trends is given the name *time value of money*.

Taking the role of inflation first, many national governments track the effect of inflation on prices by calculating an annual price index composed of the average price of a representative collection of goods. Measuring the change in the price of this collection from year to year provides an indicator of the extent to which inflation is changing prices over time. For individual consumers, a measure of the overall cost of living is of the greatest interest, whereas for manufacturers and other businesses, a measure of the cost of goods and other expenditures needed to stay in business is desired. Therefore, governments typically publish a *consumer price index (CPI)* for consumers and *producer price indices* for various industries (e.g., a construction producer price index would be most relevant for construction companies building transportation infrastructure). The CPI values for the United States for the period 1997–2014 are shown in Table 3-1, along with the CPI values for the United Kingdom for the same period for comparison. It can be observed from this table that consumer prices grew more slowly in the United Kingdom than in the United States up until 2012. (Industry-specific producer price indices are not shown, but can be obtained from relevant government agencies.)

Using the CPI value, we can make a distinction between prices that are taken at their face value in the year they are given, and prices that are adjusted for inflation. A price given in *current dollars* is worth the given value in unadjusted dollars that year. A *constant year-Y dollar* is worth the same amount in any year, and enables direct comparison of costs evaluated in different years. Comparisons made between costs given in constant dollars are sometimes said to be given *in real terms*, since the effect of inflation has been factored out. The difference is illustrated in Example 3-1.

Example 3-1   A particular model of car costs \$17,000 in 1998 and \$28,000 in 2005, given in current dollars for each year. How much are each of these values worth in constant 2002 dollars? Use the U.S. CPI values from Table 3-1.

Solution   Values from the table are used to convert to constant 2002 dollars, for \$17,000 in 1998 dollars and \$28,000 in 2005 dollars, respectively:

$$\$17{,}000\left(\frac{104.5}{94.7}\right) = \$18{,}763$$

$$\$28{,}000\left(\frac{104.5}{113.4}\right) = \$24{,}986$$

1998 dollars $\times (104.5)/(94.7) = \$18{,}763$

Thus even after factoring out *inflation*, the cost of this car model has risen considerably in real terms.

Year	United States	United Kingdom
1997	87.4	96.3
1998	94.7	97.9
1999	96.7	99.1
2000	100.0	100.0
2001	102.8	101.2
2002	104.5	102.5
2003	106.9	103.9
2004	109.7	105.3
2005	113.4	107.4
2006	117.1	109.9
2007	120.4	112.5
2008	125.0	116.5
2009	124.6	119.0
2010	126.7	123.0
2011	130.6	128.6
2012	133.3	132.2
2013	134.9	135.6
2014	137.5	137.5

Indexed to Year 2000 = 100.

Sources: U.S. Bureau of Labor Statistics (2015), for United States; U.K. National Statistics (2015), for United Kingdom.

TABLE 3-1 Consumer Price Index Values for the United States and United Kingdom from 1997 to 2014

In order to simplify the economic analysis, all costs are given in constant dollars in the remainder of this book unless otherwise stated, and we give no further consideration to the effect of inflation.[1]

Turning to the second source of time value of money, much of the funding for investment in energy projects comes from private investors or government funding sources, which are free to consider other options besides investment in energy. Each investment option, including the option of keeping money in hand and not investing it, has an *opportunity cost* associated with the potential earnings foregone from investment opportunities not chosen. There is an expectation that investments will do more than keep pace with inflation, and will in fact return a profit that exceeds the inflation rate. This expectation can be incorporated in economic calculations through *discounting*, or counting future returns, when measured in real terms, as worth less than the same amount of money invested in the present.

[1]For the interested reader, full-length texts on engineering economics such as Blank and Tarquin (2012) or Newnan et al. (2011) treat this topic in greater detail.

3-3   Economic Analysis of Energy Projects and Systems

We now turn from the question of individual prices and costs to the economic analysis of entire energy projects, which is based on a comprehensive treatment of the various cost and revenue components of the project. Since it adds complexity to the calculations, the time value of money is incorporated in some economic analyses but ignored in others, depending on the context.

3-3-1   Definition of Terms

Whether time value is considered or not, the following terms are used in economic analysis of projects:

- *Term of project:* Planning horizon over which the cash flow of a project is to be assessed. Usually divided into some number of years N.
- *Initial cost:* A one-time expense incurred at the beginning of the first compounding period, for example, to purchase some major asset for an energy system.
- *Annuity:* An annual increment of cash flow related to a project; typically these increments repeat each year for several years, as opposed to a one-time quantity such as the initial cost. Annuities can either be positive (e.g., annual revenues from the sales of energy from a project) or negative (e.g., annual expenditure on maintenance). They may also be reported as the net amount of flow after considering both positive and negative flows; any energy investment must have a net positive annuity in order to repay the investment cost. In this book, annuities are given as net positive flows after subtracting out any annual costs, unless otherwise stated.
- *Salvage value:* A one-time positive cash flow at the end of the planning horizon of a project due to sale of the asset, system, and the like, in its actual condition at the end of the project. Salvage values are typically small relative to the initial cost.

3-3-2   Evaluation without Discounting

Although the time value of money is an important element of many financial calculations, in some situations it is reasonable to deal only with face value monetary amounts, and ignore discounting. This is especially true for projects with a shorter lifetime, where the impact of discounting will be small. Two terms are considered: *simple payback* and *capital recovery factor (CRF)*.

Simple payback is calculated by adding up all cash flows into and out of a project, including initial costs, annuities, and salvage value amounts; this value is known as the *net present value (NPV)*. If the NPV of the project is positive, then it is considered to be financially viable. In the case of multiyear projects with a positive simple payback, the *breakeven point* is the year in which total annuities from the project surpass initial costs, so that the initial costs have been paid back.

The capital recovery factor (CRF), as applied to the generation of electricity,[2] is a measure used to evaluate the relationship between cash flow and investment cost.

---

[2]The definition of CRF for electric utilities is different from CRF for other types of investments, where CRF is the ratio of a capital value to its annualized amount under discounting, as in Eq. (3-4).

It can be applied to short-term investments of less than 10 years. In this case, current dollar values are used.

In order to calculate CRF, we start with NPV of the project, the yearly annuity, and the term in years N. The CRF is then

$$CRF = \frac{ACC}{NPV} \qquad (3\text{-}1)$$

where ACC is the *annual capital cost* calculated as

$$ACC = annuity - \frac{NPV}{N} \qquad (3\text{-}2)$$

In the United States, the Electric Power Research Institute (EPRI) recommends a maximum CRF value of 12%; in cases with higher values, the ratio of the annual cost of capital [as defined in Eq. (3-2)] to net value of the project is too high.

3-3-3 Discounted Cash Flow Analysis

Discounting of cash flow analysis starts with the premise that the value of money is declining over time and that therefore values in the future should be discounted relative to the present. Two additional terms are added that pertain in particular to discounted cash flow:

- *Interest rate:* Percent return on an investment, or percent charged on a sum of money borrowed at the beginning of a time horizon. Unless otherwise stated, interest is compounded at the end of each year, that is, the unit of 1 year is referred to as the "compounding period."
- *Minimum attractive rate of return (MARR):* The minimum interest rate required for returns from a project in order for it to be financially attractive, which is set by the business, government agency, or other entity that is making a decision about the investment.

Suppose there is a project that is expected to include initial cost, constant annuity, and salvage value during the course of its lifetime of N years. We can represent the planning horizon for this project in a *cash flow diagram* as shown in Fig. 3-1.

Considerations in Determining a Suitable MARR

The motivation for setting the MARR value comes from the time value of money concept discussed in Sec. 3-2-1, which states that a dollar today is worth more than a dollar tomorrow. The following thought exercise is illustrative: suppose you were offered a choice between receiving $1 today or $1 next year, the choice would be simple, as a dollar today is better. Then, suppose you were offered $1 today or $1.01 next year. The decision is still simple: the choice of $1 today is probably more attractive. If we keep increasing the value received in the future, at some point, receiving $1 today is no longer as attractive as receiving $1.xx next year. The decision has become more difficult and, if xx is large enough, you would not hesitate to wait to receive $1.xx next year. The value of xx at which you feel the decision is a toss-up is a first estimate of the MARR. If xx = $0.10, for example, the MARR equals 10% in your estimate.

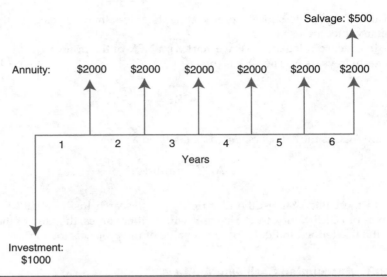

FIGURE 3-1   Example of cash flow diagram for project with initial cost, constant annuity, and salvage value.

There are three main reasons for demanding a minimum attractive rate of return on foregoing the present value of a given amount of money. The first reason is inflation, which reduces the future value of the dollar. Inflation in the United States during the past century has averaged approximately 4% per year, so during that time period a dollar in year $y + 1$ was worth 4% less than a dollar in year y, on the average. For the first decade of the twenty-first century, the total inflation was 26.7%, according to the CPI values in Table 3-1. When the MARR equals the rate of inflation, the return in real dollars is zero, but there is no actual loss of value. Logically, the MARR should not be less than the expected rate of inflation. (Deflation would make tomorrow's dollar worth more than today's dollar, but a deflation economy is an unusual and problematic situation.)

The second reason is the return possible from alternate investments. In many cases, investing in energy systems can be costly. Even though there may be incentives to invest in an energy system that goes beyond monetary return (e.g., environmental consciousness), most persons will want to see their investment pay back sufficiently to cover at least the financial outlay. If the money could be invested in a venture with a greater return, many people would choose the latter. This means that rates of return from available alternate investments determine the baseline acceptable value for the MARR (assuming it is greater than the expected rate of inflation).

The third reason, and perhaps the most important, is the *risk* associated with the investment. Any investment carries some element of risk. The risk may be low, such as a government-insured savings account in a bank. The risk may be high, such as investing in a new, unproven technology. The high-risk investment is more of a gamble, from which you would want a relatively high rate of return if all goes well to balance the possibility of losing all or most of your capital if all does not go well.

Equations for Relating Present, Annual, and Future Values

As a basis for incorporating the time value of money, it is useful to be able to make conversions between the present, annual, and future values of elements in a cash flow analysis. We begin with the conversion between present and future values of a single cash flow element. Given interest rate i, time horizon of N years, and a present value P of an amount, the future value of that amount F is given by

$$F = P(1+i)^N \qquad (3\text{-}3)$$

It is also possible to translate a stream of equal annuities forward or backward to some fixed point at present or in the future. Given a stream of annuities of equal size A over a period of N years, the equivalent present value P is

$$P = A\frac{(1+i)^N - 1}{i(1+i)^N} \qquad (3\text{-}4)$$

Also, given the same annuity stream and time horizon, the future value F of the annuity at the end of the Nth year can be calculated as

$$F = A\frac{(1+i)^N - 1}{i} \qquad (3\text{-}5)$$

Equations (3-4) and (3-5) assume a constant value of the annuity A; the case of irregular annuities is considered below.

These formulas can be inverted when needed, for example, if one knows F and wishes to calculate P, then Eq. (3-3) can be rewritten $P = F/(1 + i)^N$, and so on. Also, it is convenient to have a notation for the conversion factor between known and unknown cash flow amount, and this is written $(P/F, i\%, N)$, $(P/A, i\%, N)$, and so on, where the term P/F is read "P given F."

Example 3-2 A cash flow stream consists of an annuity of \$100,000 earned for 10 years at 7% interest. What is the future value of this stream at the end of term? What is the value of the factor $(F/A, i\%, N)$ in this case?

Solution Substituting $A = \$100,000$, $N = 10$, and $i = 7\%$ into Eq. (3-3) and solving to the nearest dollar gives

$$F = \$100,000\frac{(1+0.07)^{10} - 1}{0.07} = \$1,381,645$$

Then $(F/A, 7\%, 10) = \$1,381,645/\$100,000 = 13.82$.

Financial Evaluation of Energy Projects: Present Worth and Annual Worth Methods

The discounting equations in the previous section provide a way to compare costs and benefits of a project while taking into account the time value of money. The *present worth* method is one approach where all elements of the financial analysis of the project are discounted back to their present worth, except for initial costs since these are already given in present terms. The positive and negative elements of the cash flow are then summed, and if the NPV is positive, then the investment is financially attractive. In cases where an annuity does not begin in year 1 of a project, the stream of annuities can be discounted to their present worth in the first year of the annuity, and then this single sum can be treated as a future value, and discounted to the beginning of the project.

When evaluating financial viability using the time value of money, we incorporate the chosen MARR for the project by using this value as the interest rate for discounting annuities and future values. As MARR increases, the present equivalent of a future single sum payment or annuity decreases, and it becomes more difficult for a project to generate a sufficient return to offset initial costs.

Similarly to the present worth method, we can also calculate the *annual worth* of a project by discounting all cost components to their equivalent annual value, summing up positive and negative components, and adopting a project if the overall value is positive.

Example 3-3   A municipality is considering an investment in a small-scale energy system that will cost $6.5 million to install, and then generate a net annuity of $400,000/year for 25 years, with a salvage value at the end of $1 million. (a) Calculate the net worth of the project using simple payback. (b) Suppose they set the MARR at 5%, what is the net present worth of the project by this approach?

Solution
(a) The net worth of the project by the simple payback method is the total value of the annuities, plus the salvage value, minus the initial cost of the project, or

$$25 \times (\$400,000) + \$1,000,000 - \$6,500,000 = \$4,500,000$$

(b) Using discounting, the following factors are needed: $(P/A, 5\%, 25) = 14.09$ and $(P/F, 5\%, 25) = 0.295$. The NPV is then

$$14.09 \times (\$400,000) + 0.295 \times (\$1,000,000) - \$6,500,000 = -\$567,119$$

Thus the project is not viable at an MARR of 5%.

Discussion   The comparison of simple payback and present worth analysis using discounting illustrates the potential effect of even a relatively low MARR on financial viability over a long period such as 25 years. In this case, a project in which returns from annuity and salvage value exceed initial costs by 69% using simple payback is not viable when discounting is taken into account. This effect can be large on renewable energy projects, where the initial capital cost is often high per unit of productive capacity, relative to the annual output, so that these projects are likely to have a long payback period.

Nonuniform Annuity and Internal Rate of Return Method

In Eqs. (3-4) and (3-5) annuities are represented as remaining constant over the project lifetime. It is also possible to discount a nonuniform set of annuities to its equivalent present worth value PW by treating each annuity as a single payment to be discounted from the future to the present:

$$PW = \sum_{n=1}^{N} \frac{A_n}{(1+i)^n} \tag{3-6}$$

Here A_n is the value of the annuity predicted in each year n from 1 to N. Predicted costs that vary from year to year can be converted to present value in the same way.

Along with the NPV of a project at a given MARR, it is also useful to know the interest rate at which a project exactly breaks even at the end of its lifetime. This information can then be used to evaluate the relative attractiveness of a project, and is known as the internal rate of return (IRR) method, or alternatively as the discounted cash flow return on investment (DCFROI) method. For a project with either uniform or nonuniform annuities, it is convenient to calculate the DCFROI by setting up a table in an appropriate software package that calculates the NPV of the project as a function of interest rate, and then solving for the interest rate that gives NPV = 0. Some packages also have a built-in IRR calculation function.

Example 3-4 Calculate the IRR for the investment in Example 3-3.

Solution In order to solve using a numerical solver, set up a table in the chosen software package with the following headings: "Year," "Capital Expense or Revenue," "Net Flow," and "Discounted Net Flow," as shown. Start in year "0" and consider the capital cost of the project to be incurred in year 0. The salvage value can be considered capital revenue in year 25. The values in the table are calculated both with the initial interest rate of $i = 5\%$ from Example 3-3 and with the IRR of $i = 4.2\%$, given in units of thousands of dollars.

Year	Capital Expense or Revenue (× $1000)	Annuity (× $1000)	Net Flow (× $1000)	Discounted Net Flow @$i = 5\%$ (× $1000)	Discounted Net Flow @$i = 4.2\%$ (× $1000)
0	−6500	0	−6500	−6500	−6500
1	0	400	400	381	384
2	0	400	400	363	369
3	0	400	400	346	354
etc.	etc.	etc.	etc.	etc.	etc.
25	1000	400	1400	413	504
Total	n/a	n/a	4500	−567	0

Solving for the value that gives NPV = 0 gives $i = 4.2\%$. In other words, if we sum the values in the rightmost column in the table (note that for brevity not all values are shown, but the reader can recreate the table if desired), they will add to $0 in year 25. For comparison, if we sum the adjacent column, the cumulative NPV is −$567,000 with $i = 5\%$. The difference comes from slight differences in the annual discounted net flow after year 0, for example $384K versus $381K in Year 1, etc.

The cumulative NPV by year for the investment in Example 3-4 with interest rate $i = 4.2\%$ is shown in Fig. 3-2. The value plotted for each gives the NPV for all years up to that point. With each passing year, the value of the annuity is reduced due to discounting, so the slope of the NPV curve decreases.

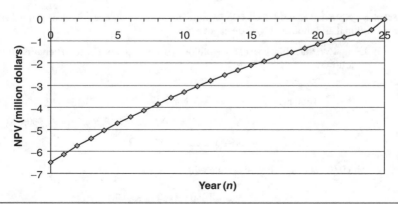

FIGURE 3-2 Net present value of energy investment at MARR = 4.2%.

Benefit-Cost Ratio Method

In addition to the PW and IRR methods, a third approach is to calculate the *benefit-cost ratio*, or ratio of all the benefits of the project to all of the costs. The B/C ratio method is especially common in public sector since it explicitly incorporates benefits accruing to different stakeholders in the benefit part of the ratio. For instance, a project to build a hydroelectric dam might provide benefits in the form of boating, fishing, and water for irrigation. The benefit of each could be estimated and included in the sum of benefits in the ratio.

In general, if the B/C ratio is greater than 1, the project is acceptable, and if it is less than 1, it is not. Where the B/C ratio is close to 1, the decision-maker should re-examine the inputs to make sure they are as accurate as possible, and to look for any additional information that might help to make a decision one way or the other.

Within the general category of B/C ratios, there are two alternatives, namely, the *conventional B/C ratio* and *modified B/C ratio*, depending on how operating costs incurred by the project operator are treated. In equation form, the distinction is the following:

$$\text{Conventional B/C} = \frac{\text{Total benefits}}{\text{Initial cost} + \text{Operating cost}} \qquad (3\text{-}7)$$

$$\text{Modified B/C} = \frac{\text{Total benefits} - \text{Operating cost}}{\text{Initial cost}} \qquad (3\text{-}8)$$

Since there is no computational advantage of one B/C method over the other in Eqs. (3-7) and (3-8) and different agencies may prefer one over the other for reasons of tradition, it is useful for the reader to be familiar with both. Given a specific project with fixed benefits, initial cost, and operating cost, both conventional and modified B/C will return the same answer, i.e., if B/C > 1 with one approach, it will be with the other. The numerical values, however, will be different. One should therefore not compare the numerical value of the B/C ratio returned by the two methods. Having a higher ratio with one method does not mean that the project is somehow better for having used that method. Examples 3-5 and 3-6 illustrate the use of the B/C ratio method.

Example 3-5   Calculate the B/C ratio for the investment in Example 3-3, (a) using the present worth of all costs and benefits, and (b) using the annual worth of all costs and benefits.

Solution   Note that because Example 3-3 does not include any operating costs, the distinction between conventional and modified B/C does not arise.

(a) Using PW: the initial cost of $6.5 million is used as given. The total benefits must be discounted, as follows:

$$\text{Total benefit} = \text{Annuity} + \text{Salvage}$$

$$= (P/A, 5\%, 25) \times \$400,000 + (P/F, 5\%, 25) \times \$1,000,000$$

$$= \$5.64M + \$295K = \$5.93M$$

The B/C ratio can now be calculated:

$$\text{B/C} = \frac{\text{Total benefit}}{\text{Total cost}} = \frac{\$5.93M}{\$6.5M} = 0.91$$

(b) Using AW: the annuity of $400,000 is used as given. Both initial cost and salvage value must be discounted. From part (a) we know that the present worth of the salvage value is $295,000. Therefore,

$$\text{Annualized initial cost} = (A/P, 5\%, 25) \times \$6.5M = \$461,191$$

$$\text{Annualized salvage value} = (A/P, 5\%, 25) \times \$295K = \$20,952$$

The B/C ratio is then calculated:

$$B/C = \frac{\text{Total benefit}}{\text{Total cost}} = \frac{\$400K + \$21K}{\$461K} = \frac{\$421K}{\$461K} = 0.91$$

Note that both using PW and AW returns the same value of B/C = 0.91, and that the result in terms of the decision about the project is the same as in Example 3-3, i.e., the project is not accepted.

Example 3-6 Now suppose the project of Example 3-3 includes an annual operating cost of $50,000, with all other values remaining the same. Calculate both the conventional and modified B/C ratios using the present worth of all benefits and costs.

Solution Since the annual worth of the operating cost is $50,000, the present worth of all annual operating costs discounted to the present is ($50,000)(P/A,5%,25) = $704,500. The two B/C ratios are therefore:

$$\text{Conventional B/C} = \frac{\text{Total benefits}}{\text{Initial cost} + \text{Operating cost}} = \frac{\$5.93M}{\$6.5M + \$704.5K} = \frac{\$5.93M}{\$7.21M} = 0.822$$

$$\text{Modified B/C} = \frac{\text{Total benefits} - \text{Operating cost}}{\text{Initial cost}} = \frac{\$5.93M - \$704.5K}{\$6.5M} = \frac{\$5.23M}{\$6.5M} = 0.805$$

Note that, as explained above, both approaches arrive at the same conclusion (project not viable) but with slightly different values for B/C.

Advantages and Disadvantages of PW, IRR, and B/C Methods

In evaluating advantages and disadvantages of different methods for economic evaluation of projects, the primary distinction is between the IRR and other methods. Other methods rely on an a priori determination of the MARR to be used, while the IRR does not. In general, IRR is more commonly used in private industry, while the B/C method is especially common in the public sector. The PW method is used in both the private and public sectors.

The primary advantage of the IRR method is transparency. Since the decision-maker is not required to set the MARR value in advance, the effort involved in evaluating MARR is avoided. To the outside observer as well, in the PW or B/C method, if one is told that the project was accepted or rejected at a certain MARR, it is not clear without further investigation how the organization considering the project arrived at the MARR, and what the impact on the decision would be if MARR were revised.

Furthermore, since IRR always arrives at a discounted cash flow where the net present value of all costs and revenues is zero, the issue of how to interpret the positive or negative worth of the net present value relative to the total initial and operating costs that would be incurred does not arise. Taking the case of Example 3-3, we know that at MARR = 5% the project has a negative worth of ~$567,000 for $6.5 million invested, but at MARR = 4.2% it breaks even. Alternatively, suppose the input values for evaluating the project changed such that at MARR = 5% the project had a positive net present value of a few hundred thousand dollars. Given the difficulty of predicting far into the future the value of the annual revenue stream for the project, at what point is the positive value high enough that the analyst can choose the project with confidence? The IRR method avoids these complexities.

The IRR method does have at least two disadvantages, however. First, since the decision-maker has not taken the effort to set an MARR, the absolute value of the IRR returned may not have meaning in and of itself. For example, one could use the IRR = 4.2% in Example 3-4 to determine that the project is not as attractive as another alternative project that had a higher IRR, providing a relative interpretation of the IRR. In the case with only the one project at IRR = 4.2%, one must have access to some absolute standard against which to measure IRR, or else further analysis is required to make a decision.

Second, depending on the cash flow pattern, the IRR method has the potential to return spurious results in cases where the cash flow changes sign more than once. In the examples so far, there was just one sign change, since the initial cash flow was negative (outlay of capital in an initial investment) and subsequent flows were positive (net positive amount of revenue or salvage value). In other cases, the cash flow may be negative, then positive, then negative again. In such a situation, there are three possible outcomes: no feasible solutions to the IRR method (in which case it is not financially attractive at any nonnegative interest rate), one solution (in which case it is treated the same as if there were only one sign change), or two solutions (in which case neither solution for interest rate is used).

In the latter case, a special *Modified IRR* method (MIRR) must be used rather than adopting one or the other of the IRR values initially returned. Presentation of the MIRR method requires considerable background and is outside the scope of this book, but is presented in any modern text on engineering economics.[3] For purposes of the present discussion, the reader should be aware in situations with two or more sign changes in the cash flow that one should test for the existence of more than one feasible root for IRR, for example by graphing net present value as a function of interest rate and observing how many times the NPV curve crosses between positive and negative. Where two or more roots exist, the MIRR must be used.

As an example relevant to energy systems, in extraction of oil from an oil field, the addition of a new well to an existing field can lead to a cash flow over multiple years with two sign changes: at first, the flow is negative due to the cost of the well, then the flow is positive due to accelerated extraction of the oil thanks to the additional well, then the flow turns negative again near the end of the investment lifetime when the value of oil extracted drops below what would have been observed with no additional well built. Example 3-7 illustrates a situation where such a cash flow leads to multiple roots and the need to apply the MIRR method.

Example 3-7   An oil company is considering adding a well to a field for the cost of $6 million. The result of this investment will be a net increase in oil production by $4.5 million in Year 1, with the net change in production declining by $1 million for each year thereafter until the end of the investment lifetime in Year 9: Year 2 = $3.5M, Year 3 = $2.5M, and so on. Note that from Year 6 onward the change in production becomes negative, i.e., Year 6 = –$0.5M, etc., and that there is a net increase in total face value of production by $4.5 million thanks to the new well. Show that the IRR method returns two values of i for which NPV = 0.

[3]See, e.g., Newnan et al., *Engineering Economic Analysis*, or Blank, L., and A. Tarquin, *Engineering Economy*. Blank and Tarquin use the term "Modified Rate of Return," or "Modified ROR," rather than MIRR.

Solution   The problem is solved by graphing the NPV of the cash flow for various interest rates, in this case starting with 0% and increasing by increments of 4% up to a large enough number to ascertain whether the problem has 0, 1, or 2 solutions, in this case out to $i = 44\%$. Example values of 0%, 4%, 16%, and 44% are shown in the table below, where the bottom row is the sum of discounted cash flow elements from each of Years 0 to 9:

Year	Interest Rate			
	0%	4%	16%	44%
	($1000s)	($1000s)	($1000s)	($1000s)
0	−$6000	−$6000	−$6000	−$6000
1	$4500	$4327	$3879	$3125
2	$3500	$3236	$2601	$1688
3	$2500	$2222	$1602	$837
4	$1500	$1282	$828	$349
5	$500	$411	$238	$81
6	−$500	−$395	−$205	−$56
7	−$1500	−$1140	−$531	−$117
8	−$2500	−$1827	−$763	−$135
9	−$3500	−$2459	−$920	−$131
Sum of flows:	−$1500	−$342	$730	−$360

From the table, it is clear that the graph of NPV as a function of i starts out negative, rises to a positive value, and returns to a negative value with increasing i. The shape of the curve is further demonstrated by graphing using other values of i for which NPV is calculated in a similar manner, as shown in Fig. 3-3.

Further investigation shows that at interest rates of approximately 5.8% and 37%, NPV = 0 for the oil well cash flow. Thus the IRR method gives two solutions, neither of which can be used to evaluate the investment. The MIRR method must be used instead.

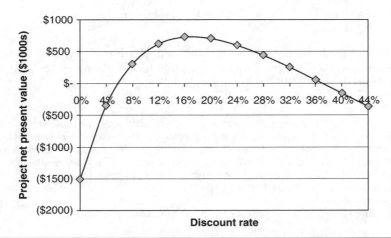

FIGURE 3-3   Net present value of Example 3-7 oil well investment as a function of discount rate from 0% to 44%.

3-3-4 Maximum Payback Period Method

In some instances, an investor is concerned that an investment pays for itself within some maximum time period, rather than its present worth, annual worth, or benefit-cost ratio over a fixed lifetime. The lifetime of the investment is still of interest, but the investment may be rejected even if it has a net lifetime positive value if the payback is insufficiently fast. The maximum payback period in years is determined by the investor based on their own subjective factors, including possibly their willingness to tolerate a longer maximum payback to support an investment that has some green benefits.

For investments that are considered without discounting, the years N to payback is simply the initial value P_{init} divided by the value of the annual return, i.e.,

$$N = \frac{P_{init}}{P_{ann}} \tag{3-9}$$

When discounting applies, the time to payback must be solved iteratively by considering the cumulative effect of the annuity from each additional year on the net present value or NPV. Let n be the number of years included in the cash flow and i the discount rate, then NPV is calculated as follows:

$$NPV = -P_{init} + \sum_{i=1}^{n} \frac{P_{ann}}{(1+i)^n} \tag{3-10}$$

As n is increased by one year to years t, $t+1$, $t+2$, etc., NPV becomes less negative until finally it changes from negative to positive, at which point the number of years to payback can be set to $N = n$. Alternatively, N can be calculated to the nearest tenth of a year. Let NPV_{t-1} and NPV_t be the net present value in the last negative and first positive year, respectively. Then the payback period to the nearest 10th of a year is:

$$N = (t-1) + \frac{|NPV_{t-1}|}{(NPV_t - NPV_{t-1})} \tag{3-11}$$

One of the limitations of the maximum payback period method is that the limited time horizon may eliminate investments that have a strong positive return in the long run. This situation is illustrated in Example 3-8.

Example 3-8   In this example, we compare two investments, both with a discount rate of 5%. Investment A is similar to the conditions of Example 3-3 (20-year life, $400,000 per year annuity), but the initial cost is now $3.5 million and there is no longer any salvage value. Investment B has an initial cost of $200,000, an annuity of $40,000, and a life of 12 years. The maximum allowable payback period is 7 years. (a) Calculate the NPV and payback period for each investment. (b) State whether each investment satisfies the payback requirement.

Solution   Part (a) NPV and payback for Investment A are the following:

$$NPV = -\$3.5M + \$400K \cdot (P/A, 5\%, 20) = -\$3.5M + \$400K \cdot (12.46) = \$1.485M$$

For payback, we create a table with face value investment and annuities, discounted values, and cumulative NPV up to year n, which we continue until the NPV is positive:

Year	Payment	Annuity	Discounted	Cumulative
0	$(3,500,000)		$(3,500,000)	$(3,500,000)
1		$400,000	$380,952	$(3,119,048)
2		$400,000	$362,812	$(2,756,236)
3		$400,000	$345,535	$ (2,410,701)
4		$400,000	$329,081	$(2,081,620)
5		$400,000	$313,410	$(1,768,209)
6		$400,000	$298,486	$(1,469,723)
7		$400,000	$284,273	$(1,185,451)
8		$400,000	$270,736	$(914,715)
9		$400,000	$257,844	$(656,871)
10		$400,000	$245,565	$(411,306)
11		$400,000	$233,872	$(177,434)
12		$400,000	$222,735	$45,301

Note that in year 11, the NPV is negative, and in year 12 it is positive. Calculating payback period to the nearest 10th of a year gives:

$$N = (t-1) + \frac{|NPV_{t-1}|}{(NPV_t - NPV_{t-1})} = 11 + \frac{177,434}{45,301 - (-177,434)} = 11 + 0.8 = 11.8y$$

For Investment B, the calculation is similar:

$$NPV = -\$200K + \$40K \cdot (P/A, 5\%, 20) = -\$200K + \$40K \cdot (12.46) = \$155K$$

	Payment	Annuity	Discounted	Cumulative
0	$(200,000)		$(200,000)	$(200,000)
1		$40,000	$38,095	$(161,905)
2		$40,000	$36,281	$(125,624)
3		$40,000	$34,554	$(91,070)
4		$40,000	$32,908	$(58,162)
5		$40,000	$31,341	$(26,821)
6		$40,000	$29,849	$3028

$$N = 5 + \frac{26,821}{3.028 - (-26,821)} = 5 + 0.9 = 5.9y$$

Part (b) Although Investment A has a much larger NPV than Investment B, it has a payback of 11.8 years, which exceeds the 7 year maximum. Investment B has a smaller NPV, but is within the limits at $N = 5.9$ years.

Example 3-8 shows how the limits to maximum payback period can reject investments that have a positive NPV in the long run. An analogy in the energy sector might be an obsolete office building that has both poor lighting and large single-pane windows in each office. Improvements to lighting may pay for themselves relatively quickly, but the annuity available and hence lifetime NPV might be modest, compared to the overall energy bill. Replacing all the windows with state-of-the-art double- or triple-pane windows with an excellent seal might significantly reduce heating and cooling requirements, but because the initial investment is so large, the time to payback is also long. The critique of the maximum payback approach is that it may prevent the investor from fully realizing the potential of green investments that save energy and reduce greenhouse gas emissions. On the other hand, it is reasonable for a private investor to assert that they should not bear the burden of having an investment in place for such a long time before it becomes profitable. Both investors and recipients of invested funds should be aware of the need to strike a balance between these two viewpoints.

3-3-5 Levelized Cost of Energy

Practitioners in the energy field may at times make cost comparisons between different options for delivering energy to customers. Where markedly different technologies compete for selection to a project, a method is needed that incorporates the role of both initial capital costs and ongoing operating costs. This is especially true in the case of the comparison between fossil and renewable technologies, since the former have a significant ongoing cost for fuel, whereas the latter have relatively low ongoing costs but high initial costs.

The *levelized cost* per unit of energy output provides a way to combine all cost factors into a cost-per-unit measure that is comparable between technologies. We will use the example of electricity generation in this case. If a predicted average output of electricity in kWh is known for a project, we can sum up all costs on an annual basis and divide by the annual output to calculate a cost per kWh value that will exactly pay for capital repayment, ongoing costs, and return on investment (ROI) at the end of the life of the project. Thus,

$$\text{Levelized cost} = \frac{\text{total annual cost}}{\text{annual output}} \quad \text{(in units of kWh)} \quad (3\text{-}12)$$

where

Total annual cost = annualized capital cost + operating cost + ROI

Here annualized capital cost is calculated using the interest rate for the project and the project lifetime applied to $(A/P, i\%, N)$. Operating cost includes cost for fuel, maintenance, wages of plant employees, taxes paid on revenues, insurance, cost of real estate, and so on. This model of levelized cost is simplified in that taxes are not treated separately from other operating costs, and ROI is treated as a lump sum. In a more complete calculation of levelized cost, taxes are levied on revenues after deducting items such as fuel, plant depreciation, and interest on capital repayment, while ROI is calculated as a percentage of net revenues from operations. The calculation is illustrated in Example 3-9.

Example 3-9 An electric plant that produces 2 billion kWh/year has an initial capital cost of $500 million and an expected lifetime of 20 years, with no remaining salvage value. The capital cost of the plant is repaid at 7% interest. Total operating cost at the plant is $25 million/year, and annual return to investors is estimated at 10% of the operating cost plus capital repayment cost. What is the levelized cost of electricity, in $/kWh?

Solution   First, annualize the capital cost using $(A/P, 7\%, 20) = 0.0944$. The annual cost of capital is then $(5 \times 10^8)(0.0944) = \$47.2$ million. Total annual cost is therefore \$72.2 million, and return to investors is \$7.22 million. Then levelized cost is

$$[(\$72.2 \text{ million}) + (\$7.22 \text{ million})] \;/\; 2 \text{ billion kWh} = \$0.0397/\text{kWh}$$

For comparison, the average electrical energy price in the United States in 2004 across all types of customers was \$0.0762/kWh, including the cost of transmission. Thus depending on the average price paid for transmission, this plant would be reasonably competitive in many parts of the U.S. market.

3-4   Direct versus External Costs and Benefits

In Example 3-9, the levelized cost of electricity was calculated based on capital repayment costs and also operating costs including energy supply, labor, and maintenance costs. These are all called *direct costs* since they are borne directly by the system operator. In addition, there may be *external costs* incurred such as health costs or lost agricultural productivity costs due to pollution from the energy system. These costs are considered external because although they are caused by the system, the system operator is not required to pay them, unless some additional arrangement has been made. On the other side of the economic ledger for the system are the *direct benefits*, which are the revenues from selling products and services. It is also possible to have *external benefits* from energy systems. For example, an unusual or interesting energy technology might draw visitors to the surrounding region or generate tourism revenue. This revenue stream is not passed along to the system operator and is therefore, a form of external benefit. External benefits from energy systems, where they exist, are small relative to direct benefits and direct or external costs, and will not be considered further here.

In cases where external costs are not borne by the system operator, the operator may make choices that are suboptimal from the perspective of society. Take the case of a plant that combusts fossil fuels to generate electricity, resulting in some quantity of pollution and significant external cost. Suppose also that the direct benefits from sales of electricity exceed direct cost, but are less than the sum of direct and external costs. Since the economics of the plant are positive in the absence of any consideration of external costs, the plant operator chooses to operate the plant, even though the economics would have been reversed had the external costs been *internalized*. Economists call this situation a "market failure" because plant operators are not seeing the true costs, and therefore the market is not functioning efficiently.

Much effort has been exerted over the years to correct for this type of market failure. One of the most common approaches is to use regulation to require emissions control technology on power plants. This policy has the effect of pushing costs and benefits closer to their true values. Costs go up for the operator, since the addition of pollution control equipment is a significant added capital cost that must be repaid over time. Also, the equipment requires ongoing maintenance (e.g., periodically removing accumulated pollutants) which adds to operating costs. At the same time, the benefits of operation are decreased slightly due to the energy losses from operating the pollution control equipment on plant output.

More recently, many governments have favored the use of pollution "markets" in lieu of "command and control" emissions regulations that require specific technologies. When participating in a pollution market system, plant operators are free to meet emissions targets in the way that is most economically advantageous for them—purchase "credits" from other operators if they fall short, or exceed their target, and sell the excess credits in the marketplace. The United States, among other countries, has since 1995 allowed for trading of SO_2 (sulfur dioxide) emissions reduction credits in the marketplace, and this approach is seen as a possible model for reducing CO_2 emissions in the future.

3-5   Intervention in Energy Investments to Achieve Social Aims

Energy investments are able to give attractive returns to investors; they also have the potential of creating significant externality costs that are borne by society. Many national and regional governments have responded by "intervening" in the energy marketplace, either legal requirements or financial incentives and disincentives, in order to give "desirable" energy options a financial advantage, or at least to lessen their financial disadvantage relative to other energy options. In this section, we refer to these options as "clean energy" alternatives since the main intention of these alternatives is to reduce pollution or greenhouse gas emissions, and thereby reduce externality costs.

There are at least two reasons to intervene on behalf of energy investments that reduce externality costs. The first is in order to give such projects credit for reducing externalities that cannot be given by the marketplace as it currently functions. For example, the operator of a project of this type may pay higher costs in order to repay capital loans, for cleaner input fuels, and so on. However, they may not be able to pass these higher costs along when they sell their energy product, or in cases where they are able to add a surcharge for a "green" energy product, the added revenue may not be adequate to cover the additional costs. Government intervention can help to reduce the costs of such an operation and thereby make them more competitive, so that overall sales of these reduced-externality options increase. From the government's perspective, it may appear that the choice is between making payments for reduced externalities or making payments for the results of the externalities themselves, such as higher health costs and reduced productivity due to pollution-related illnesses. On this basis, they choose the former option.

The second reason to intervene is to allow fledgling technologies a chance to grow in sales volume so that they can, over time, become less expensive per unit of capacity and better able to compete in the marketplace. Private investors may recognize that, over the long term, certain clean energy technologies may hold great promise; however, they may be unwilling to invest in these technologies since the return on investment may come only in the distant future, if at all. In this situation, there is a role for government to allocate part of the tax revenues (which are drawn in some measure from all members of the taxpaying population, both individual and corporate) to advance a cause in which all members of society have an interest for the future, but for which individuals find it difficult to take financial risks. With government support, pioneering companies are able to improve the new technologies and sell them in increasing amounts, and as they become more competitive, the government can withdraw the financial incentives.

3-5-1   Methods of Intervention in Energy Technology Investments

Support for Research & Development (R&D)
Governments commonly support the introduction of new energy technologies by paying for R&D during the early stages of their development. This can be done through a mixture of R&D carried out in government laboratories, sponsorship of research at universities, or direct collaboration with private industry. Once the energy technology reaches certain performance or cost milestones, government can make the underlying

know-how available to industry to further develop and commercialize. Since the end of World War II, many governments around the world have sponsored diverse energy research activities, ranging in application from nuclear energy to solar, wind, and other renewables, to alternative fuels for transportation.

Support for Commercial Installation and Operation of Energy Systems

In addition to R&D support, governments can also provide support for commercial use of energy, ranging from support for fixed infrastructure assets to support for clean fuels. Examples of this type of support include the following:

- *Direct cost support:* The government pays the seller of energy equipment or energy products part of the selling cost, so that the buyer pays only the cost remaining. This type of program can be applied to energy-generating assets, such as the case of the "million roofs" program in Japan. The Japanese government set a target of 1 million roofs with household-sized solar electric systems, and provided partial funding to help homeowners purchase the systems. Direct support can also be applied to energy products, such as the support for biodiesel in the United States. Here the U.S. government provides a fixed amount for each unit of biodiesel purchased, in order to make it more cost-competitive with conventional diesel.

- *Tax credits:* Governments can also reduce tax payments required of individuals and businesses in return for installation or purchase of clean energy. Under such a program, a clean energy purchaser is allowed to reduce their tax payment by some fraction of the value of the energy system or expenditure on clean energy resources. In cases of infrastructure purchases, government may allow the credit to be broken up over multiple years, until the full value has been credited. The government must set the credit percentage carefully in such a program. If the percentage is too low, there may be little participation; however, if it is too high, the taxpaying public may join the program in such numbers that the loss of tax revenue to the government may be too great. From the purchaser's perspective, breaking up the receipt of tax credits over multiple tax years will introduce some amount of complication in the cash flow of the project, since they may need to finance the amount of the tax credit for a period of time until they receive the credit.

- *Interest rate buydown:* In cases where the purchaser must borrow money to acquire an energy system, the government can assist the loan program by "buying down" some fraction of the interest payments. In New York, the New York State Energy Research & Development Authority (NYSERDA) has a buydown program in which they reduce the interest payment by 4% from the starting interest rate. This program reduces the monthly payment required of the purchaser and makes investments in energy efficiency or alternative energy more affordable.

Governments can also intervene in the commercial market legislatively, without providing financial support. For example, the German government provides a price guarantee for solar-generated electricity sold to the grid. The grid operator must purchase any such electricity at a fixed price, which is above the market price. This mechanism allows solar system purchasers to afford systems that otherwise would be financially unattractive. Also, in practice, since solar generating capacity is only added to the grid slowly via this program, the grid operator does not become suddenly overwhelmed with high-price electricity for which it is difficult to pass on costs to consumers. The effect of government support is shown in Example 3-10.

Example 3-10   Recalculate NPV in Example 3-3 using the following adjustments to reflect government support programs for the project:

1. A rebate of $200/rated kW, if the system is rated at 5000 kW.
2. An operating credit of $0.025/kWh, if the system produces on average 1.8 million kWh/year.
3. An interest rate buydown such that the MARR for the project is reduced to 3%.

Solution

1. The rebate is worth $200/kW × 5000 kW = $1 million. Therefore, the net capital cost of the project is now $5.5 million, and the NPV of the project is now $432,881, so it is viable.

2. The annual operating credit is worth $0.025/kWh × 1.8 × 10^6 kWh = 45,000 EU/year. The annuity is therefore increased to $445,000/year. Recalculating the present worth of the annuity gives $6.27 million, and the NPV of the project is now $67,108, so it is marginally viable.

3. With the reduced MARR, we recalculate the conversion factor for the annuity of $(P/A, 3\%, 25) = 17.41$. The present worth of the annuity is now $6.97 million, and the NPV of the project is now $942,865, so it is viable.

3-5-2   Critiques of Intervention in Energy Investments

As with other types of government subsidy program and intervention in the tax code, it is perhaps inevitable that intervention programs can be intentionally misused or send economic signals into the marketplace that turn out to be counterproductive. The purchaser of an energy system might receive the support or tax benefit from the government, but the clean energy system or product performs poorly—or worse, not at all!—so society gets little or no reduction in externality costs in return for the expenditure. Its widely cited example is the launch of wind power in California in the early 1980s, where the government provided large amounts of financial support to launch wind energy in the state. Some wind turbines installed at that time were poorly designed and did not function well, but nevertheless were sold to owners who then received generous benefits just for having purchased the systems. System owners had little incentive to maintain the device, and when the benefits program went away, these manufacturers soon went out of business. While it is true that other manufacturers made more successful products and were able to survive and in the long run make wind power a permanent presence in the state, the early stages might have gone more smoothly with a more carefully designed public funding program.

In terms of avoiding monetary waste, payments for output of clean energy supplies as opposed to installation of clean energy infrastructure may have an advantage. Payments for output, such as a production credit for the annual production of electricity from a windfarm, give the system operator an incentive to stay in the production business in order to recoup the initial investment in the asset. Payments for the installation, on the other hand, such as a rebate program for part of the cost of a small-scale renewable energy system, give an incentive for purchasing the asset, but not for using it, so the operator may in some cases not have adequate motivation to maintain the system properly or repair it if it breaks.

Even when a program is designed to motivate clean energy production as opposed to the initial purchase of clean energy assets, and the level of financial support is set carefully in order to keep out those who are seeking to exploit a tax loophole, it may not be possible to root out all possible misapplications of government funding programs, whether they are willful or not. Constant vigilance is the only recourse. Administrators of such programs should keep in communication with those receiving the support, learn how to train them to use the programs as effectively as possible, and preempt abusers wherever possible.

3-6   NPV Case Study Example

An investment in an energy system is likely to involve many cash flows over the span of its useful life. Some flows will be incomes, some will be expenses, and each may change in magnitude and direction from year to year. (Note: An avoided future expense is considered as income for the year in which it occurs.)

Consider an investment in a solar electric system for a home. The most obvious cash flow is the initial purchase of the PV panels and associated equipment, which may be in cash or may be through a loan, with or without a down payment. There may be some, or all, of the following considerations, in no particular order of importance:

- Whether to pay cash, or finance through a loan
- Expected lifetimes of system components
- Loan repayment schedule, and a possible down payment requirement
- Government or utility investment incentives, or possible grants
- Tax credits
- Yearly electricity production from the renewable source
- Local electric utility policy on net metering
- Possible scheduled maintenance, such as cleaning the panels and checking electrical connections
- Occasional major component repairs or replacements (e.g., the inverter)
- Possible property tax implications (higher property tax, or perhaps lower as an incentive)
- Electricity cost per kWh at the time of the installation
- Electricity inflation rate
- General inflation rate (not necessarily the same as the electricity inflation rate)
- Anticipated future electricity use changes (growing or shrinking family size, for example)
- Your marginal income tax rate (federal, state, and perhaps local)
- Possible removal costs at the end of the system life

This list can be extended. The rule is to think carefully about what may happen over the life of an investment and consider all reasonable possibilities.

It should also be noted that, if the investment is by a business, that business probably used the yearly cost of electricity as a tax deduction (because it is a cost of doing business). If the electricity bill is reduced, the deduction is reduced, resulting in a larger tax liability. In that case, the true yearly savings is the yearly electric bill reduction, multiplied by the complement of the marginal tax rate:

$$\text{True savings} = (\text{pretax saving}) \times (1 - \text{marginal tax rate}) \qquad (3\text{-}13)$$

As a demonstration of how NPV analysis can be applied to a renewable energy investment, consider a photovoltaic panel system, as described above, to be installed on a single-family home. In this example, the term "Present Worth (PW)" applies to the present worth of a future cash flow. The value of the NPV equals the sum of PW values through the life of the investment. In other words, the NPV is defined for the example as the sum of the present worth values of the entire series of future cash flows (incomes and expenses) related to the specific investment.

Many assumptions are necessary for a complete NPV analysis. Assumptions for this example are not intended to represent an actual situation with great accuracy, but are for illustration. Thoughtful consideration of assumptions is highly recommended for any investment.

For this example, relevant assumptions are:

- The system will have an 8 kW capacity and will be purchased with cash.
- Installation will be assumed to occur at year 0 (today).
- Assume cash flows occur yearly (end of the year) except for the initial purchase. The assumption could be monthly, or even weekly cash flows, but the resulting spreadsheet would be too long to be easily understood, and unnecessary for this example.
- Total installation cost will be $52,300, including labor, hardware, and permits.
- There are no Feed-In-Tariff (FIT) incentives.
- You will be eligible for a federal tax credit of 30% of installed cost (with no cap on the total amount), and your income is sufficient that you can benefit from the tax credit.
- You are eligible for a state energy office incentive of $4000 per kW of installed capacity for the first 5 kW, and $3500 per kW of installed capacity after the first 5 kW, up to a maximum of 10 kW. There is no added incentive for capacities greater than 10 kW.
- Utility bills for the past several years indicate your yearly electricity use averaged 7000 kWh, with no trend to use more or less than the average. You expect your family size and electricity use to remain constant, so past electricity use should predict the future. (Aggressive energy conservation may make this assumption invalid.)
- The PV panels will provide 70% of your yearly electricity, and all production has value because the local utility permits net metering for home owners over the year.
- Current electricity (production plus distribution) costs $0.16/kWh.
- You expect electricity rates to inflate by 8% yearly.
- You expect general inflation to be 4% yearly.

- You have thought seriously about the future and concluded a reasonable time value of money is 12% (this includes consideration of risk, which is an important factor discussed in more detail below).

- You expect an annual cost of operating, repairing, and cleaning the PV system to be 0.1% of the initial cost during the first year of ownership, and increase by the general inflation rate thereafter.

- After every 10 years of operation, the system will require a general overhaul that, today, costs $1200, and is a cost expected to increase with general inflation.

- The system will cease to be useful after 30 years but removal cost will be insignificant.

- You expect to remain in the house for the life of the investment (which is an important consideration).

The accompanying Table 3-2 provides cost elements used to calculate NPV in each year.

The NPV of the investment is $3804. In other words, the investment today is equivalent to receiving $3804 in today's dollars as profit, after all expenses are paid, under the assumptions considered.

Year	Purchase, One Time	Incentive, One Time	Tax Credit, One Time	kWh Saved Yearly	Dollars Saved Yearly	Maintenance, Yearly	Overhaul	Net Cash Flow, Yearly	PW Net Cash Flow, Yearly	Net Present Value
A	B	C	D	E	F	G	H	I	J	K
0	−$52,300	$30,500						−$21,800	−$21,800	−$21,800
1			$15,690	4900	$784.00	−$52.30		$16,422	$14,662	−$7138
2				4900	$846.72	−$54.39		$792	$632	−$6506
3				4900	$914.46	−$56.57		$858	$611	−$5895
4				4900	$987.61	−$58.83		$929	$590	−$5305
5				4900	$1066.62	−$61.18		$1005	$571	−$4735
6				4900	$1151.95	−$63.63		$1088	$551	−$4183
7				4900	$1244.11	−$66.18		$1178	$533	−$3651
8				4900	$1343.64	−$68.82		$1275	$515	−$3136
9				4900	$1451.13	−$71.58		$1380	$497	−$2638
10				4900	$1567.22	−$74.44	−$1,776	−$284	−$91	−$2729
11				4900	$1692.60	−$77.42		$1615	$464	−$2265
12				4900	$1828.00	−$80.51		$1747	$449	−$1817

TABLE 3-2   NPV Table for Case Study Example

Year	Purchase, One Time	Incentive, One Time	Tax Credit, One Time	kWh Saved Yearly	Dollars Saved Yearly	Maintenance, Yearly	Overhaul	Net Cash Flow, Yearly	PW Net Cash Flow, Yearly	Net Present Value
A	B	C	D	E	F	G	H	I	J	K
13				4900	$1974.25	−$83.73		$1891	$433	−$1383
14				4900	$2132.19	−$87.08		$2045	$418	−$965
15				4900	$2302.76	−$90.57		$2212	$404	−$561
16				4900	$2486.98	−$94.19		$2393	$390	−$170
17				4900	$2685.94	−$97.96		$2588	$377	$207
18				4900	$2900.81	−$101.88		$2799	$364	$571
19				4900	$3132.88	−$105.95		$3027	$351	$922
20				4900	$3383.51	−$110.19	−$2629	$644	$67	$989
21				4900	$3654.19	−$114.60		$3540	$328	$1316
22				4900	$3946.53	−$119.18		$3827	$316	$1633
23				4900	$4262.25	−$123.95		$4138	$305	$1938
24				4900	$4603.23	−$128.90		$4474	$295	$2233
25				4900	$4971.49	−$134.06		$4837	$285	$2517
26				4900	$5369.20	−$139.42		$5230	$275	$2792
27				4900	$5798.74	−$145.00		$5654	$265	$3057
28				4900	$6262.64	−$150.80		$6112	$256	$3313
29				4900	$6763.65	−$156.83		$6607	$247	$3560
30				4900	$7304.74			$7305	$244	$3804

Explanation of data and calculations in table columns:
A. Year: Year 0 (today) and the following 30 years of expected useful life
B. Purchase: Assume purchase is a one-time cash outlay to occur in Year 0
C. Incentive: (5 kW)($4000/kW) + (8 − 5 kW)($3500/kW), assumed to be received in Year 0
D. Tax Credit: 30% of purchase price, assumed to be received during Year 1, when Year 0 taxes are filed
E. kWh saved: 70% of 7000 kWh = 4900 kWh, with no change from year-to-year
F. Dollars saved: $0.16 × kWh saved Year 1, inflated by (1.08 × past year's) each year thereafter
G. Maintenance: 0.1% of purchase price in Year 1, inflated by (1.001 × past year's) each year thereafter except the final year (30) when the system ends useful life
H. Overhaul: $1200 inflated by $(1.04)^N$, where $N = 10$ and 20 (No overhaul would be scheduled for the final year of life, Year 30, of the system)
I. Net Cash Flow: Sum of cash flows for the year (rounded to nearest dollar)
J. PW of Net Cash Flow (NCF): = Net Cash Flow × 1.12^{-N}, where N is the year and 0.12 is the assumed time value of money (rounded to nearest dollar)
K. Net Present Value (NPV) = running sum of PW of NCF values (rounded to nearest dollar)

TABLE 3-2   NPV Table for Case Study Example (*Continued*)

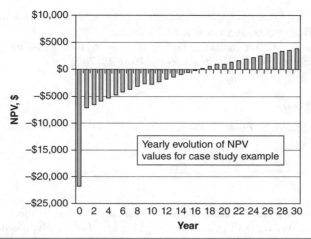

FIGURE 3-4   Yearly evolution of NPV values for case study example.

Figure 3-4 is provided as another means of visualizing the financial changes over the life of the investment. As seen from the graph, the NPV changes only a little from year-to-year in the more distant future even though the cash flow (dollars saved) grows larger and larger. Even though the savings and other cash flows are much larger in the distant future dollars, they are of lesser import in terms of today's dollars, which is what the time value of money reflects.

3-7   Summary

Two important premises of economic evaluation of energy systems are (1) the value of money depends on time, due to inflation and expectation of return on investments and (2) there are both direct and external costs from the operation of energy systems, and the engineer should be aware of the latter costs, even if they are not directly incorporated in evaluation. When analyzing energy investments, it is important to take into account the time value of money, especially for longer projects (of at least 8 to 10 years in length), since discounting of returns many years in the future can greatly reduce their value at present, and thereby alter the outcome of the analysis. Lastly, it is difficult for private individuals and private enterprises to spend money unilaterally on reducing external costs, since such investments may not be rewarded in the marketplace. Therefore, there is a role for government intervention to use financial incentives to encourage emerging energy technologies, since the financial burden associated with these incentives is spread across all those who pay taxes to the government.

References

Blank, L., and A. Tarquin (2012). *Engineering Economy*, 9th ed. McGraw-Hill, New York.

Costanza, R., R. d'Arge, R. de Groot, et al. (1997). "The Value of the World's Ecosystem Services and Natural Capital." *Nature*, Vol. 387, pp. 253–260.

Newnan, D., T. Eschenbach, and J. Lavelle (2011). *Engineering Economic Analysis*, 11th ed. Oxford University Press, Oxford.

U.K. National Statistics (2015). *National Statistics Online.* Web resource, available at www
.statistics.gov.uk. Accessed June 23, 2015.

U.S. Bureau of Labor Statistics (2015). *Bureau of Labor Statistics Website.* Web resource,
available at www.bls.gov. Accessed June 23, 2015.

Further Reading

DeGarmo, E., W. Sullivan, J. Bontadelli (1993). *Engineering Economy,* 9th ed. Macmillan,
New York.

Exercises

3-1.   A power plant operator is considering an investment in increased efficiency of their
plant that costs $1.2 million and lasts for 10 years, during which time it will save $250,000/
year. The investment has no salvage value. The plant has an MARR of 13%. Is the investment
viable?

3-2.   A municipality is considering an investment in a small renewable energy power plant with
the following parameters. The cost is $360,000, and the output averages 50 kW year-round. The
price paid for electricity at the plant gate is $0.039/kWh. The investment is to be evaluated over
a 25-year time horizon, and the expected salvage value at the end of the project is $20,000.
The MARR is 6%.
  a. Calculate the NPV of this investment. Is it financially attractive?
  b. Calculate the operating credit per kWh which the government would need to give to
      the investment in order to make it break even financially. Express your answer to the
      nearest 1/1000th of dollars.

3-3.   An electric utility investment has a capital cost of $60 million, a term of 8 years, and an
annuity of $16 million. What is the CRF for this investment?

3-4.   An energy project requires an investment in Year 0 of $8 million, and has an investment
horizon of 10 years and an MARR of 7%. If the annuities in Years 1 through 10 are as shown in the
table, what is the NPV?

Year	Annuity
1	$1,480,271
2	$1,165,194
3	$1,190,591
4	$1,286,144
5	$1,318,457
6	$973,862
7	$1,108,239
8	$1,468,544
9	$1,105,048
10	$851,322

3-5. An interest buydown program offers to reduce interest rates by 4% from the base rate. Suppose the base rate for a loan of $8000 is 8% for 10 years. What is the monthly payment before and after the buydown? In this case, use monthly compounding, that is, the term is 120 payment periods, and the interest per month is 0.667% before and 0.333% after the buydown.

3-6. One concern held by some scientists is that while investments required to slow global climate change may be expensive, over the long term the loss of "ecosystem services" (pollination of crops, distribution of rainfall, and so on) may have an even greater negative value. Consider an investment portfolio which calls for $100 billion/year invested worldwide from 2010 to 2050 in measures to combat climate change. The value per year is the net cost after taking into account return from the investments such as sales of clean energy in the marketplace. If the investments are carried out, they will prevent the loss of $2 trillion in ecosystems services per year from 2050 to 2100. Caveats: the analysis is simplified in that: (a) in 2050 the loss of ecosystem services jumps from $0 to $2 trillion, when in reality it would grow gradually from 2010 to 2050 as climate change becomes more widespread, (b) the world could not ramp up from $0 to $100 billion in investment in just 1 year in 2010, and (c) there is no consideration of what happens after the year 2100. Also, the given value of investment in technology to prevent a given value of loss of ecosystem services is hypothetical. The potential loss of ecosystem services value is also hypothetical, although at a plausible order of magnitude relative to a published estimate of the value of ecosystem services [Costanza, R. et al. (1997)].

    a. Calculate the NPV of the investment in 2010, using a discount rate of 3% and of 10%, and also using simple payback.

    b. Discuss the implications of the results for the three cases.

CHAPTER 4

Climate Change and Climate Modeling

4-1   Overview

This chapter considers one of the fundamental drivers for changing the worldwide energy system at the present time, namely, climate change due to increasing volumes of greenhouse gases in the atmosphere. Greenhouse gases benefit life on Earth, up to a point, since without their presence, the average temperature of the planet would be much colder than it is. However, as their presence increases in the atmosphere, changes to the average temperature and other aspects of the climate have the potential to create significant challenges for both human and nonhuman life on Earth. The community of nations is starting to take action on this issue through both local and national action as well as through the United Nations, even as scientists continue to learn more about the issue and are better able to predict its future course.[1]

4-2   Introduction

Climate change has been called the most pressing environmental challenge of our time, and it is one of the central motivations for the discussion of energy systems in this book. Numerous public figures, from political leaders to well-known entertainers, have joined leading scientists in calling for action on this issue. To quote President Barack Obama, "no challenge poses a greater threat to future generations than climate change."[2]

Energy use in all its forms, including electricity generation, space and process heat, transportation, and other uses, is the single most important contributor to climate change. While other human activities, such as changing land use practices, make important contributions as well, the single most important action society can take to combat climate change is to alter the way we generate and consume energy.

For the engineer, climate change poses some particularly complex and difficult problems, due to the nature of the fossil fuel combustion and CO_2 emissions process. Unlike air pollutants such as nitrogen oxide (NO_x), CO_2 cannot be converted into a more benign by-product using a catalytic converter in an exhaust stream. Also, due to the volume of CO_2 generated during the combustion of fossil fuels, it is not easy to capture and

---

[1]For a more in-depth discussion of climate and climate change, see Maslin (2013).
[2]Speaking during 2015 State of the Union Address, January 20, 2015, Washington, DC.

dispose of it in the way that particulate traps or baghouses capture particles in diesel and power plant exhausts (although such a system is being developed for CO_2, in the form of carbon sequestration systems; see Chap. 7). At the same time, CO_2 has an equal impact on climate change regardless of where it is emitted to the atmosphere, unlike emissions of air pollutants such as NO_x, whose impact is location and time dependent. Therefore, a unit of CO_2 emission reduced anywhere helps to reduce climate change everywhere.

4-2-1   Relationship between the Greenhouse Effect and Greenhouse Gas Emissions

The problem of climate change is directly related to the "greenhouse effect," by which the atmosphere admits visible light from the sun, but traps reradiated heat from the surface of the Earth, in much the same way that a glass greenhouse heats up by admitting light but trapping heat. Gases that contribute to the greenhouse effect are called *greenhouse gases* (GHGs). Some level of GHGs is necessary for life on Earth as we know it, since without their presence the Earth's temperature would be on average below the freezing temperature of water, as discussed in Sec. 4-3. However, increasing concentration of GHGs in the atmosphere may lead to widespread negative consequences from climate change.

Among the GHGs, some are *biogenic,* or resulting from interactions in the natural world, while others are *anthropogenic,* or resulting from human activity other than human respiration. The most important biogenic GHG is water vapor, which constitutes approximately 0.2% of the atmosphere by mass, and is in a state of quasi-equilibrium in which the exact amount of water vapor varies up and down slightly, due to the hydrologic cycle (evaporation and precipitation). The most important anthropogenic GHG is CO_2, primarily from the combustion of fossil fuels but also from human activities that reduce forest cover and release carbon previously contained within trees and plants to the atmosphere. Other important GHGs include methane (CH_4), nitrous oxide (N_2O), and various human-manufactured chlorofluorocarbons (CFCs). While the contribution to the greenhouse effect of these other GHGs per molecule is greater than that of CO_2, their concentration is much less, so overall they make a smaller contribution to the warming of the atmosphere.

4-2-2   Carbon Cycle and Solar Radiation

Looking more closely at the carbon cycle that brings CO_2 into and out of the atmosphere, we find that both short-term and long-term cycles are at work (see Fig. 4-1). The collection of plants and animals living on the surface of the Earth exchanges CO_2 with the atmosphere through photosynthesis and respiration, while carbon retained more permanently in trees and forests eventually returns to the atmosphere through decomposition. The oceans also absorb and emit carbon to the atmosphere, and some of this carbon ends up accumulated at the bottom of the ocean. Lastly, human activity contributes carbon to the atmosphere, primarily from the extraction of fossil fuels from the Earth's crust, which then result in CO_2 emissions due to combustion.

It is noteworthy that the fraction of the total carbon available on Earth that resides in the atmosphere is small, perhaps on the order of 0.003%, versus more than 99% that is contained in the Earth's crust. Nevertheless, this small fraction of the Earth's carbon has an important effect on surface and atmospheric average temperature. Changes that have a negligible effect on the total carbon under the Earth's surface can significantly

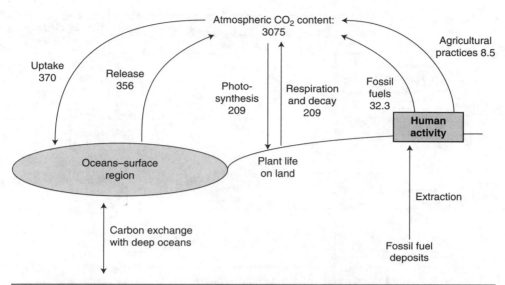

FIGURE 4-1 Flows of carbon between human activity, atmosphere, land-based plant life, and oceans, in 10^9 tonnes CO_2, with atmospheric concentration figure current as of 2012.

alter, in percentage terms, the concentration of CO_2 in the atmosphere. For example, extraction of approximately 7–8 Gt of carbon in fossil fuels per year as currently practiced is small relative to the 2×10^7 Gt total in the crust, but may potentially lead to a doubling of atmospheric CO_2 concentration over decades and centuries.

The role of the atmosphere in transmitting, reflecting, diffusing, and absorbing light is shown in Fig. 4-2. Of the incoming shortwave radiation from the sun, some fraction is transmitted directly to the surface of the Earth. The remainder is either reflected back to outer space, diffused so that its path changes (it may eventually reach the surface as diffuse light), or absorbed by molecules in the atmosphere. Light that reaches the surface raises its temperature, which maintains a surface temperature that allows it to emit longwave, or infrared, radiation back to the atmosphere. Some fraction of this radiation passes through the atmosphere and reaches outer space; the rest is absorbed by GHGs or reflected back to the surface.

4-2-3 Quantitative Imbalance in CO_2 Flows into and out of the Atmosphere

Inspection of the quantities of carbon movements in Fig. 4-1 shows that the total amount inbound to the atmosphere exceeds the amount outbound by approximately 27 Gt CO_2 per year. Indeed, the figure of 3075 Gt concentration of CO_2 in the atmosphere is not fixed but is in fact gradually increasing due to this imbalance. The imbalance between annual flows in and out is itself increasing, driven by CO_2 emissions from human activity that are rising at this time, as shown on the right side of the figure. Data presented in Chap. 1 showed that worldwide emissions increased from 18 to 32 Gt of CO_2 from 1980 to 2012; the latter quantity represents a sixteenfold increase compared to the approximately 2 Gt of emissions in 1900. Amounts of CO_2 emissions and concentration are sometimes given in units of gigatonnes of carbon rather than CO_2. In terms of mass of carbon, the atmospheric concentration was approximately 840 Gt C

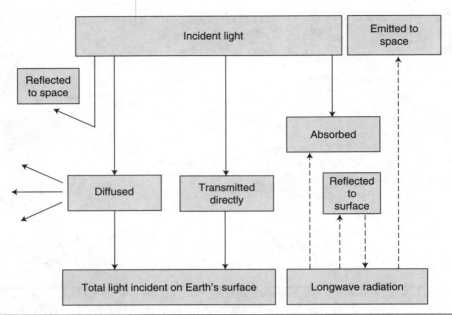

FIGURE 4-2   Pathways for incident light incoming from sun and longwave radiation emitted outward from surface of Earth (solid = shortwave from sun, dashed = longwave from Earth).

in 2012, and the annual emissions from fossil fuels are on the order of 7.5 Gt carbon. The long-term emissions trend from use of fossil fuels is shown in Fig. 4-3, along with the contribution of different sources. As shown, from approximately 1970 onward, it is the combination of emissions from petroleum products and coal that have been the largest contributor of CO_2 emissions, with a lesser, though not negligible, contribution from natural gas. As shown in Fig. 4-3, total emissions surpassed 9 Gt C for the first time in 2010, reaching 9.2 Gt C and up from 8.7 Gt C in 2009.

Historically, it has been known since the nineteenth century that CO_2 contributes to warming the surface temperature of the Earth, but it took humanity longer to realize that increasing CO_2 emissions increases CO_2 concentration and hence average surface temperature. Fourier recognized the greenhouse effect in agricultural glasshouses as early as 1827, and in 1869, Tyndall suggested that absorption of infrared waves by CO_2 and water molecules in the atmosphere affects global temperature. In 1896, Arrhenius presented a hypothetical analysis that showed that a doubling of CO_2 concentration in the atmosphere would increase average temperature by 5–6°C. For many decades thereafter it was believed that the oceans would absorb any increase in CO_2 in the atmosphere, so that increased emissions could not significantly alter temperature. Only in the 1950s did scientists conclude that the oceans did not have an unlimited capacity to absorb "excess" CO_2, and that therefore increased anthropogenic release of CO_2 would lead to higher concentration levels in the atmosphere. In 1957, Revelle and Seuss initiated the measurement of CO_2 concentration on top of Mauna Loa in the Hawaiian Islands, shown in Fig. 4-4. They chose this remote location on the grounds that it best represented the baseline CO_2 concentration, since it is far from major population centers or industrial emissions sources. These measurements have a characteristic sawtooth

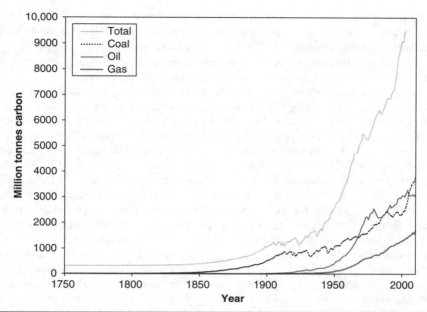

FIGURE 4-3   Carbon emissions related to fossil fuel use in tonnes carbon, 1750–2010. *Conversion:* 1 tonne carbon = ~3.7 tonnes CO_2. (*Source:* CDIAC, Oak Ridge National Laboratories. Reprinted with permission.)

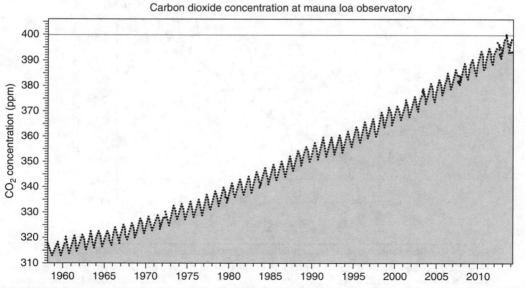

FIGURE 4-4   Concentration of CO_2 in the atmosphere in parts per million, 1957–2014. (*Source:* Scripps Institute of Oceanography.)

pattern, due to increased tree growth during summer in the northern hemisphere, when northern forests absorb CO_2 at a higher rate. Nevertheless, they show a steady upward trend, from approximately 315 parts per million (ppm) in 1960 to approximately 396 ppm in 2014. Examination of ice core samples from Vostok, Antarctica, shows that the current levels are unprecedented in the last 420,000 years.

The annual average CO_2 concentration from Fig. 4-4 is combined with temperature deviation from the 1951–1980 average global temperature to show the relative rise of both trends through the year 2010 in Fig. 4-5. CO_2 rises steadily and in fact the rate of rise is increasing because the annual rate of emissions from fossil fuel burning is increasing, thus increasing the flow of CO_2 into the atmosphere in the absence of a growing sink. The temperature deviation trend is more variable because average temperature is affected by weather patterns, precipitation, slight variability in average solar energy, and other factors. Nevertheless, although the rate of rise is uneven, the general trend is upward. The popular press has in some quarters used the general slowing in rate of rise since 1998 and variation between slightly warmer and cooler years in the period 1998–2010 as reason to question the validity of climate change. However, the running average of points in the figure (4 years on either side of every point, although it is not shown in the figure) shows a steady rise for most years compared to the value 2 years before, and in any case the warming trend is being superimposed on an underlying global average temperature value that undulated around an average value prior to the accelerating rise in CO_2 concentration since 1958 (as in Fig. 4-4).

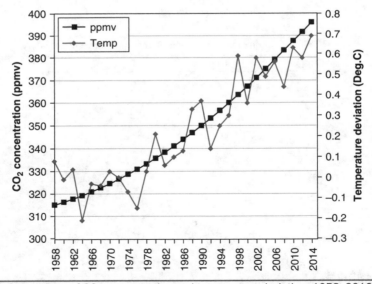

FIGURE 4-5 Comparison of CO_2 concentration and temperature deviation, 1958–2010.

Notes: CO_2 given in units of parts per million volume, and temperature deviation given in degrees Celsius compared to 1951–1980 average temperature. Sources: National Ocean & Atmosphere Administration (NOAA), for temperature; Carbon Dioxide Information & Analysis Center (CDIAC), for CO_2.

4-2-4 Consensus on the Human Link to Climate Change: Taking the Next Steps

The consensus on the connection between human activity and the increase in average global temperature reflects the work of the Intergovernmental Panel on Climate Change (IPCC), which is the largest coalition of scientists engaged to work on a single issue of this type in history, and the most authoritative voice on climate matters. According to this consensus, increasing concentration of CO_2 in the atmosphere due to human activity caused an increase in temperature of $0.74 \pm 0.18°C$ from 1900 to 2000, as illustrated in Fig. 4-6, which shows the annual and 5-year average of temperature measurements from around the world. Other phenomena, such as the level of solar activity or volcanism, have a measurable effect, but these by themselves cannot explain the upturn seen in the figure, especially since the 1960s. In January 2007, the IPCC concluded "with 90% certainty" that human activity was the cause of global warming. This was the first worldwide consensus among scientists representing all members of the community of nations on the near-certainty of this connection. A subsequent IPCC report in 2014, the Fifth Assessment Report, further strengthened this position and narrowed the statistical bounds around range of possible outcomes from temperature rise.

The strength of the scientific consensus was further bolstered by the conclusion of a random sampling of the abstracts of peer-reviewed scientific papers published between

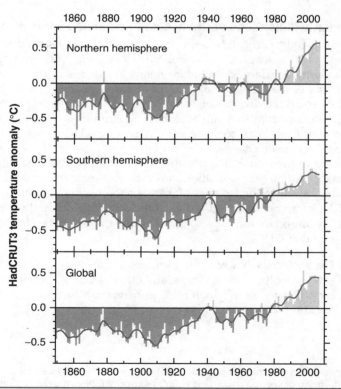

FIGURE 4-6   Variation in mean global temperature from value in 1940, for years 1850–2006. (*Source:* Climatic Research Unit, University of East Anglia. Reprinted with permission.)

1993 and 2003 that include "climate change" among their keywords (Oreskes, 2004). Out of 928 papers sampled, 75% accepted the view that climate change is connected to human activity, 25% took no position, and not a single one disagreed with the consensus position of the IPCC.[3]

Despite this confidence in the link between human activity and climate change, much uncertainty remains, both about the quantitative extent of the contribution of human activity to climate change, and especially in regard to the degree to which GHG emissions will affect the climate in the future. Some authors point to the potential benefits of increased CO_2 in the atmosphere, such as more rapid plant growth (e.g., Robinson et al., 2007). However, the prediction shared by most observers is that the net effect of climate change will be largely negative, and possibly catastrophically so. According to the IPCC report in 2014, the increase over the period from 1900 to 2100 might range from 1.1 to 7.5°C, narrowed to 2.5 to 7.5°C if mitigation steps are not taken. Values at the high end of this range would lead to very difficult changes that would be worthy of the effort required to avoid them. Because we cannot know with certainty how average global temperature will respond to continued emissions of GHGs, and because the possibility of increases of as much as 5 to 8°C at the high end have a real chance of occurring if no action is taken, it is in our best interest to take sustained, committed steps toward transforming our energy systems so that they no longer contribute to this problem.

4-2-5   Early Indications of Change and Remaining Areas of Uncertainty

There are already early indicators of climate change observed in the world, some examples of which are given in Table 4-1. They range in scope from widespread decline in the amount of surface ice in polar regions to changes in the behavior of plants and animals. Any of these events might occur in isolation due to the natural fluctuations in amount of solar activity, extent of volcanic ash in the atmosphere, or other factors. However, the fact that they are all occurring simultaneously, that they are occurring frequently, and that they coincide with increasing GHGs, strongly suggests that they are caused by climate change.

Looking to the future, it is possible to forecast possible pathways for future GHG emissions, based on the population forecasts, economic trends, and alternative scenarios for how rapidly societies might take steps to lower emissions. Even if the course of future GHGs can be predicted within some range, there are three unknowns that are particularly important for improving our ability to predict the effect of future emissions of CO_2 and other GHGs:

1. *The connection between future emissions and concentration:* Given uncertainties about how the oceans, forests, and other carbon sinks will respond to increased concentration of CO_2 in the atmosphere, the future path of CO_2 concentration is not known.

---

[3]The author points out that the sampling method may have overlooked a small number of papers published during the period under study that disagree with the consensus position. Nevertheless, the total number of such papers could not have been large, given that so many papers were sampled and none came to light that disagreed.

Effect	Description
Receding glaciers and loss of polar ice	Glaciers are receding in most mountain regions around the world. Notable locations with loss of glaciers and sheet ice include Mt. Kilimanjaro in Africa, Glacier National Park in North America, La Mer de Glace in the Alps in Europe (see Figs. 4-7 and 4-8), and the glaciers of the Himalaya. Also, the summer ice cover on the Arctic Ocean surrounding the north pole is contracting each year.
Extreme heat and weather events	In the 1990s and 2000s, extreme heat waves have affected regions such as Europe, North America, and India. A heat wave in Moscow, Russia, drove temperatures above 100°F five times during the summer of 2010, where previously they had not exceeded this level in modern record keeping. On June 1, 2010, the highest temperature ever recorded in Asia of 53.5°C (128.3°F) was reached in Pakistan. The global anomaly record of +0.66°C set in 2010 was broken in 2014, with a new record of +0.68°C. Numerous extreme weather events with growing destructive power and/or economic damage have occurred in recent years, including Superstorm Sandy in the United States in October 2012 or Typhoon Haiyan in the Philippines in November 2013.
Change in animal, plant, and bird ranges	Due to milder average temperatures in temperate regions, flora and fauna have pushed the limit of their range northward in the northern hemisphere and southward in the southern hemisphere. For instance, the European bee-eater, a bird that was once very rare in Germany and instead lived in more southern climates, has become much more common.
Coral reef bleaching	Coral reef bleaching, in which corals turn white due to declining health and eventually die, is widespread in the world's seas and oceans. Along with other human pressures such as overfishing and chemical runoff, increased seawater temperature due to global warming is a leading cause of the change.

TABLE 4-1   Examples of Early Indications of Climate Change

FIGURE 4-7   Mer de Glace glacier, near Mt. Blanc, France, summer 2007.

Thinning of the glacier at the rate of approximately 1 m/year up to 1994, with an accelerated rate of 4 m/year for 2000–2003, has been recorded.

FIGURE 4-8   As the glacier has contracted in size, operators of the Mer de Glace tourist attraction have added temporary staircases to allow tourists to descend to the edge of the glacier.

2. *The connection between future concentration and future global average temperature:* Concentration and average temperature have been correlated for the latter half of the twentieth century. Although we expect them to be correlated in the twenty-first century, so that average temperature continues to rise, reinforcing or mitigating factors that are currently not well understood may come into play, so that it is difficult to accurately predict the future course of temperature, even if future concentration levels were known.

3. *The connection between future global average temperature and future local climates:* The real environmental and economic impact from future climate change comes not from the average temperature worldwide, but from changes in climate from region to region. While some regions may experience only modest change, others may experience extreme change due to temperature rises greater than the worldwide average. This type of accelerated change is already happening at the north and south poles.

Improved modeling of the climate system can help to shed light on some of these areas of uncertainty. Section 4-3 provides an introduction to this field.

4-3   Modeling Climate and Climate Change

In the remainder of this chapter, we consider the mathematical modeling of climate, and use the concepts presented to discuss feedback mechanisms that are at work as climate changes, in order to shed light on ways in which climate change might slow or accelerate over time. As a starting point for calculating solar energy available on Earth, recall that *black-body radiation* is the radiation of heat from a body as a function of its temperature. Black-body radiation is the primary means by which a planet or star exchanges heat with its surroundings. The energy flux from black-body radiation is a function of the fourth power of temperature, as follows:

$$F_{BB} = \sigma T^4 \tag{4-1}$$

where F_{BB} is the flux in W/m², T is the temperature in kelvin, and σ is the *Stefan-Boltzmann constant* with value $\sigma = 5.67 \times 10^{-8}$ W/m²K⁴.

By treating the sun and the Earth as two black bodies, it is possible to calculate the expected temperature at the surface of the Earth in the absence of any atmosphere. The sun has an average observed surface temperature of 5770 K, which gives a value of 6.28×10^7 W/m² using Eq. (4.1). The sun's radius is approximately 7.0×10^8 m, and the average distance from the center of the sun to the Earth's orbit is approximately 1.5×10^{11} m. Since the intensity of energy flux per area decays with the square of distance, the average intensity at the distance of the Earth is

$$6.28 \times 10^7 \text{ W/m}^2 \left(\frac{7.0 \times 10^8}{1.5 \times 10^{11}} \right)^2 = \sim 1368 \text{ W/m}^2$$

This value of 1368 W/m² is known as the *solar constant,* and assigned the variable S_0. Two caveats are in order. First, although the name "solar constant" is given, the value actually fluctuates slightly over time (see Chap. 9). Second, slightly different values for S_0 are found in the literature, such as $S_0 = 1367$ W/m² or $S_0 = 1372$ W/m². For the purposes of calculations in this chapter, choice of value of S_0 will not alter the results substantially, and the reader is free to use a different value if preferred.

It will now be shown that, if there were no atmosphere present and the energy striking the Earth due to S_0 and black-body radiation of energy leaving the Earth were in equilibrium, the average surface temperature should be some 33 K (59°F) lower than the long-term average observed value over all of the Earth's surface, as follows. Suppose the Earth is a black body with no atmosphere and that its only source of heating is the energy flux from the sun. (In the case of the actual planet Earth, the only other potential source of heating, namely, heat from the Earth's core, has a negligible effect on surface temperature because the layer of rock that forms the Earth's crust is a very poor conductor.) As shown in Fig. 4-9, if we define R_E as the radius of the earth, it is observed

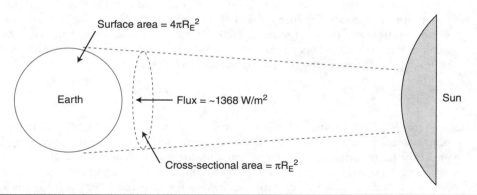

FIGURE 4-9 Schematic of solar constant traveling from sun and striking Earth.

that the cross sectional area of the energy reaching the earth is, or one-fourth of the surface area of the earth of $4\pi R_E^2$. This difference is significant when calculating the balance of energy flows to and away from the surface. In addition, some portion of the light reaching the Earth is reflected away and does not contribute to raising the Earth's temperature; the fraction of light reflected is known as the *albedo*.

The energy balance between energy reaching the Earth due to solar radiation and the energy flowing out due to black-body radiation can then be written as follows. Let F_{sun} be the energy flux from the sun and F_{BB} be the black-body radiation flux away from the earth, both in units of W/m²:

$$E_{in} = E_{out}$$

$$F_{sun}A_{disc} = F_{BB}A_{earth}$$

$$S_o(1-\alpha)\pi R_{earth}^2 = \sigma T_{earth}^4 4\pi R_{earth}^2$$

$$S_o(1-\alpha)/4 = \sigma T_{earth}^4 \qquad (4\text{-}2)$$

Here α is the albedo; the observed value for Earth is $\alpha = 0.3$. Substituting known values and solving for T_{Earth} gives

$$T_{Earth} = \left[\frac{S_o(1-\alpha)/4}{\sigma}\right]^{1/4} = 255 \text{ K}$$

This value represents the average temperature of a planet orbiting our sun at the same radius as the Earth and with the same albedo, but with no atmosphere. The observed average temperature, in the absence of recent warming in the period 1980–2000, is 288 K (i.e., the baseline temperature in Figs. 4-5 and 4-6 where temperature anomaly equals 0°C), so the difference can be attributed to the effect of GHGs.

4-3-1   Relationship between Wavelength, Energy Flux, and Absorption

The energy emitted from black-body radiation is the sum of energy of all waves emitted by the body, across all wavelengths. Energy contained in an individual wave is a function of frequency f of that wave. Let λ represent the wavelength in meters, and c represent the speed of light (approximately 3×10^8 m/s). Recall that λ and f are inversely related, with $f = c/\lambda$. Energy is released in waves as electrons drop to a lower energy state; the change in energy ΔE is proportional to f, that is

$$\Delta E = hf \qquad (4\text{-}3)$$

Here h is *Planck's constant*, with value $h = 6.626 \times 10^{-34}$ J·s. Although all black bodies emit some amount of waves at all wavelengths across most of the spectrum, bodies emitting large amounts of energy, such as the sun, release much of the energy in waves created by large energy changes, for which f will be larger and λ will be smaller. For black bodies such as the Earth that have a lower surface temperature, f is smaller and λ is larger.

For both hotter and cooler black bodies, the strength of the total energy flux varies with both wavelength and black-body temperature. For given temperature T and λ in units of meters:

$$F_{BB}(\lambda) = \frac{C_1 (\lambda \cdot 10^6)^{-5}}{\left[\exp\left(\dfrac{C_2}{(\lambda \cdot 10^6)T} \right) - 1 \right]}$$ (4-4)

Here $C_1 = 3.742 \times 10^8 \ \text{Wm}^3/\text{K}$, $C_2 = 1.4387 \times 10^4 \ \text{mK}$, and F_{BB} is given in units of W/m². Integrating Eq. (4-4) across all values of λ from 0 to 8 for a given temperature T gives Eq. (4-1). Based on Eq. (4-4), the wavelength at which F_{BB} reaches a maximum λ_{max} can also be expressed as:

$$\lambda_{max} = 2.897 \times 10^{-3}/T$$ (4-5)

Here λ_{max} is given in units of meters. Example 4-1 illustrates the application of Eq. (4-4) at different values of λ.

Example 4-1   Using a surface temperature value of 288 K, evaluate $F_{BB}(\lambda)$ at λ_{max}, and also at $\lambda = 5 \ \mu\text{m}$ (5×10^{-6} m) and 25 μm.

Solution   Solving for λ_{max} using Eq. (4-5) gives $\lambda_{max} = 1.006 \times 10^{-5}$ m. Substituting $\lambda = 1.001 \times 10^{-5}$ m and $T = 288$ K into Eq. (4-4) gives

$$F_{BB}(1.006 \times 10^{-5}) = \frac{3.742 \times 10^8 (1.006 \times 10^{-5} \cdot 10^6)^{-5}}{\exp\left[\dfrac{1.4387 \times 10^4}{(1.006 \times 10^{-5} \cdot 10^6)288} \right] - 1} = 25.5 \ \text{W/m}^2$$

Similarly, $F_{BB}(5 \times 10^{-6} \ \text{m}) = 5.49 \ \text{W/m}^2$, and $F_{BB}(25 \times 10^{-6} \ \text{m}) = 6.01 \ \text{W/m}^2$. Comparison of the three results shows the decline in energy flux away from λ_{max}. All three points are shown on the graph of flux as a function of wavelength in Fig. 4-10.

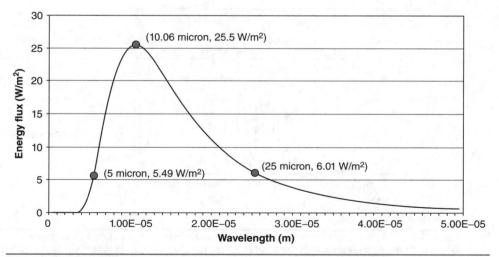

FIGURE 4-10   Energy flux as a function of wavelength for black body at 288 K.

Contribution of Specific Wavelengths and Gases to Greenhouse Effect

In Sec. 4-3-1, we established that temperature determines the range of wavelengths at which peak black-body transmission of energy will occur. To understand the extent of climate warming, we must consider two other factors: (1) the ability of different molecules to absorb radiation at different wavelengths and (2) the extent to which radiation is already absorbed by preexisting levels of gases when concentration of some gases increases.

Addressing the first point, triatomic gas molecules such as H_2O or CO_2, as well as diatomic molecules where the two atoms are different, such as carbon monoxide (CO), are able to absorb small increments of energy, due to the structure of the molecular bonds. This feature allows these molecules to absorb the energy from waves with wavelength $8 < \lambda < 30$ μm, compared to diatomic molecules where both atoms are the same, such as O_2 and N_2, which can only absorb higher energy waves. Figure 4-11 shows, in the upper half, energy transmitted as a function of wavelength for both sun and Earth; the lower half of the figure shows wavelengths at which atmospheric molecules peak in terms of their ability to absorb energy. In particular, the CO_2 molecule has a peak absorptive capacity in the range from approximately 12 to 14 μm, which coincides with the peaking of the energy flux based on the Earth's black-body temperature. H_2O has high absorption elsewhere on the spectrum, including longer wavelengths (>20 μm) and some of the incoming shorter waves from the sun. The mass percent and contribution to heating of

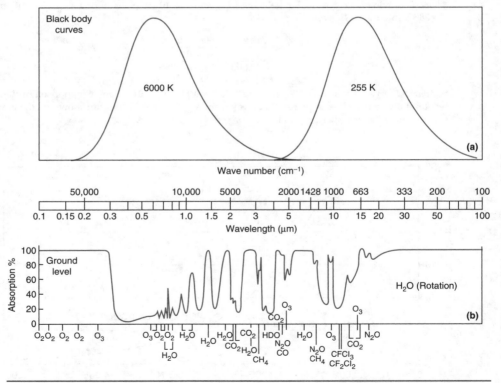

FIGURE 4-11 Distribution of relative energy of emission from sun and Earth black bodies by wavelength (a), and percent of absorption by greenhouse gases (b). [*Source:* Mitchell (1989). Published by American Geophysical Union. Image is in public domain.]

Gas	Mass Percent	Contribution to Warming (W/m²)	Wavelength Range for Peak Contribution* (10⁻⁶ m)
H_2O	0.23%	~100	> 20
CO_2	0.051%	~50	12–14
CH_4	< 0.001%	~5	3, 8
Other (For N_2O, O_3, CFC11, CFC12, and so on)	< 0.001% for each	< 2 for each	For N_2O: 16 For CFCs: 11 For O_3: Several

*As shown in Fig. 4-11.
Total atmospheric mass $= 5.3 \times 10^{18}$ kg

Note: mass percent and contribution for CO_2 shown is at 750 Gt mass of carbon in atmosphere (2750 Gt CO_2). As CO_2 concentration increases per Fig. 4-4, both mass percent and contribution to warming increase.

TABLE 4-2 Mass Percent and Contribution to Atmospheric Warming of Major Greenhouse Gases

the atmosphere of H_2O, CO_2, and other gases is given in Table 4-2. Note that the total contribution of water to climatic warming is greater than that of CO_2. Although the average concentration of water in the atmosphere may increase as a result of climate change, the percent change in atmospheric water will not be as great as that of CO_2, and hence the effect of changing concentration of water will not be as large.

Figure 4-11 also shows the fraction of energy absorbed by atmospheric molecules as a function of wavelength. At the longest wavelengths shown (>20 μm), 100% of the black-body radiation from the surface of the Earth is absorbed. In the range of the 7–20 μm peak BB radiation from the planet, however, some fraction of radiation remains unabsorbed. Therefore, increasing the amount of CO_2 in the atmosphere increases the absorption percent, leading to a warming of the average temperature of both the atmosphere and the Earth's surface.

As shown in Table 4-2, the 150 W/m² of atmospheric heating provided by H_2O and CO_2 accounts for most of the 33 K increase in surface temperature due to the greenhouse effect, with other GHGs accounting for lesser amounts of the contribution to warming. The effect of H_2O and CO_2 can be understood by considering the warming effect of each in isolation, as shown in Example 4-2.

Example 4-2 Suppose that water and CO_2 were the only GHGs in the atmosphere. Calculate the resulting temperature in each case.

Solution In the absence of the greenhouse effect, flux from the surface would be the black-body radiation value at 255 K, or:

$$F_{BB} = (5.67 \times 10^{-8})(255)^4 = 240 \text{ W/m}^2$$

For the two cases, the value of the temperature is therefore the following:

$$T_{H_2O\_only} = \left(\frac{240 + 100}{5.67 \times 10^{-8}} \right)^{0.25} = 278 \text{ K}$$

$$T_{CO_2\_only} = \left(\frac{240 + 50}{5.67 \times 10^{-8}} \right)^{0.25} = 267 \text{ K}$$

Thus water would increase temperature by 23 K by itself, and CO_2 would increase temperature by 12 K by itself. Note that these values add up to more than the total temperature change ΔT due to the greenhouse effect, which must be calculated by first calculating the total contribution to warming from all GHGs before calculating the value of ΔT.

The data in Table 4-2 are intended to explain the difference between the surface temperature with no atmosphere versus the temperature with an atmosphere composed partly of GHGs. Therefore, they do not capture the effect of *increasing* the concentration of GHGs. This point is important for the GHGs other than water whose concentrations have been increasing recently. In addition to CO_2, two gases in particular, methane (CH_4) and N_2O, have very high energy trapping potential per molecule compared to CO_2, with *Global Warming Potential (GWP)* multipliers estimated at 24.5 and 320, respectively. (GWP of methane is subject to uncertainty, with estimated values as high as 33 reported recently.) Although methane and N_2O do not contribute as much to the warming of the average temperature of the Earth as CO_2, their effect is not negligible and poses an additional climate change concern.

In comparison, the IPCC reports in its fourth assessment noted that the *anthropogenic* contribution to warming for CO_2, methane, and N_2O in 2004 were 1.7, 0.5, and 0.1 W/m^2, respectively. These values, which are all much smaller than the values reported in Table 4-2, represent the contribution made in recent times to the greenhouse effect by increasing the concentration of these GHGs compared to preindustrial levels. Although contributions in the range of 0.1 to 1.7 W/m^2 can only lead to a small temperature change relative to the total magnitude of the greenhouse effect of 33 K, even these relatively small forcings are of major climate change concern. In terms of total temperature deviation due to anthropogenic climate change as of the year 2010, the IPCC's fifth report attributed 76% of the change to CO_2, 16% to methane, 6% to N_2O, and 2% to chlorofluorocarbons.

Growing Concern about Methane (CH_4): Comparison to CO_2

Methane is the second-most significant anthropogenic greenhouse gas after CO_2, and there is growing concern about its current and potential future effect. The IPCC fifth assessment report put its contribution in 2010 to the total 49 Gt of CO_2 equivalent contribution at 16%, compared to 76% from CO_2 emissions (65% due to fossil fuel use and 11% due to agricultural/land use practices). Methane escapes into the atmosphere from the production of fossil fuels (coal, gas, and oil), from the land-filling of solid waste, and from agricultural processes, including both crops and animal husbandry. Methane concentration levels in the global atmosphere rose from approximately 600 ppb in preindustrial times to the level of approximately 1770 ppb until the year 1999, entered an undulating plateau between 1770 and 1780 ppb until 2007, and began to rise again and surpassed 1800 ppb in 2011 (at a level of 1803 ppb).

In the short run, industry can effectively mitigate climate change impacts of methane by, for instance, modifying agricultural practices so that they release less methane, or combusting landfill methane as it is vented from the landfill to convert it to CO_2, so that the GWP per mass of carbon released is greatly diminished. Landfill owners can also transport methane to nearby industrial or commercial facilities, where it can be combusted to make electricity. Such a system is in place at the Spartanburg, SC, plant of the auto manufacturer BMW, where electrical generators rated at 11 MW produce approximately 30% of the plant's electricity from landfill gas.

The fate of methane currently sequestered in the permafrost of Siberia and the Arctic poses potentially a much larger threat over the medium to long term. These sources represent 1 trillion or more tons of carbon equivalent captured as methane since the last ice age. Although there is much uncertainty about how much of this methane might eventually be released with gradually increasing global temperature, it is possible that in a more drastic scenario, the climate change impact of the resulting methane concentration in the atmosphere could surpass the heat trapping impact of anthropogenic CO_2.

Since the release of methane from fossil fuel processing, agriculture, or the thawing of the permafrost is primarily either a chemical management or natural systems process, and not directly a part of energy conversion, it is not considered further in this book. The reader should be aware, however, that because of the growing problem of methane, the study of CO_2 as in the focus of this book does not represent the full extent of the GHG problem, and that a complete understanding of the threat to the climate involves methane and other GHGs as well as CO_2.

4-3-2   A Model of the Earth-Atmosphere System

The previously introduced black-body model of the Earth-sun system shows how the Earth without an atmosphere would be colder than what is observed. In this section, we expand the model to explicitly incorporate the atmosphere as a single layer that interacts both with the surface of the Earth and with outer space beyond the atmosphere. Such a model serves two purposes:

1. It shows how the incoming solar radiation, atmosphere, and Earth's surface interact with one another, and provides a crude mathematical estimate of the effect on Earth surface temperature of adding an atmospheric layer.

2. It illustrates the function of much more sophisticated *general circulation models* (GCMs), discussed in Sec. 4-3-3, that are currently used to model climate and predict the effect of increased GHGs in the future. (A detailed consideration of the mathematics of GCMs is, however, beyond the scope of this chapter.)

The simplifying assumptions for this model are as follows. First, we are considering a one-dimensional model where energy is transferred along a line from space through the atmosphere to the Earth's surface and back out to space; the lateral transfer of energy between adjacent areas of the surface or atmosphere is not included. Second, the atmosphere is considered to have uniform average temperature T_a, and the Earth's surface is considered to have uniform average temperature T_s. Third, only radiation and conduction are considered; in the interest of simplicity, convection between surface and atmosphere is not included. The goal of the model is to calculate equilibrium temperatures of the Earth's surface and atmosphere, respectively, using an energy balance equation for each.

The flow of energy can be explained as follows. Per Fig. 4-12, inbound fluxes from the sun are shortwave, while outbound fluxes from the Earth are longwave. Radiation reaches the atmosphere from the sun, of which part is reflected, part is transmitted, and part is absorbed. The transmitted flux reaches the Earth, minus a portion that is reflected by the Earth, and heats it to an equilibrium temperature. The Earth then radiates a longwave flux back out to space, which the atmosphere again transmits, reflects, or absorbs. In addition, conductive heat transfer takes place between the Earth and atmosphere due to the temperature difference between the two. Note that the flux reaching the atmosphere is taken to be $S_0/4$, not S_0, as explained above. Black-body radiation from atmosphere to surface, as well as that from surface to atmosphere that is reflected back to the surface, is assumed to be completely absorbed by the surface, and the possibility of its reflection at the surface is ignored.

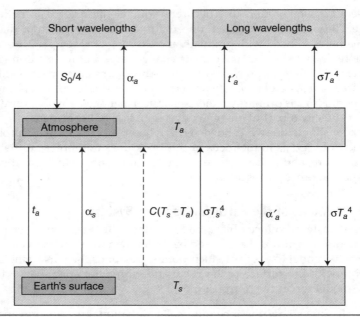

FIGURE 4-12   Diagram of simple Earth-atmosphere climate model.

In the model, variables and subscripts are defined as follows. Coefficient t corresponds to the portion transmitted through the atmosphere either inward or outward. Coefficient α is similar to the function of albedo introduced in Sec. 4-3, in that it corresponds to the portion reflected. Both the atmosphere and the surface of the Earth can reflect radiation. Subscripts a and s correspond to "atmosphere" and "surface." A prime symbol ("'") denotes flux of longwave radiation, both outward from the Earth's surface and reflected back to the surface by the atmosphere. Coefficients without prime symbol denote flux of shortwave radiation, either from the sun inward or reflected outward by the atmosphere or surface. Initial parameter values that approximate the function of the Earth-atmosphere system before increases in the level of GHGs are shown in Table 4-3. C is the conductive heat transfer coefficient in units of W/m^2K.

The intuition behind the parameter values is the following. The value of α_a is the same as the albedo of the Earth, so that incoming radiation has the same probability of being reflected by the atmosphere. More than half of the remaining radiation is then transmitted by the atmosphere through to the surface of the Earth, although some 17% is absorbed based on the values of reflection and transmission of the atmosphere. The value of α_s is less than that of α'_a since the land and water covering of the Earth is less

Inward flux:	Outward flux:	Conductive heat transfer:
$\alpha_s = 0.11$	$t'_a = 0.06$	$C = 2.5\ W/m^2K$
$t_a = 0.53$	$\alpha'_a = 0.31$	
$\alpha_a = 0.3$		

TABLE 4-3   Initial Parameter Values for Single-Layer Climate Model

likely than the atmosphere to reflect incoming radiation. For outgoing longwave radiation, probability of reflection back to Earth α'_a is slightly higher than the value of α'_a reflecting the slightly increased tendency of longwaves to be reflected. The key feature of the greenhouse effect, however, is embodied in the low value of t'_a relative to t'_a since this indicates how the atmosphere is much less likely to transmit longwave radiation than shortwave. Thus the total absorption of longwaves by the atmosphere is higher than that of shortwaves, consistent with Fig. 4-11.

The equations for the surface [Eq. (4-6)] and atmospheric [Eq. (4-7)] energy equilibrium can be written as follows:

$$t_a(1-\alpha_s)S_0/4 - C(T_s - T_a) - \sigma T_s^4(1-\alpha'_a) + \sigma T_a^4 = 0 \tag{4-6}$$

$$(1-\alpha_a - t_a(1-\alpha_s))S_0/4 + C(T_s - T_a) + \sigma T_s^4(1-t'_a - \alpha'_a) - 2\sigma T_a^4 = 0 \tag{4-7}$$

Taking the case of Eq. (4-6) for the surface, the four terms can be explained in the following way:

1. *Incoming energy flux from the sun:* The amount transmitted by the atmosphere, after taking out the amount reflected by the surface
2. *Conductive heat transfer:* Energy transferred outward due to the temperature difference between surface and atmosphere
3. *Radiative heat transfer outward:* Black-body radiation outward, after taking out the amount reflected back to the surface by the atmosphere
4. *Radiative heat transfer inward:* From atmosphere to surface

Energy flux for the atmosphere, in Eq. (4-7), can be described in a similar manner:

1. *Incoming energy flux from solar gain:* The amount retained by the atmosphere, after taking out the amount reflected by the atmosphere back to space, and also the amount transmitted to the surface, of which a portion is reflected back to the atmosphere
2. *Conductive heat transfer:* To atmosphere from surface
3. *Radiative heat transfer inward from Earth's surface:* Radiation inward, after taking out the amount reflected back to the surface and transmitted to space
4. *Radiative heat transfer outward:* From atmosphere both to surface and to space

Using the parameter values in Table 4-3, we can solve the system of equations using a numerical solver. The resulting outputs from the solver are $T_s = 289.4$ K, $T_a = 248.4$ K. Note that due to the simplicity of the model, the results do not exactly reproduce the observed surface value of 288 K and observed value in the upper atmosphere of 255 K. However, the value for the surface temperature is quite close, and the value for the atmosphere is within 7 K as well. These results show how even a simple model that incorporates a role for the atmosphere can bring predicted surface temperature much closer to the actual value. One possible refinement is to expand the atmosphere into several layers, which will tend to create a temperature gradient from the coolest layer furthest from the Earth, to the warmest layer adjacent to the Earth, as occurs in the actual atmosphere.

Given the values calculated above, it is also illustrative to map the values of energy flows measured in W/m² for the base case, as shown in Fig. 4-13. The reader can verify

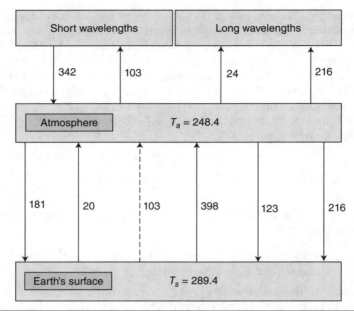

FIGURE 4-13   Single-layer climate model with representative energy flows in W/m².

that at the equilibrium temperatures, flows of 520 W/m² in and 520 W/m² out balance each other for the Earth's surface, and flows of 863 W/m² in and 863 W/m² out balance each other for the atmosphere. Note that due to the absorption of the atmospheric layer, the shortwave flux from the sun is reduced from ~240 W/m² as calculated using Eq. (4-2) to 181 W/m². However, the gain due to the black-body radiation from the atmosphere to the surface of 216 W/m² and the reflection of 123 W/m² of longwave radiation back from the atmosphere to the surface more than compensate, so that overall the surface temperature is increased.

Example 4-3   Use an appropriate change in the coefficients in Eqs. (4-6) and (4-7) from the values given in Table 4-2 to model the effect of changing characteristics of the atmosphere due to increasing GHGs on predicted T_s and T_a.

Solution   Since the predicted effect of increased GHGs is to reduce the transmission of longwave radiation, we can model this effect by lowering the value of t'_a. Changing its value to $t'_a = 0.04$ and recalculating using the numerical solver gives $T_s = 291.4$ K, $T_a = 250.5$ K. This change is equivalent to a 2.0 K increase in Earth's surface temperature. Thus a 33% decrease in the value of t'_a in the model is equivalent to the predicted effect of an increase of 200 to 300 ppm in the concentration of CO_2 in the atmosphere, which might happen between the middle and end of the twenty-first century.

4-3-3   General Circulation Models of Global Climate

General circulation models (GCMs) are the sophisticated computer models of the global climate system that are used to predict the effect of changes in GHG concentration on climate in the future. They incorporate fluid dynamics, thermodynamics, chemistry, and to some extent biology into systems of equations that model changes of state for a three-dimensional system of grid points over multiple time periods (a typical time increment in the model is 1 day). They expand on the simple concepts laid out using the single-layer model explained above by incorporating the following additional elements:

1. *Dynamic modeling:* Whereas the single-layer model is continued to be in static equilibrium with no change to the solar constant or T_s and T_a, GCMs are dynamic models where quantities of thermal or kinetic energy are retained from one time period to the next, and temperatures at specific locations in the models vary accordingly in time.

2. *Multiple vertical layers:* The atmosphere is modeled as having on the order of 20 layers, in order to track temperature gradients from the top to the bottom of the atmosphere. Also, for portions of the globe covered by oceans, these areas are modeled with 15–20 layers, so as to capture the effect of temperature gradients in the ocean on CO_2 absorption and release, and other factors.

3. *Horizontal matrix of grid points:* The globe is covered in a grid of points, so that the model can capture the effect of uneven heating of the globe due to the movement of the sun, change in reflectance based on changing ice cover in different parts of the globe, and other factors. The result is a three-dimensional matrix of grid points covering the globe.

4. *Convection:* Heat transfer between grid points due to convection is explicitly included in the model.

5. *More sophisticated temperature forcing:* In the single-layer model, the forcing function is the solar input, which is assumed to be constant. GCMs are able to incorporate historic and anticipated future variations in solar input due to variations in solar activity, as well as volcanic activity and other deviations.

6. *Biological processes:* The models capture the effect of future climate on the amount of forest cover, which in turn affects CO_2 uptake and the amount of sunlight reflected from the Earth's surface.

7. *Changes in ice cover:* Similar to biological processes, the models capture the effect of changing temperature on the amount of ice cover, which in turn affects the amount of sunlight reflected or absorbed in polar or mountain regions where ice exists.

GCMs have been developed by a number of scientific programs around the world. Some of the best-known models include those of the Center for Climate System Research (CCSR), the Hadley Centre, the Max-Planck Institute, and the National Center for Atmospheric Research (NCAR). Rather than choose a single model that is thought to best represent the future path of the climate, the IPCC, as the international body convened to respond to climate change, uses the collective results from a range of GCMs to observe the predicted average values for significant factors such as average temperature, as well as the "envelope" created by the range of high and low values predicted by the models.

In order to be used to predict future climate, GCMs must first be verified by evaluating their ability to reproduce past climate using known information about CO_2 concentration, solar activity, and volcanic eruptions. In recent years, validation of numerous GCMs over the period from approximately 1850 to 2000 has also provided evidence that solar activity and volcanism alone could not explain the increase in mean global temperature over the last several decades. Figure 4-14 shows the actual mean temperature along with the range of reproduced values from the collection of GCMs. Although each GCM does

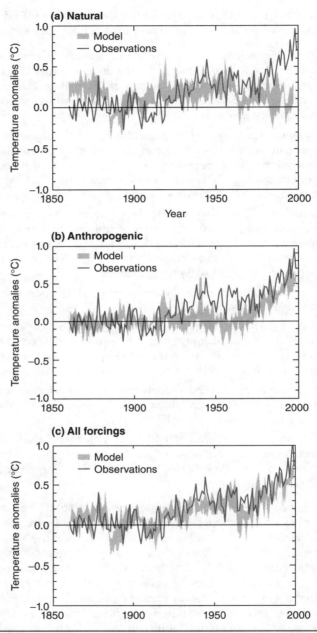

FIGURE 4-14   Observed actual global mean temperature and range of reported values from GCMs, 1850–2000. (a) Natural: Due to changes in sunspot or volcanic activity, and the like. (b) Anthropogenic: Due to increasing emissions of GHGs and other human effects on climate. (c) All forcings: Combination of natural and anthropogenic. (*Source:* Climate Change 2001: Synthesis Report, Intergovernmental Panel on Climate Change. Reprinted with permission.)

not predict the same temperature each year, the band of observed values clearly shows a turning point after approximately 1960. If the rise in CO_2 concentration is ignored and the only forcing is due to solar activity and volcanism, no such turning point emerges. GCMs have also consistently predicted that the most rapid temperature rise would be seen in the polar regions, and this prediction has been borne out in recent years.

Global Integrated Assessment Models

Global Integrated Assessment Models, sometimes also referred to as simply "integrated assessment models," extend climate modeling capacity by integrating GCMs with socio-economic models of human decisions affecting levels of GHG emissions, and climate behavior in turn affecting economies and human behavior (U.S. Department of Energy, 2015). Integrated assessment models allow policymakers to apply proposed policies to the human-climate interaction and observe outcomes in terms of future emissions and CO_2 concentration levels. These models therefore enable the evaluation of policy alternatives against each other with regard to cost and benefits, as well as working backward from a desired future climate outcome to policies that could achieve that outcome. Because of the computing power required to implement an integrated assessment model beyond the already significant needs of a GCM, work to improve the former in terms of geographic scope, level of detail, and maximum future time horizon is ongoing.

4-4   Climate in the Future

While climate change has already brought about some changes that affect the human population and the natural world, it is likely that larger changes will arise in the future. These changes may come about both because the level of GHGs in the atmosphere will continue to increase for some time, no matter what policy steps are taken immediately, and also because some of the effects of increased GHGs have a long lag time, so that past GHG emissions will have an impact some time in the future. At the same time, changing the state of the climate leads to feedback that may accelerate or slow climate change, as discussed in Sec. 4-4-1.

4-4-1   Positive and Negative Feedback from Climate Change

It is a common observation for systems, including very complicated systems like the global climate, that feedback can occur between the output state of the system and the effect of the inputs. *Positive feedback* is that which reinforces or accelerates the initial change, while *negative feedback* counteracts or dampens the change. Figure 4-15 shows the case of the feedback relationship between average global temperature, amount of ice cover, and amount of cloud cover. Ice and cloud cover have the potential to contribute to the amount of sunlight reflected back to space, but ice decreases while clouds increase with temperature, so these two factors counteract one another.

Examples of feedback include the following:

- *(Positive) Loss of ice cover:* Per Fig. 4-14, warmer average temperatures lead to loss of ice at the poles. This in turn leads to less reflection and more absorption of incoming sunlight during the summer, which further increases temperature and melts more ice. This example of positive feedback is particularly strong, and explains why the strongest effect on average temperature is being felt in polar regions. The overall effect of reducing ice coverage is to reduce the albedo

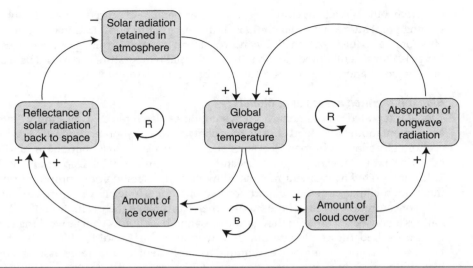

FIGURE 4-15   Causal loop diagram linking global average temperature, cloud or ice cover, and absorption of incoming solar energy.

of the Earth. The effect of reducing albedo can be approximated using Eq. (4-2), where reducing the value from $a = 0.3$ to $a = 0.27$ increases the black-body temperature by 2.5 K.

- *(Positive) Ocean absorption of CO_2:* As the oceans warm, their ability to absorb CO_2 decreases, increasing the net amount of CO_2 in the atmosphere.

- *(Positive) Release of methane from permafrost:* As permafrost melts due to warming, trapped methane is released to the atmosphere, increasing the greenhouse effect.

- *(Positive/negative) Increased forest growth:* Where it occurs, forest growth increases absorption of incoming sunlight and warms planet, but also increases absorption of CO_2 due to accelerated tree and plant growth.

- *(Positive/negative) Increased cloud cover:* Per Fig. 4-15, increased cloud cover reflects more incoming radiation to space but also increases absorption of longwave radiation.

- *(Positive/negative) Deforestation:* Where it occurs, deforestation due to climate change releases additional carbon but also increases reflectivity of the planet, since deforested areas are typically lighter colored than forests.

- *(Negative) Increase in black-body radiation from a hotter planet:* Beyond a certain point of temperature rise trend, the increase in radiation will outpace the ability of GHGs to absorb the additional radiation, so the rate of increase in global average temperature will slow.

Two important observations about feedback arise from the preceding list. First, it is of concern that positive effects outnumber the negative effects. Although much uncertainty surrounds the way in which feedback effects will play out over time, there is a strong likelihood that in the short to medium term, positive effects will outweigh negative effects, and climate change may proceed relatively quickly. Human intervention to combat climate change may be effective in slowing it or lessening its effects, but it may be difficult to reverse during this stage.

Second, in the long term, the most significant type of negative feedback that can prevent a "runaway" increase in global average temperature is the last effect, namely, longwave black-body cooling. To illustrate this feedback effect, the black-body radiation of a surface at 288 K is 390 W/m². Since radiation increases with the fourth power of temperature, the rate of energy flux increases rapidly above this temperature, to 412 W/m² at 292 K and 435 W/m² at 296 K. Thus the average temperature is unlikely to rise much beyond these values, even over centuries. Still, the potential risks to many natural systems and to worldwide built infrastructure of a long-term rise of 4–8 K are very great, so a rise of this magnitude is to be avoided if at all possible.

4-4-2   Scenarios for Future Rates of CO_2 Emissions, CO_2 Stabilization Values, and Average Global Temperature

Given the many uncertainties surrounding the way in which climate responds to the changing concentration of CO_2, it is not possible to model with certainty future global average temperature, let alone expected seasonal temperature by region, as a function of CO_2 emissions. Furthermore, government policies, economic activity, consumer choices, and so on, will determine the pathway of future CO_2 emissions, so these are yet to be determined. The role of other GHGs besides CO_2 complicates the issue further. This is a good application of the use of scenarios to bracket the range of possible outcomes, as discussed in Chap. 2.

In response, the IPCC has identified a range of possible stabilization levels for the concentration of CO_2 in the atmosphere, and "back-casted" a likely pathway of emissions per year that would arrive at the target concentration. The pathways are shown in Fig. 4-16, with matching of pathways to outcomes in terms of stabilization values in ppm of CO_2. All emissions pathways accept that the amount emitted per year will continue to rise in the short term (from 2000 onward) due to current economic activity combined with heavy reliance on fossil fuels. However, by approximately 2020, differences emerge, with the lowest pathway (450 ppm) peaking much sooner and dramatically reducing emissions by 2100, while the highest (1000 ppm) only peaking around 2075. These outcomes represent a percentage increase from the preindustrial level concentration of approximately 280 ppm of between 61% and 257%. The value of 1000 ppm may be close to the maximum possible value of CO_2 concentration, as it has been suggested that if humans were to consume all available fossil fuels, the concentration of CO_2 would not surpass 1400 ppm. Figure 4-16(a) uses a grey band around the pathways to emphasize that the exact path that would reach a given stabilization concentration is not known, and may vary up or down from the colored pathways given. Both Figs. 4-16(a) and 4-16(b) also show numbered curves in black that correspond to IPCC economic growth scenarios, namely, A1B = rapid economic growth and peaking population with balance of energy sources, A2 = continued population growth with widely divergent economic growth between regions, and B1 = similar to A1B but transition to a less materially intensive economy with cleaner energy sources.

We can use a similar approach to estimate the range of plausible outcomes for global average temperature by using predictions from the range of available GCM models.

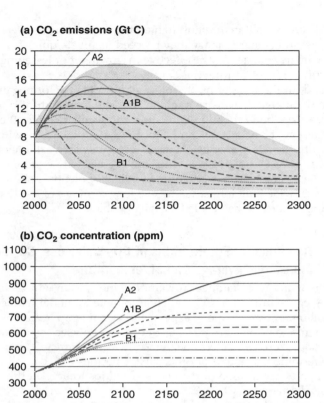

(a) CO_2 emissions (Gt C)

(b) CO_2 concentration (ppm)

FIGURE 4-16   Possible pathways for CO_2 emissions (a) and atmospheric concentration (b) for a range of stabilization values from 450 to 1000 ppm.

The numbered curves are IPCC economic growth scenarios, for example, A1B = rapid economic growth and peaking population with balance of energy sources, A2 = continued population growth with widely divergent economic growth between regions, B1 = similar to A1B but transition to less materially intensive economy, cleaner energy sources. (*Source:* Climate Change 2001: Synthesis Report, Intergovernmental Panel on Climate Change. Reprinted with permission.)

Figure 4-17 shows a temperature range over which different GCMs predict temperature to grow through the year 2100 in response to emissions scenario A2 from Fig. 4-16. In this emissions scenario, there is relatively little intervention to slow the growth of CO_2 concentration in the atmosphere. There is little disagreement among GCMs that temperature would rise by at least 2°C in response to scenario A2, but whether temperature rise might go as high as 5°C is not certain, with several GCMs predicting an outcome somewhere in the middle. Also, an alternative scenario in which CO_2 emissions were cut more aggressively would slow the rate of temperature increase, but this is not shown in the figure. The difference in effect of a temperature rise of approximately 2°C versus a rise of approximately 5°C is the subject of the next section.

Long-Term Effects of Increasing Average Global Temperature
Table 4-1 provided examples of changes currently observed in our world that are likely to be caused by climate change. It is expected that as the average global temperature

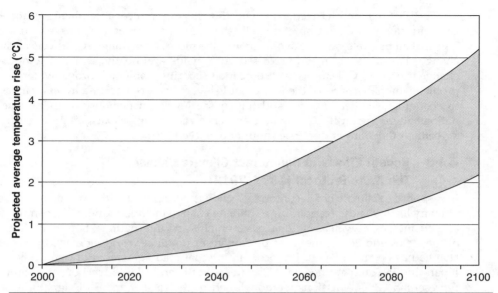

FIGURE 4-17   Predicted temperature rise during the twenty-first century in response to emissions scenario A2, "continued population growth with widely divergent economic growth between regions."

increases, the changes discussed in the table will become more intense. In addition, other long-term effects from increasing temperature and climate change may appear. For example, increasing the average temperature of the upper layers of the ocean will decrease its density, so that the level of the water rises in proportion to the amount of expansion at each layer of depth. At ambient temperature and pressure, each degree Celsius of temperature increase of seawater causes a 0.025% decrease in density, so for each 1 degree of increase of the upper 100 m layer of the ocean, the average sea level would increase by 2.5 cm. Note that this calculation is simplistic, since it does not take into account the change in density with depth, the change in coefficient of thermal expansion with increasing pressure, the effect of undersea topography on thermal expansion, nor the maximum depth to which ocean waters are affected by global warming.

Changes to average sea level due to thermal expansion are likely to be small enough that by themselves they will not create a direct threat to infrastructure that is used to keep seawater out of low-lying areas, such as the infrastructure found in the Netherlands or the Mississippi delta region in the United States. There is a risk, however, that during extreme weather events, storm surges may be more likely to overcome barriers due to their increased maximum height. Rising sea levels may necessitate the evacuation of some low-lying Pacific Island nations such as Tuvalu, which has already appealed to other nations to allow the migration of its inhabitants. The large masses of ice in Antarctica and Greenland pose an additional concern. These contain some 3×10^{16} tonnes of water, or approximately 2% of the mass of the oceans. If a significant fraction of this ice were to melt and run off into the ocean, the long-term consequences on sea level could be severe, including the abandonment of low-lying areas in many regions or the necessary investment in new barriers to keep out the sea. A combination of increased average sea temperature and changes in salinity in the future may also disrupt ocean currents such as the Atlantic Gulf Stream, affecting climate in certain regions.

The IPCC has emphasized in its reporting the importance of stabilizing atmospheric concentration of CO_2 at approximately 450 ppm and temperature increase at around 2°C, if at all possible, so as to avoid some of the more severe long-term effects of climate change. These effects include reduced crop yields around the world if temperature change exceeds 3°C, leading to chronic food security problems; chronic water security problems in many parts of the world; and elevated risk of flooding in many areas due to changing weather patterns and higher sea levels. In response to these concerns, nations attending an IPCC climate conference at Cancun, Mexico, in 2010 reached an agreement that temperature rise should not exceed 2°C.

4-4-3   Recent Efforts to Counteract Climate Change: The Kyoto Protocol (1997–2012)

In response to climate change, human society is taking, and will continue to take, steps at many levels. Those responding to climate change include individuals, communities, and businesses, as well as national governments and international bodies. In the period of the 1990s and 2000s, the most prominent example of worldwide action on climate change has been the Kyoto Protocol, an international agreement drawn up in Kyoto, Japan, in 1997 and ratified in 2005 after a sufficiently large fraction of nations and represented GHG emissions were included in the total number of signatories to the treaty. The Kyoto Protocol is an outgrowth of the United Nations Framework Convention on Climate Change (UNFCCC), which was drawn up at the Earth Summit in Rio de Janeiro in 1992 and ratified in 1994. The UNFCCC does not contain any binding targets for GHGs; instead, it calls for successive "protocols" to be created and ratified under broad guidelines in the UNFCCC.

The Kyoto Protocol was not explicitly designed to achieve one or the other of the CO_2 emissions pathways shown in Fig. 4-16. However, it can loosely be seen as trying to move the signatory nations and, to some extent, the overall community of nations away from Scenario A2 and toward Scenarios A1B or B1. The Kyoto Protocol also required no binding reductions in GHG emissions from the emerging economies. As discussed in Chap. 1, growth in CO_2 emissions from emerging economies has been out-pacing that of the industrialized economies since approximately 2000, and by the end of the period in force of the Kyoto Protocol, this group of countries has become the majority emitter of CO_2 in the world. However, from a historical perspective, the industrialized countries are responsible for most of the cumulative CO_2 emissions and the increase in concentration of CO_2 in the atmosphere, and at present their per capita CO_2 emissions are much higher. Therefore, negotiators reached a consensus that the industrialized countries should take the early lead in curbing CO_2 emissions.

Signatory nations to the Kyoto Protocol include European Union member states, Canada, Japan, Russia, and, as of 2007, Australia. The United States chose not to sign the Kyoto Protocol, on grounds that the effects would have been too detrimental to its economy, and also that it unfairly benefited the emerging economies. Some state and local governments as well as institutions such as universities and corporations in the United States responded to this decision by adopting Kyoto targets on their own.

Targets for reducing GHGs in the Kyoto Protocol were quantified in terms of percent reductions in CO_2 equivalent relative to emissions levels in 1990. Thus total impact across all GHGs, and not just CO_2, is taken into account, and nations can reach targets through a range of measures including changes to practices that affect nonfossil methane and CO_2 emissions, as well as reductions in fossil fuel consumption. The target

Kyoto Signatories:	Total Emissions, MMT:				Per Capita Tonnes CO$_2$		
	1990	2012	Change	Target	1990	2011	Change
Denmark	59	40.5	−31%	−21%	11.11	8.19	−26%
Germany*	980	788	−20%	−21%	12.48	9.63	−23%
Japan	1015	1259	+24%	−6%	8.48	9.42	+11%
United Kingdom	598	499	−17%	−13%	10.44	7.79	−25%
Australia	263	421	+60%	+8%	15.78	19.59	+24%
Nonsignatory countries							
USA	5013	5270	5%	n/a	20.19	17.60	−13%
Worldwide change:							
World total CO$_2$	21616	32310	49%	n/a	3.95	4.63	+17%

*Germany 1990 values are the sum of emissions from the former East Germany and West Germany.

Sources: Energy Information Administration, U.S. Department of Energy, for CO$_2$ from energy; European Environment Agency, for Europe Kyoto targets; Japan Environment Ministry, for Japan Kyoto targets. Note that 2012 per capita emissions values were not available from USEIA so 2011 values are used instead.

TABLE 4-4   Fossil CO$_2$ Emissions 1990–2012 and Targets for Kyoto Protocol for Sample of Signatory Countries (in million tonnes of CO$_2$)

percentages for the 1997–2012 period varied between countries (note that not all were reductions, for Australia the allowed change was +8%). Table 4-4 provides the percent change target and actual values of CO$_2$ emissions from fossil fuels (hereafter "fossil CO$_2$") for Australia, Denmark, Germany, Japan, and the United Kingdom, as well as emissions during the same period for the United States for comparison.

Note that the 1990 and 2012 emissions figures are for fossil CO$_2$ only, as these emissions represent the single largest contribution to total effective GHG emissions measured in CO$_2$ equivalent, and are also of greatest interest for the purposes of understanding energy systems. In practice, since countries can meet Kyoto obligations through a variety of GHG reductions, and the situation with emissions besides fossil CO$_2$ has often been more favorable, overall "actual" values for CO$_2$ equivalent for 2012 are closer to the "target" values shown. (For example, once non-CO$_2$ GHGs and the impact of agricultural and forestry practices on CO$_2$ are taken into account, the Australian government predicts that its 2012 CO$_2$-equivalent emissions will come in at 6% above 1990, thus fulfilling its targets.) On the other hand, it is arguably more realistic to look at the change in fossil CO$_2$ emissions, since these values in the long run will dominate the debate among the community of nations about how to address climate change. The 1990 and 2008 emissions per capita of fossil CO$_2$ are also shown, as an indication of how change in personal carbon intensity (and by implication population growth) contributed to total carbon growth. Thus, for example, Australia's growth in fossil CO$_2$ of 59% is in part due to a rise in per capita CO$_2$ emissions from 15.8 to 19.6 tonnes, but also due to a 28% rise in population, largest of any of the countries in the table.

The quantitative results in Table 4-4 show that as the protocol period drew to a close in 2012, if CO$_2$-only emissions are used in place of total CO$_2$ equivalent, the United

Kingdom and Denmark met their obligations, Germany fell slightly short, and Japan and Australia did not meet their objectives. For comparison, the nonsignatory U.S. reduced total emissions compared to Japan or Australia, but did not reduce them as much as Denmark, Germany, Japan, or the United Kingdom. Japan also started from a relatively efficient value of 8.5 tonnes CO_2 per capita in 1990, while Australia had the largest population growth among any of the countries in the table at 29%. Germany benefited from high CO_2 intensity in the former East Germany prior to reunification, which was then reduced during subsequent modernization.

4-4-4   Assessing the Effectiveness of the Kyoto Protocol and Description of Post-Kyoto Efforts

In retrospect, the Kyoto Protocol can be credited with some successes but also criticized on a number of fronts. On a positive note, the protocol spurred considerable efforts to reduce GHGs, efforts which might not have happened otherwise. This includes not only the efforts of signatories but also the efforts in nonsignatory nations such as the United States where subnational entities (states, municipal governments) have independently adopted internal targets for GHG reductions, such as the state of California in the United States. Furthermore, the Kyoto round gives the community of nations experience that can be applied to future climate negotiations and treaty enforcement.

On a negative note, a list of shortcomings of the Kyoto round includes the rather modest reductions of CO_2 emissions from the combustion of fossil fuels compared to the BAU values that might have occurred with no concerted effort to reign in GHG emissions on the part of the industrialized countries. Although the target initially was to return emissions levels to those of 1990, as a first step toward deep reductions required in the future, emissions from industrialized countries for the period 1990–2012 in fact increased by 14%, or 1.5 Gt. Emissions from emerging countries increased that much more. Because CO_2 emissions account for more than half of the global warming potential from all GHG emissions, the community of nations will need to greatly accelerate efforts in this regard if the goal is to truly reduce the threat to the climate.

The Kyoto round also did not resolve the tension between the industrialized and emerging countries over who should contribute what portion to overall GHG reductions. By design, the Kyoto agreement did not impose any limits on the emerging countries, to allow them more time to adapt to a carbon-constrained world economy. Still, negotiators might have made more progress on finding common ground so that as the Kyoto Protocol ended, a new framework could immediately be implemented that would help the emerging countries, and especially the BRIC countries (Brazil, Russia, India, and China) since their emissions are growing rapidly at present, to introduce low-carbon technologies, with sufficient support from the industrialized countries that poorer citizens of these countries would not be adversely affected.

Perhaps due to the difficulties in making real commitments to reducing fossil CO_2 emissions and to finding genuine agreement between industrial and emerging countries, a third shortcoming to Kyoto has emerged, namely, the inability to agree on a binding, global agreement at a summit meeting on climate in Copenhagen in December 2009. International negotiators had intended to agree to a "Copenhagen Protocol" that would have been the successor to the Kyoto Protocol and guided all nations (including for the first time the emerging countries) to reduce GHG emissions for the next 10 to 15 years. No such new protocol emerged.

The agreement that emerged from Copenhagen in lieu of a full protocol may provide a pattern for the function of climate negotiations in the post-Kyoto era. At the end of the Copenhagen conference, a group of large emitters including the United States along with the so-called BASIC countries (Brazil, India, South Africa, and China) put forth a multilateral (but not universal), nonbinding agreement on voluntary GHG reductions that stated a goal of keeping global temperature increase under $2°C$ and challenged all countries to put forth individual national goals for reducing emissions that would help attain this overall goal. This new paradigm encourages countries to enter into bilateral or multilateral agreements to reduce CO_2 and other GHGs without waiting for a universal agreement similar to the Kyoto Protocol. It also encourages support from rich to poor countries to help the latter adapt to climate change, protect forest resources that act as major carbon sinks, and transition to postcarbon means of energy production as quickly as possible. Subsequent conferences (Cancun, 2010; Durban, 2011; Rio +20 conference in Rio de Janeiro, 2012; Warsaw, 2013; Bonn, 2014; Paris, 2015) are intended to keep countries in discussion and eventually to find sufficient reductions to meet the $2°C$ goal.

In conclusion, if the overall targets for the post-Kyoto era are to be met, nations will need to look to much greater efficiency in the use of fossil fuels and especially to new energy technologies that either significantly reduce CO_2 emissions to the atmosphere or eliminate them entirely to achieve further cuts in emissions. In its public communications, the IPCC has stated that to reach the $2°C$ temperature stabilization goal, global CO_2 emissions should peak by the year 2020 and fall rapidly after that. Given rapid growth in global CO_2 emissions in the opposite direction as late as 2012, and the fact that in the period 1985–2012 there were only three years when total emissions were less than the year before (1991, 1992, 2009), it may be difficult to achieve this goal by 2020. Even if the 2020 target is not met, however, the need for low- and zero-carbon energy technology will persist. This shift toward emphasizing new technologies is a powerful argument for many of the energy systems presented in subsequent chapters of this book.

4-5   Summary

Human use of energy emits large volumes of CO_2 to the atmosphere due to our heavy dependence on fossil fuels as an energy source. Climate change is one of the leading concerns for the energy sector at the present time, because energy conversion is the largest emitter of CO_2 in the world. Many scientists around the world have responded to the challenge of climate change by participating in the development of computerized general circulation models (GCMs) that strive to reproduce the behavior of the atmosphere, allowing the testing of response to perturbations such as an increase in the amount of CO_2. To date, this modeling research strongly suggests that, since CO_2 is such a potent greenhouse gas, doubling or tripling its atmospheric concentration will lead to an increase in global average temperature that can have severe negative repercussions on local average temperature, amount of precipitation, frequency and intensity of extreme weather events, and other concerns. Other anthropogenic GHGs such as methane (CH_4), nitrous oxide (N_2O), and chlorofluorocarbons (CFCs) play a role as well. Some early signs of climate change have already appeared, including receding ice cover and stress on many of the world's coral reefs. Society is responding to climate change on many levels, including locally, nationally, and through international mechanisms such as the UN Framework Convention on Climate Change (UNFCCC) and the Kyoto Protocol.

References

Intergovernmental Panel on Climate Change (2007). Climate Change 2007: Impacts, Adaptation and Vulnerability. Report, IPCC, Geneva.

Intergovernmental Panel on Climate Change (2014). Summary for policymakers from fifth assessment report. IPCC, Geneva.

Maslin, M. (2013). *Climate: A Very Short Introduction*. Oxford University Press, Oxford.

Mitchell, J. (1989). "The Greenhouse Effect and Climate Change." *Review of Geophysics*, Vol. 27, No. 1, pp. 115–139.

Oreskes, N. (2004). "Beyond the Ivory Tower: The Scientific Consensus on Climate Change." *Science*, Vol. 306, No. 5702, p. 1686.

Robinson, A., N. Robinson, and W. Soon (2007). "Environmental Effects of Increased Atmospheric Carbon Dioxide." *Journal of American Physicians and Surgeons*, Vol. 12, No. 3, pp. 79–90.

U.S. Department of Energy (2015). Climate and Environmental Sciences Division (CESD): Integrated Assessment of Global Climate Change. USDOE, Office of Science. Electronic resource, available at www.science.energy.gov. Accessed June 26, 2015.

Further Readings

Berthier, E., Y. Arnaud, D. Baratoux, et al. (2004). "Recent Rapid Thinning of the 'Mer de Glace' Glacier Derived from Satellite Optical Images." *Geophysical Research Letters*, Vol. 31.

Brohan, P., J.J. Kennedy, I. Harris, et al. (2006). "Uncertainty Estimates in Regional and Global Observed Temperature Changes: A New Dataset from 1850." *Journal of Geophysical Research*, Vol. 111.

Collins, W., R. Colman, J. Haywood, et al. (2007). "The Physical Science behind Climate Change." *Scientific American*, Vol. 297, No. 2, pp. 64–73.

Congressional Budget Office (2008). Climate-change policy and CO_2 emissions from passenger vehicles. CBO, Washington, DC.

Hoffert, M., K. Caldeira, G. Benford, et al. (2002). "Advanced Technology Paths to Global Climate Stability: Energy for a Greenhouse Planet." *Science*, Vol. 298, pp. 981–987.

Hunt, J., S. Tett, and J. Mitchell (1996). "Mathematical and Physical Basis of General Circulation Models of Climate." *Zeitschrift fur Angewande Mathematik and Mechanik*, 76, Suppl. 6, pp. 501–508.

Mascarelli, A. (2009). "A sleeping giant?" *Nature Reports*, Vol. 3, April 2009, pp. 46–49.

Moore, T. (1998). *Climate of Fear: Why We Shouldn't Worry about Global Warming*. Cato Institute Press, Washington, DC.

Sterman, J., and L. Sweeney (2007). "Understanding Public Complacency About Climate Change: Adult's Mental Models of Climate Change Violate Conservation of Matter." *Climate Change*, Vol. 80, No. 3–4, pp. 213–238.

Thorpe, A. J. (2005). "Climate Change Prediction: A Challenging Scientific Problem." *Report*. Institute of Physics, London.

Warhaft, Z. (1997). *The Engine and the Atmosphere: An Introduction to Engineering*. Cambridge University Press, Cambridge.

Wigley, T. M. L., and S. C. B. Raper (1987). "Thermal Expansion of Sea Water Associated with Global Warming." *Nature*, Vol. 357, pp. 293–300.

Exercises

4-1. As reported in this chapter, the combination of water and CO_2 molecules in the atmosphere account for approximately 150 W/m² in greenhouse heating. If we start with a black body at 255 K and add 150 W/m² in energy flux out of the body, what is the new predicted temperature?

4-2. Use the model presented in Eqs. (4-6) and (4-7) to model the effect of an ice age. Suppose the combination of increased ice on the surface of the Earth and decreased CO_2 in the atmosphere changes the parameters in the model as follows: $\alpha_s = 0.15$, $t'_a = 0.12$, $\alpha'_a = 0.23$. All other parameters are unchanged. What are the new values of T_a and T_s?

4-3. The planets Mars and Venus have albedo values of 0.15 and 0.75, and observed surface temperatures of approximately 220 K and approximately 700 K, respectively. The average distance of Mars from the sun is 2.28×10^8 km, and the average distance of Venus is 1.08×10^8 km. What is the extent of the greenhouse effect for these two planets, measured in degrees kelvin?

4-4. Build a simple atmosphere-surface climate model for Venus and Mars of the type presented in Eqs. (4-6) and (4-7) in this chapter. Use values for albedo, observed surface temperature (for verification of your model), and orbital distance from Problem 3, and appropriate values of your choosing for the remaining parameters for the model. How well does the model work in each case? Is the accuracy of the outcome sensitive to your choice of parameters? Why or why not? Discuss.

4-5. Verify that the total flux across all wavelengths in Example 4-1 sums approximately to the black-body radiation value given by Eq. (4-1) by summing values for all wavelengths by increment of 1 μm up to an appropriate value of wavelength such that flux is negligible.

4-6. A general circulation model has a grid layout using 2.5 degree increments in latitude and 3 degrees in longitude. It has 19 levels in the vertical. In addition, for the 71% of the Earth's surface covered by ocean, there are layers going under the ocean, on average 10 layers deep. Approximately how many grid points does the model contain?

4-7. Scientists speculate that in a severe ice age, if the amount of ice covering planet Earth became too great, the planet would pass a point of no return in which increasing albedo would lead to decreasing temperature and even more ice cover, further increasing albedo. Eventually the Earth would reach a "white Earth state," with complete coverage of the planet in ice. At this point the albedo would be 0.72, that is, even with a completely white planet, some radiation is still absorbed. Ignoring the effect of the atmosphere, and considering only black-body radiation from the Earth in equilibrium with incoming radiation from the sun, what would the surface temperature of the Earth be in this case?

4-8. Create an "IPCC-like" stabilization scenario using the following simplified information. Starting in the year 2004, the concentration of CO_2 in the atmosphere is 378 ppm, the global emissions of CO_2 are 7.4 Gt carbon per year, the total carbon in the atmosphere is 767 Gt, and emissions are growing by 4% per year, for example, in 2005 emissions reach 7.7 Gt, 8.0 Gt in 2006, and so on. Note that concentration is given in terms of CO_2, but emissions and atmospheric mass are given in Gt carbon. The oceans absorb 3 Gt net (absorbed minus emitted) each year for the indefinite future. The change or decline of emissions is influenced by "global CO_2 policy" as follows: in year 2005, the emissions rate declines by 0.1 percentage points to 3.9%, and after that and up to the point that concentration stabilizes, the change is 0.1% multiplied by the ratio of the previous year's total emissions divided by 7.4 Gt, that is

$$\text{Chg\%}_t = \text{Chg\%}_{t-1} - \frac{(0.1\%)(\text{total emissions}_{t-1})}{7.4}$$

After concentration stabilizes, emissions are 3 Gt/year, so that emissions are exactly balanced by ocean absorption. To illustrate, (Chg%)2004 = 4%, (Chg%)2005 = 3.9%, and so on. Also, concentration can be calculated as 378 ppm × (total carbon in atmosphere/767 Gt). (a) What is the maximum value of concentration reached? (b) In what year is this value reached? (c) What is the amount of CO_2 emitted that year? (d) Plot CO_2 concentration and emissions per year on two separate graphs. Hints: Use a spreadsheet or other software to set up equations that calculate emissions rate, change in emissions, and concentration in year t as a function of values in year $t - 1$. (e) Compare the shape of the pathway in this scenario, based on the graphs from part (d), with actual IPCC scenarios, and discuss similarities and differences. (f) The scenario in this problem is a simplification of how a carbon stabilization program might actually unfold in the future. Identify two ways in which the scenario is simplistic.

CHAPTER 5

Fossil Fuel Resources

5-1   Overview

At the present time, fossil fuels such as oil, gas, and coal provide the majority of energy consumed around the world. At the same time, all fossil fuels are not the same, both in terms of their availability and their impact on climate change. The first part of this chapter compares fossil fuel alternatives in terms of their worldwide consumption rates, estimated remaining resource base, and cost per unit of energy derived. It also considers the amount of CO_2 emitted per unit of energy released, and outlines strategies for reducing CO_2 emissions. The second part of the chapter presents methods for projecting the future course of annual consumption of different fossil resources, and discusses implications of declining output of conventional fossil fuels for nonconventional sources such as oil shale, tar sands, and synthetic fossil fuel substitutes made from coal.

5-2   Introduction

As the name implies, fossil fuels come from layers of prehistoric carbonaceous materials that have been compressed over millions of years to form energy-dense concentrations of solid, liquid, or gas, which can be extracted and combusted to meet human energy requirements. Different types of fossil fuels vary in terms of energy content per unit of mass or volume, current consumption rates in different regions of the world, availability of remaining resource not yet extracted, and CO_2 emissions per unit of energy released during combustion. These differences underpin the presentation, later in this book, of many of the technologies used for stationary energy conversion and transportation applications, and are therefore the focus of this chapter.

Since the time that coal surpassed wood as the leading energy source used by humans for domestic and industrial purposes in the industrializing countries of Europe and North America in the nineteenth century, fossil fuels have become the dominant primary energy source worldwide, currently accounting for about 86% of all energy production. It is only in the poorest of the emerging countries that biofuels such as wood or dung are the major provider of energy for human needs; elsewhere, fossil fuels are the majority provider of energy for electricity generation, space heating, and transportation. Because fossil fuels are so important for meeting energy needs at present, any pathway toward sustainable energy consumption must consider how to use advanced, high-efficiency fossil combustion technology as a bridge toward some future mixture of nonfossil energy sources, or consider how to adapt the use of fossil fuels for compatibility with the goals of sustainable development over the long term.

5-2-1   Characteristics of Fossil Fuels

The main fossil fuels in use today are coal, oil, and natural gas (hereafter shortened to "gas"); in addition, there are several emerging, nonconventional fossil fuels, such as *shale oil, shale gas,* and *tar sands,* that are considered at the end of this chapter. Of the three major sources, coal cannot be combusted directly in internal combustion engines, so its use in transportation, or "mobile," applications is limited at the present time. It is used widely in "stationary" (i.e., nontransportation) applications including electricity generation and industrial applications. Coal can also be used in small-scale domestic applications such as cooking and space heating, although local air quality issues become a concern. Gas and oil are more flexible in terms of the ways in which they can be combusted, and they are used in both transportation and stationary applications. While gas generally requires only a modest amount of purification before it is distributed and sold to end-use customers, oil comes in the form of crude petroleum that must be refined into a range of end products. Each end product takes advantage of different viscosities that are made available from the refining process, as shown in Table 5-1.

Energy Density and Relative Cost of Fossil Fuels

The *energy density* of a fuel is a measure of the amount of energy per unit of mass or volume available in that resource. The energy density of a fossil fuel is important, because fuels that are less energy dense must be provided to energy conversion processes in greater amounts in order to achieve the same amount of energy output. In addition, less dense fuels cost more to transport and store, all other things being equal.

The energy density of coal is measured in GJ/tonne or million BTU/ton. Of the three main fossil fuels, coal varies the most in energy density, with coal that has been under the Earth for a shorter period of time having relatively more moisture and lower energy density than coal that has been underground longer. Energy density of oil is measured in GJ/barrel or million BTU/barrel. Energy density varies moderately between oil extracted from various oil fields around the world, depending on, among other things, the amount of impurities such as sulfur or nitrogen. For purposes of estimating output from energy conversion devices that consume petroleum, it is possible to assume a constant value for oil energy density. For natural gas, energy density is measured in kJ/m^3 or BTU/ft^3 when the gas is at atmospheric pressure, and its value can be assumed constant. In practice, gas is compressed in order to save space during transportation and storage.

Typical values for energy density are shown in Table 5-2, along with a representative range of prices paid on the world market in the 2005–2015 period in U.S. dollars. While coal varies between 15 and 30 GJ/tonne, the value of 25 GJ/tonne is used for purposes of converting cost per tonne to cost per unit of energy. Turning to costs, the

Raw gasoline	Automotive gasoline, jet fuel
Kerosene	Automotive diesel fuel, kerosene
Gas oil	Domestic heating oil, industrial fuel oil
Lubricating oil	Paraffin, motor oil for automotive applications
Heavy oil	Coke for industrial applications, asphalt

TABLE 5-1   Available End Products from the Refining of Crude Petroleum, in Order of Increasing Viscosity

Resource	Energy Density	Unit Cost	Cost per Unit of Energy ($/GJ)
Coal	15–30 GJ/tonne	$25–$100/tonne	$1.00–$4.00
Crude petroleum	5.2 GJ/barrel	$36–$150/barrel	$6.92–$28.85
Natural gas	36.4 MJ/m³	$0.11–$0.28/m³	$2.91–$7.76

Notes: Prices shown are representative values that reflect both fluctuations in market prices during the 2005–2015 period, and also differences in regional prices, especially for coal, which is subject to the greatest variability in transportation costs. Energy density of coal is highly variable with coal quality. A value of 25 GJ/tonne is used to convert coal unit cost into cost per unit of energy. Values below $25/tonne and as low as $15/tonne are reported by the Energy Information Administration; however, these prices are paid for low energy density coals and are therefore not included in the cost range figures shown.

TABLE 5-2 Energy Density and Typical 2005–2015 Cost Range for Coal, Petroleum, and Natural Gas

table shows that, in general, coal is significantly cheaper per unit of raw energy than gas or especially oil. However, individual circumstances vary from region to region. Coal is the most difficult to transport, and transportation costs may drive the actual price paid by the end user well above the values shown, especially for destinations requiring long distance movements by truck, or for customers purchasing relatively small quantities of coal (e.g., for a college campus' central-heating plant, as opposed to a large coal-fired power plant). Countries with large amounts of gas and oil resources among the emerging countries that also have relatively low GDP per capita (e.g., in the range of $5000–$10,000 per person, compared to >$30,000 per person for some of the wealthier industrialized countries) are able to sell gas and oil products to individual domestic consumers at prices below world market rates. Otherwise, in countries not possessing large fossil fuel reserves, consumers can expect to pay prices in the ranges shown.

Table 5-2 also indicates the wide variations in unit costs of fossil energy observed recently, while Fig. 5-1 shows the volatility in crude oil prices observed since 2005 compared to the relative stability in the 1990–2005 period. Oil prices have varied between $36 and $147 per barrel. While these fluctuations are concerns for both individual consumers and whole economic sectors, fossil energy prices can also be self-correcting. When energy prices rise disproportionately, they tend to choke off economic growth, leading to reduced demand for energy that allows energy prices to fall again. Over the longer term, high energy prices also allow energy-efficient technologies to become more cost-competitive, again easing upward pressure on energy demand and allowing cost growth to slow down or reverse. Changing prices have also had an impact on the relative market share of coal and gas in the United States. As shown in Fig. 5-2, coal prices rose steadily from approximately 2005 onward, while gas prices fell after 2009, so that gas market share climbed while that of coal declined.

Energy Consumption in Location, Extraction, Processing, and Distribution of Fossil Fuels

The values given in Table 5-2 for energy density reflect the energy content of the product after extraction from the ground, and in the case of crude petroleum, before refining into finished energy products. The process of extracting fossil fuels from the ground, processing

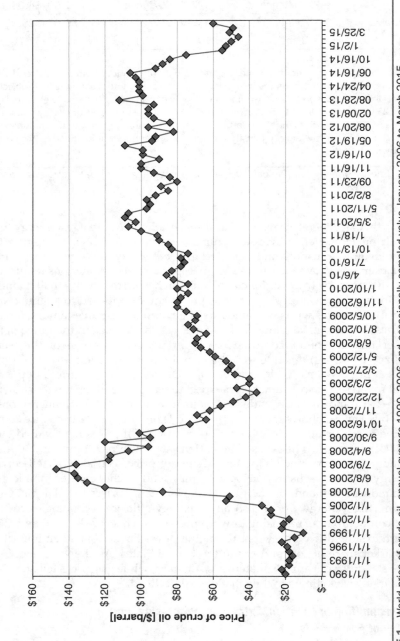

Figure 5-1 World price of crude oil, annual average 1990–2006 and occasionally sampled value January 2006 to March 2015.

Note: Change of time scale used 2005–2015 to illustrate wide swings in price during this period. Data are from Energy Information Administration (1990–2006) and occasional personal sampling of publicized values (2006 to present).

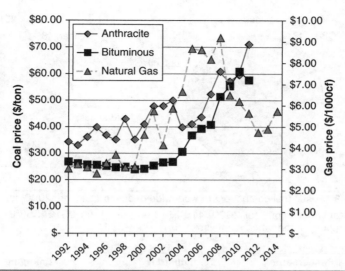

FIGURE 5-2   U.S. Price of coal and natural gas, 1992-2014.

(*Source:* U.S. Energy Information Administration. "Bituminous" and "Anthracite" are two leading types of coal. Data series for coal was not available from USEIA for years after 2011. Natural gas price shown is "city gate" price, including well-head price and transmission to local market, but not including retail price markup.)

them into finished energy products, and distributing the products to customers constitutes an "upstream loss" on the system, so that the net amount of energy delivered by the system is less than the values shown. However, this loss is usually small, relative to the larger losses incurred in conversion of fossil fuels into mechanical or electrical energy. For coal extraction, the U.S. Department of Energy has estimated that energy consumption in extraction, processing, and transportation of coal from U.S. coal mines may range from 2% to 4% of the coal energy content for surface mining or deep-shaft mining, respectively. For gas, upstream losses are on the order of 12%, while for petroleum products, they may be on the order of 17%, although this figure includes a substantial component for the refining process.[1]

From a life-cycle perspective, upstream energy losses should also include a component for energy consumed during the oil exploration process that led to the discovery of the resource in the first place. These data are not currently available in the literature; however, it can be speculated that their inclusion would not change the preceding outcome, namely, that upstream losses comprise a modest fraction of the total energy density of fossil fuels.

5-2-2   Current Rates of Consumption and Total Resource Availability

Measured in terms of gross energy content of resource consumed each year, fossil fuels account for some 82% of total world energy consumption. As shown in Fig. 5-3, in 2012 crude oil represents the largest source of energy, while coal and natural gas, the second- and third-ranked energy sources, each supply more energy than all the nonfossil

---

[1]Sources: Brinkman (2001), for petroleum product data; Kreith et al. (2002), for gas data.

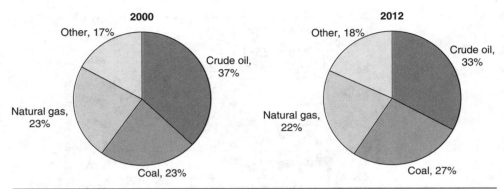

FIGURE 5-3   Fossil fuel component of worldwide energy production, 2000 and 2012.

Note: Total value of pie: for 2000, 416 EJ (394 quads); for 2012, 553 EJ (524 quads). (*Source:* Energy Information Administration.)

Resource:	Industrial		Emerging	
Year 2005	**(EJ)**	**% of Total**	**(EJ)**	**% of Total**
Crude oil	95.0	47%	83.6	39%
Natural gas	55.9	27%	55.2	26%
Coal	53.0	26%	75.9	35%
All fossil fuels	203.9	100%	214.8	100%
Year 2012				
Crude oil	85.0	43%	96.5	38%
Natural gas	57.5	29%	66.5	26%
Coal	57.5	29%	89.3	35%
All fossil fuels	200.0	100%	252.3	100%

Source: U.S. Energy Information Administration.

TABLE 5-3   Comparison of Breakdown of Energy Consumption by Fossil Fuel Type for Industrialized and Emerging Countries, 2005 and 2012

sources combined. The "other" component consists primarily of nuclear and hydropower, with smaller amounts of wind, solar, and biomass. Note also that over the 2000–2012 time period the consumption of coal has outpaced that of the other two fossil fuels, so its share of the pie has grown, while that of petroleum has declined.

Fossil fuels are used differently and in different amounts, depending on whether a country is rich or poor. Table 5-3 gives the breakdown of fossil fuel consumption among industrialized and emerging countries.[2] These data show how crude oil consumption is concentrated in the industrial countries, while the emerging countries consume a relatively large proportion of coal. There are several reasons for this distinction.

---

[2]Members of industrialized countries include North America, Europe, Japan, Australia, and New Zealand; emerging countries include all other countries. See Chap. 1.

First, the wealthier countries have greater access to motorized transportation that depends on petroleum products as an energy source, including a larger number of private motor vehicles per capita or greater volume of goods movement. Also, in and near large population centers in rich countries, the relative economic wealth enables the generation of electricity using natural gas, which costs more per unit of energy than coal, but which also can be burned more cleanly. Poor countries cannot compete in the international marketplace for gas and oil to the same extent as rich countries, and so must rely to a greater degree on coal for electricity generation and industrial purposes. The industrial countries also have greater access for electricity generation to nuclear power, which thus far has not made major inroads into the energy generation markets of the developing world (see Chap. 8).

Estimated Reserves and Resources of Fossil Fuels

A *reserve* is a proven quantity of a fossil fuel that is known to exist in a given location, within agreed tolerances on the accuracy of estimating fuel quantities. (It is of course not possible to estimate quantities of a substance that is under the ground to within the nearest barrel or cubic foot, so some variation is expected between published reserve quantities and the actual amount that is eventually extracted.) A *resource* is an estimated quantity of fuel that has yet to be fully explored and evaluated. Over time, energy companies explore regions where resources are known to exist in order to evaluate their extent, so as to replace energy reserves that are currently being exploited with new reserves that can be exploited in the future. Thus the quantity of reserves that is said to exist by an energy company or national government at any given point in time is less than the amount of the fossil fuel that will eventually be extracted.

Table 5-4 gives available reserves of coal, oil, and gas for the entire world, as well as for selected countries that are either major consumers or producers of fossil fuels. Quantities are given in both standard units used in everyday reporting for the fuel in question (e.g., trillion cubic meters for gas, and so on), and also in units of energy content, to permit comparison between available energy from different reserves. For each fuel, an additional quantity of resource is available that may be equal to or greater than the reserve value shown. For example, one estimate for the worldwide coal resource is 130,000 EJ, (Sorensen, 2002, p. 472) more than five times the 2014 value of the reserve.

Two observations arise from the table. First, even taking into account the exact energy content of proven reserves, it is clear that coal is more abundant than gas or oil. The expansion of nonconventional oil and gas extraction is changing the outlook for these two resources compared to what the relatively small reserves in Table 5-4 would suggest (See Sec. 5-3). Nevertheless, world energy supplies in the future may reach a point where coal is used as a substitute when oil and gas supplies tighten. Second, countries such as China that have sizeable crude oil demand and relatively small reserves (1.5% of the world total in the table) can expect to rely on imports for the foreseeable future. The United States was among the import-dependent countries early in the 2000s, but the rise of domestic nonconventional oil extraction has resulted in great reductions in the amount of imported oil (down from more than 10 million barrels per day in 2010 to 5 million in 2014).

The relationship between annual consumption and proven reserves is illustrated in Example 5-1.

Country:	Gas		Oil		Coal	
Year 2004	**(tril m³)**	**(EJ)**	**(bill. bbl)**	**(EJ)**	**(bill. tonnes)**	**(EJ)⁽¹⁾**
China	1.5	54.7	18.2	95	110	2750
Russia	47.6	1732.4	60.0	314	250	6250
Saudi Arabia⁽²⁾	·6.5	238.2	262.0	1373	—	—
United States	5.4	196.6	21.9	115	270	6750
Rest of World	111.0	4040.4	902.9	4731	370	9250
Total	172.0	6262.2	1265.0	6628	1000	25,000
Year 2014						
China	4.4	160.2	24.4	127.7	126.2	3155.4
Russia	47.8	1740.3	80.0	419.2	173.1	4326.8
Saudi Arabia⁽²⁾	8.2	299.8	268.4	1406.1	—	—
United States	9.6	348.8	36.5	191.4	258.6	6465.5
Rest of World	127.5	4639.6	1246.3	6530.4	421.9	10,547.1
Total	197.5	7188.7	1655.6	8674.8	979.8	24,494.8

Notes: (1) Energy values based on 36.4 MJ/m³ for gas, 5.2 GJ/bbl for oil, 25 GJ/tonne for coal. Volumetric conversion: 1 m³ = 35.3 ft³. (2) Saudi Arabia is not a significant source for coal.

Source: U.S. Energy Information Administration.

TABLE 5-4 Fossil Fuel Reserves for World and for Select Countries, 2005 and 2014

Example 5-1 Total consumption of gas, oil, and coal in the United States in 2014 measured 28.1, 37, and 18.3 EJ of energy content, respectively. (A) Calculate the ratio of reserves to energy consumption for the United States for these three resources. (B) Discuss the validity of this calculation.

Solution From Table 5-4, the total reserves for the United States are 349, 191, and 6465 EJ, respectively. Therefore, the ratios are 12.4, 5.2, and 354 for the three fuels. This calculation implies that at current rates, without considering other circumstances, the gas reserves will be consumed in about 12 years, the oil reserves in 3 years, while the coal reserves will last for more than three centuries.

Discussion The use of the ratio of reserve to consumption misrepresents the time remaining for availability of a resource in a number of ways. First, for gas and oil in this example, the calculation does not take into account the fraction of the resource that is imported, as opposed to supplied domestically. For the particular case of U.S. oil consumption, this fraction is significant. Domestic production in 2014 amounted to 5.3 billion barrels, or 27.5 EJ of energy equivalent out of 42.6 EJ of total demand. Imported oil amounting to 9.5 EJ met the remainder, a 26% share. Second, depletion of existing reserves spurs exploration to add new resources to the total proven reserves, which pushes the point at which the resource is exhausted further out into the future. Lastly, as a fossil fuel such as petroleum comes closer to depletion, the remaining resources tend to those that are more difficult and more expensive to extract. Also, awareness in the economic marketplace that the resources are scarce allows sellers to increase prices. With prices rising for both of these reasons, substitutes such as natural gas become more competitive, and displace some of the demand for oil. As a result, oil demand is driven downward, slowing the rate of depletion.

As a result of these complicating factors, the ratio of reserves to consumption is not an accurate predictor of how long a resource will last. It is, however, a relative indicator between resources of which one is likely to be exhausted first, that is, the resource with the lower reserve-to-consumption ratio. Based on the figures in this example, it is likely that the U.S. oil reserves will be exhausted first, followed by the gas reserves, and then by the coal reserves.

5-2-3 CO_2 Emissions Comparison and a "Decarbonization" Strategy

The CO_2 emissions per unit of fossil fuel combusted are a function of the chemical reaction of the fuel with oxygen to release energy. The energy released increases with the number of molecular bonds that are broken in the chemical transformation of the fuel. Fossil fuels that have a high ratio of hydrogen to carbon have relatively more bonds and less mass, and so will release less CO_2 per unit of energy released. To illustrate this point, consider the three chemical reactions governing the combustion of coal, gas, and oil in their pure forms:[3]

$$Coal: C + O_2 >> CO_2 + 30 \text{ MJ/kg}$$

$$Gas: CH_4 + 2O_2 >> CO_2 + 2H_2O + 50 \text{ MJ/kg}$$

$$Oil \text{ (gasoline)}: C_8H_{18} + 12.5O_2 >> 8CO_2 + 9H_2O + 50 \text{ MJ/kg}$$

Note that the energy released is given per unit of mass of fuel combusted, not of CO_2. The amount of CO_2 released per unit of energy provided can be calculated from these reactions by comparing the molecular mass of fuel going into the reaction to the mass of CO_2 emitted. The carbon intensity of methane (CH_4) is calculated in Example 5-2.

Example 5-2   Calculate the energy released from combusting CH_4 per kilogram of CO_2 released to the atmosphere, in units of MJ/kg CO_2.

Solution   To solve for MJ/kg CO_2, recall that the molecular mass of an atom of carbon is 12, that of oxygen is 16, and that of hydrogen is 1. Therefore, the mass of a kilogram-mole of each of these elements is 12 kg, 16 kg, and 1 kg, respectively. The mass of a kilogram-mole of CH_4 is therefore 16 kg, and the mass of a kilogram-mole of CO_2 is 44 kg, so the amount of energy released per unit of CO_2 emitted to the atmosphere is 50 MJ/kg $\times$ (16/44) = 18.2 MJ/kg CO_2.

Repeating this calculation for gasoline gives 16.1 MJ/kg CO_2, and for coal gives 8.18 MJ/kg CO_2. It is left as an exercise at the end of this chapter to carry out the calculation.

Comparison of the values for gas, oil, and coal underscores the value of carbon–hydrogen bonds in the fuel combusted for purposes of preventing climate change. Coal does not have any such bonds, so it has the lowest amount of energy released per unit of CO_2 emitted to the atmosphere. This is a problem for many countries seeking to reduce CO_2 emissions, because certain sectors, such as the electricity sector in China, Germany, or the United States, depend on coal for a significant fraction of their generating capacity. To put the relative emissions in some context, a typical power plant producing on the order of 2 billion kWh of electricity per year might require roughly 50 PJ (~50 trillion BTU) of energy input to meet this demand. Completely combusting coal with this amount of energy content would generate 5.9 million tonnes of CO_2, whereas for gas the amount is 2.7 million tonnes.

In response to this situation, some national or regional governments have pursued or advocated a policy of *decarbonization*, in which coal is displaced by oil and gas as an energy resource for electricity, industry, or other applications, in order to reduce CO_2 emitted per unit of energy produced. Decarbonization is especially appealing for countries with large gas or oil reserves, such as Canada, which had reserves of 1.7 trillion m^3

---

[3]Energy content value given for coal is for high-quality, relatively pure coal. Coal with a high moisture content or with significant impurities has lower energy content per kilogram. For oil, the formula for gasoline is used as a representative petroleum product.

(66.7 trillion ft^3) of gas and 173 billion barrels of oil in 2014, with a population of 35.5 million. As the Kyoto Protocol and other future agreements restrict the amount of CO_2 that can be emitted, these countries can use decarbonization to meet targets.

Historically, decarbonization has been taking place since the nineteenth century around the world and especially in the industrialized countries, with many sectors of the world economy replacing coal with oil and gas because they were cleaner and easier to transport and combust. The lower carbon-to-energy ratio was a convenient side effect of this shift, but it was not the motivation. As a worldwide strategy for the future; however, decarbonization is faced with (1) a relative lack of gas and oil reserves/ resources compared to those of coal and (2) rapidly growing demand for energy, especially in the emerging countries. With aggressive implementation of much more efficient energy conversion and end-use technologies, the world might be able to decarbonize by drastically cutting demand for energy services in both the industrial and emerging countries, and then using existing gas and oil resources for most or all of the remaining energy requirements. Otherwise, decarbonization will be limited to only those countries that have the right mix of large gas and oil resources relative to fossil fuel energy demand.

Already since the year 2000, a trend toward *recarbonization*, or movement back from hydrocarbons to coal, is taking place at the worldwide level, as the percent share of fossil energy from coal has been increasing. Based on Fig.5-3, between 2000 and 2012, total consumption of coal had increased by 51 EJ versus 28 EJ for both oil and natural gas. China, for example, has a rapidly growing economy and, from Table 5-4, much larger reserves of coal than oil or gas, so it is not surprising that its growing energy consumption would push the world energy breakdown toward a greater share for coal, absent a truly historic shift away from fossil fuels as a group.

5-3   Decline of Conventional Fossil Fuels and a Possible Transition to Nonconventional Alternatives

Example 5-1 showed that a linear model of fossil fuel resource exhaustion, in which the resource is used at a constant rate until it is gone, is not realistic. In this section, we consider a more plausible pathway for resource exhaustion, in which annual output declines as the remaining resource dwindles. This pathway for the life of the resource as a whole reflects that of individual oil fields, coal mines, and so on, which typically follow a life cycle in which productivity grows at first, eventually peaks, and then begins to decline as wells or mines gradually become less productive. The worldwide production of a fossil fuel follows the same pathway as that of individual fields. Near the end of the resource's worldwide lifetime, older wells or mines, where productivity is declining, outnumber those that are newly discovered, so that overall production declines.

Hereafter we focus on the particular case of crude oil, as the resource that arguably is getting the most attention at present due to concerns about remaining supply.

5-3-1   Hubbert Curve Applied to Resource Lifetime

The growth-peak-decline pattern of nonrenewable fossil resources can be seen both in the *annual* and *cumulative* production patterns of the resource. In this section, we first look at techniques to fit curves to the annual production pattern, and thereafter look at curve-fitting with the cumulative pattern.

Resource Lifetime Using Annual Output Values

The use of a curve to model the lifetime of a nonrenewable resource is sometimes called a "Hubbert curve" after the geologist M. King Hubbert, who in the 1950s first fitted such a curve to the historical pattern of U.S. petroleum output in order to predict the future peaking of U.S. domestic output. The fitting of a curve to the observed data is presented in this section using the Gaussian curve. The choice of curves is not limited to the Gaussian; other mathematical curves with a similar shape can be used to improve on the goodness-of-fit that can be obtained with the Gaussian.

The Gaussian curve has the following functional form:

$$P = \frac{Q_{inf}}{S\sqrt{2\pi}} \exp\left[-(t_m - t)^2/(2S^2)\right] \tag{5-1}$$

Here P is the output of oil (typically measured in barrels) in year t; Q_{inf} is the estimated ultimate recovery (EUR), or the amount of oil that will ultimately be recovered; S is a width parameter for the Gaussian curve, measured in years; and t_m is the year in which the peak output of oil occurs. In order to fit the Gaussian curve to a time series of oil output data, one must first obtain a value for Q_{inf}. Thereafter, the values of t_m and S can be adjusted either iteratively by hand or using a software-based solver to find the values that minimize the deviation between observed values of production P_{actual} and values of P_{est} predicted by Eq. (5-1). For this purpose, the *root mean squared deviation* (RMSD) can be used, according to the following formula:

$$\text{RMSD} = \sqrt{\frac{1}{n}\sum_{i=1}^{n}(P_{actual,\,i} - P_{est,\,i})^2} \tag{5-2}$$

Here n is the number of years for which output data are available, and the subscript i denotes the year in the time series. Example 5-3 uses a small number of data points from the historical U.S. petroleum output figures to illustrate the use of the technique.

Example 5-3   U.S. domestic output of petroleum for the years 1910, 1940, 1970, and 2000 are measured at 210, 1503, 3517, and 2131 million barrels, respectively, according to the U.S. Energy Information Administration. If the predicted ultimate recovery of oil is estimated at 223 billion barrels, predict the point in time at which the oil output will peak, to the nearest year. What is the predicted output in that year?

Solution   As a starting point, suppose that we guess at the values of the missing parameters S and t_m to calculate an initial value of RMSD. Let $S = 30$ years and $t_m = 1970$. Then for each year t, the value of P_{est} can be calculated using Eq. (5-1). The appropriate values for calculating RMSD that result are given in the following table, leading to a value of $\Sigma(P_{actual} - P_{est})^2 = 5.38 \times 10^{17}$:

t	P_{actual} (10⁶ barrel)	P_{est} (10⁶ barrel)	$(P_{actual} - P_{est})^2$
1910	210	401	3.665E+16
1940	1503	1799	8.772E+16
1970	3517	2966	3.032E+17
2000	2131	1799	1.101E+17
		Total	5.377E+17

The value of the RMSD is then calculated using Eq. (5-2):

$$\text{RMSD} = \sqrt{\frac{1}{4}(5.377 \times 10^{17})} = 3.67 \times 10^8$$

The values of S and t_m can then be systematically changed by hand in order to reduce RMSD. Here we have used a solver to find the values of S and t_m that minimize RMSD, in this case by setting up a spreadsheet to calculate RMSD based on all four values and asking the solver to find the values of S and t_m that minimize RMSD. The solver returns $S = 25.3$ years and $t_m = 1974$.

t	P_{actual} (10⁶ barrel)	P_{est} (10⁶ barrel)	$(P_{actual} - P_{est})^2$
1910	210	146	4.028E+15
1940	1503	1442	3.671E+15
1970	3517	3479	1.456E+15
2000	2131	2056	5.659E+15
		Total	1.481E+16

Based on these values, the RMSD is reduced as follows:

$$\text{RMSD} = \sqrt{\frac{1}{4}(1.481 \times 10^{16})} = 6.086 \times 10^7$$

Thus the minimum value is RMSD = 6.09×10^7. Substituting $Q_{inf} = 223$ billion, $S = 25.3$ years, and $t_m = 1974$ into Eq. (5-1), we get:

$$P = \frac{2.23 \times 10^{11}}{25.3\sqrt{2\pi}}\exp(0) = \frac{2.23 \times 10^{11}}{25.3\sqrt{2\pi}} = 3.518 \times 10^9 \text{ bbl}$$

This value is quite close to the observed maximum for U.S. petroleum output of 3.52 billion barrels in 1972.

Application to U.S. and Worldwide Oil Production Trends

The Gaussian curve can be fitted to the complete set of annual output data from 1900 to 2008, as shown in Fig. 5-4. Assuming an EUR value of $Q_{inf} = 225$ billion, the optimum fit to the observed data is found with values $S = 27.8$ years and $t_m = 1976$. Therefore, the predicted peak output value from the curve for the United States occurs in 1976, 6 years after the actual peak, at 3.22 billion barrels/year. By 2005, the actual and predicted curves have declined to 1.89 billion and 1.86 billion barrels/year, respectively. If the Gaussian curve prediction were to prove accurate, the output would decline to 93 million barrels/year in 2050, with 99.3% of the ultimately recovered resource consumed by that time. The Gaussian curve fits the data for the period 1900–2005 well, with an average error between the actual and estimated output for each year of 6%.

The same technique can be applied to worldwide oil production time series data in order to predict the peak year and output value. A range of possible projections for the future pathway of the world annual oil output can be projected, depending on the total amount of oil that is eventually recovered, as shown in Fig. 5-5. Since the world total reserves and resources are not known with certainty at present, a wide range of EUR values can be assumed, such as values from 2.5 trillion to 4 trillion barrels, as shown.

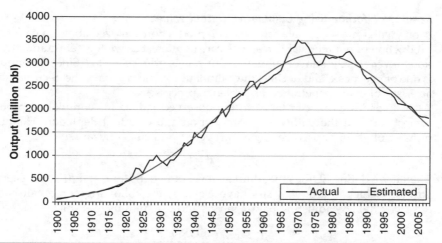

Figure 5-4 U.S. petroleum output including all states, 1900–2008. (*Source:* U.S. Energy Information Administration.)

Figure 5-5 suggests that the timing of the peak year for oil production is not very sensitive to the value of the EUR. An increase in the EUR from 2.5 trillion to 4 trillion barrels delays the peak by just 21 years, from 2013 to 2034. In the case of EUR = 3.25 trillion, output peaks in 2024. Also, once past the peak, the decline in output may be rapid. For the EUR = 4 trillion barrel case, after peaking at 35.3 billion barrels/year in 2034, output declines to 25.6 billion barrels/year in 2070, a reduction of 28% in 36 years.

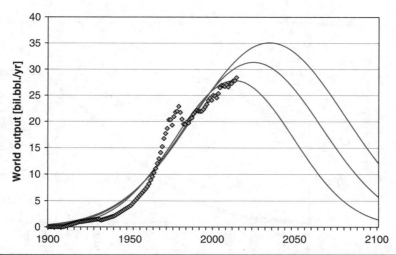

Figure 5-5 World oil production data 1900–2014, with three best-fit projections as a function of EUR values of 2.5 trillion, 3.25 trillion, and 4 trillion barrels.

(*Sources:* Own calculation of production trend and projection using Eqs. (5-1) and (5-2), and historical production data from U.S. Energy Information Administration.)

Resource Lifetime Using Cumulative Output Values

Along with time series data for *annual* oil production analyzed above, data for *cumulative* production of nonrenewable resources are of interest as well, where cumulative production for year t is the sum of annual output values from year 1 to t. Since the annual pattern is one of rise, peak, and decline, the cumulative value will have the 'S-curve' shape characteristic of many penetration or depletion processes.

A logistics curve can be fit to the observed values for cumulative production to project forward the pathway of production and eventual depletion. Define t' as the number of years elapsed between start year t_0 and current year t, i.e.,

$$t' = t - t_0 \tag{5-3}$$

Then if we define EUR as the ultimate level of penetration, and $f(t')$ as the level of penetration achieved after t' years have elapsed, the following curve can be fit to the observed data by adjusting parameters c_1 and c_2 to minimize RMSD:

$$f(t') = \frac{\text{EUR} \cdot e^{(c_1 + c_2 t')}}{1 + e^{(c_1 + c_2 t')}} \tag{5-4}$$

Either the absolute number of barrels of oil consumed or the percent of EUR can be used for fitting parameters c_1 and c_2, as illustrated in Example 5-4, which shows the application of the logistics curve to the historic pathway of U.S. oil production. The reader can refer to Fig. 5-6, which has been adapted from Fig. 5-4 to show the observed annual and cumulative production in the same graph.

Example 5-4   For an estimated ultimate recovery value of EUR = 2.23×10^{11} bbl for the United States and the production values of 1900–2008 shown in Fig. 5-4, model cumulative oil production using the logistics function by finding parameters c_1 and c_2 that minimize RMSD.

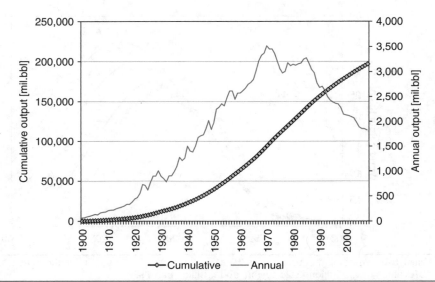

FIGURE 5-6   Annual and cumulative U.S. petroleum output including all states, 1900–2008. (*Source:* U.S. Energy Information Administration.)

Solution Set up a spreadsheet or computational software table in which the cells in the $f(t')$ column are written formulaically in terms of F, c_1, and c_2 to provide an estimated value for each year, for comparison with the actual values. A table with 109 rows, one for each year in the time series, results. An abbreviated version of the table with values for 1900, 1901, 1950, and 2008 is shown here, including calculations based on both the absolute numbers of production ("absolute approach") and percent of EUR consumed ("percent approach"). Note that absolute value of cumulative oil production is reported in millions of barrels.

Year	Cumulative Barrels (millions)			Percent of EUR		
	Observed	Modeled	(Obs − Mod)²	Observed	Modeled	(Obs − Mod)²
1900	6.36E+01	2.26E+03	4.839E+06	0.03%	1.01%	9.73E−05
1901	1.33E+02	2.40E+03	5.156E+06	0.06%	1.08%	1.04E−04
etc.	etc.	etc.	etc.	etc.	etc.	etc.
1950	39.96E+04	39.31E+04	4.165E+05	17.92%	17.63%	8.38E−06
etc.	etc.	etc.	etc.	etc.	etc.	etc.
2008	1.83E+05	1.66E+05	7.31E+05	88.3%	87.9%	1.47E−05
Sum of error terms:			2.74E+08	Sum of error terms:		5.497E−03

The solver minimizes the sum of RMSD by changing c_1 and c_2 in the model, resulting in values $c_1 = -4.58$ and $c_2 = 0.06077$ for either absolute or percent approach, giving the figures shown in the "modeled" columns in the table. Substituting the sum of error terms at the bottom of the table into Eq. (5-2), we get the following values for RMSD, respectively:

$$\text{RMSD\_Absolute} = \sqrt{(1/106)(2.73 \times 10^8)} = 1606$$

$$\text{RMSD\_Percent} = \sqrt{(1/106)(5.497 \times 10^{-3})} = 0.0072$$

Thus RMSD values are different, but both are minima for their respective approaches.

Next, we can illustrate the calculation of the modeled value in a specific year from the table by taking the year 1950 as an example, as follows:

$$t' = 1950 - 1900 = 50$$

$$f(50) = \frac{\text{EUR} \cdot e^{(c_1 + c_2 t')}}{1 + e^{(c_1 + c_2 t')}} = \frac{2.23 \times 10^{11} \cdot e^{(-4.58 + 0.06077 \cdot 50)}}{1 + e^{(-4.58 + 0.06077 \cdot 50)}} = 3.931 \times 10^{10}$$

That is, estimated value of 39.31 billion barrels, which can be compared with the observed value of 39.96 billion barrels, a difference of 650 million barrels or 1.6% of the actual value. The overall agreement between the estimated and observed curves is quite good, as can be seen from the comparison of the two curves in Fig. 5-7.

Discussion One of the benefits of fitting the penetration curve to the observed cumulative consumption is that it provides another way to project future extent of depletion. For example, if one extrapolates the curve with $c_1 = -4.580$ and $c_2 = 0.06077$ out to the year 2025 (i.e., $t'' = 125$), approximately 213 billion barrels, or 95%, of the original total amount available have been depleted, leaving 10 billion barrels to be consumed.

Also, although the annual curve undulates substantially from year to year, the cumulative s-curve shows little impact, with a maximum rate of increase achieved sometime between 1970 and 1980, and the rate of growth declining thereafter. The large magnitude of total extraction in the first 70 to 80 years explains the relative insensitivity of the cumulative curve and its smooth shape. During the 1970–1980 period, for example, annual output values vary between 3 and 3.5 billion barrels per year, but by that time, the cumulative value is growing from approximately 100 billion barrels in 1970 to 130 billion barrels in 1980, so that it is difficult to perceive swings of half a billion barrels up or down against background values in this range.

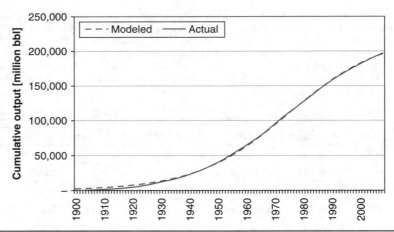

FIGURE 5-7   Comparison of observed and estimated cumulative consumption curves for U.S. conventional petroleum production 1900–2008.
(*Source:* U.S. Energy Information Administration.)

Concern about Peaking and Subsequent Decline of Conventional Oil Production

Predicting when the conventional oil peak will occur is of great interest from an energy policy perspective, because from the point of peaking onward, society may come under increased pressure to meet ever greater demand for the services that oil provides, with a fixed or declining available resource. This phenomenon has been given the name *peak oil*. For example, the world output curve with EUR = 4 trillion barrels in Fig. 5-5 suggests that output would peak around 2030 and decline by 36% over the next 40 years, even as demand for oil would continue to grow in response to increasing economic activity.

Not all analysts agree that the future world annual output curve will have the pronounced peak followed by steady decline shown in the figure. Some believe that the curve will instead reach a plateau, where oil producers will not be able to increase annual output, but will be able to hold it steady for two or three decades. Most agree, however, that the total output will enter an absolute decline by the year 2050, if not before. A similar trend is occurring with natural gas: in some countries, traditional gas resources are being depleted, creating opportunities for new extraction techniques. For this reason, it is important to consider the potential role of nonconventional fossil fuels in the future.

5-3-2   Potential Role for Nonconventional Fossil Resources as Substitutes for Oil and Gas

As mentioned in the introduction, in addition to the main fossil resources of oil, gas, and coal, there are nonconventional resources that are derived from the same geologic process but that are not currently in widespread use. There has been interest in nonconventional fossil resources for several decades, since these resources have the potential to provide a lower-cost alternative at times when oil and gas prices on the international market are high, especially for countries that import large amounts of fossil fuels and have nonconventional resources as a potential alternative. In other cases, nonconventional resources may make up for shortfalls of conventional resources if the latter dwindle in availability.

In general, the conversion of nonconventional resources to a usable form is more complex and expensive than for conventional oil and gas, which has hindered their development up until now. However, with the rapid rise in especially the price of oil in recent years, output from existing nonconventional sources is growing, and new locations are being explored.

There are four main options for nonconventional fossil resources, as follows:

1. *Oil shale:* Oil shale is composed of fossil organic matter mixed into sedimentary rock. When heated, oil shale releases a fossil liquid similar to petroleum. Where the concentration of oil is high enough, the energy content of the oil released exceeds the input energy requirement, so that oil becomes available for refining into end use products.

2. *Tar sands:* Tar sands are composed of sands mixed with a highly viscous hydrocarbon tar. As with oil shale, tar sands can be heated to release the tar, which can then be refined.

3. *Synthetic gas and liquid fuel products from coal:* Unlike the first two resources, the creation of this resource entails transforming a conventional source, coal, into substitutes for gas and liquid fuels, for example, for transportation or space-heating applications where it is inconvenient to use a solid fuel. Part of the energy content of the coal is used in the transformation process, and the potential resulting products include a natural gas substitute (with minimum 95% methane content) that is suitable for transmission in gas pipelines, or synthetic liquid diesel fuel.

4. *Unconventional gas resources from advanced extraction techniques:* In recent years, extraction techniques have been developed that allow the extraction of natural gas resources that were previously inaccessible. For example, high-pressure slipwater hydro-fracturing, or "hydro-fracking," allows the gas extractor to drill first vertically downward and then horizontally into shale layers that contain shale gas. After an initial fracturing operation to release the gas from the shale using a water-based fracturing fluid that contains a proprietary mix of chemical agents, the gas can be pumped from the well for an extended period of time.

These nonconventional fossil resources are already being extracted, converted, and sold on the world market at the present time. For example, the oil sands of Alberta, Canada, produced approximately 1 million barrels of crude oil equivalent per day in 2005. The total world resource available from oil shale and tar sands is estimated to be of a similar order of magnitude to the world's total petroleum resource, in terms of energy content, and could in theory substitute for conventional petroleum for many decades if it were to be fully exploited. The total oil shale resource is estimated at 2.6 trillion barrels, with large resources in the United States and Brazil, and smaller deposits in several countries in Asia, Europe, and the Middle East. Economically recoverable tar sand deposits are concentrated in Canada and Venezuela, and are estimated at approximately 3.6 trillion barrels.

Along with extraction and processing of nonconventional resources, conversion of coal to a synthetic gas or oil substitute is another means of displacing conventional gas and oil. Since 1984, a coal gasification plant in North Dakota, United States, has been converting coal to a synthetic natural gas equivalent that is suitable for transmission in the national gas pipeline grid. Processes are also available to convert coal into a liquid fuel for transportation, although these have relatively high costs at present and are therefore not in commercial use.

For the United States, one of the most significant developments for nonconventional fossil resources in recent years is the rise of shale gas extraction as a growing contributor to the total domestic gas supply. From a negligible quantity in the year 2000, shale gas has risen to 14% of the total gas supply in 2009 (90.4 billion cubic meters or 3.2 tcf out of 646 billion cubic meters or 23 tcf) and 44% in 2013 (323 bcm or 11.4 tcf out of 739 bcm or 26.1 tcf). The rise of shale gas also accounts for the growth over the last 5–10 years in size of gas reserves both for the United States and for the world reported in Table 5-4. For example, inclusion of nonconventional gas reserves increases the figure for the United States from 5 to 10 tcm. Furthermore, the size of these reserves is likely to continue to grow as the underlying geologic formation is better understood.

Thus the nonconventional resources show the potential to address the resource sufficiency side of the energy challenge, at least for an interim period of several decades. In some cases, nonconventional resources of one type may substitute for conventional resources of another, such as substituting nonconventional, domestic shale gas for conventional oil resources in the United States, which could provide a longer term alternative vehicle fuel and also reduce expenditures on imports. The example of growing nonconventional fossil fuel supplies is explored quantitatively in the next section.

5-3-3   Example of U.S. and World Nonconventional Oil Development

The example of U.S. nonconventional oil development can be further examined by taking the production data shown in Fig. 5-4 and extending it through the year 2014, as shown in Fig. 5-8. The latter figure shows total U.S. production, broken down between conventional and nonconventional resources. The total value given for each year is the actual total production reported by the U.S. Energy Information Administration. Conventional production is the amount predicted by the continuation of the Hubbert

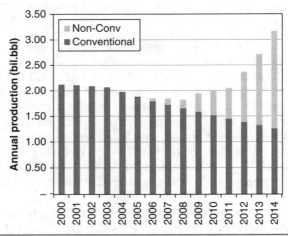

FIGURE 5-8   U.S. conventional and nonconventional oil production 2000-2014. (*Source:* U.S. Energy Information Administration, for total annual production.)

Curve from Fig. 5-4, and then the nonconventional amount shown for each year is the difference between total and conventional production.

Observation of the time series data shows how growth in nonconventional production has reversed the long-term decline in U.S. oil production. Already in the 2006–2008 period, small amounts were being produced, although total output was continuing to decline, reaching its lowest point of 1.83 billion barrels in 2008. Thereafter, nonconventional output increased rapidly, surpassing conventional production in 2013 (1.40 versus 1.33 billion barrels) and reaching 1.91 billion barrels in 2014. The 2014 total of 3.18 billion barrels is nearly as high as the earlier peak of 3.52 billion barrels of conventional oil in 1970.

The time series data for nonconventional oil for the period 2006–2014 provides another opportunity for projecting the future trajectory of oil output using the Hubbert curve technique in the form of Eqs. (5-1) and (5-2). The EUR value is highly uncertain as of 2015 because development is in early stages, but one estimate from USEIA puts the total at 46 billion barrels in the continental United States (Slaughter, 2013). Note that this EUR value is likely conservative, since the total amount available is likely to increase as the industry continues to produce and gain experience; cumulative production through 2014 already amounts to 6 billion barrels. Using the figure of EUR = 46 billion, fitting the curve to 2006–2014 production gives peak year t_m = 2020 and width parameter S = 4.6 years. The resource declines rapidly after 2020 and has been mostly consumed by 2030, as shown in Fig. 5-9.

A similar pattern of nonconventional resources substituting for conventional ones might play itself out at the world level as well. As mentioned in Sec. 5-3-2, combined oil shale and tar sand resources may be on the order of 5 trillion barrels worldwide. These resources might allow total world output of petroleum to continue to grow worldwide, even as conventional resources decline as predicted by the Hubbert Curve. For example, one scenario from Cambridge Energy Research Associates (2006) suggests that, although conventional oil production would not keep pace with demand going forward and would peak in 2040, nonconventional oil would make up the difference and postpone any pronounced decline in total availability until 2060 or later (Fig. 5-10).

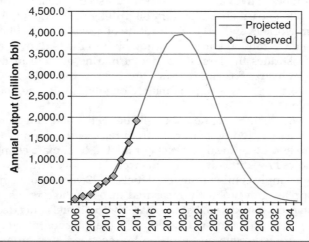

Figure 5-9 Historical pathway for U.S. nonconventional oil production and projected Hubbert Curve path, 2006–2034. *Note:* Curve assumes EUR = 46 billion barrels. See text for discussion. (*Source:* U.S. Energy Information Administration.)

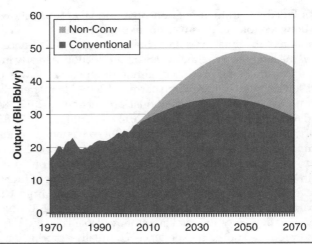

FIGURE 5-10   Projection of annual world oil output to 2070, showing impact of unconventional oil on total production over time period. [*Source:* Adapted from Cambridge Energy Research Associates (2006).]

5-3-4   Discussion: Potential Ecological and Social Impacts of Evolving Fossil Fuel Extraction

In the last few years, concerns about the ecological impact of fossil fuel extraction have come on many fronts. Two high profile disasters occurred within 15 days of each other in the year 2010. In the first incident, on April 5th, in the Upper Big Branch mine in West Virginia, USA, accumulated methane in the mine shaft exploded, killing 29 miners who were in the mine at the time. It was the worst accident in U.S. coal mining industry in terms of number of lives lost in 40 years. In the second incident, on April 20th, an explosion and fire on board an exploratory oil drilling platform in the Gulf of Mexico killed 11 platform workers and led to the rupture of a sea-floor well-head and the leaking of approximately 5 million barrels of crude oil into the Gulf over the next 90 days. This accident constitutes the largest accidental marine oil spill in the history of the U.S. petroleum industry. For comparison, the 1989 Exxon Valdez spill in Prince William Sound, Alaska, USA, was approximately one-tenth this size, measured in number of barrels spilled.

Along with risk of accidents, fossil fuel extraction and transportation to market faces other risks including both risks from release of greenhouse gases (GHGs) to the atmosphere and other local impacts not related to climate change. At the same time, Fig. 5-3 shows how difficult it is for the world to move away from the dominant share of its energy supply coming from fossil fuels. In 12 years shown in the figure, renewable energy supplies grew robustly measured in percentage terms, but total demand for energy also grew quickly, so that the share for fossil fuels only declined by one percentage point. There are no easy answers, and the question of what is the right balance between securing our energy supply and protecting the environment is the subject of a vigorous debate both among experts and the general public.

Potential Effect on Climate Change

Increasing reliance on nonconventional petroleum resources (e.g., oil shale, tar sands, etc.) is of concern from a GHG emissions point of view because of the additional heating and processing energy expenditure, which increases CO_2 emissions per unit of useful energy delivered to the end use application (e.g., motor vehicle, home-heating system, etc.). For example, from above, motor vehicles using gasoline or natural gas as a fuel incur losses on the order of 12–17% upstream of the vehicle. Substituting equivalent products from oil shale or tar sands may double this upstream loss, as an amount of energy equivalent to 10–20% of the original energy content of the extracted oil would be consumed at the extraction site in order to remove the oil from the shale or sands in the first place. Similarly, for synthetic liquids or gases made from coal, a large fraction of the energy content of the coal is used as process heat in the chemical conversion of the coal. These extra energy consumption burdens would make it even more difficult to achieve CO_2 reduction goals while using nonconventional resources, assuming that combustion by-products are vented to the atmosphere, as is done at present.

All types of fossil fuel extraction (oil, gas, and coal) have the potential for releasing incidental amounts of methane to the atmosphere. This includes shale gas, where the extra steps involved in the extraction process may lead to greater releases of methane than for conventional gas extraction and may undercut the CO_2 emissions advantage of gas over coal. To some extent, the gas industry may be able to improve this situation by investing in technologies and practices that prevent methane from escaping.

Other Potential Environmental Effects Not Related to Climate Change

Along with increasing CO_2 emissions, expanded use of oil shale, tar sands, processed coal, and shale gas as substitutes for conventional oil and gas would likely increase the amount of disruption to the Earth's surface, as large areas of land are laid bare and fossil resources removed. In the past, the record of the oil and gas industry in maintaining the natural environment in extraction sites has been mixed, and some sites have suffered significant degradation. The least impacting options are the extraction of conventional oil and gas from boreholes and wells, or the mining of coal from deep shaft mines. As we move to other extraction options and increased use of nonconventional fossil fuels, the process becomes more disruptive for the surface, mitigated somewhat by more stringent regulations and a better understanding of how to protect and restore ecosystems compared to 50 or 100 years ago.

For coal, strip mining lays bare large areas of land while the coal is extracted. A variation on this technique is *mountain-top removal*, primarily used in the Appalachian Mountains of the eastern United States, where large amounts of explosives are used to remove *overburden* or rock layers above a coal seam. Daily volume of explosives used is large, on the order of 1 to 2 kilotons per day.[4] Once the overburden is removed, the coal is extracted mechanically (as if it were a strip mine) and then the overburden is either replaced or dumped in nearby valleys. Some attempt to return natural systems (forests and wildlife) to their premining condition is required by law, but experience shows that some of the impact is irreversible. Extraction of tar sands or oil shale to extract the fossil resource is similarly disruptive to the surface.

---

[4]A *kiloton* is a measure of explosive power. One kiloton is equivalent to 1.2 GWh of energy.

For petroleum, some remaining reserves are in increasingly deep offshore locations, and the risk is that if blowouts happen, they are difficult to contain. The Gulf Oil spill occurred in 5000 ft (approximately 1500 m) deep water, where the extremely high water pressure on the sea floor hampered efforts to cap the leak and stop the spread of oil into the sea water.

Local impacts from unconventional gas extraction are multifaceted. Since the well pads for fracturing the rock and then extracting the gas are dispersed, it is not possible to reach each pad by pipeline, so hydro-fracturing fluid must be moved in and out by truck. Water for the process must be taken from local resources and then later cleaned and returned to watersheds, potentially straining both the supply and treatment systems. The fluid must be managed carefully to prevent accidental escape into the environment both during transport and from failure of well casings. Taking the Marcellus Shale formation in the eastern United States as an example, as of 2011, there have been several instances of accidental leakage of fluid, explosions at production sites, and other accidents that have led to government agencies requiring companies to clean up sites or pay fines. Because of the distributed nature of the operations, no one incident is as large as the Gulf Oil Spill of 2010. The overall effect on the entire region over time of extra traffic, water concerns, and accidents may in future years end up having a similar effect on the environment as well as the social fabric of communities, if operations are not conducted carefully.

A comparison of the quantity of current and future coal and nonconventional fuel use illustrates the way in which surface disruption might intensify. Currently, coal and nonconventionals account for on the order of 150 EJ out of some 450 EJ worldwide fossil energy use. If in the year 2050 the total fossil energy were to grow to 750 EJ, with 80% derived from coal, oil shale, tar sands, and shale gas due to the decline in availability of conventional oil and gas, the rate of extraction of these resources would quadruple. Thus regions possessing these resources could expect extraction to go forward at a much faster rate, possibly affecting livability for both human and nonhuman life in these regions.

A similar example can be developed for the United States for the case of nonconventional gas resources. The 2013 contribution to the total energy supply from this source had an energy content of approximately 11 EJ. Suppose in the future the total national energy demand stabilizes around 100 EJ: one could imagine that two-thirds or 67 EJ of this total, or about five times the current rate of production, might come from unconventional gas and be used for residential and industrial heat, electricity generation, and as a transportation fuel. This rate of gas extraction, much of it in areas of the United States with high populations such as central Texas, the lower peninsula of Michigan, or the mid-Atlantic states, would be a strain on the population.

Possible Steps to Reduce Impact

Regardless of the path fossil fuel usage takes in the future, society can pursue a number of steps now that can protect local environments and reduce GHG emissions, as follows:

1. *Heightened standards for local impacts from extraction:* As a complement to policies that encourage greater use of nonconventional fossil fuels, high standards for protecting wildlife, land and water resources, and human communities should be enacted and enforced.

2. *Curtail or discontinue mountain-top removal mining:* Because of the intense damage to local environments and waterways from this practice, it should be reduced or phased out in favor of deep shaft mining. Shaft mining typically costs more per unit of coal extracted but also creates more jobs. It is an open question whether all coal seams could be worked by shaft instead of by mountain-top removal, for some seams shaft mining may not be practical.

3. *Improve controls on methane release from fossil fuel extraction:* Due to its high global warming potential, or GWP (see Chap. 4), incidental methane releases at all stages of both conventional and nonconventional fossil fuel production should be studied, and wherever possible techniques should be implemented to reduce emissions. While CO_2 emissions from carbon combustion are inevitable for the time being, incidental methane releases do not provide any service (unlike the electricity or propulsion from energy conversion) and should be eliminated as much as possible.

4. *Increase efficiency of energy end use to slow growth in fossil fuel demand:* Improving energy efficiency helps across the board by reducing the need to extract fossil fuels of all types.

5. *Increase efforts to capture CO_2 from energy conversion from all types of fossil fuels:* In concept, fossil fuels combustion systems can be designed to capture the by-product CO_2 without allowing it to enter the atmosphere (see Chap. 7). This approach typically converts fossil fuels into a currency such as electricity or hydrogen, which can then be distributed in a grid for stationary end uses or transportation. The CO_2 is then "sequestered" in long-term deposits under the land or sea (see Chap. 7). This step does not address local environmental impacts but does reduce the climate change burden from fossil fuels.

Even were these steps robustly implemented, it is not clear whether they would adequately address environmental needs over the next two or three decades. Policies regarding climate change may shift, requiring a more rapid transition to nonfossil energy sources and rendering moot efforts to reduce the impact of fossil fuel extraction. In the short to medium term, however, the above steps provide a means of using both conventional and nonconventional fossil fuels in the most environmentally benign way possible.

5-3-5 Conclusion: The Past and Future of Fossil Fuels

Looking back at the history of fossil fuel use since the beginning of the industrial revolution, it is clear that certain key characteristics of these fuels made them highly successful as catalysts for technological advance. First, they possess high energy density, significantly higher than, for example, the amount of energy available from combusting an equivalent mass of wood. Also, in many cases, they exist in a concentrated form in underground deposits, relative to wood or fuel crops, which are dispersed over the surface of the Earth and must be gathered from a distance in order to concentrate large amounts of potential energy. Furthermore, once technologies had been perfected to extract and refine fossil fuels, they proved to be relatively cheap, making possible the use of many energy-using applications at a low cost. Indeed, it could be argued that the ability to harness fossil fuels was the single most important factor for the success of the industrial revolution.

Looking forward to the future, it is clear that while we have come to appreciate the benefits of fossil fuels, we also live in a world with a new awareness of the connection between fossil fuels and climate change. This awareness will over time transform the way fossil fuels are used.

There are two main ways to proceed with our use of fossil fuels. One way is to use them only long enough to sustain our energy-consuming technologies while we rapidly develop nonfossil energy sources that will replace them (a "long-term fossil fuel availability" scenario). Even in a very ambitious scenario, 350 EJ/year of fossil fuel consumption cannot be replaced overnight with renewable or nuclear sources, so this approach might take one century or more to complete. Thereafter, fossil fuel consumption would be restricted to use as a feedstock for the petrochemical industries, in which case fossil fuel resources might last for several millennia.

Alternatively, society might continue to use fossil fuels as the leading energy source at least into the twenty-second century, if not beyond, by developing technologies for capturing and sequestering CO_2 without releasing it to the atmosphere (a "robust fossil fuel use" scenario). Assuming that these technologies were successful and that the impact of surface mining could be successfully managed, fossil fuels might continue to hold a financial advantage over nonfossil alternatives for as long as they lasted, and society would only transition to the latter at a much later date.

Figure 5-11 shows a possible shape for both of these scenarios, with pathways beginning around the year 1800 and continuing to the right. The ultimate outcome of the robust use scenario is illustrated using a Hubbert-style curve for the case of EUR

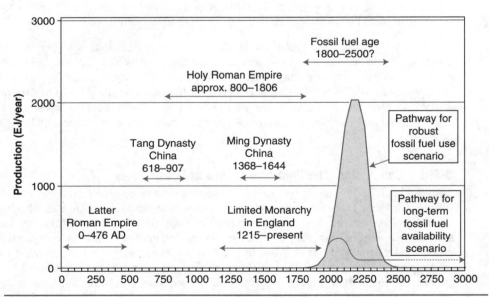

Figure 5-11 Long-term view of fossil fuel production for years AD 0–3000 with projection for "robust" world fossil fuel scenario for years AD 1800–2500 assuming EUR of 500,000 EJ for all types of fossil fuels (oil, coal, gas, oil shale, and tar sands), and projection of "long-term fossil fuel availability" scenario.

Note: Steady-state value in long-term availability scenario is to convey concept only; see text.

500,000 EJ for the total value of energy available from all fossil sources, without making a distinction among oil, coal, gas, and nonconventional sources. The curve is fitted to the historical rate of growth of fossil fuel consumption for the period 1850–2000. Assuming a stable world population of roughly 10 billion people after 2100, the peak value around the year 2200 would be equivalent to annual production of 200 GJ/capita, or on a similar order of magnitude to today's value for rich countries such as Germany, Japan, or the United States. Also shown is the long-term availability scenario, which peaks in the mid-twenty-first century and then stabilizes at a steady-state value that is much lower than the current rate of fossil fuel consumption. Note that the steady-state value shown is a representative value only; no analysis has been done to estimate what level of fossil fuel consumption might be required in the distant future in this scenario.

By using a millennial time scale from the year AD 0 to the year 3000, the figure also conveys the sense that the age of fossil fuels might pass in a relatively short time, compared to some other historical time spans, such as, for example, the period from the founding of Rome in 742 BC to the fall of the Roman Empire in AD 476. Furthermore, barring some unforeseen discovery of a new type of fossil fuel of whose existence we do not yet know, this age will not come again, once passed. What will its ultimate legacy be? Clearly, fossil fuels have the potential to be remembered some day in the distant future as a catalyst for innovation, an endowment of "seed money" from the natural world bestowed on society that served as a stepping stone on the way toward the development of energy systems that no longer depend on finite resources. On a more pessimistic note, fossil fuels may also leave as their primary legacy the permanent degradation of the planetary ecosystem, from climate change and various types of toxic pollution. The outcome depends both on how we manage the consumption of fossil fuels going forward and on how we manage the development of nonfossil alternatives.

5-4 Summary

Concern about the future availability of fossil fuels, and in particular oil and gas, is one of the other major drivers of transformation of worldwide energy systems at the present time. Among the three major conventional fossil fuel resources, coal is more plentiful and cheaper than oil or gas, but has the disadvantage of higher emissions of CO_2 per unit of energy released. A comparison of the current rate of consumption to proven reserves of a fossil fuel provides some indication of its relative scarcity. However, in order to more accurately predict the future course of annual output of the resource, it is necessary to take into account the rising, peaking, and decline of output, using a tool such as the Hubbert curve. With the awareness that conventional oil and gas output will decline over the next few decades, nonconventional resources such as oil shale, tar sands, and synthetic fuels from coal are generating more interest. These options have the potential to greatly extend the use of fossil fuels, but must also be developed with greatly reduced emission of CO_2 to the atmosphere in order to support the goal of preventing climate change.

References

Brinkman, N., M. Wang, T. Weber, et al. (2005). *GM Well-to-Wheels Analysis of Advanced Fuel/Vehicle Systems: A North American Study of Energy Use, Greenhouse Gas Emissions, and Criteria Pollutant Emissions*. Report, General Motors Corp., Argonne National Laboratories, and Air Improvement Resources, Inc.

166   Chapter Five

Cambridge Energy Research Associates. (2006) Press release 60907-9. Cambridge Energy Research Associates, Cambridge, MA.

Kreith, F., R. E. West, and B. E. Isler (2002). "Legislative and Technical Perspectives for Advanced Ground Transportation Systems." *Transportation Quarterly*, Vol. 56, No. 1, pp. 51–73.

Slaughter, Andrew. (2013) "Global supply and market impacts of U.S. unconventional oil production growth". Presentation to Energy Information Administration 2013 conference, Washington, DC, June 18, 2013. Electronic resource, available at www.eia.gov. Accessed June 1, 2015.

Sorensen, B. (2002). *Renewable Energy: Its Physics, Engineering, Environmental Impacts, Economics and Planning*, 2nd ed. Academic Press, London.

U.S. Energy Information Administration (2015). Website of the USEIA. Electronic resource, available at www.eia.gov. Accessed July 2, 2015.

Further Reading

America's Natural Gas Alliance. (2011). "ANGA Remarks on Duke Study on Methane." http://www.anga.us/media-room/links--resources/anga-on-duke-study-on-methane. Web resource, accessed June 24, 2011.

Bartlett, A. (2000). "An Analysis of US and World Oil Production Patterns Using Hubbert-Style Curves." *Mathematical Geology*, Vol. 32, Num. 1, pp. 1–17.

Greene, D., J. Hopson, and J. Li (2003). *Running Out of and into Oil: Analyzing Global Oil Depletion and Transition through 2050*. Report ORNL/TM-2003/259, Oak Ridge National Laboratories, Oak Ridge, TN.

Howarth, A., R. Santoro, and A. Ingraffea (2011). "Methane and the greenhouse-gas footprint of natural gas from shale formations: A letter." *Climatic Change*, online journal, April 12, 2011, edition. Accessed April 21, 2011.

Hubbert, M. K. (1956). *Nuclear Energy and the Fossil Fuels*. American Petroleum Institute, Drilling and Production Practices, pp. 7–25.

Osborn, S. G., A. Vengosh, N. R. Warner, et al. (2011). "Methane contamination of drinking water accompanying gas-well drilling and hydraulic fracturing." *Proceedings of the National Academy of Sciences*, Vol. 108, pp. 8172–8176.

Exercises

5-1. CO_2 intensity: Per kilogram of fuel combusted, the following fuels release the following amounts of energy: butane (C_4H_{10}), 50 MJ; wood (CH_2O), 10 MJ; gasoline (C_8H_{18}), 50 MJ; coal (pure carbon), 30 MJ. Calculate the energy released per kilogram of CO_2 emitted for these fuels.

5-2. Comparison of consumption and reserves: Research archived information on the Internet or in other sources to find published comparisons from the past of oil consumption and proven reserves in a given year, for the United States or other country, or for the whole world. Create a table using these sources, reporting the following information from each source on separate line in the table: year for which the data are provided, consumption in that year, reserves in that year, ratio of reserves to consumption. From your data, does the reserves-to-consumption ratio appear to be increasing, decreasing, or holding constant?

5-3. Effect of using synthetic fuels on CO_2 emissions: A compact passenger car that runs on natural gas emits 0.218 kg CO_2 per mile driven. (a) What is the mass and energy content of the fuel consumed per mile driven? (b) Suppose the vehicle is fueled with synthetic natural gas that is derived from coal. Forty percent of the original energy in the coal is used to convert the coal to the synthetic gas. What is the mass of coal consumed and new total CO_2 emissions per mile? Consider only the effect of using coal instead of gas as the original energy source, and the conversion loss; ignore all other factors. Also, for simplicity, treat natural gas as pure methane and coal as pure carbon.

5-4. Hubbert curve: Suppose the EUR for world conventional oil resources is 3.5 trillion barrels. Find on the Internet or other source data on the historical growth in world oil production from 1900 to the present. Then use the Hubbert-curve technique to predict: (a) the year in which the consumption peaks, (b) the world output in that year, and (c) the year following the peak in which the output has fallen by 90% compared to the peak.

5-5. An oil company refines crude oil valued at $62/barrel and sells it to motorists at its retail outlets. The price is $2.90/U.S. gallon ($0.77/L). On a per unit basis (e.g., per gallon or per liter), by what percentage has the price increased going from crude oil before refining to final sale to the motorist?

5-6. Repeat Problem 5, with oil at $100/barrel and retail outlet price at $4.00/gal ($1.06/L).

CHAPTER 6

Stationary Combustion Systems

6-1   Overview

At the present time, most of the world's electricity is produced using combustion technologies that convert fossil fuels to electricity. Since it would likely take years or even decades to develop the capacity to significantly increase the share of electricity generated from alternative sources that do not rely on the combustion of fossil fuels, there is scope for an intermediate step of implementing more efficient stationary combustion technologies, as outlined in this chapter. These technologies have an application both in reducing fossil fuel consumption and CO_2 emissions in the short to medium term, and as part of large-scale combustion-sequestration systems in the future. This chapter first explains the main technical innovations that comprise the advanced combustion systems useful for improving conversion efficiency. The second half of the chapter considers economic analysis of investment in these systems, environmental aspects, and possible future designs. Mobile combustion technologies for transportation, such as internal combustion engines, are considered later in Chaps. 15 and 16.

6-2   Introduction

At the heart of the modern electrical grid system are the combustion technologies that convert energy resources such as fossil fuels into electrical energy. These technologies are situated in electrical power plants, and consist of three main components: (1) a means of converting fuel to heat, either a *combustion chamber* for gas-fired systems or a *boiler* for systems that use water as the working fluid; (2) a *turbine* for converting heat energy to mechanical energy; and (3) a *generator* for converting mechanical energy to electrical energy (see Fig. 6-1). Since fossil fuels are the leading resource for electricity generation, the majority of all of the world's electricity is generated in these facilities. In addition, nuclear and hydropower plants use these three components in some measure, although in nuclear power the heat source for boiling the working fluid is the nuclear reaction, and in hydro power there is no fuel conversion component.

The importance of stationary combustion systems is reflected in their dominant role in U.S. electricity generation in 2004 and 2014 (Fig. 6-2). A major shift in market share

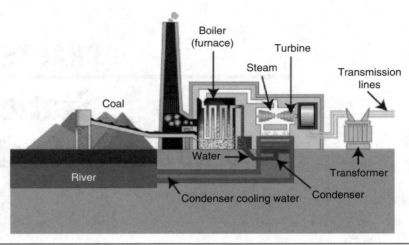

FIGURE 6-1 Schematic of components of coal-fired electric plant, with conversion of coal to electricity via boiler, turbine, and generator.

Note the presence of a large natural or human-made source of cooling water. (*Source:* Tennessee Valley Authority, U.S. Federal Government.)

occurred between electricity generated from coal and gas, as coal's share decreased from 50% to 39%, while that of gas increased from 18% to 28%. Overall, the combination of coal, gas, and oil was dominant in both years, at 71% in 2004 and 67% in 2014. If nuclear is included as an electricity source that uses steam to drive a turbine, the percentage rises to 91% and 87%, respectively. During this 10-year period, the U.S. population increased by 26 million from 293 to 319 million, while electricity consumption increased

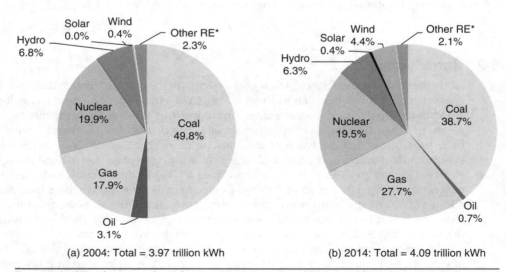

FIGURE 6-2 Mix of sources for U.S. electricity generation, 2004 and 2014. *Source:* U.S. Energy Information Administration.

by 120 billion kWh from 3.97 trillion kWh to 4.09 trillion kWh. Therefore, the country became more efficient in consuming electricity on a per capita basis, decreasing from 13,600 kWh to 12,800 kWh per person per year.

A decline in the relative cost of gas compared to coal as well as air quality and CO_2 advantages are driving the transition from coal to gas. In some cases, coal-fired plants are closed and their capacity replaced by new gas-fired plants built in other locations. In others, *repowering* is applied, in which a coal-fired plant is converted to use gas by replacing the energy conversion system and bringing in a gas supply from the national gas distribution grid. An example of repowering at Cornell University is presented in Sec. 6-5-3.

At present there exists a strong motivation to develop and install an efficient new generation of combustion-based generating systems. This new generation of systems can meet the following goals:

- *Reduce CO_2 emissions until other technologies can meet a significant share of electricity demand:* In the short to medium term, replacing obsolete, inefficient turbine technology with the state of the art can make a real difference in reducing energy consumption and CO_2 emissions. As a comparison, renewable technologies such as wind or solar power are capital intensive, and also require a large up-front capacity to build and install equipment. Even with manufacturing facilities for solar panels and wind turbines working at or near capacity at the present time, each year the total installed capacity of equipment installed from these facilities adds only a small percentage to the total quantity of electricity generated. The manufacturing capacity to build these systems is currently expanding rapidly worldwide (see Chaps. 10 and 13), which accelerates the installation of renewable energy systems, but this expansion also takes time. Manufacturing the components of a new combustion turbine takes less time than the production of the number of wind turbines or solar panels with an equivalent annual output, and the distribution system for incoming fuel and outgoing electricity is already in place. Therefore, upgrading fossil fuel combustion systems to the state of the art can have a relatively large effect in a short amount of time.

- *Enable the conversion of biofuels as well as fossil fuels to electricity and heat:* Feedstocks made from biofuel crops and forest products can be combusted in energy conversion systems in the same way that fossil fuels are combusted, but without releasing net CO_2 to the atmosphere, since the carbon in biofuels comes from the atmosphere in the first place. This resource is already being used for generating electricity in a number of countries, and its application is likely to grow in the future, bounded mainly by the need to balance providing feedstocks for energy and feeding people. The use of biofuels for electricity will depend, for the foreseeable future, both on maintaining existing combustion-based generating stations and including biofuels in the mix of fuels combusted (a process known as "co-firing"), and on adding new biofuel-powered plants, where appropriate. Heat that is exhausted from these conversion processes as a by-product can be used for other applications as well. Related examples of nonconventional fuel conversion, such as waste-to-energy and water treatment bio-waste, are discussed later in this chapter.

- *Prepare the way for conversion of fossil fuels to electricity with sequestration of by-product CO_2:* In coming years, existing power generating stations may be retrofitted or modified with equipment that can separate CO_2 generated in the energy conversion process, and transfer it to a sequestration facility (see Chap. 7). Such technology would make the use of fossil fuels viable in the long run, perhaps for several centuries, as described at the end of Chap. 5.

In all three of these cases, the electric utility sector will put a premium on having maximum possible efficiency, whether for reasons of minimizing CO_2 emissions to the atmosphere, or maximizing the amount of energy produced from biofuels, or minimizing the investment required in new sequestration reservoirs. With these motivations in mind, the rest of this chapter reviews recent developments in high-efficiency combustion technology.

6-2-1   A Systems Approach to Combustion Technology

Both "technology-focused" and "systems-focused" approaches can improve combustion technology. From a technological perspective, combustion efficiency can be improved, for example, when metallurgists develop materials that can withstand a higher maximum temperature, enabling the installation of a steam or gas turbine that can operate at a higher temperature and which is therefore more efficient. The systems perspective, on the other hand, attempts to answer questions such as

1. What is the underlying question that we are trying to answer when implementing a technology?

2. Where have we drawn boundaries around the systems design problem in the past, where perhaps changing those boundaries might lead to new solutions in the future?

3. How can we apply existing technology in new ways?

4. How can we merge the solution to the current problem with the solution to a different problem in order to create a new solution?

As opportunities for incremental improvement become harder to achieve, the use of a systems perspective to "think outside the box" will become increasingly important.

This type of systems analysis will be especially important with implementing carbon sequestration from combustion systems. A range of options will likely emerge over time for separating out CO_2 from other by-products, some more radically different from the current turbine designs than others. Electric utilities will be faced with trade-offs between retrofits that have lower capital cost but are more costly to operate per unit of CO_2 reduced, and fundamentally different technologies that sequester CO_2 at a lower cost but at the expense of a high upfront investment. In some instances, utilities may have opportunities to prepare for sequestration in the future, by upgrading power generating facilities in the near term in such a way that they are more accessible to a sequestration retrofit in the future. Such a step presents a financial risk in itself, because making a facility "sequestration-ready" entails an additional cost, and it cannot be known when or even if these adaptations will deliver a return in the form of actually capturing and sequestering carbon. In any case, the presentation of technologies in this chapter is linked to the question of both reducing CO_2 generated in the first place and preparing for sequestration.

6-3   Fundamentals of Combustion Cycle Calculation

Any discussion of combustion cycles begins with a review of the underlying thermodynamics, so in this section we first briefly review thermodynamic concepts and then present the fundamental cycles for vapor (the Rankine cycle) and gas (the Brayton cycle). Further review of this material is not covered in this book, but is widely available in other books specifically on the subject of engineering thermodynamics (e.g., Cengel and Boles, 2014; Moran and Shapiro, 2012; Wark, 1983).

6-3-1   Brief Review of Thermodynamics

According to the dictionary definition, thermodynamics is "the branch of physics that deals with the mechanical actions or relations of heat."[1] In engineering, the study of thermodynamics focuses on heat or energy-related transformations and the impact they have on substances and structures, such as the change in high pressure steam as it passes through a turbine, leading to lower pressure in the steam but greater kinetic energy in the shaft of the turbine. Thermodynamics can further be divided between *classical thermodynamics*, which views substances undergoing changes in heat or energy properties at a macroscopic level, and *statistical thermodynamics*, which brings in behavior of substances at an atomic or microscopic level to better understand phenomena observed at both the micro- and macroscopic levels. A classical thermodynamic perspective is adequate for the purposes of this book and is adopted in all subsequent discussions.

In thermodynamics, both the quantity of energy available, or enthalpy, and the quality of the energy, or entropy, are important for the evaluation of a thermodynamic cycle. Enthalpy is measured in units of energy per unit of mass, for example, kJ/kg in metric units or BTU/lb in standard units, the intuition being that containing a larger quantity of energy per unit of mass implies greater energy density or concentration. Entropy is measured in units of energy per unit of mass-temperature, for example, kJ/kg · K in metric units or BTU/lb · °R in standard units, where °R stands for degrees Rankine, and $T_{\text{Rankine}} = T_{\text{Fahrenheit}} + 459.7$.[2] For this measure, the underlying intuition is that the measure of energy per unit of mass and degrees Kelvin or Rankine evaluates the ability to contain energy for a given level of absolute temperature. In common language, "low entropy" is used to describe a highly-ordered state, and "high entropy" to describe a highly disordered state, so that on geological time scales the entire universe is undergoing a gradual transition from low to high entropy. In thermodynamics, however, the quantification of entropy results in values such that a quantity of a working fluid (e.g., water) has a higher entropy value when it has a higher value of kJ/kg · K or BTU/lb · °R. Higher values result when a fluid is at higher pressure and/or temperature, and is therefore of a higher *quality*, or ability to do work. Analysis of combustion cycles relies heavily on obtaining enthalpy and entropy from *thermodynamic tables* that are published for water, saturated steam, superheated steam, and air.

---

[1]Merriam-Webster online dictionary, available at www.m-w.com.
[2]In other words, to convert temperature in degrees Fahrenheit to degrees Rankine, add 459.7, and vice versa. For purposes of looking up values in the tables in the online Appendices, the conversion can be rounded to $T_{\text{Rankine}} = T_{\text{Fahrenheit}} + 460$.

According to the *first law of thermodynamics*, energy is conserved in thermal processes. In an energy conversion process with no losses, all energy not retained by the working fluid would be transferred to the application, for example, rotation of the turbine shaft. In practical energy equipment, of course, losses will occur, for example, through heat transfer into and out of materials that physically contain the working fluid. According to the *second law of thermodynamics*, a combustion cycle can only return from its initial state of entropy back to that state with an entropy of equal or greater value. A process in which entropy is conserved is called *isentropic*; for simplicity, examples in this chapter and elsewhere in the book will for the most part assume an isentropic cycle, although in real-world systems, some increase in entropy is inevitable.

The *Carnot limit* provides a useful benchmark for evaluating the performance of a given thermodynamic cycle. Let T_H and T_L be the high and low temperatures of a thermodynamic process, for example, the incoming and outgoing temperature of a working fluid passing through a turbine. The Carnot efficiency η_{Carnot} is then defined as

$$\eta_{\text{Carnot}} = \frac{(T_H - T_L)}{T_H} \tag{6-1}$$

or, in other words, the ratio of the change in temperature to the initial temperature of the fluid. The Carnot limit states that the efficiency value of a thermodynamic cycle cannot exceed the Carnot efficiency. This quantitative limit on efficiency points in the direction of increasing T_H and decreasing T_L in order to maximize cycle efficiency.

In subsequent sections, the objective is to evaluate the theoretical efficiency of combustion cycles, as a function of the available energy in the heat added from the fuel source, the output from the system (turbine work or process heat), and the work required for *parasitic loads*, that is, to operate pumps, compressors, and so on, that are part of the cycle. The objective, for a given amount of heat input, is to maximize output while minimizing internal work requirements, or in other words to improve the thermal efficiency η_{th}:

$$\eta_{\text{th}} = \frac{w_{\text{out}} - w_{\text{in}}}{q_{\text{in}}} \tag{6-2}$$

Here w_{out} is the work output from the system, w_{in} is the work input required, and q_{in} is the heat input. It will be seen, as we progress from simple to more complex cycles, that by adding additional components to cycles, it is possible to significantly improve efficiency. In real-world applications, these improvements of course come at the expense of higher capital cost, so the economic justification for additional complexity is important.

6-3-2   Rankine Vapor Cycle

The Rankine vapor cycle is the basis for a widely used combustion cycle that uses coal, fuel oil, or other fuels to compress and heat water to vapor, and then expand the vapor through a turbine in order to convert heat to mechanical energy. It is named after the Scottish engineer William J.M. Rankine, who first developed the cycle in 1859. A schematic of a simple Rankine device is shown in Fig. 6-3, along with the location of the four

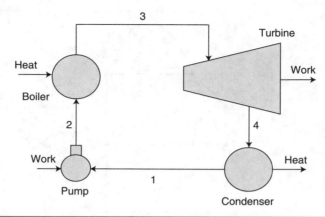

FIGURE 6-3 Schematic of components in simple Rankine device.

states in the cycle shown along the path of the working fluid. The following stages occur between states in the cycle:

1. 1-2 Compression of the fluid using a pump
2. 2-3 Heating of the compressed fluid to the inlet temperature of turbine, including increasing temperature to boiling point, and phase change from liquid to vapor
3. 3-4 Expansion of the vapor in the turbine
4. 4-1 Condensation of the vapor in a condenser

The fluid is then returned to the pump and the cycle is repeated.

The heat input and work output of the cycle are calculated based on the enthalpy value at the stages of the process. For heat input, the enthalpy change is $h_3 - h_2$. For the work output, the enthalpy change is $h_3 - h_4$. The pump work is calculated based on the amount of work required to compress the fluid from the initial to the final pressure, that is

$$w_{\text{pump}} = v_f(P_2 - P_1) \tag{6-3}$$

Here v_f is the specific volume of the fluid, measured in units of volume per unit mass.

We first consider the case of a simple Rankine cycle that is *ideal*, meaning that both compression through the pump and expansion in the turbine are isentropic. In this cycle, the pump and boiler transform water into *saturated* steam leaving the boiler and entering the turbine, that is, no part of the vapor remains as fluid. The exhaust from the turbine is then transformed into *saturated* liquid leaving the condenser, that is, no part remaining as a gas. The enthalpy and entropy values at states 1 and 3 can be read directly from the tables. The evaluation of enthalpy at the pump exit (state 2) depends on work performed during pumping. The evaluation of enthalpy at the turbine exit makes use of the isentropic process and the known value of entropy in states 3 and 4.

At the turbine exit, the working fluid exists as a mixture of liquid and vapor, which is then condensed back to a saturated liquid for reintroduction into the pump at state 1. The *quality* of this mixture, x, is the ratio of the difference between the change in entropy

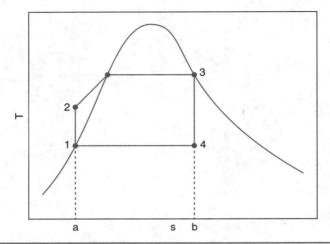

FIGURE 6-4   Temperature-entropy diagram for the ideal Rankine cycle.

from s_4 to fluid, and the maximum possible change in entropy to between saturated steam and saturated liquid at that temperature, that is

$$x = \frac{s_4 - s_f}{s_g - s_f}$$

The change in enthalpy is proportional to the change in entropy, so the value of the quality can be used to calculate the enthalpy at state 4.[3] Evaluation of the efficiency of a Rankine cycle is illustrated in Example 6-1 using metric units and Example 6-2 using standard units.[3] For both metric and standard units, refer to Fig. 6-4 for the relationship between temperature and entropy in the different stages of the cycle.

Example 6-1   Metric units: An ideal Rankine cycle with isentropic compression and expansion operates between a maximum pressure of 4 MPa at the turbine entry and 100 kPa in the condenser. Calculate the thermal efficiency for this cycle. Compare to the Carnot efficiency based on the temperature difference between extremes in the cycle.

Solution   The cycle is represented in the accompanying temperature-entropy (T-s) diagram. At the beginning of the analysis, it is useful to present all enthalpy and entropy values that can be read directly from steam tables, as follows:

$h_3 = 2800.8$ kJ/kg (enthalpy of saturated water vapor at 4 MPa)

$h_1 = 417.5$ kJ/kg (liquid water at 100 kPa)

$s_3 = s_4 = 6.070$ kJ/kg · K (due to isentropic expansion)

Pump work: The change in pressure is $4 - 0.1 = 3.9$ MPa. The specific volume of water at 100 kPa is 0.00104 m³/kg. Therefore, the pump work is

$$w_{pump} = 0.00104 \, m^3/kg \times 3.9 \, MPa = 4.0 \, kJ/kg$$

It follows that $h_2 = h_1 + w_{pump} = 417.5 + 4.0 = 421.5$ kJ/kg.

---

[3]Examples 6-1, 6-2, 6-4, and 6-5 are based on example problems in Wark (1984), Chaps. 16 and 17.

Turbine output: In order to solve for the work output from the turbine, we need to evaluate h_4. First, evaluate the quality of the steam leaving the turbine, that is, the percent of the mixture that is vapor. Use the entropy values for liquid water and saturated steam at 100 kPa:

$$s_f = 1.3028 \text{ kJ/kg·K}$$

$$s_g = 7.3589 \text{ kJ/kg·K}$$

The quality x is then

$$x = \frac{s_4 - s_f}{s_g - s_f} = \frac{6.0696 - 1.3028}{7.3589 - 1.3028} = 78.7\%$$

Using the enthalpy values from the tables for liquid water and saturated steam at 100 kPa, $h_f = 417.5$ and $h_g = 2675.0$ kJ/kg, h_4 is calculated by rearranging:

$$x = \frac{h_4 - h_f}{h_g - h_f}$$

$$h_4 = x(h_g - h_f) + h_f = 0.787(2675.0 - 417.5) + 417.5 = 2194.5 \text{ kJ/kg}$$

Overall efficiency is then

$$\eta_{th} = \frac{w_{out} - w_{in}}{q_{in}} = \frac{(2800.8 - 2194.5) - 4.0}{(2800.8 - 421.5)} = 25.3\%$$

Comparison to Carnot efficiency: From the steam tables, water that has been completely condensed at a pressure of 100 kPa has a temperature of 99.6°C, or 372.7 K. Saturated steam at a pressure of 4 MPa has a temperature of 250.4°C, or 523.5 K. The Carnot efficiency is then

$$\eta_{Carnot} = \frac{(T_H - T_L)}{T_H} = \frac{(523.5 - 372.7)}{523.5} = 28.8\%$$

This example illustrates how the Rankine cycle efficiency is somewhat less than the maximum possible efficiency dictated by the Carnot limit.

Example 6-2 Standard units: An ideal Rankine cycle with isentropic compression and expansion operates between a maximum pressure of 400 psi at the turbine entry and 14.7 psi in the condenser. Calculate the thermal efficiency for this cycle. Compare to the Carnot efficiency based on the temperature difference between extremes in the cycle.

Solution At the beginning of the analysis, it is useful to present all enthalpy and entropy values that can be read directly from steam tables, as follows:

$h_3 = 1205.0$ BTU/lb (enthalpy of saturated water vapor at 400 psi)

$h_1 = 180.2$ BTU/lb (liquid water at 14.7 psi)

$s_3 = s_4 = 1.4852$ BTU/lb · °R (due to isentropic expansion)

Pump work: The change in pressure is $400 - 14.7 = 385.3$ psi. The specific volume of water at 14.7 psi is 0.01672 ft^3/lb. Therefore, the pump work is

$$w_{pump} = \frac{(0.01672 \text{ ft}^3/\text{lb})(385.3 \text{ lb/in}^2)(144 \text{ in}^2/\text{ft}^2)}{778 \text{ ft} - \text{lb/BTU}} = 1.2 \text{ BTU/lb}$$

It follows that $h_2 = h_1 + w_{pump} = 180.2 + 1.2 = 181.4$ BTU/lb.

Turbine output: In order to solve for the work output from the turbine, we need to evaluate h_4. First evaluate the quality of the steam leaving the turbine, that is, the percent of the mixture that is vapor. Use the entropy values for liquid water and saturated steam at 14.7 psi:

$$s_f = 0.3121 \text{ BTU/lb·°R}$$

$$s_g = 1.7566 \text{ BTU/lb·°R}$$

The quality x is then

$$x = \frac{s_4 - s_f}{s_g - s_f} = \frac{1.4852 - 0.3121}{1.7566 - 0.3121} = 81.2\%$$

Using the enthalpy values from the tables for liquid water and saturated steam at 14 psi, $h_f = 180.2$ and $h_g = 1150.3$ BTU/lb, h_4 is calculated by rearranging:

$$x = \frac{h_4 - h_f}{h_g - h_f}$$

$$h_4 = x(h_g - h_f) + h_f = 0.812(1150.3 - 180.2) + 180.2 = 967.7 \text{ BTU/lb}$$

Overall efficiency is then

$$\eta_{th} = \frac{w_{out} - w_{in}}{q_{in}} = \frac{(1205.0 - 967.7) - 1.2}{(1205.0 - 181.4)} = 23.1\%$$

Comparison to Carnot efficiency: From the steam tables, water that has been completely condensed at a pressure of 14.7 psi has a temperature of 212°F (671.7°R). Saturated steam at a pressure of 400 psi has a temperature of 444.6°F (904.3°R). The Carnot efficiency is then

$$\eta_{Carnot} = \frac{(T_H - T_L)}{T_H} = \frac{(904.3 - 671.7)}{904.3} = 25.7\%$$

This example illustrates how the Rankine cycle efficiency is somewhat less than the maximum possible efficiency dictated by the Carnot limit.

Effect of Irreversibilities on Overall Performance

The pump and turbine in Example 6-1 were assumed to be 100% efficient; for example, the turbine is assumed to convert all the thermal energy available in the enthalpy change of 606.6 kJ/kg into turbine work. In practice, these devices are less than 100% efficient, and efficiency losses will affect the calculation of η_{th}. Pursuing the turbine example further, let $\eta_{turbine}$ be the efficiency of the turbine, so that the actual output from the turbine is $w_{actual} = \eta_{turbine}(w_{isentropic})$, where $w_{isentropic}$ is the work output value assuming no change in entropy. Applying the actual turbine efficiency in the analysis of the Rankine cycle affects the calculation of the exit enthalpy value, as shown in Example 6-3.

Example 6-3   Suppose the turbine in Example 6-1 has an efficiency of 85%, and that the pump operates isentropically. Recalculate the efficiency of the cycle.

Solution   Recalculating the output from the turbine gives $w_{actual} = 0.85(606.3) = 515.4$ kJ/kg. The value of the turbine exit enthalpy is $h_4 = h_3 - w_{actual} = 2800.8 - 515.4 = 2285.4$ kJ/kg. Since the value of h_4 is increased compared to the isentropic case, it should be verified the exhaust exits the turbine as wet steam, that is, $h_4 < h_g$, which holds in this case. The overall efficiency is then

$$\eta_{th} = \frac{w_{actual} - w_{in}}{q_{in}} = \frac{515.4 - 4.0}{(2800.8 - 421.5)} = 21.5\%$$

Thus the introduction of the losses in the turbine reduces the efficiency of the cycle by 3.8 percentage points.

Example 6-3 shows that taking into account efficiency losses in components such as the turbine can significantly reduce the calculated efficiency of the cycle. The effect of irreversibilities in other equipment, such as pumps and compressors, can be applied to other combustion cycles presented in this chapter. Heat and frictional losses from the steam to the surrounding pump and turbine walls, ductwork, and so on, lead to additional losses.

6-3-3 Brayton Gas Cycle

For a gaseous fuel such as natural gas, it is practical to combust the gas directly in the combustion cycle, rather than using heat from the gas to convert water to vapor and then expand the water vapor through a steam turbine. For this purpose, engineers have developed the gas turbine, which is based on the *Brayton cycle*. This cycle is named after the American engineer George Brayton, who in the 1870s developed the continuous combustion process that underlies the combustion technique used in gas turbines today.

A schematic of the components used in the Brayton cycle is shown in Fig. 6-5, along with numbered states 1 to 4 in the cycle. The following stages occur between states:

1. 1-2 Air from the atmosphere is drawn in to the system and compressed to the maximum system pressure.

2. 2-3 Fuel is injected into the compressed air, and the mixture is combusted at constant pressure, heating it to the system maximum temperature.

3. 3-4 The combustion products are expanded through a turbine, creating the work output in the form of the spinning turbine shaft.

Note that unlike the Rankine cycle, there is no fourth step required to return the combustion products from the turbine to the compressor. A gas cycle requires a continual supply of fresh air for combustion with the fuel injected in stage 2-3, and it would not be practical to separate uncombusted oxygen from the turbine exhaust for return to the compressor. Therefore, combustion products after stage 3-4 are either exhausted

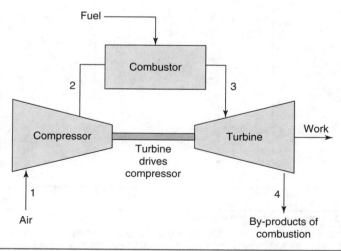

FIGURE 6-5 Components in Brayton cycle.

directly to the atmosphere, or passed to a heat exchanger so that heat remaining in the combustion products can be extracted for some other purpose. Another difference is that the amount of work required for adding pressure is much greater in the case of the Brayton cycle. In the Rankine cycle, the amount of work required to pressurize the water is small, because liquid water is almost entirely incompressible. On the other hand, air entering the Brayton cycle is highly compressible, and requires more work in order to achieve the pressures necessary for combustion and expansion.

As with the Rankine cycle, calculating the cycle efficiency requires calculation of the enthalpy values at each stage of the cycle. Alternative approaches exist for calculating enthalpies; here we use the *relative pressure* of the air or fuel-air mixture at each stage to calculate enthalpy. The relative pressure is a constant parameter as a function of temperature for a given gas, as found in the air tables in the online appendices. Since the amount of fuel added is small relative to the amount of air, it is a reasonable simplification to treat the air-fuel mixture as pure air at all stages of the cycle.

If one relative pressure value and the compression ratio in the gas cycle are known, they can be used to obtain the other relative pressure value, as follows. Suppose the gas entering a compressor has a relative pressure p_{r1}, and the compression ratio from state 1 (uncompressed) to state 2 (compressed) is P_2/P_1. The state 2 relative pressure is then

$$p_{r2} = p_{r1}\left(\frac{P_2}{P_1}\right) \tag{6-4}$$

The calculation for relative pressure leaving the turbine is analogous, except that one divides by the compression ratio to calculate the exit p_r. Both calculations are demonstrated in Example 6-4 with metric units and Example 6-5 with standard units. For both metric and standard units, refer to Fig. 6-6 for the relationship between temperature and entropy in the different stages of the cycle.

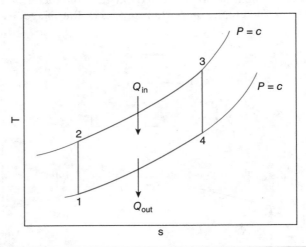

FIGURE 6-6   Temperature-entropy diagram for the ideal Brayton cycle.

Note that in a practical Brayton cycle, the expanded gas-air mixture is exhausted to the environment upon exiting the turbine, rather than being cooled at constant pressure from state 4 to 1 and reintroduced into the cycle, as suggested by the diagram.

Example 6-4 Metric units: A Brayton gas cycle operates with isentropic compression and expansion, as shown in the accompanying T-s diagram. Air enters the compressor at 95 kPa and ambient temperature (295 K). The compressor has a ratio of 6:1, and the compressed air is heated to 1100 K. The combustion products are then expanded in a turbine. Compute the thermal efficiency of the cycle.

Solution From the air tables, the values for enthalpy and relative pressure at states 1 and 3 are known, that is, $h_1 = 295.2$ kJ/kg, $p_{r1} = 1.3068$, $h_3 = 1161.1$ kJ/kg, $p_{r3} = 167.1$. In order to solve for heat added in the combustor, work output from the turbine, and work required for the compressor, we need to solve for h_2 and h_4.

Taking h_2 first, it is observed that since the compression ratio is 6:1, the relative pressure leaving the compressor is $p_{r2} = p_{r1}(6) = 7.841$. The value of h_2 can then be obtained by interpolation in the air tables. From the tables, air at T = 490 K has a relative pressure of $p_r = 7.824$, or approximately the same as p_{r2}, so h_2 is approximately the enthalpy at T = 490 K from the table, or $h_2 = 492.7$ kJ/kg.

For h_4, it is necessary to interpolate between values in the table, as follows. First calculate $p_{r4} = p_{r3}/6 = 27.85$. From the tables, air at T = 690 K has a relative pressure of $p_r = 27.29$, and air at T = 700 K has a relative pressure of $p_r = 28.8$, so we calculate an interpolation factor f:

$$f = \frac{27.85 - 27.29}{28.8 - 27.29} = 0.37$$

From the tables, enthalpy values at T = 690 and 700 K are 702.5 and 713.3 kJ/kg, respectively, so h_4 can be obtained:

$$h_4 = 702.5 + 0.37(713.3 - 702.5) = 706.5 \text{ kJ/kg}$$

Linear interpolation is commonly used in this way to obtain values not given directly in the tables. It is now possible to calculate heat input, turbine work, and compressor work:

$$q_{in} = h_3 - h_2 = 1161.1 - 492.7 = 668.4 \text{ kJ/kg}$$

$$w_{turbine} = h_3 - h_4 = 1161.1 - 706.5 = 454.6 \text{ kJ/kg}$$

$$w_{compressor} = h_2 - h_1 = 492.7 - 295.2 = 197.5 \text{ kJ/kg}$$

The overall cycle efficiency is then

$$\eta_{th} = \frac{w_{turbine} - w_{compressor}}{q_{in}} = \frac{454.6 - 197.5}{668.4} = 38.5\%$$

Discussion The value of the compressor work in this ideal cycle is 43% of the total turbine work. Furthermore, these ideal compressor and turbine components do not take into account losses that would occur in real-world equipment. Losses in the compressor increase the amount of work required to achieve the targeted compression ratio, while losses in the turbine reduce the work output. Thus the actual percentage would be significantly higher than 43%. This calculation shows the importance of maximizing turbine output and minimizing compressor losses in order to create a technically and economically viable gas turbine technology.

Example 6-5 Standard units: A Brayton gas cycle operates with isentropic compression and expansion. Air enters the compressor at 14.5 psi and 80°F. The compressor has a ratio of 6:1, and the compressed air is heated to 1540°F. The combustion products are then expanded in a turbine. Compute the thermal efficiency of the cycle.

Solution From the air tables, the values for enthalpy and relative pressure at states 1 and 3 are known, that is, $h_1 = 129.1$ BTU/lb, $p_{r1} = 1.386$, $h_3 = 504.7$ BTU/lb, $p_{r3} = 174$. In order to solve for heat added in the combustor, work output from the turbine, and work required for the compressor, we need to solve for h_2 and h_4.

Taking h_2 first, it is observed that since the compression ratio is 6:1, the relative pressure leaving the compressor is $p_{r2} = p_{r1}(6) = 8.316$. The value of h_2 can then be obtained by interpolation in the air tables. From the tables, air at T = 880°R has a relative pressure of $p_r = 7.761$, and air at T = 900°R has a relative pressure of $p_r = 8.411$, so we calculate an interpolation factor f:

$$f = \frac{8.316 - 7.761}{8.411 - 7.761} = 0.854$$

From the tables, enthalpy values at T = 880 and 900°R are 211.4 and 216.3 BTU/lb, respectively, so h_2 can be obtained:

$$h_2 = 211.4 + 0.854(216.3 - 211.4) = 215.6 \text{ BTU/lb}$$

By similar calculation, we obtain for state 4 a value of $p_r = 174/6 = 29$ and $h_4 = 307.1$.

It is now possible to calculate heat input, turbine work, and compressor work:

$$q_{in} = h_3 - h_2 = 504.7 - 215.6 = 289.1 \text{ BTU/lb}$$
$$w_{turbine} = h_3 - h_4 = 504.7 - 307.1 = 197.6 \text{ BTU/lb}$$
$$w_{compressor} = h_2 - h_1 = 215.6 - 129.1 = 86.5 \text{ BTU/lb}$$

The overall cycle efficiency is then

$$\eta_{th} = \frac{w_{turbine} - w_{compressor}}{q_{in}} = \frac{197.6 - 86.5}{289.1} = 38.4\%$$

Discussion   The value of the compressor work in this ideal cycle is 44% of the total turbine work. Furthermore, these ideal compressor and turbine components do not take into account losses that would occur in real-world equipment. Losses in the compressor increase the amount of work required to achieve the targeted compression ratio, while losses in the turbine reduce the work output. Thus the actual percentage would be significantly higher than 44%. This calculation shows the importance of maximizing turbine output and minimizing compressor losses in order to create a technically and economically viable gas turbine technology.

6-4   Advanced Combustion Cycles for Maximum Efficiency

In the preceding examples, we calculated efficiency values of 25% for a Rankine cycle and 38% for a Brayton cycle, before taking into account any losses. Not only are there significant losses within a combustion cycle, but there are also losses in boiler system that converts fuel to heat in the case of the Rankine cycle, as well as slight losses in the generator attached to the turbine (modern generators are typically on the order of 98% efficient, so the effect of these losses is limited). Given the large worldwide expenditure each year on fuel for producing electricity, and the pressure to reduce pollution, it is not surprising that engineers have been working on improvements to these cycles for many decades. While it is not possible in this section to present every possible device or technique used to wring more output from the fuel consumed, three major improvements have been chosen for further exploration, namely, the supercritical cycle, the combined cycle, and the combined heat and power system.

6-4-1   Supercritical Cycle

One way to increase the efficiency of the Rankine cycle is to increase the pressure at which heat is added to the vapor in the boiler. In a conventional Rankine cycle such as that of Example 6-1, the water in the boiler is heated until it reaches the saturation temperature for water at the given pressure, and then heated at a constant temperature in the liquid-vapor region until it becomes saturated steam. However, if sufficiently compressed, the fluid exceeds the *critical pressure* for water and does not enter the liquid-vapor region. In this case, the process is called a *supercritical cycle*, and the boiler achieves a temperature that is in the supercritical region for water, as shown in Fig. 6-7. In the supercritical region, there is no distinction between liquid and gas phases.

In the early years of the large-scale power generation, it was not possible to take advantage of the supercritical cycle, due to the lack of materials that could withstand the high temperatures and pressures at the turbine inlet. Breakthroughs in materials technology in the mid-twentieth century made possible the use of the supercritical cycle. The first fossil fuel-powered, supercritical power plant was opened in 1957 near Zanesville, Ohio, U.S., and since that time many more such plants have been built.

Example 6-6 illustrates the benefit to efficiency of the supercritical cycle.

Example 6-6   Suppose the cycle in Example 6-1 is modified so that the boiler reaches a supercritical pressure of 35 MPa and temperature of 600°C. The condenser remains unchanged. Assume isentropic pump and turbine work. By how many percentage points does the efficiency improve?

Solution   From the tables, characteristics for the turbine inlet are $h_3 = 3399.0$ kJ/kg and $s_3 = 6.1229$ kJ/kg·K. Given the much higher maximum pressure, the amount of pump work increases significantly:

$$w_{pump} = 0.00104 \, m^3/kg \times 34.9 \, MPa = 36.3 \, kJ/kg$$

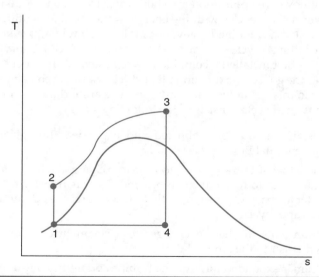

FIGURE 6-7   T-s diagram showing sub- and supercritical regions, and supercritical Rankine cycle.

Thus $h_2 = h_1 + w_{pump} = 417.5 + 36.3 = 453.8$ kJ/kg. It will be seen that the benefits to turbine output more than offset the added pump work. Using $s_3 = s_4 = 6.1229$, the quality of the steam is now $x = 79.6\%$. Therefore, the enthalpy at the turbine exit is

$$h_4 = 0.796(2675.5 - 417.5) + 417.5 = 2214.0 \text{ kJ/kg}$$

The overall efficiency has now improved to

$$\eta_{th} = \frac{(3399.0 - 2214.2) - 33.2}{(3399.0 - 453.8)} = 39.0\%$$

Therefore, the turbine output increases to 1184.8 kJ/kg, and the improvement in efficiency is 13.7 percentage points. Note that Carnot efficiency has changed as well:

$$\eta_{Carnot} = \frac{(T_H - T_L)}{T_H} = \frac{(873.1 - 372.7)}{873.1} = 57.3\%$$

Thus there is a significant gap between the calculated efficiency and the Carnot limit for the cycle given here.

Example 6-6 shows that, without further adaptation, there is a wide gap between the calculated efficiency and the Carnot limit. One possible solution is the use of a totally supercritical cycle, in which the fluid leaves the turbine at a supercritical pressure, is cooled at constant pressure to an initial temperature below the critical temperature, and then pumped back up to the maximum pressure for heating in the boiler. Narrowing the range of pressures over which the system operates tends to increase efficiency, all other things being equal.

6-4-2   Combined Cycle

Let us return to the Brayton cycle introduced in Sec. 6-4. It was mentioned that the remaining energy in the exhaust from the gas turbine can be put to some other use in order to increase overall system output. If the exhaust gas temperature is sufficiently high, one innovative application is to boil water for use in a Rankine cycle, thus effectively powering two cycles with the energy in the gas that is initially combusted. This process is called a combined gas-vapor cycle, or simply a *combined cycle*; the components of a combined cycle system are shown in Fig. 6-8. Oklahoma Gas & Electric in the United States first installed a combined-cycle system at their Belle Isle Station plant in 1949, and as the goal of maximizing plant efficiency has become more important, so has the interest in this technology in many countries around the world.

The combined cycle consists of the following steps:

1. Gas-air mixture is combusted and expanded through a turbine, as in the conventional Brayton cycle.
2. The exhaust is transferred to a heat exchanger where pressurized, unheated water is introduced at the other end. Heat is transferred from the gas to the water at constant pressure so that the water reaches the desired temperature for the vapor cycle.
3. Steam exits the heat exchanger to a steam turbine, and gas exits the heat exchanger to the atmosphere.
4. Steam is expanded through the turbine and returned to a condenser and pump, to be returned to the heat exchanger at high pressure.

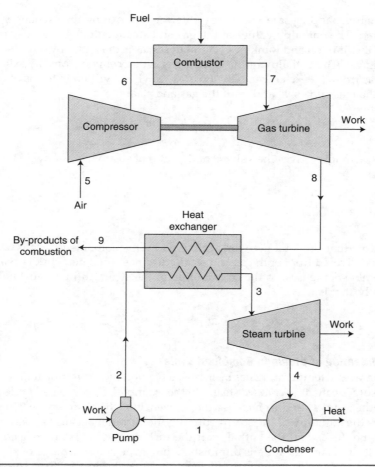

FIGURE 6-8   Schematic of combined cycle system components.

In practice, a typical combined cycle facility consists of several gas turbines installed in parallel, with usually a smaller number of steam turbines also installed in parallel, downstream from the gas turbines. The rated capacity of the combined steam turbines is less than that of the gas turbines, since the limiting factor on steam turbine output is the energy available in the exhaust from the gas turbines.

To calculate the efficiency of the combined cycle, we first calculate the enthalpy values for the Brayton and Rankine cycles in isolation, and then determine the relative mass flow of gas and steam through the heat exchanger such that inlet and outlet temperature requirements are met. Let the prime symbol denote enthalpy values in the gas cycle, for example, h'_{in} denotes the value of enthalpy at the input from the gas cycle to the heat exchanger in Fig. 6-8, whereas h_{in} denotes the value in the steam cycle at the heat exchanger entry. The mass flow is evaluated using a heat balance across energy entering and exiting the exchanger:

$$E_{in} = E_{out} \qquad (6\text{-}5)$$

Here E_{in} and E_{out} are the energy entering and exiting the exchanger, and losses are ignored. The enthalpy values of the gas and steam entering the exchanger are known from the Brayton and Rankine cycle analyses, as is the enthalpy of the steam exiting the exchanger. The enthalpy of the exiting gas is determined from the exit temperature of the gas from the exchanger. Now Eq. (6-5) can be rewritten in terms of enthalpy values and the mass flows m_g and m_s of the gas and steam:

$$m_g h'_{in} + m_s h_{in} = m_g h'_{out} + m_s h_{out} \qquad (6\text{-}6)$$

Rearranging gives the value y of the ratio of steam to gas flow in terms of enthalpy values:

$$y = \frac{m_s}{m_g} = \frac{h'_{in} - h'_{out}}{h_{out} - h_{in}} \qquad (6\text{-}7)$$

Note that the total net work delivered by the combined cycle is a combination of work delivered from both gas and steam turbines. Total net work is evaluated per unit of combustion gases, so the contribution of the steam turbine must be added proportionally, that is

$$w_{net} = w_{g.net} + y \cdot w_{s.net} \qquad (6\text{-}8)$$

Superheating of Steam in Combined Cycle

Steam whose temperature has been raised beyond the saturation temperature at a given pressure is called *superheated* steam. In the example of a combined cycle presented next, superheating is used to increase the efficiency of the steam cycle, since superheating raises the enthalpy of the steam entering the turbine. Because superheating increases q_{in} in Eq. (6-2), and the enthalpy at the turbine exit is also increased, reducing w_{out}, the effect of superheating is diminished; however, across a range of superheating temperatures, the net change in efficiency is positive. In practice, superheating steam requires a separate piece of equipment from the boiler called a *superheater*, which is installed between the boiler exit and turbine entrance, and is specifically designed to add heat to pure steam. Example 6-7 illustrates the effect of superheating in isolation. Example 6-8 presents the complete calculation of the efficiency of a combined cycle.

Example 6-7 Suppose the cycle in Example 6-1 is redesigned to superheat steam to 500°C at 4 MPa. What is the new efficiency of the cycle? Refer to the illustration in Example 6-6.

Solution From the superheated steam tables, the following values are obtained:

$h_3 = 3446.0 \text{ kJ/kg}$
$s_3 = s_4 = 7.0922 \text{ kJ/kg·K}$

Recalculating the quality of the steam gives $x = 95.6\%$, so now $h_4 = 2575.5 \text{ kJ/kg}$. The overall efficiency changes to

$$\eta_{th} = \frac{(3446.0 - 2575.5) - 4.0}{(3446.0 - 421.5)} = 28.6\%$$

Thus the overall efficiency increases by 3 percentage points compared to the original cycle.

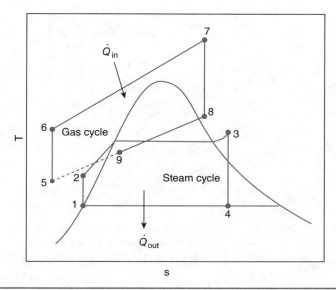

FIGURE 6-9   Combined cycle temperature-entropy diagram.
Note superheating of steam to reach point 3 in the vapor curve.

Discussion   Comparing Examples 6-7 and 6-1, the value of h_3 increases by 645.2 kJ/kg, while the value of h_4 increases by 381.0 kJ/kg. The value of q_{in} also increases by 645.2 kJ/kg, but because this increase is smaller in percentage terms compared to the original value of q_{in}, the net change in efficiency is positive.

Example 6-8[4]   Consider an ideal combined cycle with compression ratio of 7:1 in which air enters the compressor at 295 K, enters the turbine at 1200 K, and exits the heat exchanger at 400 K. In the steam cycle, steam is compressed to 8 MPa, heated to 400°C in the heat exchanger, and then condensed at a pressure of 10 kPa. Calculate the theoretical efficiency of this cycle, and compare to the efficiencies of the two cycles separately.

Solution   Refer to the accompanying T-s diagram in Fig. 6-9. Note that the gas cycle enthalpy values in the table below are numbered 5 through 8, as in Fig. 6-8. First solve for the enthalpy values for the two cycles separately. These calculations are carried out similar to Examples 6-3 and 6-5. For the gas cycle, we obtain the following values:

State	p_r	h'
5	1.3068	295.2
6	9.1476	515.1
7	238	1277.8
8	34	747.3

Let us call the enthalpy of the gas leaving the heat exchanger h_9. From the air tables, for 400 K we obtain $h_9 = 400.1$ kJ/kg. On the steam side, we calculate the following table of values:

---

[4]This example is based on Cengel and Boles (2002), pp. 545–546.

State	h	s
1	191.8	not used
2	199.9	not used
3	3139.4	6.3658
4	2015.2	6.3658

It is now possible to evaluate y for the exchanger from Eq. (6-7):

$$y = \frac{h_8' - h_9'}{h_3 - h_2} = \frac{(747.3 - 400.1)}{(3139.4 - 199.9)} = 0.118$$

In order to calculate net work and efficiency, we first need the net work for the gas and steam cycles, $w_{g.net}$ and $w_{s.net}$, respectively. From the above data, we obtain $w_{g.net} = 310.6$ kJ/kg and $w_{s.net} = 1116.1$ kJ/kg, and also $q_{in} = 762.7$ kJ/kg for the gas cycle. Total net work and overall efficiency are as follows:

$$w_{tot} = 310.6 + (0.118)1116.1 = 442.4 \text{ kJ/kg}$$

$$\eta_{overall} = \frac{w_{tot}}{q_{in}} = \frac{442.4}{762.7} = 58.0\%$$

This value compares very favorably to the efficiencies of each cycle separately. For the steam cycle, $q_{s.in} = 2938.3$ kJ/kg. Thus $\eta_s = 1116.1/2939.5 = 38.0\%$, and $\eta_g = 310.6/762.7 = 40.7\%$.

In actuality, the overall efficiency of the cycle in Example 6-8 would be less than 58% as calculated, due to the various losses in the system. However, with various modifications, efficiencies of up to 60% in actual plants are possible at the present time. Table 6-1 provides a sample of published design efficiency values for plants in various countries in Asia, Europe, and North America, including several in the 55 to 60% efficiency range.

Plant Name	Country	Efficiency
Baglan Bay	United Kingdom	60%
Futtsu (Yokohama)	Japan	60%
Scriba (New York State)	United States	60%
Vilvoorde	Belgium	56%
Seoinchon	South Korea	55%
Ballylumford	United Kingdom	50%
Avedore 2	Denmark	50%
Skaerbaek 3	Denmark	49%

TABLE 6-1 Thermal Efficiency at Design Operating Conditions for a Selection of Combined Cycle Power Plants

6-4-3   Cogeneration and Combined Heat and Power

In the combined cycle systems discussed in Sec. 6-4-2, two different types of turbine technologies, namely, gas and steam, are used for the single purpose of generating electricity. In a *cogeneration* system, a single source of energy, such as a fossil fuel, is used for more than one application. One of the most prominent types of cogeneration system is *combined heat and power* (CHP), in which a working fluid is used to generate electricity, and then the exhaust from the generating process is used for some other purpose, such as a district heating system for residential or commercial buildings located near the CHP power plant, or for process heat in an industrial process. Historically, CHP systems were popular in the early years of large-scale power generating plant in the early part of the twentieth century. Although they fell out of favor in the middle part of the century due to the relatively low cost of fossil fuels, in recent years interest has reemerged in CHP as fuel prices have become more volatile and interest in reducing emissions has grown.

Hereafter we refer to these systems as cogeneration systems. Figure 6-10 provides a schematic for the components of a cogeneration system based on a steam boiler and turbine. In addition to the pumps, boiler, turbine, and condenser seen previously, several new components are introduced:

- *A process heater:* This component extracts heat from incoming vapor flow and transfers it to an outside application, such as district heating. The fluid at the outlet is condensed to saturated water. Note that the steam extracted from the turbine for use in the process heater is partially expanded to the pressure of the heater. Steam not extracted for the heater is fully expanded in the turbine to the condenser pressure.

- *An expansion valve:* This valve permits the reduction in steam pressure so that the steam can pass directly from the boiler exit at higher pressure to the process heater at lower pressure, without expansion in the turbine. In an ideal expansion

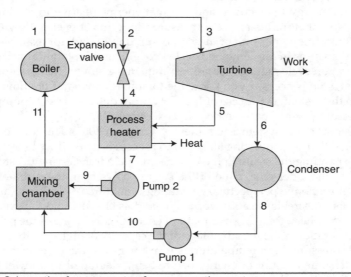

Figure 6-10   Schematic of components of a cogeneration system.

valve, the enthalpy of the steam is not changed, but the entropy is changed, due to the reduction in pressure and hence quality of the steam.

- *A mixing chamber:* This unit permits mixing of water from two different pumps before transfer to the boiler. Enthalpy of the exiting water is determined using a mass balance and energy balance across the mixing chamber.

Cogeneration can be based on gas as well as water vapor. For example, a large-scale application might entail a gas-fired combined-cycle system, where the exhaust steam from the bottom cycle is used for space heating or industrial process heat. Small-scale cogeneration from gas involves either a microturbine or a reciprocating engine used to generate electricity, with the exhaust used for water heating of domestic or service hot water. Examples of both of these applications are provided below in Sec. 6-5.

The addition in Fig. 6.10 of new pathways for the working fluid compared to the simple Rankine cycle provides maximum flexibility to deliver the appropriate mixture of electricity and process heat. For example, during periods of high demand for process heat, it is possible to turn off flow to the turbine and pass all steam through the expansion valve to the process heater. Conversely, if there is no demand for process heat, all steam can be expanded completely in the turbine, as in a simple Rankine cycle. All combinations in between of electricity and process heat are possible.

To measure the extent to which the total energy available is being used in one form or the other, we introduce the *utilization factor* ε_u defined as

$$\varepsilon_u = \frac{W_{net} + Q_p}{Q_{in}} \tag{6-9}$$

where W_{net} is the net work output from the turbine, after taking into account pump work, and Q_p is the process heat delivered. Utilization is different from efficiency in that it focuses specifically on the case of cogeneration/CHP to measure the total allocation of output energy to electrical and thermal energy, relative to the input energy in the boiler. Note that inputs are now measured in terms of energy flow (measured in kW) rather than energy per unit of mass (measured in kJ/kg), for example, heat input is Q_{in} rather than q_{in} as in the combined cycle above. This adaptation is necessary because the cogeneration cycle requires variations in the mass flow of steam depending on system demand, so it is necessary to know the total mass flow and the flow across individual links in the system. In the case of 100% of output delivered as Q_p, pump work is factored in as negative work so that ε_u does not exceed unity.

The goal of the system is to achieve ε_u at or near 100%. This level of utilization can be attained either by converting all steam to process heat, or, if we ignore pump work, partially expanding the steam in the turbine and then transferring the steam to the process heater. In the latter case, none of the steam passes from the turbine to the condenser. From an economic perspective, generating a mixture of electricity and process heat is often more desirable because work delivered as electricity to the grid is more valuable than process heat, for a given amount of energy. Indeed, one of the primary motivations for cogeneration is to take advantage of the capacity of high-quality energy sources such as high-pressure, high-temperature steam (or gas in the case of the Brayton cycle) to generate some amount of energy before being degraded to a lower quality of energy where the only viable application is process heat.

Since energy is conserved across the process heater, any incoming energy from the expansion valve or turbine not extracted as process heat must be ejected from the heater as saturated water at the system pressure. Let $m_{v.in}$ be the flow in from the valve, $m_{t.in}$ be the flow from the turbine, and m_{out} be the flow out of the heater. Using the energy balance:

$$m_{v.in}h_{v.in} + m_{t.in}h_{t.in} = m_{out}h_{out} + Q_p \qquad (6\text{-}10)$$

Rearranging terms gives the following:

$$Q_p = m_{v.in}h_{v.in} + m_{t.in}h_{t.in} - m_{out}h_{out} \qquad (6\text{-}11)$$

Analysis of the mixing chamber is similar, only there is no process heat term. Given two inlet flows from pumps 1 and 2 defined as $m_{1.in}$ and $m_{2.in}$ and outflow m_{out}, the value of h_{out} can be solved as follows:

$$h_{out} = \frac{m_{1.in}h_{1.in} + m_{2.in}h_{2.in}}{m_{out}} \qquad (6\text{-}12)$$

In cases where either the process heater or the condenser is turned off and there is no input from either pump 1 or pump 2, Eq. (6-12) reduces to an identity where the value of enthalpy in and out of the mixing chamber is constant.

Example 6-9[5] A cogeneration system of the type shown in Fig. 6-10 ejects steam from the boiler at 7 MPa and 500°C. (Use numbering notation in the figure.) The steam can either be throttled to 500 kPa for injection into the process heater, or expanded in a turbine, where it can be partially expanded to 500 kPa for the process heater, or fully expanded and condensed at 5 kPa. The function of the process heater is controlled so that the fluid leaves saturated liquid at 500 kPa. Separate pumps compress the fluid to 7 MPa and mix it in a mixing chamber before returning it to the boiler. Calculated process heat output, net electrical output, and utilization for the following two cases:

1. All steam is fed to the turbine and then to the process heater (e.g., $m_2 = m_6 = 0$).

2. Ten percent of the steam is fed to the expansion valve, 70% partially expanded and fed to the process heater, and 20% fully expanded.

Solution First, calculate all known values of enthalpy that apply to both (1) and (2)—refer to Fig. 6-11. There is no change in enthalpy across the expansion valve, so from the steam tables we have

$$h_1 = h_2 = h_3 = h_4 = 3411.4 \text{ kJ/kg}$$

Using the isentropic relationship $s_1 = s_5 = s_6$ and evaluating the quality of the steam at states 5 and 6, we obtain

$$h_5 = 2739.3 \text{ kJ/kg}$$

$$h_6 = 2073.0 \text{ kJ/kg}$$

Enthalpies h_7 and h_8 are saturated fluids at 500 kPa and 5 kPa, respectively, so from the tables $h_7 = 640.1$ and $h_8 = 137.8$ kJ/kg. In order to evaluate h_9 and h_{10}, we need to know the pump work per unit

---

[5]This problem is based on Cengel and Boles (2002), pp. 541–543.

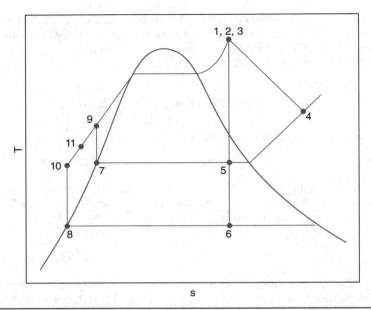

FIGURE 6-11   T-s diagram for cogeneration cycle with expansion valve, partial turbine expansion, and full turbine expansion with condenser.

of mass. Using specific volumes of $v_{@\,5\,kPa} = 1.005 \times 10^{-3}\,\text{m}^3/\text{kg}$ and $v_{@\,500\,kPa} = 1.093 \times 10^{-3}\,\text{m}^3/\text{kg}$, we calculate the corresponding values of pump work as 7.0 and 7.1 kJ/kg for the two pumps, respectively. We then obtain

$$h_9 = 640.1 + 7.1 = 647.2\,\text{kJ}/\text{kg}$$

$$h_{10} = 137.8 + 7.0 = 144.8\,\text{kJ}/\text{kg}$$

The remaining value of h_{11} depends on the specific circumstances of the mixing chamber.

Case 1. Since $m_{10} = 0$, $m_{11} = m_9$ and $h_{11} = h_9 = 647.2\,\text{kJ}/\text{kg}$. Therefore, heat input is

$$Q_{in} = 15\,\text{kg}/\text{s} \times (3411.4 - 647.2) = 41,463\,\text{kW}$$

Pump power for pump 2 is calculated as follows:

$$W_{P2} = (7.1\,\text{kJ}/\text{kg})(15\,\text{kg}/\text{s}) = 106.5\,\text{kW}$$

Given the enthalpy change for the partially expanded steam and the flow rate $m_3 = m_5 = 15\,\text{kg}/\text{s}$, power output and net work from the turbine are, respectively,

$$W_{out} = 15(3411.4 - 2739.3) = 10,081\,\text{kW}$$

$$W_{net} = W_{out} - W_{pump} = 10,081 - 106.5 = 9975\,\text{kW}$$

From Eq. (6-10), process heat is

$$Q_p = 15(2739.3 - 640.2) = 31,488\,\text{kW}$$

It can now be confirmed that utilization is 100%:

$$\varepsilon_u = \frac{W_{net} + Q_p}{Q_{in}} = \frac{9975 + 31,488}{41,463} = 1.00$$

Case 2. Based on the allocation of input steam from the boiler, we have $m_2 = 1.5$, $m_5 = 10.5$, and $m_6 = 3$ kg/s. Thus $m_{10} = 3$ and $m_9 = 10.5 + 1.5 = 12$ kg/s. Solving for h_{11} using Eq. (6-12) gives

$$h_{11} = \frac{m_9 h_9 + m_{10} h_{10}}{m_{11}} = \frac{12(647.2) + 3(144.8)}{15} = 546.8\,kJ/kg$$

Heat input is then

$$Q_{in} = 15\,kg/s(3411.4 - 546.8) = 42,969\,kW$$

Turbine work now consists of two stages, so we account for work done by $m_3 = m_1 - m_2 = 13.5$ and by $m_6 = 3$ kg/s:

$$W_{out} = 13.5(3411.4 - 2739.3) + 3(2739.3 - 2073.0) = 11,072\,kW$$

Subtracting pump work for both pumps 1 and 2 gives the net work:

$$W_{net} = W_{out} - W_{pump1} - W_{pump2} = 11,072 - 21.1 - 85.3 = 10,966\,kW$$

On the process heat side, input to the heater comes from both the expansion valve and the turbine, so

$$Q_p = 1.5(3411.4) + 10.5(2739.3) - 12(640.1) = 26,199\,kW$$

Finally, utilization in this case is

$$\varepsilon_u = \frac{10,966 + 26,199}{42,969} = 0.865$$

Discussion   Case 1 illustrates how the operator might vary allocation between the turbine and process heater to match process heat output to demand, with remaining energy allocated to electricity production. Also, although utilization decreases in case 1 compared to case 2, electricity production increases, so that, with the value of electricity often two or three times that of process heat on a per-kW basis, it may be preferable to operate using the allocation in case 2.

Limitations on the Ability to Use Process Heat

One limitation on the use of process heat is the variability of demand, whether for industrial processes or space heating, which prevents the full use of available process heat at certain times. In addition, incremental cost of building a process heat distribution system and line losses in the system limit the distance to which the heat can be transferred before the system becomes uneconomical. On the industrial side, as long as industrial facilities are located next to power plants, line losses are not a concern, but it may not be practical to transfer steam over distances of many miles or kilometers. On the residential side, existing district heating systems typically heat residences (houses or apartments) that have a sufficiently high density, in terms of number of dwellings per km^2 or mi^2, in locations adjacent to the plant. For many modern low-density residential developments in the industrialized countries, however, the retrofitting of district heating would not be effective.

6-5 Economic Analysis of Stationary Combustion Systems

Economic analysis can be applied to combustion investments at a number of levels. Perhaps the most fundamental is the decision to build a new plant on site where none existed previously. In this case, the analyst determines whether the proposed plant delivers sufficient net income (revenues from sales of electricity minus cost) to be attractive, or, if there are several alternative designs, which one is the most economically attractive. The analyst can also evaluate proposed upgrades to existing plants, which incur some additional upfront cost but pay for themselves by reducing fuel or other costs during their investment lifetime.

The prospective plant owner must make decisions in an environment of considerable uncertainty, especially for new-plant investments, where the investment lifetime may be 20 years or more, and the actual lifetime of the plant before it is decommissioned may last for decades beyond that horizon. Uncertainties around future energy costs (not only fossil fuels but also renewables such as biofuels) are one example: while the analysis may assume unchanging energy cost in constant dollars, recent history illustrates how costs in actuality fluctuate up and down. Also, demand for electricity is not known with certainty, and since many plants around the world now operate in a *deregulated* market, the price that will be paid for each kWh is not known either. Lastly, the plant *availability*, or time that it is not off line for routine maintenance or malfunctions, is not known, although with good engineering and conscientious upkeep, the amount of uncertainty here can be minimized.

The analyst has a number of tools at her or his disposal to work with uncertainties in the data available for analysis. One such tool is *sensitivity analysis* in which some previously fixed numerical input into the calculations is varied over a range of values to shed light on its effect on the bottom-line cost calculation or possibly the build versus no-build decision. Another tool is *probabilistic analysis* or the use of sampling from statistical distributions and Monte Carlo simulation to incorporate uncertainty into the quantitative solution of problems, as discussed in Chap. 2.[6] At a minimum, it should be understood that if the analyst makes a recommendation based on the assumption of fixed values, and the actual values in the future unfold differently from what was expected, then the recommendation may later prove incorrect (!).

Another dimension in economic analysis is the need to meet environmental, health, and safety regulations. Both the additional technology, especially emissions control equipment to limit or eliminate emissions to the atmosphere, and the environmental impact assessment process prior to public approval of the plant, add to the project cost in a nonnegligible way. While eliminating this layer of cost would help the profitability of the plant, societies in industrialized countries, and increasingly in emerging countries as well, have determined that the benefit to society of preventing uncontrolled emissions outweighs the extra cost of more expensive electricity, and therefore choose to establish these requirements.

---

[6]For examples of probabilistic analysis applied to the solution of energy problems, see, e.g., Schrage (1997), Chap. 10, or Dunn (2002).

6-5-1   Calculation of Levelized Cost of Electricity Production

For combustion plants, the levelized cost over the lifetime of the project can be calculated based on annual costs and expected annual output. The costs of the plant consist of three main components:

1. *Capital cost:* This item includes cost of land purchase, all mechanical and electrical equipment related to power conversion, pollution control equipment, and all required structures. The full cost includes repayment of capital debt at some specified MARR.

2. *Fuel cost:* This cost is calculated based on the amount of fuel that has the required energy content to deliver the expected annual output of electricity. Conversion losses in going from fuel to electricity must be taken into account.

3. *Balance of cost:* This item is a "catch-all" that includes all human resource wages and benefits, operations costs, maintenance costs, expenditures on outside services (e.g., payment to outside experts for nonstandard work that is beyond the skills of plant employees), and overhead. In a simple model, these costs can be treated as a fixed per annum amount, independent of the level of plant activity. Realistically, however, a plant that is producing more kWh/year can be expected to incur somewhat more of these costs, all other things being equal.

We can now modify Eq. (3-12) to incorporate these three items into the calculation of levelized cost as follows:

$$\$/kWh = \frac{\text{annual capital cost} + \text{fuel cost} + \text{balance of cost} + \text{ROI}}{\text{annual kWh}} \qquad (6\text{-}13)$$

It is also possible to calculate the levelized cost based on lifetime cost divided by lifetime expected output. If constant dollars are used and the stream of annual costs (fuel + balance) is discounted to the present instead of annualizing the capital cost, the calculation is equivalent.

The annual expenditure on fuel, or *fuel cost*, is calculated from the rated capacity of the plant and annual output of electricity as follows. The overall efficiency of the combustion cycle η_{overall} is a function of the combustion device efficiency η_c (i.e., boiler for a vapor cycle or combustor for a gas cycle), the turbine efficiency η_t, and the generator efficiency η_g. Taking the case of a vapor cycle, we have

$$\eta_{\text{overall}} = \eta_c \eta_t \eta_g \qquad (6\text{-}14)$$

Turbine efficiencies vary with the characteristics of the cycle, as described in earlier sections in this chapter. Boiler, combustor, and generator efficiencies are more uniform, with generators having very high efficiency values and combustors/boilers having somewhat lower ones; typical values are 98% and 85%, respectively. Since theoretical values for ideal turbine cycle efficiency are in the range of 25 to 60% (the highest being for combined cycle systems), the largest losses in the process occur in the conversion of thermal energy in the working fluid into mechanical energy, rather than the other components.

Next, the annual energy output of the plant E_{out} and required energy input E_{in} can be calculated as a function of the average *capacity factor* of the plant (dimensionless ratio of average output to rated output), the rated capacity of the plant in MW_e, and the number of hours per year of operation:

$$E_{out} = (\text{cap.factor})(\text{capacity})(\text{hours per year})(3.6 \text{ GJ/MWh})$$

$$E_{in} = \frac{E_{out}}{\eta_{overall}}$$

(6-15)

Average fuel cost per year is then calculated using the cost per GJ for the given fuel. Example 6-10 demonstrates the calculation of levelized cost for comparison of two investment options.

Example 6-10   Carry out a financial analysis of a coal-fired plant based on the combustion cycle in Example 6-1. The actual cycle achieves 80% of the efficiency of the ideal cycle as given, due to irreversibilities in pumping and expansion, and so on. The efficiencies of the boiler and generator are 85% and 98%, respectively.

The plant is rated at 500 MW and has a capacity factor of 70% over 8760 operating h/year. Its initial cost is $200 million. The lifetime of the analysis is 20 years, and the MARR is 8%. The balance of cost is $18 million/year. From Table 5-2, use a mid-range value of $35/tonne for the coal, delivered to the plant, and an energy content of 25 GJ/tonne. Ignore the need for ROI.

Case 1. Solve for the levelized cost per kWh.

Case 2. Now suppose a supercritical alternative is offered based on Example 6-6. The alternative plant costs $300 million and also achieves 80% of the efficiency of the ideal cycle. All other values are the same. What is the levelized cost of this option? Which option is preferred?

Solution

Case 1. From Example 6-1, the ideal efficiency is 25.3%, so the actual turbine efficiency is $(0.8)(0.253) = 20.2\%$. Multiplying by component efficiencies gives

$$\eta_{overall} = \eta_c \eta_t \eta_g = (0.85)(0.202)(0.98) = 16.9\%$$

The annual electric output is $(0.7)(500 \text{ MW})(8760 \text{ h})(1000 \text{ kWh/MWh}) = 3.07 \times 10^9 \text{ kWh}$. Annual fuel cost is calculated based on an average energy input required using Eq. (6-15):

$$E_{out} = (0.7 \times 500 \text{ MW} \times 8760)(3.6 \text{ GJ/MWh}) = 11.0 \times 10^6 \text{GJ}$$

$$E_{in} = \frac{E_{out}}{\eta_{overall}} = \frac{11.0 \times 10^6 \text{ GJ}}{0.169} = 65.5 \times 10^6 \text{ GJ}$$

This amount of energy translates into 2.62 million tonnes of coal per year, or $91.6 million cost per year.

In order to discount the repayment of the capital cost, we calculate $(A/P, 8\%, 20) = 0.102$. Therefore the annual capital cost is ($200M)(0.102) = $20.4 million.
   Combining all cost elements gives levelized cost of

$$\$/kwh = \frac{\$20.4 \text{ M} + \$91.6 \text{ M} + \$18 \text{ M} + \$0}{3.07 \times 10^9 \text{ kWh}} = \$0.0424/kWh$$

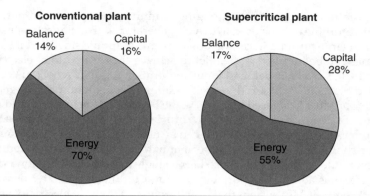

Conventional plant   **Supercritical plant**

FIGURE 6-12   Comparison of cost breakdown for conventional and supercritical plants.

Case 2. Using data from Example 6-6, the new actual turbine efficiency is $(0.8)(0.39) = 31.2\%$. The overall efficiency is therefore 26.0%, and from repeating the calculation in case 1, the annual expenditure on coal is reduced to $59.6 million. The annualized capital cost has increased, however, to ($300 M)(0.102) = $30.6 million. Recalculating levelized cost gives

$$\$/\text{kwh} = \frac{\$30.6\,\text{M} + \$59.6\,\text{M} + \$18\,\text{M} + \$0}{3.07 \times 10^9 \text{kWh}} = \$0.0353/\text{kWh}$$

Therefore, the supercritical plant is more economically attractive over the life cycle, although it is more expensive to build initially.

Discussion   The plant in case 1 is given for illustrative purposes only; its efficiency is poor and it would never be built in the twenty-first century. The plant in case 2 is more in line with some plants that are currently in operation in the world, but the overall efficiency of new supercritical plants has now surpassed 40%, thanks to additional efficiency-improving technology that is added to the cycle. Such plants are, however, more expensive, costing on the order of $600 million in the United States for a 500-MW$_e$ plant. Using the assumptions in this exercise, a supercritical plant with $\eta_{\text{overall}} = 40\%$ and $600 million capital cost has a levelized cost of $0.0384/kWh.

Inspection of the cost breakdown between the two plants shows that although for both plants, energy costs are the leading component, the improvement in efficiency noticeably reduces the proportion allocated to energy costs (see Fig. 6-12). In the case 1 plant, energy consumes 70% of the total, whereas in case 2 this has been reduced to 55%.

6-5-2   Case Study of Small-Scale Cogeneration Systems[7]

Example 6-10 above considered the costs and benefits of energy efficiency on a large scale, equivalent to the electricity output of a typical modern power plant. Here we consider a smaller-scale example, namely, a cogeneration system that is sized to supply electricity and DHW to an apartment complex.

This type of cogeneration system is growing in popularity for commercial and institutional uses. Units as small as 30 kW$_e$ of capacity can be installed in facilities such as primary and secondary schools or apartment buildings where natural gas is being used for space heating or domestic hot water (DHW). At this size, reciprocating engines

---

[7]This case study appeared previously in Vanek and Vogel (2007).

(e.g., diesel cycle combustion engines running on compressed natural gas) as well as microturbines can be used to drive a generator, with exhaust heat used to heat water. As with large turbines in central power plants, multiple small turbines or reciprocating engines can be installed in parallel to achieve the desired total electric output as dictated by the available demand for water heating. A public school in Waverly, New York, United States, uses five 75-kW reciprocating cogeneration units to generate electricity and by-product heat (see Fig. 6-13), while Pierce College in Woodland Hills, California, United States, uses six 60-kW microturbines for the same purpose (see Fig. 6-14). In both cases, the by-product heat can be used for DHW, space heating, or heat supply to air-conditioning chillers. Some manufacturers have further reduced micro-scale cogeneration so that it can be used for a single household. The Honda Corporation, for example, has adapted its portable generator technology into a Micro Combined Heat and Power (MCHP) system that provides up to 1.2 kW of electricity and 12,000 BTU/h (3.5 kW) of by-product heat for space heating or DHW.

The economic analysis compares the proposed new cogen system to an existing system which uses a stand-alone water heater for DHW and purchases all electricity from the grid. Financial payback time rather than levelized cost is used to evaluate the proposed investment. The example is based on a feasibility study for an actual multifamily housing complex with approximately 500 residences in Syracuse, New York, U.S. The facility has an estimated average of electrical and DHW load of 7000 kWh and 23,250 gal/day, respectively. The goal of the cogen system is to produce as much electricity as possible while not exceeding demand for DHW. Any electricity not supplied by the system will be purchased from the grid, and the existing boiler that previously delivered all DHW requirements will make up any shortfall in DHW

Figure 6-13   375-kW cogen installation using five Tecogen reciprocating generator units in Waverley, New York.

Piping loops (emerging from front of each unit) connect hot water by-product to DHW, hydronic heaters, or absorption chillers. (*Source:* Tecogen, Inc. Reprinted with permission.)

(a)

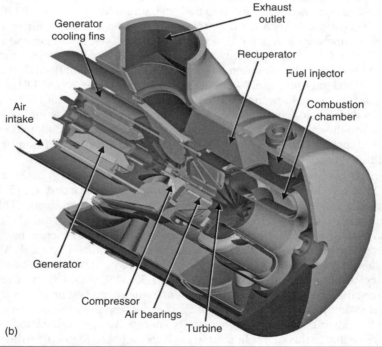

(b)

FIGURE 6-14   Capstone microturbine 30 kW (left), 65 kW (center), and 65 kW with integrated CHP for hot water supply (right) units in Fig. 6-14(a); Fig. 6-14(b) is cutaway of 65 kW turbine components.

Module on top of right unit recovers heat from turbine to generate hot water for space heating or DHW. (*Source:* Capstone Turbine Corp. Reprinted with permission.)

Number of units	2
Total cost (materials + labor)	$170,000
Daily electricity out, kWh	2160
Daily DHW out, gallons	20,800
Reduction in elec. cost, year	$82,782
Increase in gas cost, year	$27,620
Additional maintenance cost, year	$15,768
Net savings per year	$39,394
Payback period, years	4.3

TABLE 6-2   Summary of Cost and Savings for Example
Cogen Project

output. For cost purposes, a value of $0.105/kWh is used for the avoided cost of electricity from the grid, and $7.58/GJ is used for natural gas. The analysis assumes that the local utility cooperates with owners who wish to install cogeneration systems, and ignores the effect of peak electrical charges that are typically charged to large commercial customers in proportion to their maximum kW load in each billing period.

The cost breakdown for the project is shown in Table 6-2. Either microturbine or reciprocating engine technology can be used as the electrical generation system. Two 60-kW cogen units are selected so that, running 18 h/day (except for midnight–6 a.m. when DHW demand is expected to be low) the exhaust heat from the turbines can generate 20,805 gal/day, or 89% of the total demand. Based on the DHW output and the heat input required to generate this amount, the electrical output is 2160 kWh/day. In a region with a cold winter climate such as that of Syracuse, it would be possible in winter to use the exhaust heat for space heating as well as DHW, allowing the units to run continuously and improving their utilization and total annual output. However, this additional benefit is not included in the analysis.

Each turbine costs $85,000 including all materials and labor, for a total cost of $170,000. In addition, the project anticipates a maintenance cost of $15,768/year, on the basis of a maintenance contract calculated at $0.02/kWh produced. This calculation is conservative, in the sense that the existing gas-fired boiler would be used less in the new system and therefore should incur less maintenance cost, but these potential savings are not included. Electrical costs have been reduced by nearly $83,000 thanks to on-site generation of electricity, while gas costs have only increased by $28,000, so that the net savings from the project are about $39,000/year. On the basis of these cost figures, the savings pay for the investment in the cogen system in a little over 4 years. It is left as an exercise at the end of the chapter to use energy requirements for the turbine and the existing water heater to quantify energy savings payback for a similar system.

Note that in this example, simple payback is chosen to calculate the payback period. This choice is acceptable, because the payback period is short enough that the difference between simple and discounted payback for MARR values up to 10% would not change significantly, although discounting would make the result more accurate. Also, the capital equipment has a lifetime of 10 years or more without a major overhaul, so barring unforeseen quality problems with installation or maintenance, the investment will pay for itself over its lifetime.

6-5-3   Case Study of Combined Cycle Cogeneration Systems

In general, plant operators build all-new cogeneration systems on previously undeveloped sites, but in some situations, they may retrofit an existing single-output facility (electricity or heat) to provide cogeneration to improve overall plant utilization. Such is the case with the central heating plant at Cornell University, where the Utilities & Energy Management (UEM) department has retrofitted a district heating system to provide electric power, and more recently converted to a gas-fired combined-cycle system (see Fig. 6-15), thus creating a system that incorporates both combined cycle combustion and combined heat and power.

Historically, the campus district heating system was originally built in 1922 to consolidate the function of delivering steam to buildings and laboratories in one location. The facility operated continuously in this form until the 1980s, when utility managers recognized the economic benefit of generating some fraction of the campus's electric demand from the steam generated to offset electricity purchases from the grid, and two turbines were installed downstream from the boiler for this purpose. A district heating system with electric generation of this type is different from a typical power generating operation in that there is no condenser. Instead, any steam expanded in the turbine must then enter into the district heating lines of the campus to eventually transfer heat to buildings and return to the cogeneration plant as condensate. (Effectively, the district heating network is the condenser for the turbines.) Campus demand for steam, which is higher in winter than in summer, therefore limits electric output.

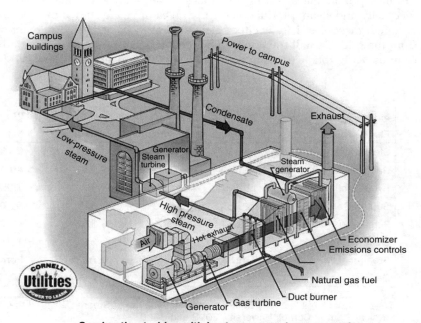

Combustion turbine with heat recovery steam generator

FIGURE 6-15   Schematic of Cornell combined heat and power project with gas turbine, steam turbine, and district heating system for campus buildings.

(*Source:* Cornell University Utilities & Energy Management. Reprinted with permission.)

With the growing awareness in recent years of the threat posed by greenhouse gas emissions, UEM planned and implemented a transition from coal to gas as the fuel for the power plant, bringing on line in December 2009 a combined-cycle system on the site of the central heating plant [the "Cornell Combined Heat and Power Plant" (CCHPP)], and phasing out the burning of coal at the site by the middle of 2011. This transition from coal to gas is part of the overall Cornell Climate Action Plan, approved by the board of trustees of the university in 2009, which seeks to make the campus fully carbon neutral by the year 2050. A key feature of this upgrade is that the steam loop to and from the campus and the steam turbines are maintained, and a gas turbine and gas supply system is added in such a way that the exhaust heat from the gas turbine creates the steam for the steam turbine and steam loop, thus creating a combined cycle.

In this section, we present a case study of the conversion of the system to relying more on combined heat and power and less on electricity purchases from the grid, and from coal to gas combustion, by comparing annual performance before and after the launch of the CCHPP facility, specifically in years 2008 and 2010, since 2010 is the first year in which CCHPP was operational for an entire year. Note, however, that some amount of primary energy still came from coal in 2010, as the transition was only completed in 2011. The focus of this case study is on generation of electricity in GWh, generation of steam for heating and other processes measured in trillion BTU of thermal content, and CO_2 emissions both on-site and derived from electricity purchased from the grid. With CCHPP, the maximum output from the two turbines is 30 MW_e[8] as compared to an observed peak electrical load for the campus of 35 MW_e. The two steam turbines put in place during the 1980s provide an additional 8 MW_e of capacity. Electricity and steam production for use on campus are shown in Table 6-3. Note that although most electricity is either generated at the central energy plant or purchased from the grid, a small fraction is generated at a run-of-the-river plant adjacent to the campus. Also, total electricity generation for 2010 on site was 125 GWh from CCHPP

Source	2008		2010	
Electricity:	GWh	Percent	GWh	Percent
Steam turbine	26.7	10.7%	23	9.4%
Gas turbine	0	0.0%	99	40.4%
Hydroelectric	3.1	1.2%	3	1.2%
Subtotal	29.8	11.9%	125	51.0%
Purchases from grid	220.1	88.1%	120	49.0%
Electricity total	249.9	100%	245	100%
Steam:	Tril. BTU	Percent	Tril. BTU	Percent
Delivered to buildings	1.32	100%	1.09	100%

TABLE 6-3   Mix of Electricity Sources and Total Volume of Steam Delivered to Campus Physical Plant at Cornell University, 2008 and 2010

[8] By convention, the notation MW_e, kW_e, and so on is used to refer to electrical power output, as distinct from MW_t, kW_t, etc., for thermal power output.

and the hydro plant. An additional 13 GWh was generated at CCHPP and sold to the grid at times when the plant could profitably generate steam for the campus and sales to the grid, but this amount of electric generation does not appear in the table. Steam usage on campus declines between 2008 and 2010, but this change is within normal year-to-year fluctuations and does not reflect any change in the way steam is delivered in the system.

Note that electricity is broken out by source, including both internally generated and purchased from grid ("net purchases") measured in GWh, but total annual volume of steam is treated as a single unit without breakdown measured in heat content (trillion BTU).[9] Also, hydroelectric production figures are lower than the typical long-term average of 5 GWh/year.

Since the transition from purchase of electricity from the grid to on-site generation at the central plant increases CO_2 emissions at the site but decreases total emissions embodied in grid electricity purchases, it is the total emissions from coal and gas combustion as well as emissions from grid-generated electricity purchased that indicate the overall effectiveness of launching CCHPP. On-site and grid-generated emissions are shown in Table 6-4 for years 2008 and 2010. Also, observed emissions rates from both fuel combustion and average grid generation vary from year to year, so these values are provided in Table 6-5. Emissions from electricity purchased from the grid are assumed to occur at the NYS average rate, which are recorded for each year by Cornell UEM and used as a basis for calculating overall emissions generated by the campus.

Energy Type	2008		2010	
	Volume	1000 t. CO_2	Volume	1000 t. CO_2
Grid electricity (GWh)	220.1	91.0	120	46.0
Coal (tril. BTU)	1.64	168.2	0.19	20.0
Gas (tril. BTU)	0.12	8.0	1.92	114.0
Total	n/a	267.2	n/a	180

TABLE 6-4   Volume of Grid Electricity, Coal, and Steam, and Corresponding Volume of CO_2 Emissions, for Cornell Campus, 2008 and 2010

Emissions Rates	2008	2010
Grid electricity (tonsCO$_2$/GWh)	413	383
Coal use (tonsCO$_2$/bil. BTU)	102.6	105.3
Gas use (tonsCO$_2$/bil. BTU)	66.2	59.4

TABLE 6-5   Average CO_2 Emissions Rates in Units of U.S. Customary tons for Electricity Purchased from Grid and for Combustion of Coal and Gas at Central Plant, 2008 and 2010.

[9]Original data for 2008 and 2010 electricity and steam generation and sales, as well as emissions, are available at http://energyandsustainability.fs.cornell.edu/file/FY_2010_ENERGY_FAST_FACTS.pdf, respectively. Accessed November 16, 2015.

The large shift in combustion and emissions from coal to gas between 2008 and 2010 is explained by the launch of CCHPP. Once the new facility is on line, most electricity and steam used on campus is generated from gas combustion—coal consumption is reduced by 88% and gas use increases by a factor of 16. This change is in line with the university's goal of reducing CO_2 emissions from steam and electricity consumption. Further reductions will be made possible by complete elimination of coal usage from 2011 onward.

The bottom line of the case study as shown in Table 6-4 is that overall CO_2 emissions for combined electricity and heat delivery to meet campus needs decline by 33%. As a simplification, the study looks only at 2008 and 2010 electricity and steam consumption and CO_2 emissions as reported, and does not consider what emissions would be if the 2008 configuration of mostly purchasing electricity from the grid and mostly generating steam from coal had been used to deliver the 2010 demand for steam and electricity. Even so, the emissions decline shown in the bottom line of Table 6-4 is large enough that it cannot be explained by changes in emissions factors in Table 6-5, nor can it be explained by reductions in demand for electricity and steam, supporting the finding of real GHG emissions reductions, thanks to CCHPP. Note that this evaluation is based on CO_2 emissions from the conversion stage of the fuel life cycle, in other words the combustion of fuels to create electricity or steam; other life cycle emissions from the extraction and transportation of coal and gas are not considered.

6-5-4   Integrating Different Electricity Generation Sources into the Grid

Along with economic analysis of individual types of generating systems, it is also worthwhile to understand how different types of systems work together to meet the varying load requirements of the grid. Type of fuel source, flexibility in stopping and starting, and fixed versus variable cost characteristics all influence the role of different generating assets. Initially we will focus strictly on fossil fuel options, and then discuss nonfossil options at the end of this section. The discussion of grid integration here is of necessity brief and simplified, for more detail the reader is referred to full-length works on the topic.[10]

Fossil-fired generating asset types that provide electricity for the grid can be broadly divided into three types, as follows:

1. *Baseline plants:* Typically these plants have high fixed capital cost but low variable cost due to low cost of fuel and high efficiency. Until recently the main fuel for these plants was coal, but gas may also be competitive for use in baseline plants going forward if recent low prices for this resource continue. Since these plants operate for many or most of the hours of the year, the ability to start, stop, or vary output quickly is not as important (see discussion of load duration curve below).

2. *Load-following plants:* Typically these plants have lower fixed capital cost but higher variable cost than baseline plants; on the other hand, capital cost is higher than peaking plants because additional investments in efficient operation are cost-effective. For example, a load-following plant might be combined cycle rather than single cycle. As the name implies, the ability to vary output with

---

[10]See for example Grainger and Stevenson (1994) or Wood and Wollenberg (1996).

load is emphasized; therefore, a short start-and-stop-time is desirable. Gas is the fuel of choice among fossil alternatives.

3. *Peaking plants:* These plants minimize fixed capital cost at the expense of high variable cost because they operate the fewest hours per year when demand is at or near its peak. Short start-and-stop time is emphasized. Fuel may be gas or oil.

For planning purposes, all the stakeholders in the grid, including the electrical generators, grid owner, electrical supply companies, independent system operator (or ISO, whose role is discussed further below), and public utilities commission of the government have an interest in ascertaining that there is adequate grid capacity to meet demand. To understand demand over the course of a year, analysts create a *load duration curve*, in which for the most recent year available the hours of the year are arranged in order of decreasing load, measured in GW or MW. A fictitious load curve is shown for illustration in Fig. 6-16, in which peak loads between 11 and 18 GW occur for 2000 h/year, and the remaining 6760 h/year have lower loads between 6 and 11 GW. A load duration curve of this size might represent the electricity market for an entire single-state or multi-state region with millions of customers served by a single utility. Note that the curve presented in the figure is linear to simplify analysis of the load; an actual load duration curve would be continuously decreasing but have a nonlinear path from highest to lowest hour of the year. The "spike" on the left side of the curve for the 500 or 1000 highest load hours of the year is typical, and this is a significant concern for grid operation, since it means that a measurable fraction of the overall generating capacity will be used for very few hours per year. In any case, stakeholders can use the load duration curve to ascertain that the right mix of baseline, load-following, and peaking plants will be available to meet anticipated demand, with some extra capacity available for contingencies.

The future load duration curve of course cannot be predicted with certainty, so grid operations are not scheduled for an entire year in advance as the load duration curve might suggest. Instead, the grid relies on a mix of advanced and instantaneous (or "spot") markets in which plant owners offer to generate electricity with a certain cost and maximum capacity, and then the ISO chooses the mix of generators that meets anticipated demand. The ISO seeks to meet as much of the demand as possible with the advance market (often called the "day-ahead" market) to lock in a favorable price, with the spot market used to fill in gaps in the moment, albeit at a higher average cost per kWh.

Once the available generators are known, thanks to the function of the electricity market, the ISO oversees the match between instantaneous load and electricity being fed into the grid to meet demand. On a continuing basis, the ISO signals load-following plants to vary their output so that supply balances demand and voltage in the grid is maintained within an acceptable range. Of course, instantaneous input from generators does not balance exactly with output to all types of loads (for instance, industrial users, institutions, households, and so on) down to the electron, but all electricity-using devices can tolerate to some extent variations in input voltage. To further respond to ongoing rise and fall of total demand, the ISO can also call on electrical storage facilities, such as pumped storage or large stationary batteries. These assets are at present expensive per kWh of charge stored, and therefore used sparingly (see below). As a last resort, if the ISO cannot meet demand, for example, if a generating asset suddenly fails on a hot summer day and a dangerous drop in voltage is threatened, the grid will go down, incurring the consequences of customers suddenly losing supply rather than risking permanent, irreversible damage to many billions of dollars of electrically powered equipment. Consequences of large-scale

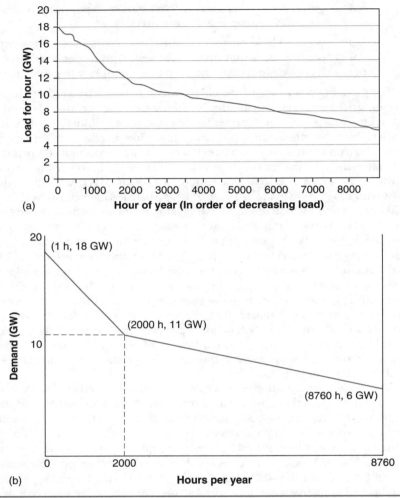

FIGURE 6-16 Load duration curve for a fictitious utility market having a minimum of 6 GW of demand for all 8760 h in the year, and a maximum of 18 GW of demand in the highest-demand hour of the year: (a) raw hourly values, (b) smoothed curve with demand as piecewise linear function of hour.

grid failures can be dire: for instance, the Northeast Blackout of August 14, 2003 affected 55 million customers in Canada and the United States and is estimated to have had an economic cost of $6 billion (U.S. dollars).

Example 6-11 illustrates calculations involving the load duration curve and generation cost information to predict a mix of generating assets that can meet demand at minimized cost.

Example 6-11 A mix of generating plants with arbitrary sources of fossil fuel is used to meet the demand represented in the load duration curve in Fig. 6-16. The table below gives fixed and variable cost characteristics for the plants. What mix of baseline, load-following, and peaking plant capacity in GW could exactly meet the demand at minimal cost? Ignore the need for spare capacity to meet contingencies. Note also that for the variable cost the units should be interpreted as follows: for example, in the case of baseline plants, "one kW generated for one hour incurs $0.025 of cost," and so on.

Plant Type	Fixed Cost ($/kW)	Variable Cost ($/kWh)
Base	$180.00	$0.025
LF	$120.00	$0.040
Peak	$90.00	$0.065

Solution Approach: we can start with the cost information given in the above table to work out breakeven hours of function for asset types in comparison to one another, independent of the information in the load duration curve. Once breakeven values are known, we can apply them to the particular demand pattern in question.

Taking the case of the baseline versus load-following plant as an example, at the breakeven number of hours per year, generating 1 kW of electricity will cost the same with either asset. Therefore, the following equality holds:

$$\$180+\$0.025(\text{Hours}) = \$120+\$0.04(\text{Hours})$$

$$\text{Hours} = \frac{\$180-\$120}{\$0.04-\$0.025} = \frac{\$60}{\$0.015} = 4000 \text{ h}$$

Thus for more than 4000 h/year, it is cheaper to generate with the baseline plant, and for less than 4000, it is cheaper using the load-following plant. Setting up a similar equation for the number of hours for comparing peaking and load-following plants:

$$\text{Hours} = \frac{\$120-\$90}{\$0.065-\$0.04} = \frac{\$30}{\$0.025} = 1200 \text{ h}$$

Turning to the load duration curve, we need to know the demand for the 1200th and 4000th hours of the year to calculate capacity. The two segments of the curve have the following equations:

$$\text{Demand}_{0-2000} = (-0.0035)\text{Hour}+18$$

$$\text{Demand}_{2000-8760} = (-0.0007396)\text{Hour}+12.48$$

Plugging in the demand for the two breakeven points gives the following:

$$\text{Demand}(1200 \text{ h}) = (-0.0035)(1200)+18 = 13.8 \text{ GW}$$

$$\text{Demand}(4000 \text{ h}) = (-0.0007396)(4000)+12.48 = 9.52 \text{ GW}$$

From these figures, we infer that for at least 4000 h/year the demand will be between 6 GW and 9.52 GW, and that this demand can be most economically served by the baseline plants because the fixed cost can be defrayed over a sufficiently large number of hours to keep overall cost to a minimum. For at least 1200 h/year the demand will be at or less than 13.8 GW. The demand above 9.52 GW and below 13.8 GW can most economically be met by the load-following plants. Peaking plants should meet remaining demand above 13.8 GW and up to 18 GW. Accordingly, the allocation of the total necessary 18 GW of capacity:

> Baseline: 9.52 GW
> Load following: 13.8 − 9.52 = 4.28 GW
> Peaking: 18 − 13.8 = 4.2 GW
> Total: 9.52 + 4.28 + 4.2 = 18 GW

Discussion   Two important simplifications are evident in this example. First, actual available capacity to meet demand going forward into the future would need to be larger than 18 GW, both because peak demand is variable and might exceed 18 GW in a subsequent year, and because when load is at or near peak, the grid needs a margin of reserve in case one generating asset fails unexpectedly. Suppose instantaneous demand were at 17.5 GW already, and a large 1-GW unit suddenly goes off-line. The grid would have only 500 MW in reserve, so the ISO would either need to quickly cut off a total of 500 MW of load (often prearranged with large industrial customers in return for discounted rates), or else a blackout would result.

Second, the actual grid might resemble the allocation of capacity to baseline, load-following, and peaking assets; for example, 11.4 GW/5.1 GW/5.0 GW, respectively, would represent the values calculated in the example with a 20% margin added to each. (Actual values would of course vary slightly from these numbers, since plant sizes are discrete and might not add up exactly to 11.4 GW for baseline capacity, and so on.) However, the function of the electricity markets and not the a priori planning exercise would determine allocation of supply. Suppose demand were exactly 13.8 GW: the above calculation implies that 69%, or 9.5 GW, would be generated by the baseline plants, and the remaining 4.3 GW or 31% by load-following plants. If the load-following plants had offered more favorable prices, their share might be larger than 31% under the assumption of up to 20% extra capacity, and vice versa.

Short-Term Storage Options for Extra Electricity Generated

Although most of the electricity loaded into the grid is used in real time, there is a small and growing role for electrical storage. Electrical storage can be used to achieve price "arbitrage" (buying at night when rates are low and then selling during the day when rates are at their peak), to improve reliability in remote locations in the grid where blackouts are a concern, or as a buffer for intermittent renewable sources.

Options for storage with relevant examples include the following:

1. *Pumped hydroelectric storage:* The principle of hydroelectricity can be used in reverse, with electrical energy expended to raise water into a reservoir, and then released again into the grid by allowing the water to run out of the reservoir through a typical hydroelectric turbine. New York Power Authority (NYPA) applies this principle at their 1160 MW_e Blenheim-Gilboa pumped storage facility in the Catskill Mountains, where power stored at night can be used to relieve pressure on electricity rates for the greater New York City region during the day, especially during the peak summer cooling season.

2. *Compressed air energy storage:* In a similar concept to pumped hydro storage, air can be compressed in a natural or manufactured reservoir, and later allowed to expand through an air turbine to make electricity. Where available, natural reservoirs can greatly reduce system construction costs. A 150-MW facility is proposed for an abandoned salt mine underneath Watkins Glen, NY, in western New York State, where the underground cavity left by salt extraction provides a natural, leak-proof reservoir for storing compressed air.

3. *Stationary battery storage:* New materials used for charge storage can gather electricity in large quantities at a fraction of the historical cost of consumer-sized or mobile batteries used in electronics and vehicles. The city of Presidio, TX, is pioneering the use of a stationary battery storage system due to difficulty of providing reliable grid power to their remote location along the Rio Grande River in western Texas. A 4-MW_e sodium-sulfur battery system provides temporary power for up to 8 h when long-distance transmission is disrupted or must be taken off line for repair work.

4. *Flywheels:* New materials and precision manufacturing have enabled the fabrication of flywheel systems that store large quantities of energy in a compact space, have minimal friction so as to reduce losses over time as the flywheel spins, and maintain environmental safety in case of any accident involving collision of other objects with the flywheel assembly. Beacon Power Corporation has installed a 20-MW_e flywheel storage station in Stephentown, NY, near the Albany Capital region, to store and release electricity in a network of flywheels.

Role of Nonfossil Generating Assets in Meeting Grid Electrical Demand

The role of nonfossil (nuclear or renewable)-generating assets depends very much on characteristics of the asset in question. First, nuclear-generating facilities are similar to large-scale coal-fired plants: they have high capital cost, low operating cost (even lower than coal due to high energy-generating potential per dollar spent on fuel), and very long start and stop times. They are therefore ideal baseline plants.

Large-scale hydropower plants can vary output, thanks to the water impounded upstream from the dam, so long as rain- or snowfall is sufficient and other needs such as irrigation have not excessively drained available water. Hydropower dams therefore function well as load-following plants and provide a CO_2-free alternative to gas-fired ones. Most hydropower dams are available as required by the grid, which is ideal for load-following, but some are "controlled release" on a once-per-day or once-per-week schedule and are therefore less flexible.

Unlike large-scale hydropower, other large renewable resources such as wind and solar are intermittent and therefore their output cannot be predicted in advance, so they do not fit in either the category of baseline or load-following. When wind- or solar-generated electricity is available, it is generally desirable to make full use of it as long as there is sufficient demand in the grid, since the variable cost is negligible. Therefore, where sufficient quantities of centrally generated wind or solar exist (i.e., not distributed generation such as household PV systems), they can be conveniently treated on the "load" rather than the "supply" side of the balancing equation. The stakeholders in the grid can think of an "adjusted load" equivalent to the actual load minus the sum of all wind or solar electricity available. The adjusted load is the amount that must be met by the remaining combination of fossil, nuclear, and large hydropower assets. Other renewables such as biomass-fired or steam-driven geothermal power plants are "dispatchable" rather than being intermittent, so the same logic does not apply.

The detailed characteristics of nuclear and renewable systems presented in Chaps. 8 through 14 can be used to further understand how they are integrated into the grid alongside fossil-generated electricity.

Role of Microgrids in Meeting Demand for Energy and Resiliency

Discussion of grid distribution thus far has assumed a "macro-scale" grid, dominated by large centralized generating assets that then transmit power over intercity distances at high voltages before they are stepped down for final distribution. One disadvantage of this approach is that in the event of a grid failure large numbers of customers can lose power, and it is incumbent on individual consumers to have in place a backup system in the event of a failure. For critical facilities such as hospitals or water treatment plants, these systems are a matter of course, but for others it can be complex and expensive to fulfill this requirement or else disruptive if they must persevere through an extended failure without power.

The *microgrid* is a response to this problem in which a community, business park, campus, etc., is still connected to the macro-grid, but also has at its disposal several forms of local or distributed generation, including solar and wind, CHP units, or small power plants (perhaps on the order of 10–30 MW, in contrast with the 500–1000 MW of a typical generating station). Further, the microgrid is built with equipment so that in the event of a regional power failure each microgrid in the region can isolate from the long-distance transmission system and continue to support local loads for the duration of the disruption. Depending on its generating and in some cases storage capacity, the microgrid may not have sufficient capacity to support all typical loads, in which case some prioritization system may be in place to cut off loads that can be postponed until normal grid power returns.

6-6   Incorporating Environmental Considerations into Combustion Project Cost Analysis

Environmental considerations affect combustion project decisions both at the level of regulatory compliance and of market opportunity. As mentioned earlier, a combustion plant is usually required to comply with emissions regulations, so pollution control equipment adds to the capital cost and to the levelized cost per kWh. Where more than one alternative technology for emissions compliance exists, the operator can reduce costs and increase net revenue by choosing the technology that has the lowest life-cycle cost.

In recent years, environmental choices have also provided a market opportunity to improve the economics of plant operation, as regulators have created an *emissions credit market* that allows operators to either reduce emissions below what is required and sell the "extra" emissions credits on the market, or buy credits on the market if they fail to meet targets for their plant. For example, a system for SO_2 emissions credits instituted by the 1990 Clean Air Act in the United States helped plant operators to reduce total emissions below a government target in a shorter time than what was required by law. Under this system, an operator may observe the market price of emissions credits and decide that it can reduce emissions for a lower cost per tonne than the value of the credit. The difference between earnings from credit sales and payment for the reductions then adds to the profits of the plant.

Such a regimen is especially interesting for reducing CO_2 emissions, because, unlike what was done in the past for emissions that are harmful to breathe such as NO_x or SO_2, it is unlikely that any governmental body will in the future legislate exactly how much CO_2 a given plant can emit. Instead, plant operators, governments, businesses, and others will either be forced to comply with mandatory emissions requirements or adopt them voluntarily, and then have flexibility about how much emissions are achieved in which location. In this economic environment, it is very likely that carbon markets that have already formed will come into widespread use. For example, since 2009, 10 states in the northeastern and mid-Atlantic United States have entered into the binding Regional Greenhouse Gas Initiative (RGGI). RGGI aims to cap total CO_2 emissions from power generation in the region and then reduce them by 10% by the year 2018.

One possible scenario is that combustion power plant emissions in a given region will fall under a CO_2 emissions cap, that is, an absolute limit that must not be exceeded. A regulator for that region might then assign an emissions target to the various plants, and require each plant to reduce emissions from the current level in order to achieve its targets. Plants that find it economically difficult to reach their target might reduce what they can, and buy credits from other plants that surpass their target in order to make up the shortfall. Also, if the region as a whole were unable to achieve its overall target, it might buy credits from outside.

Some choices might help an operator to achieve CO_2 reductions targets in line with the cap, while at the same time adding value to the electricity produced. For instance, use of biofuels as a substitute for fossil fuels in combustion systems is a way to avoid CO_2 taxes that also provides a "green" power source for which some consumers are willing to pay extra.

If the ability to sell credits in the market is the "carrot" for reducing CO_2 emissions, then the concept of a tax on carbon emissions is the "stick." Carbon taxes have been implemented by a small number of countries, notably in Scandinavia, and are under consideration in a number of others. For example, the government of Sweden charges a tax of \$150/tonne of carbon emitted to the atmosphere to most emitters, with industrial firms receiving a 75% discount, and some energy-intensive industries not required to pay any tax, for fear that the tax would drive the industry out of business. The revenue stream generated from the carbon tax can be used to invest in efficiency across the economy, reduce other taxes such as income taxes, or assist poor countries that do not have the means to adapt to climate change.

The potential effect of a carbon tax on the price of electricity can be quite significant, as shown in Example 6-12.

Example 6-12 Recalculate the levelized cost of electricity for the two plants in Example 6-10 using a carbon tax like that of Sweden. Assume that power plants, as large operations, receive a 75% discount, and that the coal used is 100% carbon (i.e., ignoring noncarbon impurities).

Solution Solving for the basic plant first, and applying a discount factor of 0.25 to the \$150/tonne tax, we obtain the dollar value of the tax imposed:

$$(2.62 \times 10^6 \text{ tonnes/year})(0.85 \text{ carbon content})(\$150/\text{tonne})(0.25) = \$83.5 \text{ million/year}$$

Therefore, the total cost incurred per year is \$130 M + \$83.5 M = \$213.5 million/year. Dividing by 3.07×10^9 kWh gives a levelized cost of \$0.0696/kWh.

Repeating this calculation for the supercritical plant gives a tax value of \$54.4 million/year and a new levelized cost of \$0.0531/kWh.

Discussion The effect of the carbon tax in this case is to further skew the economics in favor of the more efficient plant. The percentage increase in cost for both plants is large, approximately 64% for the basic plant and 33% for the supercritical plant. Even if such a tax were phased in gradually, the effect on the competitiveness of either plant would be severe, so the operators would likely respond by upgrading to the most efficient technology possible. In this way, the carbon tax would achieve its desired effect of reducing carbon emissions over the long term. Operators might also consider shifting some production to natural gas, which emits less carbon per unit of energy released and therefore incurs a lower carbon tax.

6-7 Reducing CO_2 by Combusting Nonfossil Fuels or Capturing Emissions

This section focuses on four opportunities to reduce CO_2 emissions through alternative means of combusting fuels to generate electricity. The first three involve combusting fuels other than fossil fuels, namely, waste-to-energy conversion (WTE), biomass combustion, and waste water energy recovery (WWE). These applications are explained using case studies in use or under consideration in the vicinity of Cornell University in New York State, but they are highly replicable elsewhere. For instance, virtually all urban areas in industrialized countries and many in emerging countries as well generate a solid waste stream that can be tapped for WTE. Similarly, waste water treatment systems provide a stream of biological material that can be extracted, gasified, and combusted to make electricity. The fourth opportunity is more long-term, entailing redesign of the fossil fuel combustion system from the ground up so that all CO_2 is diverted without being emitted to the atmosphere. Along with lower CO_2 emissions from electricity generation, these four systems can create local industry, use up local waste streams to avoid landfilling, and meet electricity demand at reduced cost.

6-7-1   Waste-to-Energy Conversion Systems

Waste-to-energy (WTE) conversion entails gathering municipal solid waste after separating noncombustible recyclable materials (e.g., metals, glass), combusting them in a specialized power generation plant, recovering any residual metals for recycling, and then safely disposing of ash. As an energy source, WTE has the advantage of reducing the volume of space required for landfilling, and generating electricity at a relatively low cost, since the tipping fee that would otherwise go to the landfill goes to the plant operator. In other words, instead of paying for fuel, the operator is paid to receive the fuel from the disposer. WTE requires specialized pollution controls, and is not strictly a zero-CO_2 source, because some of the fuel is in fact derived from hydrocarbons that would otherwise end up in the landfill instead of the atmosphere. (Note that some decomposing hydrocarbons in landfills generate methane, an extremely potent greenhouse gas or GHG, although we do not attempt to quantify the benefit of avoided CH_4 emissions by combusting fuels using WTE here.)

WTE generated 21.3 billion kWh in the United States in 2014, or about 0.5% of the total electricity supply generated, according to the U.S. Energy Information Administration. WTE facilities are distributed around the country, such as the 39-MW Jamesville plant outside of Syracuse, NY. No new WTE facilities have been opened in the United States in approximately the last 20 years due to disappointing performance on pollution controls and resulting public controversy from an earlier generation of WTE technology. Peer countries in Europe and Asia, notably Japan and Denmark, have in the meantime forged ahead to reduce overall emissions from WTE combustion, and today the general opinion of WTE in those countries, especially as an alternative to landfilling solid waste, is much higher than in the United States. In Copenhagen fully 97% of the solid waste stream is either recycled or combusted using WTE, with only 3% going to landfills. WTE facilities in Denmark operate in close proximity to residential neighborhoods.

Example of a 50-MW Regional WTE Facility

The feasibility of a 50-MW WTE plant in the region around Cornell University was studied as part of a larger economic study of replacing a coal-fired electric plant that was no longer economically viable for generating base load electricity. The feasibility study is summarized here. All figures used in the study are taken from government sources such as the USEIA, or from similar facilities currently in operation. The study assumes the federal government discount rate of 7%, a 20-year investment lifetime for the plant, and a 60% capacity factor, which is approximately that of the coal-fired plant that is being phased out.

The capital cost of the WTE plant is $2060 per kW, or $103 million initial cost. Discounting this value over 20 years gives an annual capital cost of $9.72 million. The plant has an overall efficiency from energy content in the waste to electricity going out of 28.4%. This relatively low value reflects the presence of moisture and noncombustibles in the fuel. The favorable economics of the incoming waste stream compensate for this low value, however. The waste has an energy content of 13.9 GJ/tonne, and the plant charges $55 for each tonne of waste delivered.

Operating cost including labor, maintenance, and specialized procedures for cleaning up combustion, removing pollutants, and landfilling any remaining ash for which there is no market value amount to $0.0751/kWh, or $19.7 million per year. The plant produces 262.8 million kWh per year, which has an equivalent energy content of 946 TJ. Based on the plant efficiency the required energy input is 3331 TJ per year, equivalent to 239,200 tonnes per year. This intake results in a revenue stream of $13.2 million per year.

The results are shown in Table 6-6. The plant has a relatively high capital cost and especially operating cost compared to a natural-gas-fired plant, due to the complexity of combusting solid waste and the relatively small capacity of the plant, compared to natural-gas-fired plants that may be 1000 MW in size or more. However, it appears that in countries such as Japan or Denmark a network of small, decentralized WTE plants is favored so that the hinterland for obtaining waste for any one plant is not too large, and it is likely that WTE plants will remain relatively small (40–50 MW typical maximum capacity). The revenue from solid waste deliver fees lowers total annual cost resulting in a levelized cost of $0.062/kWh. The analysis also does not include a potential small revenue stream that might be derived from selling by-product metals that are left after combusting waste, although this figure would not substantially change the levelized cost calculation. Overall, WTE has the benefits of relatively low cost per kWh, low price volatility thanks to a steady supply of municipal waste, partially avoided CO_2 emissions from combusting waste not derived from fossil fuels, and avoided use of space for landfilling. The maximum scope for WTE is limited, however. Based on typical per capita rates of waste generation and maximal recovery of waste possible, the most electricity that could be generated from WTE would be on the order of 3% of national demand.

Technical Components		
Annual output	262.8	million kWh/y
Plant efficiency	28.4%	
Waste energy input	3331	TJ/y
Waste required	239,200	tonnes/y
Economic Components		
Capital cost	$9.72	million/y
Operating cost	$19.73	million/y
Fuel revenue	$(13.16)	million/y
Total cost	$16.29	million/y
Levelized cost	$0.062	per kWh

TABLE 6-6   Summary of Technical and Economic Figures for 50-MW WTE Plant Feasibility Study

6-7-2 Electricity Generation from Biomass Combustion

Burning of biomass such as wood is another alternative to burning fossil fuels already in use around the United States, such as the 50-MW Joseph McNeil generating station outside of Burlington, Vermont (USA). In 2014, 43 billion kWh or about 1% of the U.S. electricity supply came from wood. Often wood-fired plants have a backup fuel source such as coal since wood supplies can be seasonal and in certain months the full supply needed to meet a given demand level may not be available.

As part of the same feasibility study for WTE discussed in Sec. 6-7-1, an economic analysis was conducted for a plant in the vicinity of Cornell University. Many of the assumptions are the same as for the WTE study; however, the plant has higher efficiency at 35%, and the energy content of wood is also higher at 21.6 GJ/tonne. However, unlike municipal waste, the wood fuel is a net cost to the operator at $35/tonne including delivery. Capital cost of a wood plant is higher at $4300 per kW, so that total cost for a 50-MW plant is $215 million, or $20.3 million per year when annualized. Operating cost is estimated at $0.0245/kWh, or $6.45 million per year for the proposed 50-MW plant. The fuel requirement is 125,100 tonnes per year, for a cost of $4.38 million per year.

The results are summarized in Table 6-7. The plant has an overall higher levelized cost at $0.118/kWh, although the output is now 100% carbon free at the conversion stage since unlike WTE no fossil fuels are combusted. Like the 50-MW WTE plant, the plant size keeps the hinterland required to supply sufficient wood small, compared to fossil-fired plants that draw from a national delivery grid for coal or gas, where fuels can come from hundreds of miles away. In summary, the feasibility of the plant depends on whether the community or region is willing to pay a premium for power generation from wood, in return for 100% CO_2 reduction and the creation of local jobs. Another concern is local air pollution, since some U.S. plants have been cited for excessive emissions and encountered local opposition on this point before.

6-7-3 Waste Water Energy Recovery and Food Waste Conversion to Electricity

Waste water energy recovery consists of separating biological materials from treated water at a waste water treatment facility (WWTF), preparing those materials in some

Technical Components		
Annual output	262.8	million kWh/y
Plant efficiency	35.0%	
Fuel energy input	2703	TJ/y
Fuel required	125,100	tonnes/y
Economic Components		
Capital cost	$20.30	million/y
Operating cost	$6.45	million/y
Fuel cost	$4.38	million/y
Total cost	$31.13	million/y
Levelized cost	$0.118	per kWh

TABLE 6-7 Summary of Technical and Economic Figures for 50-MW Wood Biomass Plant Feasibility Study

Location	Plant Average Flow (MGD)	Energy Capacity (kW$_e$)	CHP Type	Installation Date
Albert Lea, MN	5	120	Turbine	2004
Chippewa Falls, WI	2	60	Turbine	2003
Fairfield, CT	9	200	Fuel cell	2005
Flagstaff, AZ	3.5	290	Reciprocating	2008
Ithaca, NY	5.5	260	Turbine	2006*
Lewiston, NY	2	60	Turbine	2001
Santa Maria, CA	7.8	300	Turbine	2009

TABLE 6-8   Representative U.S. Waste Water Treatment Facilities Currently Using CHP to Convert Waste to Electricity. *Note:* MGD = million gallons per day of waste water flow. *Prior to 2006 this facility used a reciprocating engine system for CHP. *Source:* U.S. Environmental Protection Agency.

way, and combusting them to make electricity and process or space heat (combined heat and power, or CHP). A typical preparation for the biological materials is to process them in an anaerobic digester to make biogas (methane or CH_4). After leaving the biodigester, the gas must be processed to remove CO_2 and trace contaminants (e.g., hydrogen sulfide) so that it can be combusted in a reciprocating engine or turbine without damaging the equipment.

A typical WWTF has a strong motivation to capture energy contained in the by-products of water treatment, since they can both reduce the cost of landfilling sludge and reduce electricity and in some cases gas purchases from the grid. Table 6-8 shows a sample of U.S. waste water plants that are using CHP to generate part of their electricity requirement internally, with microturbines, reciprocating engines, or fuel cells. For many WWTFs, the total energy available in sewage and contents of septic tanks delivered by trucks known as *septage* does not generate enough electricity to make the plant self-sufficient for electricity. Therefore, the biodigestion facility within the plant can be adapted to receive other outside waste (e.g., food waste from food processing facilities, waste streams from institutions that may be high in waste food content, etc.) to increase the total potential for electricity production, so that the plant can become self-sufficient and even a net generator of electricity for the surrounding load. Some WWTFs are seeking to rebrand themselves as "energy centers" for which the waste water stream is just one among several energy sources.

Case Study of Ithaca Area Waste Water Treatment Facility Energy Conversion Program

The Ithaca Area Waste Water Treatment Facility (IAWWTF) in New York State serves approximately 100,000 people with a maximum capacity in peak periods of 30 million gallons per day (MGD), although average flow is smaller at 5.5 MGD. Starting with waste water treatment at a newly constructed facility in the 1980s, the IAWWTF has recently been expanding its capacity to receive "truck waste" (waste food and other waste biological materials) so as to reduce the regional waste flow to landfills and also reduce grid electricity purchases. The plant has used electricity generation from its inauguration, but recently the original reciprocating Cummins engine generators were replaced with four 65-kW Capstone microturbines, for a capacity of 260 kW.

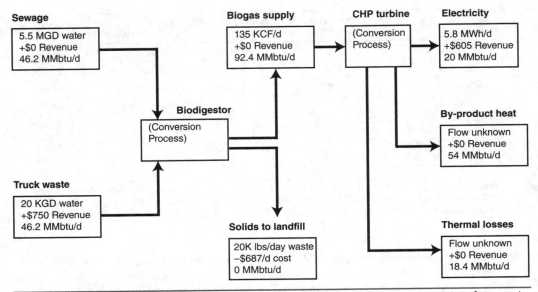

Sewage

5.5 MGD water
+$0 Revenue
46.2 MMbtu/d

Biogas supply

135 KCF/d
+$0 Revenue
92.4 MMbtu/d

CHP turbine

(Conversion Process)

Electricity

5.8 MWh/d
+$605 Revenue
20 MMbtu/d

Biodigestor

(Conversion Process)

By-product heat

Flow unknown
+$0 Revenue
54 MMbtu/d

Truck waste

20 KGD water
+$750 Revenue
46.2 MMbtu/d

Solids to landfill

20K lbs/day waste
−$687/d cost
0 MMbtu/d

Thermal losses

Flow unknown
+$0 Revenue
18.4 MMbtu/d

FIGURE 6-17   Schematic of material and energy flows in waste water energy recovery system from waste water and truck waste to electricity and heat generation.

The goal of this case study is to cost the ancillary equipment needed to generate electricity from waste at the IAWWTF and calculate the levelized cost. Figure 6-17 shows the flow of waste into and energy products out of the system, where each bubble in the schematic includes if possible physical flow volume, energy flow, and revenue of cost (if any). In the configuration shown, the energy that is eventually converted in the microturbines arrives 50% from sewage and 50% from "truck waste" (including septage, food processing waste, portable toilet waste, and animal processing waste). Although the energy contribution is equal the physical volume of the truck waste is much smaller, since the waste water flow is mostly water (20,000 gallons per day of truck waste, or 20 KGD, versus 5.5 MGD). Energy content passes through the biodigester and enters the CHP system as part of the biogas supply, and is allocated to electricity, heat supply, or thermal losses in proportions similar to those of the apartment complex case study of Table 6-2. Some energy content is of course lost with the residual solids (also known as "cake") sent to the landfill, but these are not quantified here. Also, the byproduct heat has $0 value in this case because it is used to heat the biodigester, but in another context it might be used to reduce grid gas purchases, creating another revenue stream.

In terms of cost or revenue, the truck waste is charged approximately $0.038 per gallon, which generates $750 in revenue for 20,000 gallons per day of flow. This amount is slightly more than the daily cost of disposing of landfill solids, which amount to $687 per day at a typical rate of $68.70 per ton. The electricity output of 5800 kWh per day amounts to approximately 2.1 million kWh per year, which is equivalent to a 92% capacity factor. At an avoided cost of $0.105/kWh (currently paid by the IAWWTF in this case), the output is worth ~$605/day. The net revenue per day is then $668 per day, or ~$244,000 per year.

For evaluation of life cycle cost and levelized energy, the capital cost of the system is estimated at $2883 per kW, or $749,000 for the 260-kW system. The cost per kW is higher than that of Table 6-2, since additional equipment is required as part of the bio-digester-CHP system. The system is discounted at 7% over 12 years, for an annualized cost of $113,600. An additional maintenance cost of $42,000/year applies. If we consider only the net revenue of charging for truck waste but paying for landfill solids of $23,000, the result is a levelized cost of $0.063/kWh, as shown:

$$\text{LevCost} = \frac{\text{CapCost} + \text{OpCost} - \text{Rvn}}{\text{Output}} = \frac{\$113,600 + \$42,000 - \$23,000}{2.1\,\text{MkWh}} = \$0.063/\text{kWh}$$

This figure is well below the purchase cost of electricity of $0.105/kWh (note that the plant must pay for both generation and distribution of purchased electricity), which shows the value of generating electricity from biological waste streams on site.

Extensions of the system in Fig. 6-17 are possible. Waste water flow is essentially capped unless the size of the community grows, but truck waste could be increased to make the plant a net generator of electricity. If so, surplus electricity could be sent to the grid. Another concept is to make the IAWWTF the core of a district heating and electricity generation system, where a district heating loop from the plant would extend to surrounding homes or businesses, which would also receive some of the electricity generated. The thermal content of the waste water passing through the plant represents another potential energy source for a low-temperature heat pump system that would remove heat from the water flow and distribute it to the plant or surrounding heat loads. (See Chap. 17 for an explanation of the function of heat pumps.) A system in Whistler, British Columbia, Canada takes advantage of low-cost hydroelectricity to operate this type of heat pump system at their waste water plant.

6-7-4 Zero-Carbon Systems for Combusting Fossil Fuels and Generating Electricity

We return now from nonfossil fuels to the question of existing plants that use fossil fuels. Initially, under a carbon trading or carbon tax regime, operators might respond by improving efficiency at existing coal-fired plants, rather than fundamentally changing the combustion technology used. Looking further into the future, however, it is clear that under tightening emissions restrictions or increasing carbon taxes, a technology that converts coal to electricity and other forms of energy without any CO_2 emissions to the atmosphere would be highly desirable. Such a technology is the goal of a number of research programs around the world.

One leading consortium that is pursuing zero-carbon energy technology is the FutureGen Industrial Alliance, a public-private consortium that includes private firms, representatives from various national governments such as India and South Korea, and, until February 2008, the U.S. Department of Energy (USDOE). The Alliance has been developing a prototype 275-MW facility that will verify the technical and economic feasibility of generating electricity from coal with no harmful emissions to the atmosphere and with the sequestration of by-product CO_2 in a long-term reservoir, with the goal of bringing the proposed facility online in the year 2012.

The USDOE withdrew from the Alliance in early 2008 because it had concluded that the technology was already sufficiently mature, and did not require continued financial support at the level of building an entire demonstration facility. This decision has been criticized by the leadership of the Alliance as well as some of the researchers in the field of carbon capture and sequestration (CCS) as being premature. Without a commercial-scale demonstration plant, it may be difficult for the coal combustion with carbon sequestration program to develop quickly enough to significantly reduce CO_2 emissions from coal in the short to medium term. In the meantime, the USDOE is providing other support to private firms that are interested in this technology. In 2011, plans for the construction of a pilot FutureGen plant in the state of Illinois were indefinitely postponed, and the technology currently awaits a new site for a pilot project.

As shown in Fig. 6-18, the first step in the conversion process is the gasification of coal to create a gasified carbon-oxygen mixture. This mixture is then fed to a water-shift reactor, which produces high temperature hydrogen and CO_2 as by-products; the CO_2 is then transported out of the system for sequestration. Some of the hydrogen released is kept in that form for use in hydrogen-consuming applications, which in the future might include fuel cell vehicles. The rest is reacted in a high-temperature fuel cell to produce electricity. Since the by-product of the hydrogen-oxygen reaction in the fuel cell is steam at high temperature, this output can be expanded in a turbine to create additional electricity, effectively resulting in a fuel-cell and turbine combined cycle generating system. Steam exhausted from the turbine is returned to the water-shift reactor for reuse. The structure required to house this energy conversion operation is shown in Fig. 6-19.

As a first step toward coal combustion with carbon sequestration, plant operators are considering integrated gasification combined cycle (IGCC) plant technology that first gasifies coal, removes impurities, and then combusts the coal in a combined cycle of the type presented earlier in this chapter. While coal-fired plants with gasification cost more to build and operate than those with conventional coal-fired boilers, implementing the gasification technology now would provide a first step toward eventually transforming plants to be able to separate and sequester CO_2 in the future.

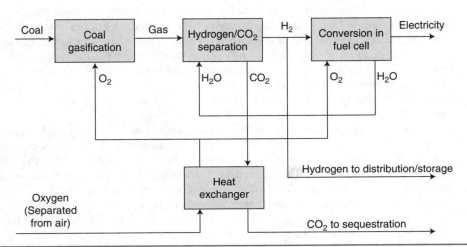

Figure 6-18   Schematic of advanced coal-fired power plant concept with hydrogen generation and CO_2 separation.

FIGURE 6-19   Architectural rendering of a zero-emission "FutureGen" coal-fired plant, with by-product CO_2 sequestered locally in reservoir or transported off site for sequestration. The facility is located next to a body of water for ready access to cooling water, and possibly also for delivery of coal if the body of water is navigable. Note the absence of the tall stacks that are characteristic of today's fossil-fuel-fired power plants. [*Source:* Courtesy of USDOE and LTI (Leonardo Technologies, Inc.). Reprinted with permission.]

6-8   Systems Issues in Combustion in the Future

To conclude this chapter, we return to the systems approach to combustion technology that was discussed at the beginning, with a view toward some of the challenges that await in the future. All evidence suggests that the bulk of the changes over the next several decades will involve major transformation of technology rather than incremental improvements. These types of transformations include major changes in core technology (e.g., from single to combined cycle or from CO_2 release into atmosphere to carbon capture and sequestration), changes in fuel (e.g., increasing the use of biofuels), or changes in end use application (e.g., increased use of cogeneration with new applications for by-product heat). In this environment, systems engineering and systems thinking will have an important role to play. Here are several examples of systems issues:

- *Bridge versus backstop technologies:* A *backstop* technology is one that fully achieves some desired technological outcome, for example, eliminating CO_2 emissions to the atmosphere. Thus an advanced combustion plant of the type discussed in Sec. 6-3 is an example of a bridge technology, and the FutureGen zero-emissions plant in Sec. 6-7-4 is a backstop technology. When making investment decisions, utility owners must consider the level of advancement of bridge and backstop technologies, and in some cases choose between them. In the short run, if the backstop is not available, the owner may have no choice but to adopt the bridge technology in order to keep moving forward toward the goal of reducing CO_2 emissions. Eventually, if the backstop technology proves successful, the owner would need to carefully weigh whether to continue building even very efficient versions of the bridge technology, or whether to transition to the backstop technology.

- *Importance of advance planning for cogeneration and other multiple use applications:* Decision-makers can help to expand the use of cogeneration by considering possible cogeneration applications at the time of building new power plants or renovating existing ones. This time period provides a window of opportunity to simultaneously develop process heat or space heating applications, for example, by bringing new industries to the plant site that can take advantage of the available process heat. Otherwise, once the plant is built, operators may face the problem of *retrofit syndrome* in that they find it technically impossible or economically prohibitive to add an application for by-product heat after the fact.

- *Maintaining use of infrastructure when transitioning between energy sources:* Along with end-use energy applications and the energy resources themselves, energy systems include a substantial investment in infrastructure to convert and distribute energy products. It is highly desirable to continue to use this infrastructure even after discontinuing the original energy product involved, so as to maintain its value. For example, many commercial and residential properties have pipes that bring in gas from the natural gas distribution grid. With modifications, this infrastructure might be some day used to deliver a different gaseous product that has a longer time horizon than gas or is renewable, such as compressed hydrogen, or a synthetic natural gas substitute made from biofuels.

- *Pushing technology toward smaller-scale applications:* One way to expand the positive effect of an energy-efficient technology is to expand its potential market by replicating it for use in smaller-scale applications. Thus the cogeneration technology that began at the 500 MW power plant level has become established at the 100-kW building complex level. Furthermore, household size cogeneration systems that provide enough hot water for a house and generate electricity at the 1-kW level are penetrating the market. Given the number of single-residence water heating systems in use around the world, this technology might greatly expand the total amount of electricity produced from cogeneration if it proves reliable and cost-competitive.

- *Creating successful investments by coupling multiple purposes:* In some cases, the business case for a transformative technology can be improved if the technology addresses more than one energy challenge at the same time. In the case of the FutureGen zero-emission plant concept, the technology addresses one issue requiring a major transformation but also promising great benefit in the future, namely, carbon capture and sequestration (CCS). At the same time, it addresses another fundamental need, namely, for a clean energy resource for transportation, by generating hydrogen as a by-product of the water-shift reaction. The success of this application of hydrogen requires the development of a satisfactory hydrogen fuel cell vehicle (HFCV) technology (See Chap. 15.). It is possible that the FutureGen plant with sequestration and the HFCV technology are more attractive as a package than they are separately.

6-9 Representative Levelized Cost Calculation for Electricity from Natural Gas

The following calculation of levelized cost per kWh for a typical large natural gas powered generating asset is based on values and discussion throughout this chapter. It represents a reduced-CO_2 option (325 g CO_2/kWh) compared to the typical current U.S. energy mixture [424 g CO_2/kWh in 2014 (USEIA)]. Figures calculated will be compared to similar calculations for nuclear (Chap. 8), solar (Chap. 10), and wind energy (Chap. 13), and again in concluding Chap. 17. Throughout the comparison, all options use an investment lifetime of 20 years, a discount rate of 7% (U.S. federal government typical rate), and no salvage value, which are thought to be conservative values.

We assume a capital cost of $1050/kW for a 1-GW plant, resulting in a $1.05 billion capital cost. The required discounting factor is (A/P, 20 years, 7%) 0.0944, which will be used for this and other energy options, giving a capital cost per year of $99 million. We assume a capacity factor of 65%, resulting in 5694 GWh/year of output. Dividing capital cost among output gives a value of $0.0174/kWh, which is in line with a 2013 U.S. Department of Energy survey of capital cost in U.S. gas-fired power plants. For fuel cost, we use an energy cost of $6/GJ and 55% average efficiency of the combustion process, resulting in $0.039/kWh cost for fuel. The final component is nonfuel operating cost, where we again use a USDOE average value of $0.008/kWh.

The final value is $0.0644/kWh, as summarized in the calculations below:

$$Output = (1\,GW)(8{,}760\,h/y)(0.65) = 5{,}694\,GWh/y$$

$$CapCost = \frac{(\$1.05B)(0.0944)}{5{,}694\,GWh/y} = \frac{\$99M}{5694} = \$0.0174/kWh$$

$$EnergyCost = (1\,kWh)(3.6\,MJ/kWh)\left(\frac{1}{0.55}\right)(\$0.006/MJ)$$

$$= \$0.039/kWh$$

$$TotCost = CapCost + EnergyCost + OpCost = \$0.0174 + \$0.039 + \$0.008$$

$$= \$0.0644/kWh$$

6-10   Summary

Combustion technologies that convert energy in fuel resources into electrical energy are crucial for the energy systems of the world, since they generate the majority of the world's electricity. Many of these technologies are built around two combustion cycles, namely, the Rankine cycle for vapor and the Brayton cycle for gaseous fuels. The basic Rankine and Brayton cycles convert only 30 to 40% of the energy in the working fluid into mechanical energy, with additional losses for combustion and electrical conversion, so over time engineers have achieved a number of technological breakthroughs to improve the overall efficiency of power plants in order to increase efficiency and minimize operating costs. Some of the most important of these breakthroughs are the result of a radical redesign of the overall energy conversion system: (1) the combination of a Rankine and Brayton cycle in combined cycle systems and (2) the addition of other applications for by-product exhaust energy in the case of cogeneration systems. A systems approach to combustion technologies will be useful in the future to help adapt and expand the ecologically conscious use of combustion technologies in an environment of changing costs and energy resources.

References

Cengel, Y., and M. Boles (2002). *Thermodynamics: An Engineering Approach*, 4th ed. McGraw-Hill, Boston.

Dunn, A. (2002). "Optimizing the Energy Portfolio: An Illustration of the Relationship between Risk and Return." *Journal of Structured & Project Finance*, Vol. 8, No. 3, pp. 28–37.

Grainger, J., and W. Stevenson (1994). *Power Systems Analysis*. McGraw-Hill, New York.

Schrage, L. (1997). *Optimization Modeling with Lindo*. Duxbury Press, Pacific Grove, CA.

U.S. Department of Energy (2006). *FutureGen: A Sequestration and Hydrogen Research Initiative*. Project update. USDoE, Washington, DC. Available at http://www.fe.doe.gov/programs/powersystems/futuregen/Futuregen_ProjectUpdate_December2006.pdf. Accessed May 15, 2007.

Vanek, F., and L. Vogel (2007). "Clean Energy for Green Buildings: An Overview of On- and Off-Site Alternatives." *Journal of Green Building*, Vol. 2, No. 1, pp. 22–36.

Wark, K. (1984). *Thermodynamics*, 4th ed. McGraw-Hill, New York.

Wood, A., and B. Wollenberg (1996). *Power Generation, Operation, and Control*, 2nd ed. John Wiley & Sons, New York.

Further Reading

Biello, D. (2008). "'Clean,' Coal Power Plant Canceled—Hydrogen Economy, Too." *Scientific American*, February 6, 2008 Internet version. Available at www.sciam.com. Accessed March 18, 2008.

Carnot, S. (1824). *Reflections on the Motive Power of Fire*. Self-published monograph, Paris.

Harvey, D. (2010). *Energy and the New Reality 1: Energy Efficiency and the Demand for Energy Services*. Earthscan, London.

Harvey, D. (2010). *Energy and the New Reality 2: Carbon-free Energy Supply*. Earthscan, London.

Kelly, H. and C. Weinberg (1993). "Utility Strategies for Using Renewables," Chap. 23. In: S. Johannson, ed., *Renewable Energy: Sources for Fuels and Electricity*. Island Press, Washington, DC.

Oak Ridge National Laboratory (1999). *An assessment of the Economics of Future Electric Power Generation Options and the Implication for Fusion*. Report, U.S. Dept. of Energy, Oak Ridge National Laboratory, Oak Ridge, TN.

Progressive Management (2003). 21st Century Complete Guide to Biofuels and Bioenergy: Compilation of U.S. Government Research on CD-ROM. Progressive Management, Washington, DC.

Sorensen, B. (2002). *Renewable Energy: Its Physics, Engineering, Environmental Impacts, Economics & Planning*. Academic Press, Amsterdam.

Tester, J., E. Drake, and M. Driscoll (2005). *Sustainable Energy: Choosing Among Options*. MIT Press, Cambridge, MA.

Williams, R. H. (1998). "A Technology for Making Fossil Fuels Environment- and Climate-Friendly." *World Energy Council Journal*, pp. 59–67.

Yergin, D. (2012). *The Quest: Energy, Security, and the Remaking of the Modern World*. Penguin, London.

Exercises

6-1. A Brayton cycle operates with isentropic compression and expansion, and air entering the compressor 27°C. The pressure ratio is 8:1, and the air leaves the combustion chamber at 1200 K. Compute the thermal efficiency of the cycle, using data from a table of ideal gas properties of air.

6-2. Suppose a power plant using the cycle in Problem 6-1 produces 700 million kWh/year of electricity. Ignore losses in the compressor and turbine. If the generator is 98% efficient and combustor is 88% efficient, and gas costs $10/GJ, calculate:
  a. The total gas energy content required per year
  b. The annual expenditure on gas
  c. The mass of CO_2 emitted per year from the plant
  d. The mass of CO_2 emitted per million kWh produced

6-3. Compute the thermal efficiency of a Rankine cycle with isentropic compression and expansion for which steam leaves the boiler as saturated vapor at 6 MPa and is condensed at 50 kPa.

6-4. Suppose a power plant using the cycle in Problem 6-3 produces 2.1 billion kWh/year of electricity. Ignore losses in the boiler and turbine. If the generator is 98% efficient and boiler is 84% efficient, and coal costs $30/tonne, calculate:
  a. The total coal energy content required per year
  b. The annual expenditure on coal
  c. The mass of CO_2 emitted per year from the plant
  d. The mass of CO_2 emitted per million kWh produced

6-5. A theoretical supercritical Rankine cycle has a maximum pressure of 30 MPa, a maximum temperature of 700°C, and a condenser pressure of 10 kPa. What is its thermal efficiency? What is its Carnot efficiency?

6-6. The top cycle of a combined cycle power plant has a pressure ratio of 12. Air enters the compressor at 300 K at a rate of 12 kg/s and is heated to 1300 K in the combustion chamber. The steam in the bottom cycle is heated to 350°C at 10 MPa in a heat exchanger, and after leaving the turbine is condensed at 10 kPa. The combustion gases leave the heat exchanger at 380 K. Assume that all compression and expansion processes are isentropic. Determine the thermal efficiency of the combined cycle. (Refer to Fig. 6-8.)

6-7. Suppose an actual combined cycle plant, in which air enters the compressor at 300 K, with compression ratio of 12 to 1, at a rate of 443 kg/s and is heated to 1427 K in the combustion chamber, has a thermal efficiency of 52% (not considering losses in the combustor and generator). The ratio of air to fuel is 50:1. The steam in the bottom cycle is heated to 400°C at 12.5 MPa in a heat exchanger, and after leaving the turbine is condensed at 15 kPa. The combustion gases leave the heat exchanger at 400 K. Determine the ideal efficiency of (a) the top cycle, (b) the bottom cycle, (c) the combined cycle, (d) the total heat input to the top cycle in kW, (e) the total work output from the combined cycle in kW, and (f) by how many percentage points the losses decrease the thermal efficiency between the ideal cycle and the actual cycle that is in use in the plant.

6-8. Consider a combined heat and power (CHP) system similar to the system in Fig. 6-10 in which steam enters the turbine at 15 MPa and 500°C. An expansion valve permits direct transfer of the steam from the boiler exit to the feedwater heater. Part of the steam is bled off at 800 kPa for the feedwater heater, and the rest is expanded to 20 kPa. The cycle uses a generator that is 98% efficient. The throughput of steam is 20 kg/s. Assume isentropic compression and expansion.

    a. If 40% of the steam is extracted for process heating and the remainder is fully expanded (i.e., the mass flow through the expansion valve is $m = 0$ kg/s), calculate the rate of heat output for process heat and electrical output, both in kW.

    b. Calculate the utilization factor ε_u in this mode of operation.

    c. If this type of operation is the average setting and the plant runs continuously, how many kWh of electricity and process heat does it produce in a year?

6-9.   One potential benefit of a cogen system like the one in Problem 6-8 is that it can reduce CO_2 emissions compared to generating the equivalent amount of process heat and electricity in separate facilities. Suppose the cogen system is coal fired, with a boiler efficiency of 85%, and that for both alternatives, coal has an energy content of 25 GJ/tonne and a carbon content of 85%. Take the output from the system required in Problem 6-8a, with all energy coming from coal combustion. Compare this facility to a process heat plant that generates heat with an 82% conversion efficiency from the energy content of the coal to the process heat, and a separate coal-fired electric plant that expands steam at 6 MPa and 600°C in a turbine and condenses it at 20 kPa. The stand-alone electric plant has a boiler efficiency of 85% and a generator efficiency of 98%. How many tonnes of CO_2 does the cogen facility save per hour?

6-10.   Make a comparison of simple cycle and combined cycle gas turbine economics by comparing two plants based on the cycles in Examples 6-2 and 6-6. In both cases, the actual cycle achieves 80% of the efficiency of the ideal cycle as given, due to irreversibilities in compressors, turbines, and the like. The efficiencies of the combustor and generator are 85% and 98%, respectively. Both plants are rated at 500 MW and have a capacity factor of 75% over 8700 operating hours per year; for the remaining 60 h, the plants are shut down. The simple cycle initial cost is $400 million and the combined cycle is $600 million. The lifetime of the analysis is 20 years, and the MARR is 6%. The balance of cost is $15 million/year for both plants. From Table 5-2, use a mid-range value of $9.75/GJ for gas delivered to the plant. Ignore the need for ROI.

    a. Show that the combined cycle plant is more attractive on a levelized cost basis.

    b. As a gas-fired power plant operates fewer and fewer hours per year, the extra investment in combined cycle becomes less and less viable. At what number of hours per year, assuming that each plant is operating at rated capacity, does the simple cycle plant become more attractive? Note that the capacity factor is different from part (a).

6-11.   Economic analysis of small-scale cogeneration. You are considering an investment in three 60-kW microturbines, which generate both electricity and DHW, for a large apartment complex. The microturbines will displace output from the existing natural-gas-fired boiler for DHW such that 100% of the demand is met by the microturbines. The microturbines start running at 5 a.m. in order to generate DHW for the morning peak; they then run continuously at full power until they have made enough DHW for the total day's demand. Thus the cogen system supplies 100% of the DHW need. It can be assumed that when the system turns off for the night, there is enough DHW in storage to last until it comes back on the next morning. The prices of gas and of electricity are $10.75/GJ and $0.105/kWh, respectively. The turbines cost $85,000 each, with an additional $0.02/kWh produced maintenance contract cost; the project life is 20 years with an MARR of 3%. The turbines convert 26% of the incoming energy to electricity, and 55% of the incoming energy to DHW. The existing boiler system transfers 83% of the incoming energy to DHW. Questions: (a) To the nearest whole hour, what time does the system turn off? (b) Determine the net present value and state whether or not the investment is economically viable. (c) If the complex uses 8000 kWh of electricity per day, what percentage of this total demand do the turbines deliver? (d) Suppose that for 5 months of the year (assume 30 days/month), the system can provide heat for space heating with no additional upfront capital cost, and is therefore able to run 24 h/day. How does the answer in part (b) change?

6-12. An example of a tradeoff in electric generation sources is between low and high capital cost natural-gas-fired power plants. In this problem, you are to supply an electricity market with two types of gas-fired plant technology. The description of the load duration curve for this market is the following: the curve has a peak portion where the highest demand hour per year has a demand of 15 GW and a curve of form $y = -0.003*x + 15$, where x is the hour of the year and y is demand in GW. The curve then has a lower demand portion with a curve of form $y = -0.0004427*x + 8.628$. Hint: in the lowest demand or 8760th hour per year, the demand according to this curve is approximately 4.75 GW. The values for fixed cost as a function of capacity in kW and variable cost per kWh are given in the table below.

    a. Draw the load duration curve for this electricity market. What is the number of the hour in the load duration curve where the "peak" and "lower demand" curves intersect?

    b. What is the combination of low and high capital cost plants that can meet the demand for electricity at minimum cost?

    c. What is the total fixed cost paid for this capacity? Note: you can ignore the variable cost payments for purposes of this calculation.

    d. If chosen plant capacity is discounted at 7% over 25 years, what is the overall levelized cost per kWh in the system?

	Hi-cap	Lo-cap
Fixed [$/kW]	140	120
Variable [$/kWh]	0.0238	0.0336

Carbon Sequestration

7-1   Overview

Carbon sequestration is the key to long-term sustainable use of fossil fuels. Without sequestration, it will not be possible to make significant use of these fuels while at the same time stabilizing the concentration of CO_2 in the atmosphere. In the short term, it is possible to sequester some carbon currently emitted to the atmosphere by augmenting existing natural processes, such as the growth of forests. For the longer term, engineers are currently developing technologies to actively gather CO_2 and sequester it before it is emitted to the atmosphere. Both short- and long-term options are considered in this chapter.

7-2   Introduction

In 1994, worldwide CO_2 emissions, measured in tonnes of carbon equivalent, surpassed the 6 gigatonnes (Gt) per year mark for the first time; by 2012 they had surpassed the 8.8 Gt/year mark. Many countries are reducing their rate of growth of CO_2 emissions per year. Some countries have even reversed their growth and are reducing emissions per year. It is clear, however, that the amount of carbon emitted per year, from use of energy as well as other sources, will remain significant for the foreseeable future. Therefore, there is great interest in finding ways of preventing CO_2 that results from fossil fuel combustion from entering the atmosphere. The term for any technique that achieves this end is *carbon sequestration*, or keeping carbon in a separate reservoir from the atmosphere for geologically significant periods of time (on the order of thousands, hundreds of thousands, or even millions of years).

Sequestration falls into two general categories. *Direct* (or active) sequestration entails the deliberate human-controlled separation of CO_2 from other by-products of combustion and transfer to some nonatmospheric reservoir for permanent or quasi-permanent storage. *Indirect* sequestration, by contrast, does not require human-controlled manipulation of CO_2; instead, natural processes, such as the uptake of CO_2 by living organisms, are fostered so as to accumulate CO_2 at a greater rate than would otherwise have occurred.

Three main methods for sequestration are presented in this chapter, in the following order:

1. *Indirect sequestration:* The primary means of indirect sequestration currently in use is the growth of forests.[1] For many individual and institutional consumers seeking a way to sequester carbon, it is the cheapest option currently available; however, it is also limited by available land area for forest plantations.

---

[1]Grasses and other nonforest plants may also be used for sequestration.

2. *Geological storage of CO_2:* The potential volume of CO_2 that might be stored using this option is much larger than is available from indirect sequestration, and may exceed the volume of all CO_2 that might be generated if all available fossil resources available on Earth were extracted and combusted. One possible disadvantage is that geologic storage leaves the CO_2 underground such that it may remain as CO_2 for very long lengths of time, with the potential for reemerging into the atmosphere at a later date and contributing to climate change for generations in the distant future. This risk is sometimes termed a "carbon legacy" for future generations, since the carbon is not deliberately converted to some other more inert substance (although this may happen due to natural forces during geologic storage). Geologic storage is currently practiced on a small scale in certain locations worldwide, but engineering research and development work remains in order to perfect geologic storage as a reliable and cost-effective means of sequestering carbon on the order of gigatonnes of CO_2 per year.

3. *Conversion of CO_2 into inert materials:* This option is the most permanent, since the CO_2 is not only separated from the stream of by-products but also converted to some other material that is geologically stable. It is also the most expensive due to the additional step of carbon conversion, although the additional cost may be offset if the resulting product can be used as a value-added material, for example, in the construction industry. Conversion of CO_2 is not as technologically developed as geologic sequestration at the present time.

Two further observations are merited. First, there are a great number of alternatives for carbon separation and carbon storage currently undergoing experimental work in laboratories around the globe. It is beyond the scope of this chapter to present each variation in detail, let alone to determine which alternatives are the most likely to emerge as the dominant technologies of the future.[2] The intention of the alternatives presented here is to give the reader a sense of the range of possible technologies that might become available in time. Second, the technologies presented here are applicable only to stationary combustion of fossil fuels, due to the large volume of CO_2 emissions concentrated in a single location. These technologies are not practical for dispersed combustion such as that of transportation, although in the future it may be technologically and economically feasible on a large scale to convert fossil fuels to a currency (electricity or hydrogen) in a fixed location, sequester the resulting CO_2, and distribute the currency to the individual vehicles for consumption in transportation networks.

7-3   Indirect Sequestration

Indirect sequestration is presently carried out primarily using forests, both in tropical and temperate regions of the planet. Experimental work is also underway to use smaller organisms that carry out photosynthesis in order to sequester carbon. Researchers have in the past also considered techniques to accelerate the uptake of CO_2 in oceans by altering the chemical balance, for example, by fertilizing the oceans with iron or nitrogen. These efforts have been largely abandoned out of concern for possible unintended negative side effects, however, and are not considered further here.

---

[2]For an in-depth look at recent developments in carbon separation and storage, the reader is referred to White et al. (2003).

Forest-based sequestration can be carried out either through *afforestation*, or planting of forests in terrain where none previously existed (in recent history), or through *reforestation*, or replanting of forests that have been recently lost. Hereafter both are referred to as forestation. During a forestation project, some additional CO_2 is released from fossil fuels used for transportation and equipment used in planting and maintaining forests, so it is the net absorption of CO_2 that is of interest for sequestration purposes. Once the forest begins to grow, carbon is sequestered as trees grow larger and accumulate mass of woody tissue in their trunks and limbs. When the forest reaches maturity, the total amount of CO_2 sequestered in it reaches its maximum value. Thereafter, the amount of carbon contained in the total number of trees in the forest reaches a steady state in which old trees die and release CO_2 to the atmosphere as they decompose, to be offset by young trees that grow to take their place. Thus it is important to maintain the mature forest in approximately its ultimate size and degree of biological health in order to consider the carbon to be truly sequestered. Some forest managers may also be able to harvest and sell a small fraction of the wood from the forest for other applications, so long as the total mass of carbon in the trees does not change significantly.

Sequestration in forests is taking place at the present time against a backdrop of overall loss of forest coverage worldwide. As shown in Fig. 7-1, total forest cover around the world declined by 0.18% per year for the period 2000 to 2005, or a total loss of nearly 1% for the 5-year period, according to the United Nations Food and Agriculture Organization (FAO). It is estimated that loss of forest cover contributes some 20% to the total net emissions of CO_2 to the atmosphere, as carbon that was previously contained in cut trees is released. On the positive side, the rate of forest loss is declining at present, both due to natural extension of existing forests and active human efforts to increase planting, according to the FAO. Also, the rate of forest loss varies widely by region, with Europe and Asia actually increasing forest cover over the period in question. Individual countries such as the United States and China are experiencing a net increase in total forest cover at present as well. In any case, forest sequestration efforts cannot only fight

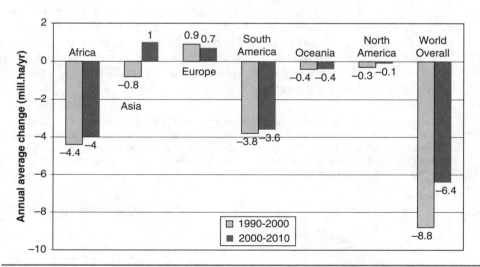

FIGURE 7-1   Average annual net change in forest area, 1990 to 2010. (*Source:* United Nations Food and Agriculture Organization, 2010.)

climate change but also bring other biological benefits, especially in Africa and Latin America where forest loss is currently happening at the fastest rate.

7-3-1 The Photosynthesis Reaction: The Core Process of Indirect Sequestration

All indirect sequestration involving plants depends on the photosynthesis reaction to convert airborne CO_2 along with water into energy that is stored in plant tissue as glucose ($C_6H_{12}O_6$) or other compounds for later use. Oxygen is released from the plant as a by-product. Although a detailed view of photosynthesis entails the division of the reaction inside the plant into several more complex reactions that occur in sequence, the net reaction can be reduced to the following:

$$6\,CO_2 + 6\,H_2O + energy \rightarrow C_6H_{12}O_6 + 6\,O_2 \qquad (7\text{-}1)$$

The energy comes to the organism in the form of sunlight. The amount of energy required for photosynthesis per molecule of CO_2 is 9.47×10^{-19} J. If one compares the energy requirement for the photosynthesis reaction to the total available energy from incident sunlight on the Earth's surface, at first glance it might appear that the sequestration of carbon might take place at a very high rate, perhaps equivalent to the rate at which CO_2 is currently emitted from worldwide fossil fuel combustion. However, a number of factors greatly reduce the rate at which carbon is actually absorbed and retained by forests around the world. First, recall from Chap. 4 that light arrives from the sun in a range of wavelengths, each with different energy values. Plant and tree leaves absorb only light waves in the visible light range (approximately 400 to 700 nm), so that energy in waves outside these wavelengths is not available to the organism. Furthermore, due to the quantum nature of the photosynthesis reaction, the leaf can only make use of energy in the arriving photon up to a maximum value, so additional energy in shorter wavelength (higher energy) photons is also lost. In addition, the coverage of the plant or tree leaf canopy is not uniform across the surface of areas where vegetation is growing, so much of the arriving light energy is either absorbed by the Earth's surface or reflected away. Lastly, a large fraction of carbon converted to glucose in photosynthesis is subsequently consumed by the organism for various cellular functions (growth, repair of cells, and so on) and is therefore not retained permanently in the organism. The effects of these factors on conversion of sunlight to sequestered carbon is illustrated in Example 7-1.

Example 7-1   A given species of tree can sequester 2 kg of CO_2/year·m^2 of planted area, in growing conditions where average sunlight makes available 100 W/m^2, taking into account the night/day and seasonal cycles, as well as variability in the weather. (a) How much energy is required to convert the CO_2 via photosynthesis into carbon that is sequestered? (b) What percentage of annual available energy is used for this conversion?

Solution

(a) First calculate the number of molecules of CO_2 in 2 kg, using *Avogadro's number*, or 6.022×10^{23}:

$$\frac{6.022 \times 10^{23}\ \text{molecules/g·mol}}{44\ \text{g/g·mol}\ CO_2}\left(1000\frac{g}{kg}\right)(2\ kg) = 2.74 \times 10^{25}$$

The energy requirement is then $(2.74 \times 10^{25})(9.47 \times 10^{-19}\ \text{J}) = 2.59 \times 10^7$ J, or 25.9 MJ.

(b) The rate of energy transfer from the sun of 100 W/m² is equivalent to 100 J arriving each second. Therefore, the energy available per year is:

$$100\,J/s \times \left(3600\frac{s}{h}\right)\left(8760\frac{h}{year}\right) = 3.15 \times 10^9\,J$$

The amount used in conversion of carbon of 25.9 MJ is therefore 0.822% of the arriving energy.

Discussion   Note that some additional energy use in photosynthesis will have taken place to convert CO_2 to glucose for later use by the tree. Because of the difficulties in accurately measuring in real time the rate of photosynthesis of trees in their natural habitat, it is difficult to measure exactly what percent of available energy in sunlight is converted to glucose in photosynthesis. Some estimates put the amount at 10 to 11%. Also, because of the relatively small percentage of energy that is actually channeled to sequestration, one fruitful area of research may be the development of trees, plants, and microorganisms that convert CO_2 more rapidly, and retain more of the carbon that they convert.

7-3-2   Indirect Sequestration in Practice

Indirect sequestration is carried out both by national governments seeking to meet carbon emissions reduction objectives, and NGOs that administer forestation projects on behalf of both institutions and individuals. Rates of sequestration may be on the order of 1 to 3 kg CO_2/m² · year during the growth period of the forest, with the highest rates of sequestration occurring in tropical forests.

Sequestration of carbon in trees is, in many instances, a relatively low-cost alternative; costs as low as $1/tonne of CO_2 removed have been reported, although costs in the range of $20 to $30/tonne may be more typical. In addition, this option leads to a number of ancillary benefits to natural ecosystems and to society that enhance its value. In some regions, forests can prevent flooding and reduce topsoil runoff, since they retain rainwater more easily than bare ground. They are also capable of moderating extremes in local climate in the vicinity of the forest. With careful management, forests can provide sustainable harvesting of forest products such as specialized timbers, edible products, or resins and other extracts that can provide a revenue stream from a project without reducing its ability to sequester carbon. Both the planting and management of forests provide employment for local populations, which is especially beneficial in some emerging countries where job opportunities are currently inadequate.

While these advantages hold promise in the short run, comparison of the total land area that might be available for sequestration with the expected rate of total CO_2 emissions over the coming decade shows that this alternative will be limited in its ability to play a major role in solving the carbon emissions problem over the long term. Approximately 4 billion hectares (4×10^{13} m²) or 30% of the Earth's land surface is currently forested. Indirect sequestration in forests might yield 1 to 2 kg of CO_2 sequestered per m²/year, for 20 to 30 years, before forests matured and the intake and release of CO_2 reached a steady state. Even if the countries of the world were able to restore the 2.7 billion hectares that are estimated to have been deforested since the Neolithic revolution—a Herculean task, given the current human population and pattern of the built environment—the total amount sequestered would be on the order of 500 to 1600 Gt. At the 2012 emissions rate of 32.3 Gt CO_2/year (8.8 Gt of carbon equivalent per year), this is equivalent to 16 to 50 years of world CO_2 emissions. Therefore, if widespread use of fossil fuels is to continue for the long term and sequestration is sought as the solution, additional techniques will be necessary.

An interesting innovation that has entered the sequestration market is the selling of forestation "offsets" to individual consumers to counteract their use of electricity, transportation, or other applications that emit CO_2. Sellers in this market are typically not-for-profit organizations that estimate the amount of CO_2 sequestered by their operations that is equivalent to a kilometer of flying or driving, or a kWh of electricity. Consumers then estimate their usage of these services for a period of time (e.g., a vacation, or an entire year) and then pay the organization the appropriate amount that will result in the requisite number of trees planted. These same organizations may also invest in other activities that offset carbon emissions, such as installation of renewable energy systems or upgrading systems so that they are more efficient. Example 7-2 provides an example of this practice.

Example 7-2   A family consumes 400 kWh/month of electricity and wishes to buy sequestration offsets from an organization that plants trees in Africa. The organization charges $19.00 per tree, and estimates that each tree will on average absorb 730 kg CO_2 net over its lifetime after accounting for life cycle emissions associated with transportation, planting activities, and so on. The electricity consumed emits CO_2 at the 2014 U.S. national average rate, based on the energy mix given in the table below, as published by the U.S. Energy Information Administration. The tree planting organization allows consumers to purchase fractions of trees. (a) How much must the family spend to offset emissions for 1 year? (b) Suppose each tree is planted in a stand of trees where it covers a cross-sectional area with 6 m diameter, and takes 20 years to mature. What is the rate of CO_2 sequestration in kg/m² year?
Data on emissions by source:

Source:	Output (bil. kWh)	Share	Emit Rate (gCO₂/kWh)
Coal	1585.7	38.7%	850
Oil	30.5	0.7%	661
Gas	1133.5	27.7%	325
Nuclear	797.1	19.5%	0
Hydro	258.7	6.3%	0
Solar	18.3	0.4%	0
Wind	181.8	4.4%	0
Other RE*	87.3	2.1%	0

*Note: Figure for other renewable energy ("other RE") includes biomass, geothermal, waste to energy conversion, and so on.

Solution

(a) The average emissions per MWh of electricity is the sum of the emissions rates for each source multiplied by the fraction allocated to each source. Only sources with nonzero emissions rates need to be considered. Thus the average is (38.7%)(850) + (0.7%)(661) + (27.7%)(325) = 424.2 gCO₂/kWh, or 0.4242 tonne CO_2 per MWh. The annual consumption of the household is 4800 kWh, or 4.8 MWh. This amount of electricity is equivalent to (4.8)(0.4242) = 2.04 tonnes emitted, or 2.8 trees, which costs approximately $53.

(b) The total area covered by the tree is $\pi(3)^2 = 28.3$ m². The sequestration rate per unit area is therefore

$$730 \, \text{kg} \times \left(\frac{1}{28.3 \, \text{m}^2}\right) \times \left(\frac{1}{20 \, \text{years}}\right) = 1.3$$

or 1.3 kg/m² · year.

Discussion   The cost per tonne of sequestration in this case is $26.03, which is on the high end of projected sequestration cost from forestation. An institutional investor would expect to be able to significantly lower its cost of sequestration by purchasing trees in large volumes.

7-3-3   Future Prospects for Indirect Sequestration

Indirect sequestration is a straightforward and economically viable option for sequestering carbon that governments, institutions, and individuals can use today to offset their emissions. Its future success depends on avoiding pitfalls that may arise, so a sustained commitment to high standards of planting and protecting forests is key. The following issues are relevant:

- *Competence and integrity of planting organizations:* Estimates of carbon sequestered from planting forests assume that stands are planted correctly and managed well. Investigations have already found some examples of sequestration projects where trees were lost or net carbon content was not as high as expected. Third party certification of forest carbon credits can help hold those planting the trees to a high standard.

- *Need to maintain forests in equilibrium over the long term:* Sequestration of carbon in forests comes in the form of a one-time benefit from the initial growth of the forest. Thereafter, as long as the forest remains healthy and the total carbon mass in the trees remains roughly constant, sequestration is taking place. It will clearly be a challenge to maintain this type of very long-term commitment to maintaining forests through whatever political and economic swings the future may hold. However, there are also historical precedents, such as the *bannwald*, or forbidden forests, of the European Alps that for centuries have been maintained to protect communities from landslides and avalanches. It is conceivable that, in the distant future, if both nonfossil sources of energy and other types of carbon sequestration (as discussed in sections below) are flourishing, sequestration forests might gradually be released for other uses. At present, it will benefit sequestration forests if they serve multiple purposes, such as protecting local ecology, preventing erosion, landslides, and avalanches, or other uses.

- *Increasing cost of sequestration over time:* Since the growing of trees anywhere on the planet can sequester carbon, organizations offering this service will tend to choose locations where costs are relatively low in order to maximize CO_2 sequestered per unit of money spent. If the investment in forests for sequestration continues to grow over the coming years and decades, prime locations will be exhausted, and forestation efforts will move to more marginal lands, driving up the cost.

- *Uncertain effect of future climate change on the ability of forests to retain carbon:* As the climate warms, the amount of CO_2 retained by sequestration forests may change, although the direction of the change is not clear at the present time. Some studies suggest that warming soil in these forests will release more CO_2, while other studies suggest that forest undergrowth will expand more rapidly, accelerating the uptake of CO_2. In any case, the actual amount of CO_2 sequestered may be different from the amount predicted at present, and there is a risk that it may fall short of expectations. Further research may allow biologists to better understand this relationship, permitting more accurate evaluation in the future of the lifetime CO_2 sequestered per tree planted.

7-4 Geological Storage of CO$_2$

Geological storage of CO$_2$ (also called *geologic storage* in the literature) as an option for sequestration entails (1) removal of CO$_2$ from the stream of combustion by-products and (2) accumulation in some stable location other than the atmosphere. In geological storage, there is no human-controlled conversion of CO$_2$ to other carbon-containing materials, although CO$_2$ may be transformed into other compounds that are potentially more stable through interaction with other minerals found in the geologic storage reservoir. Geologically sequestered CO$_2$ may also "leak" from the underground reservoir back into the atmosphere, but as long as this leakage rate is kept to a minimal amount, the capacity for reservoirs to sequester large amounts of CO$_2$ may be very large.

7-4-1 Removing CO$_2$ from Waste Stream

Any geological storage scheme begins with the separation of CO$_2$ from the stream of combustion by-products. Separation techniques should strive to remove CO$_2$ under conditions that are as close as possible to the existing flue gas conditions, so as to minimize any energy penalty involved in manipulating or transforming the flue gases. A gas stream that is rich in CO$_2$ facilitates the collection of CO$_2$. The stream from a conventional coal-fired power plant may be on the order of 10 to 15% CO$_2$ by volume—uncombusted nitrogen is the dominant component—so an alternative with higher concentration of CO$_2$ is desirable. Emerging combustion technologies such as the FutureGen concept (see Chap. 6) incur an energy penalty upstream from the combustion process, for example, for oxygen-blown coal gasification, in order to create a stream of by-products that consists largely of CO$_2$ and is therefore easier to separate.

Many CO$_2$ separation processes rely on either *absorption* or *adsorption* to isolate CO$_2$. In absorption, the substance whose removal is desired is imbibed into the mass of the absorbing agent. In adsorption, the substance adheres to the surface of the agent, which can be either solid or liquid.

Figure 7-2 shows a typical CO$_2$ absorption-separation process, in which the flue gas stream is reacted with an absorbing agent in an absorption chamber. The absorbing agent is then transferred to a regeneration chamber, where the CO$_2$ and the agent are separated; the agent returns to the absorption chamber and the CO$_2$ is removed from the cycle for sequestration. The primary energy burden in this cycle is in the regeneration chamber, where a significant amount of energy must be expended to separate the agent and the CO$_2$. Carbon separation processes based on this cycle are currently in use in pilot sequestration projects in Europe and North America, and also for commercial separation of CO$_2$ for sale to industrial customers, such as the carbonated beverage industry. A fossil-fuel-fired electric power plant in Cumberland, Maryland, United States, uses the absorption-regeneration cycle, with Monoethanolamine (MEA) as the absorption agent, to remove CO$_2$ and sell it to regional carbonated beverage bottlers.

The cycle used at the Cumberland plant is a mature technology for separating relatively small volumes of CO$_2$ (compared to the total worldwide emissions from stationary fossil fuel combustion) and selling it as a product with commercial value in the marketplace. However, the energy penalty for this cycle as currently practiced is too large for it to be economically viable for the full inventory of power plants around the world. Experimental work is ongoing to find variations on the cycle using either absorption or adsorption that are effective in removing CO$_2$ but with a lower energy penalty. Another potential approach is the development of a membrane for separating CO$_2$; this

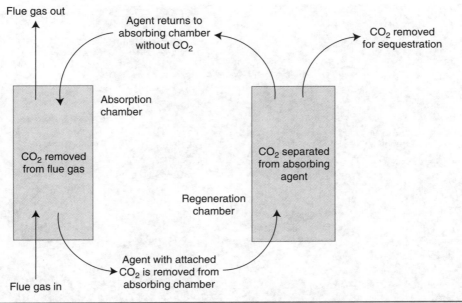

FIGURE 7-2   Absorption-regeneration cycle for removing CO_2 from flue gas stream.

alternative could potentially use less energy and require less maintenance than the absorption-regeneration cycle. Researchers are exploring the use of cycles from the natural world as well. The natural enzyme carbonic anhydrase, which exists in both plants and animals as an absorber of CO_2, might be synthesized and used in an engineered CO_2 absorption facility. Alternatively, photosynthesis might be exploited; simple organisms such as algae might be maintained in a facility with controlled conditions where they would be situated in a stream of by-product CO_2 and absorb it as it passed.

7-4-2   Options for Direct Sequestration in Geologically Stable Reservoirs

Geological storage of CO_2 implies accumulation in some type of *reservoir*. The term "reservoir" in this case includes any geological formation underground or under the sea floor where CO_2, once injected, might remain for significant lengths of time. It also includes the bed of accumulated carbonaceous material at the bottom of the oceans, which, though not contained by rock strata, are unable to escape to the atmosphere due to the high pressure of ocean water at these depths.

Injection into Underground Fossil Fuel Deposits, with or without Methane Recovery

Among various reservoir options, CO_2 can be injected into partially depleted oil or gas deposits, thus occupying space made available by oil and gas extraction (see Fig. 7-3). For some time the petroleum industry has practiced *enhanced oil recovery* (EOR), in which CO_2 is pumped to an oil field and then injected into the deposit, increasing the output of crude oil. This technique can also be applied to natural gas fields. Although the invention of the technique predates interest in carbon sequestration, its use may be expanded in the future as requirements for carbon sequestration grow. In order for this

FIGURE 7-3   Great Plains Synfuels Plant, Beulah, North Dakota, United States.
By-product CO_2 from the plant, which converts coal into methane for distribution in the national gas grid, is transported by pipeline to Canada for sequestration and enhanced oil recovery in Canada. (*Source:* Basin Electric Power Cooperative. Reprinted with permission.)

technique to make a useful contribution to reducing CO_2 in the atmosphere, care must be taken that the amount of CO_2 sequestered equals or exceeds the amount of CO_2 released from combustion of the oil or gas that results from the CO_2 injection process.

Enhanced coal bed methane extraction (ECBM) is a related technique in which CO_2 is injected into unmineable coal seams, driving out methane that is adsorbed to the coal. Figure 7-4 illustrates this process, in which the CO_2 injection and CH_4 extraction are in geographically separate locations, and water is removed at the CH_4 wellhead, leaving a pure gas stream that can be transferred to the long-distance distribution grid. The commercial value of the gas in the energy marketplace thus helps to finance the function of sequestering CO_2.

Table 7-1 gives both sequestration and gas recovery for select major coal fields from around the world that are estimated to have the highest potential for ECBM. Note that the total sequestration potential of approximately 4.4 Gt is small relative to the projected eventual CO_2 sequestration requirement; an additional 5 to 10 Gt of sequestration might be available in smaller fields, but the overall amount sequestered using ECBM will likely be small relative to other options that may eventually be used. However, this option is also economically attractive, so it would be sensible to exploit it in an initial round of geological sequestration efforts before moving on to other options.

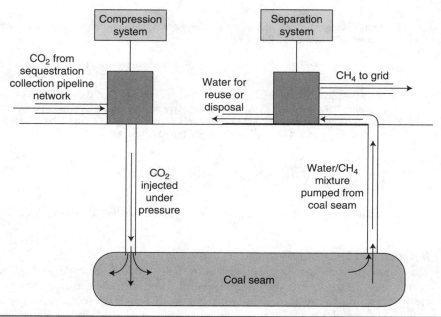

Figure 7-4   System for methane production from coal seam using CO_2 injection.

Example 7-3 illustrates sequestration potential and value of gas output from an ECBM operation.

Example 7-3   Suppose the CO_2 output from the plant in Example 6-9 in the previous chapter is piped to the Sumatra coal field in Indonesia (see Table 7-1), where it is 100% injected into the field for sequestration and to generate methane. Assume that natural gas production without CO_2 injection has been exhausted prior to the project, and that methane is worth $10.50/GJ. (a) For how long will the coal field be able to sequester the CO_2? (b) What is the annual value of the methane production?

Name of Region	Country	CO_2 Sequestration Potential (Gt)	CH_4 Recovery (bil. m³)	Ratio (m³/tonne CO_2)
San Juan	USA	1.4	368	263.0
Kuznetsk	Russia	1	283	283.2
Bowen	Australia	0.87	235	270.2
Ordos	China	0.66	181	274.6
Sumatra	Indonesia	0.37	99	267.9
Cambay	India	0.07	20	283.2

Note: Projected methane recovery shown is the amount recovered after methane without ECBM is complete.
Source: White et al. (2003).

Table 7-1   Representative World Coal Fields with ECBM Potential

Solution

(a) From Example 6-9, the annual coal consumption is 2.26 million tonnes/year. With a carbon content of 85%, the resulting CO_2 is

$$(2.26 \times 10^6)(0.85)(44/120) = 7.04 \times 10^6 \text{ tonnes } CO_2$$

Since the total capacity of the field is 370 million tonnes CO_2, there are 52.5 years of capacity available. This is similar to the lifetime of a typical coal-fired plant built today.

(b) The conversion rate of the field is 267.9 m³ CH_4 per tonne CO_2, and from Chap. 5, the energy content of gas is 36.4 MJ/m³. Accordingly, the gas output is

$$(7.04 \times 10^6)(267.9)(36.4)\left(\frac{1\,GJ}{1000\,MJ}\right) = 6.87 \times 10^7 \text{ GJ}$$

At \$10.50/GJ, the value of the gas output is approximately \$721 million/year.

Injection into Saline Aquifers

This alternative for geologic sequestration takes advantage of large bodies of saline water that are present under many parts of the Earth. Due to its salinity, this water resource is not of interest for other purposes such as irrigation or drinking water. Many existing power plants are located near saline aquifers; in the United States, for example, the majority of power plants are located on top of saline aquifers, so the requirement for long-distance overland transmission of CO_2 to an aquifer would be minimized.

The available capacity for sequestration from saline aquifers is much larger than that of ECBM, and may be 10,000 Gt CO_2 or more, which is similar in magnitude to the total emissions of CO_2 that would be produced from combusting all known fossil resources. This technique is already in use today, albeit on a relatively small scale. Since 1996, the Norwegian company Statoil has been sequestering 1.1 million tonnes of CO_2/year, which is removed from natural gas as it is extracted from the Sleipner field in the North Sea (Fig. 7-5), in a saline aquifer below the sea floor. The CO_2 must be removed in order for the gas product to meet maximum CO_2 requirements for the gas market, and by sequestering the CO_2, Statoil is able to avoid the Norwegian carbon tax. The CO_2 removal technique used at this facility is a variation on the absorption-regeneration cycle shown in Fig. 7-2. Since 2005, development of carbon capture and sequestration (CCS) has moved from energy extraction platforms such as the example of StatOil to application at coal-combusting power plants. Table 7-2 gives representative examples from several countries. The typical development strategy is to start with small-scale pilot project development, with the goal of scaling technology up to larger volumes of CCS once it has been proven at the pilot level. Countries represented include Canada, China, Germany, Italy, and Poland. Note that in the short- to medium term with the exception of Boundary Dam in Canada the technologies in use aim to capture a fraction of the total CO_2 produced from coal combustion, but not yet capture 100% of it. (The Boundary Dam plant consists of several generating units, of which one unit has all CO_2 diverted to sequestration.). This choice reflects the incremental nature of carbon reduction targets in many countries such as those of the European Union: since CCS rates at or close to 100% are technologically difficult and expensive, it is simpler to achieve partial CCS in the interim while developing 100%-effective technologies for the longer term.

Figure 7-5   Sleipner gas platform sequestration process. (*Source:* StatoilHydro. Reprinted with permission.)

As shown in Fig. 7-6, the CO_2 that is sequestered in an aquifer either remains trapped as a gas pocket, or dissolves into the saline fluid. In either case, long-term entrapment of the CO_2 depends on a nonpermeable caprock layer to prevent the CO_2 from returning to the surface. Fluid in a saline aquifer moves at a very slow pace (1 to 10 cm/year) in response to a hydrodynamic gradient that is present, and the sequestered CO_2 that dissolves in the fluid will move with it, away from the injection well. Depending on the extent of the caprock, however, it may take on the order of tens of thousands to millions of years for it to move to a location where it could return to the surface. Some CO_2 may also react with minerals present in the aquifer, resulting in the precipitation of carbonate molecules out of the solution, including calcite ($CaCO_3$), siderite ($FeCO_3$), and magnesite ($MgCO_3$). This type of precipitation reaction improves the overall reliability of the sequestration operation, as the resulting minerals are very stable.

Plant Name	Country	Sequestration Rate 1000 tonnes/yr	Year Commissioned
Brindisi Pilot project	Italy	8	2011
Porto Tolle	Italy	< 1000	2015
Belchatow	Poland	100	2015
Vattenfall	Germany	< 600	2014
Luzhou*	China	60	2011
Boundary Dam, Sask	Canada	1000	2014

*Carbon captured at Luzhou plant but used in industrial process rather than being sequestered.

Table 7-2   Early Examples of Carbon Capture and Sequestration (CCS) Projects in Select Countries

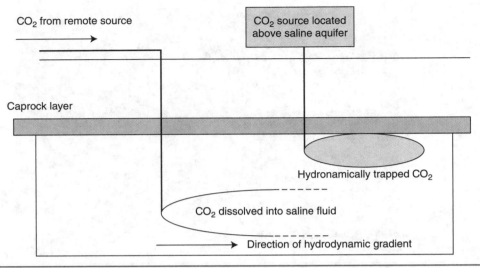

FIGURE 7-6 Two options for injection of CO_2 into saline aquifer.

On the left, the CO_2 arrives from a source located at a distance, and dissolves into the surrounding saline fluid. On the right, the CO_2 originates from a source directly atop the aquifer, and remains as a separate trapped pocket of gas, which is held in place by the pressure of the aquifer against the nonpermeable caprock. A hydrodynamic gradient in the aquifer gradually moves the CO_2 away from the well shaft.

The capacity of a reservoir to sequester CO_2 at subcritical pressures (less than approximately 7.4 MPa) can be expressed as a first-order approximation as a function of the porosity f and the water saturation s_w, or fraction of reservoir volume filled with water, as follows. Let V_r be the total volume of the reservoir and V_g be the volume displaced by the gas sequestered, both in units of cubic meter. The volume of gas in the reservoir is

$$V_g = \phi V_r (1 - s_w) \tag{7-2}$$

Recall from the ideal gas law that the number of moles n of a gas in a volume V_g is quantified $n = (V_g P)/(RT)$, where P is the pressure in cubic meter, T is the temperature in degrees kelvin, and $R = 8.3145$ J/mol·K. Using the ideal gas law to substitute in Eq. (7-2) gives

$$n = \frac{\phi V_r (1 - s_w) P}{RT} \tag{7-3}$$

The intuition behind Eq. (7-3) is that water saturation makes some fraction of the reservoir volume unavailable for sequestration. Of the remaining volume, the fractional value of the porosity is available for sequestration of the injected gas. In all such calculations, the reader should be aware that, while V_r is presented as an exactly known figure, the exact volume of an aquifer is subterranean and therefore subject to some imprecision when it is estimated with underground sensing techniques. This approach is also not appropriate for injection of CO_2 to depths below approximately 800 m, where the CO_2 becomes a supercritical fluid and the maximum volume that can be sequestered is greatly increased. Lastly, the approach does not consider the effect of carbonate precipitation on the total volume that can be sequestered.

The full cost of developing CO_2 extraction and injection into saline aquifers is not known at present, so its effect on the life-cycle cost of fossil fuel electric production is not clear. Given the size and complexity of the sequestration system, its cost will not be negligible; however, it may be low enough to keep fossil fuel electricity competitive with other alternatives. The sequestration literature suggests that an additional $0.01/kWh may be a feasible target.

Example 7-4 tests for adequacy of a saline aquifer for sequestration over the lifetime of a power plant operation, and illustrates the effect of sequestration technology cost on life-cycle electricity costs.

Example 7-4   For the supercritical plant in Example 6-8, suppose that the CO_2 from 15 such plants is piped to a saline aquifer where it is injected. The aquifer is approximately 1000 km long and 500 km wide, with an average depth of 1000 m. All of the aquifer's volume is available for dissolution of CO_2, and the porosity is 10%. The pressure in the aquifer is 500 kPa on average, and the temperature is 325 K. (a) If the plants function for 75 years, is there sufficient capacity in the aquifer to sequester all the CO_2 generated? (b) If the total cost per tonne CO_2 sequestered, including repayment of capital cost and all operating costs, is $25/tonne, by what percent does the cost of electricity increase?

Solution

(a) Since the full volume of the reservoir is available, there is no part that is saturated with water, and $s_w = 0$. The volume of the reservoir is 5×10^{14} m^3. Therefore, using the pressure and temperature values given, the maximum number of moles is

$$n = \frac{(5 \times 10^5 \text{ N/m}^2)(0.10)(5 \times 10^{14} \text{ m}^3)}{8.314(325)} = 9.25 \times 10^{15} \text{ mol}$$

At 44 g/g·mol for CO_2, this converts to 4.07×10^{11} tonnes. From Example 6-8, the CO_2 generated per plant is 1.47×10^6 tonnes/year. The amount of CO_2 generated by 15 plants over 75 years is 1.66×10^9 tonnes. Therefore, there is adequate capacity in the aquifer.

(b) If the cost per tonne is $25.00, then the cost per year is $36.9 million. Spreading this cost among 3.07 billion kWh generated per year gives $0.0120/kWh. Since the original cost was $0.0353/kWh, the increase is $(0.0120)/(0.0353) = 34\%$.

Other Geological Options

Producers of CO_2 from fossil fuel combustion may also be able to sequester it on the ocean floor, in such a way that the additional CO_2 cannot return to the surface and does not affect the chemical balance of the ocean at large. In one scenario under development, CO_2 is transported by pipeline to a depth of 1000 to 2000 m below the ocean surface, where it is injected at pressure into the surrounding water. From there, the CO_2 falls until it reaches the layer of carbonaceous material at the bottom of the oceans sometimes referred to as the "carbon lake," which may contain on the order of 30,000 to 40,000 Gt of carbon. Other options for transporting CO_2 to the ocean bottom are under consideration as well.

Finding Suitable Locations for Underground Sequestration Reservoirs

Finding a suitable location for a sequestration reservoir depends first on using available information about underground geological strata. Both saline aquifers and fossil fuel deposits have been studied extensively as part of ongoing geologic activities and, in the latter case, as part of efforts to estimate the value of fossil fuel resources awaiting extraction. Based on this information, the analyst then estimates the capacity of the reservoir to sequester CO_2, since this quantity will determine whether the reservoir is sufficiently large to be an attractive site, and if so, to which CO_2 sources (e.g., fossil-fuel-powered electric plants) it should be connected. The site must also be carefully surveyed for indications of possible leakage points, such as formations above the reservoir that may allow CO_2 to pass, or passages left by previous fossil fuel extraction activities.

7-4-3   Prospects for Geological Sequestration

Sequestration research up to the present time indicates that between subterranean and sea-bottom options, there is adequate *capacity* for sequestration of all foreseeable CO_2 emissions from fossil fuel production, provided that we invest sufficiently in facilities to separate and sequester it. The *risk* associated with sequestration, both from leakage under ordinary conditions and from seismic activity, is a more complex issue, and is considered in greater detail here.

Risk of leakage can be divided into acute leakage, where a hazardous amount of CO_2 is released suddenly, and chronic leakage, where continuous low-level leakage leads to detrimental effects over the course of time. The negative effects of chronic leakage include possible detrimental effects on human health and on atmospheric CO_2 levels. Carbon sequestration has received a negative image in some quarters because of the Lake Nyos disaster that occurred in Cameroon in August 1986, in which an underground cloud of CO_2 erupted to the surface suddenly and killed 1700 people. This event was in all likelihood unique, because the underground geologic formation first allowed a large amount of CO_2 to accumulate at high pressure, and then allowed the CO_2 to come to the surface all at once (a system for venting CO_2 so that it cannot build up in lethal concentrations has since been installed at this lake). For geological sequestration of CO_2, engineers will have the option of choosing aquifers or reservoirs with more favorable formations for keeping CO_2 underground, and will also be able to limit the maximum concentration of CO_2 in any one location in the reservoir. Therefore, conditions that might lead to acute leakage can be avoided.

Risks from chronic leakage are a greater concern. Leaking CO_2 may contaminate groundwater or leak into homes; it may also leak back into the atmosphere at a high enough rate to undercut the objective of stabilizing atmospheric CO_2 concentration. CO_2 sequestered in an underground reservoir might, in conjunction with surrounding materials, react with the caprock and increase its permeability, leading to higher than expected rates of leakage. A sequestration reservoir may leak a small amount of CO_2 per unit of time without compromising climate change objectives. However, some research suggests that a maximum of 0.01% of injected CO_2 per year should be the limit for this leakage.

Other risks must also be considered. Previous experience with the underground injection of hazardous waste has shown that these practices carried out in certain locations can increase the frequency of earthquakes. The same principle may apply to CO_2. Along with negative effects on the built environment, seismic activity might accelerate the leakage of CO_2. There are also risks inherent in the transmission of CO_2 away from the combustion site. Because of the large volumes of gas involved, it is not practical to construct intermediate reservoirs for temporary holding of CO_2; instead, the gas must go directly to the final point of injection. In case of disruption to the transmission system, operators would need to build in redundancy or else have in place a permit system for the temporary venting of CO_2 to the atmosphere.

Thus far, experience with initial sequestration projects such as the Sleipner field have been positive, which provides some indication that the risks outlined above can be managed on the path toward creating a successful geological sequestration system. However, geological sequestration is not yet in widespread use, and CO_2 may behave in unexpected ways when injected into different reservoirs around the world. The research community should continue experimental work, as well as careful observation of new sequestration operations as their numbers grow, to ascertain that there are no unforeseen adverse effects from injection of CO_2 into various types of reservoirs. Historically, scientists have accumulated only limited previous experience with CO_2 injection, and conditions in a reservoir are difficult to reproduce in the laboratory or using computer simulation. To address this issue, existing CO_2 deposits, for example, in the states of New Mexico and Colorado in the United States, are being used to calibrate computer models of CO_2 in underground formations. Once these models can success-fully reproduce conditions in real-world reservoirs, they can be used as test beds for modeling the effect of sequestration procedures.

7-5   Sequestration through Conversion of CO_2 into Inert Materials

In this approach to sequestration, captured CO_2 is reacted in engineered facilities with mineral such as calcium or magnesium to form carbonate molecules that are stable over long periods of time and can therefore sequester carbon. Carbonate forming reactions occur naturally, and they continually precipitate out some amount of carbon from the oceans and from subterranean reservoirs, although the amount is small relative to the large volumes of net carbon emissions each year to the atmosphere from anthropogenic sources. As discussed above, these compounds can precipitate as a by-product of geo-logical sequestration of CO_2 in saline aquifers, where the necessary minerals are present. In this section, we consider actively transforming carbon into carbonate materials in an industrial process, at rates sufficient to keep pace with CO_2 production from energy generation.

One favorable characteristic of this process is that the reactions are exothermic, and therefore can occur without a large energy input during the transformation of CO_2 to carbonate materials. Taking the case of magnesium oxide and calcium oxide as examples, we have

$$CaO + CO_2 \rightarrow CaCO_3 + 179\,kJ/mol \tag{7-4}$$

$$MgO + CO_2 \rightarrow MgCO_3 + 118\,kJ/mol \tag{7-5}$$

However, at ambient conditions the reaction occurs at too slow a pace to be practi-cal, so the temperature or pressure must be elevated to accelerate the transformation of CO_2, incurring an energy penalty. Even after expending energy to accelerate the rate of carbonate production, the overall balance is favorable when one takes into account the energy initially released from combusting fossil fuels. Taking coal as an example, the reaction is:

$$C + O_2 \rightarrow CO_2 + 394\,kJ/mol \tag{7-6}$$

Thus for a coal-based energy life cycle in which coal is first combusted and then the resulting CO_2 is reacted with CaO to form $CaCO_3$, there is a net gain of 573 kJ/mol, from which energy for the sequestration operation including transmission of CO_2 and heating of carbonate materials can be drawn, and the remainder used for energy end uses.

In the natural world, silicate materials that contain magnesium and calcium are much more common than pure calcium or magnesium oxides, and the energy penalty incurred in preparing these oxides for use in sequestration may be excessive. Instead, a more commonly found mineral might be used, even if it has a lower energy yield per unit of CO_2 reacted. Taking the case of forsterite (Mg_2SiO_4), at 515 K we have

$$\frac{1}{2}Mg_2SiO_4 + CO_2 \rightarrow MgCO_3 + \frac{1}{2}SiO_2 + 88\ kJ/mol \tag{7-7}$$

In addition, the reaction requires the supply of 24 kJ/mol of thermal energy to heat the forsterite from ambient temperature to the temperature required for the reaction.[3] Similarly, the energy released by reacting CaO and MgO at high temperatures needed for a more rapid reaction is reduced:

$$(\text{At } 1161\ K)\ CaO + CO_2 \rightarrow CaCO_3 + 167\ kJ/mol \tag{7-8}$$

$$(\text{At } 680\ K)\ MgO + CO_2 \rightarrow MgCO_3 + 115\ kJ/mol \tag{7-9}$$

The energy required to heat CaO and MgO from ambient conditions to the required temperature is 87 kJ/mol and 34 kJ/mol, respectively.

Example 7-5 illustrates the material requirements and energy availability for a sequestration process using this material.

Example 7-5   Suppose that the CO_2 output from one of the plants in Example 7-4 is sequestered by reaction with MgO to form $MgCO_3$, rather than injection into an aquifer. In order to sequester CO_2 at a sufficient rate, the reactants must be heated to 680 K, consuming 34 kJ/mol CO_2; the reaction itself releases 115 kJ/mol CO_2. (a) What is the mass of MgO that is required each year to sequester the CO_2 stream? (b) What is the mass of $MgCO_3$ produced? (c) Ignoring any losses in the process, what is the net energy flux during $MgCO_3$ formation, in MW?

Solution

(a) The atomic masses, rounded to the nearest whole number, of magnesium, oxygen, and carbon are 24, 16, and 12, respectively. Therefore, the mass of MgO is 40. Since the amount of CO_2 sequestered is 1.47×10^6 tonnes/year, the amount of MgO is $(40/44)(1.47 \times 10^6$ tonnes/year) $= 1.34 \times 10^6$ tonnes/year.

(b) The mass of $MgCO_3$ is 84, so the amount of $MgCO_3$ is $(84/44)(1.47 \times 10^6$ tonnes/year) $= 2.81 \times 10^6$ tonnes/year.

(c) The net energy released per mole is $115 - 34 = 81$ kJ/mol. The number of moles of CO_2 per year is

$$\frac{1.47 \times 10^6\ \text{tonne}}{44\ \text{g/mol}}\left(\frac{10^6\ \text{g}}{\text{tonne}}\right) = 3.34 \times 10^{10}$$

Therefore, the net energy released per year is $(3.34 \times 10^{10})(81) = 2.71 \times 10^{12}$ kJ. Dividing by the number of seconds per year, or 3.1536×10^7 s, gives 8.58×10^4 kW or 85.8 MW.

---

[3]By convention, the input of 24 kJ/mol does not appear on the left-hand side of the reaction in Eq. (7-7), although this input is implied since it is a requirement for achieving the net release of 88 kJ/mol shown in the reaction.

An engineered system for converting CO_2 to inert carbonates would include a CO_2 delivery network, an extraction operation to mine the required raw material, and a system for accumulating the carbonate material on site, or possibly a supply chain to deliver the material to the marketplace as a raw material. Assuming the material produced in the reaction is stored in place, the operator would likely minimize cost by siting the sequestration facility at the location of the required mineral deposit and piping the CO_2 to the facility, rather than moving the mineral to the CO_2. Large-scale removal of minerals from under the Earth raises issues of scarring the natural landscape during extraction. However, given that the total mineral resource available for CO_2 conversion is many times larger than the amount needed to sequester the remaining fossil fuel reserve, it is likely that, as long as the conversion process itself is economically viable, the necessary facilities can be cited in locations that are not sensitive to either natural habitat disruption or detrimental effect on other human endeavors.

7-6   Direct Removal of CO_2 from Atmosphere for Sequestration

Current sequestration R&D activities for the most part presuppose that CO_2 must be prevented from entering the atmosphere if at all possible, since it is not feasible to remove it in large quantities once it has accumulated there. (As discussed in Sec. 7-3, some CO_2 can be removed indirectly through the forestation process, but the total potential for this option is likely to be small relative to the amount of CO_2 that might accumulate in the atmosphere over the next 50 to 100 years.) If, however, a reliable and economical device could be developed to capture CO_2 from the air for sequestration in reservoirs or carbonate materials, new possibilities for offsetting current carbon emissions and even removing past buildup of CO_2 would emerge. This technology would of course not only need to operate in a cost-effective way compared to other options for mitigating climate change, but also incur only a limited CO_2 emission penalty, that is, for each unit of CO_2 generated in the life-cycle energy used in the capture process, many units of CO_2 should be captured.

One approach to designing such a device is to locate it in terrain with high average wind speeds, where CO_2 can be extracted from the air as it passes through the facility without requiring the mechanical suction of the air, thus reducing the energy consumption requirement of the facility. Long-term sequestration options should also be located nearby, if possible, so as to minimize CO_2 transmission cost. One vision of the CO_2 capture device is a tunnel with a lining containing some adsorbing material. As the air passed through the tunnel, the CO_2 would adsorb to the lining, to be removed by some secondary process and transferred to the sequestration site. A CO_2 capture facility would consist of a field of such tunnels, perhaps colocated with a field of wind turbines to provide carbon-free energy for the operation, given that the site is chosen to have high wind speeds. Alternatively, the energy might be taken from the grid, with CO_2 from the required energy sequestered elsewhere.

Assuming the capture technology can be successfully developed, a comprehensive system might be built up around the sequestration or possibly recycling of captured CO_2, as shown in Fig. 7-7. In this case, the assumed energy source for carbon capture is a fossil-fuel-powered energy plant which delivers both energy and CO_2 to the capture and sequestration site. The sequestration facility then injects the CO_2 into an underground reservoir. Alternatively, the energy generation facility might emit the CO_2 to the atmosphere, to be offset by the capture facility. Also, the sequestration facility

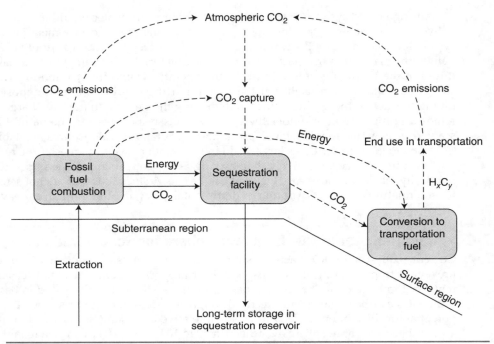

FIGURE 7-7 CO_2 cycle with combustion, capture, and sequestration.
Solid lines show essential parts of cycle, including the use of fossil fuel energy to capture and sequester CO_2, with CO_2 from combustion sent directly to sequestration facility without release to atmosphere. Dashed line shows optional components, including release and recapture of CO_2 to and from atmosphere, and conversion of captured CO_2 to hydrocarbon fuel for transportation applications.

might transmit some fraction of the CO_2 to a transportation energy conversion plant, which would transform the CO_2 into a synthetic transportation fuel H_xC_y (i.e., the hydrocarbon combustion reaction in reverse), with the carbon emitted to the atmosphere during transportation end use and returned to the sequestration facility by the capture device.

7-7   Overall Comparison of Sequestration Options

Sequestration options are presented in this chapter in increasing order of complexity, but also in increasing order of potential for reliably and permanently sequestering carbon in unlimited quantities. Geological sequestration may provide the optimal balance of high potential volume of sequestration and manageable life-cycle cost, and at the time of this writing, it is also the option receiving the greatest research effort. All discussion in the next several paragraphs assumes that our R&D efforts can achieve mature sequestration technologies that function reliably and that meet required cost-per-tonne CO_2 targets.

Indirect sequestration is limited by the amount of Earth surface area available for planting of trees or possibly exploitation of smaller organisms that perform photosynthesis. It may be possible to increase the yield of CO_2 sequestered per square meter of area by using genetic engineering or creating some process to assist with the conversion of plant or tree activity into sequestered carbon, but the potential is limited. Nevertheless, in the short run it should be exploited vigorously, not only because of low cost but also because of ancillary benefits from forests.

Some questions remain at the present time about the permanence of geological CO_2 sequestration, including the possibility of leaks from some reservoirs if the practice is used on a large scale, and the concern over contamination of drinking water or increases in seismic activity. If these issues are resolved favorably, then there may not be a need to pursue CO_2 conversion. On the other hand, the issue of the "carbon legacy" lasting for thousands or even millions of years is difficult to rule out with absolute certainty. If CO_2 conversion is sufficiently cost-competitive, it may be desirable to prioritize it in order to completely eliminate any concern about negative consequences of geological sequestration that would be impossible to detect until long after human society had committed itself to the latter.

Energy penalties are important for all sequestration options. Due to high energy content in fossil fuels per unit of carbon combusted, it is often possible to sustain a significant energy penalty during the sequestration life cycle and still create an energy system that delivers electricity to the consumer at a competitive cost per kWh with no net CO_2 emissions to the atmosphere. If, however, energy penalties are excessive, too much of the energy available in the fossil fuel is eaten up in the sequestration process, and fossil fuels become uncompetitive with nuclear or renewable alternatives.

The role of nonfossil resources in carbon sequestration raises interesting possibilities. In some cases it may be possible to use nonfossil energy sources to power any component of the sequestration process (air capture, separation, transmission, or injection of CO_2, as well as recycling of CO_2 into hydrocarbon fuels for transportation). On the other hand, depending on the efficiency and cost of the processes involved, it may be preferable to use the nonfossil resources directly in the end-use applications, rather than directing them toward the carbon-free use of fossil fuels. Also, per the preceding discussion, CO_2 must be sequestered in real time as it is generated, while many renewable options function intermittently. Therefore, use of these intermittent resources for carbon sequestration would require backup from some dispatchable resource. We pose these questions about fossil versus nonfossil energy resources here without attempting to answer them; a quantitative exploration is beyond the scope of the book.

In all consideration of options, the "end game" for carbon sequestration should be kept in mind. In the future, the maximum concentration of CO_2 in the atmosphere, and the length of time over which that maximum exists, will determine the extent and effect of climate change, and not the absolute volume of CO_2 that is emitted over the remaining lifetime of all fossil fuels on Earth. Therefore, imperfect or "leaky" sequestration options may still be useful if they can help to lower the peak over the coming decades and centuries.

Lastly, carbon sequestration, and especially geological sequestration, may be applicable first in the emerging economies, where utility operators are rapidly adding new power generating facilities to keep up with burgeoning demand. Each new power plant represents an opportunity to launch the separation and sequestration of CO_2 without needing to retrofit plants that are currently operating, especially in the industrialized countries, at large cost and part way through their life cycle. These sequestration projects might provide a very good opportunity for the clean development mechanism (CDM) under UNFCCC agreements going forward in the post-Kyoto era., in which industrialized countries meet their carbon reduction obligations by financing the sequestration component of the new plants. Timing is important—for geological sequestration to be useful for the CDM, our R&D efforts must perfect the technology and make it available commercially before the current phase of power plant expansion is complete. After this time, carbon sequestration will require retrofitting of plants, which would greatly increase the cost per tonne of CO_2 sequestered.

7-8   Summary

Carbon sequestration, as overviewed in this chapter, is the key technology for making fossil fuel use sustainable over the long term; without some means of preventing CO_2 from reaching the atmosphere, an increase in its concentration is inevitable if we are to continue to use fossil fuels at current rates. Options for carbon sequestration are evolving and will change over time. In the short run, the planting of forests is a viable alternative, along with sequestration in underground reservoirs on a demonstration basis in order to sequester some CO_2 while learning more about its long-term behavior in these reservoirs. Enhanced recovery options, such as enhanced oil recovery (EOR) or enhanced coal basin methane recovery (ECBM), provide a means of sequestering CO_2 while generating a revenue stream from recovered oil or gas that is otherwise unrecoverable. Gradually, as opportunities for forestation or enhanced recovery are exhausted, geological sequestration in saline aquifers or in the carbon pool at the bottom of the ocean may expand in order to increase the volume of CO_2 sequestered, provided earlier results are satisfactory in terms of minimizing leakage and preventing negative side effects. Geological sequestration may eventually sequester all CO_2 generated from the combustion of remaining fossil resources. Further in the future, conversion of CO_2 to inert materials or air capture of CO_2 to reduce atmospheric concentration may emerge, although these techniques are still in the laboratory. For all sequestration options, we have an imperfect ability to measure how much CO_2 will be sequestered and its long-term fate once sequestration has started, so further research is necessary to clarify the extent to which sequestration can safely and effectively be used to prevent climate change.

References

United Nations Food and Agriculture Organization (2010). State of the World's Forests. Report, UNFAO, Rome.

White, C., B. Strazisar, E. Granite, et al. (2003). "Separation and Capture of CO_2 from Large Stationary Sources and Sequestration in Geological Formations—Coalbeds and Deep Saline Aquifers." *Journal of the Air & Waste Management Association*, Vol. 53, pp. 645–713.

Further Reading

Bruant, R., A. Guswa, M. Celia, et al. (2002). "Safe Storage of CO_2 in Deep Saline Aquifers." *Environmental Science and Technology*, Vol. 36, No. 11, pp. 240–245.

Fanchi, J. (2004). *Energy: Technology and Directions for the Future*. Elsevier, Amsterdam.

Herzog, H., B. Eliasson, and O. Kaarstad (2000). "Capturing Greenhouse Gases." *Scientific American*. Feb. 2000 issue, pp. 72–79.

Holloway, S. (1997). "An Overview of the Underground Disposal of Carbon Dioxide." *Energy Conversion and Management*, Vol. 38, No. 2, pp. 193–198.

Lackner, K., C. Wendt, D. Butt, et al. (1995). "Carbon Dioxide Disposal in Carbonate Materials." *Energy*, Vol. 20, No. 11, pp. 1153–1170.

Lackner, K., and H. Ziock (2001). "The US Zero Emission Coal Alliance Technology." *VGB PowerTech*, Vol. 12, pp. 57–61.

Lackner, K. (2002). "Can Fossil Carbon Fuel the 21st century?" *International Geology Review*, Vol. 44, No. 12, pp. 1122–1133.

Lawrence Berkeley National Laboratory, Lawrence Livermore National Laboratory, Oak Ridge National Laboratory, Stanford University, University of Texas Bureau of Economic Geology, Alberta Research Council (2004). *GEO-SEQ Best Practices Manual—Geologic Carbon Dioxide Sequestration: Site Evaluation to Implementation.* Technical report LBNL-56623 Earth Sciences Division, Ernest Orlando Lawrence Berkeley National Laboratory. Web resource, available at http://www.osti.gov/bridge/servlets/purl/842996-Bodt8y/native/842996.PDF. Accessed Nov. 5, 2007.

New York State Energy Research and Development Authority (NYSERDA) (2003). *Patterns and Trends: New York State Energy Profiles 1989–2003.* Annual report, available on website at http://www.nyserda.org/publications/trends.pdf. Accessed Aug. 10, 2005.

Parsons, E., and D. Keith (1998). "Fossil Fuels without CO_2 Emissions." *Science*, Vol. 282, No. 5391, pp. 1053–1054.

Wigley, T., R. Richels, and J. Edmonds (1996). "Economic and Environmental Choices in Stabilization of Atmospheric CO_2 Concentrations." *Nature*, Vol. 379, No. 6562, pp. 240–243.

Winters, J. (2003). "Carbon Underground." *Mechanical Engineering*, Feb. 2003 issue, pp. 46–48.

Exercises

7-1. For a country of your choice, create a rudimentary sequestration program for CO_2 emissions from fossil-fuel-powered electricity generation. Assume that geological sequestration technology is mature enough for widespread use, and that options for forest sequestration are also available. What is the total annual inventory of CO_2 by fossil fuel type? How is this production of CO_2 to be allocated among available sequestration options? Document the availability of the chosen carbon sinks using information from the literature or Internet. Is the carbon to be sequestered domestically within the country, or will some of the sequestration requirement be met outside the borders?

7-2. China has 9.6 million km² of area. Suppose the government embarks on a comprehensive sequestration program in which 5% of this area is targeted for reforestation with a species of tree that can absorb 1.4 kg carbon/m²·year net for 20 years, after accounting for carbon emissions associated with executing the project, for example, transporting tree seedlings, equipment for planting, and so on. In order to develop the reforested area evenly, the project is divided into two phases. In the first phase, the reforestation project starts from a small scale in the year 2010 and follows a triangle curve in reaching 100% of annual capacity to plant trees over 10 years. In the second phase, planting continues at a constant rate of 50,000 km²/year until all the area has been reforested. Over the life of the project, emissions from the Chinese economy are projected to start at 7 Gt/year in 2010 and grow linearly by 0.27 Gt/year. As a simplification, assume for the purposes of solving the problem that lifetime sequestration from trees planted occur in the year they are planted, although in reality there is a lag since each tree takes 20 years to sequester its full quota.

a. How many years does it take to plant the entire area with trees, to the nearest whole year?

b. Over the lifetime of the project, what is the percentage of total of carbon emissions from the economy sequestered by the project?

c. Comment on your answer from part (b).

7-3. Design an ECBM system with methane recovery, electricity generation, and carbon sequestration for a coal field with 17 billion m^3 methane capacity and a yield of 280 m^3/tonne CO_2 sequestered. The methane (assume it is pure CH_4 and ignore impurities) is delivered to an advanced technology-generating plant where it is mixed with a stream of pure oxygen to deliver electricity for the grid, CO_2 for sequestration, and water as a by-product. Of the energy available in the methane, 40% is delivered to the grid in the form of electricity. The remainder is dissipated in losses in the plant and to supply energy needed for the operation (methane, CO_2 and water pumping, CO_2 injection, and so on). All CO_2 from the generating plant is injected into the coal field, so that there are no emissions of CO_2 to the atmosphere. Additional CO_2 is transported by pipeline from other sources located at a distance. The plant is planned to operate for 40 years at a constant output, after which all the methane will have been withdrawn.

a. How much electricity does the plant produce per year?

b. What is the net amount of CO_2 sequestered, after taking into account the fraction that is derived from carbon that was under the ground in the first place?

7-4. An advanced coal-fired electric power plant uses 1.34 million tonnes of coal to generate 3 billion kWh of electricity per year. All of the CO_2 output from the plant is separated and sequestered in a saline aquifer. The aquifer has a volume of 1.76×10^{12} m^3, average pressure of 300 kPa, average temperature of 350 K, water saturation of 0.22, and a porosity of 14%. How many years would it take for the CO_2 from the plant to fill the aquifer, to the nearest year?

7-5. Suppose that the plant in Problem 4 requires a 1000-km pipeline to move the CO_2 from the plant to the injection site. The project has a financial life of 25 years and an MARR of 6%. There is also an operating cost of $5.00/tonne to transport the CO_2, and a fixed capital cost of $100 million for the facilities to transmit CO_2 from the plant and to inject it into the aquifer. What is the additional levelized cost per kWh to sequester the CO_2:

a. If the pipeline costs $500,000/km?

b. If the pipeline costs $5,000,000/km?

7-6. Sequestration in carbonate materials. You are to build up a sequestration process using forsterite, using data on energy required for heating reactants and energy released during carbonate formation found in the chapter. Use of electricity in the problem can be assumed to emit CO_2 at the rate of 0.12 tonne of CO_2 per GJ. Answer the following questions:

a. How many tonnes of forsterite are required per tonne of CO_2 sequestered? How many tonnes of magnesium carbonate ($MgCO_3$) are produced?

b. How much net energy is released per tonne of CO_2 sequestered?

c. The sequestration facility is located on top of a layer of rock strata that is 40% forsterite by weight, which is to be mined for the sequestration process. The volume available to be mined is 500 million tonnes. The operation is to work for 20 years. How many tonnes of CO_2 can be sequestered per year?

d. The total energy consumption per tonne of rock mined to extract the rock, prepare the forsterite, and power the plant is 1.0 GJ/tonne of forsterite yielded. Suppose 3/4 of the energy released during the sequestration reaction can be usefully made available to the plant either as electricity or as process heat. What is the net energy required from outside sources per tonne of CO_2 sequestered?

e. Suppose the energy requirement in part (d) is made available in the form of electricity from a fossil-fuel-burning power plant. What is the net amount of CO_2 sequestered per ton of forsterite processed, taking into account this CO_2 "loss"?

f. Suppose the CO_2 for the process is to come from an air capture device. A device is available that has a swept area of $20 \, m \times 20 \, m$ and is 50% efficient in terms of capturing the passing CO_2 (ignore any CO_2 emissions resulting from powering the device, for this part). The average air speed is $3.6 \, m/s$. How many of such devices are needed to supply the CO_2 for sequestration in part (c)?

g. Suppose that instead of being captured from the air, the CO_2 is delivered from a power plant. Also, the output from the sequestration process is not accumulated on site: instead it is sold as a building material, similar to gravel, at a value of $8.00/tonne. The facility, including CO_2 delivery, material mining, and carbon transformation, has a capital cost of $3 billion. The energy requirement calculated in part (d) is delivered in the form of electricity, and the cost of the electricity is $0.10/kWh; all other operating costs are $5,000,000/year. If the project lifetime is 25 years and the MARR is 5%, what is the net additional cost per kWh for electricity whose CO_2 emissions are offset using this facility?

CHAPTER 8

Nuclear Energy Systems

8-1 Overview

Nuclear energy is the first of several energy sources to be discussed in this book that can deliver significant amounts of energy without emitting greenhouse gases, and without consuming finite fossil fuel resources. At present, its future is uncertain. On the one hand, a reliable and dispatchable (nonintermittent, unlike solar or wind) electricity source that emits no greenhouse gases and also no air pollutants can help the community of nations address both climate change and ongoing problems with air quality. On the other hand, questions remain about the vulnerability of nuclear power plants to accidents, and the possible repercussions of those accidents, not only on human health and the natural environment but also on investors' capital and on electric-grid reliability. Regardless of how the debate about nuclear energy is eventually resolved, electricity generation from existing plants will continue to play a role in the short to medium term, since countries such as China, France, India, Japan, and Korea have recently inaugurated new plants, and other plants in these countries and elsewhere are nearing completion, at a time when demand for electricity continues to grow. The further expansion of the use of nuclear energy, or its use over a much longer time span, however, will require a more solid worldwide consensus on the impregnability of plant designs and the ability to contain any malfunctions. Furthermore, the total availability of the primary fuel for nuclear energy, uranium-235, is relatively small compared to other possible resources, so long-term growth of nuclear energy will require the development of new uranium resources, new fission reactor designs capable of fast fission, the successful development nuclear fusion, and a more permanent solution for radioactive waste products. These issues are discussed in this chapter.

8-2 Introduction

The term "nuclear energy" means different things in different professions. To a nuclear physicist, nuclear energy is the energy in the bonds between subatomic particles in the nucleus of an atom. To an energy systems engineer, nuclear energy is the controlled release of that energy for application to human purposes, most commonly the generation of electricity for distribution in the electric grid.[1]

---

[1] For a full-length work on the principles of nuclear energy and nuclear reactor design, see, for example, Glasstone and Sesonske (1994), Volumes 1 and 2.

Nuclear energy can be obtained either through *fission*, or splitting of a nucleus of one heavy atom resulting in two or more smaller nuclei, or *fusion*, which is the joining together of two light nuclei into a larger one. Energy released during these nuclear reactions is transferred to a working fluid such as pressurized water or steam, and then the thermal energy is converted into electrical energy in a conversion device similar to those studied in Chap. 6 for fossil fuels. Because nuclear bonds are extremely strong, the amount of energy released by a nuclear reaction is very large relative to a chemical reaction, and nuclear power plants can therefore deliver very large amounts of power per mass of fuel consumed.

In an age of worldwide concern about climate change and other environmental issues, nuclear energy has some unique positive attributes, and also some unique concerns. Nuclear energy emits no CO_2 during the end-use stage of its life cycle, and because the facilities built have a small footprint relative to the amount of energy they are capable of producing, nuclear energy also emits relatively small amounts of CO_2 during the infrastructure construction stage. There are also no air pollutants emitted from the consumption of nuclear fuels in the plant. There are, however, radioactive by-products created at the fuel extraction and fuel consumption stages, as well as at the disposal stage of a reactor. Managing radioactivity requires management techniques that are unique to nuclear energy, including the shielding of reactors, the careful movement of by-products in the transportation network, and the management of high-level nuclear waste that includes some isotopes that remain radioactive for thousands of years.

It is perhaps also the presence of concentrated radioactivity, unfamiliar as it was to the general public before the invention of nuclear weapons and the first construction of nuclear power plants in the mid-twentieth century, which has made nuclear energy one of the most controversial energy options at the present time. Of course, large-scale generation and distribution of energy resources has caused controversy in many forms, ranging from debates over the extent to which power plant emissions should be regulated, to the effect of inundating land with reservoirs impounded behind hydroelectric dams, to the visual and auditory effect of wind turbines. Still, it is hard to think of an energy source where two sides have such diametrically opposing views. In the words of one author, "...one side is certain that nuclear energy will have to expand in the twenty-first century to meet energy demand, whereas the other side is equally certain that this energy form is too dangerous and uneconomical to be of longer-term use."[2] What is clear is that, in the short run, nuclear energy provides a significant fraction of the world's electricity, and since this output cannot easily be replaced, even if that were the intention of the community of nations, nuclear power will certainly be with us for many decades. It is less certain whether in the long-term fission power will be retained as a permanent option for electricity, or whether it will be phased out in favor of some other option.

8-2-1   Brief History of Nuclear Energy

The origin of nuclear power begins with the development of nuclear weapons through the use of *uncontrolled fission* of fuels to create an explosion. Fission in a nuclear weapon is uncontrolled in the sense that, once the chain reaction of the weapons fuel begins, a substantial fraction of the nuclear fuel is consumed within a fraction of a second, leading to an extremely powerful and destructive release of energy.

---

[2] Beck (1999).

Many of the scientists involved in the development of nuclear weapons advocated the application of nuclear energy toward nonmilitary applications such as electricity generation. Several countries shared the early milestones in its peacetime development. In 1951, scientists at the National Reactor Testing Station in Idaho, U.S., the present-day Idaho National Laboratories, generated electricity for use inside a building for the first time. In 1954, a nuclear reactor with a 5-MW capacity at Obninsk in the erstwhile USSR delivered power to the surrounding electric grid for the first time in history. The first commercial nuclear plant in the world was the 50-MW Calder Hall plant at Sellafield, England, which began operation in 1956. The first commercial nuclear power plant in the United States was the 60-MW unit at Shippingsport, Pennsylvania, U.S., which began to deliver electricity to the grid in 1958.

From the late 1950s and early 1960s, the number of nuclear plants grew rapidly in North America, Japan, various countries in Europe, and the former Soviet Union, as did the size of individual reactors. In 1974, the first plant with a 1000-MW capacity, Zion 1, came on line in the United States near Warrenville, Illinois. Since that time, the size of the largest plants has remained approximately constant around 1 to 1.5 GW, although some utilities have concentrated the total nuclear generating capacity in specific locations by installing multiple plants adjacent to one another. In New York State in the United States, for example, three plants with a total capacity of approximately 3 GW are colocated along Lake Ontario near Oswego, New York: Nine Mile Point 1, Nine Mile Point 2, and the James Fitzpatrick reactor.

The other main application of nuclear energy besides electricity generation and explosive weapons is in propulsion for naval vessels. Nuclear reactors are ideal for warships because a relatively small mass of fuel in the on-board reactor allows aircraft carriers, submarines, and other vessels to function for months or years at a time without needing refueling, an asset for military applications. On the civilian side, in the 1960s and 1970s the Japanese and U.S. governments sponsored the development of prototype nuclear-powered merchant ships for carrying commercial cargo, namely, the Mutsu Maru in Japan and the N.S. Savannah in the United States. These vessels eventually proved not to be cost-competitive with conventional diesel-powered merchant ships and were taken out of service. There are no nuclear-powered merchant ships in use at the present time.

Three highly publicized accidents involving nuclear energy since the 1970s have slowed its growth around the world. In the first incident, in 1978, a partial meltdown of one of the reactors at Three Mile Island (TMI) near Harrisburg, Pennsylvania, U.S., led to the permanent loss of use of the reactor, although the adjacent reactor in the facility was not damaged and continues to be used at present.[3] After TMI, many contracts for the construction of nuclear plants were cancelled, and since that time no new plants were ordered in the United States until an announcement of an agreement for a new plant at the Bellefonte site in the state of Alabama in 2007, and the continuation of the construction of the Watts Bar 2 plant in Tennessee, which had been discontinued in 1988. Work at Bellefonte was later discontinued, while work at Watts Bar 2 continues.

In the second accident, in 1986 a reactor malfunction and fire triggered by a poorly designed and executed experiment at Chernobyl in the Ukraine led to the complete destruction of a reactor and its surrounding structure, as well as the loss of life of 28 firefighters and the evacuation of a large geographic area around the plant. Although the accident could not

[3] A full report on the accident was published by the President's Commission, also known as the "Kemeny Commission Report" (United States, 1979). For an account of the sequence of events in the Chernobyl accident, see Cohen (1987). The sequence of events in the Fukushima accident is reported in the United States NRC Fukushima Accident Task Force Report (Nuclear Regulatory Commission, 2011).

have happened if the plant operators had not violated several key safety regulations, the accident revealed the vulnerability of the design to the release of radioactivity in the case of a fire or meltdown (see discussion of reactor designs, Sec. 8-4). At the time of the Chernobyl accident, many European countries were moving toward fossil-based electricity generation due to relatively low fuel costs, and the Chernobyl accident served as further discouragement for the construction of new plants. Italy, for example, passed a referendum in 1987 to phase out domestic nuclear power production, and closed its last plant in 1990. Other countries, notably France and Japan, continued to expand nuclear capacity; the former brought on line four 1450-MW$_e$ plants between 1996 and 2003.

The third high profile incident was an accident at the Fukushima Dai-ichi, or Fukushima Number One, nuclear power plant in Fukushima Prefecture, Japan, in March 2011. In this accident, a tsunami from the 9.0 Magnitude Great Northeast Japan Earthquake of 2011 overtopped a tsunami protection wall adjacent to the plant, destroying both the plant's connection to the power grid and backup generation system. In the ensuing station blackout (SBO), the loss of coolant to the reactor cores led to both a partial meltdown and explosions that to varying degrees damaged the containment structures and buildings of the four reactors at the site. Although there were no deaths among the workforce caused by direct, acute radiation exposure, the government was forced to institute a 19-mile or 30-km diameter evacuation zone around the plant to protect residents from radiation-related risks. As a result of the accident, the Japanese government closed all nuclear plants in the country for inspection and upgrading until the first plants were allowed to reopen in 2015.

8-2-2   Current Status of Nuclear Energy

The memory of the three incidents previously mentioned continues to influence nuclear policy around the world as nations struggle with the challenge of providing sustainable energy for the twenty-first century. At the same time, operators of existing nuclear plants have improved efficiency and also availability of nuclear plants, so that once capital costs have been amortized, the plants operate for the largest percent of hours per year of any energy generation technology, and also produce electricity for the grid at some of the lowest production costs per kWh of any technology. In countries with deregulated electricity markets, operators of these plants can sell electricity for whatever price the market will bear while producing it at low cost, leading to attractive profits.

Among the industrialized nations, at present some countries are expanding the use of nuclear energy, some countries are keeping output constant, and some countries are attempting to phase out nuclear energy entirely. As shown in Table 8-1, countries with high reliance on nuclear energy, in terms of percent of total kWh produced, include France with 77%, Belgium with 48%, and Switzerland with 38% in 2014. Japan had a 30% share in 2006 but 0% in 2014 due to the national shutdown. As mentioned earlier, the United States, along with the United Kingdom and a number of other European countries, has not recently been adding new plants, although the operators of the plants continue to increase total nuclear power production by improving the performance of existing plants. In the United States, new plant construction is currently underway in the form of new units being added to existing nuclear power plants. The Watts Bar 2 plant mentioned above (near Chattanooga, Tennessee) is nearing completion and is expected to come on line in 2016. Other plants actively under construction are Alvin Vogtle 3 and 4 (Waynesboro, Georgia) and Virgil Summer 2 and 3 (Jenkinsville, South Carolina). Countries with the stated intention of phasing out nuclear energy include Belgium and

Sweden. Taking the example of Sweden, the government responded to the accident at TMI by making the commitment in the early 1980s to phase out nuclear energy within the country by 2010. The actual value in 2014 was 41% of the total consumption, down from 48% in 2006, but still far from a complete phase-out as shown in Table 8-1.

Another major effort has been the upgrading of nuclear plants in the former Soviet Union and former Eastern European bloc from obsolete Soviet designs to state-of-the-art technology. For example, a plant at Temelin in the southern Czech Republic was

Country	2006		2014	
	2006 Output (bill. kWh)	Share (%)	2014 Output (bill. kWh)	Share (%)
Armenia	2.4	42.0	2.3	30.7
Belgium	44.3	54.4	32.1	47.5
Bulgaria	18.1	43.6	15.0	31.8
Canada	92.4	15.8	98.6	16.8
Czech Republic	24.5	31.5	28.6	35.8
Finland	22.0	28.0	22.6	34.6
France	428.7	78.1	418.0	76.9
Germany	158.7	31.8	91.8	15.8
Hungary	12.5	37.7	14.8	53.6
Japan	291.5	30.0	0.0	0.0
Korea	141.2	38.6	149.2	30.4
Russia	144.3	15.9	169.1	18.6
Slovakia	16.6	57.2	14.4	56.8
Slovenia	5.3	40.3	6.1	37.2
Spain	57.4	19.8	54.9	20.4
Sweden	65.1	48.0	62.3	41.5
Switzerland	26.4	37.4	26.5	37.9
Ukraine	84.8	47.5	83.1	49.4
United Kingdom	69.2	19.9	57.9	17.2
United States	787.2	19.4	798.6	19.5
Subtotal	2492.6	n/a	2145.9	n/a
Other	168.4	n/a	218.1	n/a
Total	2661	16.0	2364	12.0

Note: "Share" is based on percent of total kWh/year generated by nuclear reactors. "Subtotal" is sum of output from countries listed in the table, and "Total" is worldwide production of electricity from all countries. Note that although Japan had a 0% share in 2014 due to the national shutdown it is included in the table due to historically having a large share. *Share figures for combination of all countries included in Subtotal and for countries not included are not available.

Source: International Atomic Energy Agency (2015).

Table 8-1 Output of Nuclear-Generated Electricity, and Nuclear's Share of Total Electric Market, for Countries with More than 15% Share of Electricity from Nuclear Energy, 2006 and 2014

partially completed with Soviet technology at the time of the changing of the regime in the former Czechoslovakia in 1990. The Czech operators of the plant subsequently entered into an agreement with a U.S. firm, Westinghouse, to complete the plant using technology that was compliant with western safety standards. The way in which the project was completed was of great interest to the Austrian government, which had originally expressed serious concern about the operation of a plant with suspect technology some 100 km from the Austrian border. The plant began delivering electricity to the grid in 2002.

The use of nuclear energy is concentrated in the industrialized countries, and is relatively less of a market player in emerging countries—just 2% of primary energy consumption in emerging countries of the world is met with nuclear energy, versus 10% in industrialized countries. For some of the emerging countries, the combination of a need for a large electricity market to justify the size of a nuclear plant, the high capital cost of the plant, and the level of technological sophistication are out of reach (although this might change in the future if smaller, modular designs become available). On the other hand, emerging countries in many parts of the world, including Argentina, Mexico, Pakistan, and South Africa, make use of nuclear energy to meet some of their electricity demand, so it cannot be said that nuclear power is exclusively the domain of the richest countries. At the present time, countries such as India and China, with rapidly growing economies and concerns about the pollution impacts of heavy reliance on coal, are expanding the use of nuclear energy. China in particular has in the period from 2002 to 2015 brought on line 23 plants with a total capacity of 20.9 GW.

8-3   Nuclear Reactions and Nuclear Resources

As a starting point for investigating the fuels and reactions used in nuclear energy, recall that the *atomic number* Z of one atom of a given element is the number of protons in the nucleus of that atom, which is also the number of electrons around the nucleus when the atom has a neutral electrical charge. The *atomic mass* A is the sum of the number of protons and neutrons in the nucleus of that atom, with each proton or neutron having a mass of 1 atomic mass unit (AMU). In precise terms, the mass of a neutron is 1.675×10^{-27} kg, while that of a proton is 1.673×10^{-27} kg; the value of 1 AMU is 1/12 of the mass of the most common carbon nucleus, which has 6 protons and 6 neutrons. The number of protons identifies a nucleus to be that of a specific element. Nuclei with the same number of protons but different numbers of neutrons are *isotopes* of that element. For example, Uranium 233, or U-233, is an isotope of Uranium with 92 protons and 141 neutrons, U-238 is an isotope with 92 protons and 146 neutrons, and so on.

The binding energy of the nucleus is a function of the ratio of Z to A at different values of A across the spectrum of elements. It suggests which isotopes of which elements will be most useful for nuclear energy. Using Einstein's relationship between energy and mass:

$$E = mc^2 \tag{8-1}$$

we can calculate the mass of a nucleus with neutrons and protons (N, Z) and binding energy B(N, Z) using

$$M(N,Z) = Nm_n + Zm_p - \frac{B(N,Z)}{c^2} \tag{8-2}$$

In other words, the mass of a nucleus is the mass of the individual particles within the nucleus, minus the mass equivalent of the binding energy which holds the nucleus together. It is this binding energy which is released in order to exploit nuclear energy for useful purposes.

From experimental observation, the following semiempirical formula for binding energy B of a nucleus as a function of A and Z has been derived:

$$B(A,Z) = C_1 A + C_2 A^{0.667} + C_3 \frac{Z(Z-1)}{A^{0.333}} + C_4 \frac{(A-2Z)^2}{A} \qquad (8\text{-}3)$$

From empirical work, the following coefficients have been found to fit observed behavior of nuclear binding energy: $C_1 = 2.5 \times 10^{-12}$ J, $C_2 = -2.6 \times 10^{-12}$ J, $C_3 = -0.114 \times 10^{-12}$ J, and $C_4 = -3.8 \times 10^{-12}$ J.

Figure 8-1 shows the binding energy per unit of atomic mass for the range of observed atomic mass values, for values from AMU = 1 to 244 (equivalent to the range from hydrogen to plutonium), as predicted by Eq. (8-3). At each value of A, a representative value of Z is chosen that is typical of elements of that size. The resulting values of B/A are therefore slightly different from the observed values when elements at any given value of A are analyzed. However, the predicted values of binding energy are reasonably close. Table 8-2 gives the predicted and actual binding energy for a sample of five elements from Al to U, measured in electron volts or eV (1 eV = 1.602×10^{-19} J). The error values are small, ranging from 2 to 5%.

The ratio Z/A that results in the lowest energy state for the binding energy of the nucleus, and hence is predicted to be the most likely value for a given value of A, can be approximated by a related, empirically derived formula:

$$Z/A = \frac{1}{2C_3 A^{0.667} + 4C_4} \left[-(m_n - m_p)c^2 + \frac{C_3}{A^{0.333}} + 2C_4 \right] \qquad (8\text{-}4)$$

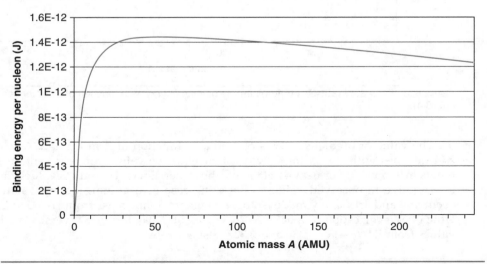

Figure 8-1 Predicted binding energy per nucleon as a function of number of AMU, using Eq. (8-3).
Note: One nucleon equals one proton or one neutron.

	A	*Z/A*	Predicted (MeV)	Observed (MeV)	Error %
Al	27	0.481	8.74	8.33	5%
Cu	64	0.453	9.03	8.75	3%
Mo	96	0.438	8.85	8.63	3%
Pt	195	0.400	8.08	7.92	2%
U	238	0.387	7.73	7.58	2%

Note: 1 MeV = 1 million eV, or 1.602×10^{-13} J.

TABLE 8-2 Predicted Binding Energy per Nucleon Using Eq. (8-3) versus Observed for a Selection of Elements

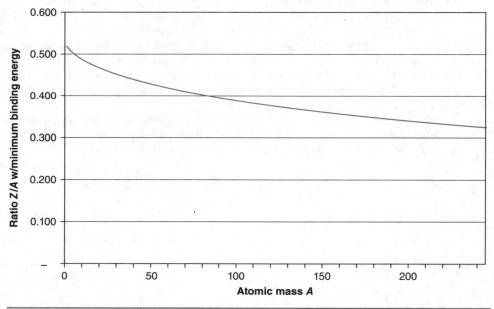

FIGURE 8-2 Estimated ratio Z/A that minimizes binding energy as a function of A, according to Eq. (8-4).

The value of predicted Z/A is plotted as a function of A in Fig. 8-2. As shown, Z/A declines with increasing A, consistent with the values for representative elements in Table 8-2. For heavier elements, however, Eq. (8-4) predicts values that are low relative to observed values. Taking the case of uranium, the main naturally occurring and artificially made isotopes range in atomic mass from U-230 to U-238, which have corresponding actual Z/A values of 0.400 and 0.387, versus predicted values from Eq. (8-4) of 0.330 and 0.327, respectively.[4]

---

[4]Isotopes of uranium with atomic number as low as 218 and as high as 242 are possible, but some of these synthetic isotopes have a very short half-life.

In summary, the binding energy of the nucleus tells us first that for light atoms, the number of neutrons and protons in the nucleus will be approximately equal, but for heavier atoms, the number of neutrons will exceed the number of protons. Second, atoms with the strongest nuclear binding energy per nucleon occur in the middle of the periodic table, while those with the weakest binding energy occur at either the light or heavy end of the element spectrum. Nuclei with the lowest binding energy per nucleon will be the easiest to either fuse together or break apart. Therefore, the elements that are most suited for use in nuclear energy are either very light or very heavy.

8-3-1 Reactions Associated with Nuclear Energy

The most common reaction used to release nuclear energy for electricity production today is the fissioning of uranium atoms, and in particular U-235 atoms, since fissioning of U-235 creates by-products that are ideal for depositing energy into a working fluid and at the same time perpetuating a chain reaction in the reactor. U-235 is also sufficiently common that it can be used for generating significant amounts of energy over the lifetime of a nuclear power plant. The chain reaction for splitting U-235 uses a "thermal" neutron, which, after being released by the fissioning of a heavy nucleus, is slowed by collisions with other matter around the reaction site, reducing its energy. The most probable energy E_p of the thermal neutron is a function of temperature T in degrees kelvin, and can be approximated as:

$$E_p = 0.5\,kT \tag{8-5}$$

where k is Boltzmann's constant, $k = 8.617 \times 10^{-5}\,\text{eV/K}$. So, for example, if the temperature around the reaction site is 573 K (300°C), the most probable energy of the thermal neutron is 0.0247 eV.

When a U-235 nucleus is struck by a thermal neutron, the neutron is absorbed and then one of two reactions follows, as shown in Eqs. (8-6) to (8-8):

$$U_{235} + n \rightarrow U_{236} \tag{8-6}$$

$$U_{236} \rightarrow Kr_{36} + Ba_{56} + 2.4n + 200 \text{ MeV (in a typical reaction)} \tag{8-7}$$

$$U_{236} \rightarrow U_{236} + \gamma \tag{8-8}$$

where γ is an *energetic photon*, or high-energy electromagnetic particle. Most of the energy comes out of the nucleus in the form of gamma rays with an energy of at least $2 \times 10^{-15}\,\text{J}$ and maximum wavelength on the order of $1 \times 10^{-10}\,\text{m}$. In Eq. (8-7), the splitting of U-236 into Krypton and Barium is a representative fission reaction, though it is not the only one; other products of fission are possible. Also, the exact number of neutrons released varies with the products of fission; the production of 2.4 neutrons represents the average number of neutrons released, and values as low as 0 and as high as 8 are possible. In Eq. (8-8), the excess nuclear energy that normally drives fission of the U-236 is eliminated by radioactive decay and γ-ray emission, and the atom remains as U-236. Splitting of the U-236 nucleus into smaller nuclei [Eq. (8-7)] occurs in about 85% of cases, while in the remaining 15% of cases, the atom remains as U-236.

The output of somewhat more than two neutrons per fission reaction is useful for nuclear energy, because it compensates for the failure to release neutrons in 15% of reactions, as well as the loss of neutrons to other causes, and allows the chain reaction to continue. A process in which on average each reaction produces 1 neutron that eventually fissions another U-235 atom according to Eqs. (8-5) and (8-6) is called *critical*. This state allows the nuclear chain reaction to continue indefinitely, since the number of reactions per second remains constant, as long as a supply of unreacted fuel is available. A process wherein the average is less than 1 fission producing neutron in the next generation is *subcritical* and will eventually discontinue. In a *supercritical* process, the average fission producing neutron per reaction is more than 1 and the rate of fission will grow exponentially. Nuclear reactors control systems are designed to regulate the reaction and prevent the process from becoming either subcritical or supercritical during steady-state operation.

Uranium found in nature is more than 99% U-238, and about 0.7% U-235. For certain reactor designs, it is necessary to have uranium with 2 to 3% U-235 reaction in order to achieve a critical reaction process. Therefore, the uranium must be *enriched*, meaning that the U-235 percent concentration in the mixture is increased above the naturally occurring 0.7%, a process which entails removing enough U-238 so that the remaining product achieves the required 2 to 3% concentration of U-235. The most common enrichment process currently in use is the gas centrifuge technique. Uranium fuel for military applications such as ship propulsion is enriched to concentration levels of U-235 much higher than 3%, so as to reduce the amount of fuel that must be carried and extend the time between refueling stops.

U-238 present in nuclear fuels also provides a basis for a fission reaction, in several steps, starting with the collision of a neutron with the U-238 nucleus. The reactions are the following:

$$U_{238} + n \rightarrow U_{239} \tag{8-9}$$

$$U_{239} \rightarrow Np_{239} + \beta \tag{8-10}$$

$$Np_{239} \rightarrow Pu_{239} + \beta \tag{8-11}$$

Here the term β represents beta decay, in which a neutron in the nucleus is converted to a proton, releasing a *beta particle*, which is a loose electron in Eqs. (8-10) and (8-11) that has been released from a nucleus. (There are also $\beta+$ decays that produce a *positron*, or positively charged particle that has the same mass as an electron.) In the beta decay in this case, a negatively charged beta particle is released, because the change of charge of the remaining particle inside the nucleus is from neutral to positive.

The plutonium atom that results from this series of reactions is fissile, and will fission if struck by a neutron, releasing energy. In a conventional nuclear reaction that focuses on the fission of U-235, some quantity of U-238 will transform into Pu-239 after absorbing a neutron, and the Pu-239 will in turn undergo fission after being struck by another neutron. Although these reactions contribute to the output of energy from the reactor, most of the U-238 present in the fuel remains unchanged. Alternatively, the reaction can be designed specifically to consume U-238 by maintaining the neutron population at a much higher average energy, on the order of 10^5 to 10^6 eV. This process requires a different design of reactor from the one used for electricity generation at present; see discussion in Sec. 8-4.

Release of Energy through Fusion of Light Nuclei

While splitting of the nuclei of very heavy atoms such as uranium and plutonium leads to the net release of energy, at the other end of the spectrum, the fusion of very light nuclei also releases energy. In the case of our sun and other stars, it is the conversion of

4 atoms of *protium*, or the isotope of hydrogen that has 1 proton and no neutrons in the nucleus, into a helium atom that leads to the release of energy. This conversion happens by two different dominant pathways, depending on the mass of the star. This reaction is deemed not to be replicable on Earth on a scale practical for human use. However, two other reactions involving other isotopes of hydrogen, namely, *deuterium* (1 proton, 1 neutron), abbreviated *D*, and *tritium* (1 proton, 2 neutrons), abbreviated *T*, are thought to be more promising. These reactions are:

$$D + T \rightarrow \text{He}_4 + n + 17.6\,\text{MeV} \tag{8-12}$$

$$D + D \rightarrow \text{He}_3 + n + 3.0\,\text{MeV} \tag{8-13}$$

$$\text{or} \quad \rightarrow T + p + 4.1\,\text{MeV}$$

While these reactions do not yield as large an amount of energy per atom reacted as the fission of uranium in Eq. (8-6), the yield is still very large per mass of fuel by the standard of conventional energy sources. It is also equal to or larger than the yield from fission on a per mass of fuel basis.

Deuterium occurs naturally in small quantities (approximately 0.02% of all hydrogen), and must be separated from protium in order to be concentrated for use as a fuel. Tritium does not occur naturally, and must be synthesized if it is to be made available as a fuel.

Comparison of Nuclear and Conventional Energy Output per Unit of Fuel

To appreciate the extent to which nuclear energy delivers power in a concentrated form, consider the comparison of uranium and a representative fossil fuel, namely, coal, in Example 8-1.

Example 8-1   Compare the energy available in 1 kg of U-235 to that available in 1 kg of coal.

Solution   From Eq. (8-6), 1 atom of U-235 releases approximately 200 MeV of energy. Suppose we react 1 kg of pure U-235, and take into account that 15% of collisions with neutrons do not yield fission, the amount of energy available is

$$(1\,\text{kg})(4.25\,\text{mol/kg})(6.022 \times 10^{26}\,\text{atoms/kg·mol})(200\,\text{MeV/atom})(0.85) = 4.36 \times 10^{32}\,\text{eV}$$

This amount is equivalent to 69.8 TJ. Recall that from Eq. (7-6), oxidation of 1 mol of carbon releases 394 kJ, so 1 kg of carbon (pure coal with no moisture content or impurities) would release 32.8 MJ/kg. Thus the ratio of energy released per kilogram is approximately 2 million to 1.

Even taking into account the fact that reactor fuel is only 2 to 3% U-235, and also accounting for losses in the conversion of nuclear energy into thermal energy in the working fluid, the increase in concentration going from conventional to nuclear power is very large. It is little wonder that, at the dawn of the age of commercial nuclear power in the 1950s, this source was projected to be "too cheap to meter" once it had come to dominate the electricity market. Proponents of nuclear energy early in its history looked only at the operating cost savings of purchasing uranium versus purchasing other fuels such as coal or oil, and underestimated the impact that high capital cost would have on the final levelized cost per kWh for nuclear energy. While existing nuclear plants operating today are able to produce electricity at very competitive prices (see Example 8-2), they are not orders of magnitude cheaper than the most efficient fossil-fuel-powered plants, as had originally been predicted.

8-3-2   Availability of Resources for Nuclear Energy

The availability of fuel for nuclear energy, relative to the annual rate of consumption, depends greatly on the reaction used and the maximum allowable cost of the resource that is economically feasible. Focusing first on uranium as the most commonly used nuclear fuel at present, Table 8-3 shows the total available uranium estimate published by the International Atomic Energy Agency, as well as amounts available in individual countries. This table does not consider the economic feasibility of extracting uranium from various sources, so that depending on the price for which the uranium can be sold, the amount that is ultimately recovered may be less. For example, for the United States, at a cost of \$65/kg (\$30/lb), the recoverable amount is estimated at 227,000 tonnes, while at \$110/kg (\$50/lb), the amount increases to 759,000 tonnes. Either amount is less than the full amount, including resources that are the most expensive to extract, 1.64 million tonnes, shown in the table.

Based on the amounts shown in the table, and taking into account that additional resources may be discovered in the future, it is reasonable to project that resources should last into the twenty-second century, or possibly as long as several centuries, at current rates. If the world were to choose a "robust" nuclear future, for example, if nuclear electricity were used to replace most fossil-fuel-generated electricity, then the lifetime of the uranium supply would be shorter. It is likely that increasing cost of uranium will not deter its extraction, even up to the maximum currently foreseen cost of \$260/kg (\$118/lb), since the cost of the raw uranium is a relatively small component of the overall cost of nuclear electricity production.

The above projections assume that the world continues to rely mainly on the U-235 isotope for nuclear energy. If emerging nuclear technologies can be made sufficiently reliable and cost-effective, the lifetime of fuel resources for nuclear energy might be extended by up to two orders of magnitude. Recall that in the current uranium cycle, 98 to 99% of the

Australia	2,232,510
Brazil	190,400
Canada	824,430
Kazakhastan	910,975
Niger	259,382
Russian Fed.	737,985
United States	1,642,430
All other	2,312,348
Total	9,110,460

Source: IAEA (2007).

TABLE 8-3   Available Uranium Resource for World and from Major Source Countries, from All Extraction Techniques, in tonnes, 2007

uranium, in the form of U-238, is left unchanged by the reaction. If this resource could be converted to plutonium and then used as fuel, as in Eqs. (8-9) to (8-11), the lifetime of the uranium supply would be extended to several millennia. Nuclear fusion holds the promise of an even longer lifetime. Given the amount of water in the ocean, and the fact that approximately 1 in 6500 hydrogen atoms is deuterium, a $D + D$ reaction could be sustained for billions of years on available hydrogen. The $D + T$ reaction is more limited, since tritium must be synthesized from other atomic materials that are in limited supply. Tritium can be produced by colliding either a Lithium-6 or Lithium-7 atom with a neutron, yielding one tritium and one Helium-4 atom (in the case of Lithium-7, an additional neutron is also produced). Reactors based on the $D + T$ reaction could supply all of the world's electricity for several millennia before available Li-6 resources would be exhausted.

8-4   Reactor Designs: Mature Technologies and Emerging Alternatives

Of the more than 400 nuclear power plants operating in the world today, most use water as a working fluid, either by using nuclear energy directly to create steam (boiling water reactor, or BWR), or by heating high-pressure water to a high temperature, which can then be converted to steam in a separate loop (pressurized water reactor, or PWR). National programs to develop nuclear energy in various countries have, over the decades since the 1950s, developed along different paths, and in some cases developed other technologies that are currently in use, though not in the same numbers as the BWR and PWR designs.

Two factors are currently driving the evolution of new reactor designs: (1) the need to reduce the life cycle and especially capital cost of new reactors, so that electricity from new reactors can be more cost-competitive in the marketplace and (2) the desire to make use of new fuels other than U-235, including U-238, plutonium, and thorium.

8-4-1   Established Reactor Designs

The PWR and BWR reactors in most instances use "light" water, that is, H_2O with two protium atoms combined with an oxygen atom, as opposed to "heavy" water that uses deuterium instead, or water with the molecular makeup D_2O. The nuclear industry uses the nomenclature *light-water reactor* (LWR) and *heavy-water reactor* (HWR) to distinguish between the two. In this section, we focus on the BWR design; the PWR design is similar, but with the addition of a heat exchanger from the pressurized water loop to the vapor loop which connects the reactor to the steam turbine.

The *reactor core* houses the nuclear reactions, which take place continuously, while being monitored and controlled so as to remain in the critical range (see Fig. 8-3). Within the reactor core, three key components interact to allow the reactor to operate steadily:

1. *Fuel rods:* Many thousands of fuel rods (labelled "fuel assemblies" in Fig. 8-3) are present in the reactor when it is functioning. Fuel in the form of UO_2, where the uranium has previously been enriched in an enrichment facility, is contained within the fuel rods. The rods are encased in a zirconium alloy. Rods are typically 1.2 cm (1/2 in) in diameter. Length varies with reactor design, but a length of approximately 4 m (12.5 ft) is typical.

2. *Moderating medium:* Because the neutrons from fission of U-235 are emitted at too high an energy to split other U-235 atoms with high probability before being lost from the reactor core, it is necessary for a moderating medium to

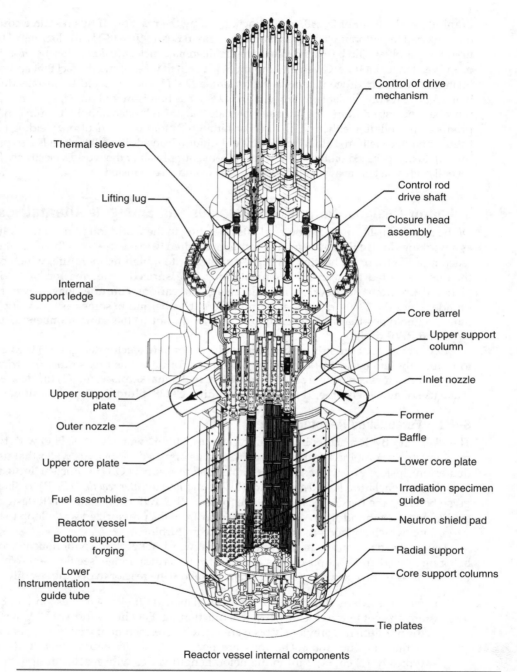

Thermal sleeve

Lifting lug

Internal support ledge

Upper support plate

Outer nozzle

Upper core plate

Fuel assemblies

Reactor vessel

Bottom support forging

Lower instrumentation guide tube

Control of drive mechanism

Control rod drive shaft

Closure head assembly

Core barrel

Upper support column

Inlet nozzle

Former

Baffle

Lower core plate

Irradiation specimen guide

Neutron shield pad

Radial support

Core support columns

Tie plates

Reactor vessel internal components

FIGURE 8-3 Westinghouse MB-3592 nuclear reactor. (*Source:* Westinghouse Electric Company. Reprinted with permission.)

reduce the energy of the neutrons to the thermal energy corresponding to a temperature of a few hundred degrees Celsius. The water that circulates through the reactor core serves this purpose; in other designs, graphite and water are used in combination. Not all neutrons are uniformly reduced to thermal energy levels. Some are lost to collisions with other materials besides U-235 and others leave the reactor core. These losses drive the need for more than 2 neutrons released per fission, when only 1 neutron is needed to perpetuate the chain reaction.

3. *Control rods:* A typical reactor design has on the order of 200 control rods that can be inserted or removed from the reactor during operation through the control rod drive shaft as shown in Fig. 8-3, so as to maintain the correct rate of fission. Control rods can be made from a wide range of materials that are capable of absorbing neutrons without fissioning, including silver-indium-cadmium alloys, high-boron steel, or boron carbide. In addition to providing a moderating effect on the neutrons, the control rods also provide a means to stop the reaction for maintenance or during an emergency. When the control rods are fully inserted, the chain reaction goes subcritical and quickly stops.

Major components outside the reactor core include the *pressure vessel*, the *containment system,* and the balance of system. The pressure vessel is a high-strength metal vessel that contains the core, and the combination of water at the bottom of the vessel and steam at the top. Reactors also incorporate a comprehensive containment system that is designed to cool the contents of the pressure vessel in the event of an emergency, and also to prevent the escape of radioactive materials to the environment. Part of the containment system is the containment vessel, a reinforced domed concrete structure that surrounds the pressure vessel, which, along with the large cooling towers, often provides a familiar landmark by which one can identify a nuclear power plant from a distance. Reactors also usually contain some type of containment spray injection system (CSIS); in an emergency, this system can spray water mixed with boron inside the containment structure so as to both cool the contents and absorb neutrons. The balance of system includes primarily the turbine and the generator, which are like those in a fossil-fuel-fired plant and are situated outside the outer concrete structure.

Thermal and Mechanical Function of the Nuclear Reactor

In the largest reactors currently in operation around the world, energy is released at a rate of approximately 4 GW during full power operation. At the given rate of energy release per atom split, this rate of energy transfer is equivalent to approximately 1.25×10^{20} reactions per second. Energy from the fission reaction is transferred primarily to the fission fragments, and to a lesser extent to the gamma photons and neutrons that result from fission. The fragments collide with many other particles within a short distance from the fission site, and the kinetic energy of the fragments is transformed into heat energy, which eventually reaches the working fluid through heat transfer. The heat from the reactions generates steam, which is transported out of the containment structure and into the turbine. In the case of the PWR reactor, a heat exchanger located within the containment structure transfers heat from the pressurized water to water that is allowed to boil in order to generate steam.

To date, nuclear power plants have been limited to operating at steam pressures that are below the critical pressure (22.4 MPa), in order to address safety concerns. The loss of additional efficiency that might come from a supercritical pressure plant is not as significant for nuclear power as it is for fossil fuels, since the cost of fuel is not as dominant a concern, and there are no regulated air pollutant or greenhouse gas emissions from nuclear plants. The release of water used as a coolant and working fluid in a nuclear reactor is strictly limited so as to avoid release of radioactivity to the environment.

Exchange of Fuel Rods

Under normal operating conditions, the fuel rods remain stationary within the reactor core as the available U-235 is gradually consumed. After a period of between 3 and 6 years, enough of the U-235 has been consumed that the reduction in available fissile material begins to affect the output, and fuel rods must be replaced. Plant operators must then shut down the reactor to replace the rods. Fuel rod replacement typically takes place every 12 to 24 months, with one-fourth to one-third of all rods replaced in each refueling operation, so that, by rotating through the fuel rods, all the fuel is eventually replaced.

The operator can take a number of steps to minimize the effect of this downtime. Where more than one reactor is situated in a given site, fuel rod replacement can be scheduled so that when one reactor is taken down, others continue to function. Similarly, an operator that owns more than one nuclear power plant may rotate the fuel rod replacement operation among its multiple plants so that the effect on total electrical output from combined assets is minimized. Lastly, minimizing the time required to replace fuel rods is a means of achieving the best possible productivity from a plant. In the United States, for example, the industry median length of time for refueling has declined from 85 days in 1990 to 37 days in 2014, with durations as short as 15 days achieved during recent years.

Costing of Nuclear Power Plant Operations

The total cost of operating a nuclear plant includes the fuel cost, nonfuel operating cost, and, during the first part of the plant's life when capital must be repaid, capital cost. The role of these components in determining the overall cost of producing electricity is best illustrated in a worked case study, as shown in Example 8-2.

Example 8-2 A nuclear power plant, whose capital cost has been fully repaid, operates with a total of 155,000 kg of uranium (or 175,800 kg of UO_2) in its reactor core, which has a U-235 concentration of 2.8%. Each year one-third of the reactor fuel is replaced at a cost of $40 million, which has built into it $0.001/kWh to be paid to a fund for the management of nuclear waste. The facility also incurs $118 million/year in nonfuel operating and maintenance cost.

The plant has a rated capacity of 1.2 GW_e. The working fluid is compressed to 10 MPa, and then heated to 325°C, after which it is expanded in the turbine to 10 kPa, condensed, and returned to the pump for return to the reactor core. The generator is 98% efficient. Assume that all energy released in nuclear fission is transferred to the working fluid, and that, due to various losses, the actual Rankine cycle used achieves 90% of its ideal efficiency. Ignore the contribution of fission of Pu-239 that results from transformation of some of the U-238 in the fuel.

(a) Calculate the cost per kWh of production for this plant.

(b) Suppose instead that the plant is newly built at a cost of $8000 per kW to be amortized over 20 years at a 5% discount rate. What is the new cost per kWh?

Solution

(a) We solve this problem by first calculating the thermal efficiency of the cycle and the rate of fuel consumption, and then the annual electrical output and the components of total cost.

From the steam tables, at 10 kPa we have $h_f = 191.8$ kJ/kg, $h_g = 2584.7$ kJ/kg, $s_f = 0.6493$ kJ/kg $\cdot$ K, and $s_g = 8.1502$ kJ/kg $\cdot$ K. Using the standard technique for analyzing an ideal Rankine cycle from Chap. 6 gives $w_{pump} = 10.1$ kJ/kg and the following table of temperature, enthalpy, and entropy values at states 1 to 4:

State	Temp. (C)	Enthalpy (h) (kJ/kg)	Entropy (s) (kJ/kg · K)
1	45.8	191.8	0.6493
2	*	201.9	*
3	320	2809.1	5.7568
4	*	1821.1	5.7568

*Value is not used in the calculation, and therefore not given.

Based on the enthalpy values $h_1 - h_4$, we obtain $q_{in} = 2607.2$ kJ/kg, and $w_t = 988.0$ kJ/kg. Accordingly, the actual efficiency of the cycle is the ideal efficiency multiplied by the cycle and generator losses:

$$\eta = \frac{988.0 - 10.1}{2607.2}(0.9)(0.98) = 33.1\%$$

Next, regarding the fuel, since one-third of the fuel is replaced each year, the annual fuel consumption is one-third of the total, or 51,700 kg. Taking into account the proportion of the fuel that is U-235, the consumption rate m is

$$m = \frac{(51.7 \times 10^6 \text{ g})(0.028)}{3.154 \times 10^6 \text{ s}} = 0.0459 \text{ g/s}$$

Given an energy release of 69.8 TJ/1 kg or 69.8 GJ/g of fissioned U-235, the consumption of fuel is transformed into electrical output as follows:

$$(\text{input}) \times (\text{efficiency}) \times (\text{h/year}) = \text{output}$$

$$(0.0459 \text{ g/s})\left(69.8\frac{GJ}{g}\right)(0.331)\left(1\frac{GW}{GJ/s}\right)\left(8760\frac{h}{year}\right) = 9279 \text{ GWh/year}$$

The average output of the plant is 1.059 GW$_e$. The average output is less than the peak output of the plant of 1.2 GW$_e$, taking into account downtime and the fluctuations in demand for output.

Total cost per kWh is calculated from the fuel and O&M cost components. Fuel cost is ($40 million)/$(9.255 \times 10^9$ kWh$)$ = $0.00431/kWh. O&M cost is ($118 million)/$(9.255 \times 10^9$ kWh$)$ = $0.0127/kWh. Thus the total cost is $0.0170/kWh.

(b) At capital cost of $8000 per kW, the total construction cost of the plant is $9.6 billion. Since the term is 20 years and MARR = 5%, the appropriate conversion factor for calculating the annual cost of capital is $(A/P, 5\%, 20)$ = 0.08024. Thus the new levelized cost value over the 20-year investment lifetime is

$$\text{Levelized cost} = \frac{\$9.6 \times 10^9 (0.08024)}{9.255 \times 10^9 \text{kWh}} + \$0.0170 = \$0.0831 + \$0.0170 = \$0.1001/\text{kWh}$$

Thus levelized cost is approximately $0.10/kWh.

Discussion The result of the calculation in part (a) shows the cost-competitiveness of a fully amortized nuclear power plant. The cost of fuel and O&M are at or very close to the U.S. average values of $0.0045/kWh and $0.0127/kWh, published by the U.S.-based Nuclear Energy Institute. At $65/kg, the cost of the raw uranium constitutes only a small fraction of the total cost of the fuel resupply, so even a substantial increase in this price would not appreciably change the overall cost of electricity production. Actual plants might vary somewhat from the value given, for example if the maximum pressure is different, or if one takes into account the contribution from fissioning of Pu-239, especially toward the end of a fuel rod's time in the reactor core when its concentration is highest. Also, not all of the losses of efficiency have been detailed in this simplified problem, so the electrical output per unit of fuel consumed is slightly overstated. However, these modifications to the calculation would not appreciably change its outcome, namely, that the cost of electricity production is approximately $0.02/kWh.

A more substantial impact on electricity cost comes from the recovery of capital investment in part (b) of the problem. The relatively high cost per unit of capacity shows how the resulting levelized cost can make new nuclear plants uncompetitive with alteratives such as coal or gas discussed in Chap. 6. The resulting electricity in this example is CO_2 free, but 83% of the levelized cost is for capital repayment. Calculation with a lower capital cost made possible by more efficient design and economies of scale in manufacturing appears as an end-of-chapter exercise. Up until the rise in fossil fuel prices that started around 2005, it was difficult to justify the capital cost of nuclear power, despite the large amount of energy that could be produced per dollar spent on fuel. In a number of countries with large fossil fuel resources such as the United States, investors have been reluctant in recent years to invest in new plants for this reason. A country such as China also has large fossil resources in the form coal but as of 2015 has been investing significantly in nuclear energy (24 plants under construction with a total output of 24 GW) due to significant air quality problems from coal combustion. Some countries with few domestic energy resources, such as France and Japan, had continued to build nuclear plants up to the time of the Fukushima accident in 2011, judging that the energy security benefit for a country completely dependent on imported fossil fuel outweighed any additional cost for nuclear compared to fossil. As of 2015, France and Japan have one and two plants under construction, respectively, so continued investment in nuclear exists but is not significant.

Cost of Nuclear Plant Decommissioning

Nuclear plant operators and the government agencies that regulate them anticipate an additional cost to decommission the plant and its site and the end of its useful lifetime. The decommissioning process has several steps, including removal of spent fuel rods, the reactor core, and any cooling water to appropriate waste storage facilities, and the physical sealing of the facility (e.g., using concrete) to prevent any contamination of the surrounding area.

Operators are required to charge a cost per kWh to accumulate a fund to cover the cost of decommissioning. The eventual cost of the process and the adequacy of funds that are accumulating is a topic of controversy. For example, in the United States, official estimates project the cost of decommissioning at $300 to $500 million per plant on average, or $30 to $50 billion for all plants currently in use. However, because the industry is still young, few plants have actually been decommissioned, so there is relatively little actual data on the true cost. Critics express concern that the true cost may be higher than the initial estimates, and if these concerns are realized, that the true life-cycle cost of nuclear energy will turn out to be higher than the price at which it was sold in the marketplace over the lifetime of the plant. With the oldest generation of plants in North America, Europe, and Japan due to retire over the next 10 to 20 years, the true costs will become more transparent, as the industry develops a track record with the decommissioning process. The estimated cost of decommissioning can be compared to the cost of cleaning up the destroyed reactor at TMI, which cost approximately $1 billion and took 12 years.

8-4-2   Alternative Fission Reactor Designs

In addition to the BWR/PWR reactors discussed in the previous section, a number of other reactor designs are either in use or under development at the present time. Some are remnants from the early experimental years in the evolution of nuclear energy, while others have emerged more recently in response to the need for lower capital cost or a more diverse fuel base.

To improve the understanding of the evolution of nuclear reactor designs, the World Nuclear Association has created a nomenclature of generations to make distinctions. Generations I and II were, respectively, the set of early experimental and the first mature generation of commercial, large-scale nuclear reactors, notably the PWR and BWR reactors that came to dominate the market. Generation III and III+ are a newer generation of reactors currently being deployed that primarily use U-235 as a fuel and light water as a moderator, but improve safety and affordability through design improvements. Finally, Generation IV reactors are future designs that move away from U-235/light water to further improve economic viability and long-term access to fuel supply. Generation IV reactors are anticipated to enter the market between 2021 and 2030.

Alternative Designs Currently in Full Commercial Use

In a fraction of the reactors in use today, gases such as helium or carbon dioxide, rather than water, are used to transfer heat away from the nuclear reaction. Early development of gas-cooled reactors took place in the United Kingdom and France. In the United Kingdom, some gas-cooled reactors known as *advanced gas-cooled reactors (AGRs)* are still in use. Although engineers developed these reactors in the pursuit of higher temperatures and therefore efficiency values than were possible with water-cooled reactors, these designs eventually lost out to the water-cooled BWR and PWR designs, which then became dominant in most countries with nuclear power. In the current designs, heat in the gas is transferred to steam for expansion in a steam turbine. However, in future designs, the gas may be used to drive the turbine directly, with the aim of providing a more cost-effective system that can compete with water-cooled designs.

In another variation, heavy water (D_2O) is used as a coolant rather than light water (H_2O). This approach is embodied in the CANDU (CANadian Deuterium Uranium) reactor used by all nuclear power plants in Canada, as well as some plants in other countries including Argentina, China, and India. Since heavy water absorbs fewer neutrons than light water, it is possible to maintain a chain reaction with natural-grade uranium (~0.7% U-235), thus eliminating the need for uranium enrichment. However, these plants require the production of heavy water from naturally occurring water. For example, a now-decommissioned heavy-water plant at Bruce Bay in Ontario, Canada, produced up to 700 tonnes of heavy water per year, using electrolysis and requiring 340,000 tonnes of freshwater feedstock for every 1 tonne of heavy water. CANDU plants also house the nuclear fission process in a network of containment tubes, rather than in a single large steel reactor vessel, thus requiring less of an investment in high-precision steel machining ability.

Arguably, the Soviet-designed RBMK (Reaktor Bolshoy Moshchnosti Kanalniy, meaning "high-power channel-type reactor") reactor has proven to be the least successful alternative design. Although the RBMK design requires neither enriched uranium nor heavy water, its combination of graphite reaction moderation and water cooling has proven to have positive feedbacks at low-power output levels. That is, loss of coolant flow at low output will actually lead to accelerating reaction rates, creating a serious safety hazard of the type that caused the Chernobyl disaster. RBMK reactors as built in the former Soviet bloc also do not have the same combination of steel and reinforced concrete containment systems that are common to other designs, posing a further safety risk. As mentioned in Sec. 8-2-2, in the case of the Temelin plant in the Czech Republic, a major effort is underway to replace Russian designs with safer alternatives.

Both the CANDU and RBMK reactor designs allow at-load refueling of the system (i.e., adding new fuel without needing to power down the reactor), an advantage over PWR/BWR designs. At-load refueling is theoretically possible with existing gas-cooled designs as well, but earlier attempts to carry out this procedure led to problems with the reactor function, and the practice is not currently used in these plants.

Emerging Designs that Reduce Capital Cost (Generations III and III+)

One goal of the next generation of nuclear power plants is to reduce capital costs by simplifying the plant design. For example, an *advanced pressurized water reactor (APWR)* design developed by Westinghouse in the United States uses the concept of *passive safety* to simplify the need for a containment system around the reactor. In a passive safety approach, a major aberration from the normal steady-state operating condition of the nuclear reaction leads to the stopping of the reaction, so that the types of accidents that are possible in current generation PWR/BWR plants can no longer happen. This design has been developed in both 1000 MW and 600 MW versions. Westinghouse anticipates that the levelized cost per kWh for this design will be lower than that of current-generation plants, thanks to the simpler containment system.

Another approach is to build smaller, more numerous reactors that can be mass-produced in a central facility, thus reducing overall cost per MW of capacity. An example of this technology is the *pebble bed modular reactor (PBMR)*. Each commercial-scale PBMR unit has a capacity of 120 to 250 MW, and the reactor fuel is in the form of small spheres rather than fuel rods, with each sphere, or "pebble," clad in a graphite casing. The PBMR simplifies the system design by using high-temperature gas to drive a Brayton cycle, rather than transferring heat to steam for use in a vapor turbine. This system is thought to be simpler to engineer and also capable of achieving higher temperatures and hence efficiencies than current PWR/BWR technology, further reducing costs. PBMR reactors are currently under active development only in China, where a 10-MW demonstration system is in use at Tsinghua University, and the government has plans to scale up PBMR at several new nuclear plant sites in the future. South Africa has been the other country actively developing PBMR in recent years, but discontinued its efforts in 2010 due to commercial difficulties.

Emerging Designs that Extend the Uranium Fuel Supply (Generation IV)

Both the APWR and PBMR reactors in the previous section assume the continued use of U-235 as the main nuclear fuel, either in a once-through or reprocessed fuel cycle. Emerging *breeder reactor* designs provide the potential to "breed" or convert U-238 to plutonium and then fission the plutonium, thus greatly expanding the amount of energy that can be extracted from the uranium resource. In order for breeding to take place effectively, neutrons must have a much higher energy at the moment of absorption by a uranium nucleus than in the current reactor technology, on the order of 10^5 to 10^6 eV per neutron. Therefore, the moderating medium in the reactor must exert relatively little moderation on the neutrons that emerge from the fissioned atoms of uranium or plutonium. Breeder reactors have the ability to generate large amounts of plutonium over time, so that a future network of breeder reactors might produce enough plutonium early in its life cycle that subsequent breeder reactors would no longer require uranium as fuel, instead consuming the plutonium stockpile for many decades.

In one leading breeder reactor design, the *liquid metal fast breeder reactor (LMFBR)*, liquid sodium at a high temperature is used as a coolant for the breeding reactor, because it exerts much less moderating effect compared to water. The liquid sodium

Reactor Type	Main Countries	Number	
		2004	**2015**
PWR	CN, FR, JP, KR, US	252	229
BWR	JP, KR, US	93	77
Gas-cooled	UK	34	15
Heavy water	AR, CD, CN, KR	33	49
Breeder*	JP, FR, RU	3	4
Other*	Various	14	69
Total		429	443

*Note: Change in number of "other" reactors is partially due to recategorization of some reactors.

Code: AR = Argentina, CD = Canada, CN = China, FR = France, JP = Japan, KR = S. Korea, RU = Russia, PWR = Pressurized Water Reactor, BWR = Boiling Water Reactor.

TABLE 8-4   Distribution of World Reactors by Type, 2004 and 2015

transfers heat to water in a heat exchanger, and the steam thus generated then operates a turbine. Experimental-scale reactors of this type are in use in France, Japan, and Russia. It has proven difficult to use liquid sodium on an ongoing basis without encountering leaks in the reactor system. Furthermore, the high capital cost of breeder reactors compared to other reactor alternatives presently is not acceptable, given the relatively low cost and widespread availability of uranium resources at the present time. However, some governments are continuing to develop this technology, in anticipation that it will eventually become necessary when supplies of U-235 become more scarce. High-temperature gas cycle breeder reactor designs are also under development. The Generation IV International Forum, a consortium of 11 countries led by the United States, has designated a total of six technologies for development in this area, including breeder reactors using liquid sodium or high-temperature gas.

In summary, the distribution of reactors by type around the world is shown in Table 8-4. PWR/BWR technology is dominant at present, but as the current generation of reactors is retired at the end of its lifetime, it is likely that other reactor types, including APWR, pebble bed, heavy water, and possibly breeder reactors will gain a larger share of the market in the future. Also, it is the intention that the new designs will reduce the capital cost of nuclear energy, and thereby make it more competitive with other alternatives. It remains to be seen how much the capital costs will be reduced, what will happen with competing energy sources, and in the end the extent to which demand for new reactor designs around the world will grow in the years to come.

8-5   Nuclear Fusion

To take advantage of the most abundant nuclear fuel, namely, deuterium (and also tritium synthesized from lithium), scientists and engineers will need to develop a reactor that can sustain a fusion reaction at high enough levels of output to produce electricity at the rate of 1000 MW or more. The leading design to solve this problem is the *tokamak*, or toroidal magnetic chamber that uses a carefully designed magnetic field configuration and a high vacuum chamber to sustain deuterium and tritium in a high-temperature plasma where fusion can take place. The magnetic field keeps the plasma away from the walls of toroid, and by-products of fusion (neutrons and alpha particles) leaving the plasma collide with the walls of the chamber and a thermal breeding blanket, transferring heat to a working fluid which can then power a turbine cycle.

Advantages of fusion in regard to radioactive by-products are twofold. First, the main product of the fusion process is helium, so there is no longer a concern about high-level nuclear waste as in nuclear fission. Second, if materials development efforts are successful, the materials that are used in the structure of the reactor and that are activated by continued exposure to high-energy neutrons will have a short half-life, so that their disposal at the end of the fusion reactor's lifetime is not a long-lasting concern.

Current and recent research activities related to developing fusion are distributed among several sites around the world. The Tokamak Fusion Test Reactor (TFTR) at Princeton University operated from 1983 to 1997, and was designed on a 10-MW scale. The Joint European Toroid (JET) in Oxfordshire, England, was operated collaboratively by European governments from 1983 to 1999, and since then by the U.K. Atomic Energy Agency. JET is scaled to be able to achieve 30 MW of output, and is currently the world's largest tokamak. The Japan Atomic Energy Agency carries out fusion research at its JT-60 facility, which is of a similar size to JET. Other tokamak research facilities exist as well in other countries.

In 2006, representatives of the European Union, along with the governments of China, India, Japan, Russia, South Korea, and the United States, signed an agreement to build the next-generation *International Thermonuclear Experimental Reactor (ITER)*, in France. The goal of this project is to successfully demonstrate all aspects of fusion reactor science and some of the engineering aspects. This stage will be necessarily followed by a demonstration commercial reactor that will use the experience of ITER to adapt fusion for demonstration of all engineering science required for electric power production, and deliver electricity to the grid by 2045. Objectives for achieving interim milestones include the first creation of plasma in 2019 and initial operation using a deuterium-tritium reaction in 2027. Although output from this first grid-connected reactor is not expected to be cost-competitive with alternatives at that time, as the technology is further developed and replicated, costs should come down to competitive levels sometime in the latter half of the twenty-first century.

Although the majority of funding for nuclear fusion research at this time is dedicated to the development of the tokamak design, as pursued by ITER (see Fig. 8-4), other designs are also being researched, as follows:

- *Stellarator:* Like the tokamak, the stellarator concept seeks to create a sustained fusion reaction in a toroidal plasma whose shape is controlled by magnetic fields. Instead of the symmetrical torus shape used in the tokamak, the stellarator uses twisted coils whose configuration precisely confines an asymmetrical torus shape without requiring a current to be passed through the toroid, as is the case with the tokamak. In theory, this approach should allow physicists to define the shape of a plasma field that is optimal for fusion, and then design the coil configuration that will create the desired shape.

- *Reverse-field pinch:* A reversed-field pinch (RFP) is similar to a tokamak, but with a modification of the magnetic field that has the potential to sustain the plasma configuration with lower fields than are required for a tokamak with the same power density. The magnetic field reverses in the toroidal direction as one moves outward along a radius from the center of the toroid, giving this configuration its name.

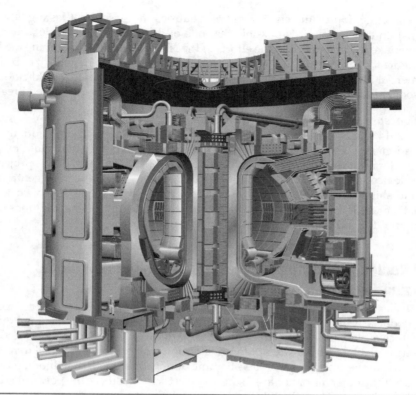

FIGURE 8-4 Tokamak design under development by ITER. (*Source:* ITER Organization. Reprinted with permission.)

- *Field-reverse configuration:* A field-reverse configuration is a magnetically confined plasma that leads to a compact toroid without requiring an evacuated chamber with a torroidal shape, as is the case with the tokamak. Thus the engineered structure for this configuration has only an outer confinement while leading to potentially lower costs than the tokamak.

- *Inertial-confinement fusion:* Inertial-confinement fusion is an approach that does not use a plasma confined by a magnetic field. Instead, the deuterium and tritium fuel is compressed to 100 times the density of lead in a controlled space to the point where fusion can take place. Current experiments aim to use lasers and a quantity of fuel a few millimeters in diameter held at a "target" point to test the feasibility of creating a fusion reaction and releasing net positive amounts of energy.

As with tokamak research, laboratory facilities in several countries are pursuing one or another of these approaches to fusion. One of the largest efforts is the National Ignition Facility (NIF) at the Lawrence Livermore National Laboratories in California, which was inaugurated in 2009 for the purpose of conducting inertial-confinement fusion experiments. Since its inauguration, the NIF has been scaling up its capability to conduct these experiments, and in June 2011, the facility set a record for productivity, generating approximately 200 trillion neutrons in a laser shot lasting 20 nanoseconds (ns). In conclusion, it is possible that one of these variations on nuclear fusion might overtake the tokamak and the efforts of the ITER consortium to become the leading candidate for nuclear fusion for the future. Even if this were to happen, it appears that commercial electricity from nuclear fusion at a competitive price would not arrive much sooner than projected by the ITER timetable, barring some unexpected breakthrough.

Concept for Small-Scale Nuclear Fusion

In 2014, the U.S. engineering firm Lockheed Martin announced an initiative to develop and market a smaller-scale (100-MW) nuclear fusion device that could be produced in greater numbers and at lower levelized cost per kWh. The goal of this project is to create a device that is 7 m long in its longest dimension, so that it can be centrally produced and shipped to its location of installation. Along with economies of scale, a potential advantage is that small size allows the firm to go through testing iterations more quickly than ITER and improve the design more rapidly.

8-6 Nuclear Energy and Society: Environmental, Political, and Security Issues

Nuclear energy is a unique energy source, in terms of both its fuel source and its by-products; it has therefore posed a number of interesting and challenging questions with which society has grappled for the last five decades. Nuclear energy is also a pioneer in terms of a large-scale attempt to use a resource other than fossil fuel to meet electricity needs in the era after fossil fuels became dominant in the first half of the twentieth century. (The other major nonfossil source of electricity, hydroelectric power, has been a part of electricity generation since the beginning of the electric grid.) Therefore, many of the questions posed by the attempt to use nuclear power to move away from fossil fuels are relevant to other alternatives to fossil fuels that are currently emerging, or may emerge in the future.

8-6-1 Contribution of Nuclear Energy to Reducing CO_2 Emissions

In recent years, nuclear power has surpassed hydroelectric power as the largest source of nonfossil electricity in the world. Therefore, the use of nuclear power makes a significant contribution to slowing global climate change, since electricity generated might otherwise come from the combustion of fossil fuels, resulting in increased CO_2 emissions.

The use of nuclear power can also help to lower the per capita carbon emissions of countries with a high percentage of electricity from nuclear generation. Taking the case of France in 2014, some 77% of France's total demand for electricity of 544 billion kWh are from nuclear power plants, according to Table 8-1. France's per capita CO_2 emissions are relatively low compared to other peer countries, at 5.7 tonnes per person per year.[5] If the nuclear-generated electricity were replaced with fossil-generated electricity at the U.S. average of 0.61 tonnes CO_2/MWh, French per capita emissions would increase to 9.7 tonnes per person per year. Figure 8-5 shows how per capita emissions for France have been consistently low compared to those of peer European countries, and also of the United States, for the period 1990–2011. In the period 2004–2011, the four comparison countries saw a decline in per capita CO_2 emissions of between 8%

[5]As a means of providing background, the per capita energy consumption values in 2011 for France, Germany, Netherlands, U.S., and U.K. were 175, 174, 258, 330, and 141, respectively.

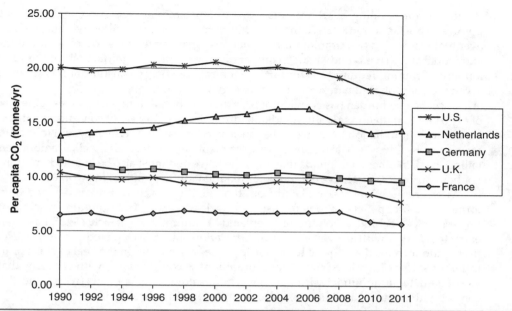

FIGURE 8-5   Comparison of per capita CO_2 emissions 1990–2011 for France and peer countries (Germany, the Netherlands, and the United Kingdom), and also the United States. (*Source:* U.S. Energy Information Administration.)

and 19%, while France's declined by 15%. France's CO_2 emissions per capita remain 2.1 tonnes/year lower than the nearest other country, namely, the United Kingdom. High share of nuclear power is not the only factor contributing to low CO_2 intensity, since land use patterns, distribution of transportation modes, mixture of industries that contribute to the economy of a country, and other factors, all play a role. Nevertheless, it is clear that use of nuclear power instead of fossil fuels for generating electricity helps France to hold down carbon emissions.

8-6-2   Management of Radioactive Substances during Life Cycle of Nuclear Energy

Radioactive by-products that result from the nuclear energy process include both materials from the nuclear fuel cycle and equipment used in the transformation of nuclear energy that becomes contaminated through continued exposure to radioactive fuel, impurity elements, and cooling media. Although these waste products, especially high-level radioactive wastes, are moderately to very hazardous, they are also produced in volumes that are much smaller than other types of waste streams managed by society (e.g., municipal solid waste, wastewater treatment plant effluent, and so on). Therefore, it is easier to retain these waste products in temporary storage while engineers devise a permanent solution for the long term, than it might be if they were produced in the same order of magnitude as other wastes.

Radioactivity from Extraction Process

The extraction of uranium for use in nuclear power plants involves the removal of material covering the resource, or "overburden"; the removal of the uranium ore; the milling of the ore to isolate the uranium; and the enrichment of the uranium to a grade that is suitable for use in a reactor.

Removal of mining overburden is like any other mining operation, and does not involve exposure to radioactive materials. In terms of the removal of overburden and eventual restoration of a uranium mine when resource extraction is complete, uranium extraction may have an advantage compared to open-pit coal mining, since the total amount of mining required to support a given rate of energy consumption is smaller, given the concentrated nature of nuclear energy in uranium ore.

Once the overburden has been removed, extraction of uranium ore involves some risk of exposure to radiation. Although radiation from uranium ore is not concentrated in the same way that radiation from high-level nuclear waste is, long-term exposure to the ore can elevate health risks. The milling process is potentially hazardous as well; uranium mill tailings contain other compounds that are radioactive and must also be handled carefully. For both processes, protective equipment is required for the health of the worker.

After uranium has been isolated from the ore, it is enriched if necessary and then formed into fuel pellets or rods for use in a nuclear reactor. The uranium enrichment process leaves behind a volume of "depleted" uranium from which U-235 has been extracted, that is, with an even greater concentration of U-238. Depleted uranium is useful for applications that require high density materials, such as in the keel of ships or in munitions. Use of depleted uranium in munitions is controversial, with concerns that use of depleted uranium in warfare exposes both soldiers and civilians (who come into contact with the uranium after hostilities end) to excessive health risk.

Radiation Risks from Everyday Plant Operation

Radiation risk from everyday plant operations consists of risks to employees, and risks to the general public both in the immediate vicinity of the plant and further afield. Risks to the employees at the plant are managed through protective equipment and clothing, and through the use of radiation badges. Tasks at nuclear power plants are designed to keep employees well below the maximum allowable threshold for radiation exposure, and continuous wearing of the radiation badge gives the plant worker assurance that health is not at risk.

The radiation risk to the public from modern nuclear plants with state-of-the-art safety systems is extremely small. This is true because the release of harmful amounts of radiation requires not only that the reactor experience a major failure (reactors are strictly prohibited from releasing radiation during normal operating conditions) but that the containment system designed to prevent radiation release must fail as well. The chances of both events happening are infinitesimally small in countries with strong safety standards such as the United States or France. A comparison of failures at different plants illustrates this point. In the case of TMI, the failure of the reactor was a very low probability event that nevertheless cost the utility, and its customer base, a significant financial amount in terms of both the loss of the reactor and the resulting cleanup requirements. Any operator in the nuclear energy marketplace has a very strong incentive to design and operate its plants so that such an event will not be repeated. However, the containment system did keep the contaminated fuel and coolant from escaping, and there was no perceptible health effect on the public in the subsequent months and years. Operators did in the days after the accident release contaminated steam and hydrogen to the atmosphere, but after making the judgment that the released material would not pose a health risk. Furthermore, after TMI, plant operators made changes to control equipment to further reduce the risk of a repeat accident of this type.

The Chernobyl accident had the unfortunate effect of persuading many members of the public that nuclear power is inherently dangerous in any country. However, one of the key failures of the Chernobyl design was precisely the lack of a comprehensive containment system that succeeded at TMI to prevent the escape of large amounts of radioactive material. RBMK designs of the Chernobyl type are being phased out or upgraded to international safety standards, and once this task is complete, it will no longer be accurate to make assessments of the safety of nuclear power on the basis of the Chernobyl accident.

More recently, the accident at the Fukushima plant in Japan in March 2011 showed that, although the containment structure combined with the ability to institute a mandatory evacuation worked adequately to protect both workers and residents in the region around the plant from immediate, acute harm, the risks from disruption and economic loss are the primary weakness of nuclear energy, especially in a seismically active region. It is difficult to anticipate all possible low-probability but high-consequence events. Indeed, the designers of the Fukushima plant had anticipated the risk of a tsunami and had built a sea wall around the plant. The earthquake that took place, however, was of a magnitude that might occur only once every 1000 years, so the sea wall proved to not be high enough. Other sources of energy, such as renewable energy systems with the distributed nature of energy conversion assets such as wind turbines or solar collectors, may have an advantage in terms of their ability to withstand or recover from such an event. To summarize, the Fukushima accident showed that radiation risks from operating a nuclear plant can be minimized even in the face of an extremely challenging natural disaster, although given the disruption to the Fukushima workforce, the residents in the region around the plant, the Tokyo Electric Power Company (or TEPCO, which owns the plant), the function of the Japanese electric grid, and the Japanese economy in general, this point may provide little consolation.

Management of Radioactive Waste

Radioactive waste from nuclear energy comes in two types, namely, *high-level waste*, primarily made up of spent fuel from reactors, and *low-level waste*, which includes reactor vessels at the end of their life cycle, cooling water from nuclear reactors, garments that have been used by nuclear plant employees, and other artifacts. The main objective in management of low-level waste is to keep these moderately radioactive materials separate from the general waste stream. Facilities for storing low-level waste therefore tend to be widely distributed. The management of high-level waste is more complex, and is the focus of the rest of this section.

At the end of their useful life in the reactor, fuel rods are removed by remote control (due to high levels of radiation) from the reactor and initially placed in adjacent pools of water inside the containment of the reactor structure. After a period of several months, the fuel rods have cooled sufficiently that they can be moved into longer-term storage outside the reactor. To prevent radiation risk once the fuel is outside the reactor, and to protect the waste during shipment, they are placed in storage containers made of lead and steel. Storage facilities for holding the waste containers may be on the site of the nuclear plant, or in some off-site location.

Unlike other types of wastes, high-level wastes require ongoing monitoring for two reasons:

1. *High levels of radiation:* Facilities must be carefully monitored to protect the health of the public.
2. *Use of wastes for unlawful purposes:* High-level wastes could provide potential materials for illicit weapons (see Sec. 8-6).

A useful goal is therefore to have high-level waste concentrated in relatively fewer facilities, located far from population centers. Although a number of governments are working toward permanent storage facilities that address these needs, at the present time there are none in operation, and all high-level waste storage is considered to be temporary and awaiting further technological developments.

The movement of high-level wastes from nuclear plants to distant storage sites naturally requires a carefully devised system for moving the wastes with minimal risk. The worldwide nuclear energy industry has developed a system for safe waste movement that includes extremely secure waste movement containers, routing of wastes over the transportation network to minimize risk to the public, and a network of recovery facilities in key locations to provide support in case of emergencies. To date, there have been no major accidents involving the large-scale leakage of nuclear waste during shipment, and although the occurrence of such an event in the future is theoretically possible, its probability is very low.

One alternative to long-term storage of spent nuclear fuel is *reprocessing*, in which useful components are extracted from spent fuel in order to be returned to nuclear reactors in *mixed-oxide pellets*, containing both uranium and plutonium. Levels of reprocessing vary around the world. France, Japan, and the United Kingdom carry out significant amounts of reprocessing, with major reprocessing centers at Sellafield in the United Kingdom, at La Hague in France, and at Tokaimura in Japan. Reprocessing has been banned in the United States since the 1970s as a precaution against nuclear weapons proliferation, so U.S. plant operators instead use uranium on the *once-through cycle* from extraction, to single use in the reactor, to long-term storage of the used fuel rods as high-level waste. An effort is underway to lift the ban so as to accommodate an anticipated rise in the amount of spent nuclear fuel generated in the future, and it is possible that the ban will be lifted. As of 2015, however, the ban had not been overturned.

Another concept for managing nuclear waste is to "transmute" long-lived isotopes in the waste stream into shorter-lived ones by inserting them into an appropriately designed reactor. The long-lived isotopes would be separated from others in the waste stream and then positioned in the reactor so that they collide with neutrons and fission or otherwise are transformed into isotopes with shorter half-lives.

Yucca Mountain: A Debate about the Best Strategy for High-Level Waste

One of the leading efforts made to date to develop a permanent storage facility for high-level nuclear waste is the underground facility at Yucca Mountain, some 150 km northwest of the city of Las Vegas, in the state of Nevada in the United States. In this proposed facility, tunnels are to be dug into igneous rock deep below the surface; waste containers are then placed in the tunnels, which are sealed once they are filled. The location has been chosen to minimize risk of seismic activity or seepage of water into the waste. The tunnels are partially completed as of 2015. Although construction was indefinitely suspended after 2007, projected cost of the final facility is $60 billion, should it eventually be completed.

Yucca Mountain has been the subject of a protracted legal battle between the U.S. federal government, which is its principal sponsor, and the state of Nevada, which opposes the facility. In 2004, federal courts ruled in favor of the federal government, which appeared to open the door to eventual completion of the project. At that time, a timetable was set that would have allowed the facility to begin receiving waste in the year 2017. However, with the change in the leadership of the U.S. Senate in 2007, political leaders in the legislative branch were able to greatly reduce funding for the project, effectively stopping its continued development for the indefinite future. It is not clear if work were to resume how long it would take to complete the project, or if instead the project will be permanently abandoned and the site of the first permanent disposal facility in the United States will move elsewhere.

Assuming Yucca Mountain eventually opens and operates as planned, it should achieve the goals of permanent storage, namely, that the high-level waste will be removed from the possibility of harming people, and theft of the waste for illicit purposes would be virtually impossible. The facility design could then be reproduced in other geologically stable formations around the world, with the hope that the lessons learned at Yucca Mountain site would reduce the cost per tonne disposed at other sites.

Even as the construction of the facility has been halted, the debate continues over its value. Proponents claim that Yucca Mountain will allow nuclear plant operators to take responsibility on behalf of both present and future generations to permanently solve the waste disposal problem. They point out that the cost of the facility, while large in absolute terms, is small relative to the total value of electricity represented by the waste that it will eventually contain. Depending on the amount of funds eventually gathered from surcharges on nuclear electricity that are dedicated to waste disposal, it may be possible to cover the cost entirely from these funds, though this cannot be known until the project is finally completed.

Detractors claim that it is both expensive and premature to put nuclear waste beyond reach in an underground facility, when other less costly options (including not moving the waste at all for the time being) are available. For example, a well-guarded ground-level facility, in a remote location and with sufficient capacity to store large amounts of waste, would achieve the same aims for a fraction of the cost. Such a location might be found in the United States in one of the western states, which contain large expanses of land that are for the most part sparsely populated. [As an indication, the average population density of the state of Nevada outside of the great Las Vegas metropolitan area is 7 persons per square mile (sq. mi.), compared to 1196 per sq. mi. for the state of New Jersey.]. Alternatively, the government might maintain the status quo, in which high-level waste is in temporary distributed storage at nuclear plants across the United States. Ground-level storage, either in situ in current plant locations or in a future centralized facility, would also allow for the high-level waste someday to be "neutralized" in a permanent way, should researchers eventually develop the necessary technology. Potential options include the transmutation to isotopes with shorter half-lives mentioned earlier, or reuse of the waste for some other purpose that is yet to be identified. Skepticism has also been raised that underground storage, even in a very stable location such as Yucca Mountain, can truly be "guaranteed" to be safe for the lengths of time envisioned, and that some future unintended consequence might emerge that would be difficult to mitigate, once the wastes had been sealed underground in large quantities. Lastly, because the project cost has grown so significantly since its inception, the waste disposal fund may eventually prove inadequate to finance the entire facility, in which case the U.S. taxpayer would be forced to cover the shortfall.

In response to these objections, proponents might argue that the development of such a use or superior disposal method for high-level waste is highly unlikely, and that it is therefore better to take responsibility now for putting the waste into permanent underground storage, rather than postponing. They might also argue that a ground-level facility implies the presence of human staff in some proximity to the stored waste (for guarding and monitoring), and that this staff is then put at unnecessary risk of exposure to radiation.

The debate over Yucca Mountain illustrates the complexities of designing a permanent storage solution for high-level waste, many of which revolve around the long-lasting nature of the hazard. Other efforts to develop permanent solutions involving deep geologic burial are proceeding slowly in a number of other countries, including China, Germany, Sweden, and Switzerland. If the United States or any of these countries can eventually complete and bring into operation a geological storage facility, this achievement would likely accelerate development in the other countries. However, many of the same concerns about the long-term safety and viability of underground permanent disposal are being raised in these other countries as in the United States, so it may be some time before any such facility is completed, if at all.

Total Volume of Nuclear Waste Generated

The problem of the disposal of high-level nuclear waste requires an eventual solution, but at the same time does not pose an imminent threat such that the nuclear industry must act overnight. This is true because the total amount of high-level waste accumulates at a very slow rate. Example 8-3 illustrates this point.

Example 8-3 Uranium dioxide, a common nuclear fuel, has a density of 10.5 g/cm^3. Assuming that the fuel used is pure uranium dioxide, and ignoring the effect of changes to the composition of the fuel during consumption in nuclear reactor, estimate the total volume of waste generated each year by the world's nuclear reactors.

Solution From Table 8-1, the world's total nuclear electricity output in 2006 was 2661 billion kWh. Using the values from Example 8-2 as an approximation of the average rate of UO_2 consumption in nuclear plants, each year a plant generating 9.26 billion kWh yields 58,600 kg in spent fuel. On this basis, nuclear plants around the world would generate (58,600 kg) × (2661)/(9.26) = 1.68 × 10^7 kg of waste. This amount is equivalent to 1.60 × 10^9 cm^3 of volume, or approximately 1600 m^3. To put this volume in visual terms, if it were laid out on the playing area of a standard tennis court (32 m × 16 m), it would fill the area to a depth of approximately 3 m. This calculation assumes no reprocessing of the fuel.

The actual footprint of land area required to store high-level waste at ground level is actually higher than the value given in the example, since the volume of the containment vessel is not included, and space must be allowed between each unit of stored waste for access and heat dissipation. Nevertheless, the total area required is still small, compared to other types of waste streams. Fuel reprocessing has the potential to further reduce the total area occupied by waste storage, since 80 to 90% of the waste stream is diverted into new fuel pellets.

In conclusion, should we be concerned about finding a permanent solution for waste storage? Given the recent growth in concern about other environmental challenges such as climate change or protection of the world's supply of clean water, it is not surprising that nuclear waste as a global issue has been pushed out of the media spotlight, except in regions that are directly affected by disposal plans, such as in the region around Yucca Mountain. There are at least, however, two compelling reasons why society should continue to push toward a solution in a timely fashion. First, the total demand for nuclear energy may grow significantly in the future. Under a "robust" nuclear future scenario, total demand for electricity and the total share of electricity generated by nuclear power may both grow rapidly, greatly accelerating the rate at which high-level waste is generated. Under such a scenario, the logic that the increment of waste added each year is small, may no longer hold, especially if the amount of reprocessing does not also increase. Second, the generation of waste without providing a solution over the long term violates the "intergenerationality" principle of sustainable development, since future societies may be left to solve a problem that they did not create, or suffer adverse consequences if they are not able to do so.

8-6-3   Nuclear Energy and the Prevention of Proliferation

The relationship between commercial use of nuclear energy to generate electricity, and military use of nuclear technology to develop weapons, complicates the international status of the nuclear energy industry. Uranium enrichment technology that is used to increase the concentration of U-235 for use in commercial reactors can also be used to enrich uranium to the point where it can be used in a weapon. Commercial energy generation and nuclear weapons production can be located in the same facilities. It is difficult to separate the two.

In order to prevent *nuclear proliferation,* or the spread of nuclear weapons capabilities to states not currently possessing them, the world community has ratified and put into force the Nuclear Nonproliferation Treaty (NPT). The NPT makes a distinction between nuclear states and nonnuclear states. Major provisions include the following:

1. Nuclear states are prohibited from transferring nuclear weapons technology to nonnuclear states.

2. Nuclear states are prohibited from using, or threatening the use of, nuclear weapons in aggressive (i.e., nondefensive) military action.

3. Nonnuclear states agree not to develop nuclear weapons, and to subject their nuclear energy facilities to inspection by the International Atomic Energy Agency (IAEA), to verify that they are meeting this requirement.

The nuclear states that adhere to the treaty include China, France, Russia (previously the USSR was party to the treaty), the United Kingdom, and the United States. Most of the remaining nations of the world are signatories to the NPT as nonnuclear states. There are some exceptions as well: India and Pakistan, for example, are not signatories to the NPT and have both tested nuclear weapons in the past.

While the NPT has been fairly effective in preventing nuclear proliferation, it has inherent loopholes and weaknesses that are difficult if not impossible to entirely eliminate. First, there is no mechanism within the NPT to prevent a nation from withdrawing from the treaty in order to develop nuclear weapons. Second, the NPT does not specifically limit the ability of nonnuclear parties to develop dual-use technologies such as uranium enrichment. If other parties suspect that an NPT signatory nation is developing enrichment for the purposes of weapons production, the treaty does not provide any means to address the situation effectively. Two recent experiences with NPT members and former members illustrate these weaknesses:

- *North Korea:* In 2003, the government of North Korea formally withdrew from the NPT, signaling its intentions to develop nuclear weapons. By late 2006, the North Korean military was able to successfully test a small nuclear device. Although the international community attempted to block this development through diplomatic pressure in the intervening 3 years, it was unable to do so.

- *Iran:* At present a conflict between Iran (an NPT signatory nation) and other UN member states is ongoing, regarding the nature of Iran's uranium enrichment efforts. The Iranian government claims that its enrichment program is for peaceful purposes only, and is necessary for it to develop an alternative to the generation of electricity from fossil fuels. Out of fear that the ability to enrich uranium will eventually lead to Iran obtaining a nuclear weapons capability, the UN demanded that the enrichment program be halted, and in response to Iran's refusal, passed a series of limited sanctions. These include the June 2010 imposition of restrictions on weapons purchasing, financial transactions, and diplomatic travel. At the time of writing in 2015, Iran and other parties to a 2-year negotiating process had reached tentative agreement on a treaty to restrict Iran's nuclear program in return for sanctions relief. If the treaty is ratified and goes into effect, it would resolve a long-standing disagreement.

Without delving into details of either of these conflicts, the underlying issue is that in an era of international mistrust, much of which stems from a previous history of armed conflict (e.g., on the Korean peninsula or in the Middle East), it is impossible to create an international agreement that is 100% certain to prevent proliferation. For the foreseeable future, the possibility exists that other states might feel compelled to leave the NPT and develop nuclear weapons, claiming the need for self-defense. We can hope that, many decades from now, these sorts of international tensions might dissipate to the point that nonnuclear states would develop their commercial nuclear programs without other states fearing that they will also develop weapons, or indeed that the nuclear states might completely dismantle their arsenals, thus eliminating the distinction between nuclear and nonnuclear states. Leading senior world diplomats such as George Schultz and Henry Kissinger have exhorted governments to take steps in this direction (see, e.g., Schultz et al., 2011). In the short to medium term, however, a commitment to preventing proliferation through ongoing monitoring and diplomacy is an essential part of the peaceful use of nuclear energy. It must therefore be seen as part of the cost of nuclear-generated electricity. Furthermore, if the amount of nuclear electricity generation grows in the future and the state of international relations does not improve, the need for monitoring will intensify, and the risk that a nonnuclear state may secretly develop a weapon may increase as well.

8-6-4   The Effect of Public Perception on Nuclear Energy

Many energy technologies, nuclear energy among them, are hampered by a lack of understanding or trust by the public, which slows their deployment and use. While engineers and other professionals in related technology fields may dismiss problems of perception as not being "real" problems, and therefore not relevant to their line of work, the power of these problems should not be underestimated. If enough people believe incorrect information about a technology having some serious flaw, and base their opinions and decisions about that technology on this information, then the "perceived" flaw in the technology is as large of a problem as if it were an actual flaw.

A comparison of nuclear and fossil energy illustrates this point. In 50 years of commercial use of nuclear energy, acute exposure to radiation in nuclear-related facilities has led to very few radiation deaths, those of firefighters at Chernobyl and of workers at an accident at a reprocessing facility in Tokaimura, Japan, in 2009 being the two exceptions. It is possible that there have been a larger number of premature deaths (i.e., not stemming from emergency circumstances) from long-term exposure to radiation brought about by nuclear energy, including both workplace and mining exposure, and exposure caused by the Chernobyl accident. Controversy surrounds this issue, since it is difficult to isolate this contribution from the background rate of cancer and other chronic, life-threatening disease. In the most conservative estimate, nuclear energy might contribute to hundreds or thousands of premature deaths per year, but not more, and probably less. Industrial accidents in the nuclear energy supply chain may also claim some number of lives, though these risks are associated with any energy source and are not particular to nuclear energy. For example, use of heavy equipment, whether in uranium or coal mines, in large-scale energy conversion plants, or in the installation and maintenance of wind turbines, entails some risk, but these risks can be minimized with proper safety equipment and practices, and there is no evidence that the nuclear energy industry is worse than any other.

By comparison, the effect of fossil fuel combustion on human health is orders of magnitude worse. This is especially true in the emerging countries, where inadequate emissions control leads to severe air quality problems that cause literally hundreds of thousands of premature deaths per year. Emissions controls are better in industrial countries, and the state-of-the-art scrubbing technology on fossil-fuel-powered plants can eliminate all but a small fraction of harmful emissions, but many premature deaths still occur in these countries as well, especially among those who are vulnerable to poor air quality (e.g., those suffering from asthma and other respiratory diseases), because there remains a large stock of antiquated power plants and vehicles that contribute to smog. There are additional effects on quality of life due to poor health, on lost productivity due to workplace absenteeism, and on nonhuman living organisms that also suffer from poor air quality. Yet, despite this toll of death and suffering, there is no widespread outcry that might dramatically change the way fossil fuels are used.

Thus it appears that nuclear energy has been "handicapped" in the debate over energy choices that has taken place in the "court of public opinion," so that the world's energy resources priorities have not been those that statistics on mortality and health might dictate. Some of the contributing factors include:

1. *A different standard for high-consequence, low-risk events, versus chronic risk:* Social scientists have long noted that many people weigh the risk of death in a large-scale disaster more heavily than death in an event that happens on an individual basis. Thus flying in airplanes induces profound fear in some consumers compared to driving on the highway, even though, statistically, flying is much safer than driving, per unit of distance traveled. An analogy can be made between the steady-state and continuous loading of the atmosphere with harmful emittants from fossil fuel combustion, and the hypothetical loading of the atmosphere with highly poisonous reaction by-products in a nuclear accident. In the latter scenario, according to folk wisdom, a nuclear reactor meltdown might spew clouds containing dangerous radioactive elements, which would come into contact with the human population once fallen to Earth, leading to thousands of poisoning deaths. This scenario is in fact close to impossible, with modern reactor designs other than the RBMK reactors, which had no enclosure to prevent the escape of radioactive by-products. Even a complete meltdown of a modern reactor vessel would be contained within the containment system. Nevertheless, fears such as these lead many consumers to consider nuclear energy to be more dangerous than fossil fuel energy, so that they prefer to tolerate compromised air quality in order to avoid the perceived greater "risk" from nuclear energy.

2. *The association of controlled nuclear fission in energy generation with uncontrolled fission in nuclear weapons:* There is a misperception in some quarters of the general public that the controlled reaction in a nuclear plant could, if control systems were to fail, degenerate into an uncontrolled, accelerating chain reaction, leading to a nuclear explosion. According to this logic, a terrorist attack on a nuclear plant might provide the catalyst for the explosion, for example, the impact of a missile or airplane might provide a source of heat or impact that would accelerate the reaction and at the same time disable the control systems that normally function, triggering an explosion. Alternatively, a reactor core might melt down with such force that it would melt through the entire containment structure and engulf the surrounding area in a runaway nuclear reaction. Theoretical calculations and empirical experience both show that either scenario is either virtually or completely impossible. For example, the partial meltdown at TMI barely penetrated the 13-cm (5-in) thick steel casing of the reactor vessel, and crash-testing of aircraft at high speeds into structures that replicate a reactor containment system has shown that the structure of the aircraft collapses without destroying the building. Furthermore, by the laws of physics, a nuclear reaction cannot consume nonfissile material outside the reactor. Nevertheless, some members of the public continue to believe that a nuclear plant poses a risk of a nuclear explosion in case of a catastrophic failure or terrorist attack.

3. *The association of the nuclear energy industry in nuclear states with secretive nuclear weapons development programs:* In the nuclear states, the development and maintenance of nuclear weapons is by necessity a closely guarded secret; among other reasons, governments in these states are committed to preventing proliferation, including through theft of plans or materials by agents of nonnuclear states or terrorist organizations. The civilian nuclear power industries in many of these states began as spin-offs from nuclear weapons programs, so that early nuclear power operations inherited a "culture of secrecy" from their military predecessors. In time, public opinion came to view both military and civilian applications of nuclear energy with suspicion, and it has proven difficult to overcome this history of mistrust. In particular, in some countries, the nuclear industry downplayed both the time required and cost of permanent disposal of high-level waste (e.g., with the Yucca Mountain project), and when initial projections of both time and cost proved overly optimistic, the industry was faulted.

Public perception factors such as these have hindered the growth of nuclear energy in the past. In recent years, nuclear energy industries in a number of countries have taken steps to address this obstacle by becoming more open about explaining the function of nuclear energy, and sharing data about its operation, including rates of resource consumption and waste generation, as well as the cost of generating nuclear electricity. Industry representatives have also made the case for the safety of nuclear power and for its contribution to prevent climate change. It is difficult to judge the overall effectiveness of these efforts. It appears that in the United States and some of the European countries prior to the 2011 Fukushima accident, criticism of nuclear power was more muted and part of the population was more open to supporting some construction of new nuclear plants to meet growing demand for electricity while limiting CO_2 emissions. In the aftermath of Fukushima, the pendulum of public opinion swung in the other direction and enthusiasm for new plants was dampened. In any case, lessons learned from the experience of trying to introduce a new, large-scale energy industry, including mistakes made, are valuable for other emerging energy sources, such as large-scale wind energy, which is also facing opposition in some regions.

8-6-5 Future Prospects for Nuclear Energy

The future of nuclear energy can be divided into two phases: a short- to medium-term phase, approximately to the middle of the twenty-first century, and a long-term phase that lies beyond.

In the short to medium term, the world can expect electricity generation from nuclear energy on the order of trillions of kWh per year, as shown in Table 8-1, for at least two or three decades more. Some countries such as Germany and Sweden may be curtailing nuclear power generation going forward, but others such as China and India have recently added plants, with the IAEA listing 67 plants around the world in various stages of construction in 2015. In the United States, the situation is mixed, with the Bellefonte project going on hold in 2009 but the Watts Bar 2 project continuing with an expected completion in 2016. The ongoing challenge with reducing world annual CO_2 emissions will lend support to the continued operation of existing plants, and the completion of at least some of those under development. A key variable is the resolution of the Fukushima accident. If the community of nations concludes that the problems of Fukushima can be avoided in the future through changes in operating practices or improvements in design going forward, new plants may be started. If not, enthusiasm for new plant starts will wane. For any scenario of plant development based on fission of U-235, fuel supplies will be sufficient through 2050 to prevent any problems with price rises or supply scarcity.

Beyond the middle of this century, long-term continued use of nuclear energy requires that two problems be solved. First, new fission and/or fusion technologies must evolve to levels of technical and economic performance so that the total amount of nuclear electricity that can be produced, based on available resources of deuterium, tritium, U-238, or other nuclear fuel, is greatly expanded. (Alternatively, the horizon for current fission reactor technology might be greatly extended by developing unconventional sources of U-235, such as extracting it from the oceans.) Second, a satisfactory solution for the storage of high-level waste must be found that is not only safe and cost-effective, but that also gives society confidence that future generations will not be adversely affected in a way inconsistent with the ideals of sustainable development.

8-7 Representative Levelized Cost Calculation for Electricity from Nuclear Fission

The following calculation of levelized cost per kWh is based on values and discussion throughout this chapter. It represents the construction of a newly built nuclear reactor in the United States, and does not assume any cost savings from economies of scale since such a reactor would be one of only a handful built since the 1970s. The chosen value for capital cost is thus conservative, and its implications are discussed further as part of the summary discussion in Chap.17.

We assume a capital cost of \$8000/kW for a 1 GW plant, resulting in an \$8 billion capital cost, as was discussed in a conservative case earlier in the chapter. The required discounting factor is (A/P, 20 years, 7%) = 0.0944 (see Chap.6, Sec. 6-8). This factor gives a capital cost per year of \$755 million. We assume a capacity factor of 88% (a typical value from the Nuclear Energy Institute), resulting in 7709 GWh/year of output. Dividing capital cost among output gives a value of \$0.098/kWh. For fuel and nonfuel operating cost, we use the typical values mentioned earlier of \$0.0043/kWh and \$0.017/kWh, respectively.

The final value is \$0.1193/kWh, as summarized in the calculations below:

$$\text{Output} = (1 \text{ GW})(8760 \text{ h/y})(0.88) = 7709 \text{ GWh/y}$$

$$\text{CapCost} = \frac{(\$8B)(0.0944)}{7709 \text{ GWh/y}} = \frac{\$755M}{7709} = \$0.098/\text{kWh}$$

$$\text{TotCost} = \text{CapCost} + \text{EnergyCost} + \text{OpCost} = \$0.098 + \$0.0043 + \$0.017$$

$$= \$0.1193/\text{kWh}$$

8-8   Summary

Nuclear energy is the conversion of energy available in nuclear bonds to other forms of energy that are useful for human purposes, most commonly electricity. Formulas that quantify the energy in the nuclear bond or the ratio of atomic number to mass that minimizes binding energy show that the most promising materials for use in nuclear fusion are at either the light or heavy ends of the atomic spectrum. Scientists and engineers have therefore pursued the harnessing of nuclear energy using either uranium or other very heavy elements in fission reactions, or deuterium and tritium in fusion reactions. An analysis of the function of today's nuclear reactors with typical efficiency values confirms that, based on current cost of nuclear fuels, existing plants are capable of producing electricity at a very low cost per kWh. Capital costs for new nuclear plants are high; however, current efforts are focused on new designs that can reduce the cost of nuclear power per installed kW, by making the designs more uniform. Breeder or fusion reactors may someday greatly expand the fuel available for use in generating nuclear energy.

Since nuclear plants produce electricity without emitting CO_2, there is a strong motivation for maintaining or expanding the use of nuclear energy systems. At the same time, issues surrounding financial risks from premature loss of plants due to accidents, the long-term disposal of nuclear waste, and the potential for nuclear weapons proliferation pose ongoing concerns for the nuclear industry. Long-term use of nuclear energy will require a more permanent solution for the waste problem, a system for preventing proliferation that is effective regardless of the number of reactors in use for civilian purposes, and an expansion of the fuel supply beyond U-235 obtained from mining, which supplies most nuclear reactors at present.

References

Beck, P. (1999). "Nuclear Energy in the Twenty-First Century: Examination of a Contentious Subject." *Annual Review of Energy & Environment*, Vol. 24, pp. 113–137.

Cohen, B. (1987). "The Nuclear Reactor Accident at Chernobyl, USSR." *American Journal of Physics*, Vol. 55, No. 12, pp. 1076–1083. Available online at http://adsabs.harvard.edu/. Accessed Dec. 1, 2007.

Glasstone, S., and A. Sesonske (1994). *Nuclear Reactor Engineering: Reactor Design Basics, Volume 1*. Springer, New York.

Glasstone, S., and A. Sesonske (1994). *Nuclear Reactor Engineering: Reactor Systems Engineering, Volume 2*. Springer, New York.

Nuclear Regulatory Commission (2011). *Recommendations for Enhancing Reactor Safety in the 21st Century: The Near-term Task Force Review of Insights from the Fukushima Dai-ichi Accident*. Technical report, available at www.nrc.gov. Accessed Aug. 5, 2011.

Schultz, G., W. Perry, H. Kissinger, et al. (2007). "A World Free of Nuclear Weapons." Guest editorial, *Wall Street Journal*, pp. A15, Jan. 4, 2007.

United States (1979). *President's Commission on the Accident at Three Mile Island*. The need for change, the legacy of TMI: report of the President's Commission on the Accident at Three Mile Island (also known as the "Kemeny Commission report"). President's Commission, Washington, DC. Available online at www.threemileisland.org. Accessed Dec. 1, 2007.

Further Reading

Chapin, D., K. Cohen, and E. Zebroski (2002). "Nuclear Power Plants and Their Fuel as Terrorist Targets." *Science*, Vol. 297, No. 5589, pp. 1997–1999.

Fanchi, J. (2004). *Energy: Technology and Directions for the Future*. Elsevier, Amsterdam.

Feiveson, H. (2003). "Nuclear Power, Nuclear Proliferation, and Global Warming." *Physics & Society*, Vol. 32, No. 1, pp. 11–14.

Fewell, M. P. (1995). "The Nuclide with the Highest Mean Binding Energy." *American Journal of Physics*, Vol. 63, No. 7, pp. 653–659.

Greenpeace (1997). *Nuclear Energy: No Solution to Climate Change—a Background Paper*. Greenpeace, Washington, DC. Report available online at http://archive.greenpeace.org/comms/no.nukes/nenstcc.html#6. Accessed Dec. 1, 2007.

International Atomic Energy Agency (2007). *INFCIS—Integrated Nuclear Fuel Cycle Information System*. Website, available at www.iaea.org. Accessed Jun. 25, 2007.

Lamarsh, J. (1983). *Introduction to Nuclear Engineering*, 2nd ed. Addison-Wesley, Reading, MA.

List, G. F., B. Wood, M. A. Turnquist, et al. (2006). "Logistics Planning under Uncertainty for Disposition of Radioactive Wastes." *Computers & Operations Research*, Vol. 33, No. 3, pp. 701–723.

Nuclear Energy Institute (2015). *Nuclear Statistics*. Website, available at www.nei.org. Accessed Jul. 14, 2015.

Rhodes, R., and D. Beller (2000). "The Need for Nuclear Power." *Foreign Affairs*. Vol. 79, No. 1, pp. 30–44.

Ristinen, R., and J. Kraushaar (2006). *Energy and the Environment*, 2nd ed. John Wiley & Sons, New York.

Turnquist, M. A., and L. K. Nozick (1996). "Hazardous Waste Management," Chap. 6. In: Revelle, C. and A. E. McGarity, eds., *Design and Operation of Civil & Environmental Engineering Systems*, pp. 233–276. Wiley-Interscience, New York.

U.S. Energy Information Agency (2006). *Current Nuclear Reactor Designs*. Website, available at http://www.eia.doe.gov/cneaf/nuclear/page/analysis/nucenviss2.html. Accessed Aug. 18, 2007.

Wald, M. L. (2004). "A New Vision for Nuclear Waste." *MIT Technology Review*, Dec. issue, pp. 1–5.

World Nuclear Association (2010). *International Standardization of Nuclear Reactor Designs*. Technical report, WNA, London. Electronic resource, available at http://www.world-nuclear.org. Accessed Aug. 6, 2011.

Exercises

8-1. For thorium-234 (atomic number 90), calculate the percentage difference between the actual ratio Z/A and the ratio Z/A for this atomic number that results in the lowest energy state.

8-2. Revisit Example 8-2. (a) If the cost of raw uranium is $65/kg, what percent of the cost of $40 million worth of fuel rods does this amount constitute? (b) If the cost increases to $260/kg, and all other cost components of the refueling remain the same, what is the new cost of the refueling each year?

8-3. Repeat the calculation of levelized cost per kWh for electricity in Example 8-2 applied to a new nuclear plant with a total capital cost of $2200/kW, an output of 1000 MW_e, an investment lifetime of 20 years, and an MARR of 5%. This is the cost of one of the most recently completed nuclear plants in the United States, at Seabrook, NH, according to one estimate. Include fuel, O&M, and capital cost in your calculation.

8-4. Carry out the rudimentary design of a small modular nuclear reactor which delivers 100 MW_e. The reactor is based on a Rankine cycle where the steam is expanded in a turbine from 12.5 MPa and 350°C to condensed water at 5 kPa. Assume that the generator is 98% efficient, the actual cycle achieves 90% of the efficiency of the ideal cycle, and 90% of the thermal energy released in the nuclear reactions is transferred to the working fluid. (a) If the reactor uses uranium as a fuel which is enriched to 3% U-235, what mass of fuel is consumed each year? (b) What is the flow rate of steam in kg/s required when the reactor is running at full power?

8-5. Suppose in Problem 8-4 that the reactor has a capital cost of $1500/$kW_e$, costs $5 million/year for refueling, and $6 million/year for nonfuel maintenance costs. If the project lifetime is 25 years and MARR = 6%, what is the levelized cost of electricity per kWh?

8-6. A reactor has a rated capacity of 1 GW_e and produces 7.9 billion kWh/year. It is fueled using fuel rods that contain 2.5% U-235 by mass. The electric generator is 98% efficient, the thermal cycle of the plant is 33% efficient and 90% of the thermal energy released in fission is transferred to the working fluid. (a) What is the capacity factor of the plant? (b) What is the rate of U-235 consumption in g/s in order to deliver the average output? (c) What is the total mass of fuel rods consumed each year? (d) Based on the amount of energy released in fission reactions, what is the reduction in mass of fuel each year?

8-7. Refueling the reactor in Problem 6 costs $35 million/year, and other maintenance costs are $90 million/year. What is the levelized cost of electricity: (a) if the plant is fully amortized, (b) if the capital cost of the plant is $2 billion and must be repaid over 30 years at MARR = 6%?

8-8. Suppose that the Yucca Mountain facility discussed in the chapter ends up costing $60 billion, and stores 70,000 tonnes of nuclear waste. Ignoring transportation costs for waste, and the effect of the time value of money, what is the surcharge that is required on each kWh of electricity to exactly pay for the facility, based on the amount of electricity production that results in this volume of waste? State your assumptions regarding the amount of waste generated per unit of electricity produced.

8-9. From the internet or other source, gather information on the rated capacity in MW of several nuclear reactors and the year of commissioning. Plot the rated capacity as a function of year. Does your graph agree with the claim in the chapter that reactor size has reached a plateau since the 1970s? Discuss.

8-10. For a country of your choice, calculate the amount of CO_2 emissions per year avoided by using nuclear power to generate electricity, based on total output from nuclear reactors and average emissions of CO_2 from fossil fuel sources.

8-11. Analyze the proposed small scale nuclear fusion reactor from Sec. 8-5 as follows. Assume capital cost of $4000/kW thanks to mass production, and assume an investment lifetime of 20 years at 7% with no salvage value. Assume that the value for fuel is similar to that of uranium fuel rods in Example 8-2, i.e., $800/kg, and further assume that for simplicity all of the deuterium will be reacted and converted to helium. Thereafter 85% of the energy released by the fusion reaction is transferred to the surrounding water vapor cycle, and 50% of the transferred energy leaves the device as electricity. Nonfuel operating cost is $0.01/kWh. The capacity factor for the device is 90%. What is the levelized cost per kWh?

CHAPTER 9

The Solar Resource

9-1 Overview

This chapter explains how light from the sun becomes available to do work on Earth, and acts as a background to determine how sunlight can be turned into useful electrical or thermal energy. Topics covered here include the measurement of energy available from the sun, the effect of the atmosphere on light transmission, the impact of angular geometry between sun, Earth, and the surface of a solar device, instantaneous versus integrated values, and variations of real data from averaged values.

9-1-1 Symbols Used in This Chapter

Many symbols are used in this chapter and recalling their definitions can become a nuisance. To help, Table 9-1 is provided. Angles can be used as degrees or radians, depending on the requirements of the calculator or computer software being used and the equation in which they are embedded or defined.

9-2 Introduction

Solar power and solar energy resources on Earth are enormous, nonpolluting, and virtually inexhaustible. Moreover, solar energy is the driving mechanism behind other renewable energy sources such as wind, hydropower, biomass, and animal power. Until the past two centuries and the exploitation of coal, oil, and natural gas, civilization grew and developed almost entirely based on solar energy in its various manifestations.

Solar intensity is more than 1 continuous kW/m^2 outside the Earth's atmosphere, while on the surface of the Earth the average daily interception is nearly 4 kWh/m^2. The solar energy intercepted by Earth in less than 1 month is equivalent of all the energy originally stored in the conventional energy resources of coal, petroleum, and natural gas on the planet. This chapter reviews techniques for measuring and understanding the availability of the solar resource, as background for understanding the devices explained in Chaps. 10, 11, and 12.

9-2-1 Availability of Energy from the Sun and Geographic Availability

The intensity of energy arriving from the sun in space just outside the Earth's atmosphere is approximately 1368 W/m^2, called the *solar constant*. Although it is termed a "constant," it varies over time. Solar flares and sun spots change the value slightly. The distance from the sun to the Earth is a greater factor, varying the value during the year.

Symbol	Definition, with Units (SI) Where Appropriate (Overhead Bar Signifies Averaged)
a_0, a_1, k	Empirical parameters, see Eq. (9-21)
a_i, b_i: $i = 1...4$	Empirical coefficients, see Eq. (9-3)
AB, CB	See Fig. 9-7
AM	Air mass, optical path length through the Earth's atmosphere, equals unity at sea level for insolation from the zenith, see Eq. (9-17)
csc	Cosecant of an angle, the inverse of the sin
ET	Equation of time, decimal minutes, see Eq. (9-3)
F	Cumulative distribution function, see Eq. (9-23)
H_A	Thickness of the atmosphere, see Fig. 9-8
H_{day}	Daily insolation integral (energy per m²) outside the Earth's atmosphere on a plane parallel with the Earth's surface at the same latitude and longitude
$\overline{H}$	Global daily insolation sum, thermal energy m⁻²
$\overline{H_d}$	Diffuse component of daily insolation sum, thermal energy m⁻²
$\overline{H_\beta}$	Global insolation on a tilted surface, thermal energy m⁻²
$\overline{H_\tau}$	Monthly averaged total daily insolation on a tilted surface, thermal energy m⁻²
$I_{\beta, b}$	Beam insolation on a tilted surface, W m⁻²
$\overline{K_\tau}$	Monthly average clearness index, ratio of solar irradiation on horizontal surface to solar irradiation outside the atmosphere on a horizontal plane directly above the surface
L	Local latitude, angle north or south of the equator, positive if north
Long	Local longitude, angle east or west of the prime meridian, positive if west
m	Air mass number, ratio of the virtual distance a solar ray travels to reach the surface of the Earth, to the distance it would travel if the sun were directly overhead, see Eq. (9-19)
N	Day number of the year, January 1 = 1, December 31 = 365 if not a leap year
$\overline{R}$	Ratio, total insolation on tilted surface to that on an equivalent horizontal surface
$\overline{R_b}$	Ratio, beam insolation on tilted surface to that on an equivalent horizontal surface
R_E	Radius of the Earth, see Fig. 9-8
α	Solar altitude angle, measured up from the horizon
α_{max}	Solar latitude at solar noon (when it is maximum for the day), see Eq. (9-5)
β	Tilt angle of a surface, measured from the back
γ	Surface azimuth angle, relative to true south
γ_s	Solar azimuth angle, relative to true south, negative if morning
γ	Clearness index shape parameter, see Eq. (9-24) and Example 9-4
δ	Solar declination angle, relative to the equator, see Eq. (9-4)
θ_i	Incidence angle between direct beam insolation and surface direct normal vector

TABLE 9-1   Symbols Used in Chap. 9

Symbol	Definition, with Units (SI) Where Appropriate (Overhead Bar Signifies Averaged)
θ_z	Solar zenith angle, measured down from directly overhead
ρ	Insolation reflectivity of a surface
τ	Scaled time of the year $= 2\pi N/365$
$\overline{\tau}_{atm}$	Transmittance of atmosphere to global solar spectrum, see Eq. (9-20)
γ_b	Transmittance of atmosphere to direct beam solar radiation, see Eq. (9-21)
Ψ	Solar altitude angle plus $\pi/2$ radians (or 90 degrees), see Eq. (9-19)
ω	Solar hour angle, $(\pi/6)*$(solar time $-$ 12), negative if morning
ω_s	Solar hour angle at sunset

TABLE 9-1   Symbols Used in Chap. 9 (*Continued*)

The daily solar constant, I_0, can be calculated from the following equation, where N is the day number of the year (e.g., January 1 is day 1; December 31 is day 365). N is also called the "Julian date," from the Julian calendar. A Julian date calendar is provided, for convenience, as an appendix at the end of the book.

$$I_0 = 1368 \{1 + 0.034 \cos [2\pi (N - 3)/365]\} \qquad (9\text{-}1)$$

The factor of $(N - 3)$ in Eq. (9-1) arises because January 3 is currently the day of the solar perihelion (closest approach of the Earth to the sun).

Average *insolation*, or solar energy reaching a given location on Earth, will be lower than the amount available outside the atmosphere due to absorption and diffraction of sunlight in the atmosphere, changing weather, loss of sunlight at night, and so on. Table 9-2 shows that worldwide average values for some representative cities, taking all these factors into account, range between 100 W/m² for Glasgow, Scotland, and 280 W/m² for Cairo, Egypt.

Figures 9-1 and 9-2 give national average values for representative countries. These figures indicate that countries with large areas of arid land, such as the United States and Australia, have relatively high values of average energy gain and hours of sunshine per day. The figures for the United States must be interpreted carefully, of course, as certain regions (e.g., Pacific Northwest and parts of the Northeast) have much lower

City	Insolation	City	Insolation
Seattle	125	Naples	200
El Paso	240	Cairo	280
Rio de Janeiro	200	Johannesburg	230
Glasgow	100	Mumbai	240
Tokyo	125	Sydney	210

Source: Data obtained from Trewartha & Horn (1980).

TABLE 9-2   Average 24-Hour Insolation in W/m² for Selected World Cities

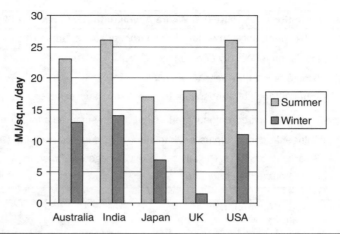

FIGURE 9-1 Average seasonal solar energy gain per square meter for select world countries. [*Source:* Norton (1992).]

values than those shown. In general, winter values are lower than those in the summer, especially in the case of high-latitude countries such as the United Kingdom (between 46° and 54° north latitude), due to shorter days and a more oblique angle with which direct sunlight strikes horizontal surfaces. In contrast, for India, proximity to the equator gives seasonal values that are closer together in the case of energy gain, and reversed for average daily sunshine compared to cities in the temperate countries, because much of the summer energy gain is in the form of diffuse rather than direct solar energy due to high humidity and monsoon rains.

It is likely that global climate change in the twenty-first century will lead to changes in regional values given here, although with the long data series required to establish

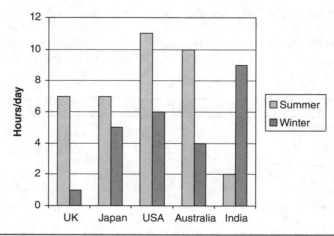

FIGURE 9-2 Average seasonal daily hours of direct sun for world countries. [*Source:* Norton (1992).]

average annual values, such changes will only become apparent over the course of years if not decades.

Insolation varies greatly during a day, during a year, and as a function of latitude. Tables 9-3(a) and (b) show this variation for two latitudes. Data are for direct normal

Month	6 a.m. 6 p.m.	7 a.m. 5 p.m.	8 a.m. 4 p.m.	9 a.m. 3 p.m.	10 a.m. 2 p.m.	11 a.m. 1 p.m.	Noon
Jan	0	4/0	640/102	848/278	931/430	966/524	977/556
Feb	0	352/30	771/200	905/401	963/556	990/651	998/684
Mar	0	582/100	820/314	914/515	958/666	980/762	986/795
Apr	210/23	648/191	804/408	876/593	913/735	931/824	937/854
May	374/67	665/256	787/460	848/632	881/766	898/846	903/872
Jun	412/89	662/279	773/476	831/642	863/770	879/847	884/871
Jul	357/70	640/257	761/457	823/626	856/757	873/836	878/861
Aug	187/25	599/192	756/403	830/583	869/721	888/808	894/837
Sep	0	514/97	758/303	857/498	905/644	927/737	934/769
Oct	0	312/30	723/198	862/394	923/546	951/639	959/671
Nov	0	5/0	619/102	828/277	912/427	948/521	959/552
Dec	0	0	556/69	811/228	907/375	948/468	960/499

(a)

Month	6 a.m. 6 p.m.	7 a.m. 5 p.m.	8 a.m. 4 p.m.	9 a.m. 3 p.m.	10 a.m. 2 p.m.	11 a.m. 1 p.m.	Noon
Jan	0	0	116/7	584/79	754/174	823/244	842/269
Feb	0	11/1	568/78	780/210	869/330	908/408	919/435
Mar	0	481/64	743/214	852/371	905/493	931/568	939/593
Apr	340/46	646/189	778/359	846/506	883/618	902/688	908/712
May	510/125	689/287	781/448	833/584	863/689	879/755	883/785
Jun	544/162	693/324	774/479	822/610	849/711	864/774	868/796
Jul	492/130	666/290	757/448	809/581	839/684	854/749	859/771
Aug	311/51	599/193	732/358	801/501	839/610	858/678	864/702
Sep	0	414/62	678/206	792/358	848/476	875/549	883/573
Oct	0	12/1	521/79	734/208	825/326	866/403	878/428
Nov	0	0	115/7	565/80	734/175	803/244	822/269
Dec	0	0	0	442/42	675/119	765/181	789/204

(b)

Note: Times of day in column headings indicate hours for which insolation is equal in the morning and afternoon, since insolation is symmetric around noon.

Source: Handbook of Fundamentals, ASHRAE, 2001.

TABLE 9-3   Direct Normal/Horizontal Surface Clear Sky Insolation, W/m^2, as a Function of Solar Time, (a) at 32° North Latitude and (b) at 48° North Latitude

radiation from a clear sky, local solar time, and for the twenty-first of each month in the northern hemisphere. The clearness of the sky can be measured using the "sky clearness number," where the value 1.0 represents a perfectly clear sky, and values less than 1.0 represent some degree of obstruction. Thus the data in the table assume sky clearness number = 1.0.

9-3   Definition of Solar Geometric Terms and Calculation of Sun's Position by Time of Day

In this section, we introduce several terms used to discuss solar geometry, and then chart the sun's position in the sky as a function of various parameters. Measures used in this section capture the effect of changing tilt of the Earth relative to the sun, changing distance from the sun as a function of time of year, and the changing angular distance traveled by the sun across the sky, as a function of the time of year and time of day.

Solar angle calculations are done in solar time, which is defined by solar noon being when the sun is due south of the observer. Solar time is synchronous with local standard time but leads or lags it because of two factors. One factor is the difference between the local longitude and the standard longitude value that defines the local time zone. One degree of longitude difference is equivalent to 4 min of solar time difference. The second factor is the equation of time, ET, which exists due to the Earth's path around the sun being an ellipse rather than a circle, while the Earth's rotation on its axis does not change through the year. The angle swept by the Earth's path is less per 24-h cycle when the Earth is farthest from the sun. The reverse is true during the other half of the year.

Solar and standard time are related by the following equation, where time is formatted as hour:minute and $Long_{std}$ and $Long_{loc}$ are standard and local longitudes (degrees), respectively.

$$\text{Solar time} = \text{Standard time} + 0{:}04(Long_{std} - Long_{loc}) + ET \tag{9-2}$$

Standard time zones are defined in 15° longitude steps, with the 0° longitude passing through Greenwich, England. For example, within the North American continent, Atlantic, Eastern, Central, Mountain, and Pacific standard longitudes (time zones) are defined as 60°, 75°, 90°, 105°, and 120°, all west of the zero meridian. These are the standard longitudes but actual time zones have ragged borders to accommodate various geographic areas and cities, and some time zones around the world have half or quarter hour offsets from their standard definitions.

The equation of time is calculated from a fitted equation and may be expressed in Fourier series form. One expression is the following, where $\tau = 360N/365$ in degrees, or $\tau = 2\pi N/365$ in radians, and scales the year through one trigonometric cycle. Note, Eq. (9-3) provides ET in decimal minutes, which should be converted to hour:minute:second format when used in Eq. (9-2).

$$ET = a_1 \sin(\tau) + b_1 \cos(\tau) + a_2 \sin(2\tau) + b_2 \cos(2\tau) + a_3 \sin(3\tau) + b_3 \cos(3\tau)$$
$$+ a_4 \sin(4\tau) + b_4 \cos(4\tau) \tag{9-3}$$

$a_1 = -7.3412$	$b_1 = +0.4944$
$a_2 = -9.3795$	$b_2 = -3.2568$
$a_3 = -0.3179$	$b_3 = -0.0774$
$a_4 = -0.1739$	$b_4 = -0.1283$

The value of ET makes the sun appear to be "fast" during part of the year and "slow" during other parts. The effect is small (plus or minus a quarter of an hour at most) but leads to unexpected observations such as the following. In the northern hemisphere, the shortest day of the year is the winter solstice, usually December 21, but the earliest sunset is during the first week of December and the latest sunrise is during the first week of January.

We now define several terms used to calculate solar geometry. The first two are *zenith angle,* θ_z, and *solar altitude,* α. Zenith angle is the angle formed between a line from the sun and a line to a point directly overhead; solar altitude is the angle between the line from the sun and a line to the horizon. Thus $\theta_z + \alpha = 90°$, and the maximum value for either is 90°.

Next, consider the surface of some solar device, for example, a solar photovoltaic panel, which may be rigidly fixed to the Earth's surface and therefore keep a certain orientation relative to the compass directions. With this, we also introduce the *surface azimuth angle,* γ, the angle between the orientation of the surface and due south, and the *solar azimuth,* γ_s, which is the angle between the direction of the sun and due south, as shown in Fig. 9-3.

For both γ and γ_s, angles to the west of due south are positive and angles to the east are negative (e.g., in Fig. 9-3, $\gamma_s > 0$, $\gamma < 0$), and at solar noon (the time when the sun is the highest in the sky), $\gamma_s = 0$. For a solar panel that is installed facing due south, $\gamma = 0$ as well. In Fig. 9-3, the surface of the solar panel is tilted toward the south with some positive *tilt angle* β, which is the angle at which the surface is tilted from horizontal. A flat solar panel (i.e., tilt angle = 0) is also possible. This tilt is usually in a southern direction, especially at higher northern latitudes, in order to increase the energy flux reaching the surface. Conversely, at higher southern latitudes the surface is tilted in a northern direction to increase the flux.

We must also consider the *declination* δ based on the time of year. Declination is the angle between the line of the sun and the plane of the equator; as shown in Fig. 9-4 for the case of a point on the equator. At the equinox, the line of the sun is in the plane of the equator, and the declination is zero. At the solstices, the north pole is either inclined toward or inclined away from the sun, so the declination is +23.45° or −23.45° (the maximum possible variation from zero) in (northern hemisphere) summer or winter, respectively.

The declination on any given day of the year N, for example, $N = 1$ on January 1, $N = 365$ on December 31, and so on, can be calculated using the following formula, where the value is in units of degrees:

$$\delta = 23.45 \sin\left[\frac{360(284 + N)}{365}\right] \tag{9-4}$$

It is also possible to calculate the solar altitude at solar noon, α_{max}, using this formula, where again the value is given in degrees:

$$\alpha_{max} = 90 - L + \delta \tag{9-5}$$

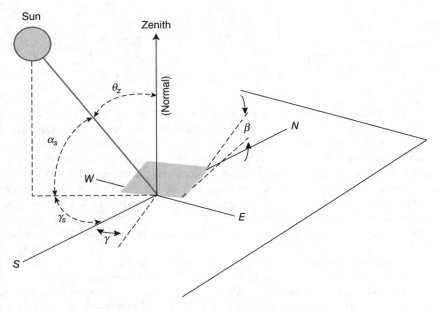

Legend for symbols	
Greek symbol	**Definition**
α_s	Solar altitude
β	Slope angle (of device)
γ	Surface azimuth angle
γ_s	Solar azimuth angle
θ_z	Zenith angle

FIGURE 9-3   View of geometric relationship between the sun's position, orientation of raised solar device, and compass directions, showing sun striking device at an oblique angle. Explanation: the device is raised by an angle β and faces east of due south by an angle γ; the sun is positioned in the sky above the horizon at an angle α_s and west of due south by an angle γ_s.

Note: Diagram is shown for northern hemisphere, in southern hemisphere, orientation would be reversed, with sun positioned to the north of the east-west line.

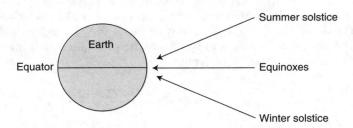

FIGURE 9-4   Declination at summer solstice, equinox, and winter solstice, from northern hemisphere perspective.

For example, at 40° north latitude, the sun will reach $\alpha = 26.55°$ ($= 90 - 40 - 23.45$) at solar noon on the winter solstice and $\alpha = 73.45°$ ($= 90 - 40 + 23.45$) at solar noon on the summer solstice. Example 9-1 illustrates the calculation of the declination value for a given day of the year.

Example 9-1 Verify that the declination reaches its lowest value at the winter solstice.

Solution The winter solstice occurs on December 21, or the 355th day of the year. Substituting $N = 355$ into the equation gives the following:

$$\delta = 23.45 \sin\left[\frac{360(284 + 355)}{365}\right]$$

$$= 23.45 \sin(630°)$$

$$= 23.45 \sin(3.5\pi \text{ radians})$$

$$= 23.45(-1) = -23.45$$

Thus δ is in fact at its lowest possible value of $\delta = -23.45°$.

Once declination is known, zenith angle and solar altitude can be calculated as a function of latitude, declination, and *hour angle, ω*:

$$\cos \theta_z = \sin \alpha = \sin \delta \sin L + \cos \delta \cos L \cos \omega \qquad (9\text{-}6)$$

where ω can be calculated by subtracting or adding 15° for each hour before or after solar noon, respectively (e.g., $\omega =$ hour $\times 15 - 180$, where hour is given in 24 h decimal time, so that 10 a.m. $= -30°$, 2 p.m. $= +30°$, 10:30 a.m. $= -22.5°$, and so on). The solar hour angle expresses the degrees of Earth rotation before or after solar noon at a given longitude. (Note that the solar hour angle is the same at any latitude along a line of constant longitude.)

The value of ω is different from that of the *solar azimuth γ_s*, with which it may be confused when first encountered but which is calculated as a function of δ, ω, and α, as follows:

$$\sin \gamma_s = \cos \delta \sin \omega / \cos \alpha \qquad (9\text{-}7)$$

The calculation in Table 9-4 of ω and γ_s on the summer solstice at 42° north latitude from 10 a.m. to 2 p.m. illustrates the difference:

Time	Hour Angle (deg)	Solar Azimuth (deg)
10:00 a.m.	−30	−62.8
11:00 a.m.	−15	−38.6
12:00 p.m.	0	0.0
1:00 p.m.	15	38.6
2:00 p.m.	30	62.8

TABLE 9-4 Representative Comparison of Hour Angle and Solar Azimuth

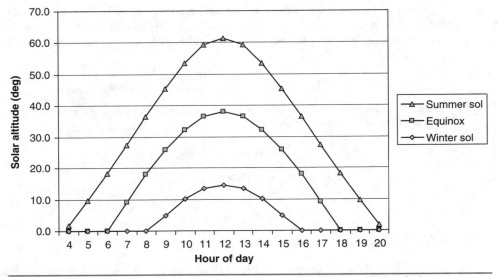

FIGURE **9-5**   Solar altitude as a function of time of day at 52° north latitude (equivalent to The Hague, The Netherlands).

Calculating α through the hours of the day for different declination values at different times of year gives us the figure shown for the case of 52° north latitude (Fig. 9-5). Note that when the formula predicts a negative value for the solar altitude, for example, before 8 a.m. at the winter solstice, the value is given as 0°, since the sun is below the horizon at this point. Repeating this exercise for a tropical location, such as 5° north as shown in Fig. 9-6, gives a range of solar altitudes in which the altitude around solar noon

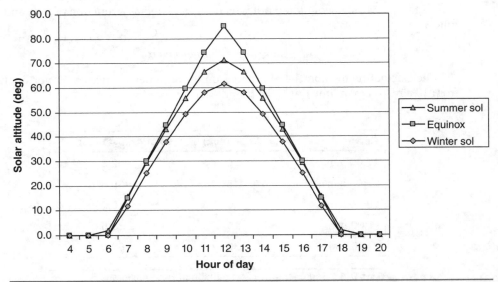

FIGURE **9-6**   Solar altitude as a function of time of day at 5° north latitude (equivalent to Abidjan, Ivory Coast).

consistently reaches high values (50° to 75°) year round, but without the variation in length of day over the course of the year (such as 4 a.m. sunrise on the summer solstice at 52° latitude). The first point explains why many of the sunniest locations in the world (Mumbai, Cairo, and so on), as shown in Sec. 9-2-1, are in regions near the equator.

9-3-1   Relationship between Solar Position and Angle of Incidence on Solar Surface

Ideally, the surface of a solar device would always face the sun (i.e., the plane of the surface is orthogonal to the direction of the beam); this can be accomplished with a *tracking mechanism* that moves the device to face the sun as it moves through the sky during the course of the day. In practice, not all devices are able to face the sun. For example, local conditions may dictate attaching the device to a fixed surface that is oriented away from due south or due north (which are ideal in the northern and southern hemispheres, respectively) by some surface azimuth angle γ, and tilted up from horizontal by some angle β. Here we solve for the *angle of incidence* θ_i between the beam of sunlight and the direction of the surface, as a function of solar altitude, tilt angle, hour angle, and surface azimuth angle, in the following way.

Suppose we have two unit vectors, one parallel to the incoming solar rays and the other normal to the device surface. Then the value of θ_i is given by the following relationship:

$$\cos \theta_i = S \cdot N \tag{9-8}$$

where S and N are the parallel and normal vectors, respectively, and "·" is the dot product operator from vector calculus. It can be shown using the Cartesian axes of the site that θ_i is given by

$$\cos \theta_i = \sin \alpha \cos \beta + \cos \alpha \sin \beta \cos (\gamma - \gamma_s) \tag{9-9}$$

This formula can be simplified in certain situations. For a flat surface, $\beta = 0°$ and

$$\cos \theta_i = \sin \alpha \tag{9-10}$$

For a vertical surface, $b = 90°$ and

$$\cos \theta_i = \cos \alpha \cos (\gamma - \gamma_s) \tag{9-11}$$

Lastly, for a south-facing surface, $\gamma = 0°$ and

$$\cos \theta_i = \sin \alpha \cos \beta + \cos \alpha \sin \beta \cos \gamma_s \tag{9-12}$$

Knowing the relationship between incident sunlight and position of the device surface gives us the proportion of incoming energy available to the device. Suppose we know the hourly beam radiation normal to the sun $I_{n,b}$ (i.e., the component of global solar radiation that arrives directly to the Earth's surface, as opposed to by the diffuse pathway). We can solve for the component normal to a horizontal surface, defined as I_b, using the following relationship:

$$I_b = I_{n,b} \cos \theta_z \tag{9-13}$$

We can also solve for the value of $I_{\beta,b}$ from the relationship

$$I_{\beta,b} = I_{n,b}\cos\theta_i \tag{9-14}$$

The ratio R_b of beam radiation received by the tilted surface to radiation received by an equivalent horizontal surface is of interest for solar installations, and can be solved from

$$\overline{R_b} = \frac{I_{\beta,b}}{I_b} = \frac{I_{n,b}\cos\theta_i}{I_{n,b}\cos\theta_z} \tag{9-15}$$

9-3-2   Method for Approximating Daily Energy Reaching a Solar Device

In this section, we take the formulas from the previous two as a basis for a means of approximating the total amount of energy reaching the surface of a solar device, as a function of the amount of energy in the direct solar beam, the orientation of the surface, the time of day, and the time of year. It is possible to estimate this quantity of energy by continuously calculating the energy flux from sunrise to sunset. However, for brevity, we will limit the discussion to an approximation in which a single measure each hour is used to represent the average value for that hour.

The amount of energy reaching the device surface is a function of the angle of incidence between the beam of light and the normal to the surface. The process of calculating the angle of incidence includes the following steps:

1. Calculate the hour angle based on the hour of day in solar time.
2. Calculate the solar altitude based on latitude, declination, and hour angle.
3. Calculate the solar azimuth from the declination, hour angle, and solar altitude.
4. Calculate the angle of incidence from the solar altitude, the solar azimuth, the surface angle β, and the surface azimuth.

Care must be taken that the correct values are used in each step of this process. First, the analyst must verify that the beam of light is above the horizon for the time of day in question; negative values of α imply that the sun is below the horizon and flux at this time should not be included in the energy estimate. Second, the beam of light must strike the front of the solar device and not the back of it. It is possible (especially at higher latitudes between the spring and autumn equinoxes), when solar altitude is low and solar azimuth is large, for the outcome of step 4 to be a value of θ_i greater than 90°, that is, the sun is striking the back of the device, and therefore is assumed not to be contributing useful energy to its function! Last, the solar azimuth must be interpreted correctly from the arcsin value in step 3, especially at high latitudes and at early and late hours of the day. Refer, for example, to Fig. 9-5, which shows that on the summer solstice at 52° north latitude, the sun rises at approximately 4 a.m. At this time, the solar azimuth is −127°, that is, the sun is rising at a point that is north of due east. At this point, the value of the arcsin from step 3 is 0.794, for which $\gamma_s = 53°$ is also a solution. If care is not taken, the result may lead to a spurious result in which the value of γ_s starts out with a value less than 90° at sunrise, increases to 90° by 9 or 10 o'clock, then decreases to 0° at solar noon. This problem arises because calculators and computer programming languages, by default, may return inverse trigonometry values in the first and fourth quadrants only.

One way to avoid the quadrant problem is to consider the date for the calculation. If the date is between the first day of autumn and the first day of spring, accept the inverse trig value as given. If the date is between the first day of spring and the first day of autumn, you may need to adjust the inverse trig value by 180° (or π radians depending on the value returned) manually for times early in the morning or late in the afternoon. As a general rule, be sure you know how the calculator or computer you are using handles inverse trigonometry functions.

Example 9-2   A normal beam of radiation $I_{n,b}$ has a flux value of 350 W/m². What is the flux reaching a solar surface facing due south and tilted south by 25° at 11 a.m. on the summer solstice at 42° north latitude?

Solution   The surface faces south, so the value of surface azimuth is $\gamma = 0$. On the summer solstice, the declination is $\delta = 23.45°$, and at 11 a.m., the hour angle is −15°. Solving for solar altitude:

$$\sin\alpha = \sin\delta\sin L + \cos\delta\cos L\cos\omega$$

$$\sin\alpha = \sin(23.45)\sin(42) + \cos(23.45)\cos(42)\cos(-15) = 0.9903$$

$$\alpha = 67.6$$

Solving for solar azimuth on the basis of ω, δ, and α gives:

$$\sin\gamma_s = \frac{\cos\delta\sin\omega}{\cos\alpha} = \frac{\cos(23.45)\sin(-15)}{\cos(67.6)} = -0.7387$$

$$\gamma_s = -38.6$$

Solving in turn for angle of incidence as a function of α, β, γ, and γ_s gives

$$\cos\theta_i = \sin\alpha\cos\beta + \cos\alpha\sin\beta\cos(\gamma - \gamma_s)$$

$$\cos\theta_i = \sin(67.6)\cos(25) + \cos(67.6)\sin(25)\cos(0 + 38.6) = 0.2705$$

$$\theta_i = 15.5$$

Thus the energy flux onto the device is

$$(350\text{ W/m}^2)\cos(15.5) = 337\text{ W/m}^2$$

Using the calculation in Example 9-1 as a basis as well as the insolation values from Table 9-6(a), we can create a complete table of results for an entire day, this time for 32° north latitude, as shown in Table 9-5. Clear sky conditions are assumed for the entire day. Recall that the value of the normal insolation changes as a function of time of day, from a low of 412 W/m² at close to sunrise and sunset, to 884 W/m² at solar noon.

The hourly values in Table 9-5 can be interpreted as taking the flux measured on the hour (8, 9, 10 o'clock, and so on) as representing the average flux from 30 min before to 30 min after. For example, the value at 8 a.m. represents the energy per unit area from 7:30 a.m. to 8:30 a.m. Since the chosen unit for estimation is 1 h, the total energy received for that hour can be directly converted from flux in W/m² to hourly energy in Wh/m². Using the preceding example, the flux is 389 W/m² on average from 7:30 a.m. to 8:30 a.m., so the energy received during that hour is 389 Wh/m². Summing the values in the column entitled "Actual Flux" gives a total energy value of 6.27 kWh/m². This value can be compared to the sum of the "Available Flux" of

Hour	Available Flux* (W/m²)	Hour Angle (deg)	Solar Altitude (deg)	Solar Azimuth (deg)	Incident Angle (deg)	Actual Flux (W/m²)	Fraction
6	412	−90	12.2	−110.2	87.2	20	0.05
7	662	−75	24.3	−103.4	73.5	188	0.28
8	773	−60	36.9	−96.8	59.7	389	0.50
9	831	−45	49.6	−89.4	46.2	575	0.69
10	863	−30	62.2	−79.7	33.2	722	0.84
11	879	−15	74.2	−60.9	21.9	816	0.93
12	884	0	81.5	0.0	16.5	848	0.96
13	879	15	74.2	60.9	21.9	816	0.93
14	863	30	62.2	79.7	33.2	722	0.84
15	831	45	49.6	89.4	46.2	575	0.69
16	773	60	36.9	96.8	59.7	389	0.50
17	662	75	24.3	103.4	73.5	188	0.28
18	412	90	12.2	110.2	87.2	20	0.05

*Values from Table 9-6(a) for June 21.

TABLE 9-5 Estimate of Energy per m² Received by the Surface of a Solar Device ($L = 32°$ N, $\delta = 23.45°$, $\beta = 25°$, $\gamma = 0°$).

9.72 kWh/m²; 36% of the available energy is lost due to the surface not being normal to the sun's position changes.

Four-Quadrant Method for Calculating Solar Azimuth

The problem of ambiguous inverse trigonometric functions mentioned above can be addressed using a "four-quadrant method" that explicitly takes into account the location of the solar azimuth γ_s in the four quadrants surrounding the device (Jain, 1997). Recall the definition of γ_s, where $\gamma_s = 0°$ is due south and values to the east and west of $\gamma_s = 0°$ are negative and positive, respectively.

The location of the sun in the four quadrants is explicitly quantified in the following way. Let x be the relative position of the sun in the east-west direction, and y be the position in the north-south direction. Values for x and y are calculated as follows:

$$x = -\sin \omega \cos \delta \tag{9-16a}$$

$$y = -\cos \omega \cos \delta \sin L + \cos L \sin \delta \tag{9-16b}$$

The value of γ_s is then found from x and y as follows. For times before solar noon, the following equation gives the value:

$$\gamma_s = -\arctan\left(\frac{y}{x}\right) - 90 \tag{9-16c}$$

For times after solar noon, the equation changes as follows:

$$\gamma_s = 90 - \arctan\left(\frac{y}{x}\right) \qquad (9\text{-}16\text{d})$$

Note that at solar noon when $\omega = 0$, $\sin(\omega) = 0$ and $x = 0$ so that the value returned by Eqs (9-16c) and (9-16d) is unbounded. The actual value in all cases is $\gamma_s = 0°$.

Example 9-3 illustrates the use of the four-quadrant method.

Example 9-3 Use the four-quadrant method to calculate the solar azimuth at 5 a.m. on the summer solstice at 42° north latitude.

Solution First, it can be confirmed that the sun is in fact above the horizon at this time, at a solar altitude of approximately 5°. Since the day is the summer solstice, the declination is 23.45°. At 5 a.m., the hour angle is −105°. Substitution gives the following:

$$x = -\sin\omega\cos\delta = -\sin(-105)\cos(23.45)$$

$$= 0.886$$

$$y = -\cos\omega\cos\delta\sin L = -\cos(-105)\cos(23.45)\sin(42)$$

$$= 0.159$$

$$\gamma_s = -90 - \arctan\left(\frac{y}{x}\right) = -90 - \arctan\left(\frac{0.159}{0.886}\right) = -90 - 10.2$$

$$=\sim -100.2$$

Thus the solar azimuth is $\gamma_s = -100.2°$.

9-4   Effect of Diffusion on Solar Performance

There are a number of possibilities for loss of beam energy as the sun's rays travel from the edge of the atmosphere to the Earth's surface, such as absorption by atmospheric particles or diffusion in different directions. One factor that affects the amount of diffusion is the distance that the beam must travel through the atmosphere, which is in turn a function of the angle relative to vertical at which the sunlight reaches the Earth. Another factor is the amount of humidity, pollution, or other particles in the atmosphere.

9-4-1   Direct, Diffuse, and Global Insolation

Light traveling from the sun can reach the surface of the Earth through either direct or diffuse transmission. In the direct case, it is transmitted through the atmosphere without interference. In the diffuse case, light is diffused by refraction and aerosols in the atmosphere, and some portion of the diffused light continues to travel forward toward the Earth and strikes the surface. This diffuse light can strike the Earth at a range of angles, depending on how it was diffused. The diffuse and direct light together make up the total amount of light reaching the surface, also called "global" insolation. The energy flux embodied in the light can be measured in units of W/m^2, so energy flux in the global insolation is the sum of the direct and diffuse components.

To put the pathways in the figure in context, we can consider the energy flux available on a clear day at or close to midday, when the sun is highest in the sky. Even in this situation, the full energy flux in solar radiation striking the Earth will not be available, as some portion is reflected or absorbed by the atmosphere. Thus, for purposes of estimating available solar energy, practitioners use values of 900 or 1000 W/m² to represent the maximum value possible [see Tables 9-3(a) and (b)].

Note that the calculated values in the table do not include the slight variation of solar intensity during the year due to the elliptical shape of the earth's path around the sun, which is beyond the purview of this text. The earth is closest to the sun in early January, and farthest in early July. The magnitude of this effect can be seen in Eq. (9-7), where the solar constant varies by +/−3.4% (the coefficient of the cosine term.) This variation leads to asymmetry of values for some months that would otherwise be symmetric relative to the solstices.

A simple model to calculate insolation intensity is based on the air mass number, which functions to quantify the reduction of solar power passing through the atmosphere and attenuated by aerosols and air molecules. The air mass number can be interpreted as the ratio between the direct beam path length divided by the path length were the sun directly overhead. In its simplest form (see Fig. 9-7), the air mass number, AM, is calculated from Eq. (9-17), where θ_z is the zenith angle, the angle from directly overhead, described later in this section. Curved paths due to light refraction is assumed to be negligible in this model (may be ignored up to approximately 75° from normal.)

$$AM = \frac{CB}{AB} = \frac{1}{\cos(\theta_z)} = \sec(\theta_z) \tag{9-17}$$

Equation (9-17) becomes incorrect for large values of the zenith angle because it predicts an infinitely large air mass number when the sun is at the horizon and the

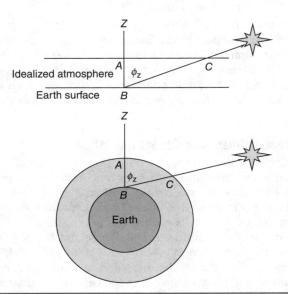

FIGURE 9-7   Definition of air mass; air mass = CB/AB. Top sketch is idealized atmosphere as a constant thickness layer. Bottom sketch greatly exaggerates the atmosphere thickness.

zenith angle is 90°. To account for the Earth's curvature, the following somewhat more complex equation can be used when the sun is near the horizon.

$$AM = \frac{1}{\cos(\theta_z) + 0.50572(96.07995° - \theta_z)^{-1.6364}} \qquad (9\text{-}18)$$

The next way to calculate air mass is by using the distance from where the light enters the atmosphere, the radius of the Earth R_E, and the geometry between the two lines thus defined, as shown in Fig. 9-8. The height of the atmosphere is denoted by H_A in this approach. As shown in the figure, a beam travels from the sun to the surface of the Earth, denoted by the inner concentric circle, and strikes the surface at point B. Therefore the distance from point B to the center of the Earth is $H_A + R_E$. The distance the beam travels through the atmosphere is along the line from point B to point C on the outer edge of the atmosphere, denoted by the outer concentric circle. The angle Ψ is formed by the beam and the radius at point B, hence $\Psi = \alpha + 90$, where α is the *solar altitude*, or angle of the sun above the horizon.

We next calculate the ratio m of the distance over which the beam travels to the distance over which it would travel if it were exactly overhead. For a given Ψ, m can be calculated using the law of cosines as

$$m = \sqrt{1 + 2R_E/H_A + (R_E \cos \Psi)^2/H_A^2} + R_E \cos \Psi/H_A \qquad (9\text{-}19)$$

The derivation of this value is given as a homework problem at the end of the chapter. To calculate representative values for m, we can use standard values of $R_E = 6372$ km and $H_A = 152.4$ km. The effect of flattening of the Earth at the poles has a negligible impact on the value of m and can be ignored. When the value of Ψ is 90° (i.e., at

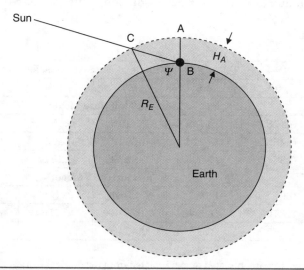

FIGURE 9-8   Relationship between angle Ψ between Earth radius and incoming ray of sun, and length of path CB through atmosphere.

the horizon), 135° (half way between horizon and overhead), and 180° (directly over-head), the value of m is 9.2, 1.4, and 1.0, respectively. Given that this ratio is up to 9 to 1, depending on the hour of the day, one would expect that the insolation at early or late hours of the day would be diminished, which is consistent with the findings in Table 9-3(a) and (b).

As an approximation for clear sky (pollution-free) atmosphere transmittance, Eq. (9-20a) can be used to within approximately 5% accuracy.

$$\overline{\tau}_{atm} = 0.5(e^{-0.65AM} + e^{-0.095AM}) \qquad\qquad (9\text{-}20a)$$

Models of varying complexity exist in the meteorological and solar energy literature to calculate direct, diffuse, and total solar radiation as functions of atmospheric condi-tions and other factors. An early and relatively simple model to estimate "clear sky" beam radiation was proposed by Hottel (1976), based on solar altitude, local elevation above sea level, and atmospheric parameters. Atmospheric transmittance (τ_b, beam radi-ation transmittance only) is calculated from Eq. (9-20b).

$$\tau_b = a_0 + a_1 \exp[-k\,csc(\alpha)] \qquad\qquad (9\text{-}20b)$$

Parameters a_0, a_1, and k depend on local elevation above sea level and atmo-spheric conditions. These parameters are available for a variety of situations, such as in Table 9-6. The 23 km haze model, for example, applies to atmospheric condi-tions of sufficient clarity to see as far as 23 km, in contrast to the 5 km haze model that refers to a clear sky condition but with sufficient atmospheric haze that visual distances are limited to 5 km. The data in Table 9-6 are used to calculate the example data in Table 9-7, which contains calculated atmospheric transmittance values for a range of solar altitudes, the 5 km haze model, and an elevation of 0.5 km above sea level. Direct normal insolation is the product of atmospheric transmittance and the solar constant. Horizontal surface insolation values are based on the law of cosines

	Altitude above Sea Level, km					
	0	**0.5**	**1.0**	**1.5**	**2.0**	**2.5**
	23 km Haze Model					
a_0	0.1283	0.1742	0.2195	0.2582	0.2915	0.320
a_1	0.7559	0.7214	0.6848	0.6532	0.6265	0.602
k	0.3878	0.3436	0.3139	0.2910	0.2745	0.268
	5 km Haze Model					
a_0	0.0270	0.063	0.0964	*0.126*	0.153	0.177
a_1	0.8101	0.804	0.7978	*0.793*	0.788	0.784
k	0.7552	0.573	0.4313	*0.330*	0.269	0.249

Italicized figures are interpolated or extrapolated from the original work.

TABLE 9-6 Coefficients for the Hottel Clear Sky Solar Transmittance Model

Solar Altitude	Atmosphere Transmittance	Beam Radiation, Direct Normal	Beam Radiation, Horizontal Surface
10	0.0927	126.7	22.0
20	0.2135	291.9	99.8
30	0.3186	435.5	217.8
40	0.3927	536.8	345.1
50	0.4435	606.3	464.5
60	0.4779	653.2	565.7
70	0.5000	683.4	642.2
80	0.5123	700.4	689.7
90	0.5163	705.8	705.8

TABLE 9-7   Example Results, 5 km Haze Model, 0.5 km above Sea Level, Assuming 1,368 W/m² outside the Atmosphere

to take into account the spreading of irradiation that is not normal to the surface, $I_{horizontal} = I_{normal} \cos(\theta_z)$.

Weather changes from day to day, more so at some locations of the Earth than at others. Seldom is there frequent ideal atmospheric transmittance. Figure 9-9 shows an example of substantial variability. The figure contains 14 years (1983 to 1996) of integrated daily kWh/m² on a horizontal surface in Ithaca, New York, U.S. (latitude 42° 24′ N and 76° 30′ W). The upper line in the figure is the variation of insolation calculated for directly outside the atmosphere on a plane horizontal to the Earth's surface, directly above Ithaca. The daily integral outside the atmosphere, H_{day}, can be calculated from the following equation where L is the local latitude, δ is the solar declination, and γ_s is the solar azimuth at sunset.[1]

$$H_{day} = I_0 \cos(L)\cos(\delta)\sin(\gamma_s)$$

$$= 1368\left(\frac{24}{\pi}\right)\left[1+0.034\cos\left(\frac{2\pi N}{365}\right)\right]\cos(L)\cos(\delta)\sin(\gamma_s) \qquad (9\text{-}21)$$

Figure 9-10 is included to show the trend of clear sky *atmospheric transmittance*, which is defined as the fraction of insolation available outside the atmosphere that is transmitted to the Earth's surface. Atmospheric transmittance values in the figure are

[1]Note that Eq. (9-21) is slightly simplified by centering the values around January 1 rather than approximately 2 days later when the true perihelion is usually reached. Also note that solar data such as equinoxes, solstices, perihelions, and aphelions, can vary by date from year to year due to the solar year being approximately 365.25 days long rather than 365. There can also be differences due to the local time zone at the exact instances of the events. Also, the variable earth/sun distance, with the associated variable angular velocity of the earth's path, leads to the observation that the winter half of the year (northern hemisphere) is approximately 5 days shorter than the summer half of the year (as measured from the spring and autumnal equinoxes).

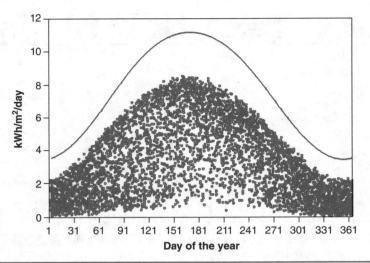

FIGURE 9-9   Fourteen years of integrated kWh/m² for Ithaca, New York, United States.

based on total received daily insolation on the best day of each date of the 14 years of data, divided by the expected value outside the atmosphere. The scatter, although not large, arises from almost no days being perfectly clear during the 14-year period. Transmittance is lower during winter when the solar altitude is lower, and decreases as summer proceeds and the atmosphere becomes dustier and more humid.

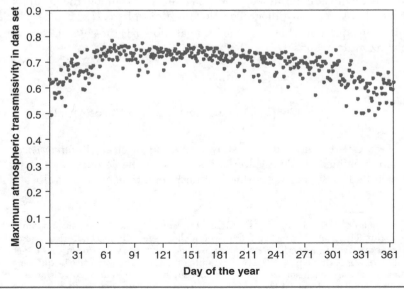

FIGURE 9-10   Maximum averaged daily atmospheric transmittance for data in Fig. 9-9.

The average of the real data in Fig. 9-9 over the year is 3.29 kWh/m² · day (11.8 MJ/m² · day) but the highest values are between one and two orders of magnitude greater than the lowest values regardless of the time of year. This magnitude of variation becomes an important consideration when analyzing and designing solar energy systems and their associated thermal energy and/or electricity storages and the location of the installation is characterized by variability such as in Fig. 9-9.

9-4-2 Climatic and Seasonal Effects

The amount of atmospheric diffusion of sunlight can be quantified using empirical observations under different climatic and seasonal conditions. First, we introduce the variable H to represent daily integrated insolation, in contrast to I used in previous sections that represents hourly insolation. We make a distinction between the global value of insolation, $\overline{H}$, and the diffuse component of that insolation, $\overline{H}_d$. The more clear the air, the smaller the ratio $\overline{H}_d/\overline{H}$. An additional indicator of diffusion is the *clearness index K*, which is the ratio of the insolation on a horizontal surface to the insolation on an extraterrestrial horizontal surface in the same location. From observation, the following relationship for the ratio $\overline{H}_d/\overline{H}$ holds, assuming isotropic sky conditions:

$$\overline{H}_d/\overline{H} = 1.39 - 4.03\overline{K_T} + 5.53\overline{K_T}^2 - 3.11\overline{K_T}^3 \qquad (9\text{-}22)$$

where $\overline{K_T}$ is the monthly clearness index, or average value of K over a month. Numerous similar equations can be found in the solar engineering literature to calculate $\overline{H}_d/\overline{H}$.

If the yearly average clearness index, $\overline{K}$, is known then it is possible to provide a projection of the likelihood of a given value for the daily value of the clearness index, K. Plotting the distribution of K for a range of values of $\overline{K}$ gives Fig. 9-11, where $\overline{K} = 0.7$ might represent a sunny region, $\overline{K} = 0.3$ a cloudy region, and $\overline{K} = 0.5$ a region in between. As an example of how to read the curves, taking $\overline{K} = 0.7$, we can expect $K \leq 0.63$ for 20% of the days of the year, $K \leq 0.78$ for 80% of the days of the year, and so forth.

Let $\overline{K}_T$ represent the average clearness index for each month. If $\overline{K}_T$ is known, then it is possible to approximate a distribution for the likelihood that the value of $\overline{K}_T$ is less than or equal to a given value $K_{T,i}$. Taking $K_{T,\text{min}}$ and $K_{T,\text{max}}$ as the lowest and highest values of the clearness index, respectively, the cumulative distribution function for K_T can be written

$$F(K_T) = \frac{\exp(\gamma K_{T,\text{min}}) - \exp(\gamma K_T)}{\exp(\gamma K_{T,\text{min}}) - \exp(\gamma K_{T,\text{max}})} \qquad (9\text{-}23)$$

The value of γ must be found by taking the average annual clearness index K_{avge} (i.e., average of the 12 monthly values) and finding γ such that Eq. (9-24) holds:

$$\overline{K} = \frac{(K_{T,\text{min}} - 1/\gamma)\exp(\gamma K_{T,\text{min}}) - (K_{T,\text{max}} - 1/\gamma)\exp(\gamma K_{T,\text{max}})}{\exp(\gamma K_{T,\text{min}}) - \exp(\gamma K_{T,\text{max}})} \qquad (9\text{-}24)$$

The use of these equations is illustrated in Example 9-4, as follows.

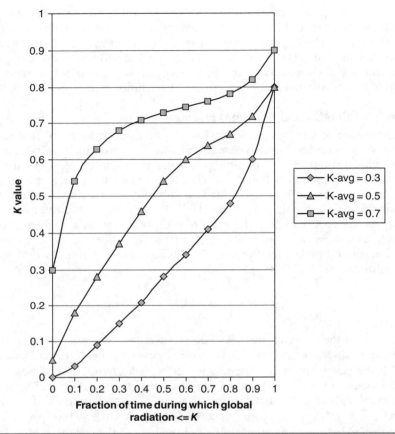

FIGURE 9-11   Distribution of daily clearness ratios K for a range of annual average clearness values K_{avg}.

Example 9-4   Suppose you are given the following values for K_T for a location in the midwestern United States:

Jan	0.41	Jul	0.68
Feb	0.45	Aug	0.67
Mar	0.51	Sep	0.66
Apr	0.55	Oct	0.59
May	0.61	Nov	0.44
Jun	0.67	Dec	0.38

Calculate γ and graph $F(K_T)$ as a function of K_T.

Solution   The values of $K_{T,min}$, $K_{T,max}$, and K are 0.38, 0.68, and 0.55, respectively. The best-fit value for γ can be solved iteratively using software; solving to two decimal places, in this case using the Solver function in MS Excel, gives either $\gamma = -1.12 \times 10^{-7}$ or $\gamma = 3$. The appropriate value must then be chosen based on observed conditions. For the case of $\gamma = -1.12 \times 10^{-7}$, there is a 50% chance that the clearness

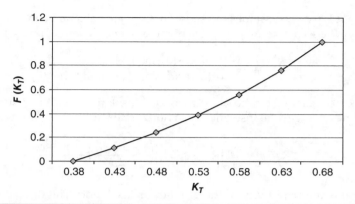

N

FIGURE 9-12 Plot of $F(K_T)$.

will be less than or equal to 0.53, which implies relatively cloudy conditions. In the case of $\gamma = 3$, the value is only 39%, so that conditions are relatively sunny. From the given data, we see that 8 out of 12 months have values of K_T in the range of 0.5 to 0.68, so $\gamma = 3$ is the correct value. Plotting $F(K_T)$ then gives the curve shown in Fig. 9-12.

9-4-3 Effect of Surface Tilt on Insolation Diffusion

The preceding formulas consider only the case of a flat, horizontal surface. Next, we adapt the calculations for tilted surfaces that are common to many installations of solar collector arrays and photovoltaic panels. The global insolation on a surface tilted at an angle β is

$$H_\beta = (H - H_d)R_{b,\beta} + H_d\left(\frac{1 + \cos\beta}{2}\right) + H\left(\frac{1 - \cos\beta}{2}\right)\rho_g \qquad (9\text{-}25)$$

Here $R_{b,\beta}$ is the ratio of insolation on a tilted surface to insolation on a flat surface, and ρ_g is the reflectivity of the ground. Note that in the case of perfect transmission of light through the atmosphere, $H_d = 0$, and $H_\beta = HR_{b,\beta} + H[(1-\cos\beta)/2]\rho_g$. Also note that Eq. (9-25) assumes a south-facing device.

The procedure to calculate the daily integral of insolation on a south-facing tilted surface begins with Eq. (9-22). The next step is to calculate the sunset hour angle on the surface. From the fall to spring equinox, the sunset hour angle on the tilted surface is the same as for a horizontal surface (the sun rises south of east, and the like). Between the spring and fall equinoxes, the calculation must be adjusted because the sun rises north of east and sets north of west. Equations (9-26a) and (b) apply to autumn to spring equinoxes, and spring to fall equinoxes, respectively.

$$\omega_s = \cos^{-1}[-\tan(L)\tan(\delta)] \qquad (9\text{-}26a)$$

$$\omega'_s = \min\{\omega_s, \cos^{-1}[-\tan(L - \beta)\tan(\delta)]\} \qquad (9\text{-}26b)$$

The next step is to calculate $\overline{R}_b$, the ratio of average direct beam insolation on the tilted surface to the corresponding average direct beam insolation on a horizontal surface.

Note that Eq. (9-27) assumes the sunset hour angle value to be in radians. If in degrees, multiply the second term in both the numerator and denominator by $\pi/180$.

$$\overline{R}_{b,\beta} = \frac{\cos(L-\beta)\cos(\delta)\sin(\omega_s') + \omega_s'\sin(L-\beta)\sin(\delta)}{\cos(L)\cos(\delta)\sin(\omega_s') + \omega_s'\sin(L)\sin(\delta)} \tag{9-27}$$

Then calculate the ratio of average total insolation on a tilted surface to the corresponding average total insolation on a horizontal surface

$$\overline{R} = \left(1 - \frac{\overline{H}_d}{\overline{H}}\right)\overline{R}_{b,\beta} + \left(\frac{\overline{H}_d}{\overline{H}}\right)\left(\frac{1+\cos(\beta)}{2}\right) + \frac{\rho[1-\cos(\beta)]}{2} \tag{9-28}$$

where ρ is the reflectivity of the ground for insolation. The three terms on the right-hand side in Eq. (9-28) are for the direct, diffuse, and reflected components of insolation on the tilted surface.

The final step is to calculate $\overline{H}_T$, the average daily total insolation on the tilted surface, as illustrated in Example 9-5

$$\overline{H}_T = \overline{H}\,\overline{R} \tag{9-29}$$

Example 9-5   Consider February in Ithaca, New York, U.S. Calculate the expected daily insolation to fall on a south-facing solar collector panel having a tilt angle of 45°. The ground reflectance is 0.7 (snow more than 1 in deep). Solar data for February in Ithaca are $\overline{H} = 8.61$ MJ/m² and $\overline{K}_T = 0.435$; these parameters are derived from the analysis of local insolation data over time[2]. Assume isotropic sky conditions as a first approximation.

Solution   From geographic data sources, the latitude of Ithaca is 42.3° north. The middle of February is day number 45. The solution sequence is to calculate solar declination, the sunset hour angle, and then the terms as defined through Eqs. (9-22), (9-26), (9-27), (9-28), and (9-29).

$$\delta = -13.6° = -0.237 \text{ rad}$$

$\omega_s = 77.3°$, equivalent to $77.3/15 = 5$ h and 9 min after solar noon and $\omega_s' = \omega_s$

$\overline{R}_{b,\beta} = 1.973$, or 97.3% more beam insolation than on the horizontal surface

$\overline{R} = 1.598$, or 59.8% more total insolation than on the horizontal surface

$$\overline{H}_T = 8.61 \times 1.598 = 13.5 \text{ MJ/m}^{-2}$$

Components of H_T are

    direct: $1.130 \times 8.61 = 9.73$ MJ/m²

    diffuse: $0.365 \times 8.61 = 3.14$ MJ/m²

    reflected: $0.103 \times 8.61 = 0.89$ MJ/m²

---

[2]One approach is to approximate local insolation data by looking at national weather data websites such as http://rredc.nrel.gov/solar/old_data/nsrdb/bluebook/state.html and estimating local values using nearby cities. In this case, nearby cities such as Syracuse, NY, and Binghamton, NY, could be used.

9-5   Summary

Solar energy, which arrives at an approximately constant intensity at the outer edge of the Earth's atmosphere, is affected by geographic location on Earth, weather, and time of year, in terms of the amount of energy available at a given location. The total energy available at a given location, or "global" insolation, is the sum of direct insolation, which arrives unimpeded from the sun, and diffuse insolation, which is diffracted as it passes through the atmosphere.

The location of the sun in the sky at a given time of day, and hence amount of energy available to a solar device, can be determined based on latitude and time of year. The orientation of the device relative to due south and the angle of inclination of the device also impact the amount of energy available. An appreciation for the day-to-day variations of insolation is important in understanding the temporal dynamics of any solar energy gathering system.

The clearness index, or ratio of energy available on the Earth's surface to energy available in space, can be measured on a daily, monthly, or yearly basis. Observation has shown that if the average clearness index in a location is known, the distribution of days or months with a given amount of diffusion can be approximated for a wide range of locations on Earth.

References

ASHRAE (2001). *Handbook of Fundamentals*. American Society of Heating, Refrigerating, and Air-conditioning Engineers, Atlanta, GA.

Hottel, H. C. (1976). "A Simple Model of Transmittance of Direct Solar Radiation through Clear Atmospheres." *Solar Energy*, Vol. 18, pp. 129–134.

Jain, A. (1997). Solar Concepts. Electronic resource, available at www.usc.edu. Accessed November 29, 2015.

Norton, B. (1992). *Solar Energy Thermal Technology*. Springer-Verlag, London.

Reddy, T. (1987). *The Design and Sizing of Active Solar Thermal Systems*. Clarendon Press, Oxford.

Trewartha, G., and L. Horn (1980). *An Introduction to Climate*, 5th ed. McGraw-Hill, New York.

Further Reading

Duffie, J. A., and W. A. Beckman (2006). *Solar Engineering of Thermal Processes*. John Wiley & Sons, New Jersey.

Fanchi, J. (2004). *Energy: Technology and Directions for the Future*. Elsevier, Amsterdam.

Goswami, D. Y., F. Kreith, and J. F. Kreider (2000). *Principles of Solar Engineering*, 2nd ed. Taylor & Francis, New York.

Johannson, J. (1993). *Renewable Energy: Sources for Fuels and Electricity*. Island Press, Washington, DC.

Kasten, F. and A. T. Young (1989). "Revised Optical Air Mass Tables and Approximation Formula." *Applied Optics*, Vol. 28, No. 22, pp. 4735–4738.

Kreith, F., and J. F. Kreider (1978). *Principles of Solar Engineering*. McGraw-Hill, New York.

Leckie, J., G. Masters, H. Whitehouse, and L. Young. (1981). *More Other Homes and Garbage*. Sierra Club Books, San Francisco.

Liu, B., and R. Jordan (1960). "The Interrelationship and Characteristic Distribution of Direct, Diffuse and Total Solar Radiation." *Solar Energy*, Vol. 4, pp. 1–19.

McQuiston, F., J. Parker, and J. Spitler. (2004). *Heating, Ventilation, and Air Conditioning: Analysis and Design*. Wiley, New York.

National Renewable Energy Laboratories (2005). *National Solar Radiation Data Base: 1991-2005 Update, Typical Meteorological Year 3*. Electronic Resources, available at http://rredc.nrel.gov/solar/old_data/nsrdb/1991-2005/tmy3/. Accessed Jun. 22, 2015.

Sorensen, B. (2000). *Renewable Energy: Its Physics, Engineering, Environmental Impacts, Economics and Planning*, 2nd ed. Academic Press, London.

Exercises

9-1. A solar panel is mounted facing due south in a location at 48° north latitude, with a tilt angle of 30°. If the sun shines on the panel for the entire day on the equinox with clear skies, (a) calculate the total energy incident per square meter on the panel during the course of the day in kWh, (b) suppose there is an obstruction on the site such that from 2:30 p.m. onward on this day, the obstruction blocks the sun from reaching the panel. What percent of energy is lost due to the obstruction, compared to the amount that would reach the panel if it were not there, on the day being studied?

9-2. A house at 42° north latitude has a roof that faces due south, and is elevated to an angle of 23°. A solar panel is mounted to the roof. What is the angle of incidence between the sun and the array at 10 a.m. on May 5th?

9-3. A solar device is located at a point that is blocked by an adjacent building for part of the morning (see illustration). The building has a west wall that faces due west, and is located 10 m east of the point. The south wall of the building extends 8 m south of the point. The building is located in Edinburgh, Scotland, at 56° north latitude. On the summer solstice, does the building block the sun from shining on the device at 9 a.m. solar time? You can assume that the building is sufficiently tall such that it is not possible for the sun to reach the device by shining over the top, it must shine in past the southwest corner of the building.

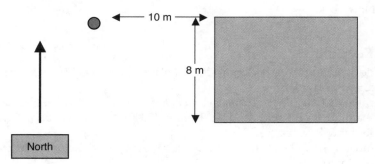

9-4. Use the law of cosines [$c^2 = a^2 + b^2 - 2ab\cos(c)$] to derive Eq. (9-19) in this chapter.

9-5. As a thought experiment, determine where on the Earth, and on which days, the solar azimuth can be the same value all day as the solar hour angle. Then prove this equality by calculations.

9-6. You are at 76° west longitude, 37° north latitude, and observe the sun to rise at 05:10:20 hours local clock time. Your friend is at 87° west longitude and 49° north latitude. What will be your clock time when she sees the sun rise?

9-7. You are at 94° west longitude in the Mountain time zone of the United States. The date is July 14, and daylight savings time is in effect. If your watch reads 13:55:23 hours, what is the corresponding local solar time?

9-8. On July 17 (daylight savings time is in effect), what is the value of the equation of time, and is the sun "fast" or "slow"? What will be the clock time when the sun is exactly due south? What are the answers for November 17 (daylight savings time is not in effect)? Assume both the local and standard longitudes are 90 W.

9-9. Use Eq. (9-3) to graph the equation of time as a function of the day of the year, and identify those periods of the year when the sun is "fast" and those periods when it is "slow."

CHAPTER 10

Solar Photovoltaic Technologies

10-1 Overview

This chapter discusses the internal function and application of solar photovoltaic technology. In the first part, we discuss how photovoltaic cells convert sunlight to electricity, and how to quantify their performance relative to objectives for acceptable efficiency values. In the second part of the chapter, we discuss practical issues related to photovoltaic panel arrays, including design considerations, output in different regions, economic payback, and environmental considerations.

10-1-1 Symbols Used in This Chapter

Many symbols are used in this chapter and recalling their definitions can become a nuisance. To help, Table 10-1 is provided. Angles can be used as degrees or radians, depending on the requirements of the calculator or computer software being used and the equation in which they are embedded or defined.

10-2 Introduction

The *photovoltaic cell*, or PV cell, is a technology that converts energy in sunlight to electricity, using an adaptation of the electrical semiconductor used in computers and other types of information technology. PV cells are designed to transfer the energy contained in individual photons penetrating the panel to electrons that are channeled into an external circuit for powering an electrical load. This function is achieved through both the layout of molecular structure at the microscopic level, and the arrangement of conductors at the edge of the cell at the macroscopic level.

A *PV panel* consists of multiple photovoltaic cells (on the order of 50 to 120 cells) connected together in an electrical circuit that can be connected to an exterior circuit at a single point. Since the individual PV panel has a limited output compared to typical residential or commercial loads (maximum output on the order of 80 to 400 W per panel), a PV system usually combines a large number of panels together in an *array*, or integrated system of panels, so as to deliver the electricity produced at a single connection point and at the desired voltage.

Although other systems for converting sunlight to electrical energy exist, such as systems that generate steam to drive a turbine of the type discussed in Chap. 6, PV

Symbol	Definition, with Units (SI) Where Appropriate (Overhead Bar Signifies Averaged)
a, b	Empirical parameters, see Eqs. (10-8) and (10-9)
A	Area of PV panel, m^2
c	Speed of light, 299,792,458 m/s
C	Rated capacity of PV panel system, kW
C_f	Optimum panel tilt angle correction factor, see Eq. (10-24)
E	Electrical charge per electron, 1.602E−19 J/V
E	Average monthly PV panel system output, kWh/month/kW of capacity for simple approach (Sec. 10-4-2), and kWh/month for extended approach [Sec. 10-4-3 and Eq. (10-16)]
E_G	Band gap energy for an electron in PV material, e.g., joules
F	Cumulative distribution function of the Weibull equation
FF	Fill factor, see Example 10-2
h	Planck's constant, reflecting quanta step, 6.626E−34 J·s
$\overline{H_d}$	Average insolation on tilted PV panel for a given month
I	Net current in a PV cell, see Eq. (10-4)
I_D	Dark current in a PV cell, see Eq. (10-3)
I_i	Total insolation for an hour, kWh/m^2
I_{NOCT}	Efficiency test standard insolation at NOCT, normally 800 or 1000 W/m^2
I_o	Saturation current in a PV cell
I_L	Light current in a PV cell, see Eq. (10-2)
I_M	Electrical current at maximum power output
I_{sc}	Short circuit current, see Fig. 10-6
$\overline{K_T}$	Atmospheric clearness index, see Eq. (9-22)
k	Boltzmann constant, 1.38E−23 J/K
M	Parameter depending on PV cell operating voltage, see Eq. (10-3)
N	Number of days
NOCT	Nominal Operating Cell Temperature, see Eq. (10-25)
P	Power, volts × amperes
RF	Regional factor, see Eq. (10-12)
S	Solar flux, photons m^{-2} s^{-1}
S	Actual PV panel tilt angle for a month (from horizontal)
S_M	Optimum PV panel tilt angle for a month (from horizontal)
T	Temperature, kelvin
T_a	Monthly average air temperature during daylight hours
T_{ai}	Ambient (air) average temperature for collection period (typically one hour)
T_c	Monthly average cell temperature
T_{ci}	Cell average temperature for collection period (typically one hour)
T_m	Mean ambient (air) temperature for the month of analysis

TABLE 10-1   Symbols Used in Chap. 10

Symbol	Definition, with Units (SI) Where Appropriate (Overhead Bar Signifies Averaged)
T_r	Cell temperature at rating condition
U_L	Heat loss coefficient, based on panel area, often assumed as 20 W m^{-2} K^{-1}
V	Volts
V_M	Voltage at maximum power output
V_{OC}	Open circuit voltage, see Eq. (10-5)
W	Cell width
α	Absorption coefficient, see Eq. (10-1)
β	Cell temperature coefficient of PV efficiency (C^{-1})
Γ	Cell insolation coefficient of PV efficiency, see Eq. (10-15)
η	Efficiency
η_{col}	Collection efficiency
η_i	Cell efficiency for specific hour, see Eq. (10-15)
η_r	Cell efficiency at rating conditions
λ	Wavelength of light
ρ	Reflection coefficient, see Eq. (10-1)
τ	Transmission coefficient, see Eq. (10-1)

TABLE 10-1 Symbols Used in Chap. 10 (*Continued*)

systems are the largest producers of electricity directly from solar energy[1] in the world at this time, in terms of kWh produced per year.

The PV panel takes advantage of the *photovoltaic effect*, discovered by Henri Becquerel in 1839, in which sunlight striking certain materials is able to generate a measurable electrical current. It was not until 1954 that scientists were able to generate currents using the photovoltaic effect that could be applied to an electrical load in a laboratory setting. Thereafter, application in the field followed relatively quickly, with PV panels first used commercially in a remote telecommunications relay station and on the U.S. satellite Vanguard I by the end of the 1950s. PV cells were first used in applications such as these because the high added value of an independent remote energy source justified the high cost per kW of rated panel output of the original PV panels. In the case of telecommunication systems, the high added economic value of the relay station system combined with the remote location of some of these facilities made it attractive to install PV. For orbiting satellites, PV panels proved to be an ideal energy technology, due to stronger insolation values in space, lack of cloud cover or weather-related deterioration, long lives with little output change, and ability to function except when in the shadow of the earth. PV panels have been used for both of these types of applications ever since.

While the cost per unit of output matters relatively little in telecommunications or space applications, where the cost of the core technology will dominate the total cost of the system in any case, it is clear that the technology must be affordable for use by mainstream electricity consumers in the commercial and residential sectors. It is here that

---

[1]Wind energy, which has a larger annual output worldwide than PV panels, is an indirect form of solar energy since it originates from uneven heating of the earth's surface by the sun; see Chap. 13.

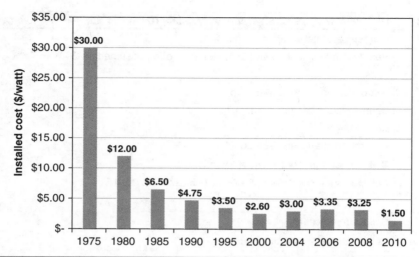

FIGURE 10-1   Cost per installed watt for solar panels, 1975–2010.

(*Source:* U.S. Energy Information Adminstration.)

R&D from the 1970s onward has lowered cost per installed watt and opened up growth potential in the world market for PV panels. Figure 10-1 shows the reduction in cost of photovoltaic panels for the period 1975 to 2010. Costs fell from $30/W in 1975 to approximately $4/W by 1995; thereafter they fluctuated between $2.50 and $3.50/W up to 2005–2007, when a shortage of silicon raw material for production led to a rise in price. From 2007 onward as the silicon shortage eased, prices fell to $1.50/W by the end of 2010, and as low as $1/W by 2015, although it is not shown in Fig. 10-1. Note that these costs do not include balance of system cost, installation labor, and overhead. Whole systems cost on the order of $1.70-$2.00/W when installed at a multi-megawatt scale, as discussed in Sec. 10-4-4.

Note that cost per watt, as used by the solar energy industry, is the total cost of a panel divided by the output at ideal conditions at which it is rated (full sun, perpendicular to the sun's rays, and so on). Because panels operate much of the time at less than optimal conditions (e.g., with partial cloud cover or at oblique angles to the sun), or do not operate at all at night, economic analysis of the panel system must be calculated based on actual and not ideal operating conditions, as discussed later in the chapter.

The U.S. solar PV market responded to the breaking of the $3/W barrier around 2000 by growing substantially, from 79 MW in that year to 2400 MW in 2010 (Fig. 10-2). It then grew much more quickly after the further reduction in price from $3/W to $1.50/W between 2008 and 2010, reaching ~20 GW at the end of 2014, or about eight times the capacity at the end of 2010. The period from 2010 to 2014 saw not only growth in distributed residential and commercial PV arrays but also the entry of some multi-megawatt solar power stations that supply power directly to the grid. As shown in Fig. 10-3, cumulative installation in 2013 amounted to 138.8 GW, with large capacity in world-leader Germany as well China, the U.S., and Italy. Growth in installed capacity in some of the leading countries has been dramatic. As recently as 2001, Germany had 75 MW of PV panels installed in grid-connected systems, out of a world total of 1900 MW; in 2013 this figure was 35 GW. The growth from 2010 to 2013 was

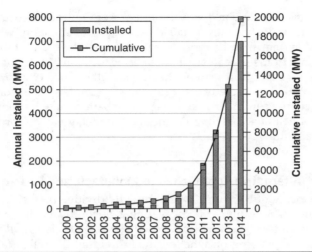

FIGURE 10-2   U.S. annual and cumulative solar PV installations, 2000–2014.

(*Source:* Solar Energy Industries Association.)

remarkable as well, consisting of 98.5 GW or 244% of the cumulative capacity installed as of the end of 2010. China had 2% of world installed PV capacity at the end of 2010 but grew to 18.6 GW or 13% of the world capacity in 2013, thanks to large ongoing investments in renewables generally and an ambitious national commitment to invest in PV. On the production side, solar PV manufacturing is distributed among many countries as well, including Germany, Japan, China, the United States, and Norway, among others.

Included in the 24% (33.2 GW) "ROW" category in Fig. 10-3 are several hundred megawatts of small PV systems installed in rural areas of relatively poor tropical

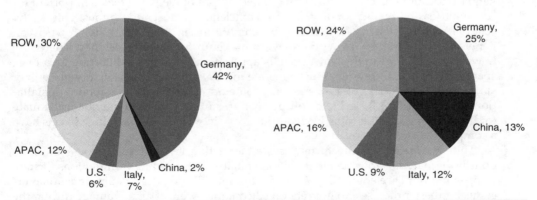

FIGURE 10-3   World cumulative installed solar PV capacity, 2010 and 2013.

Notes: Total world capacity in 2010 and 2013 was 40.3 and 138.8 GW, respectively. APAC = Asia-Pacific, not including China. ROW = rest of world.

(*Source:* European Photovoltaic Industry Association.)

countries, as are found in, for example, parts of Central America or Sub-Saharan Africa. These systems are typically installed in off-the-grid situations where the cost of a stand-alone PV system with battery storage compares favorably to the cost of bringing in electric lines from the nearest grid-connected location. While homeowners in the United States might expect a system on the order of 2000–5000 W in size to meet a substantial fraction of load, in these tropical locations, a simple system consisting of a 50–100 W panel, a charge controller, and a deep cycle battery may be enough to provide electric light at night and access to radio and the Internet, greatly improving the quality of life for a rural family.

10-2-1   Alternative Approaches to Manufacturing PV Panels

Manufacturing of PV panels can be broadly divided between *crystalline* and *thin-film* approaches. The former entails creating a silicon wafer that harnesses the photovoltaic effect and can be further divided between *monocrystalline* and *multicrystalline* solar cells, depending on whether the individual PV cell is fabricated from a single crystal or multiple crystals. In either case, the crystals are first grown and then thin slices of silicon are cut from them and placed on the PV panel. Monocrystalline cells are more efficient than multicrystalline in terms of converting sunlight to electrical output, but they are also more difficult to manufacture and hence more expensive per unit of surface area covered.

Thin-film panel manufacturing entails a fundamentally different process, namely, laying down a thin layer of photovoltaic effect capable materials, either *amorphous silicon* or some nonsilicon combination of metals, on a backing material. Thin-film cells use a mass-production process that is easier to scale up than the growing and cutting of crystals, and are therefore the cheapest to produce but have the lowest efficiency per unit of surface area. Both manufacturing cost per unit area and average efficiency have been improving in recent years. Thin film panels constituted approximately 10% of the world PV market in 2013.

Photovoltaic materials with lower output per square meter can compete in the marketplace because it is the cost per unit of rated output of the panel that is important for economic feasibility, and not the maximum efficiency, in most applications. This is why cost per watt of capacity is used as a benchmark in Fig. 10-1: the system owner can generally cover more area with panels to make up for lower efficiency. There are, however, exceptions to this rule, such as applications with limited surface area (e.g., spacecraft), where the system must generate the maximum possible energy per unit of surface area, regardless of cost. Also, for some applications such as rooftop installations, with thin-film PV cells the output per area of roof covered may be inadequate for the system owner's annual kWh output goals, in which case crystalline panels have a competitive edge.

Thin-film panel manufacturing uses either a thin layer of amorphous silicon or nonsilicon materials to generate the photovoltaic effect. Leading formulations include Copper Indium Diselenide (CIGS) and Cadmium Telluride (CdTe). Since manufacturers using these processes do not rely on silicon, they were at an advantage during the silicon shortage of 2006–2008 observed in the price "spike" in Fig. 10-1. On the other hand, use of heavy metals such as cadmium may in the future create a hazardous waste challenge both at the manufacturing and disposal stages of the life cycle, so the industry will need to proceed carefully.

PV manufacturers are pursuing several developments either to increase efficiency or to reduce manufacturing cost:

1. *Concentrating PV (CPV):* CPV is an adaptation of the conventional PV cell in which a light concentrating system is added to the panel to increase output per unit of area of PV surface. Thus an additional cost is added for the concentrating system, but expense for PV cells per unit of rated output is reduced. Since CPV focuses sunlight, the panel system must incorporate a sun-tracking capability over the course of the day. An example of a CPV system is the 330-kW installation that opened near Tarragona, Spain, in 2009.

2. *Multi-junction PV:* In a traditional *single-junction* PV panel, the photovoltaic effect takes place at a single layer within the PV cell. Depending on the material used, the cell is optimized for a specific energy level of incoming photon, with silicon crystal cells optimized for lower energy photons, and nonsilicon cells typically optimized for higher energy levels (see Sec. 10-3-1). The *multi-junction* PV cell uses layers of junctions one underneath the other to at first convert higher energy (shorter wavelength) photons and then lower energy ones, increasing maximum potential efficiency. There is of course a cost premium for this more complex manufacturing process, so the goal of multi-junction panel makers is to bring costs down so that additional output compensates for additional cost per square meter. In the short term, multi-junction cells are ideal for applications such as space flight that put a premium on maximizing output.

3. *Nanotechnology:* A number of thin-film PV makers in the United States, Japan, China, and Germany are using nanotechnology [i.e., manipulating components in the cell at a scale of 1 nanometer (nm) or 1×10^{-9} m] to increase maximum cell efficiency and reduce cost per rated watt. Fabricating cells at a nanoscale level allows the engineer to design the desired structural qualities of the cell more precisely, so as to withdraw as much energy as possible out of the incoming photon, minimizing the amount that escapes as heat.

10-3   Fundamentals of PV Cell Performance

The objective of the PV cell is to convert as much of the energy in the incoming stream of photons into electricity as possible. In this section, we will describe the phenomena that are occurring as the individual photons are either converted to electrical charge, or not, depending on the circumstances. We will also quantify the effect of rates of conversion and rates of loss on the overall efficiency of the cell.

In order to encourage the conversion of photons to electrons using the photoelectric effect, the two layers of silicon that constitute a silicon-based PV cell are modified so that they will be more likely to produce either (1) loose electrons, or (2) holes in the molecular structure where electrons can reattach. Recall that the silicon atom has 4 valence electrons in its outer shell. In one PV cell design, the upper or n-type layer is doped with phosphorus, with 5 valence electrons, while the lower or p-type layer is doped with boron, which has 3 valence electrons (see Fig. 10-4). This type of molecular structure is known as a *p-n junction* and is common to most designs of PV cells. The imbalance of electrons across the cell (due to migration of some free electrons to open holes) creates a permanent electrical field which facilitates the travel of electrons to and from the external circuit and load to which the PV system is attached.

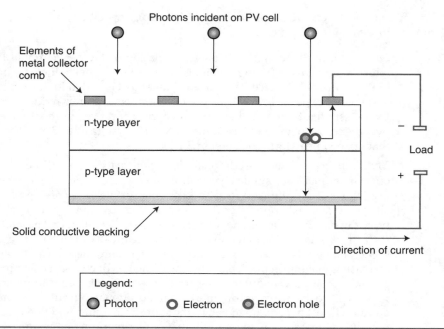

Figure 10-4   Cross section of PV cell.

In Fig. 10-4, the upper layer is an n-type silicon doped with phosphorus, which has an excess of electrons; the lower layer is a p-type silicon doped with boron, which has extra holes for absorbing electrons. On the right-hand side, an incoming photon breaks loose an electron within the cell structure. The electron then travels toward the collector comb, while the electron hole travels toward the conductive backing, contributing to the current flow to the load outside the cell.

An arriving photon possessing an energy value equal to or greater than the *bandgap energy* E_G is able to break loose an electron from the structure of the PV cell. The value of E_G depends on the type of panel used; recall that energy is inversely related to wavelength, that is, $E = hc/\lambda$, where h is the Planck's constant and c is the speed of light. Therefore, photons with a wavelength greater than the *bandgap wavelength* λ_G do not have sufficient energy to convert to an electron. Electron holes can also move through the cell structure, in the opposite direction to that of the electrons, in the sense that as an electron moves to fill a hole, it creates a hole in the position previously occupied.

Electrons leave the structure of the PV cell and enter the exterior circuit via conductive metal collectors on the surface of the cell. Typical collectors are laid out in a "comb" pattern and are readily observed on the surface of many PV cells. The collector networks, where present, block the incoming sunlight from entering the PV cell. Therefore, design of the collectors involves a trade-off between having sufficient area to easily collect as many electrons as possible, but not so much area that the ability of the sun's energy to enter the cell is greatly reduced.

10-3-1   Losses in PV Cells and Gross Current Generated by Incoming Light

If it were possible to convert all the energy in the incoming light into electrical current, then the magnitude of this current would be equivalent to the energy available in the light per unit of surface area. For applications outside the earth's atmosphere, this flux averages 1.37 kW/m², while for applications at the earth's surface, in full sun, the flux is reduced to approximately 1 kW/m² due to refraction and absorption in the atmosphere. However, there are numerous sources of loss within the cell itself that reduce the amount of energy actually available, as follows:

- *Quantum losses:* As mentioned above, photons with $E < E_G$ are unable to produce the photoelectric effect, so the energy they contain is not available to the system, constituting one type of loss. Also, each photon can only produce 1 electron, regardless of the energy of the photon, so the "excess" energy is also not available for producing current.

- *Reflection losses:* Photons arriving at the surface of the PV cell are subject to a fractional reflection loss $\rho(E)$, where $0 < \rho < 1$; the amount of reflection varies with the energy of the photon. PV cells are typically coated with an antireflective coating to reduce $\rho(E)$.

- *Transmission losses:* Because PV cells possess finite width (i.e., thickness of the cell in the direction of travel of the photon), it is possible for a photon to pass through the cell without colliding with an atom in the structure. In this circumstance, the photon has been "transmitted" through the cell, leading to transmission losses. The amount of these losses is a function of both the width of the cell and the energy of the photon, and is written $\tau(E, W)$, where $0 < \tau < 1$. One possible way to make better use of all available energy is to position a solar thermal application, such as a solar water heater, underneath the cells of a PV panel; photons that pass through the cell can then be used to accumulate thermal energy.

- *Collection losses:* In a real-world PV cell, some fraction of the electrons released will permanently reabsorb into the structure before they have a chance to leave the cell, so that the collection efficiency, denoted by $\eta_{col}(E)$, will be less than 100%. For a range of photons with energy E such that $E_G < E < E^*$, where E^* is some approximate upper bound that is specific to the design of cell in question, $\eta_{col}(E)$ is close to unity for many types of cells. However, especially at very high photon energy values, empirical observation shows that electrons released are very likely to reabsorb, and $\eta_{col}(E)$ falls to nearly 0. Thus collection losses constitute another type of significant loss in the cell.

The gross current generated by the PV cell, also known as the *light current* I_L since it occurs only when the cell is illuminated, can now be calculated by taking into account the full range of losses. To simplify notation, define the *absorption coefficient* $\alpha(E, W)$ to take into account reflection and transmission losses as follows:

$$\alpha(E,W) = 1 - \rho(E) - \tau(E,W) \tag{10-1}$$

Also define the function $S(E)$ of frequency of arriving photons as a function of photon energy. In practice, the curve for wavelength of arriving photons is observed to have a shape approximated by an appropriate statistical function with a single peak and zero probability of negative values, such as a Weibull or Gamma distribution. I_L is now calculated by integrating over the range of photon energy values from E_G upward:

$$I_L = e \int_{E_G}^{\infty} \eta_{col}(E)S(E)\alpha(E,W)\, dE \qquad (10\text{-}2)$$

Here I_L is given per unit of area (e.g., mA/cm²), and e is the unit charge per electron.

Typical values of losses vary widely between PV cell designs, but in all cases the total effect is significant. Depending on the design, band gap energy losses may amount to up to 23%, while quantum losses from photon energy greater than electron energy, combined with collection losses, may amount to up to 33%. Absorption losses are typically on the order of 20%. Therefore, a device that achieves on the order of 15 to 25% overall efficiency is considered to be relatively efficient. Some commercially available devices with efficiency values lower than this range can compete in the marketplace if their cost per unit of area is acceptable. Also, there are inherent limits on the extent to which losses can be eliminated: while there is no theoretical lower bound on the reduction of absorption losses possible, through advances in technology, quantum losses are an integral part of the photoelectric effect and cannot be avoided.

Example 10-1 illustrates the effect of various types of losses on overall current production and device efficiency.

Example 10-1   A solar cell has a constant collection efficiency of 95% between energy values of 1.89×10^{-19} and 5.68×10^{-19} J, and 0% outside this range. The absorption coefficient is 80% for all energy values at the given width of the solar cell. Photons arrive at the surface of the panel at a rate $S = 2.5 \times 10^{21}$ particles/s·m². Photon energy values E have a Weibull ($\alpha = 3$, $\beta = 3$) distribution with E measured in units of 10^{-19} J. Calculate I_L.

Solution   From the data, we are given that $\eta_{col}(E) = 0.95$ over the range in which the device operates, and $\alpha(E,W) = 0.8$ throughout. Therefore, we can integrate Eq. (10-2) between 1.89 and 5.68×10^{-19} J, and take the η_{col} and $\alpha(E,W)$ terms outside the integral since they are both constant:

$$I_L = e \int_{1.89}^{5.68} \eta_{col}(E)\alpha(E,W)S \cdot f(E)$$

$$= e\eta_{col}(E)\alpha(E,W)S \int_{1.89}^{5.68} f(E)$$

The Weibull (3, 3) probability density function, or pdf, has the form shown in Fig. 10-5 over the interval from $E = 1 \times 10^{-19}$ J to $E = 6 \times 10^{-19}$ J, which contains the desired interval 1.89×10^{-19} J $< E < 5.68 \times 10^{-19}$ J. From the figure, the peak frequency of photons occurs around $E = 2.5 \times 10^{-19}$, and over 99% of the photons incident have an energy less than the upper bound of $E = 5.68 \times 10^{-19}$ J. To illustrate the application of the pdf, take the case of $E = 1 \times 10^{-19}$ J:

$$f(x) = \frac{\alpha}{\beta^{\alpha}} x^{\alpha-1} \exp[-(x/\beta)^{\alpha}]$$

$$f(1) = \frac{3}{3^3} 1^{3-1} \exp[-(1/3)^3] = 0.107$$

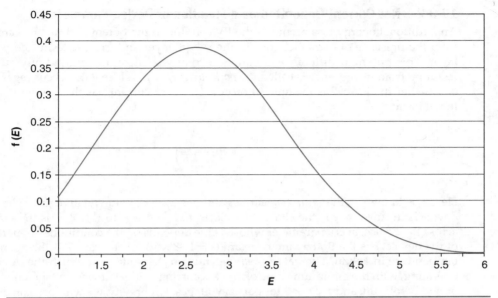

FIGURE 10-5 Distribution of photon frequency as a function of energy 10^{-19} J.

To solve for I_L, we can convert the integral into the probability that E falls in the desired interval. For this purpose, we use the cumulative distribution function $F(E)$, which for a Weibull distribution has the following form:

$$F(E) = 1 - \exp[-(E/\beta)^\alpha]$$

The desired probability is then calculated as follows:

$$P(1.89 < X < 5.68) = F(5.68) - F(1.89)$$
$$F(5.68) = 1 - \exp[-(5.68/3)^3] = 0.999$$
$$F(1.89) = 1 - \exp[-(1.89/3)^3] = 0.221$$
$$P(1.89 < X < 5.68) = 0.999 - 0.221 = 0.778$$

The number of electrons released is then the probability of being in the correct energy range multiplied by the rate of photon arrival, by η_{col}, and by α:

$$(0.778)(2.5 \times 10^{21})(95\%)(0.8) = 1.476 \times 10^{21} \text{ e/s} \cdot \text{m}^2$$

Since 1 A of current flow is defined as 6.241×10^{18} electrons per second, we can then calculate I_L per square meter of surface:

$$I_L = \frac{1.476 \times 10^{21} \text{ e/s} \cdot \text{m}^2}{6.241 \times 10^{18} \text{ e/s} \cdot \text{A}} = 236.4 \text{ A/m}^2$$

or 236.4 A/m^2. Note that the inputs into this calculation are simplistic, because the actual values of $\eta_{col}(E)$ and $\alpha(E, W)$ are likely to vary with E, and the chosen value of $S(E)$ is a simplified approximation of an actual spectral distribution.

10-3-2   Net Current Generated as a Function of Device Parameters

An additional energy loss occurs in the PV cell due to the potential difference created from the presence of extra electrons in the n-type layer and extra holes in the p-type layer. This potential difference is common to any semiconductor, and any current resulting from its presence is called the *dark current*, or *ID*, since it can occur regardless of whether the PV cell is in sunlight or not. The dark current has the following functional form:

$$I_D = I_0 \left[\exp\left(\frac{eV}{mkT}\right) - 1 \right] \tag{10-3}$$

Here I_0 is known as the *saturation current* and has a small value relative to I_L when the device is in full sun, e is the charge per electron, or 1.602×10^{-19} J/V, V is the voltage across the device, m is a parameter whose value depends on the conditions of operation of the device, k is the Boltzmann constant ($k = 1.38 \times 10^{-23}$ J/K), and T is the operating temperature in kelvin. The value of m depends on the voltage at which the device is operating, with the minimum value of $m = 1$ occurring at ideal device voltage under full or nearly full sun, and $1 < m < 2$ for partial voltages. For brevity, we will focus on device operation under ideal conditions with voltage near the maximum value possible, and use $m = 1$ through the remainder of the chapter.

Because I_D flows in the opposite direction from I_L, the net current from the PV cell is the difference between the two, that is

$$I = I_L - I_D$$

$$= I_L - I_0 \left[\exp\left(\frac{eV}{mkT}\right) - 1 \right] \tag{10-4}$$

Plotting I as a function of V for representative parametric values for a cell gives the function shown in Fig. 10-6. At $V = 0$, the amount of current produced is the *short-circuit current* I_{SC}, which is equal to the light current since $\exp(eV) = 1$ and I_D makes no contribution, that is, $I_{SC} = I_L$. Due to the exponential nature of I_D, its effect on I grows rapidly above a certain value, so that at the *open-circuit voltage* V_{OC}, the device ceases to produce a positive current.

It is now possible to solve for V_{OC} by setting $I = 0$ in Eq. (10-4) since, by definition, there is no current flow in an open circuit:

$$I_L - I_0 \left[\exp\left(\frac{eV_{OC}}{mkT}\right) - 1 \right] = 0$$

$$V_{OC} = \frac{mkT}{e} \ln\left[\frac{I_L}{I_0} + 1\right] \tag{10-5}$$

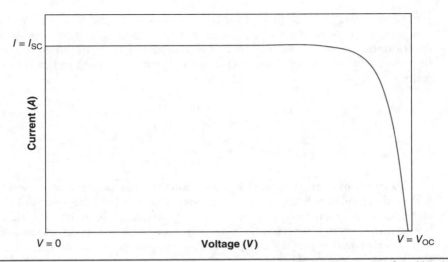

FIGURE 10-6 Current as a function of voltage in ideal PV cell for values between $V = 0$ and $V = V_{oc}$.

We are now able to characterize the function of the cell, since we know the range of voltages over which it operates, $0 \leq V \leq V_{OC}$, and we can calculate the current at any value V given fixed device parameter I_0, value of m at the operating condition, and temperature T.

Calculation of Maximum Power Output

Now that I can be written in terms of fixed device parameters I_{SC} and V_{OC}, and also independent variable V, we wish to know the values of I_M and V_M that maximize the power output from the device. Recall that power P is the product of current and voltage, that is, $P = I \times V$. From inspection of Fig. 10-6, it is evident that P will at first increase with V, but that at some value of V close to V_{OC}, P will peak in value and thereafter decline due to falling I. We can solve for values of I_M and V_M by taking the derivative and setting equal to zero:

$$P = IV$$
$$dP/dV = IdV + VdI = 0 \tag{10-6}$$
$$dI/dV = -I_M/V_M$$

In addition, the values of I_M and V_M must satisfy Eq. (10-4):

$$I_M = I_L - I_0\left[\exp\left(\frac{eV_M}{mkT}\right) - 1\right] \tag{10-7}$$

We now have two equations for two unknowns, I_M and V_M. The solution to this system of equations does not exist in analytical form; however, an approximate solution that gives satisfactory results is available. Define parameters a and b such that $a = 1 + \ln(I_L/I_0)$ and $b = a/(a + 1)$. I_M and V_M can be written in terms of a and b as follows:

$$I_M = I_L (1 - a^{-b}) \tag{10-8}$$

$$V_M = V_{OC}\left(1 - \frac{\ln a}{a}\right) \tag{10-9}$$

Once we can calculate I_M and V_M, we would also like to know how good the values achieved are, relative to some maximum possible value. Looking again at Fig. 10-6, it is clear that a device with $I_M = I_{SC}$ and $V_M = V_{OC}$ would achieve the maximum possible power output from the device. Such a result cannot be achieved in an actual PV cell, so a measure known as the *fill factor* (FF) is used to evaluate the actual maximum value of P relative to this upper bound:

$$FF = \frac{I_M V_M}{I_L V_{OC}} \tag{10-10}$$

Substituting Eqs. (10-8) and (10-9), we can rewrite FF in terms of a and b:

$$FF = (1 - a^{-b})\left(1 - \frac{\ln a}{a}\right) \tag{10-11}$$

In actual devices, fill factor values of $FF \geq 0.7$ are considered to be an acceptable result. Example 10-2 illustrates the performance evaluation for a representative device.

Example 10-2   Consider a PV cell with light current of 0.025 A/cm², saturation current of 1.0×10^{-11} A/cm², open circuit voltage of 0.547 V, and operating temperature of 20°C. For simplicity, assume a value of $m = 1$ and constant saturation current across its operating range. What is the combination of voltage and current that maximizes output from this device, and what is its fill factor?

Solution   From the description above, we have $I_L = 0.025$ A/cm², $I_0 = 1.0 \times 10^{-11}$ A/cm², and $V_{OC} = 0.547$ V. Since V_{OC} is known, we can proceed directly to calculating I_M and V_M:

$$a = 1 + \ln(0.025/1 \times 10^{-11}) = 22.64$$

$$b = a/(a+1) = 0.958$$

$$I_M = 0.025(1 - 22.64^{-0.958}) = 0.0237$$

$$V_M = 0.547\left(1 - \frac{\ln 22.64}{22.64}\right) = 0.471$$

By plotting I as a function of V across the operating range of the cell, it is possible to verify by inspection that the values of I_M and V_M from Eqs. (10-8) and (10-9) are in good agreement with the peak shown in the curve in Fig. 10-7.

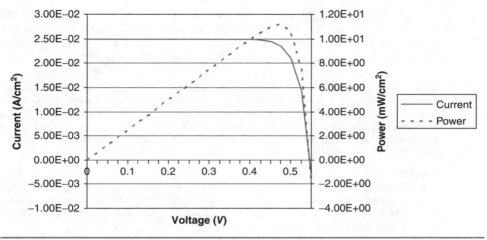

FIGURE 10-7   Current and power as a function of voltage for PV cell.

The FF is then calculated

$$FF = \frac{I_M V_M}{I_L V_{OC}} = \frac{(0.0237)(0.471)}{(0.025)(0.547)} = 0.82$$

Thus, the device is considered to have acceptable output.

10-3-3   Other Factors Affecting Performance

A number of factors affect the actual performance of a PV cell, causing it to vary from the theoretically derived values above. Notable among these

- *Temperature:* Temperature affects the output of the cell, with output at first increasing as temperatures increase from cryogenic temperatures close to 0 K, peaking in the range of 50 to 100°C, and decreasing thereafter. Current I increases with rising temperature, but V_{SC} falls; up to the peak output value, the increase in I drives the change in the power output, but thereafter the decrease in V dominates.

- *Concentration:* It is possible to increase the output from a PV cell by *concentrating* the amount of light striking the cell relative to ambient insolation, up to a point. Concentration is achieved using flat or curved mirrors to focus additional light onto the cells. Up to 15 to 20 times concentration will increase output from a cell; however, above this range of values resistance losses dominate, so that output decreases with increasing concentration.

Other factors that affect performance include resistance in series and in parallel with the PV cells in a panel, as well as the age of the device, for performance degrades gradually with age.

10-3-4   Calculation of Unit Cost of PV Panels

The calculated maximum output per unit of area from a PV cell leads directly to the cost to build a complete PV panel, if the cost per unit of cell area is known. In addition to the individual cells, the panel consists of a backing material to support the cells, a transparent laminate to protect the cells from above, a structural frame for mounting the panel, and all necessary wiring to connect the cells into a circuit. Most of the cost of the panel is in the cost of the cells, as illustrated in Example 10-3.

Example 10-3   Suppose that the cell from Example 10-2 is to be incorporated into a panel with rated output of 120 W at peak conditions. Total cost for manufacturing the cell material including labor, and so forth, is $0.009/cm². The cells are approximately square, with 15 cm on each side. The balance of the panel, including all materials and assembly labor, adds 30% to the total cost of the panel. (a) What is the cost per watt for this panel? (b) Approximately how many cells are required to be fitted to the panel to achieve the desired output?

Solution

(a) To calculate the cost of the panel, we first calculate output per unit of area in peak conditions, and then calculate the amount of area required to meet the target. From the calculation of I_M and V_M in Example 10-2, the output is $(23.7 \text{ mA/cm}^2)(0.471 \text{ V}) = 11.2 \text{ mW/cm}^2$. Therefore, the required area is

$$\frac{120 \text{ W}}{11.2 \text{ mW/cm}^2}(1000 \text{ mW/W}) = 10{,}700 \text{ cm}^2 = 1.07 \text{ m}^2$$

For this amount of area, the cost of the cell material is $(10{,}700 \text{ m}^2)(\$0.009/\text{cm}^2) = {\sim}\$96$. Adding 30% for balance of panel adds ~$29, so that the total cost is $125. Thus the cost per watt is ($125)/(120 W) = $1.04/W. This value is in line with typical current wholesale market values in the 2014–2015 time frame.

(b) Each cell is 15 cm on a side, so its total area of 225 cm² will produce 2.52 W in rated conditions. Therefore, the number of cells required is $(120)/(2.52) = 47.6 = {\sim}48$ cells. Since panels typically have an oblong shape, a suitable pattern for arranging the cells is 8 cells long by 6 cells wide, which gives 48 cells total. The resulting panel dimensions considering only the area occupied by the cells is 120 cm long by 90 cm wide (48″ L × 36″ W in standard units); an actual panel would be slightly larger to take into account the frame around the perimeter, area occupied by wires in the interior, and so forth. (See Fig. 10-7 for an example of an array in which each panel has cells laid out in an 8 by 6 pattern.)

10-4   Design and Operation of Practical PV Systems

The design of a PV system requires both understanding of the possible components that can be incorporated in a system, and a means of estimating the future output per year from the system, taking into account both the available solar resource in the region and the conversion efficiency of the PV panels. Using known capital cost values and projected future output, it is possible to forecast the economics of the investment. PV systems come in a great range of sizes, from those sized to meet the needs of a single family-sized residence, to multimegawatt commercial systems (see Figs. 10-8, 10-9, and 10-10). The size of the system will influence design and component choices.

10-4-1   Available System Components for Different Types of Designs

Any PV system will require the purchase of the panels themselves (the single most-expensive component) and some sort of structure to mount them in place. In addition, since most systems generate dc power so as to serve ac loads, an *inverter* is required to transform the current from dc to ac, although some new panels incorporate the conversion to ac into the panel itself, avoiding the need for a separate, detached inverter. Also, roofing shingles with integrated PV cells are an emerging technology that blends into

(a)

(b)

FIGURE 10-8   Solar PV power stations: (a) 10 MW station in Muehlhausen, Germany, (b) 80 MW station in Sarnia, Ontario, Canada.

[*Source:* For image (a) courtesy of SunPower Corporation. Reprinted with permission.]

the roofline of the building and also avoids the need for a structural rack for mounting PV panels. This technology may enable the implementation of solar electric generation in situations where aesthetics are a primary concern and the owner is unwilling to install PV panels on racks for aesthetic reasons.

Whether and how the PV system is connected to the electrical grid will affect what other equipment is necessary. An off-the-grid system requires a battery bank to store excess production during sunny times, to be used at night or during cloudy weather. If there are no batteries, then some other outlet must be provided for excess production, which otherwise poses a hazard. For grid-connected systems, the grid provides the sink

FIGURE 10-9   Ground-mounted PV array. Ground mounting permits seasonal adjustment of array angle (higher angle in winter, lower angle in summer).

(*Source:* Philip Glaser, 2007, philglaserphotography.com. Reprinted with permission.)

FIGURE 10-10   Roof-mounted PV system on a multifamily housing complex near Boston, MA.
(*Source:* Paul Lyons, Zapotec Energy. Reprinted with permission.)

for excess production. In such cases, the system owner must have in place an agreement to *interconnect* with the grid, giving the owner the right to supply power to the grid. In addition, it is highly desirable to have an agreement for *net metering*, whereby excess production is counted against electricity bought from the grid, so that the system owner gets the economic benefit of the excess production.

The effect of net metering is illustrated by comparing a sample of daily output from a PV system with a typical consumption pattern, as shown in Fig. 10-11. The output data were gathered on June 15, 2011, on a day with clear skies and slight haze and humidity,

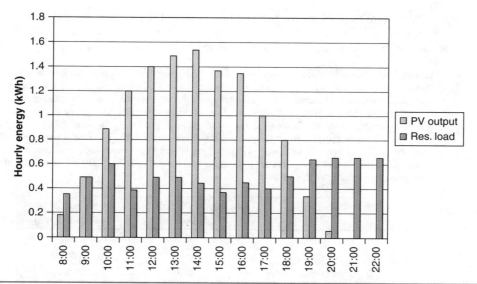

FIGURE 10-11 Comparison of daily 2.24 kW PV system output and residential load on June 15, 2011, for household with grid-connected array in Ithaca, New York.

Note: values shown are either array production or household consumption for the previous hour (e.g., from 7 to 8 a.m. array produced 0.18 kWh and residence consumed 0.35 kWh, and so on). Prior to the 7–8 a.m. block, the array had no output. PV output data are accurate to nearest watt-hour, thanks to display on inverter; accuracy of residential load data is limited by accuracy of utility meter outside residence. See text.

thus at or near the top output that would be expected for this system.[2] The hourly electricity consumed by the residence ("Res. Load") is deduced from current flow from the array and past the utility meter; although it is crude due to the limitations on the meter accuracy, it shows some consumption through the day and a peak in the late afternoon and evening, as one or more residents were in the building throughout the day. The residence on which the system is installed uses more electricity per year than the system producing, so with net metering it is possible for any excess output on a given day to be used to offset purchases from the grid at some other time of the year. Production exceeds load from 9 a.m. to 6 p.m., and for the whole period of 7 a.m. to 10 p.m., the array produces 12.1 kWh and the residence consumes 7.3 kWh, for a net surplus of 4.8 kWh. The array output can be divided between 5.2 kWh that is consumed directly by the load and the remaining 6.9 kWh that is excess and is delivered to the grid.

With net metering, this surplus is counted against electricity purchases from the grid; at $0.13/kWh, it is worth $0.90. Without net metering, the owner does not get credit for this production, only the 5.2 kWh or $0.67 when load was equal to or greater than array output. While the loss on this one day is not significant, the percentage loss is. Over the course of an entire year, these types of losses would noticeably diminish the value of the investment. One caveat: if the building total electric load is large and

---

[2]For readers interested in making comparisons between solar applications, data from tests of hot water and solar cooking technologies performed on the same day and under the same conditions are presented in Chap. 11 "Active Solar Thermal."

the PV system is small, there may be few hours of the year when PV output exceeds consumption, in which case it may not be as important to obtain net metering status. Example 10-4 illustrates the benefit of net metering on an annual basis.

Example 10-4   A residential grid-connected solar PV system is equipped with a two-way utility meter, that is, the meter tracks kWh flowing from the grid into the house and from the PV array onto the grid in two separate channels. The meter readings for a 12-month period from summer solstice to summer solstice are as given in the table below. Meter readings #1 and #2 are the two directions of the two-way utility meter. Meter #3 is the meter recording flow on the load side of the inverter, in other words, current produced by the array that either flows into the residential wire network to help meet the load requirement, or flows out to the grid if array output exceeds total demand inside the residence. (a) What is the net electricity consumption for the residence? (b) What percent of the total energy was generated by the PV array?

Meter	June 21, 2009 (Lifetime kWh)	June 21, 2010 (Lifetime kWh)
1. Incoming from grid	6757	8258
2. Outgoing to grid	4387	5460
3. PV production meter	10,954	12,846

Solution

(a) The first step is to measure annual change in kWh from the meter readings, as follows:

$$\text{Meter #1: } 8258 - 6757 = 1501 \text{ kWh}$$

$$\text{Meter #2: } 5460 - 4387 = 1073 \text{ kWh}$$

$$\text{Meter #3: } 12{,}846 - 10{,}954 = 1892 \text{ kWh}$$

The underlying logic of the grid-tied system is the following: any current produced by the array that does not flow out to the grid (Meter #2) must be consumed by the electrical load of the residence. Therefore, the net electricity consumption over the 12-month period is:

$$\text{Net Consumption} = \text{Array Output} + \text{Incoming From Grid} - \text{Outgoing To Grid}$$

$$= 1892 + 1501 - 1073 = 2320 \text{ kWh}$$

(b) The percent of net electricity consumption coming from the array is

$$\frac{\text{Array Output}}{\text{Net Consumption}} = \frac{1892}{2320} = 82\%$$

As a comment on this finding, the residence uses electricity quite efficiently, as 2320 kWh per year is well below typical U.S. figures, which are on the order of 5–6 MWh per year, making it possible for the array to cover most of the net consumption.

For owners of larger residential units such as apartment complexes, it may be beneficial to have in place a *master meter* with *submetering*, whereby the owner pays the utility for electricity purchased at the master meters and then charges individual tenants for consumption at their submeters. The owner may then be able to benefit from economies of scale in purchasing a larger single PV system for the entire complex, rather than several smaller systems that are each attached to individual meters.

The above options assume that the building owner, whether residential or commercial, owns the PV system as well, but it is also possible for the utility to own the system and install it on behalf of the building owner. The owner then purchases electricity produced by the system when available, and the utility manages the excess production from the system, when it is available. Utilities may also own and operate stand-alone PV power stations that produce electricity for the grid like any other generating asset.

To connect safely to the grid, a grid-connected (or "intertied" or "synchronous") inverter must be able to sense the ac oscillations of the electricity being delivered by the grid, and synchronize its ac output panels with those oscillations. This capability adds complexity to the design of the inverter beyond the inverter designs used for remote production of ac power, for example, producing ac electricity from the battery in a car or other vehicle to run household appliances. The standard grid-tied inverter requires a pulse from the grid in order to operate; during a grid failure, or "blackout," the system cannot produce output and the system owner loses power just like other grid customers. Alternatively, the owner may install a modified inverter along with a battery backup system, which allows the owner to isolate from the grid during a blackout and to continue to produce and use electricity from the system. To remain connected to the grid during a power outage can endanger persons working on the grid lines.

Options for Mounting PV Panels

For all PV systems associated with either a residential or commercial building (i.e., other than PV power stations), the mounting options are either on the roof of the building or on the ground adjacent to it. In many instances, roof-mounted systems are favored because they do not take up additional space (especially for larger commercial systems in the size range of 50 to 200 kW, where the required surface area is very large), they are less susceptible to incidental damage, and they reduce the building heat load in summer since they convert or reflect thermal energy that might otherwise penetrate the building. However, ground-mounted systems are advantageous in some situations, especially for residential system owners who may have system sizes on the order of 2 to 5 kW. Ground mounting may allow more options for finding a shade-free location; it also simplifies cleaning and snow removal from the panels, and may allow seasonal adjustment of the tilt angle of the panels in response to the changing declination of the sun (see Fig. 10-12).

Inverter Sizing Considerations

Strings of PV panels are fed into inverters in series, and since each additional panel in the string increases the expected voltage arriving at the inverter, the PV string must be sized correctly for the range of acceptable voltage values for the inverter. If this is not possible, a larger or smaller inverter should be specified that is compatible with the string size. On the one hand, the string must be able to generate a "wake-up" voltage that is large enough to start the inverter when sunlight first begins to fall on the array. At the other extreme, the maximum achievable string voltage must not exceed the maximum acceptable voltage for the inverter, so as to avoid permanently damaging the internal circuitry. For example, the array studied in Fig. 10-11 consists of 16 panels each rated at 24 V of dc output. At this voltage, the system inverter is capable of taking input from strings with 7 to 10 panels per string, so the array is configured with two strings of 8 panels each feeding the inverter. An increasingly popular way to avoid the constraints of a centralized inverter is to use panels with built-in "microinverters" so that current is transmitted from the panel already in an AC form.

Auxiliary Components That Enhance Value

Some variations in the components of the array are possible that add to the cost but also improve the output. PV systems can be mounted on a "tracking" frame that follows the sun as it moves across the sky; depending on local factors, the additional output from

FIGURE 10-12   Ground-mounted 2.8-kW PV array near Mecklenburg, New York, January 2007. Ground mounting allows snow removal, and snow has just been swept from the panels at the time the picture was taken. Note that due to the overcast conditions, snow removal had only marginal effect at the time: with a thin cover of snow, the system produced 60 W, while in the condition shown it produced 120 W.

the array may offset the extra cost of the tracking system. Similarly, it is possible to add a mirrorized concentrating system that focuses light on the panels so as to increase output. Roof-mounted systems are often mounted flat to the roof so that they assume the tilt angle of the roof, but it is possible to add a structure that raises the tilt angle of the panels above that of the roof (see Fig. 10-13). Any such additional investment should be evaluated on the basis of the additional capital cost and the projected increase in output from the system.

Operation and Maintenance Considerations for Design of PV Systems
PV panels have no moving parts and a long expected life, so there are relatively few operations and maintenance considerations. PV panels typically fail by gradually losing their ability to produce the expected output for a given amount of insolation, rather than abruptly ceasing output altogether. Therefore, although panels typically are under warranty for 25 years, they are expected to function for 35 or 40 years before the output is so diminished that they would merit replacing. Inverters have a shorter expected life, on the order of 15 to 20 years for the current generation of inverters, so the system owner should plan for at least one inverter replacement over the system lifetime.

Washing or removal of snow is beneficial for the output of the panels, but weather events such as rain and wind also remove dust, snow, or other fouling that blocks light from penetrating the panels. Here panels that are installed flat are at a disadvantage compared to those installed at a raised angle, since wind or rain cannot remove fouling as effectively in the former case. As long as the weather can adequately "clean" the panels, a design for access to the panels for washing is helpful, but not essential. PV

FIGURE 10-13 Roof-mounted PV array, Ithaca, New York, January 2007. Compare to mounting the PV rack flush to a gently sloped roof (see Fig. 10-10); the owner has incurred an extra cost to raise the angle of the panels, in order to improve productivity.

systems should allow for easy replacement of individual panels in case of damage or failure, without affecting surrounding panels.

10-4-2   Estimating Output from PV System: Basic Approach Using PV Watts

As a first approximation, it is possible to estimate the output from a proposed PV system by starting with the rated capacity of a system, system azimuth angle γ, and system elevation relative to horizontal β, and obtaining an estimated conversion factor from the National Renewable Energy Laboratory's PV Watts database (NREL, 2015). The PV Watts database uses climate data from a nearby weather station and the estimated conversion efficiency from panel to AC current leaving the system to predict monthly and annual output. Where the estimate is made in a location that does not have a weather station, a nearby location can be used. For example, PV Watts does not include data for Ithaca, NY, but because predictions from nearby weather stations in Binghamton, Rochester, and Syracuse are all similar to each other, an average value can be used, or one of the three nearby weather stations can be chosen to represent Ithaca. Another resource is Sunny Portal (www.sunnyportal.com), which reports archived production data for numerous participating PV arrays.

The basic approach is best for small systems (e.g., single-family residences) where the investment risk is limited. For larger systems where a larger capital expenditure is required, the comprehensive approach in the next section is appropriate. The use of a regional conversion factor also assumes that solar access is not obstructed, for instance by trees or tall buildings; if this assumption does not hold, then the method will overestimate output.

Sample Location	Regional Factor (kWh/kW/y)	
	Before Losses	**w/ 15% Losses**
Syracuse, NY	1445	1228
Chicago, IL	1508	1282
Atlanta, GA	1751	1488
San Diego, CA	1948	1656

Note: Table assumes south-facing array raised to latitude angle of location (e.g., 42° tilt for Chicago, 43° tilt for Syracuse, etc.).

TABLE 10-2 Representative Values of Regional Productivity per Unit of Installed Capacity for Select Locations Using PV Watts

Suppose the rated capacity of a proposed system in kW is some value C. The regional factor (RF) is given in units of kWh/kW/year of capacity, on an average annual basis. Use of the online PV Watts calculator gives the predicted system output and not RF, but RF can be calculated by comparing predicted output to a perfect value of 8760 kWh/kW/year. Representative values for RF obtained from PV Watts are given for several cities in Table 10-2, assuming a south-facing array raised to the latitude angle of the location, with no obstructions and in good working order. The array conversion efficiency η_{array} must also be included to take into account losses in conversion to ac and other losses in the system between the panels and the load, so Table 10-2 also gives an adjusted value for each location with η_{array} = 85% efficiency. The expected annual output E in kWh is then

$$E = C \times RF \times \eta_{array} \qquad (10\text{-}12)$$

Equation (10-12) can be rewritten to calculate the required capacity to achieve a desired yearly output, for example, in a case where an owner wishes to produce approximately the same amount of electricity that is consumed in an average year:

$$C = \frac{E}{RF \times \eta_{array}} \qquad (10\text{-}13)$$

Note that this method does not estimate the output for a particular time of year, say in the summer or winter. However, since in many situations it is the annual output that is desired, in order to know how much the system will produce over the course of many years, this limitation may not matter.

To give an example we compare predicted and observed output for a 2.24 kW array (16 panels rated at 140W each) oriented 15 degrees west of south, and raised at a 25 degree angle. The expected average output for an ideal array in the region around Syracuse, New York, is 1445 kWh/kW/year before losses. However, because the array azimuth and elevation angles are different from optimal, the expected output is reduced to 1416 kWh/kW/year. The actual output for 4 years from January 2011 to December 2014 for a 2.24-kW system installed in this region in Ithaca, New York, is shown in Fig. 10-14a, along with the average for each month for 4 years. Average monthly output varies widely, from a low of

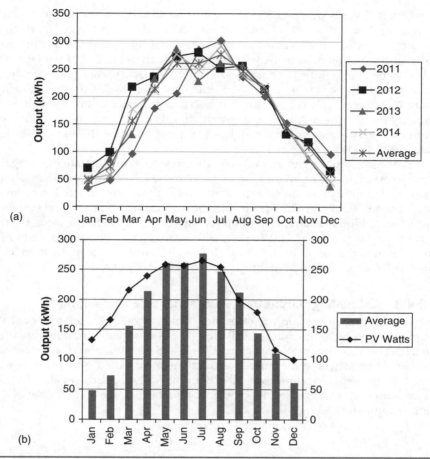

FIGURE 10-14 Output from 2.24 kW array by month in Ithaca, New York, January 2011 to December 2014, with comparison PV Watts estimate.

Notes: Average value in Fig. 10–14a sums to annual value of 2065 kWh. The same average curve appears in Fig. 10-14b, in comparison with PV Watts annual value of 2379 kWh.

48.3 kWh in January to a high of 276.8 kWh in July. Total observed output for the system averaged over 4 years is 2065 kWh. Figure 10-14a shows how monthly output in this location varies widely between summer and winter months, with the former having monthly output on the order of five times as large as the latter. It also shows that output for any chosen month is consistent from one year to the next, e.g., all January values are between 35 and 75 kWh, all July values between 250 and 300 kWh, and so on.

Losses in the array under study are known to be higher than the 15% figure used in Table 10-2, so an array efficiency value of 75% is used instead. According to Eq. (10-12), the predicted output for the system is

$$E = 2.24\,\text{kW} \times 1416\frac{\text{kWh}}{\text{kW}\cdot\text{month}} \times 0.75 = 2379\,\text{kWh}$$

Thus PV Watts overpredicts the output by 15% compared to the 4-year average output of 2065 kWh/year. Figure 10-14b shows the monthly comparison between predicted and observed values, and suggests that the prediction is more accurate in the high-insolation months (April–September) than in the rest of the year. The shortfall may be due to mitigating factors not captured by PV Watts, such as the presence of snow on the panels that blocks production for some of the time during winter months. The result also illustrates that, as a first approximation, the method is able to roughly predict the output of the system. In this case, since the system is small, the difference between actual and predicted output is worth \$41/year at the local utility rate of \$0.13/kWh.

By the same token, the example could also be used to illustrate the danger of using a crude estimate of output for a larger system used on a large commercial or government building. Suppose an error of similar magnitude was incurred in estimating output from a system with 2.2 MW of capacity, which is an appropriate size for such an application. With the same assumptions about output and electric rates, the cost of the error has now grown to \$41,000/year, which could have serious financial implications for recouping the cost of the investment. This risk provides the motivation for the comprehensive approach to estimating output in the next section.

10-4-3   Estimating Output from PV System: Extended Approach

A more comprehensive approach to system design should account for how PV cell output is affected by temperature, insolation flux density, suboptimum panel slope, and characteristics of the local solar climate as it varies throughout the year. This section provides parameters and example calculations for detailed estimation of output, for which software packages more detailed than PV Watts are also available. Based on insolation summed (perhaps hourly) over a month, the average daily output of a panel, E, can be approximated by

$$E = \frac{A \Sigma \eta_i I_i}{N} \tag{10-14}$$

The value of E (taken here to be for one month) is the averaged (daily) energy output (kWh) for the period. Panel area is A (m²), η_i is average hourly panel efficiency, I_i is the integrated insolation for the hour (kWh/m²), N is the number of days in the integration period, and the sum is hourly (for the daylight hours) for the month.

Although weather can change from day to day (and more frequently in many locations), the method outlined below assumes weather within a month will be similar to the average for the month, but changes from month to month are sufficiently large that months cannot be averaged, even into 2-month blocks of time.

Hourly cell efficiency, η_i, is a function of cell and array design, cell temperature, and insolation intensity. It can be characterized by the following equation where η is the rated efficiency, β is the temperature coefficient of efficiency (C^{-1}), T_{ci} is the cell average temperature for the hour, T_r is the temperature at which the cell efficiency was rated, and γ is the insolation flux density coefficient of cell efficiency. All temperatures are in Celsius.

$$\eta_i = \eta_r [1 - \beta(T_{c,i} - T_r) + \gamma \log_{10}(I_i)] \tag{10-15}$$

If these two equations are combined and algebraically rearranged, T_{ai} is defined as the average ambient air temperature for the hour and T_M is defined as the mean air temperature for the month, we find

$$E = \left(\frac{\eta_r A}{N}\right)[\Sigma I_i - \beta\Sigma(T_{ci} - T_{ai})I_i - \beta\Sigma(T_{ai} - T_M)I_i - \beta\Sigma(T_M - T_r)I_i + \gamma\,\Sigma I_i \log_{10}(I_i)] \quad (10\text{-}16)$$

From a different and simpler perspective, it would be preferable to calculate E from the following equation, where the monthly average efficiency, η, is used in preference to hourly values.

$$E = \frac{\eta A}{n}\Sigma(I_i) \quad (10\text{-}17)$$

For appropriate definitions of $(T_c - T_a)$, $(T_a - T_M)$, $(T_M - T_r)$, and $\gamma \log(I_i)$, an expression for average efficiency that makes the two equations for E comparable is

$$\eta = \eta_r[1 - \beta(T_c - T_a) - \beta(T_a - T_M) - \beta(T_M - T_r) + \gamma \log_{10}(I_i)] \quad (10\text{-}18)$$

The monthly average cell efficiency of conversion, η, can be calculated if the various terms on the right-hand side of the above equation can be determined. The following equivalencies must exist if the two equations are comparable:

$$(T_c - T_a)\Sigma I_i = \Sigma(T_{ci} - T_{ai})I_i \quad (10\text{-}19a)$$

$$(T_a - T_M)\Sigma I_i = \Sigma(T_{ai} - T_M)I_i \quad (10\text{-}19b)$$

$$\log(I\Sigma I_i) = \Sigma I_i \log(I_i) \quad (10\text{-}19c)$$

From intuition, the term $(T_c - T_a)$ in Eq. (10-19a) can be viewed as representing the monthly "average difference" between the cell temperature and ambient temperature during daylight hours. Additionally, intuition suggests the term $(T_{ci} - T_{ai})$ should strongly depend on insolation on the cell and thermal loss from the array to the environment. These ideas are developed next.

Equating solar gains in the PV array to the sum of electrical output and thermal loss yields the following equation, where U_L is the unit area thermal loss coefficient (or thermal conductance based on panel area, $kW/m^2 \cdot K$), and α and τ are the PV cell surface absorptance and array protective covering transmittance for insolation, respectively,

$$\alpha\tau I_i = \alpha\tau\eta_i I_i + U_L(T_{ci} - T_{ai}) \quad (10\text{-}20)$$

If we assume electrical generation is small relative to heat generation (an approximation for today's commercial PV units but becoming less true as PV cells become more efficient), we find

$$(T_{ci} - T_{ai}) \approx \frac{\alpha\tau I_i}{\Sigma I_i} \quad (10\text{-}21)$$

Equation (10-21) can be substituted into Eq. (10-19a) to yield Eq. (10-22). Hourly insolation data are available for many meteorological sites in the United States. Thus, the right-hand side of Eq. (10-22) can be quantified for the local site of interest.

$$\frac{U_L(T_c - T_a)}{\alpha\tau} = \frac{\Sigma I_i^2}{\Sigma I_i} \tag{10-22}$$

The right-hand side of Eq. (10-22), taken on a monthly time step, should correlate with the corresponding monthly value of $\overline{K}_T$. Evans (1981) analyzed hourly solar data for seven cities in the United States, found the correlation to be linear with relatively little scatter of the data about the regression, and determined the following equation by regression.

$$\frac{\Sigma I_i^2}{\Sigma I_i} = 0.219 + 0.832\,\overline{K}_T \tag{10-23}$$

Equation (10-23) was based on the PV panel being at the optimum tilt angle for each month, which changes from month to month. A tilt correction factor for $(T_c - T_a)$ was determined by computer analysis and simulation to be approximated by Eq. (10-24), where S is the actual panel tilt angle and S_M is the optimum tilt angle for the month.

$$C_f = 1.0 - 0.000117(S_M - S)^2 \tag{10-24}$$

The optimum tilt angle depends, of course, on major factors such as the local latitude and month of the year, and minor factors such as ground reflectivity (unless the panels are installed horizontally on a flat roof). Additional computer simulations suggested optimum tilt angles on a monthly basis as listed in the following table, where L is local latitude and tilt is expressed in degrees. The tilt factor, C_f, is factored with $(T_c - T_a)$ on a month-by-month basis, as will be seen in an example later in this section.

Month	Jan	Feb	Mar	Apr	May	Jun	Jul	Aug	Sep	Oct	Nov	Dec
Opt. tilt	L+29	L+18	L+3	L−10	L−22	L−25	L−24	L−10	L−2	L+10	L+23	L+30

The next step is to estimate a value of U_L, based on manufacturer's data from standard test protocols. NOCT symbolizes the nominal operating cell temperature (normally 20°C) and I_{NOCT} is the insolation during the test with the cell held (by mechanical cooling) at the NOCT (I_{NOCT} is normally equal to 800 or 1000 W/m²). The reduction of cell temperature caused by electricity output during the test instead of heat generation is neglected, leading to

$$\frac{U_L}{\alpha} = \frac{I_{NOCT}}{NOCT - T_{air,NOCT}} \tag{10-25}$$

where $T_{air,NOCT}$ is ambient air temperature that existed during the calibration test and which permitted the cell to be held at NOCT. Equation (10-25) is, in effect, a thermal balance on the cell. Test conditions normally include an ambient wind speed of 1.0 m · s⁻¹. Although Eq. (10-25) can be used in PV system output analysis, consistency of manufacturing suggests $U_L = 20$ W · m⁻² · K⁻¹ is a reasonable value that may be assumed to apply in the general case.

The next value to obtain is $(T_a - T_M)$. Rearranging Eq. (10-19b) leads to

$$(T_a - T_M) = \frac{\Sigma T_{a,i} I_i}{\Sigma I_i} - T_M \tag{10-26}$$

Equation (10-26) is, in essence, a measure of the difference between the average air temperature during daylight hours of the month, and the mean monthly (24 h) temperature. Typical Meteorological Year (TMY) data exist for many cities (see, e.g., NREL 2005). The seven cities used to obtain Eq. (10-23) were considered and $(T_a - T_M)$ was found to be, typically, within the range 3 ± 2 K. This is a relatively large range compared to the actual value, but the magnitude of the difference anywhere within the range is small compared to $(T_c - T_a)$, making any inaccuracy of limited concern. However, the increasing availability of weather data on the Internet makes it possible to calculate a value for $(T_a - T_M)$ for most large cities, if one wishes.

The difference $T_M - T_r$ is found readily from weather data and the manufacturer's data for T_r, the rating test cell temperature.

The final term to be explored is Eq. (10-19c), which can be rearranged to

$$\log(I) = \frac{\Sigma I_i \log(I_i)}{\Sigma I_i} \tag{10-27}$$

Insolation data for the seven cities were used to calculate values for the right-hand side of Eq. (10-27), which were found to correlate with $\bar{K}_T$ linearly.

$$\log(I) = 0.640 - 0.732 \bar{K}_T \tag{10-28}$$

The coefficient, γ, in Eq. (10-19c) may be assumed to be sufficiently small in magnitude that the entire term can be dropped as a good approximation. The effect of insolation intensity within its normal range does not influence PV cell efficiency in a significant way.

Practical considerations introduce small de-rating factors that reduce PV ideal panel system output in realistic applications. There are 11 factors, each with an assumed default value, as follows:

1. PV module nameplate dc rating may not be achieved, 0.950
2. Losses in the inverter and transformer, 0.920
3. Mismatch of panels and inverter, 0.980
4. Losses in diodes and connections, 0.995
5. Resistance losses in dc wiring, 0.980
6. Resistance losses in ac wiring, 0.990
7. Soiling of the panel surface, 0.925
8. System availability, 0.980
9. Shading, 0.950
10. Sun tracking, 1.000
11. Aging, 1.000

The 11 factors multiply, providing an adjustment factor that de-rates the expected output of the integrated PV system (panels and associated hardware). The product of the default values equals 0.71, or a 29% reduction of output from the ideal.

All of the above has led to a group of equations that permit monthly estimates of the electrical output of a PV system. The process can be best illustrated by an example for a single month. Extension to every month of a year is obvious.[3]

Example 10-5   Assume a PV array tilted at 45° to the south at a location where the latitude is 42° north. Determine the total May kWh output from the array based on the following parameter values, and the standard de-rating values:

$$U_L = 0.02 \text{ kW} \cdot \text{m}^{-2} \cdot \text{K}^{-1} (\text{standard assumption})$$

$$\alpha = 0.88$$

$$\tau = 0.95$$

$$\eta_r = 0.15$$

$$T_r = 20°C$$

$$\beta = 0.0045°C^{-1}$$

$$K_T = 0.502$$

$$\overline{H}_d = 4.97 \text{ kWh m}^{-2} \text{day}^{-1} (\text{irradiating the tilted PV array})$$

$$T_M = 15.3°C$$

Solution   From Eqs. (10-22) and (10-23),

$$\frac{U_L(T_c - T_a)}{\alpha\tau} = 0.219 + 0.832(0.502) = 0.637 \text{ kW} \cdot \text{m}^{-2}$$

From Eq. (10-24), for May, $S_M = 42 - 22 = 20°$ and

$$C_f = 1.0 - 0.000117(20 - 45)^2 = 0.927$$

Leading to

$$(T_c - T_a) = \frac{(0.927)(0.637)(0.88)(0.95)}{(0.02)} = 24.7 \text{ K}$$

Assume

$$(T_a - T_M) = 3 \text{ K}$$

From the data provided

$$(T_M - T_r) = 15.3 - 20 = -4.7 \text{ K}$$

The average cell efficiency for the month is calculated from Eq. (10-18) as follows:

$$\eta = 0.15[1 - 0.0045(24.7 + 3 - 4.7)] = 0.134(13.4\%)$$

The daily average insolation integral is 4.97 kWh · m^{-2}, and May has 31 days, thus

$$E_{May} = 0.134(31)(4.97) = 20.6 \text{ kWh} \cdot \text{m}^{-2}$$

The final step is to include the 11 de-rating factors, which equal a default value of 0.71 as shown above:

$$E_{May} = 0.71(20.6) = 14.6 \text{ kWh} \cdot \text{m}^{-2}$$

Extending this solution to an entire year requires weather data for a year and perhaps a spreadsheet or computer language such as Matlab to expedite calculations.

---

[3]Typical Meteorological Year (TMY) data are available at, e.g., National Renewable Energy Laboratories (2005).

Calculations in the example above have provided an estimate of the electrical energy a sample PV panel could provide during a single month. Actual delivery of electrical energy to an end use depends on other components of the complete system. When PV electricity is used directly, such use is likely to be as ac power. In such applications, the inefficiency introduced by an inverter that converts dc to usable ac power must be acknowledged. If, for example, the inverter component has an efficiency of 0.92, the delivered energy will be 0.92 (20.6 kWh · m^{-2}), or 19.0 kWh · m^{-2} for the month in the example above. If, instead, electricity is delivered first to battery storage and then to an inverter for later use, the energy delivered will also be reduced by the inefficiency of the battery component. For example, if storage and later retrieval of electricity from a battery bank component has an overall efficiency of 75%, 0.75 (19.0 kWh · m^{-2}), or 14.2 kWh · m^{-2} will be finally delivered for the month in the example above. If the PV system provides some power for use during daylight collection hours, and the rest of its power for storage and night use, the two components should be treated separately to determine the total benefit of the PV system. Finally, sizing electrical wires becomes another issue with regard to efficiency of transport. When wire lengths are long and their diameters small, the conversion of electrical power to heat creates yet another inefficiency to be accounted for. These factors are important from a systems viewpoint and should be considered when sizing a system (larger) to meet a predetermined energy need, as detailed in Examples 10-6 and 10-7.

Example 10-6 Extended Case Study. A rural township located near Ithaca, New York (42.4° N), wishes to install a PV system on their new Town Hall. Panels will be installed on the roof surface at the roof's tilt angle. The rated capacity of the system chosen, based on the roof area, is 10.8 kW. Relevant data are below. Determine the yearly kWh that can be expected to be produced by the system.

Basic Data, PV Panel		Basic Data, Location	
Rated efficiency, decimal	0.181	Latitude, degrees	42.4
Rating temperature, T_c, C	25	Latitude, radians	0.740
Temperature coefficient, β, K^{-1}	0.0038	Fixed tilt offset (from latitude)	−19.8°
U_L, kW m^{-2} K^{-1}	0.02	Fixed tilt angle, degrees	22.6
$\tau\alpha$	0.85	Fixed tilt angle, radians	0.395
γ	0 (assumed)	Area per panel, m^2	1.244
$T_a - T_M$, K	3	Number of panels	48
		Roof rise, ft	5
		Roof run, ft	12

Note: The roof angle is the arc tangent of the rise over the run, which was also the tilt angle for the panels (mounted flat on the south-facing roof). Assume default values of the eleven de-rating factors.

Solution The first step is to determine the expected insolation on the roof for each month of the year. Ground reflectivity assumptions were 0.2 for March through November, 0.3 for December, and 0.5 for January and February. A ground reflectivity value of 0.2 indicates thick turf, and a value of 0.7 indicates more than one inch (25 mm) snow depth.

Insolation calculations are followed by monthly calculations for electricity output, patterned after Example 10-5. The results are in the following table. The sum of 10,925 kWh per year is very close to the value estimated by the system installer, using a commercial PV chart program and a slightly different weather data set.

Mo	K_T	$\bar{H}$ MJ m^{-2}	T_m, C	Opt. tilt, deg.	Tilt Factor	Solar, kWh m^{-2}d^{-1}	U_L*etc.	$T_c - T_a$, K	Effic.	Days mo^{-1}	kWh m^{-2} mo^{-1}	Derated kWh mo^{-1}
J	0.351	4.95	−2.67	71.4	0.72	1.98	0.37	15.67	0.187	31	11.47	488
F	0.435	8.61	−3.06	60.4	0.83	3.23	0.48	20.57	0.184	28	16.62	707
M	0.450	12.26	2.22	45.4	0.94	4.02	0.56	23.69	0.178	31	22.23	945
A	0.428	15.08	9.11	32.4	0.99	4.41	0.57	24.17	0.173	30	22.93	975
M	0.502	20.28	15.33	20.4	1.00	5.56	0.64	27.04	0.167	31	28.76	1223
J	0.538	23.10	20.50	17.4	1.00	6.11	0.66	28.24	0.163	30	29.83	1268
J	0.554	23.15	23.28	18.4	1.00	6.21	0.68	28.84	0.160	31	30.88	1313
A	0.530	19.80	22.17	32.4	0.99	5.65	0.65	27.73	0.162	31	28.35	1206
S	0.497	15.05	17.89	40.4	0.96	4.75	0.61	25.89	0.166	30	23.63	1005
O	0.465	10.47	12.00	52.4	0.90	3.73	0.54	23.08	0.172	31	19.90	846
N	0.324	5.31	5.28	65.4	0.79	1.99	0.38	16.32	0.181	30	10.80	459
D	0.337	4.90	−1.33	72.4	0.71	1.99	0.35	15.07	0.187	31	11.54	491
Yearly sum												10,925

Example 10-7 Extend Example 10-6 to five other cities in the United States, chosen to provide a variety of climates. The cities are:

1. Miami, Florida, United States: semi-tropical, and sunny, latitude 25.78° N
2. Grand Junction, Colorado, United States: moderately cold, and sunny, latitude 39.12° N
3. Phoenix, Arizona, United Stated: Desert climate, latitude 33.43° N
4. Seattle, Washington, United States: More cloudy than Ithaca, with more moderate temperatures, latitude 47.45° N
5. Winnipeg, Manitoba, Canada: cold winters, and sunny, latitude 49.90° N

Solution Relevant weather data[4] are in the following table, with units converted as appropriate.

City	Data	Jan	Feb	Mar	Apr	May	Jun	Jul	Aug	Sep	Oct	Nov	Dec
1	K_T	0.604	0.616	0.612	0.600	0.578	0.545	0.552	0.549	0.525	0.534	0.559	0.588
	T_a, C	22.0	22.2	23.2	25.0	26.6	28.3	28.9	29.2	28.5	26.8	24.2	22.6
	$\bar{H}$	14.43	17.72	20.85	23.03	23.58	22.70	22.72	21.56	18.77	16.38	15.06	13.49
2	K_T	0.597	0.633	0.643	0.632	0.643	0.704	0.690	0.650	0.705	0.654	0.590	0.621
	T_a, C	−2.8	1.7	7.0	13.2	19.1	24.3	28.1	26.4	21.9	14.6	5.6	−0.3
	$\bar{H}$	8.38	12.94	18.00	21.39	22.51	27.01	23.98	19.48	19.56	13.82	8.84	7.42
3	K_T	0.650	0.691	0.716	0.728	0.753	0.745	0.667	0.677	0.722	0.708	0.657	0.652
	T_a, C	12.3	14.9	18.2	22.3	27.1	31.8	34.8	33.6	30.8	24.3	17.6	13.7
	$\bar{H}$	12.84	17.27	22.42	27.23	30.89	31.71	27.94	26.22	24.30	19.25	14.71	11.87
4	K_T	0.266	0.324	0.423	0.468	0.477	0.459	0.498	0.511	0.459	0.372	0.284	0.269
	T_a, C	3.8	6.1	8.3	11.1	14.5	17.1	19.6	19.3	16.4	12.2	7.6	5.3
	$\bar{H}$	2.87	5.38	10.46	15.68	18.98	19.65	20.58	18.43	12.87	7.27	3.71	2.49
5	KT	0.601	0.636	0.661	0.574	0.550	0.524	0.587	0.567	0.504	0.482	0.436	0.503
	T_a, C	−16.0	−13.8	−5.9	4.9	13.3	18.5	22.2	20.8	14.8	7.6	−3.8	−12.2
	$\bar{H}$	5.57	9.52	15.44	18.71	21.71	22.37	24.21	20.08	13.57	8.75	5.07	3.93

$\bar{H}$ converted from BTU ft^{-2} as listed in footnote reference, MJ m^{-2} d^{-1}; BTU ft^{-2} * 0.0113565 => MJ m^{-2}.

Results of analysis: Table of total kWh produced by month and city

City	Jan	Feb	Mar	Apr	May	Jun	Jul	Aug	Sep	Oct	Nov	Dec	Total
Ithaca	488	707	945	975	1223	1268	1313	1206	1005	846	459	491	10,925
Miami	1157	1145	1332	1277	1259	1136	1184	1181	1091	1096	1094	1082	14,033
Grand J.	1009	1123	1391	1418	1509	1564	1540	1431	1440	1269	1024	979	15,696
Phoenix	1163	1248	1544	1566	1645	1539	1427	1431	1448	1400	1215	1107	16,732
Seattle	271	422	812	1250	1168	1121	1212	1161	889	597	327	245	9475
Winnipeg	790	977	1362	1281	1351	1271	1410	1273	970	793	565	563	12,603

[4]See Leckie et al. (1981).

The results of the calculations are not a surprise in terms of values relative to each other. However, several observations may be made.

1. The lowest output (Seattle) is approximately 57% of the highest output (Phoenix). This is perhaps a bit more than might be expected *a priori*.

2. Although the solar data favors Phoenix over Grand Junction, their yearly outputs are only a few percent different. This is attributable to the high Phoenix air temperatures, which significantly reduce PV cell efficiency. The same is true of the Miami results.

3. Ground reflectivity is one parameter that must be estimated. For this case study, the values for Ithaca were assumed to apply to Grand Junction and Winnipeg, also. A constant value of 0.2 was assumed for every month for the other cities. This value is for turf, which may not be appropriate for a desert location such as Phoenix. Changing that value to 0.5 for each month (which applies to bright, bare ground) increases the calculated yearly output to 16,904 kWh, which is only a 1% change and well within the level of accuracy to which other parameters are known. With PV panels on the roof of a building, the effect of ground reflectivity is essentially negligible.

4. Sensitivity analyses are often useful to gain an appreciation of the influence of the various input parameters as they are varied individually. Based on data for Ithaca, New York, U.S., as used above, calculations show the following effects:
   a. Increase the cell temperature parameter (β) from 0.0038 to 0.005 and yearly kWh decreases to 10,701.
   b. Increase $T_a - T_M$ from 3 K to 10 K and yearly kWh changes to 10,615.
   c. Change to a latitude 10°C further north and yearly kWh increases to 11,933 (due to longer summer days).
   d. Change to a latitude 10°C further south and yearly kWh decreases to 10,281.
   e. Increase every monthly temperature by 10°C and yearly kWh decreases to 10,483.
   f. Decrease every monthly temperature by 10°C and kWh yearly increases to 11,357.
   g. Increase ground reflectivity for each month to 1.0 and yearly kWh increases to 11,217.
   h. Decrease ground reflectivity for each month to 0.0 and yearly kWh decreases to 10,838.

10-4-4   Year-to-Year Variability of PV System Output

As was already suggested by Fig. 10-14a, PV systems have relatively low variability of output for any given month, and in addition peaks and valleys in relative monthly output tend to balance out over time. Therefore variability in annual as opposed to monthly output is especially low from year to year. Annual production for the example array from Fig. 10-14 averaged 2,006 kWh per year over nine years from 2006 to 2014, with a coefficient of variation of 4.4% (Fig. 10-15). All individual years were within 10% of the average except 2012, which was a drought year during which almost no rain fell in the summer, resulting in an annual value 11% higher than the average. Low variability of annual output is common to arrays large and small, so long as the operator remains vigilant about any possible malfunctions and keeps the array in proper working order.

10-4-5   Economics of PV Systems

The following economic analysis of PV systems assumes an initial capital cost for the purchase of the system and then an annual revenue stream based on the value of displacing electricity purchases from the grid with production from the system. The analysis therefore assumes that the system owner has access to net metering, so that all production will count toward the revenue stream at the retail rate. Net metering is very important for receiving the maximum possible value of production, since otherwise electricity that is worth the equivalent of $0.10 to $0.15/kWh retail may only earn the wholesale rate of $0.03 to $0.05/kWh, or even $0.00/kWh, at times when system production is exceeding consumption in the building.

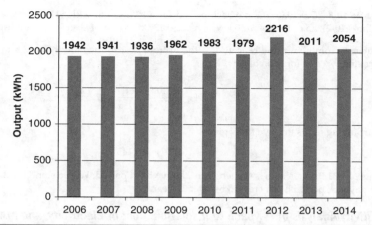

FIGURE 10-15   Annual output from example 2.24-kW PV array in Ithaca, NY, 2006–2014.
Notes: 9-year average output 2,003 kWh, standard deviation = 88.6 kWh, coefficient of variation
$CV = \sigma/\mu = (88.6)/(2003) = 4.4\%$.

As a starting point, we use a 5 kW system with $RF = 1588$ kWh/kW/y (a midrange value from Table 10-2) and 85% efficiency as a case for economic analysis, with a resulting annual production of 6750 kWh. The components of total cost of this system, as well as the tax credits and rebates applied, are shown in Table 10-3. The largest component of the cost is that of the panels, with additional cost for the inverter, balance of system, and labor. The rebates and credits are substantial in this case, reducing the cost by 50%.

Suppose the system maintains 6750 kWh/year, down stream from the inverter, over the long term. At $0.13/kWh in constant 2015 dollars, this output is worth approximately $877/year. Thus the simple payback for the system is ($11,000)/($877/year) = ~12.5 years. With a long simple payback such as this, it is clear that any use of discounted cash flow will result in a long payback period indeed, as shown in Example 10-8.

Item	Cost/W	Total Cost
Panels	$2.00	$10,000
Inverter	Fixed	$2000
Balance of hardware	$0.60	$3000
Labor	$1.00	$5000
Design	$0.40	$2000
Total before credits	$4.40	$22,000
Federal incentive	−25%	$(5500)
State incentive	−25%	$(5500)
Net cost	$2.20	$11,000

TABLE 10-3   Cost Components and Rebate for
Representative Solar PV System (2015 Dollar Costs)

Example 10-8   Calculate the payback period for a PV system with $11,000 cost after incentives (as in Table 10-2), an output of 6750 kWh/year, $0.13/kWh cost of electricity, and discount rate = 7%.

Solution   From above, the ratio P/A of the upfront cost to the annuity is 12.5 to the nearest tenth. Recalling P/A as a function of i and N:

$$P/A = \frac{[(1.07)^N - 1]}{[0.07(1.02)^N]} = 12.5$$

Rearranging terms to solve for N gives

$$(1.07)^N = 8.145$$

Thus $N = 31.0$; 31.0 years is the length of the payback period. With a standard discount rate of 7%, the payback period has increased by over 17 years.

Sensitivity Analysis of Payback to Insolation, Cost of Electricity, and MARR

As suggested by the preceding discussion, the effect of changing factors such as the amount of insolation can be examined through the use of a sensitivity analysis, using a base PV system with a total installed cost of $2.20/W after incentives, as in Table 10-2. In Table 10-4, payback is calculated as a function of output per unit of installed capacity, measured in kWh/kW/year after accounting for inverter losses, and price paid for electricity. Both cases of simple payback and discount rate = 7% are given, to show the effect of discounting at the U.S. federal government standard discount rate. The range of output values shown is from 900 kWh/kW/year, representing locations such as the British Isles that are relatively cloudy, to 1700 kWh/kW/year, representing locations such as Arizona or California that are relatively sunny (see Chap. 9). The range of electric rates, from $0.10 to $0.16 per kWh, is typical of regions in the United States or Europe where the cost of electricity favors the pursuit of alternative sources.

Table 10-4 provides a number of lessons about the sensitivity of economic payback to the various input factors. First, payback varies greatly from the most favorable to the most

kWh/kW/year:	900	1300	1700
(a) Simple payback			
$0.10	24.4	16.9	12.9
$0.13	18.8	13.0	10.0
$0.16	15.3	10.6	8.1
(b) Discount rate = 7%			
$0.10	>50	>50	34.9
$0.13	>50	35.8	17.6
$0.16	>50	19.9	12.3

*Cost per kWh in the left column is the retail cost of grid electricity which the PV system displaces.

TABLE 10-4   Payback Time of PV System, in Years, as a Function of Productivity Measured in kWh/kW/year, Retail Electric Prices in $/kWh, and Discount Rate (Case a: Simple Payback or Case b: Discount Rate = 7%)

unfavorable locations, with simple payback values from 8 to 24 years shown in the table. In locations with good sun and high retail cost of electricity, such as the states of California or Hawaii in the United States, these systems are relatively attractive, as represented by the combination of 1700 kWh/kW/year and $0.16/kWh in the table. On the other hand, if either of these conditions does not hold, then simple payback may lengthen to 20 years or more. Furthermore, there are a number of regions where the price paid for electricity is below the lower bound on electric rate of $0.10/kWh shown in Table 10-4. For example, in the United States, as of 2014, 35 out of 50 states had residential electric rates lower than this threshold, although prices have in general been rising both before and since.

Second, introducing a 7% discount rate changes the economic picture substantially. For the most attractive region, payback increases from 8 to 12 years. For the least attractive four combinations, the payback is over 50 years. Since the time period is longer than the system will last in even the most optimistic circumstances, the result is that, effectively, the system can never pay for itself.

The assumption of constant retail price of electricity (in constant dollars) can work either for or against the viability of the investment. If carbon taxes or disruptions in energy supplies over the lifetime of the PV system raise the retail price of electricity, the length of the payback period might be shortened. On the other hand, if inexpensive, large-scale energy sources that both emit no carbon and possess an abundant supply of resource (e.g., large-scale wind or nuclear) become more prevalent during the lifetime of the PV system, retail prices might instead drop, lengthening the payback period.

Large-scale PV Installations: An Illustrative Example

Along with residential size PV systems discussed above, large-scale "solar farm" operations have been another major source of the overall growth in installed PV capacity in recent years. In this section, we will look at a representative solar farm case study based on the Sarnia Photovoltaic Station in Sarnia, Ontario, Canada. This plant has a rated capacity of 80 MW of thin-film panels, and was the largest plant by rating in North America at the time of its completion in October 2010. Since 2010, the Sarnia farm has been surpassed in size by larger farms such as the 550-MW Desert Sunlight farm that opened near Riverside, California in 2015.

Through bulk purchasing and streamlining of the installation process, the solar farm is able to cut cost per installed watt roughly in half, to $3/watt, as shown in Table 10-5.

Component	Cost/W ($)	Total (million $)
Panels	1.00	80
Inverter	0.30	24
Hardware	0.20	16
Labor	0.30	24
Design, permits, etc	0.20	16
Total cost	2.00	160

Acknowledgment: Thanks to Rob Garrity, President of Finlo Energy Systems, for his input into the choice of cost values shown.

TABLE 10-5 Components of Cost Per Watt and Total Cost for 80 MW Solar Farm (2015 Prices in U.S. Dollars)

Chapter Ten

Another factor is uniform ground mounting of the panels, which reduces cost compared to roof mounting. A solar farm in the region around Sarnia is thought to generate approximately 1300 kWh per kW of capacity, assuming proper maintenance and good operating condition, for a capacity factor of 14.8%, so the 80-MW farm is thought to generate ~104 GWh/year. Once the actual Sarnia plant has accumulated several years of operation, it should be possible to compare the representative figure to the actual annual output.

Based on rated capacity and capacity factor, the annual output is estimated at 104 million kWh per year. Suppose we build an 80-MW farm in Sarnia at today's prices and adopt a 20-year lifetime and discount rate of 7% to calculate levelized cost of production, and a figure of $0.004/kWh to cover all operating cost: repairs and maintenance, insurance, and so on. The needed discounting factor is $(A/P, 7\%, 20) = 0.0944$. Levelized cost is calculated as follows:

$$ACC = (0.0944)(\$160\ M) = \$15.1\ M/year$$

$$Operating\ Cost = (104\ million\ kWh)(\$0.004) = \$416,000/year = \$1.13\ M/year$$

$$Levelized\ Cost = \frac{\$15.1M + \$0.416M}{104\ GWh} = \$0.149/kWh$$

The Ontario Provincial Government's support for photovoltaic electricity in the form of a feed-in tariff requirement made the project economically viable at the time of construction. The tariff requires that any solar photovoltaic produced electricity be purchased by the grid operator at rates up to $0.45/kWh (Canadian dollars), depending on conditions at the time. The adoption of this tariff policy was a key factor in the feasibility of such a large PV power station, whereas early North American stations of the 2000–2005 period were on the order of 5 to 10 MW. Another consideration is the capacity factor of the plant. The reader can work out as an exercise that if the productivity is increased to 1700 kWh/kW (CF = 19.4%), the levelized cost declines to $0.114/kWh.

The Role of Government and Individual Support for PV Systems

As demonstrated by the incentives in Table 10-3 or tariff policy in the solar farm example, purchase of PV systems can benefit greatly from the use of policy instruments to encourage investment. These benefits are available in Europe, Japan, and many states in the United States, among other locations. Governments justify these incentives through the perceived ability to dampen peak demand periods (peak periods of demand during the middle of summer days are the times when PV systems have their highest output) and the anticipation that they will incubate a PV industry that will over time achieve cost reductions through R&D and economies of scale, therefore requiring less financial support in the future. The impact of these incentives is measured in Example 10-9, by considering the levelized cost of electricity from a residential PV system using current technology *without* incentives.

Example 10-9 Compute the levelized cost of electricity for the PV system in Table 10-3 with a 20-year lifetime, at MARR = 7%, without incentives. Assume the system has negligible salvage value at the end of its lifetime. Ignore the cost of replacing the inverter midway through the project life span. Carry out the calculation for output rates of (a) 900 kWh/kW/year, (b) 1300 kWh/kW/year, or (c) 1700 kWh/kW/year, all downstream from the inverter.

Solution Since there is no incentive in this case, the capital cost is $P = \$22,000$. The annual value of this cost is

$$A = P(A/P, 7\%, 20y) = (\$22,000)(0.0944)$$

$$= \$2077/\text{year}$$

Next, the levelized cost is calculated on the basis of 2.2 kW capacity and the given conversion factors for each of the scenarios 1 to 3:

1. Annual output = (900)(5) = 4500 kWh/y; cost per kWh = ($2077)/(4500) = $0.46/kWh
2. Annual output = (1300)(5) = 6500 kWh/y; cost per kWh = ($2077)/(6500) = $0.32/kWh
3. Annual output = (1700)(5) = 8500 kWh/y; cost per kWh = ($2077)/(8500) = $0.24/kWh

Discussion The above calculation shows that, at a total system cost of around $4/W and without incentives, electricity prices from PV are very high compared to grid prices, even with very favorable sun conditions. In remote off-the-grid locations where PV must compete against the cost of bringing in a line from the grid, the above prices are more competitive (although an additional cost for battery storage would be incurred). In such a case, the building owner is typically building with an understanding that their electricity prices will be substantially higher than the $0.08 to $0.16/kWh paid by grid customer. However, where PV must compete with the grid, there is presumably a significant set of middle-class consumers who might consider the system with tax rebate, but be unwilling to consider a system that delivers electricity at a cost of up to $0.46/kWh.

Economic analyses presented so far in this section show that in most circumstances, PV systems as investments in green energy have a relatively long payback period, except where the retail cost of electricity is high and/or the solar resource is excellent. Province of Ontario or German laws requiring that excess electricity production from PV systems be bought from the producer at a premium feed-in tariff can make PV systems economically quite attractive. Otherwise, for many consumers in other countries, purchasing carbon-free electricity from the grid from large-scale sources, for example, wind-generated electricity from commercial wind farms, is a more cost-effective means of achieving the goal of reducing CO_2 emissions from electricity. Nevertheless, both individual and institutional consumers continue to purchase these systems, resulting in the accelerating growth in manufacturing and installation from 2005 to 2014 shown in Figs. 10-2 and 10-3. In some cases, the owner of the system may derive an indirect economic benefit from having the system directly attached to the building in which they live or work, such as protection against grid unreliability. They may also view the system as a mixture of economic investment and donation to a cleaner environment, similar to making a charitable contribution to an environmental nonprofit organization. Lastly, even if they are expensive as investments, in many cases the total cost of a PV system relative to the cost of the building which it powers is small (often between 3% and 15%), so that the decision about the PV system is combined with the larger set of decisions about size of rooms, quality of floor, wall, and window finishes, and so on, and is therefore seen by the consumer as an item within the total "cost envelope" of the building.

The concept of economic decision making with "flexibility" is illustrated in Fig. 10-16. In case 1, the decision is made strictly on economic grounds, with investments in the black zone (i.e., net cost above $0) rejected because they do not break even. In case 2, there is a "grey area" where an investment with positive net cost is

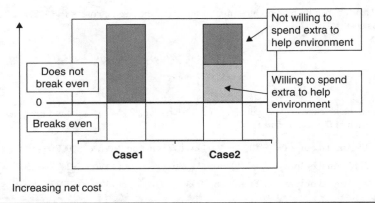

Does not break even

0

Breaks even

Not willing to spend extra to help environment

Willing to spend extra to help environment

Case1   Case2

Increasing net cost

FIGURE 10-16 Conceptual graph of two approaches to economic break-even analysis, comparing "case 1," based purely on grounds of cost, and "case 2," based on cost combined with consideration of environmental benefit.

still accepted due to its combination of environmental benefit and net cost not exceeding some positive amount, whose value is up to the subjective judgment of the decision maker. Once an investment has a net cost that is in the black zone for both cases 1 and 2, it is rejected in both cases.

Importance of Energy Efficient Appliances

There is an adage in the PV industry that "the first parts of the PV system that the customer should buy are the energy-efficient appliances that it will run." From a cost-effectiveness standpoint, this saying holds true, since it is in general cheaper to upgrade appliances to a more efficient level than to add extra panels to run less efficient appliances.

For example, suppose a 2.5-kW system is installed on a newly built 2000 ft^2 home in a region with an average output-to-capacity ratio of 1200 kWh/kW/year. Thus the output would be 3000 kWh/year. Since the home is not overly large, it might require 5000 kWh/year for a family of four, gas instead of electric dryer, and median efficiency in other major appliances (e.g., refrigerator, dishwasher, and washing machine). However, upgrading just these three appliances to high-efficiency Energy Star models, plus some additional compact fluorescent light bulbs, might trim the annual electric consumption by 1000 kWh to 4000 kWh, all for a cost differential of $1000 to $1500. This is much less than the approximately $3,000 more it would cost, including labor and materials, to add enough additional capacity to generate the additional 1000 kWh/year needed. (The additional capacity can be installed for slightly less than $6/watt because it is an incremental cost payment, e.g., the same inverter may work for the larger system, so this cost will not be incurred.) In other words, it helps to see the PV array and the collection of lights, appliances, and other devices in the house as a "complete system," rather than making decisions about the size of the PV array in isolation from other decisions about the building.

10-5   Life-Cycle Energy and Environmental Considerations

Along with economic return on investment, PV technology must achieve an adequate energy return on investment in order to qualify as a truly green technology. PV panels are made using a number of different processes under differing conditions, so there is no one standard for measuring the energy input required per panel. Also, recouping the energy investment in embodied energy in the panel will depend on the location where it is installed, with sunny locations being more favorable than cloudy ones. Thus there is not yet consensus on the value of PV energy return on investment or energy payback at this time. For conventional silicon technology, estimates vary, with numbers as low as 3 years and as high as 8 years reported. A study of building-integrated photovoltaic (BIPV) technology found that the difference in energy payback was nearly 2 years for a sampling of U.S. cities using this technology, ranging from Phoenix (3.4 years) to Detroit (5.1 years) (Keoleian and Lewis, 2003). Low energy consumption in manufacturing may provide an advantage for thin-film PV technology, with payback as low as 2 years anticipated for this technology, once it matures.

Another question is whether or not the possible toxic emissions from manufacturing PV panels and the material throughput from their disposal at the end of the lifetime undercuts the environmental benefit of reducing CO_2 emissions through their installation and use. A particular concern is PV manufacture in China, which has become a world leader in the industry but has shown evidence of neglecting the resulting pollution (Smith, 2015). As with EROI, there is no one standard for PV production, and emissions data from manufacturing tied directly to units of panel output are not easily obtained, so it would be difficult to make any definitive statement about emissions per unit of PV capacity. Also, the productive capacity of PV panels produced worldwide each year at present is small relative to total world electricity consumption. For example, at an average of 300 W per panel, the 33,000 MW of capacity installed on average between 2010 and 2013 from Fig. 10-3, would represent approximately 110 million panels manufactured worldwide. The dollar value or number of units is small relative to annual output of other mass-produced consumer electronic products such as laptop computers or mobile phones.

It is therefore not current emissions so much as future emissions that are of interest, were PV technology to grow into a major player. Future emissions are, however, difficult to predict. Because the technology is changing rapidly, projections about future total emissions or materials requirements based on today's technology are not likely to be accurate. As an alternative to projecting toxic emissions or solid waste from a robust global PV industry in the future, we instead estimate an order-of-magnitude number for the volume of panels that would be turned over each year in such a scenario in Example 10-10.

Example 10-10   Suppose that at some future time several decades from now, PV panels have become the dominant producer of electricity in the world, so that they produce as much as all sources combined did in the year 2000, namely, approximately 17 trillion kWh/year, equivalent to 2400 kWh for every one of the approximately 7 billion human beings currently living. This calculation assumes that electricity consumption per capita becomes much more efficient in the future, thanks to advances in the efficiency of end uses. The remaining electricity at that future time, assuming that total demand has grown in line with increasing population and average per capita wealth, is produced from some other complementary electricity source, most likely one that can generate electricity on demand, since

PV is intermittent. Suppose that each panel lasts on average 50 years before requiring replacement, and that the PV technology produces on average 1200 kWh/year per 1 kW of capacity. How many panels must the world roll over each year in order to maintain the required electric output from PV panels?

Solution   Based on the ratio of output-to-capacity, the required total capacity for 17 trillion kWh/year is 14 billion kW. At an average of 300 W per panel, this amount is equivalent to 46 billion panels installed. At a lifetime of 50 years, 1/50th of the panels are replaced each year, or ~1 billion panels per year. This quantity is around 6 times as many panels as were produced in 2013.

The result of Example 10-10 is crude, but it does give a sense of what a robust PV future might entail. Large-scale manufacturing of PV panels of this magnitude would create a challenge for preventing escape of hazardous materials into the environment, unless they have been completely eliminated from the product. Also, the large volume of panels would require either safe disposal or else dismantling and recycling, which would not be a trivial task. The implication of the exercise is that as 1 billion new panels are installed, 1 billion discarded panels must be broken down and recycled as much as possible. Since most of the panels installed as of 2015 have not yet reached their end of life, the launch of this recycling system has not yet even begun.

The actual future impact of the PV industry is of course sensitive to the assumptions made. For example, new technologies might emerge that are much more durable than the 50-year-lifetime projected, reducing throughput per year. Also, the PV industry might not grow as large as shown. New technologies may also emerge that take advantage of the photovoltaic effect without requiring a panel of the dimensions currently used, which might reduce the amount of material throughput. In conclusion, there are many uncertainties about the future growth potential, environmental benefits, and environmental burden associated with the production, use, and disposal of PV technology. If anything can be said with certainty at this time, it is that R&D aimed at reducing energy and materials requirements (and especially toxic materials such as heavy metals) per unit of PV generating capacity can only help to ensure that this technology, which today has an environment-friendly reputation, will not in the future become an environmental liability.

10-6   Representative Levelized Cost Calculation for Electricity from Solar PV

The following calculation of levelized cost per kWh is based on values and discussion throughout this chapter. It represents the construction of a newly built multi-megawatt solar PV plant built in the United States at recent costs. We assume a capacity factor of 17%, similar to that of Atlanta (see Table 10-2) and thus a midrange value for the entire country: more productive than New York, but not as much as California.

We assume a capital cost of $2000/kW for a 100-MW plant, resulting in a $200 million capital cost, on par with several recent U.S. multi-megawatt facilities. The required discounting factor is (A/P, 20 years, 7%) 0.0944 (see Sec. 6-8). This factor gives a capital cost per year of $18.9 million. The chosen capacity factor results in 149 GWh/year of output. Dividing capital cost among output gives a value of $0.1268/kWh. Noncapital operating cost is set at a small but nonzero amount of $0.004/kWh, derived from a survey of recent typical plant operating cost values, reflecting the need to occasionally inspect and maintain plants.

The final value is \$0.1308/kWh, as summarized in the calculations below:

$$\text{Output} = (100 \text{ MW})(8760 \text{ h/y})(0.17) = 149 \text{ GWh/y}$$

$$\text{CapCost} = \frac{(\$200M)(0.0944)}{149 \text{ GWh/y}} = \frac{\$18.9M}{149} = \$0.1268/\text{kWh}$$

$$\text{TotCost} = \text{CapCost} + \text{EnergyCost} + \text{OpCost} = \$0.1268 + \$0 + \$0.004$$

$$= \$0.1308/\text{kWh}$$

10-7   Summary

Photovoltaic technologies consist of a family of devices that take advantage of the photovoltaic effect to transform solar energy into electricity, without using a thermal working fluid or a mechanical conversion device such as a turbine. A PV panel is made up of many PV cells wired together to produce a rated power output in full sun; multiple panels can be joined in a PV array to produce a total output per month or year that matches the required demand of the system owner.

A PV cell produces current over a range of voltages from 0 V to the open-circuit voltage V_{OC}. Because the cell does not produce current at $V = V_{OC}$, the maximum power output from the cell occurs at some combination of maximum current I_M and maximum voltage V_M, where $V_M < V_{OC}$. Engineers design cells to have a fill factor (FF), or ratio of actual maximum power to theoretically possible maximum power if the cell were to produce the short-circuit current I_{SC} at V_{OC}, as close as possible to FF = 1.

In response to global concern about CO_2 and pollutant emissions from fossil fuel combustion, as well as government incentives that promote PV systems, PV sales worldwide have been growing rapidly in recent years. Relatively high levelized cost per kWh as well as uncertainties about supplies of key materials may limit future growth of PV, so research activities are currently focusing on technologies that produce cells with lower material requirements and at a lower cost per watt of capacity.

Analysis tools exist to predict the monthly/yearly electrical output of a PV panel array, with reasonable accuracy and sensitivity to local climates and the statistical variation of weather and load. The method is based on extensive research at the University of Wisconsin and parallels the *f*-chart method for active solar thermal panel array analysis (see Chap. 11). The PV analysis method is sufficiently simple to implement using a spreadsheet, and has been institutionalized in commercially available programs, under names such as PV chart and PV *f*-chart.

References

Keoleian, G., and G. Lewis (2003). "Modeling the Life Cycle Energy and Environmental Performance of Amorphous Silicon BIPV Roofing in the US." *Renewable Energy*, Vol. 28, No. 2, pp. 271–293.

Leckie, J., G. Masters, H. Whitehouse, et al. (1981). *More Other Homes and Garbage*. Sierra Club Books, San Francisco. Available on line through Cornell University at http://hdl.handle.net/1813/1006. Accessed Aug. 9, 2011.

National Renewable Energy Laboratories (2005). *National Solar Radiation Database: 1991-2005 Update Typical Meteorological Year 3*. Electronic resource, available at rredc .nrel.gov. Accessed July 29, 2015.

National Renewable Energy Laboratories (2015). *PV Watts Calculator*. Electronic resource, available online at pvwatts.nrel.gov. Accessed July 28, 2015.

Smith, R. (2015) China's Communist-Capitalist Ecological Apocalypse. Real-world economics review, Vol. 71.

Further Reading

Beckman, W. A., S. A. Klein, and J. A. Duffie (1977). *Solar Heating Design by the f-chart Method*. John Wiley & Sons, New York.

Bullis, K. (2007). "Solar Power at Half the Cost." *MIT Technology Review*, May 11, 2007.

European Photovoltaic Industry Association (2013). Global Market Outlook for Photovoltaics, 2014-2018. Annual report, EPIA, Brussels.

Evans, D. L. (1981). "Simplified Method for Predicting Photovoltaic Array Output." *Solar Energy*, Vol. 27, No. 6, pp. 555–560.

Fanchi, J. (2004). *Energy: Technology and Directions for the Future*. Elsevier, Amsterdam.

Fraunhofer Institute (2014). Photovoltaics report. Fraunhofer Institute for solar energy systems, Freiburg, Germany.

German Solar Energy Society (2013). *Planning and Installing Photovoltaic Systems: A Guide for Installers, Architects, and Engineers*. Earthscan, London.

Lorenzo, E. (1994). *Solar Electricity: Engineering of Photovoltaic Systems*. Progensa, Sevilla.

Stecky, N. (2004). "Renewable Energy for High-Performance Buildings in New Jersey: Discussion of PV, Wind Power, and Biogas and New Jersey's Incentive Program." *ASHRAE Transactions*, 2004, pp. 602–610.

Stone, J. (1993). "Photovoltaics: Unlimited Energy from the Sun." *Physics Today*, Sep. Issue, pp. 22–29.

Vanek, F., and Vogel, L. (2007). "Clean Energy for Green Buildings: An Overview of On- and Off-site Alternatives." *Journal of Green Building*, Vol. 2, No. 1, pp. 22–36.

Exercises

10-1. Consider an improved version of the photovoltaic cell in Example 10-1, which achieves a 97% collection efficiency for photons in the range of energy values from 1.5×10^{-19} J and 6.00×10^{-19} J. The absorption coefficient is also improved, to 82%. Using the same distribution for frequency of photons as a function of energy, calculate I_L for this device.

10-2. Given a PV cell with light current 0.035 A/cm^2 and saturation current 1.5×10^{-10} A/cm^2, calculate maximum voltage, current, and fill factor if the ambient temperature is 32°C. Use parameter $m = 1$. Also plot current and power as a function of voltage from $V = 0$ to $V = V_{OC}$ to verify that the approximation method for maximum current and voltage is in good agreement with the plotted values.

10-3. This problem is to estimate the yearly kWh provided by 1 m² of a photovoltaic panel installed on a building at latitude 39.4° north. The panel has the following specifications:

- Temperature coefficient for cell efficiency: 0.0055°C⁻¹
- Permanent tilt angle at 40° south
- Efficiency at rated (1000 W · m⁻², 20°C) conditions: 0.16
- $\alpha\tau$ value of 0.85
- Use will be direct, through an inverter with an efficiency of 93%

Assume values of any other parameters that might be needed. Climate data are

Month	Jan	Feb	Mar	Apr	May	Jun
$\bar{H}$,MJ·m⁻²	6.15	8.95	13.05	16.43	20.53	22.87
$\bar{K}_T$	0.40	0.43	0.48	0.48	0.52	0.55
$\bar{t}_{air}$,C	−1.0	0.0	4.0	11.0	17.0	22.0
Month	Jul	Aug	Sep	Oct	Nov	Dec
$\bar{H}$,MJ·m⁻²	22.67	20.32	16.94	12.25	7.36	5.44
$\bar{K}_T$	0.56	0.56	0.57	0.54	0.44	0.39
$\bar{t}_{air}$,C	24.0	23.0	19.0	13.0	5.0	0.0

As a second part of this exercise, quantify the advantage gained by operating the panel at the optimum tilt angle, with monthly adjustments.

10-4. This exercise is to explore the influence of photovoltaic panel temperature coefficient for cell efficiency on the kWh produced from the panel for a month where $\bar{H} = 12$ MJ·m⁻², $\bar{K}_T$ and $\bar{t}_{air} = 18$°C. Graph the output for the range of coefficient from 0.001 to 0.01, in steps of 0.001. Assume values of any other parameters that might be needed.

- Permanent tilt angle at 38°; the panel is located at latitude 43° north
- $\alpha\tau$ value of 0.85
- Temperature coefficient for cell efficiency: 0.0055°C⁻¹
- Efficiency at rated (1000 W · m⁻², 20°C) conditions: 0.16

10-5. Show the equivalence of Eq. (10-16) to Eqs. (10-14) and (10-15).

10-6. You are to evaluate a solar PV investment in a sunny location where each 1 kW of capacity generates on average 115 kWh/month year round. The system size is 5 kW rated, and electricity in the region costs $0.12/kWh. What is the value of the output from the system per year?

10-7. Now consider the cost of the system in Problem 6. You will finance the entire cost of the system with a loan, and thanks to an interest rate buy down, you are able to obtain the loan at 4% interest. There is a fixed cost of $5000 for the system for which no rebates or tax incentives are available; however, there are rebates available for the PV panels as a function of the number of watts of capacity installed. The cost of the panels is $6/W. Assume an investment lifetime of 20 years, and that the price of electricity remains constant, as measured in constant dollars. What is the dollar amount of rebate required per watt for the system to exactly break even over its lifetime?

10-8. Evaluate the viability of investment in PV in your own location. Obtain data on expected output from PV systems per kW of capacity, or use data on average insolation in your location from the Internet and assume that the system can return 10% of this value in electricity. Obtain local cost estimates for a system, as well as the current price paid for retail electricity. If the investment lifetime is 25 years, what is the NPV of a system (a) if the price of electricity in constant dollars remains the same, (b) if it increases on average by 3% per year due to rising fossil fuel costs, constraints on grid electric supply, and the like?

10-9. Recalculate the expected yearly kWh for Example 10-6, assuming the panels are re-oriented every month to be at the optimum tilt angle.

CHAPTER 11

Active Solar Thermal Applications

11-1 Overview

Solar power, when absorbed by a surface, converts to either thermal or electrical power, or both. This chapter focuses on conversion to thermal power (or energy), primarily for water and space heating. Solar collectors are described, along with their design, testing, and performance, leading into a presentation of the *f*-chart method for characterizing expected flat-plate solar collector output. The *f*-chart method is presented as an analysis tool, and then expanded to demonstrate an ad hoc approach to use the method to optimize system design. Included in the *f*-chart method is the interaction of the collectors with the total system, and how thermal storage influences collector system efficiency.

11-2 Symbols Used in This Chapter

Many symbols are used in this chapter and recalling their definitions can become a nuisance. To help, Table 11-1 is provided.

11-3 General Comments

As mentioned early in Chap. 9, the amount of energy reaching the Earth from the sun over the period of 1 day or 1 year is very large and is, in fact, much larger than the energy used by humans for an equivalent period. Although the solar energy resource is large, it is not immediately useful at large scale at most locations on Earth. It must be captured at a useful temperature, transported, and perhaps stored for later use. Solar energy installations are obviously better suited to locations with the best solar availability, but well-designed systems can be useful even in locations with significant cloudiness and cold weather during parts of the year.

The most obvious use of solar energy is for direct heating. Solar heating has been applied to crop drying in agriculture, space heating of homes and businesses, heating water, growing food (and other crops) in greenhouses, producing salt from evaporating seawater, cooking food, and driving heat engines in power cycles, as examples. The diffuse and temporally variable availability of solar energy requires large collection

Symbol	Definition, with Units (SI) Where Appropriate (Overhead Bar Signifies Averaged)
A_{ap}	Solar collector absorber plate area, m^2
A_c	Solar collector total area, including frame, m^2
c_p	Specific heat, $J\ kg^{-1}\ K^{-1}$
C	Fluid mass flow thermal capacitance, mc_p, $J\ s^{-1}\ K^{-1}$, or $W\ K^{-1}$
CF	Capacity factor, ratio of actual output to idealized maximum output
C_{min}	Smaller of the two thermal capacitance values (hot and cold sides) in a heat exchanger, $W\ K^{-1}$
C_c	Marginal cost of adding one more unit area of solar collector
C_{UA}	Marginal cost of adding one more unit of UA, the heat exchanger conductance
f	Fraction of heat need met by solar collector system, see Eqs. (11-15) and (11-16)
F_R	Collector heat removal factor, dimensionless, see Eq. (11-4)
F_R'	Collector heat removal factor, corrected for thermal storage effect, see Eq. (11-12)
$\overline{H_T}$	Integral of insolation on collector for a month, $J\ m^{-2}$
G	Mass flow rate of solar collector fluid per unit collector area, $kg\ s^{-1}\ m^{-2}$
I	Direct normal insolation, $W\ m^{-2}$
L	Thermal (heating) load for a month, J
M	Fluid mass flow rate, $kg\ s^{-1}$
N	Number of seconds in a month
q_u	Useful rate of heat gain, W
SR	Sunrise
SS	Sunset
t_a	Ambient air temperature, °C
t_{ap}	Average temperature of collector absorber plate, °C
t_i	Transport fluid temperature at inlet to collector, °C
t_o	Transport fluid temperature at outlet from collector, °C
t_{ref}	Reference temperature for collector heat loss, assumed to be 100°C
T	Absolute temperature, K
U_L	Collector thermal conductance, normalized to solar collector plate area, $W\ m^{-2}\ K^{-1}$
X	Intermediate value, ratio of heat lost from collector/heat need, see Eq. (11-10)
Y	Intermediate value, ratio of absorbed solar heat/heat need, see Eq. (11-11)
α	Solar collector absorptance for solar irradiation, dimensionless
$\beta_0, \beta_1, \beta_2$	Regression coefficients, see Eqs. (11-6) and (11-7)
ε	Heat exchanger effectiveness, see Eqs. (11-8) and (11-9)
η	Solar collector thermal efficiency, dimensionless
μm	Micrometer, or micron, 1E-6 m
τ	Transmittance of solar collector cover plate for solar energy, dimensionless
Φ	Intermediate variable, see Eq. (11-12)

TABLE 11-1   Symbols Used in Chap. 11

areas to capture the energy and convert it to heat, suitable storage means to hold the heat until it is needed, well-insulated distribution methods, and effective controls to avoid wasting the energy after it is collected.

Any method of solar energy capture and use must be viewed within the context of the system of which it is a part, as well as the stochastic nature of weather. Flat-plate solar collector systems to heat water, for example, must be designed based on actual need, be compatible with the energy storage capability of the water heating system, and work in concert with a conventional backup system.

Solar energy is diffuse, more available in some areas than others, and is nowhere continuously available or predictable, except in space. This chapter provides an overview of some of the technologies available today that are in various states of maturity, and provides analysis tools useful for designing water and space heating systems of near optimality based on economic return. Economic return is the evaluation metric to which most people can relate. Life-cycle analysis and assessment can provide a different metric that balances embodied energy against captured energy over the life of the renewable energy system, but such analyses are beyond the scope and level of detail of this book.

11-4   Flat-Plate Solar Collectors

11-4-1   General Characteristics, Flat-Plate Solar Collectors

Flat-plate solar collectors form the heart of most solar-powered living space and domestic water heating systems. The concept of a collector is simple—provide a dark surface to absorb as much solar energy as practical and include a means to transport the collected energy without serious loss for either immediate needs elsewhere or storage for later use.

Components of a solar collector include some or all of the following:

1. a surface (typically a metal sheet) that is black to absorb nearly all the incident solar energy (insolation),

2. one or more glazing sheets to transmit insolation readily to the absorber plate while intercepting and reducing thermal radiation and convection heat loss to the environment,

3. tubes or ducts to transport a fluid through the collector to accumulate the solar heat and transport that heat out of the collector,

4. structure (basically, a box) to hold and protect the components and withstand weather, and

5. insulation placed on the sides and behind the absorber plate to reduce parasitic heat loss.

In Fig. 11-1 are section views of three common flat-plate collectors, one that uses liquid as the transport medium (typically water with an antifreeze additive), a second that uses air as the transport medium, and a third, an unglazed collector using water as the transport medium and representing collectors commonly installed to heat swimming pools. Figure 11-2 shows a flat-plate collector installed on a rooftop for heating domestic hot water (DHW) for a residence.

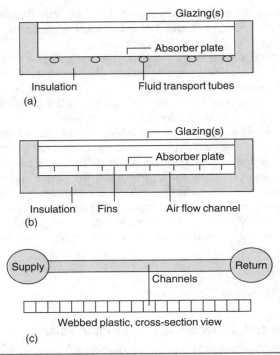

FIGURE 11-1 Section views of three flat-plate solar collector types: (a) water-type, (b) air-type, and (c) unglazed water (swimming pool) type.

FIGURE 11-2 Flat-plat solar hot water collector, Ithaca, NY, 2015.

Glass is typically the material of choice for solar collector glazing, although plastic may be used. Glass withstands weather better than plastic and does not lose transmittance due to aging—often seen as yellowing and surface degradation. Glass with low iron content (e.g., 0.02% to 0.10% Fe_2O_3) transmits a greater insolation fraction, particularly in the solar infrared spectrum up to approximately 3 μm, which can increase collector efficiency by several percent. Careful economic analysis is needed to determine whether the added capital cost of low-iron glass is compensated by the value of the additional energy collected over the life of the collector system. In large systems, the greater efficiency allows installation of fewer panels and may more than compensate for the added cost of low-iron glass, especially in cold climates with limited access to solar energy, where numerous collectors are needed to capture sufficient solar energy.

11-4-2   Solar Collectors with Liquid as the Transport Fluid

Water as a heat transport fluid has several advantages. The first is its relatively high volumetric heat capacity and high specific heat (4186 kJ m^{-3}K^{-1} and 4.186 J kg^{-1}K^{-1}, respectively, at 25°C). The second is its relative incompressibility. A third is its relatively high mass density, permitting use of small pumps, tubes, and pipes for transport.

However, water freezes well within the winter temperature range of colder climates; freezing water will damage a solar collector and its piping system. One option to avoid freeze damage is to use a drain-down collector system that empties the collectors and pipes as soon as the solar input drops below some critical insolation level. The system remains empty until the next sunny morning. Drain-down systems may not drain completely at night if pipe slope is insufficient (e.g., less than 2%) or if an air valve sticks, creating a vacuum within the plumbing and leaving behind a pocket of water that can freeze, causing damage. Drain-down sensors can fail to signal the need for drainage. Refilling a solar collector system may be incomplete because of air pockets that block water flow, reducing system efficiency. These potential problems are reasons to avoid drain-down collector systems for most applications in cold climates.

The alternate, preferred, strategy is to add antifreeze to the water. The typical antifreeze fluid is either ethylene glycol (which is toxic, requiring double-walled, closed loop systems) or propylene glycol, mixed with water. Either fluid must be adjusted to the proper concentration for adequate freeze protection. Antifreeze can degrade over time and lose effectiveness. Manufacturer's instructions strongly recommend replacement at least every five years, perhaps creating a disposal problem.

11-4-3   Solar Collectors with Air as the Transport Fluid

Solar collectors that use air as the transport fluid are usually better suited for space heating (or heating ventilation air) and drying crops in agriculture, but less well suited to water heating. Although most applications will require a fan to move the air, carefully designed collectors can be integrated into building systems to provide passive movement of the warmed air (by thermal buoyancy) to the heated space, with a return flow of cooler air, sometimes coordinated with a Trombe wall (see Chap. 12) for a hybrid form of passive solar space heating. Heat transfer from a solid to air by convection is significantly less vigorous than heat transfer from a solid to a liquid. Fins in air-based collectors add surface area and facilitate convective heat transfer, as shown in the second sketch in Fig. 11-1.

11-4-4   Unglazed Solar Collectors

Certain solar heating applications require only low temperatures, perhaps even below afternoon air temperatures in hot climates. Examples include heating swimming pools, preheating (only) water, warming water for aquaculture facilities, and heating water for showers in camping facilities. Heating water for swimming pools may be restricted to only summer months for outdoor pools, or possibly year round for indoor pools. Unglazed solar collectors have found wide acceptance in these, and comparable, applications.

Unglazed collectors typically are fabricated of channelized panels of black plastic, as shown in the third sketch in Fig. 11-1. They can be highly efficient because there is no loss of solar energy passing through a glazing and ambient air can be a source of secondary heat when outdoor air is warmer than the solar collector. Solar collection efficiencies greater than unity have been obtained when heat is gained from the air in addition to the solar gain. Unglazed collectors are significantly less expensive than conventional collectors are and their total installed area may be the majority of installed solar collectors in the United States.

11-4-5   Other Heat Transfer Fluids for Flat-Plate Solar Collectors

Refrigerants and other phase-change liquids have the advantage of a low boiling point and high energy density through the phase-change process, making them useful where physical restrictions require small volumes of fluid to transfer large amounts of heat. Phase-change fluids can respond quickly to rapid collector temperature changes and are available in a wide selection of refrigerants that do not freeze. Hydrocarbon oils and silicones can also function as heat transfer fluids in collectors but have lower specific heats than water, and higher viscosities, which increases pumping volume, energy, and cost.

11-4-6   Selective Surfaces

Absorber plates should be thin sheets of metals having excellent thermal conductivity (copper or aluminum, for example.) The upper surface is blackened for maximum solar absorption. Solar absorptance values of such collectors are typically greater than 0.95. Unfortunately, thermal radiation emittance values are also near 0.95 for such surfaces, increasing heat loss by thermal radiation and reducing the transport fluid temperature. Surfaces having a high value of absorptance for solar radiation and low value of emittance for longwave thermal radiation do not exist as natural materials.

Special fabrication techniques produce surfaces that do exhibit such behavior, however, and are termed "selective surfaces." Selective surfaces make use of the phenomenon that thermal radiation properties of a surface are determined by the surface properties to a depth of only several wavelengths of the involved radiation. Solar radiation peaks at approximately 0.6 μm. Longwave (earth temperature) thermal radiation peaks at 9 to 10 μm. The peak intensity wavelength of emitted thermal radiation can be determined using Wien's law, where the absolute temperature of the emitting surface, T, is surface temperature in kelvin and the wavelength is in microns.

$$\text{wavelength, } \mu\text{m} = 2898/T \qquad (11\text{-}1)$$

An absorbing layer that is only several micrometers thick will almost completely absorb solar radiation having wavelengths peaking near 0.6 μm but be nearly invisible to longwave thermal radiation having wavelengths peaking near 10 μm. The thermal radiation emittance value of a material can be proven to equal, numerically, its absorptance

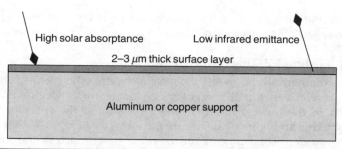

FIGURE 11-3   Section view of absorber plate with a selective surface.

value *at the same electromagnetic radiation wavelength* by making use of the second law of thermodynamics (meaning that if a sufficiently thin layer of a material absorbs a small fraction of incident longwave thermal energy, it emits an equally small fraction of the maximum possible longwave thermal radiation.) Thus a surface coating that is 2 to 3 μm thick absorbs solar radiation readily but re-radiates only a little thermal (IR) energy. (Note: It cannot emit shortwave thermal radiation because it is not at the sun's temperature!)

The surface needs support, of course, which can be provided by a thin, polished metal sheet (e.g., copper or aluminum). Polished metal exhibits a low value of absorptance and, subsequently, emittance for longwave thermal radiation. A composite made of a sheet of polished metal with a thin, visually black, surface layer (e.g., chrome oxide, nickel oxide, or aluminum oxide with nickel) can be deposited uniformly (typically applied galvanically) and creates a surface that absorbs solar energy but does not emit a significant amount of thermal radiation (see Fig. 11-3). Solar energy absorption is maximized and parasitic loss by re-radiation is minimized.

11-4-7   Reverse-Return Piping

Flat-plate solar collectors typically operate in parallel. An important goal in connecting the collectors in parallel is to have the heat transport fluid flowing through each collector at the same rate to balance thermal and flow characteristics and collector efficiencies. The method to achieve comparable flows is to connect the collectors to create the same flow path length for each. This is termed "reverse-return" piping, shown in the sketch of Fig. 11-4. Balance is assured because the first collector plumbed to the supply pipe is the last collector plumbed to the return pipe. The length of flow path and the number of fittings in each path through the collector array are the same, creating the same flow

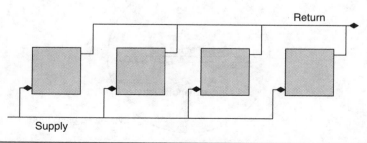

FIGURE 11-4   Reverse-return piping of a solar collector array.

resistance and transport fluid flow rate in each path. If return flow were to be toward the left of the sketch, flow paths would not be comparable and neither would be the flows.

11-4-8   Hybrid PV/Thermal Systems

Today's commercial photovoltaic panels are more efficient than in the past, but still convert a relatively small percentage of insolation to electricity. The remainder is absorbed in the form of thermal energy, warming the collector and reducing cell efficiency. Hybrid photovoltaic/thermal panels (PV/T or PVT) that produce electricity from the incident solar power, and then transport much of the concomitant thermal energy (waste heat) to thermal storage, are available in today's market. The thermal part of the collector uses metal piping attached to the back of the PV panel, through which a transfer fluid flows (typically water with antifreeze in cold climates) as in typical solar thermal panels.

11-4-9   Evacuated-Tube Solar Collectors

Evacuated-tube solar collectors, fabricated from arrays of one or two concentric glass tubes and connected in parallel, are housed within a protective structure for physical protection and insulation. The tubes resemble common fluorescent bulbs in shape and size. The evacuated tubes block convective heat loss from the absorber. Efficiency is generally higher than for typical flat-plate solar collectors and the collection temperature is higher (e.g., from 50°C to near the boiling point of water). The higher collection temperature permits useful applications for process heat in commercial applications and sufficiently high temperature to drive absorption refrigeration systems for solar air conditioning, as examples.

Evacuated-tube collectors are also suitable for domestic hot water and may be preferred for northern climates where very cold winters and the greater resulting heat loss make flat-plate solar collectors less efficient. The evacuated tube reduces heat loss from the absorber plate by thermal convection, making this design possibly useful even on cloudy days when normal flat-plate collectors may not reach a temperature adequate for useful collection. On the other hand, the greater thermal insulation value of an evacuated tube solar collector slows the rate of snowmelt from the collector panel, reducing collection efficiency on days following snowfall.

The simplest evacuated-tube design uses a single tube that houses a flat or curved metal plate (generally aluminum, see Figs. 11-5 and 11-6). The plate typically has a

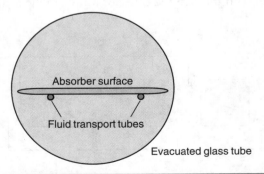

FIGURE 11-5   Section view of glass/metal evacuated-tube solar collector.

(a)

(b)

FIGURE 11-6 Evacuated tube collector system, 2007: [Fig. 11-6(a)] front view of ground-mounted system, [Fig. 11-6(b)] back view of roof-mounted system, showing insulated pipes leading to/from the collector.

(*Source:* © Philip Glaser, 2007, philglaserphotography.com.)

selective surface to enhance collection efficiency. Although these collectors are efficient and relatively simple, problems can arise in the seal between the glass and metal at the ends due to differential thermal expansion. Material selection and fabrication techniques are, thereby, critical issues.

11-4-10   Performance Case Study of an Evacuated Tube System

Performance data for an actual evacuated tube system in Ithaca, New York, with 16 tubes and a total collection area of 3.8 m² are shown in Fig.11-7. Both the annual production for 4 years (2011–2014 inclusive) in kWh of thermal energy gathered for heating water, broken down by month, and hourly output for a representative day are

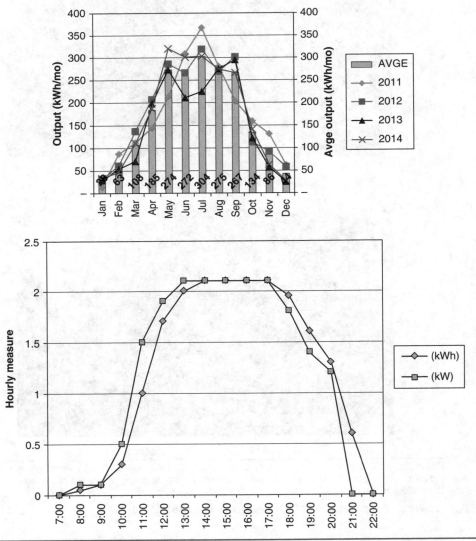

FIGURE 11-7   Annual output 2011–2014 (a) and performance for representative day (b) for 3.8 m² available collection area solar hot water system in Ithaca, New York.

Note: instantaneous output in kW is the reading given by the pump control system on the hour; hourly production in kWh is the arithmetic average of the hour shown and the previous hour, that is, average power in kW × 1 h = measured kWh. Annual output values for 2011–14 are 2095, 2142, 1835, and 2085 kWh, respectively. Numbers in (a) = monthly averages.

provided. The daily test took place on June 15, 2011, a day near the summer solstice with clear skies (essentially no cloud cover) and slight haze due to airborne humidity. The system is roof-mounted with a sealed fluid loop (water and glycol mix due to freezing temperatures in winter) that transfers heat from the collector to the storage tank in the basement of the residence. The rate of energy transfer is measured by the sensors in the pumping and control subsystem for the overall system, factoring temperature difference between cold water entering and hot water leaving the collector, multiplied by the volume of water transferred in liters per minute. Continual monitoring of energy transfer rate in kW is used to estimate cumulative energy delivered in kWh. On the daily performance side, we can observe from Fig. 11-7(b) that peak performance is shifted into the afternoon hours, with the system reaching peak output of 2.1 kW at 1 p.m. and then maintaining this output until 5 p.m., with output dropping to 1.8 kW at 6 p.m., and so forth.

We can calculate an approximate capacity factor (CF) for the device by comparing the observed 4-year average annual output of 2039 kWh [the sum of the 12 average monthly values in Fig. 11-7(a)] with the productivity that would be observed were ideal conditions maintained continually. According to the hot water system manufacturer, the system should be able to produce 37 $MJ/m^2/day$ under ideal conditions of 20°C and clear skies. Instantaneously, this rate is equivalent to 428 W/m^2. Taking into account the size of the collector and multiplying by 8760 hours per year, we can see that with 2039 kWh/y average the CF is 14.3% of what would be produced under continuously ideal conditions. Thus the CF is similar to and consistent with the CF for the solar PV system in the same location discussed in Chap. 10—since the two systems are seeing the same weather and climate conditions, it makes sense that their CF values should be comparable. Note the wide variation in monthly productivity of the system, from a low of 15 kWh in January 2011 to a high of 368 kWh in July 2011. Also, the productivity is an inferred measure from temperature difference and fluid flow, unlike the electric meter used by the solar PV array, so it is subject to greater inaccuracy, which may explain some of the year-to-year variability in monthly figures in Fig. 11-7.

We can also evaluate the relative contribution of the system to expected demand for hot water. A conservative estimate for U.S. households is 100 L/person/day, requiring 16.7 MJ/person/day using standard input and output hot water temperature levels (see Example 11-3 below). Therefore, the requirement for a household of four persons is 66.8 MJH per day, or 24.4 GJ per year. Conversion of the units for annual production from the device from kWh to MJ gives 7.34 GJ per year (6.96 million BTU), or 30% of the total expected demand. Several comments are in order:

1. The low output in the months nearest the winter solstice compared to those nearest the summer solstice (45 kWh/month average for D/J/F versus 283 kWh/month average for M/J/J) hurts the overall productivity of the system. In a different part of the United States, with more winter insolation (such as the Southwest,) year-round productivity would be more balanced.

2. If it were not for the space constraints on the rooftop collector system, the system owner would ideally add more collector units to offset more of year-round hot water demand. The cost incurred to make this change is only the marginal cost of more evacuated tubes—the tank, pump, controls, and plumbing are already in place and require no additional investment.

3. Conservation measures can bring the need closer to the output. Well-insulated hot water pipes and thermal storages, Navy showers, careful dishwashing, and short distances from the water heater to the major use point are examples of measures worth considering.

11-5   Concentrating Collectors

11-5-1   General Characteristics, Concentrating Solar Collectors

Flat-plate solar collectors, even with selective surfaces, are generally limited to collection temperatures below the boiling point of water. Applications of the collected heat are, thereby, somewhat limited in utility. Concentrating solar collectors address this issue and can provide heat for food processing, industrial processes, absorption chilling, and solar air conditioning, and even for generating electricity in Rankine, Brayton, or Stirling cycle power stations, as examples. Concentrating solar collectors follow three main designs: parabolic troughs and dishes, nonimaging solar concentrators, and central receivers (or power towers).

11-5-2   Parabolic Trough Concentrating Solar Collectors

Parabolic collectors function on the principle that light rays aligned with the axis of a parabolic surface reflect to a single focal point or line, as shown in Fig. 11-8. The focal line is where a pipe carrying a heat transport fluid should be placed in a two-dimensional trough. Heating to temperatures near 300°C is common. A characteristic of parabolic troughs is that the focal line is unchanged if the solar radiation is oblique to the face of the collector, in the direction of the trough axis. This is an advantage because solar tracking needs only a single degree of freedom—the south-facing trough can rotate to maintain a constant focal line. Of course, as the degree of obliqueness increases, total solar power collected is reduced by the cosine of the angle away from perpendicular.

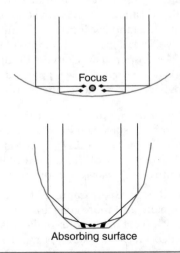

FIGURE 11-8   Cross-section schematics of two forms of concentrating solar collectors.

The second sketch in Fig. 11-8 shows an alternate configuration for concentrating insolation using parabolic shapes in a nonimaging solar concentrator. In this version, solar energy entering the aperture of the collector reflects (perhaps with multiple reflections) to the bottom, where there is an absorber plate. Solar concentration is less intense but the accuracy of focus is less critical and the collector does not need to track the sun as precisely, leading to a collector that can increase collection temperature without high cost. In simpler versions, the sides are not curved because straight but shiny sides will reflect (specular reflection) much of the insolation to the absorber plate. Tilt angle adjustment can be monthly or seasonal to aim the collector at the midday sun. Such concentrators are a more recent development than parabolic concentrating collectors that focus solar energy to a line or a point. They provide sufficient power concentration that fluid heating for process purposes, and even power generation are possible at lower cost of system manufacture and less precision of control. The absorber is smaller, creating savings of both cost and embodied energy.

11-5-3   Parabolic Dish Concentrating Solar Collectors

Parabolic dish collectors are similar to parabolic trough concentrators except the concentrator is bowl shaped and focuses solar energy on a single point instead of a line. This greater intensity of concentration achieves much higher temperatures. In theory, the focal point of a bowl-shaped, concentrating collector of perfect manufacture can reach the sun's effective surface temperature. A more typical maximum temperature may be 2000°C. Collector tracking to follow the sun must have two degrees of freedom and the collection area is limited by the maximum concentrator diameter that is practical to manufacture and move to track the sun's movement throughout the day. Difficulties such as these have limited, to date, application of dish concentrators to research and demonstration projects.

11-5-4   Power Tower Concentrating Solar Collectors

Rankine, Brayton, and other thermodynamic cycles used to power heat engines are more efficient when higher source temperatures are available. Concentrating solar collectors yield high temperatures but it is currently impractical to manufacture parabolic dishes and perhaps troughs large enough to capture sufficient solar energy to make grid power generation competitive. Instead, large fields of individually movable flat mirrors are able to reflect solar energy onto a stationary solar receiver atop a high tower where the concentrated power is absorbed and transferred to heat a (closed cycle) steam boiler, or to heat air in an open Brayton cycle and drive a turbine (see Fig. 11-9). The field of mirrors is termed a "heliostat." The mirrors are computer controlled and track with two degrees of freedom to follow the sun, and can create focus temperatures well above 1000°C. They are costly to build today but the cost is shrinking as technology improves. Several power towers operate today in the world. An early model operated in southern California for many years. Newer and larger units operate today in the United States Southwest as well as the Sanlucar la Mayor site near Seville in southern Spain. As of 2015, the largest power tower facility in the world is the Ivanpah station on the border between California and Nevada in the United States, which opened in 2014 and has a nameplate capacity of 392 MW. Simpler and smaller central receiver solar concentrators have used mirrors to reflect solar energy to a Stirling engine for uses such as water pumping in remote areas.

FIGURE 11-9 Solar One solar furnace project developed by Sandia National Laboratories in Barstow, California, United States.

(*Source:* NREL PIX No 00036. Reprinted with permission.)

Although heat transfer to the electricity generation cycle may be direct, variable solar altitude during the day, dust collection due to locations in dry desert regions, and intermittent cloudiness, greatly varies the collected solar power. To buffer the changes, an approach has been to transfer the reflected solar heat to create molten salt for use immediately, or transfer to a highly insulated storage for later in the day when solar input lessens but high demand continues for electrical power (and even during the night). A useful salt is a mix of 40% potassium nitrate and 60% sodium nitrate, which starts to melt at approximately 700°C and is molten at 1000°C. The phase change process has a high energy density, reducing storage volume compared with sensible heat storage methods.

11-5-5 Solar Cookers

Energy for cooking food on a daily basis is in limited supply in many parts of the world. Wood, the traditional cooking energy source, is renewable but woodland regrowth may not keep pace with need when population density is high, leading to deforestation and a need to spend many hours a day (and long walking distances) to find sufficient wood to meet daily needs. This burden frequently falls on women and children. In addition, burning wood, particularly burning it indoors when weather is cold, can lead to unhealthy concentrations of carbon monoxide and other noxious fumes. Health data have shown that exposure to such air pollutants leads to chronic obstructive pulmonary disorder (COPD) and chronic bronchitis/asthma. In regions, such as western China, where indoor cooking and heating with wood are common, severe eye damage is common among

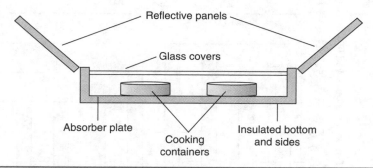

FIGURE 11-10 Schematic of solar cooker based on side reflectors.

older residents. Solar cookers provide a low-tech and low-cost alternative source of heat for cooking. The collectors are a form of concentrating solar collector.

A solar cooker in its simplest form is typically a well-insulated box with a glazing as the top (see Figs. 11-10 and 11-11) and perhaps as one or more sides. Through use of reflectors (mirrors) or lenses, solar energy is concentrated to pass into the oven, creating a usable temperature for a sufficient length of time to cook a daily hot meal. The floor of the cooker is an absorber, so cooking occurs from underneath the cooking containers, as done on a conventional stove. The glass covers and cooker insulation help assure the absorber surface retains its collected heat. Warming food to at least 100°C for at least 30 min is one definition of sufficient heating.

Solar energy is most available during midday and the largest meal in many societies typically is not until evening. Thermal mass and insulation in the solar cooker, and use of slow cooking, can delay the end of cooking until later in the day and synchronize

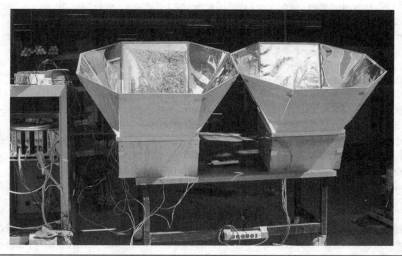

FIGURE 11-11 Solar ovens with four reflective sides undergoing parametric testing at Cornell University, May, 2004.

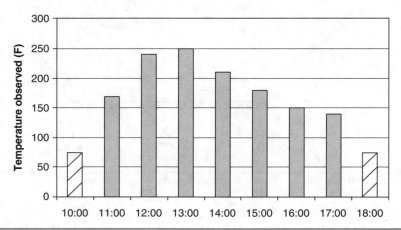

FIGURE 11-12   Air temperature measurements during test solar cooking of 12-pound gross weight casserole on June 15, 2011, in Ithaca, New York.

Note: Hash-marked bars indicate when the oven is not in use and temperatures have returned to ambient, 75°F.

need with availability. Additionally, repositioning the cooker (or its reflectors) during the day assures that it will face the sun for most of the afternoon to facilitate afternoon heat collection. Many designs have been proposed and tested and an Internet search will locate a great deal of information about solar cooking around the world. The schematic in Fig. 11-10 shows only one configuration. Reflective panels may be on one, two, three, or four sides of the cooker in actual application. Other, more complex, configurations use parabolic mirrors or Fresnel lenses as solar energy concentrators.

Figure 11-12 shows measured hourly temperature values for an example solar cooker tested on June 15, 2011, under the same conditions as described earlier for the solar hot water test (see Sec. 11-4-10). Being a simpler solar technology, the solar cooker did not have the same access to digital energy flow measurement, so hourly oven temperature was used as a proxy for more accurate measures of performance.

The cooker used for the test was similar in shape to those in Fig. 11-11, with a maximum aperture open to the sun of 38"L × 34"W, or 1292 in² (0.83 m²). Sunlight entered directly or was focused into a glass cover with dimensions 20"L × 15"W, or 300 in² (0.19 m²). The cooker had walls 2" thick for base dimensions 24"L × 19"W. The vertical depth of the oven was 12". The resulting combined outer area of the base and four walls was 1488 in² that was as well insulated as possible, in this case using high-temperature fiberglass, to retain the solar gain that entered through the glass cover. The oven was tilted at a fixed angle $\beta = 30°$ for the duration of the test, and was refocused to face the sun each hour on the hour when temperature was measured. In other words, tracking was one-dimensional only; the surface azimuth angle, γ, was adjusted, but not the slope angle.

The oven sustained temperatures at or above 150°F from 11 a.m. to 4 p.m. At 10:30 a.m., a casserole (net weight 10 lbs of food plus 2 lb of pot with flat black exterior) was inserted into the cooker. From 11 a.m. to 1 p.m., the temperature rose from 170°F to 250°F as the casserole heated through. Prime cooking happened between 12 noon and 2 p.m. when the oven was at or above the nominal boiling point of water of 212°F;

incoming solar gain maintained the food at a cooking temperature and catalyzed the chemical process of cooking. In the later afternoon, the oven continued to lightly cook the food and keep it warm; the dish was removed fully cooked at 5 p.m. Although taste is subjective, to some palates, the slow-cooked dishes produced by a solar oven at temperatures in the range 180–250°F have superior flavor.

Year-on-Year Productivity of Solar Cooking

As a means of assessing the ability of solar cooking to contribute significant quantities of cooked food to the overall diet of a household, solar cooking productivity was tracked from spring 2012 to fall 2014 for a household with access to two solar cooking ovens. One device was a mass-produced oven from Sun Ovens International of Elburn, Illinois, and the other was a hand-built solar oven. The hand-built oven had larger interior dimensions but the mass-produced oven reached a higher temperature and cooked foods faster, so each brought advantages to the productivity test. (Performance comparison was outside the scope of the test, however.) The test was carried out in Ithaca, NY, at 42° north latitude, thus in a temperate climate where solar cooking is practical between early April and mid-October. Outside these times some individuals continue to use solar cooking devices, but in general inclement weather and lack of sufficient daylight hours render solar cooking difficult.

During the testing period, each use of either oven for any single dish was recorded as a solar cooking event or "run." Thus, near the solstice with sufficient daylight hours it is possible to carry out two runs per oven, for up to four runs in a single day. In general, though, either oven performed either zero or one run per day. Also, no attempt is made to interpret the amount of energy converted based on the length of cooking time: an omelet that takes 20 minutes or a casserole that takes a day to cook both count as a single run. The number of runs per month and average monthly value are shown in Fig. 11-13. Note that during the test the apparatus was not available for month of July 2014, so that no production is shown in Fig. 11-13, and the July average figure is for July 2012 and July

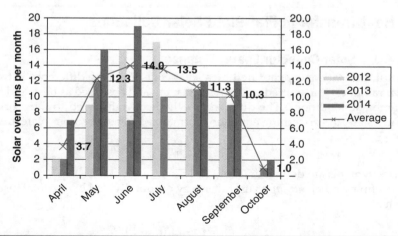

Figure 11-13 Number of cooking runs per month during cooking season (April–October) for two-oven cooking test base in Ithaca, NY, 2012–2014.

Note: "Average" curve is 3-year average of runs per month, except for July which includes 2012 and 2013 only.

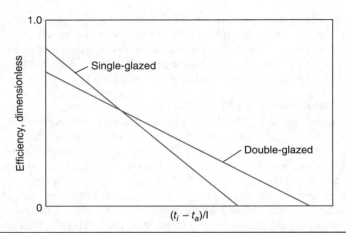

FIGURE **11-14** Example efficiency graphs for single- and double-glazed, flat-plate, solar collectors.

2013 only. Also, although a large fraction of days had insufficient sun for cooking, on another substantial portion of the days the ovens were not used even though the sun was adequate because there was no demand for additional cooked food. These limitations notwithstanding, solar cooking productivity follows the familiar pattern for solar PV arrays (Fig. 10-14) or solar hot water systems (Fig. 11-7). Output starts relatively small in the spring, peaks around the summer solstice, and declines again into the fall. Since during the summer season the diet in this region and others like it shifts to increased uncooked, fresh foods (e.g., fresh salads) the remaining cooking requirement is reduced, and with the sum of the average monthly output in Fig. 11-13 across 7 months equal to 65–70 dishes cooked, the contribution can indeed be substantial.

11-6 Heat Transfer in Flat-Plate Solar Collectors

11-6-1 Solar Collector Energy Balance

Useful thermal energy captured by a solar collector is the difference between incident solar radiation transmitted through the glazing and absorbed as heat, and heat lost back to ambient conditions. The difference is the heat transferred to the working fluid. In steady state, the governing equation, normalized to the absorber plate area, is

$$q_u = \tau \alpha I A_{ap} - U_L A_{ap}(t_{ap} - t_a) = mc_p(t_o - t_i) \tag{11-2}$$

where symbols are defined in Table 11-1.

Defining efficiency, η, as useful energy output divided by the total insolation (IA_{ap}), we have

$$\eta = \tau \alpha - U_L(t_{ap} - t_a)/I \tag{11-3}$$

The difficulty in using the above equations is imprecise knowledge of the absorber plate temperature in a real installation. Moreover, absorber plate temperature is not likely to be uniform and a means must be found to average temperature over the plate.

ASHRAE[1] proposed an empirical adjustment factor, called the heat removal factor (F_R), to permit using the inlet transport fluid temperature (which can be readily measured, or specified as thermal storage temperature) to substitute for absorber plate temperature.

$$q_u = F_R(\tau\alpha)IA_{ap} - F_R U_L A_{ap}(t_i - t_a) \qquad (11\text{-}4)$$

$$\eta = F_R(\tau\alpha) - F_R U_L(t_i - t_a)/I \qquad (11\text{-}5)$$

This adjustment, quantified empirically as part of collector efficiency testing, can be viewed as the ratio of the net heat actually collected to the net heat that would have been collected were the collector absorber plate entirely at the transport fluid entering temperature, t_i. A value for F_R is not obtained by itself, and is not needed. Efficiency data are collected over a range of $(t_i - t_a)/I$ and regression is used to obtain a value for the intercept $(F_R\tau\alpha)$ and the slope $(F_R U_L)$, which are used for analysis and system design, as covered later in this chapter.

The general form of the efficiency curve is in Fig. 11-14. Single-glazed collectors transmit more solar radiation to the absorber plate but are insulated less well against heat loss to the environment; thus their usual efficiency curves have higher intercepts but steeper slopes. The reverse is, of course, true for double-glazed collectors.

The performance model to describe the efficiency curve follows one of the following forms, where values of β_n are determined by least-squares regression using measured data to determine the best fit and decide whether the second-order term is statistically significant and provides a higher coefficient of fitness [Eq. (11-7)]. If not, Eq. (11-6) is used. Linearity is typically an adequate assumption.

$$\eta = \beta_0 + \frac{\beta_1(t_i - t_a)}{I} \qquad (11\text{-}6)$$

$$\eta = \beta_0 + \frac{\beta_1(t_i - t_a)}{I} + \beta_2\left[\frac{(t_i - t_a)}{I}\right]^2 \qquad (11\text{-}7)$$

When solar irradiation is weak and ambient air is cold, efficiency can become negative (operation below the x-axis intercept). Heat is lost from the collector more rapidly than it is collected, which bleeds heat from the thermal storage. This situation can occur early in the morning when the sun is low in the sky, at a time of low air temperature, and late in the afternoon as the sun wanes and thermal storage temperature is high after a day of collection. The concept of "critical insolation" applies. Critical insolation is the insolation value at which heat gain balances heat loss.

Figure 11-15 shows the concept of how critical insolation varies over the day. As a physical interpretation of the critical insolation graph, the area under the insolation curve is the total energy available for collection during the day. The area above the critical insolation line during collection hours is the net amount available for collection and the area below the critical insolation line represents the amount not available (lost back to the ambient conditions).

In practice, the critical insolation line is not straight. The line depends on ambient air temperature, among other factors, which is likely to be highest in the middle of the afternoon, reducing the critical insolation value during those hours. The critical insolation line bows down during the middle of a typical day. However, critical insolation is

---

[1]ASHRAE = American Society of Heating, Refrigerating, and Air-Conditioning Engineers.

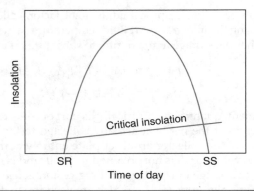

FIGURE 11-15   Example of critical insolation during an idealized solar day.

generally a concern only at the beginning and end of the collection day. As an observation, increased collector/system insulation lowers the critical insolation line, providing two benefits. Less energy is lost during the day, so net collection is greater, and the collection day can start earlier and end later.

11-6-2   Testing and Rating Procedures for Flat-Plate, Glazed Solar Collectors

Standard testing procedures are based on instantaneous efficiency measurements (generally in steady state, or nearly steady state) taken over a range of ambient temperatures. Energy available for capture is the integrated solar irradiance. The product of fluid mass flow rate, specific heat, and temperature change from inlet to outlet of the collector quantifies the energy captured. Efficiency is the energy captured divided by the irradiating energy imposed by the test facility.

Additional data are recorded, reflecting thermal shock resistance, static pressure characteristics, fluid pressure drop through the collector, and the thermal time constant of the collector, using randomly selected collectors from a production line. Collectors with water as the transport fluid are typically tested under ISO standards. Collectors with air as the transport fluid are typically tested under ASHRAE standards. Rating agencies exist that collect and publish rating data, and most collector manufacturers can provide efficiency curves for their units.

11-6-3   Heat Exchangers and Thermal Storages

Water is the typical heat storage medium in a liquid-based solar collector system and pebble beds frequently provide thermal storage for air-based systems. The typical liquid storage design is for fluid flow on one side (the collector) and no flow on the other (the storage). Figure 11-16 shows a liquid system heat exchanger. The UA value is the overall thermal conductance of the exchanger (A is the exchange area; U is the average unit area thermal conductance). A typical exchanger surface would be made of a coiled copper pipe, perhaps with fins to provide sufficient area of exchange. Manufacturer specifications generally include a UA value, usually as a function of transport-fluid flow rate.

Water from the collector enters the storage warm or hot, and cools as it passes through the exchanger. In the ideal situation, the circulating fluid leaves the storage at the lowest storage temperature—all the possible heat has been removed. In practice, this is impossible without an indefinitely large heat transfer area or residence time.

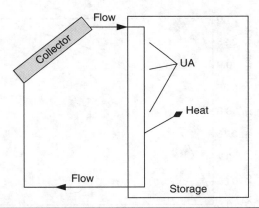

FIGURE 11-16 Sketch of heat exchanger, liquid system.

The effectiveness, ε, is the measure of how closely the heat exchange has approached the ideal. If the storage temperature is uniform at t_s and the circulating fluid enters at temperature t_i and leaves at temperature t_o, the effectiveness can be calculated as

$$\varepsilon = \frac{t_i - t_o}{t_i - t_s} \qquad (11\text{-}8)$$

For a heat exchanger with flow on only one side, the effectiveness is based on Eq. (11-9).

$$\varepsilon = 1 - \exp\left(\frac{-\mathrm{UA}}{mc_p}\right) \qquad (11\text{-}9)$$

Equations are available to determine effectiveness values for other types of heat exchangers and may be found in heat transfer books. Example 11-1 is proved to show how effectiveness is influenced by exchanger UA value.

Example 11-1 Graph the heat exchanger effectiveness with transport liquid flow on one side (such as shown in Fig. 11-16). The mass flow rate through the exchanger is 0.15 kg s^{-1} and the transport fluid heat capacity is 3500 J kg^{-1}K^{-1}. Explore the UA range from 500 to 2500 W K^{-1}.

Solution A simple approach to solve this kind of problem is to write a short Matlab script file such as the following:

```
m = 0.15;  % fluid mass flow rate, kg/s

cp = 3500; % capacitance, J/kg-K

UA = 500:100:2500;   % thermal conductance range, W/K

e = 1 - exp(-UA/[m*cp]); % exchanger effectiveness

plot(UA, e, 'k')

xlabel('UA of exchanger, W/K')

ylabel('Exchanger effectiveness, dimensionless')
```

The script file creates the graph shown in Fig. 11-17. As a perspective, a mass flow rate of 0.15 kg/s is typical of that required for a 10 m^2 solar collector array, and a fluid heat capacitance of 3500 J Kg^{-1} represents a solution of antifreeze and water.

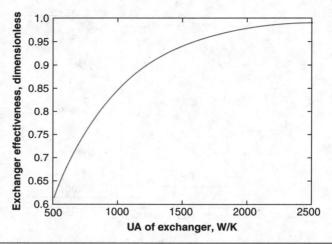

FIGURE 11-17 Heat exchanger effectiveness as a function of UA, with fluid flow on one side of the exchanger. Mass flow rate is 0.15 kg/s and fluid specific heat is 3500 J kg^{-1} K^{-1}.

11-6-4   *f*-Chart for System Analysis

Solar angles, insolation, and average weather will change throughout the year in most climate zones, often from day to day. Time-dependent, long-term, solar heating system analysis must be sensitive to these changes but not overly sensitive. One calculation to represent an entire year is unlikely to be accurate. In contrast, minute-by-minute analysis is unlikely to be more accurate than analyses using longer time bases, such as hourly or daily.

Computer simulations of systems have shown calculations on a month-by-month basis are suitably accurate without incurring excessive calculation time. That is, historical averages of insolation and air temperature are suitably consistent on a monthly basis but are not likely to be so on shorter time periods (such as weekly). This observation, based on detailed computer simulations of example systems, has led to a practical solar heating system analysis process called the "*f*-chart" method. The parameter *f* represents the predicted *fraction of the heating load* that will be provided by the thermal solar system for a given month. The method applies to both liquid-based and air-based systems.

The calculation of *f* requires two dimensionless parameters. One parameter considers the solar energy collected for the month. The other considers the heat lost from the collector for the month. (The difference represents, conceptually, the energy captured, although it is not used as such.) Dividing each by the heating load for the respective month normalizes the values and provides a perspective regarding their importance relative to the heat need. With appropriately consistent units, each parameter is dimensionless.

$$X = \frac{F_R{'} U_L A_C \Delta t(t_{ref} - \overline{t_a})}{L} \tag{11-10}$$

$$Y = \frac{F_R{'} \overline{\tau \alpha} A_c \overline{H_T}}{L} \tag{11-11}$$

where Δt is change in time and other symbols are defined above in Table 11-1.

$F_R'U_L$ and $F_R'\,\tau\alpha$ data for the collector *system* are found starting with the efficiency rating data describing the solar collector by itself ($F_R\tau\alpha$ and F_RU_L), which are typically provided by the collector manufacturer or published by an independent rating agency. F_R' is calculated starting from the value of F_R and modifying the value based on characteristics of the heat exchanger that connects the collector and its associated thermal storage. If the heat exchanger is separate from but links the collector and storage, F_R' is equal to the following:

$$F_R' = F_R\varphi = F_R\left[1 + \frac{F_RU_L}{Gc_p}\left(\frac{A_cGc_p}{\varepsilon C_{min}} - 1\right)\right]^{-1} \tag{11-12}$$

where symbols are defined above in Table 11-1.

If the heat exchanger is located within the thermal storage and there is flow only from the collector through a (finned) pipe located within the storage (described in the following section), the calculation for F_R' is simplified to the following:

$$F_R' = F_R\varphi = F_R\left[1 + \frac{F_RU_L}{Gc_p}\left(\frac{1-\varepsilon}{\varepsilon}\right)\right]^{-1} \tag{11-13}$$

The factor φ may be assumed to have a default value of 0.95, but this is an unnecessary approximation unless a collector system is being evaluated without knowledge of the thermal storage to be used.

The value of $\tau\alpha$ must also be corrected. Solar collector rating curves are based on direct normal irradiation. Unless a collector is mounted in a tracking device with two degrees of freedom, irradiation will seldom be normal to the absorber plate. Glazing transmittance and the collector plate absorption coefficient will be somewhat lower in value when irradiation arrives at an oblique angle. A simple but relatively accurate approximation is to multiply the y-intercept of the collector efficiency curve by 0.93 to obtain a value that represents the average over a day.

$$F_R'\,\overline{\tau\alpha} = 0.93F_R'\tau\alpha \tag{11-14}$$

Extensive computer simulations using the computer program TRNSYS (Transient System Simulation) provided values of f over the ranges of Y from 0 to 3 and X from 0 to 18. Regression analysis for f as a function of X and Y led to separate equations for liquid and air systems (valid only within the regression ranges of X and Y).

Liquid: $$f = 1.029Y - 0.065X - 0.245Y^2 + 0.0018X^2 + 0.0215Y^3 \tag{11-15}$$

Air: $$f = 1.040Y - 0.065X - 0.159Y^2 + 0.00187X^2 - 0.0095Y^3 \tag{11-16}$$

Fluid flow rate and thermal storage size affect collector performance. In theory, high flow rates of transport fluids make collectors more efficient. However, the advantage of a high flow rate diminishes as the flow rate exceeds approximately 0.015 L s^{-1} m^{-2} in a water-based system, based on a unit area of absorber plate. Flow through air-based

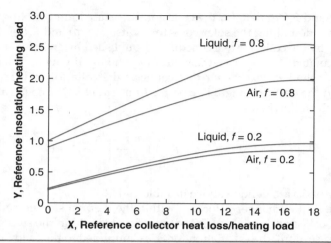

FIGURE 11-18   *f*-Value comparisons between liquid and air solar collectors.

collectors is typically best at approximately 10 L s^{-1} m^{-2}. Collector rating data should include the actual flow rate that applies to the collector efficiency curve.

Optimum storage size shows a broad plateau with little change that ranges from 50 to 100 L of water, or 0.25 m^3 of pebbles for an air-based system, each based on a unit collector area, m^2.

The graph in Fig. 11-18 compares the two *f*-value equations for low and high values of *f* and liquid and air collectors, where each has the same values of the various parameters (F_R, G c_p, and so on). It is unlikely that a liquid and an air collector will have the same parameter values for similar operating conditions and, thus, the graph is not suitable to compare efficacies of the two types of collectors. Rather, the graph acts to show only comparative behaviors of the two *f*-value equations. As one observation, for the same values of all parameters, the collection potential of collector systems using air as a transport fluid is less than the collection potential of collector systems using liquid.

The equations for *f*-values are useful for analyses using spreadsheets and computer programs directly. Alternately, the equations, as graphed in Figs. 11-19 and 11-20, require separate calculations of *X* and *Y* and the corresponding value of *f* is then read directly from the appropriate graph at the intersection of the *X* and *Y* values.

Example 11-2 Calculate the thermal energy you could expect to collect from a south-facing, liquid-based, solar collector tilted at 45° from horizontal during March, with climate variables having the following values: $\overline{t}_a$ = 6°C, $\overline{H}$ =12 MJ m^{-2}, $\overline{K}_t$ = 0.62, ground reflectivity = 0.2, and local latitude is 38° north. Heat exchanger/solar collector parameters have the following values: Heat exchanger effectiveness = 0.70, transport fluid specific heat, c_p = 3500 J kg^{-1} K^{-1} , transport fluid flow rate, G = 0.014 kg s^{-1} m^{-2}, $F_R U_L$ = 3.5 W m^{-2} K^{-1}, $F_R \tau\alpha$ = 0.88, and collector area = 10 m^2.

The monthly heating load (domestic hot water) is calculated to be 2500 MJ. Day number 75 is March 16 (or March 15 in a leap year), and can represent the month on the average.

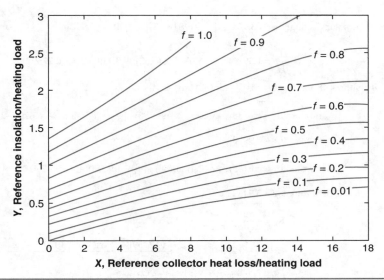

Figure 11-19   Liquid systems *f*-chart.

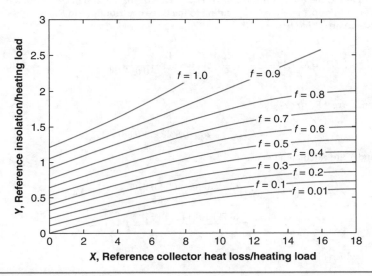

Figure 11-20   Air systems *f*-chart.

Solution   As a first step, based on the method of calculation presented in Chap. 9, estimate the average daily insolation on the tilted collector in the following sequence of steps, where two intermediate variables, numerator and denominator, are created for convenience. Angles are expressed in degrees in this example.

$$\delta, \text{ solar declination} = 23.45° \sin\left[\frac{360}{365}(284+75)\right] = -2.4°$$

$$\frac{\bar{H}_d}{\bar{H}} = 1.39 - 4.03(0.62) + 5.53(0.62)^2 - 3.11(0.62)^3 = 0.2759$$

$$h_s, \text{ sunrise hour angle} = \cos^{-1}[-\tan(38)\tan(-2.4)] = 88°$$

$$\text{numerator} = \cos(38-45)\cos(-2.4)\sin(88) + \left(\frac{\pi}{180}\right)(88)\sin(38-45)\sin(-2.4)$$

$$\text{numerator} = 0.9990$$

$$\text{denominator} = \cos(45)\cos(-2.4)\sin(88) + \left(\frac{\pi}{180}\right)(88)\sin(45)\sin(-2.4)$$

$$\text{denominator} = 0.7469$$

$$\bar{R}_b = \frac{\text{numerator}}{\text{denominator}} = 1.3375$$

$$\bar{R} = (1-0.2759)(1.3375) + \frac{(0.2759)[1+\cos(45)]}{2} + \frac{(0.2)[1-\cos(45)]}{2} = 1.2332$$

$$\bar{H}_t = \bar{H}\bar{R} = (12)(1.2332) = 14.8\,\text{MJ}\,\text{m}^{-2}$$

As an aside, the three components of $\bar{H}_t$, (direct, diffuse, and reflected insolation) are 11.6, 2.8, and 0.4 MJ m^{-2}, respectively, corresponding to the three contributions to $\bar{R}$.

Next, correct *FRUL* to *FR'UL*.

$$F_R'U_L = F_R U_L \left[1 + \frac{F_R U_L}{Gc_p}\left(\frac{1-\varepsilon}{\varepsilon}\right)\right]^{-1} = 3.5\left[1 + \left(\frac{3.5}{0.014*3500}\right)\left(\frac{1-0.7}{0.7}\right)\right]^{-1} = 3.5(0.97) = 3.4$$

Then, correct $F_R'\tau\alpha$ to $F_R'\overline{\tau\alpha}$.

$$F_R'\tau\alpha = 0.93(0.97)(0.88) = 0.79$$

X and Y are calculated next.

$$X = \frac{(3.4)(10)(86,400)(31)(100°-6°)}{2500E6} = 3.42$$

$$Y = \frac{(0.79)(10)(14.8E6)(31)}{2500E6} = 1.46$$

From which f is calculated,

$$f = (1.029)(1.46) - (0.065)(3.42) - (0.245)(1.46^2) + (0.0018)(3.42^2) + 0.0215(1.46^3) = 0.84$$

The average thermal energy collected daily for this example equals (0.84) (2500E6 J) = 2100 MJ.

This has analyzed the system and predicted the energy benefit. A continuation would be to complete this analysis for every month of the year, assign a monetary value to the energy, and compare that yearly benefit to the yearly cost of owning the solar collector system. Such a step takes the process from a snapshot to system design. Example 11-3 (below) is such an analysis and system design example.

Net present value (NPV) analysis is suggested as an economic evaluation approach that considers costs, benefits, and the time value of money over the lifetime of an investment. Although not as accurate or nuanced, calculating ownership cost based on yearly amortization is another approach, and is used in Example 11-3 for simplicity, leaving the more detailed economic analysis to the reader.

11-6-5 *f*-Chart for System Design

Numerous rules of thumb exist to determine the best size of a solar collector system to meet expected heat needs. None can be applied widely because of differences of climate, heat need, cost, and system component characteristics. Expanding the method of analysis using the *f*-chart approach provides insight into the best system for specific, local conditions. The method can be demonstrated through an extended example. The example considers domestic hot water as the only heating load, although space heat delivered through a hydronic heating system or by using a heat exchanger and forced hot air could be included.

Example 11-3 Design an optimally sized solar collector system for Sacramento, California, and estimate how much useful energy the system will provide in an average year, and its monetary value. The solar heating system will provide domestic hot water only. You are considering a solar collector with rating data as follows: $F_R U_L = 4.57$ W m^{-2} K^{-1} and $F_R \tau\alpha = 0.737$. The collector tilt angle will be the latitude plus 5°, often chosen for overall yearly best performance, or 43° from the horizontal for Sacramento. The area of each individual collector is 2 m^2.

The expected effectiveness of the heat exchanger in the heat storage tank you plan to use is 0.75. Use $G = 0.014$ kg s^{-1} m^{-2} and $c_p = 3500$ J kg^{-1} K^{-1}, as representing the heat transfer fluid. Storage will be 50 L m^{-2}, based on collector area. This is on the low end of the suggested range of 50 to 100 L m^{-2} but the solar climate in Sacramento is relatively consistent from day to day, making a large thermal storage less preferred.

Tentative estimates of costs include $750 per collector module, $2500 for pumps, piping, controls, and so on, and $2.00/L of thermal storage, with the heat exchanger included. Conventional water heating is by electricity at $0.16/kWh.

Design for a family of four, with one additional adult who will live in the home from June through September. Make a reasonable assumption of daily water use of the household. Water specific heat is 4180 J kg^{-1} K^{-1}. The home is located at latitude 38.5° north. Meteorological data for Sacramento are in Table 11-2. Snow is rare in Sacramento and it seldom freezes; assume ground reflectivity of 0.2 for all months.

Dealing with an NPV analysis is outside the scope of this example. As a simplified alternative, assume the installation cost will be amortized at a yearly cost of 12% of the installed cost. That is, when loan repayment, property taxes, maintenance, and operating costs are considered, they sum to 12% of the installed cost. Additionally, assume (for simplicity purposes) that tax credits/deductions, and other rebates, have been factored into the 12% value.

Month	Jan	Feb	Mar	Apr	May	Jun
$\bar{H}$, M J m^{-2}	6.92	10.7	15.6	21.2	25.9	28.3
$\bar{K}_t$	0.42	0.49	0.54	0.60	0.65	0.67
$\bar{t}_a$,°C	7.3	10.4	12.0	14.6	18.5	22.0

Month	Jul	Aug	Sep	Oct	Nov	Dec
$\bar{H}$, MJ m^{-2}	28.6	25.3	20.5	14.5	8.63	6.24
$\bar{K}_t$	0.70	0.68	0.66	0.60	0.48	0.42
$\bar{t}_a$,°C	24.3	23.9	21.9	17.9	11.8	7.4

Source: WBAN Data, National Renewable Energy Laboratory.

URL: http://rredc.nrel.gov/solar/old_data/nsrdb/bluebook/data/23232.SBF. Accessed Aug., 2011.

TABLE 11-2 Meteorological Data for Sacramento, California, United States

Month	Jan	Feb	Mar	Apr	May	June
GJ	2.76	2.50	2.76	2.68	2.76	3.34
Month	**Jul**	**Aug**	**Sep**	**Oct**	**Nov**	**Dec**
GJ	3.46	3.46	3.34	2.76	2.68	2.76

TABLE 11-3   Monthly Heating Load to Provide Hot Water for Example 11-3, GJ

Solution   Two preliminary calculations determine the heating load for each month and the monetary value of the collected energy. Heating load: Assume each person will use 100 L of hot water on a daily basis, which is heated from 15 to 55°C. Water main average temperature is typically equal to the average yearly air temperature of the region, which is *ca*. 15°C in Sacramento. Water heating set point temperature is often 60°C, but this is above the pain threshold level for humans. If the temperature is lower (such as 50°C), thermal losses will be less and collector efficiency will be slightly greater. Assume 55°C as a compromise. Each person requires the following for hot water every day.

$$(100 \text{ L})(1 \text{ kg L}^{-1})(4180 \text{ J kg}^{-1} \text{ K}^{-1})(55°C - 15°C) = 16.7\text{E6 J}$$

Thermal storage loses some of the collected energy through parasitic loss. Assume 25% of the collected heat will be lost this way, from pipes and storage. Thus each person will require a gross collection of 16.7 E6 J/0.75 = 22.3 E6 J day^{-1} for hot water. Monthly heat needs, based on these assumptions, are given in Table 11-3.

Conventional energy cost: Each kWh of purchased (assume electrical) energy will cost $0.16 and provide 3.6 MJ. Assume electric heating is 100% efficient (resistance heating). Thus the value of each GJ of heat (conventional or solar) is (1E9/3.6E6)($0.16) = $44.44.

The data provided above suffice to calculate the *f*-value and energy savings for every month of the year if the collector area is known, but the optimum area is unknown. To find it, a sequence of areas will be analyzed and the best one chosen. A closed form solution is not available. The process can be accomplished best using a computer language such as Matlab, or a spreadsheet, but is presented here as a series of tables that could be mimicked in a spreadsheet.

The first step is to determine the average daily insolation on the collectors for each month of the year. The procedure detailed in Chap. 9 is used, based on the data in Table 11-2, to calculate the following insolation values on a south-facing collector installed at a 43° tilt angle (and other assumptions listed above) to obtain the data in Table 11-4.

Values for X and Y depend on collector area. For simplicity of presentation, X/A and Y/A are calculated separately and are listed in Table 11-5. For each collector area, values in Table 11-5 are separately multiplied by the corresponding area, m^2, to find X and Y values, which then are used to calculate monthly *f*-values. For example, for a system using four collectors, the total area is 8 m^2 and the corresponding X, Y, and *f* data are as listed in Table 11-6.

Table 11-6 contains data for a single collector array—4 collectors with a total area of 8 m^2. Yearly energy supplied by the collectors is the sum of the monthly GJ, or 23.66, at a value of $44.44/GJ, which is a total value of $1050. Comparable calculations should be made for other areas. For example, considering 2 . . . 8 collectors, with areas of 4 . . . 16 m^2, leads to the data in Table 11-7.

Month	Jan	Feb	Mar	Apr	May	June
GJ m^{-2} day^{-1}	11.49	15.63	19.08	21.34	22.53	22.88
Month	**Jul**	**Aug**	**Sep**	**Oct**	**Nov**	**Dec**
GJ m^{-2} day^{-1}	23.74	23.88	23.61	20.63	14.25	10.91

TABLE 11-4   Average Daily Insolation, $\bar{H}_T$, on a South-Facing Solar Collector Tilted at 43° in Sacramento, California, MJ/m^2

Month	Jan	Feb	Mar	Apr	May	June
X/A	0.398	0.385	0.378	0.367	0.350	0.268
Y/A	0.086	0.116	0.142	0.159	0.168	0.136
Month	Jul	Aug	Sep	Oct	Nov	Dec
X/A	0.260	0.261	0.268	0.353	0.379	0.398
Y/A	0.142	0.142	0.141	0.154	0.106	0.081

TABLE 11-5 Intermediate Calculation Values for Example 11-3, X/A and Y/A

Month	Jan	Feb	Mar	Apr	May	June
X	3.19	3.08	3.02	2.93	2.80	2.14
Y	0.685	0.932	1.14	1.27	1.34	1.09
f	0.408	0.581	0.705	0.782	0.824	0.728
GJ saved	1.13	1.45	1.95	2.09	2.28	2.43
Month	Jul	Aug	Sep	Oct	Nov	Dec
X	2.08	2.09	2.15	2.82	3.03	3.18
Y	1.13	1.14	1.13	1.23	0.850	0.651
f	0.755	0.758	0.748	0.766	0.530	0.383
GJ saved	2.61	2.62	2.50	2.12	1.42	1.06

TABLE 11-6 Monthly Values of X, Y, f, and GJ Provided by 8 m² of Collector Area for Example 11-3

Area, m²	Yearly GJ	Yearly Value
4	13.84	U.S. $615
6	19.21	U.S. $854
8	23.66	U.S. $1050
10	27.29	U.S. $1210
12	30.15	U.S. $1340
14	31.83	U.S. $1410
16	32.52	U.S. $1450

TABLE 11-7 Energy Saved, and the Corresponding Monetary Value, for each of the Solar Collector Arrays Considered in Example 11-3

Final calculations are for the yearly cost of the collector array, to be used to compare to the benefits listed in Table 11-7. Considering the assumed costs of the problem description, amortized at 12% yearly, cost data in Table 11-8 apply.

The final column in Table 11-8 shows the most important data—the net yearly benefit. The data can also be graphed, as in Fig. 11-21. There is a definite optimum collector array size (12 m², or 6 collectors) resulting from the diminishing returns behavior as the collector area increases, combined with steadily increasing costs.

Area, m²	Collectors	Storage	Other	Total	Yearly	Net
4	U.S. $1500	U.S. $400	U.S. $2500	U.S. $4400	U.S. $528	U.S. $87
6	U.S. $2250	U.S. $600	U.S. $2500	U.S. $5350	U.S. $642	U.S. $212
8	U.S. $3000	U.S. $800	U.S. $2500	U.S. $6300	U.S. $756	U.S. $295
10	U.S. $3750	U.S. $1000	U.S. $2500	U.S. $7250	U.S. $870	U.S. $343
12	U.S. $4500	U.S. $1200	U.S. $2500	U.S. $8200	U.S. $984	U.S. $356
14	U.S. $5250	U.S. $1400	U.S. $2500	U.S. $9150	U.S. $1098	U.S. $317
16	U.S. $6000	U.S. $1600	U.S. $2500	U.S. $10,100	U.S. $1212	U.S. $233

TABLE 11-8   System Costs, Amortized at 12% Yearly, and Net Benefit for Each of the Solar Collector Arrays Considered in Example 11-3

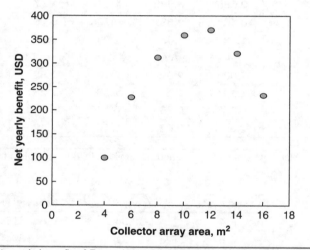

FIGURE 11-21   Net yearly benefit of Example 11-3 solar collector system design.

Final thoughts: Although the example of solar house design was presented as a relatively straightforward process, such is seldom the case in practice. Important caveats include the following considerations:

1. Weather varies from year to year and the parameter values used in the example come from historical data. There is no guarantee they will apply in the future. Localized climate change may improve or diminish benefits. Local development that adds air pollution and changes insolation values can occur (or, to be optimistic, stricter local and regional air pollution controls may reduce air pollution and increase insolation), surrounding trees may grow tall and shade the collectors for some time during the day, tall buildings may be constructed nearby and do the same.

2. The assumed quantity of hot water use per person per day is based on using only half of the current average in the United States. Simple water savings measures can reduce use even further. An unexpected result of conservation may be that a solar collector array is no longer attractive based on economic return in today's situation. This may be the case for any renewable energy scenario based on taking conservation measures very seriously. If the need is small,

savings, of necessity, will be small and the monetary advantage of adding a renewable system may be lost. This is not a negative outcome, for energy is saved either way. Conservation is often the most renewable of renewable energy resource, with the least embodied energy required.

3. The assumed costs were used for this example only and must not be assumed to apply necessarily to today or the future. Advancing technologies, economies of scale in manufacture, applicable tax deductions and credits, rebate programs, and the changing global economy can bring large differences in equipment and labor costs. However, the analysis processes used in the example will continue to apply.

4. The example considered a single set of assumptions. A more revealing analysis could use Monte Carlo simulation whereby the value each parameter is allowed to vary randomly within an assumed range and the results are recalculated. Results from a large number of simulations (e.g., at least 100; 500 would be better) can be graphed as a cumulative probability distribution function to show the most probable benefit and the range of possible benefits, with their associated probabilities of occurrence. Obviously, this is better done using a computer program rather than a spreadsheet.

5. The assumed cost of the heat exchanger increased monotonically and steadily with size. This will not be the case. Each available thermal storage/heat exchanger unit is likely to be appropriate for several array sizes and the cost will be more of a step function. Additionally, the cost of installation is not likely to remain constant and there may be a quantity discount available as the number of collector units increases.

6. As the electric utility industry moves to time-of-day rate structures, the cost of electrically heating water for domestic use becomes much more difficult to determine. Larger hot water storages may become more attractive, due less to improved solar heating system performance than to the ability to heat water using conventional energy at times of lowest cost and storing that water to compensate for times of higher costs, or days and times of reduced solar capture.

7. The design example is centered on the economic value of the installed solar collector system. Most people can relate to this metric. An alternative would include the embodied energy in the solar heating system, the yearly input of energy (as for operating pumps and maintaining the system) and the saved energy, and reach a design decision based on life-cycle net energy rather than yearly net money saved. This is a more complex analysis approach, of course, but life-cycle analysis is an established field of knowledge that could apply.

8. The same procedure used in the design example can apply to air-based solar collector systems and water-based systems used for space heating. For space heating analysis, monthly heat loss values must be calculated, based perhaps on local heating degree-day data.

11-6-6 Optimizing the Combination of Solar Collector Array and Heat Exchanger

Matching the solar collector array to the thermal storage and heat exchanger for liquid-based systems depends on their relative costs. If collectors are relatively expensive compared to thermal storages, the optimum combination will include a heat exchanger with a high UA value to enhance heat exchange. The converse is also a possibility. An approach to find the best balance is based on the marginal costs of the collectors and the heat exchange system. If C_c is the marginal cost of adding collector area and C_{UA} is the marginal cost of adding more heat exchange capacity, the following has been proposed to estimate the optimum ratio between the UA for heat exchange and collector area, where symbols are as used previously:

$$\frac{UA}{A_c} = \sqrt{\frac{F_R'U_L C_C}{C_{UA}}} \qquad (11\text{-}17)$$

This can be applied to the solar thermal design example (Example 11-3) to illustrate the method. The candidate collectors cost US $750 for each module and each has an area of 2 m². Thus, C_c = $375 m⁻². The value of $F_R'U_L$ was 4.57 W m⁻² K⁻¹. Assume manufacturer's charges for a typical thermal storage is $600 for a UA value of 1500 W K⁻¹, making C_{UA} = $600/1500 = $0.40/W K⁻¹. These data result in a UA/A_c ratio of

$$\frac{\text{UA}}{A_c} = \sqrt{\frac{(4.57)(375)}{\$0.40}} = 65 \text{ W K}^{-1} \tag{11-18}$$

A collector array of 12 m² would be optimized with a heat exchange capacity, UA, = (12)(65) = 780 W K⁻¹.

The value of U is relatively constant among exchangers of different sizes but the same design, so the exchange area must increase as collector array size increases. Storage volume also increases. However, product lines are frequently limited as to the number of available units, so exact matches cannot be expected for all collector array areas being investigated in a design exercise.

11-6-7   Pebble Bed Thermal Storage for Air Collectors

Air collectors are typically used for space heating, but times of solar heat availability may not coincide with times of need, especially during autumn and spring when heat may be needed only during the middle of the night. Thermal storage is required to accommodate the asynchronicity of availability and need.

Air is a poor heat storage medium. Although water is a good storage medium, transferring heat from air to water through a heat exchange surface is inefficient on the air side and a large exchange area is needed. A widely adopted compromise is to store solar heat in a well-insulated box containing uniformly sized stones (Fig.11-22). The boxes can be wood-framed, concrete, steel, concrete block, or other sturdy materials. The solar-warmed air is pushed down through the rock mass (vertical air flow) by a fan. A plenum is included between top of the box and the top of the stones. As air flows through the stones, it gives up its heat and exits from a second plenum through a grill

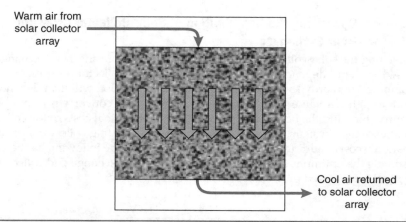

Warm air from solar collector array

Cool air returned to solar collector array

FIGURE 11-22   Cross-sectional view showing vertical air flow through a pebble thermal storage.

at the bottom of the storage. At night, air flow is reversed and cool room air is drawn up through the stones, recovering the stored heat.

The volumetric heat capacity of stone is approximately half that of water, but stone surfaces can provide a large heat exchange area if the stones are sufficiently small. (Recall that a heat exchanger is characterized by its UA value.) In the past, mention of "fist-sized rocks" for thermal storage was frequent. However, smaller stones provide significantly greater heat exchange area for the same thermal mass capacity and permit creation of a sharp heating/cooling front. The smaller the stones, the greater will be the surface area. When the stones are of uniform size, porosity near 30% is provided by small stones. Stone sizes follow standards where, for example, #1 stone is typically 6 to 12 mm, with no dimension greater than 25 mm; and # 2 stone is typically 12 to 25 mm with no dimension greater than 50 mm. Graded stones must be used to assure sufficient porosity for air to move through the pebble bed storage. Rounded river stones are preferred over crushed stone for the same reason—they provide greater porosity. If there is dirt or sand on the stones, they should be washed and sieved to remove small particles.

A typical pebble bed for thermal storage is 1.2 to 2.4 m tall, with stones resting on a metal grid above the bottom to create a plenum, with a second plenum at the top, each sized to equal approximately 10% of the bed depth. Recommended bed volume should be chosen to store approximately two-thirds of the solar heat collected and delivered on a good collection day during the heart of the heating season (which assumes the remaining third that is collected is used directly for space heating). To keep air velocity through the bed sufficiently low, limit the air face velocity to no more than 0.1 m/s.

The pressure required to push air through a pebble bed is typically less than pressure drops in the supply and return ducts and fittings. For example, the pressure drop through uniformly sized pebbles, as a function of air face velocity at the top of the bed, is as follows for two recommended stone sizes (ASHRAE, 1988).

Air Face Velocity, m/s	Pressure Loss, 19 mm Pebbles	Pressure Loss, 38 mm Pebbles
0.05	7 Pa/m	2 Pa/m
0.08	14 Pa/m	7 Pa/m
0.10	23 Pa/m	12 Pa/m
0.13	38 Pa/m	19 Pa/m

Using small stones, and moving the warmed air down from the top, encourages creation of a sharp and stable heating front. There are several advantages of a sharp heating front. Virtually all the solar heat is removed; air exits from the bottom at the lowest bed temperature. The solar heat is stored in a limited volume of the pebble bed (starting at the top). The temperature of the top volume of pebbles is maximized because heat is not spread over the entire depth of bed, warming all the pebbles only a little. Stored heat can be recaptured by reversing the air flow direction to remove the heat collected on a partially sunny day, and the air will be adequately warmed for comfort if there is a sharp heating front, with the pebbles at the top heated to the greatest extent.

Example 11-4 Consider an air collection system expected to capture 10 MJ/day per m² of collector area. Design for a 20°C temperature rise in the pebble bed storage. You have measured the porosity of the pebbles to be 30% and the pebble density is 1800 kg/m³. The pebbles have a specific heat capacity of 0.9 kJ kg⁻¹ K⁻¹. How large should the thermal storage be, normalized to the solar collector size, if the heat storage box will hold pebbles? (Note: These data are examples and may not represent locally available materials.)

Solution   Assume two-thirds of captured thermal energy will be stored on good collection days. Per square meter of collector,

$$\text{pebble mass} = \frac{\left(\frac{2}{3}\right)\left(10{,}000\,\frac{\text{kJ}}{\text{m}^2}\right)}{(0.9\,\text{kjkg}^{-1}\text{K}^{-1})(20\text{K})} = 370\ \text{kg m}^{-2}$$

Common practice is to add an additional 10% of storage volume to the bottom of the storage as a safety factor to prevent heat completely filling the storage and being then directed back to the collectors. Accordingly, total pebble mass should be *ca.* 410 kg/m² of gross collector array area. With a porosity of 30%, specific density of 1800 kg/m³, and top and bottom plenums each equal to 10% of the bed volume, total storage volume can be estimated.

$$\text{pebble volume} = \left(\frac{410\ \text{kg m}^{-2}}{0.7}\right)\left(\frac{1}{0.8}\right)\left(\frac{1}{1800\ \text{kg m}^{-3}}\right) = 0.4\ \text{m}^3/\text{m}^2$$

Naturally, the air flow rate is determined by the collector array requirements. The 0.1 m/s face air velocity limit leads to a minimum area of the bed top surface, and the calculated storage volume of 0.4 m³/m² leads to the minimum bed depth.

11-7   Summary

Solar energy can be useful even in the middle of winter when insolation intensity is weakest. Whenever there is some direct sunlight, it can be captured to heat space or provide domestic hot water. Collection can be through separate collectors with the heat then carried to the places where heat is needed. Modern analysis procedures have been developed that can be expanded to design processes. The *f*-chart method is available for sizing solar collector systems. However, the procedure is most valuable when heat collection and distribution processes are viewed within the context of a total system. The system viewpoint should start with enforced energy conservation to prevent extravagant use of collected solar energy, and to minimize the size and cost of the solar heating system. After a reasonable degree of conservation has been imposed, the other components of the system must be evaluated. This is particularly important with active solar heating systems where the thermal storage and associated heat exchanger can significantly affect overall system behavior.

Fortunately, modern personal computers can mechanize the analysis and design process, to include iterative solution procedures that permit interactions of system components to be quantified. The methods outlined in this chapter can be readily implemented using spreadsheets or programming tools such as Matlab, whereupon true design can start by asking the "what if" questions that lead to exciting innovation.

References

ASHRAE (1988). *Active Solar Heating Systems Design Manual*. American Society of Heating, Refrigerating, and Air Conditioning Engineers. Atlanta, GA.

ASHRAE (2011). *HVAC Applications*. American Society of Heating, Ventilating, and Air-Conditioning Engineers, Atlanta, Georgia.

Beckman, W. A., S. A. Klein, and J. A. Duffie (1977). *Solar Heating Design by the f-chart Method.* John Wiley & Sons, New York.

Leckie, J., G. Masters, H. Whitehouse, et al. (1981). *More Other Homes and Garbage: Designs for Self-sufficient Living.* Sierra Club Books, San Francisco.

Lund, P. (1983). *Storage Heat Exchanger: Simplified Design. Solar Age*, April, pp. 49–50.

Further Reading

Butti, K. and J. Perlin (1980). *A Golden Thread: 2500 Years of Solar Architecture and Technology.* Cheshire/Van Nostrand Reinhold Books, New York.

Duffie, J. A., and W. A. Beckman (2006). *Solar Engineering of Thermal Processes.* Wiley-Interscience, New York.

Kreith, F., and J. F. Kreith (1978). *Principles of Solar Engineering*, 3rd ed. Hemisphere Publishing Corporation, New York.

Gordon, J. (2001). *Solar Energy: The State of the Art.* James & James, London.

Exercises

11-1. This is a thought experiment. Based on the definition of F_R, should its value normally be greater than unity, or less? What physical reasoning led to your answer?

11-2. Contrast two solar collectors, one liquid based and one air based, with the following conditions. Which collector will yield the largest quantity of net energy for the month?
 a. $F_R'U_L = 1$ BTU h^{-1} ft^{-2} °F^{-1} (or 5.7 W m^{-2} K^{-1}) for each and $F_R'(\overline{\tau\alpha}) = 0.75$ for each.
 b. You want to heat the water and the air to 140°F (or 60°C).
 c. Average outdoor air temperature of the month being considered is 65°F (or 18°C).
 d. Insolation for the month will be 50,000 BTU ft^{-2} (or 570 MJ m^{-2}).
 e. The heating load for each collector system will be 2E6 BTU (or 2E3 MJ).
 f. The month is June.
 g. Collector area for each collector type is 50 ft^2 (or 5 m^2).

11-3. You plan to install an active, liquid-based solar heating system for hot water. There are four candidate collector systems. Your calculations (*f*-chart) provided the following data for the month of June. Which of the four collector systems will provide the most collected energy for the month?

X	Y
2.870	0.960
3.466	0.998
3.229	1.080
5.525	1.094

11-4. Graph heat exchanger effectiveness with transport fluid flow on one side only (such as shown in Fig. 11-14) as a function of the transport fluid heat capacity, mc_p. The UA value of the exchanger is 1000 W K^{-1}. Explore the mc_p range from 100 to 1500 J kg^{-1}K^{-1}.

11-5. Compare two liquid-based solar collectors to provide hot water for a home. System A is characterized by a corrected slope of the collector efficiency graph of 6.8 W m^{-2} K^{-1} and corrected intercept of 0.70. System B is characterized by a corrected slope of the collector efficiency graph of

7.9 W m^{-2} K^{-1} and corrected intercept of 0.75. Which collector system will provide the most energy for the month? Assume:

a. Water will be heated to 60°C.

b. The average outdoor air temperature for the month of June is 18°C.

c. Monthly insolation on the collector for June is expected to be 450 MJ m^{-2}.

d. Heat need for the month is expected to be 2 GJ.

e. Collector area is 5 m^2.

11-6.   A rudimentary Solar Electric Generating Station (SEGS) system consists of 2500 heliostats, each of 10 m^2, focusing on a central tower. The tower-heliostat system is able to transfer 50% of the incoming solar energy to incoming water, which then boils and is transferred to the turbine. Ignore other losses between the heliostats and expansion in the turbine. The plant averages 10 h/day of operation year round, and during that time the incoming sun averages 400 W/m^2. Ignore losses of insolation due to the heliostats not facing in a normal direction to the sun. Before entering the tower, the water is compressed to 20 MPa. In the tower it is then heated to 700°C. The superheated steam is then expanded through the turbine using a basic Rankine cycle, and condensed at 33°C. Assume that the actual efficiency achieved by this cycle is 85% of the ideal efficiency. The generator is 98% efficient. What is the annual electric output in kWh/year?

11-7.   The cost of the facility in the previous problem is $15 million (assuming that mass production has brought the cost down from the current cost of such experimental facilities), and it has an expected life of 25 years with no salvage value. If the MARR is 5%, what is the levelized cost of electricity from the system in $/kWh? Ignore the portion of electricity used onsite (also known as parasitic loads), and ignore maintenance and other costs.

CHAPTER 12

Passive Solar Thermal Applications

12-1 Overview

Passive solar heating describes solar energy collection systems where heat moves by passive means—transport without a fan or pump or other, mechanically based, energy inputs. Passive movement can be by thermal radiation, natural (buoyancy-driven) convective air circulation, or sensible heat conduction through a solid. Passive solar heating can be used to heat water but the most common application in the United States is for thermal comfort conditioning of indoor air. This is not necessarily the situation everywhere. Passive solar domestic water heating systems are used widely throughout the world.

12-2 Symbols Used in This Chapter

Many symbols are used in this chapter and recalling their definitions can become a nuisance. To help, Table 12-1 is provided.

12-3 General Comments

Passive solar building design is not a new technology. Villages in Ancient Greece were planned and constructed so that homes faced south to gain winter solar heat, but were designed to reject excess solar heat in summer. Ancient desert cultures understood the importance of a building's thermal mass in equalizing the extremes of day and night temperatures. Passive solar designs have proven to be reliable, require little or no maintenance, can be based on locally available materials, and do not use motive energy to transport collected heat (or coolness) from place to place in a building.

Although passive solar energy methods are simple in concept, they are complex to analyze. Conduction, convection, and radiation heat transfer modes are likely to operate simultaneously, some in parallel and some in series (and sometimes in conjunction with phase-change heat storage). Material properties can vary widely from place to place within passive solar components. Thermal mass creates thermal time constants as long, or longer, than the diurnal temperature cycle, making computational analysis complex. As computers became more powerful, numerical analysis methods (such as finite difference and finite element algorithms) came to be applied to passive solar systems and analysis methodologies became systematized. This chapter presents such a methodology

Symbol	Definition, with Units (SI) Where Appropriate (Overhead Bar Signifies Averaged)
A_p	Projected area of passive solar glazing
base	Base temperature for HDD calculations, historically taken as 65°F (18.3°C)
BLC	Building load coefficient, see Eq. (12-9)
c	Specific heat, BTU lb^{-1} F^{-1}
c_d	Coefficient of discharge for air flow through an opening
CF	Conservation factor, see Eq. (12-11)
D	Derivative of SSF with respect to the inverse of the LCR, see Eq. (12-11)
F_{HL}	Building heat loss factor, BTU h^{-1} °F^{-1} or W K^{-1}
factor	Adjustment factor to convert HDD to a different temperature base, see Eqs. (12-4) and (12-7)
G	Height of solar glazing, see Fig. 12-11
g, g_c	Force to mass conversion factor used in the IP system of units, see Eq. (12-18)
HDD	Heating degree day
h	Elevation difference, ft or m, see Fig. 12-11
h	Vertical distance between top of solar glazing and shading overhand, see Eq. (12-21)
k	Thermal conductivity, BTU h^{-1} ft^{-1} °F^{-1}
LCR	Load:collector ratio
NRL	Net reference load, see Eq. (12-10)
O	Distance shading overhang extends from wall, see Eq. (12-20)
P	Air pressure, lb ft^{-3}, or Pa
Q	Heat, BTU or J or kWh
R	Thermal resistance, h ft^2 F^2 BTU^{-1}, or m^2 kW^{-1}
S	Summer solar altitude angle at noon
SLR	Solar load ratio
SSF	Solar savings fraction, fraction of heating need provided by passive solar heating installation
T	Absolute temperature
$\bar{t}_o$	Average outdoor dry-bulb temperature
UA	Heat loss factor (BTU h^{-1}°F^{-1} or W K^{-1}) for a building's wall section of total area A
V	Air velocity, ft min^{-1}, or m s^{-1}
W	Winter solar altitude angle at noon
α	Solar altitude above horizon
ρ	Air density, lb ft^{-3} or kg m^{-3}

TABLE 12-1   Symbols Used in Chap. 12

and shows how it can be used to optimize passive solar heating systems for buildings, and touches briefly on other aspects of passive environmental modification, including natural ventilation based on the well-known "chimney problem" in fluid mechanics.

12-4   Thermal Comfort Considerations

Passive solar thermal heating is most generally (but not entirely) applied to space heating to create thermal comfort. Thermal comfort is a subjective state that can be described quantitatively. Although stress itself (thermal or otherwise) is difficult to define precisely, an accepted thermal comfort definition is "that condition of mind that expresses satisfaction with the thermal environment.[1]" The degree of thermal comfort is influenced by many factors, including thermal radiation, air temperature, humidity partial pressure, age, activity, gender, diet, metabolic rate, acclimation, and clothing. The effects of many of these factors have been quantified and are described in detail in the *ASHRAE Handbook: Fundamentals* (2009).

Thermoregulation acts to achieve body temperature homeostasis (and thermal comfort) through two primary mechanisms: vasoconstriction and vasodilation. Vasoconstriction is a narrowing of blood vessels (primarily capillaries) near the body surface, which reduces heat loss from the body. Vasodilation is the opposite phenomenon where capillaries near the surface expand, bringing more blood flow near the skin surface and increasing heat loss as long as the effective environmental temperature is below body temperature. This mechanism permits the body to maintain a consistent core temperature over a range of environmental conditions.

When vasoconstriction reaches its maximum extent, the corresponding environmental effective temperature is termed the "lower critical temperature." When vasodilation is maximized, the "upper critical temperature" has been reached. The temperature range between the lower and upper critical temperatures is termed the thermoneutral region. It must be emphasized that lower and upper critical temperatures represent the environment, for body temperature remains constant over the thermoneutral region, and beyond. The thermoneutral region shifts depending on clothing and other factors listed above. Moreover, the length of an exposure period affects the perception of thermal comfort—a cool room may be comfortable for an hour but grow uncomfortable after several hours of relative inactivity.

Some renewable energy systems, such as passive solar methods for space heating, have been criticized for ignoring the fundamental desire of people for thermal comfort. Some people will happily don another sweater during a period of limited solar availability and lower temperature. Others will not. This consideration becomes central to the design, control, and backup of solar heating systems for human occupancy. Moreover, the environmental parameter important to comfort is more than simply air temperature. The effective environmental temperature depends also on water vapor partial pressure in the air, air movement velocities, and the mean radiant temperature (magnitude and symmetry) of the thermal environment. As an aside, a frequent cold season recommendation is that humidifying indoor air can balance lowered air temperature. Data suggest a 15% increase of air relative humidity compensates for a 1°F air temperature decrease. Whether this truly saves heating energy depends on the

---

[1]ASHRAE 1992. Thermal environmental conditions for human occupancy. Standard 55-1992.

rate of air infiltration into the building and the subsequent need to evaporate water at a greater rate when the infiltration rate is high.

12-5 Building Enclosure Considerations

Three aspects of a building are critical for suitable passive solar heating. They are: (1) south-facing glazing of sufficient, but not excessive, area, (2) appropriately sized thermal mass, and (3) excellent thermal insulation of the building to make best use of the captured solar heat. A fourth aspect is necessary for adequate passive cooling by natural ventilation: a sufficient elevation difference between low and high vent areas to encourage vigorous air flow by thermal buoyancy. Note that the solar wall azimuth need not be exactly south. An acceptance angle within 30° of true south can often provide acceptable solar gain. (Use of "south" in this chapter should be interpreted as "north" in the southern hemisphere.) Moreover, at latitudes between the Tropics of Cancer and Capricorn (±23.45°, as defined by the extreme values of solar declination), solar geometry is more complex because the sun is in the northern sky at solar noon for part of the year and in the southern sky at solar noon for the remainder of the year. However, passive solar heating is seldom useful near the equator except perhaps at high altitudes.

Awnings and permanent overhangs for shade can be placed above and to the sides of solar glazings to permit winter sunlight full access to the glazing but exclude the summer sun. Knowledge of solar angles can help guide placement of awnings and shades to limit entry of summer sunlight. Deciduous trees and vines on trellises can be of some help for summer shade. However, a bare mature shade tree in winter may cast as much as 50% shade. If such a tree is to the south of a passive solar home, and close enough to shade the solar glazing when the sun is low in the sky even during midday, the effectiveness of passive solar heating will be noticeably reduced.

12-6 Heating Degree Days and Seasonal Heat Requirements

The concept of "heating degree days (HDD)" arises from the experiences of companies delivering heating fuel to rural homes as the conversion from coal to heating oil and liquid forms of natural gas became common in the United States during the first half of the twentieth century. The companies did not wish to deliver too frequently, which would have been an inefficient use of time and delivery trucks. Neither did they want to deliver too infrequently, which would have risked alienating customers.

Data showed the rate of fuel use for a home or other heated building was proportional to the difference between 65°F and the average daily outdoor temperature, $\bar{t}_a$, where only positive values were considered.

$$HDD_{65} = (65 - \bar{t}_a)^+ \tag{12-1}$$

Each house was calibrated (based on the experience of multiple deliveries) and when ΣHDD_{65} for that house reached a value known to correlate to a need for more fuel, a fuel delivery was scheduled. Note that the temperature scale when the method evolved was Fahrenheit and inch-pound (I-P) units were used for calculations, which is the system of units used generally in what follows in this chapter.

The total heat need, Q, corresponding to a HDD_{65} summation over time, can be calculated from the following equation, where F_{HL} has units of BTU h^{-1} °F^{-1}. The factor

24 converts BTU h^{-1} °F^{-1} to BTU day^{-1} °F^{-1}, or BTU per HDD in the I-P system of units. In the SI system, and with F_{HL} expressed in W K^{-1}, the same equation supplies Q in daily watt-hours, or kJ when further multiplied by 3.6.

$$Q = 24 F_{HL} \, HDD \tag{12-2}$$

Heating degree data have been accumulated for many cities of the world. The temperature base in Fahrenheit has been converted in various ways to representative bases in Celsius, including the simple and direct conversion of 65°F to a base of 18.3°C (divide HDD in F by 1.8 to find HDD in C).

The original base temperature of 65°F represents houses of the early twentieth century in the United States when insulation was minimal and indoor air temperatures were approximately 75°F (24°C). The 10° difference (temperature "lift") was a consequence of internally generated heat (cooking, lights, etc.) and solar gain through windows. Today's homes are generally insulated to a higher level, control temperatures may be lower than 75°F, and the number of electrical devices, with their associated parasitic electricity drains, has increased markedly. The 10°F lift may no longer be appropriate but to determine a new value is not a straightforward calculation. One approach would be to calculate the F_{HL} of the building as if it had no wall insulation, minimum ceiling insulation, and single-glazed windows, and then calculate a second F_{HL} value corresponding to the building designed to modern insulation standards. The lift is estimated by determining the ratio of the two values, multiplied by 10°F (5.6°C). For example, if a poorly insulated building is calculated to have an F_{HL} value of 500 W K^{-1} and an insulated version an F_{HL} value of 250 W K^{-1}, lift = (500/250)(5.6) = 11°C (or 20°F). A higher value of lift leads to a lower value of HDD by definition. This conversion does not consider today's greater electricity use, which can further increase the lift, of course. A more complete energy balance on the building is one approach to include that effect.

One should note the compounding advantage of insulation. A building with more insulation has a lower F_{HL} value and a lower effective HDD value. Heating load is proportional to the product of the two values *and the benefit is compounded*. This is an important insight for any heated building and particularly for designing a passive solar home, for the optimum area of solar glazing and associated thermal mass is likely to be significantly smaller and less expensive when the rest of the building is well insulated and the heating load is moderated.

12-6-1   Adjusting HDD Values to a Different Base Temperature

During cold months, when the average outdoor air temperature is below the HDD base temperature every day, the HDD sum for a month is reduced by the number of days in the month multiplied by the difference between the standard base temperature and the adjusted base temperature. For example, if the HDD base temperature for January in a cold climate is 15.3°C instead of 18.3°C, the number of HDD is reduced by 31(18.3 – 15.3) = 93°C days. The calculation is more complicated during spring and fall months. Some days that are included under the standard base temperature are omitted under the adjusted base temperature [e.g., a day with an average outdoor temperature of 16.3°C contributes 2 HDD to the monthly total at the standard base (of 18.3), but none at the adjusted base of 15.3°C]. This nonlinearity can be approximated by the following equations, one for temperature in Fahrenheit and one for temperature in Celsius. The parameter, base$_{new}$, is the modified base for calculation. For example, the new base might be 12°C for a well-insulated building controlled to an indoor air temperature of perhaps 22°C (or a new base of 55°F instead of 65°F for the same house).

For I-P units and monthly conversions
If $HDD_{65} < 50(65 - base_{new})$

$$HDD_F = HDD_{65} - days_{month} (65 - base_{new}) (1 - factor) \qquad (12\text{-}3)$$

$$factor = \left[1.05 - 0.01 (65 - base_{new}) - \frac{HDD_{65}}{60 (65 - base_{new})}\right]^{1.9} \qquad (12\text{-}4)$$

Else

$$HDD_F = HDD_{65} - days_{month} (65 - base_{new}) \qquad (12\text{-}5)$$

For SI units and monthly conversions
If $HDD_{18.3} < 50(18.3 - base_{new})$

$$HDD_C = HDD_{18.3} - days_{month} (18.35 - base_{new}) (1 - factor) \qquad (12\text{-}6)$$

$$factor = \left[1.05 - 0.01 (18.3 - base_{new}) - \frac{HDD_{18.3}}{60 (18.35 - base_{new})}\right]^{1.9} \qquad (12\text{-}7)$$

Else

$$HDD_C = HDD_{18.3} - days_{month} (18.3 - base_{new}) \qquad (12\text{-}8)$$

Note: These conversions must be applied on a month-by-month basis to heating degree-day data. Days within a month are somewhat similar and may be considered together; mid-winter days are so unlike spring days that they cannot be grouped and averaged. The two conversion processes can be graphed, and are shown in Figs. 12-1 and 12-2 for Fahrenheit and Celsius temperatures.

As a confirmation of the HDD adjusting method, predictions based on the method can be compared to actual data. Daily averaged air temperatures for Ithaca, New York, for 2006 were used to generate the graph in Fig. 12-3. Temperature bases of 6.3, 9.3, 12.3, 15.3, and 18.3°C were included. Each daily average air temperature was subtracted from each base temperature (as done in the definition of HDD) and the differences

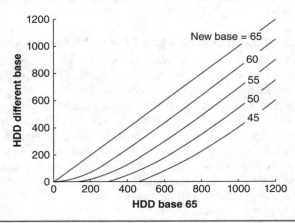

Figure 12-1 Adjusted heating degree-days for several base temperatures and indoor air temperatures, Fahrenheit temperatures.

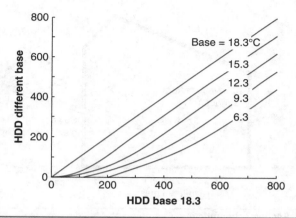

FIGURE 12-2   Adjusted heating degree-days for several base temperatures and indoor air temperatures, Celsius temperatures.

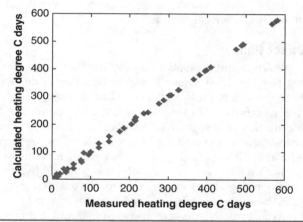

FIGURE 12-3   Correlation of measured and adjusted monthly heating degree data for Ithaca, New York, United States (2006 weather data).

summed monthly to obtain values of ΣHDD_{base} for each of the five base temperatures. Each monthly value at the standard base was used to calculate a corresponding adjusted HDD_c, using the appropriate equation above. The two sets of HDD values are graphed in Fig. 12-3 and show good correlation to the 1:1 line, with little scatter.

12-7   Types of Passive Solar Heating Systems

Passive solar space heating systems can be classified as *direct gain*, *indirect gain*, or *isolated gain*. Direct gain is based on solar radiation being absorbed within the heated space. Indirect gain is based on absorbing solar energy into a separate thermal mass designed to provide later release. Isolated gain describes situations where solar energy is absorbed to heat a separate space for later release, if desired, into the living space. Thermal mass is required to buffer large and rapid temperature swings and store heat

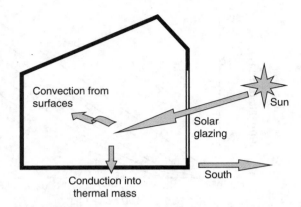

FIGURE 12-4   Section schematic of direct gain solar building for winter heating, northern hemisphere (arrow points north in southern hemisphere).

for later, and can be water, concrete, stone, adobe, or other material with a high volumetric specific heat and high thermal conductivity.

12-7-1   Direct Gain

South-facing glass transmits insolation directly into the living space in a direct gain passive solar building (see Fig. 12-4), where it is absorbed. Part of the absorbed heat is released immediately to the indoor air by convective heat transfer from the heated surfaces into the space. The rest moves into the thermal mass by conduction for release back to the indoors late in the day, or at night. An advantage of a direct gain passive system is its relatively low construction cost and rapid warm-up in the morning. Disadvantages can include overheating during midday, excessive light glare in the living space, and accelerated solar degradation of materials near the solar glazing (wood and fabrics, for example). Additionally, carpets and furniture tend to prevent solar gain from reaching the thermal mass, exacerbating overheating and reducing thermal mass effectiveness.

Several practical rules are suggested when planning a direct gain solar home:

- Thermal mass elements such as the floor should not be more than 15-cm (6 in) thick.
- Keep floors and the living space near the solar glazing as free of carpets and furniture as functionally and aesthetically reasonable.
- Spread thermal mass throughout the living space rather than concentrating it in one element (such as only in the floor).
- If concrete blocks are used for thermal mass (such as for walls), fill the cores with concrete.
- Use dark colors for floors and walls most directly irradiated by the sun—to absorb insolation intensity and reduce glare.

12-7-2   Indirect Gain, Trombe Wall

The most common form of indirect gain passive solar system is the Trombe wall (see Fig. 12-5). Insolation is absorbed by the outer surface of the thermal mass wall.

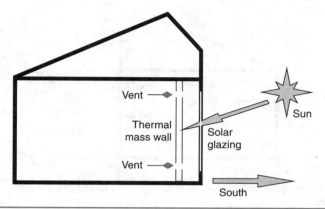

The distance between the glazing and wall is narrow (e.g., 6 in or 15 cm). As the wall
outer surface heats, warm air is created within the space. The heated air can move by
thermal buoyancy if vent areas are provided near the floor and ceiling. Vents must be
closed at night to prevent reverse siphoning of cold air into the living space, of course.

As the outer surface of the wall warms, heat conducts as a thermal wave through
the mass and toward the living space side, typically traveling at a velocity of 1 to 1.5 in
h^{-1} (2 to 3 cm h^{-1}). The wall is typically 12- to 16-in (30 to 40 cm) thick so the thermal
wave and, thereby, the living space side surface temperature, peaks approximately
12 h after peaking at the outer surface (which generally occurs during early afternoon).
Timing is important, so more heat is delivered to the living space during the night when
outdoor air temperature is lowest. Heat transfers to indoor air by convection and to
other surfaces within the space by thermal radiation. A warm wall, radiating to people,
can compensate for an air temperature several degrees lower than normally perceived
as comfortable.

Movable insulation can be deployed between the wall and glass at night (or exterior
to the glass) to limit heat loss back to outdoors and keep the mass from cooling too
rapidly at night. If movable insulation is installed, a system of automated control to
open and close it is best to make most effective use of the system.

The thermal wall, alternately, may hold water in vertical tubes with a diameter of
approximately 15 cm. In such a system, if the tubes are translucent and the water is
dyed, the wall can function strictly as an indirect gain system if the dye is opaque, or a
combination of direct (with daylighting) and indirect passive gain when a useful portion
of insolation is transmitted (e.g., 50%).

Hybrid direct gain and indirect gain passive solar systems can be integrated to create a compromise and limit the disadvantages of direct gain systems, but achieve faster
warming of the living space in the morning than would be provided by an indirect gain
system, as well as providing natural lighting (daylighting). Natural light is usually
more acceptable for visual comfort and natural lighting uses no conventional energy.
A hybrid system would have vision strips or windows placed in the Trombe wall at
appropriate locations for vision and lighting. A room generally requires glazing area
equal to at least 5% of its floor area for adequate vision lighting during the day. A hybrid

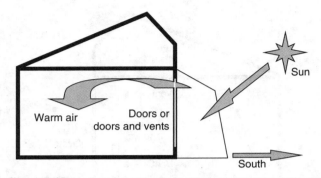

FIGURE 12-6   Section schematic of sunspace on a solar building for winter heating, northern hemisphere.

direct/indirect gain system may require glazing to be somewhat more than 5% of the floor area to provide sufficiently rapid warm-up during mornings, but not so much as to create overheating and visual glare.

12-7-3   Isolated Gain

A sunspace (sun room, or solar greenhouse, see Fig. 12-6), combined with a convective loop to the attached house, permits isolation from solar gain when heat is not needed but gives access to solar heat during cold days and nights. The wall between the sunspace and living space may be thermally massive to act somewhat like a Trombe wall. Although a thermally massive wall may buffer the heat from the sunspace somewhat, unless it is constructed with the thickness and volumetric heat capacity of a Trombe wall, there will be less thermal wave delay than provided by a Trombe wall. If the thermal wave arrives at the living space side during the afternoon, for example, overheating and wasted energy could result.

A sunspace can overheat on hot and sunny days and must be vented when it is used as a greenhouse as well as a means to provide solar heat. Plants are generally stressed at temperatures above 35°C. Solar greenhouses must also be heated during winter if plants are to be kept alive. Additionally, sunspaces may be sources of insects, other pests, and mildew spores that move into the living space. Plants with large leaf canopies can add a considerable amount of humidity to the air, which may be an advantage or a disadvantage. One disadvantage is that transpiration and evaporation reduce the quantity of sensible solar heat that can be collected. However, a sunspace can provide a very pleasant addition to a building for relaxation and entertainment.

Roof ponds may be another isolated gain passive solar heating system. Fifteen to thirty centimeters of water is contained in a large plastic or fiberglass container on a flat roof, with the container covered with a glazing. Water within the container warms during the day and the stored heat conducts downward at night through the roof to the living space, where the major heat transfer path to occupants is thermal radiation. Movable insulation over the container is required to retain the heat at night. Conversely, operation can be reversed to permit the water to cool at night to provide a form of passive air conditioning the next day. Roof ponds require drainage systems and a structural support to support the weight of the water. Few roof ponds are in active use today. Note: A roof pond in the ceiling of a home would be classified an indirect gain system, but an isolated gain system if on the roof of a nearby or attached garage. The precise classification, of course, is not very significant.

12-8 Solar Transmission through Windows

Passive solar homes require considerable attention be paid to windows, solar or otherwise. Windows frequently provide two functions—the first is to provide access to outdoor views and the second is to permit solar entry without excessive heat loss back to the outdoors. These functions may be required of the same window, or the total glass area for a building may be separated into separate solar glazing and other windows. (Windows provide other functions, of course, such as ventilation and daylighting.)

There is little reason for all windows in a building to be of the same type. Nor is there need for all to be operable. Operable windows typically experience greater unwanted air infiltration in addition to being more expensive. North-facing windows should be chosen to limit heat loss with little concern for excluding excessive solar heat. East- and west-facing windows may be limited in size and chosen for solar exclusion to limit solar heat gain during the summer when intense sunlight in mid-morning and mid-afternoon can exacerbate overheating and increase the air conditioning load. South-facing windows may be selected to emphasize solar admission over thermal resistance to heat loss, although perhaps coupled with operable night insulation.

Most windows today are double glass. The glass itself does not add greatly to thermal resistance of the combination. Rather, the air space, and its associated convection and radiation heat transfer coefficients, creates most of the resistance (see Fig. 12-7).

The surface heat exchange coefficient at surface one is a combination of convection and thermal radiation. At surfaces two and three, convection and thermal radiation each are present but thermal radiation is a relatively more important factor because surfaces two and three are not likely to be close in temperature. Thermal radiation flux depends on the difference between the fourth powers of the absolute temperatures of the two surfaces, and the effective cavity emittance (refer, e.g., to the thermal radiation section of any comprehensive heat transfer text). Wind typically leads to dominance of convective heat transfer over thermal radiation at surface four.

Inert gas, such as argon, between the two glass layers increases thermal resistance because it conducts heat less readily than air. Unless there is physical damage during installation, even if there is leakage, it is slight and reduces the benefit by only a few percent over a decade. However, if the window unit is manufactured at a low elevation above sea level and installed in a building high above sea level, the gas pressure difference may lead to seal failure and inert gas loss, and noticeably reduced efficiency.

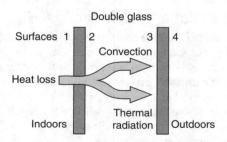

FIGURE 12-7   Heat loss paths through a double-glazed window.

Thermal radiation is an important path of heat loss through windows. Glass by itself has a thermal radiation emittance close to unity. Thin surface coatings on glass can change surface emittance to low values, such as 0.1 or 0.2 instead of 0.95. This is similar to the action of selective surfaces on solar collector absorbers, described earlier in Chap. 11, except absorption of insolation is not part of the process. Clear silver or tin oxide layers that are vacuum deposited on surfaces provide the property of reducing thermal radiation emission but permitting visible light to pass. The surface coatings are generally not able to withstand much physical abrasion and, thus, are usually protected by being placed on either surface two or three. In cold climates, surface two is the preferred side to reduce thermal radiation heat loss to surface three. In hot climates, surface three is preferred to limit heat gain from the outdoors.

Window units in the United States are energy rated by the National Fenestration Rating Council (NFRC). (A fenestration is defined as an opening through a wall into a building, such as a window or a door.) Rated windows bear an NFRC seal containing several items related to window energy performance. The U value (thermal conductance) in I-P units is provided and values generally fall between 0.2 and 1.20 BTU h^{-1} ft^{-2} $°F^{-1}$. The second item is the solar heat gain factor (SHGF) that measures the fraction of incident insolation transmitted through the window unit as a whole. Values are bounded by 0 and 1, with higher values better for solar gain (e.g., on south-facing windows) and lower values better for reducing cooling loads in summer (e.g., on east- and west-facing windows). As a rule, an SHGF < 0.4 is preferred to reduce air conditioning loads and an SHGF > 0.6 is preferred for solar gain/heating. The U value and SHGF are related because a low U value (e.g., lower than 0.32 BTU h^{-2} ft^{-2} $°F^{-1}$) generally means a high value of SHGF cannot be achieved. Measures taken to reduce U, such as special glass coatings, generally reduce solar transmittance. Certain window units even have a third layer, a thin plastic film layer between the glass layers, with a low-emittance surface coating for infrared radiation to create even better resistance to heat loss but, obviously, also less transmittance of insolation.

Visible transmittance is the third item provided and is the fraction of visible light transmitted by the window unit. The fourth item is air leakage, a measure of resistance to unwanted air infiltration, which contributes to heat loss. The fifth item is condensation resistance, which measures the ability of the window unit to resist formation of condensation on the interior surface of the glass. The value ranges between 0 and 100 (highest resistance) and provides a relative measure of condensation resistance. However, other factors such as air tightness of the building, local climate and weather, and moisture production within the building also affect condensation.

12-9   Load:Collector Ratio Method for Analysis

A widely accepted tool to analyze passive solar heating systems is the load:collector ratio (LCR) method. [The method is called the solar load ratio (SLR) method when applied on a month-by-month basis.] The LCR method is sufficiently simple that hand calculations can suffice, although computer programs that include a database needed for the method are preferred. The method is based on extensive hour-by-hour computer simulations that were tested against experiments to validate the results. The method applies to three generic passive solar heating methods: direct gain, indirect gain (Trombe wall and water wall), and attached sunspace systems.

The method starts with a calculation of the building thermal characteristic, called the building load coefficient (BLC). By definition, the BLC is the building heat loss per heating degree-day for all but the solar glazing. The solar glazing is omitted because the computer simulation results on which the method is based provide the solar glazing *net* benefit, not the total benefit. Heat loss through the solar glazing to the outdoors is included in the simulation results and should not be counted twice.

The original computer simulations on which the LCR method is based were done using I-P units of measure, which is what will be used here. When the heat loss factor (F_{HL}) of the building is calculated (BTU h^{-1} °F^{-1}), the BLC is calculated from the following equation, where the inclusion of the number 24 converts heat loss per degree hour to heat loss per degree day.

$$\text{BLC} = 24(F_{HL} \text{ for all but the solar wall}) \tag{12-9}$$

The net reference load (NRL) of the building is the calculated need for heat for a year for all but the solar glazing, as calculated from the following equation.

$$\text{NRL} = (\text{BLC})(\Sigma\text{HDD}) \tag{12-10}$$

The HDD should be calculated for the nearest city or weather station and should be adjusted on a month-by-month basis to a new temperature base as appropriate, as described earlier in this chapter.

The final step is to relate the calculated value of the LCR to the solar-savings fraction (SSF). This can be done using data tables developed by extensive computer simulations undertaken in the early 1980s. The simulations were based on standardized material and other properties, which are listed in Table 12-2. More than 50 passive system design combinations were simulated and extensive data comparable to the small selection presented below for Billings, Montana, are available in Jones et al. (1982) as well as Balcomb et al. (1984). A subset of their data is below and in Appendices B and C at the end of this book.

Four types of passive systems were analyzed, abbreviated as WW for water wall, TW for Trombe wall, DG for direct gain, and SS for sunspace. Within the types, further definitions delineate properties such as availability of movable night insulation, use of either double or triple glazing for the solar window, thermal mass thickness, application of a normal or selective surface on the absorbing surface (Trombe walls), and other type-specific parameters. As an example, and for later use in examples and exercises, Table 12-3 contains the reference design characteristics for vented Trombe walls.

More than 200 cities in the United States and Canada were analyzed for each of the four types of systems, such as for each of the design designations for vented Trombe walls listed in Table 12-3. Results of the simulations were presented for solar savings fractions of 0.1 to 0.9, in tenths. Table 12-4 contains resulting combinations of LCR and SSF data for a representative city, Billings, Montana, selected as an example because of its cold winters (7265 base 65°F heating degree-days) but reasonably good winter solar climate.

Example 12-1 With data in Table 12-4, a specific installation can be analyzed. Consider a home with a heat loss factor (if it were to be built without the solar wall) of 225 BTU h^{-1} °F^{-1}. The insulated wall that the solar glazing will replace has thermal insulation equal to R-25, I-P units. Architectural

Masonry properties	
Thermal conductivity, sunspace floor	0.5 BTU h^{-1} ft^{-1} °F^{-1}
Thermal conductivity, all other masonry	1.0 BTU h^{-1} ft^{-1} °F^{-1}
Density	150 lb ft^{-3}
Specific heat	0.2 BTU lb^{-1} °F^{-1}
Solar absorptances	
Water wall	0.95
Masonry Trombe wall	0.95
Direct gain and sunspace	0.8
Sunspace lightweight common wall	0.7
Sunspace, other lightweight surfaces	0.3
Infrared emittances	
Normal surface	0.9
Selective surface	0.1
Glazing properties	
Transmission	Diffuse
Orientation	Due south
Index of refraction	1.526
Extinction coefficient	0.5 in^{-1}
Thickness of each pane	1/8 in
Air gap between panes	0.5 in
Control	
Room temperature	65 to 75°F
Sunspace temperature	45 to 95°F
Internal heat generation	None
Thermocirculation vents (when used)	
Vent area (upper and lower)	6% projected area
Height between upper and lower vents	8 ft
Reverse flow	Prevented
Movable night insulation (when used)	
Thermal resistance	R9 (I-P units)
Deployed, solar time	1730 to 0730
Solar radiation assumptions	
Shading	None
Ground diffuse reflectance	0.3

TABLE 12-2   Reference Design Characteristics of Systems Listed in LCR Tables

Designation	Thermal Storage Capacity, BTU ft^{-2} °F^{-1}	Wall Thickness, in	ρck^2, BTU^2 h^{-1} ft^{-4} °F^{-2}	Number of Glazings	Wall Surface	Night Insulation
TW A1	15	6	30	2	Normal	No
TW A2	22.5	9	30	2	Normal	No
TW A3	30	12	30	2	Normal	No
TW A4	45	18	30	2	Normal	No
TW B1	15	6	15	2	Normal	No
TW B2	22.5	9	15	2	Normal	No
TW B3	30	12	15	2	Normal	No
TW B4	45	18	15	2	Normal	No
TW C1	15	6	7.5	2	Normal	No
TW C2	22.5	9	7.5	2	Normal	No
TW C3	30	12	7.5	2	Normal	No
TW C4	45	18	7.5	2	Normal	No
TW D1	30	12	30	1	Normal	No
TW D2	30	12	30	3	Normal	No
TW D3	30	12	30	1	Normal	Yes
TW D4	30	12	30	2	Normal	Yes
TW D5	30	12	30	3	Normal	Yes
TW E1	30	12	30	1	Selective	No
TW E2	30	12	30	2	Selective	No
TW E3	30	12	30	1	Selective	Yes
TW E4	30	12	30	2	Selective	Yes

$^2\rho$ is density, c is specific heat, and k is thermal conductivity.

TABLE 12-3 Definitions of Reference Design Designations for Vented Trombe Wall to be Used in LCR Method of Passive Solar Heating System Analysis

considerations suggest a solar glazing of 200 ft^2. Calculate the expected energy savings if a Trombe wall, type C3, is included in the design of the home for Billings, Montana. Assume the heating degree-day value for Billings is adjusted to 5000°F-days due to keeping the indoor air temperature at 70°F and insulating the house to well above minimum standards.

Solution As a first step, the F_{HL} is adjusted. This is a small change for this house, but is done for completeness. The UA value of the wall to be replaced by solar glazing is A/R, where R is the inverse of U and is the thermal resistance, and A is the glazing area.

$$\Delta F_{HL} = \Delta UA = \frac{\Delta A}{R} = \frac{200}{25} = 8$$

Designation	SSF = 0.1	0.2	0.3	0.4	0.5	0.6	0.7	0.8	0.9
TW A1	460	46	19	5	2	–	–	–	–
TW A2	161	52	27	16	10	6	4	2	1
TW A3	130	53	29	18	12	8	5	3	2
TW A4	113	52	30	19	13	9	6	4	2
TW B1	225	46	21	11	6	4	2	–	–
TW B2	132	49	26	16	10	7	4	3	1
TW B3	117	48	27	17	11	7	5	3	2
TW B4	109	45	25	16	10	7	5	3	2
TW C1	152	44	22	13	8	5	3	2	1
TW C2	119	43	23	14	9	6	4	3	1
TW C3	115	41	22	13	9	6	4	3	1
TW C4	123	38	19	12	8	5	3	2	1
TW D1	74	23	8	–	–	–	–	–	–
TW D2	138	63	37	24	16	11	8	5	3
TW D3	141	69	42	27	19	13	9	6	4
TW D4	145	79	50	34	25	18	13	9	6
TW D5	140	79	51	36	26	19	14	10	6
TW E1	168	83	51	34	23	16	12	8	5
TW E2	157	82	51	35	24	18	13	9	6
TW E3	172	101	67	47	34	25	19	13	9
TW E4	156	93	62	44	32	24	18	13	8

Note: Entries designated by a dash are not recommended combinations.

TABLE 12-4   LCR Table for Billings, Montana, and the 21 Passive Solar Heating System Designations Listed in Table 12-3

The F_{HL} value after the change is $225 - 8 = 217$ BTU h^{-1} °F^{-1}, from which the BLC and NRL can be calculated as follows:

$$BLC = 24F_{HL} = (24)(217) = 5208 \text{ BTU HDD}^{-1}$$

$$NRL = (BLC)(HDD) = (5208)(5000) = 26.0\text{E}6 \text{ BTU} = 7.6\text{E}3 \text{ kWh} = 27.5 \text{ GJ}$$

The LCR for the house is calculated as follows, where A_p is the (vertically) projected area of the passive glazing:

$$LCR = \frac{BCL}{A_p} = \frac{5208}{200} = 26.0$$

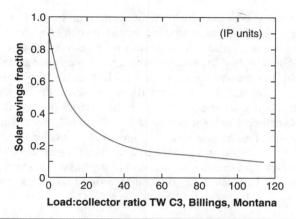

FIGURE 12-8 Solar savings fraction as a function of LCR for passive solar wall type TW C3 in Billings, Montana, used for Example 12-1.

From Table 12-4, LCR and respective SSF values can be interpolated to find an SSF value for a desired wall type, corresponding to an LCR value of 26.5. The relationship is not linear but linear interpolation is likely to suffice between adjacent values in the table. The graph in Fig. 12-8 shows the SSF:LCR data combinations and can also be used for interpolation.

   An LCR value of 26.5 corresponds to an SSF of approximately 0.28, or a 28% savings of conventional heat for the house for a year. A savings of 28% is an energy savings of 0.28×NRL, or 7.4E6 BTU per year. It must be noted that Billings is in an area of cold winters and the Trombe wall of the example had no night insulation. Were the type changed to D4, which would be a similar design but with night insulation, the SSF rises to nearly 0.50. Depending on the source of conventional heat for the house, movable night insulation may be well justified as an installation expense. Note in Table 12-4 that several designs permit SSF values near 0.5 at an LCR value of 26.5, but much higher SSF values for this LCR are not found in the table for this city. In large part, this is because of the high heating load (large HDD value) compared to any reasonable size of passive glazing.

12-10   Conservation Factor Addendum to the LCR Method

Optimizing passive solar heating systems requires a balance between energy conservation and the added construction cost of solar heating. If insulation is expensive, emphasis is put on the passive system. If insulation is inexpensive, the decision is to insulate heavily, so a smaller passive system can be installed.

   A fundamental maxim for solar energy is "Insulate first, then add solar." This maxim is true but the question is, which combination is optimum? In part, this can be answered by understanding the diminishing returns aspect of adding insulation. If a wall begins at a low thermal resistance value of, for example, R-10 (I-P units) and is redesigned for R-20, heat loss is reduced to half, which is an incremental gain of 50%. Increasing to R-30 reduces heat loss to a third, but with an incremental gain of 17% over half. Increasing to R-40 reduces heat loss by a factor of four, but with an incremental gain of 8% over R-30. The marginal cost of insulation increases more or less linearly with increasing R but the incremental benefit shrinks with each additional insulation step. Additionally, as insulation thickness increases, framing dimensions and other construction details will need to be changed, probably in steps, adding to the cost. Depending on the balance between construction costs and energy costs, there is an optimum from an economic viewpoint.

A similar analysis of marginal benefit change can be made to assess the value of the passive solar heating system. A first increment of solar glass helps heat the house for the entire heating season. A next increment may be excessive for some part of the season, with less marginal benefit than the first increment but linearly (or greater) increasing construction cost. A third increment has even less benefit over the heating season, and so on.

The perspective remains the same if one considers the balance as between embodied energy and collected energy, although the balance points may be rather different.

The LCR method has been extended to include a "conservation factor" (CF) used to balance insulation cost and passive solar area cost. The CF is the result of a mathematical balancing of conservation and passive solar and has been reduced to a set of simple equations that are based on the incremental cost of the passive system, the incremental cost of insulation, and a calculated CF specific to individual cities (and climate). Table 12-5

Designation	SSF = 0.1	0.5	0.8
TW A1	1.8	4.7	–
TW A2	1.6	2.9	4.9
TW A3	1.5	2.5	4.0
TW A4	1.5	2.4	3.6
TW B1	1.8	3.7	8.6
TW B2	1.6	2.7	4.3
TW B3	1.6	2.6	3.9
TW B4	1.6	2.6	3.9
TW C1	1.8	3.2	5.2
TW C2	1.7	2.8	4.3
TW C3	1.8	2.9	4.3
TW C4	1.9	3.1	4.6
TW D1	2.2	–	–
TW D2	1.3	2.0	2.9
TW D3	1.2	1.9	2.7
TW D4	1.1	1.6	2.2
TW D5	1.1	1.5	2.1
TW E1	1.1	1.7	2.3
TW E2	1.1	1.6	2.2
TW E3	1.0	1.3	1.8
TW E4	1.0	1.4	1.8

Note: Entries designated by a dash are not recommended combinations.

TABLE 12-5   Conservation Factor Table for Billings, Montana, and the 21 Passive Solar Heating System Designations Listed in Table 12-4 (Balcomb et al. 1984)

contains CF values for Billings, Montana, as an example to parallel the data of Table 12-4. As a general equation to calculate CF based on LCR:SSF data, the following equation can be used, where symbols are as defined previously, and $D = d(SSF)/d(1/LCR)$.

$$CF = \sqrt{24\left[\frac{1}{LCR}+(1-SSF)\right]/D} \qquad (12\text{-}11)$$

Three applications of the CF process are presented here to calculate the optimum wall and ceiling R-value, optimum slab floor perimeter R-value, and optimum heated basement wall R-value.

Walls:

$$R_{opt} = CF\sqrt{\frac{\text{passive system cost, \$ per unit area}}{\text{insulation cost, \$ per unit area } R \text{ unit}}} \qquad (12\text{-}12)$$

Perimeter:

$$R_{opt} = 2.04CF\sqrt{\frac{\text{passive system cost, \$ per unit length}}{\text{insulation cost, \$ per unit length } R \text{ unit}}} - 5 \qquad (12\text{-}13)$$

Heated Basement:

$$R_{opt} = 3.26CF\sqrt{\frac{\text{passive system cost, \$ per unit area}}{\text{insulation cost, \$ per unit perimeter length per } R \text{ unit}}} - 8 \qquad (12\text{-}14)$$

Example 12-2 Consider possible installation of a passive solar heating system of type TW type D4 that will cost \$10 ft^{-2} more than the insulated wall it is to replace. Wall insulation will cost \$0.025 R^{-1} ft^{-2}. Calculate the optimum R-value for a home in Billings, Montana, if the expected solar savings fraction (SSF) is to be 0.50.

Solution By data in Table 12-5, for SSF = 0.5 and type TW C3, CF = 2.9. Using the wall equation above,

$$R_{opt} = 1.6\sqrt{\frac{10}{0.025}} = 32$$

This R-value, in I-P units, describes a wall that contains fiberglass insulation approximately 10-in (25 cm) thick, with typical inside and outside wall sheathing, and siding. A wall insulated this well would be classified as "super insulated" but well within the realm of possibilities. The value is so high because of the cold winters in Billings and the assumption that half the heating need would come from the sun. Solar capture is not greater with this much insulation, but the heat loss is reduced to the point that half of the remaining heating load can be met by the passive solar thermal system.

Note that insulating to this level may suggest a recalculation of the modified heating degree-days, possibly leading to a smaller HDD value and revised calculations for the design. Iterative calculations are often part of optimized design. The SSF assumed in the above example may not be the best. If a building design is assumed, the LCR method can be used to find a first estimate of the SSF. This value can be used to estimate a better wall design using the CF equations and the LCR calculations redone to find a refined value of SSF. With several iterations, the values converge to a consistent set and the calculated best design has been achieved.

12-11   Load:Collector Ratio Method for Design

The analysis procedure in the above example can be extended to find the best combination of passive solar area and building insulation. One possible method is an extension of the method to consider a range of passive solar areas, areas that fit architecturally and reasonably into the south wall of the building. Each building system is analyzed and the net solar benefit is found. The passive area with the greatest net benefit becomes the design optimum.

Example 12-3   This case study has three parts:

a. Determine the optimum area of a vertical, glazed, south-facing, vented Trombe wall. You are considering the system for your solar home to be constructed near Baltimore, Maryland. The candidate Trombe wall you are considering will have the following characteristics:
   - Wall thickness of 12 in and thermal storage capacity of 30 BTU/ft² °F
   - A selective absorption surface
   - Double glazing
   - Movable night insulation
   - $\rho ck = 30$

   You expect the Trombe wall will cost more than the well-insulated wall it will replace by $5/ft² of glazing area. You have also concluded the yearly cost of the wall will be 5% of the installed cost, based on the assumption that the installation cost will be part of the mortgage, the system will last many years, the interest on the mortgage is a tax deduction on both the federal and state returns, and you are in the highest marginal tax bracket.

   The Heat Loss Factor (HLF), were the house to be built without a solar wall, would be 265 BTU h^{-1} °F^{-1}. The walls will be R-25 (IP units). You calculated the insulation benefit and found that you should reduce the heating degree-days to a temperature base of 50°F. Your conventional heat source is natural gas at $2.15/therm (a therm is 100,000 BTU) with annual fuel utilization efficiency (AFUE) of 92%.

b. What is the CF for your design? For your design, installed insulation costs $0.04/ft² per unit increase of R-value, and $5/ft² of solar wall installed cost. What would be the optimum R-value of the insulated walls and ceiling of the house? (Do not redo the Trombe wall optimization using this new value, however.)

c. By how much would the design change if the Trombe wall were triple-glazed, with no night insulation and a normal surface, but costing $9 more per square foot of glazing area? Assume other conditions and properties remain the same.

Solution

(a) As a first step, convert HDD values from base 65°F[3] to a lower base. This house is super insulated; assume the effective HDD base is 50°F. Monthly HDD values are in the following table. Note the large change when the base is lowered by 15°F.

Month	HDD Base 65	HDD Base 50	Month	HDD Base 65	HDD Base 50	Month	HDD Base 65	HDD Base 50
Jan	936	471	May	90	0	Sep	48	0
Feb	820	400	Jun	0	0	Oct	264	0
Mar	679	226	Jul	0	0	Nov	585	167
Apr	327	15	Aug	0	0	Dec	905	440
Sum HDD$_{65}$	4654							
Sum HDD$_{50}$	1719							

---

[3]HDD data source: Leckie et al. (1981).

Unit energy cost is found from the AFUE, fuel heating value, and fuel unit cost:

$$\frac{\$}{MBTU} = \frac{\left(1,000,000\,\frac{BTU}{MBTU}\right)\left(unit\ cost\,\frac{\$}{therm}\right)}{\left(100,000\,\frac{BTU}{therm}\right)(AFUE)} = \frac{(1,000,000)(\$2.15)}{(1,000,000)(0.92)} = \frac{\$23.37}{MBTU}$$

The description of the Trombe wall corresponds to type TW E4 in Table 12-3. The optimum area can be found by repeated analyses, with the glazing area ranging from 50 to 500 ft². Sequential calculations to determine the SSF as in Example 12-1 above are needed, but are not repeated here in detail. Other calculations are:

- Wall cost = (wall area)(cost per unit wall area)(5% yearly amortization)
- BLC, no passive wall = (24)(265) = 6360
- Yearly energy, MBTU = (BLC)(HDD)(1E-6)
- Yearly heating cost (no passive wall) = (Yearly energy using BLC = 6360)($/MBTU fuel cost)
- Yearly heating cost (with passive wall) = (Yearly heating cost, no wall)(1 – SSF)
- Total cost = Wall cost + Yearly heating cost
- $$ saved = Yearly heating cost (no passive wall) – (Yearly heating cost)(with passive wall)
- Net $$ saved = Yearly $$ saved – Yearly wall cost
- The peak value of net $$ saved suggests the optimum solar wall area. The minimum value of total yearly cost suggests the same

The (rounded) results are in the following table:

Wall Area	HLF	BLC	LCR	SSF	Yrly Wall Cost	Yrly Htg Cost	Yrly Total Cost	Yrly Htg $$ Saved	Net Yrly $$ Saved
0	265	6360	–	0	$0	$256	$256	$0	$0
50	263	6312	126.2	0.154	$13	$217	$229	$39	$27
100	261	6264	62.6	0.313	$25	$176	$201	$80	$55
200	257	6168	30.8	0.547	$50	$116	$166	$140	$90
300	253	6072	20.2	0.683	$75	$81	$156	$175	$100
400	249	5976	14.9	0.782	$100	$56	$156	$200	$100
500	245	5880	11.8	0.855	$125	$37	$162	$219	$94

Inspection of the data shows the optimum area to be *ca.* 350 ft² if the criterion is to maximize net savings. For the TW-E4 wall, the optimum SSF is somewhat higher than 0.7. It is noted that the relatively mild winter climate of Baltimore, a movable night curtain, a selective surface on the solar absorber, and a super-insulated house, permit a large area of solar glazing to be optimum. (However, 350 ft² may not fit the house architecturally, e.g., a glass wall 6.5 ft tall by 54 ft long would be required. A compromise is necessary in the architectural design, the degree of super insulation, and/or the type of Trombe wall to install.)

(b) Using Appendix C at the end of this book Table of CF values for Baltimore, an SSF = *ca.* 0.74, and interpolating, CF = 1.7 (rounded). Using Eq. (12-12) from the text,

$$R_{opt} = 1.7(5.00/0.04)^{0.5} = 19 \text{ (I-P units)}$$

This is near the value of 25 that we started with, and may be acceptable without further calculations. (Of course, to be sure, the problem should be re-examined with this new value.) The modest HDD values and moderate winter solar values make the solar wall more effective and greater wall insulation is not needed.

(c) The description of the revised Trombe wall corresponds to type TW D2 in Table 12-3. Repeating the sequential calculations as above, the (rounded) results are:

Wall Area	HLF	BLC	LCR	SSF	Yrly Wall Cost	Yrly Htg Cost	Yrly Total Cost	Yrly Htg $$ Saved	Net Yrly $$ Saved
0	265	6360	–	0	$0	$256	$256	$0	$0
50.00	263	6312	126.2	0.122	$23	$225	$248	$31	$8
100.00	261	6264	62.6	0.213	$45	$201	$246	$55	$10
150.00	259	6216	41.4	0.295	$68	$180	$248	$76	$8
200.00	257	6168	30.8	0.366	$90	$162	$252	$94	$4
250.00	255	6120	24.5	0.419	$113	$149	$262	$107	–$6
300.00	253	6072	20.2	0.456	$135	$139	$274	$117	–$18

For the TW-D2 wall, the optimum passive wall area is much smaller, more costly per unit area, and the optimum SSF is reduced to *ca.* 0.213. From the relevant CF table, this corresponds to CF = 1.5 (rounded) and $R_{opt} = 1.5(9.00/0.04)^{0.5} = 23$, which is close to the original design assumption. Note how the passive heating system is less effective here, making it important to consider increasing insulation in spite of the modest HDD value and decent solar environment. The benefit of installing type TW-E4 wall, using night insulation with a selective surface, reduces night heat losses through the solar glazing, and permits a larger Trombe wall (part A above) to be installed as the optimum size. However, it must be emphasized that assumptions of costs included in this example are only values to illustrate the method and do not necessarily represent actual costs today. However, when real cost estimates exist, continuing this example and exploring other passive solar wall types can lead to selection of the "best" selection of type.

12-12   Passive Ventilation by Thermal Buoyancy

Passive ventilation by thermal buoyancy can be vigorous when there are two or more openings of suitable size in the exterior walls of a building and there is no internal obstruction to air movement through the air space from the inlet to the outlet. An example is sketched in Fig. 12-9. The rate of ventilation can be calculated if indoor and outdoor air

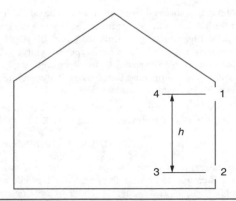

Figure 12-9   Sectional view of building ventilated passively by thermal buoyancy.

temperatures, vent opening areas, and elevation differences between the vents are known. In the simplest situation for calculation, indoor and outdoor air temperatures are assumed constant, there are two vents, and vertical vent dimensions are small relative to the elevation difference between them. Indoor and outdoor absolute air temperatures are uniform and denoted by T_4 and T_2, respectively, and the corresponding air densities are denoted by ρ_4 and ρ_2. Air velocities are V_4 through the upper (outlet) vent and V_2 through the lower (inlet) vent.

The sketch in Fig. 12-9 is conceptual only. Opening an exit vent on the downwind side of a building and an intake vent on the upwind side enhances the ventilation effect. The opposite suppresses airflow and even a slight breeze can block the thermosyphon effect. For this reason, operable vents to take advantage of the direction of evening breezes can be an advantage.

Principles of fluid statics and dynamics can be applied to quantify pressure changes in a loop starting at point 1, passing through points 2, 3, and 4, and returning to point 1. Points 1 and 4 are at the same elevation, as are points 2 and 3. Traversing the loop from 1 to 2 and 3 to 4 is a fluid statics problem related to fluid density and elevation difference. Traversing the loop from 2 to 3 and 4 to 1 is a fluid dynamics problem that can be analyzed using the Bernoulli equation.

$$P_1 - P_2 = -\frac{\rho_2 g h}{g_c}$$

$$P_2 - P_3 = \frac{\rho_2 V_2^2}{2 g_c}$$

$$P_3 - P_4 = \frac{\rho_4 g h}{g_c} \tag{12-15}$$

$$P_4 - P_1 = \frac{\rho_4 V_4^2}{2 g_c}$$

The factor g_c is the force to mass units conversion factor that is equal to 32.2 in the I-P system when mass is in pounds, and 1.0 in the SI system when mass is in kilograms.

The pressure change loop is closed by summing the four pressure changes, which must sum to zero.

$$2gh(\rho_2 - \rho_4)/g_c = \rho_2 V_2^2 + \rho_4 V_4^2 \tag{12-16}$$

The principle of mass flow continuity applies, where c_d is the coefficient of discharge, often taken as 0.65 for openings such as windows.

$$\rho_2 C_{d2} A_2 V_2 = \rho_4 C_{d4} A_4 V_4 \tag{12-17}$$

When the continuity equation is combined with the pressure loop equation, and the perfect gas law is used to relate air temperature and density, one of the airflow velocities can be determined and the continuity equation used to solve for the other, if desired, and for the ventilation rate using either half of the mass flow continuity equation.

$$V_4 = \left[\frac{2gh(T_4 - T_2)/g_c}{T_2 + \left(\dfrac{T_2^2}{T_4}\right)\left(\dfrac{C_{d4}A_4}{C_{d2}A_2}\right)^2} \right]^{0.5} \tag{12-18}$$

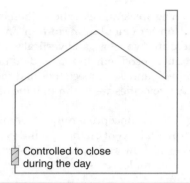

Controlled to close
during the day

FIGURE 12-10 Sectional view of building ventilated passively by a thermosyphon chimney.

The form of the solution equation suggests two important physical insights. First, the vigor of the natural ventilation process is proportional to the square root of the indoor to outdoor air temperature difference. Thus regions that tend to cool more at night make better use of passive ventilation. However, passive ventilation can be effective at even modest temperature differences [e.g., a 10°C difference is better than a 5°C difference by $\sqrt{2}$ (41% better), not double the benefit]. The second observation is that the vigor of the ventilation process is proportional to the square root of the elevation difference between the two vents, leading to the same relative insight as for the temperature difference—doubling the elevation difference increases ventilation by the square root of 2.

If a situation should arise where an elevation difference is insufficient for adequate ventilation, a thermosyphon chimney can be added, as sketched conceptually in Fig. 12-10. If the cross-section area of the chimney is sufficiently large that airflow is unimpeded, the equations above apply. If the chimney adds significant drag to the moving air, the pressure drop due to the drag can be expressed as a function of flow velocity and added to the pressure loop equation. The solution remains a straightforward application of fluid mechanics and reduces, again, to finding one unknown velocity using one equation. (Hint: The mass flow continuity equation permits the velocity in the chimney to be related directly to the velocity through either vent.)

Thermal buoyancy calculations become more complex when there are more than two vents, vents are large relative to the elevation difference between them, wind-induced ventilation is present, or the indoor temperature changes with the ventilation rate. These complexities are addressed in Albright (1990), to which the reader is referred for details. Solar chimneys for ventilation are of frequent interest, which is a situation where the ventilation rate and air temperature difference are closely, but inversely, related. The value of the temperature, T_4, in Eq. (12-18), in this case, is a function of solar heating and ventilation rate, which complicates the details but the applicable equations remain similar and the solution becomes iterative.

12-13   Designing Window Overhangs for Passive Solar Systems

Ideal design of a passive solar space heating system provides full solar exposure during the coldest days of winter and full shade of the solar window during summer. Fortunately, the critical time of day for each is solar noon. The sun is highest in the sky at

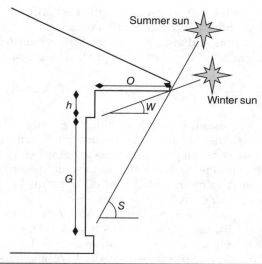

FIGURE 12-11   Sketch of overhang for summer shade and winter sun.

solar noon in winter and lower in the sky for the rest of the day. The sun is farthest south at solar noon from the spring to the fall equinox (north of the Tropic of Cancer). The summer (S) and winter (W) solar angles at solar noon, and relevant dimensions of the solar window installation, are shown in the sketch in Fig. 12-11. The distance the over-hang (or roof) extends beyond the solar wall is denoted by O, the distance down from the overhang to the top of the window is h, and the height of the window is G. At solar noon, the simple equation to calculate solar altitude, α, can be used:

$$\alpha = 90 - \text{latitude} + \text{solar declination} \tag{12-19}$$

Unfortunately, need for passive solar heat is not symmetric around the summer solstice. Passive heating can usually be useful later in the spring than would be acceptable at the symmetric date in the late summer or autumn. April 21 is not the same as August 21. Thus, when designing a fixed shade for a solar window, a compromise is necessary to specify the dates of full shade and full sun. Does the design require full summer shade until the fall equinox to prevent overheating? If so, full shade must start at the spring equinox when weather may still be sufficiently cool that a degree of pas-sive heating would be useful. The same problem arises when deciding the dates for full solar exposure. Full solar exposure through late winter might be preferred but that is accompanied by full solar exposure starting early in the autumn. If a compromise will not be satisfactory, a movable shade system (or awning) will be required.

Once suitable dates for full shading and full solar exposure have been decided and the respective solar declinations and solar noon solar altitudes (S and W) calculated, the distance O can be found if either G or h is known. A value for G is typically known, based on dimensions of available window units. The other two dimensions can be determined from the following equations.

$$O = \frac{G}{[\tan(S) - \tan(W)]} \tag{12-20}$$

$$h = O \tan(W) \tag{12-21}$$

Additional shades may be placed to the east and west of the solar window if the overhang does not extend sufficiently far past the two edges of the window and some sun enters before and after solar noon. Trigonometric calculations and the solar angle equations can be used in the design process to size such shades.

12-14   Summary

Solar energy can be useful even in the middle of winter when insolation intensity is weakest. Whenever there is some direct sunlight, it can be captured using passive means to heat space or even provide domestic hot water. Collection can rely on structural features to provide passive solar heating of space. For greatest effectiveness, energy conservation using insulation and other means becomes a critical matter for successful passive solar energy installations.

Extensive computer analyses, and experimental tests, 30 years ago at the Los Alamos National Laboratory have provided a method, called the load:collector ratio (LCR) on which a designer can choose from an extended list of passive solar designs for a building, and optimize the combination of building and passive solar heating method. Combinations of several passive solar space heating methods installed in a single house operate in parallel and can be evaluated using the same tools. Computer-based analysis procedures have been developed, based on analysis methods, that can be applied to create design processes.

The LCR method can be applied to the process of designing passive solar heating systems for buildings of various types. The systems viewpoint should begin with enforced energy conservation to prevent extravagant use of collected solar energy, and minimize the extent of structural modifications needed to create successful passive solar heating systems. Increased insulation for a building leads to two advantages. First, the heat loss factor for the building is smaller in magnitude. Second, the applicable monthly and seasonal heating degree-day sums are smaller. The required space heat for comfort is the product of the two factors. Thus adding insulation creates a compounded advantage, for each factor is made smaller by adding insulation. After a reasonable degree of conservation has been imposed, the other components of the system must be evaluated. A systems viewpoint, with an iterative design approach, can lead to the optimum combination of conservation and solar heating.

References

Albright, L. D. (1990). *Environment Control for Animals and Plants*. Am. Soc. Agr. and Bio. Engrs, St. Joseph, MI.

ASHRAE (2003). *ASHRAE Handbook: HVAC Applications*. Am. Soc. Htg., Refrig. and Air Cond. Engrs. Atlanta, Georgia.

ASHRAE (2005). *ASHRAE Handbook: Fundamentals*. Am. Soc. Htg., Refrig. and Air Cond. Engrs. Atlanta, Georgia.

Athienitis, A. K., and M. Santamouris (2002). *Thermal Analysis and Design of Passive Solar Buildings*. James & James, London.

Balcomb, J. D., R. W. Jones, R. D. McFarland, et al. (1984). *Passive Solar Heating Analysis: A Design Manual*. Am. Soc. Htg., Refrig. and Air Cond. Engrs. Atlanta, Georgia.

Duffie, J. A., and W.A. Beckman (2006). *Solar Engineering of Thermal Processes,* 3rd ed. Wiley-Interscience, New York.

Goswami, D. Y., F. Kreith, and J. F. Kreider (2000). *Principles of Solar Engineering,* 2nd ed. Taylor and Francis, New York.

Jones, R. W., J. D. Balcomb, C. E. Kosiewicz, et al. (1982). *Passive Solar Design Handbook, Volume Three: Passive Solar Design Analysis.* NTIS, U.S. Dept. of Commerce, Washington, D.C.

Leckie, J., G. Master, H. Whitehouse, and L. Young (1981). *More Other Homes and Garbage: Designs for Self-sufficient Living.* Sierra Club Books. Available online at http://hdl .handle. net/1813/1006 for no charge. Accessed Aug. 2011.

Exercises

12-1. You have recorded air temperatures (°C) for one week and have calculated the following average air temperatures for the week:

Day 1	Day 2	Day 3	Day 4	Day 5	Day 6	Day 7
8	12	10	16	18	21	4

Calculate the number of heating °C-days contributed by this week to the yearly total for temperature base 18 and temperature base 12.

12-2. Explain why, as a house becomes more insulated (or super insulated), the base temperature for calculating heating degree-days is lower for the same indoor air temperature.

12-3. Determine the effect of adding window screens to two windows used for thermal buoyancy ventilation in a house near sea level. Each window is 1 m² and the elevation difference between them is 4 m. Outdoor air temperature is 15°C and indoor air temperature is 24°C. Assume window screens reduce the effective flow areas through windows by 50%. What is the volumetric flow rate in each case, and by how much do the screens reduce the flow rate?

12-4. You are considering adding a Trombe wall solar window to your home. Architectural considerations suggest three possibilities for the window area, 80, 150, and 200 ft² will fit the current south-facing wall. Construction cost increments of the three possibilities are $850, $1200, and $1500, respectively. The heat loss factor (F_{HL}, I-P units) of the house has been calculated as 275 BTU h^{-1} °F^{-1} and the wall into which the solar window can go has an R-value of 20 (I-P units). Your local (adjusted) heating °F-day value is 5000 and your conventional heating energy is expected to cost $42 MBTU^{-1}. If the installation cost is amortized at 10% yearly, which is the most economically beneficial window area if the Trombe wall system you are considering is represented by the following SSF:LCR data?

SSF	0.1	0.2	0.3	0.4	0.5	0.6	0.7	0.8	0.9
LCR	135	64	32	20	13	9	5	2	–

12-5. Choose LCR/CF data for one of the cities as listed in an appendix and use the data that applies to the home described in Problem 12-4 above, and other data as described in Example 12-1. For a Trombe wall of type D2, calculate the recommended R-value for the walls and ceiling.

12-6. Choose three of the passive solar wall types and their corresponding LCR data as listed in the appendix. If the house in Problem 12-4 above is to be constructed in Denver, Colorado, which has a yearly heating °F-day value of 4292, calculate and contrast the passive solar wall areas for a solar savings fraction of 0.1, 0.3, and 0.5.

12-7.   Repeat Problem 12-6 for Edmonton, Alberta, Canada, which has a yearly heating °F-day value of 10,645.

12-8.   Complete Problems 12-6 and 12-7 above and write a descriptive explanation for the differences you found.

12-9.   Start this exercise by finding the monthly heating °F-day values for Cleveland, Ohio (or other city of your choice). By what percentage is the yearly heat need reduced if the HDD base is 50°F?

CHAPTER 13

Wind Energy Systems

13-1   Overview

This chapter introduces the wind turbine technology and the fundamentals of understanding the wind resource on which it is based. Estimation of available wind energy depends on an understanding of the frequency and duration of different wind speeds, which can be measured empirically or modeled using statistical functions. Knowledge of the available wind speeds can be converted to a projection of the average annual output from a specific turbine using a "power curve" for that turbine. The function of the turbine can be modeled with either a simple theoretical model known as an "actuator disc" or a more sophisticated "strip theory" model that takes into account the shape of the wind turbine blades. Both analytic solutions to blade design and approximate solutions that analyze performance at a finite number of points along the blade are presented. Lastly, the economics of both small- and large-scale wind systems are considered, in terms of their ability to repay investment through sales of electricity or displacement of electricity purchases from the grid.

13-2   Introduction

Use of wind energy for human purposes entails the conversion of the kinetic energy that is present intermittently in the wind into mechanical energy, usually in the form of rotation of a shaft. From there, the energy can be applied to mechanical work or further converted to electricity using a generator. As was said of the solar resource in Chap. 9, the amount of energy available in the wind around the planet at any one moment in time is vast. However, much of it is too far from the earth's surface to be accessible using currently available technology, and the energy that is accessible requires investment in a conversion device, such as a windmill or wind turbine. Furthermore, no location is continuously windy, and the power (i.e., rate of energy flow) in the wind is highly variable, requiring provision both for alternative energy supplies during times of little or no wind, and means of protecting the wind energy conversion device from damage in times of extremely high wind.

While the use of wind energy for mechanical applications such as grinding grain dates back many centuries, the use of wind power for the generation of electricity, which is the focus of this chapter (specifically at a utility scale as opposed to residential scale), dates back to the end of the nineteenth century. Early machines were based on windmill designs for water pumping, and produced limited output. Throughout the twentieth century, the design of the small-scale wind turbine evolved toward fewer blades, lighter weight, and faster rotational speeds. Although some attempts were made at utility-scale wind devices in the mid-twentieth century, these efforts did not bear fruit, so that through the 1930s and 1940s, wind turbines were developed on the scale of tens of kilowatts, especially for rural locations, prior to national electrification efforts. The world's first megawatt-scale wind turbine, a 1.25 MW unit erected in 1941 at Castleton, Vermont, USA, failed prematurely due to the lack of sufficiently advanced materials to build a durable device at that scale. As of the 1960s, the only commercially available devices capable of generating electricity were sized for powering households or farms, especially in remote locations that were not easily reached by the electric grid.

The evolution of the modern utility-scale wind turbine began with experimental devices in the 1970s that tested many possible design variables: horizontal or vertical axis; one, two, or three blades; or blades upwind or downwind of the tower. By the early 1980s, turbines began entering the U.S. market for grid electricity, especially in California, and the total installed capacity of these devices in the United States grew from a negligible amount in 1980 to approximately 2500 MW in the year 2000 (Fig. 13-1). Thereafter the U.S. market entered an accelerated growth phase, reaching some 9100 MW by the end of 2005 and 65,900 MW by the end of 2014. Added capacity often varies greatly from year to year due to tax policy and other factors. Therefore, a 5-year "Smoothed" running average is included in the figure as well (2013 and 2014 data points use 4- and 3-year averages, respectively). The smoothed curve suggests that the typical rate has been 6–8 GW added per year in recent years.

World use of wind energy has been growing rapidly as well in recent years. In 2002, there were 32,400 MW of large-scale wind power in use by utilities around the world; this figure reached 59,100 MW by 2005, 195,000 MW by 2010, and 369,600 GW by 2014. Global leaders in terms of share of the 2014 total installed capacity include China (31%),

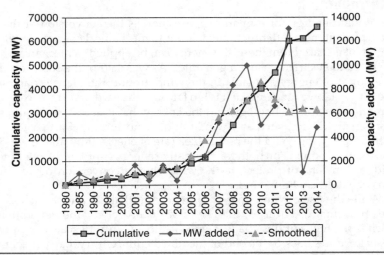

FIGURE 13-1 Growth in annual and cumulative installed capacity of utility-scale wind capacity in USA, 1980–2014.

(*Source:* American Wind Energy Association.)

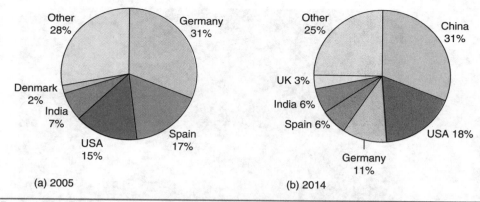

FIGURE 13-2   Total installed capacity: 59.1 GW (2005), 369.6 GW (2014). Small-scale wind turbines are not included, but including these turbines would not noticeably change the percentage values.

(*Source:* Global Wind Energy Consortium.)

United States (18%), Germany (11%), Spain (6%), and India (6%) (see Fig. 13-2). Note that statistics on total installed capacity should be seen in context. In the year 2013, installed capacity of fossil fuel powered plants in the United States alone totaled 870,000 MW.

The success of the large-scale wind turbine emerged from advances involving several engineering disciplines. First, an improved understanding of the fluid dynamics of wind moving past the device enabled improved design of the turbine blade. Second, improved materials allowed for the fabrication of large turbines that were both light and strong enough to perform robustly and efficiently, and also increasing the maximum *swept area* (i.e., circular area formed by the rotational span of the turbine blades) of the device, which helped to reduce the cost per kWh. These larger turbines also rotate more slowly, reducing the risk to birds that tarnished the image of some of the early utility-scale devices. Lastly, improvements in power electronics made it possible for large numbers of turbines, operating intermittently, to transmit electric power smoothly and reliably to the grid, as well as to optimally control total output from an entire wind farm in real time, as exemplified by the Madison wind farm in Fig. 13-3. In this chapter, we will focus on the energy available in the wind resource and the design of the turbine blade, which corresponds to the first point. However, the other two points are significant as well, and the reader is referred to full-length works on wind energy engineering for further reading.[1] Improvements in large-scale wind energy have also helped small-scale turbines (see Fig. 13-4). As the technology improves, wind energy becomes cost-effective not only for the windiest sites, but also for those with less strong winds, as the turbines cost less and produce more electricity and hence more revenue per year, assuming the cost of electricity remains the same.

---

[1]For a full-length work on wind energy systems, the reader is referred to Manwell et al. (2010). Also, the electrical power control dimensions of wind energy, including generator design and integration of wind turbine electrical output with grid ac oscillation frequency, are outside the scope of this chapter. The reader is referred to, for example, Grainger and Stevenson (1994) or Wood and Wollenberg (1996), as well as the "Electrical Aspects" chapter in Manwell et al.

FIGURE 13-3 Madison utility-scale wind farm with seven 1.5-MW turbines near Utica, New York. (*Source:* Chris Milian/photosfromonhigh.com. Reprinted with permission.)

(a)

(b)

FIGURE 13-4 (a) Bergey 50-kW turbine on free-standing tower outside a retail center in Aurora, CO. (b) A 10-kW Weaver Wind Energy Model 5 turbine erected on tower.

[*Source:* For image (a) Bergey Windpower Co.; for image (b) Weaver Wind Energy. Reprinted with permission.]

13-2-1   Components of a Turbine

The horizontal axis wind turbine (HAWT) system consists of blades attached to a central hub to form a rotor that rotates when force is exerted upon them by the wind. The hub is in turn attached to a driveshaft that transmits rotational energy to the interior of the *nacelle*, a central enclosure that sits atop the turbine tower and is rotated by a "yaw mechanism" on a vertical axis to face the wind from any direction. Here "yaw" is defined as the angular orientation of the nacelle and rotor around its vertical axis. The nacelle contains the bearing for the driveshaft, the transmission, generator, mechanical brake, and gears and drives to change both the orientation of the nacelle and the pitch of the turbine blades. Because it is difficult to access the nacelle, controls and monitors are installed inside the base of the tower, when possible. Major components of the turbine system are shown in Fig. 13-5; Figs. 13-6 and 13-7 give a sense of the size of these

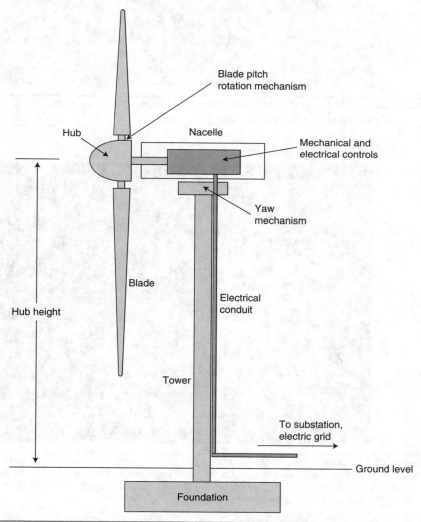

FIGURE 13-5   Main parts of a utility-scale wind turbine. Not to scale.

FIGURE 13-6   Ellie Weyer demonstrates the size of the hub and blade ends.
(*Source:* Ellie Weyer, 2007. Reprinted with permission.)

FIGURE 13-7   Access to the turbine nacelle via doorway and stairwell inside the turbine tower. Additional control equipment is housed inside the bottom of the tower at ground level for ease of access, since the nacelle can only be reached by climbing a ladder inside the tower using climbing equipment, a time-consuming process.

(*Source:* Ellie Weyer, 2007. Reprinted with permission.)

devices. The mechanical and electrical controls inside the nacelle include a rotor brake, a mechanical gearbox, a generator, and electrical controls. The yaw mechanism rotates the nacelle relative to the vertical axis of the tower, using a system of motors and gears, so that the turbine can face into the wind.

Most utility-size designs have now settled on three blades as the optimal number (enough blades to take sufficient energy out of the wind and to function in a stable way in changing wind conditions, but not so many that the turbine would become excessively heavy or costly). In terms of function, the blades and hub of a turbine are similar to an airplane propeller or helicopter rotor operating in reverse, that is, the device takes energy out of the air instead of putting energy into it like those of aircraft. Hence, the process of designing the shape of the turbine blade resembles the design of a propeller.

In the case of a utility-scale wind park with multiple turbines feeding the grid, power is transformed from mechanical to electrical inside the nacelle, and then transmitted via cables down the *tower* of the turbine to a substation, where the voltage is stepped up for long-distance transmission over the grid. A successful wind park project must have access to the grid, so the park must either be built adjacent to a transmission line, or else the developer must secure agreement with local governments, and the like, for the construction of a high-voltage extension to the nearest available line.

13-2-2 Comparison of Onshore and Offshore Wind

Onshore wind comprises any wind installation that is installed on land, as opposed to *offshore wind*, which is installed in bodies of water adjacent to land. Offshore wind is more expensive per MW of installed capacity, due to the complexity of providing an underwater foundation for the tower and turbine, but it has several advantages as well. Since the approach to the construction site for offshore wind is by water, and components are brought in by ship, the maximum size of each individual turbine is larger compared to onshore sites, where road widths and other factors limit the maximum size of components. Maximum size offshore turbines have therefore reached 5 and 6 MW, and may eventually grow to 10 MW. The open water around an offshore wind site can also provide higher average windspeeds, offsetting some of the installation cost disadvantage. The waters off the U.S. Atlantic coast from Virginia to New England, for example, provide a higher average year-round wind energy density than the parallel onshore regions at or near the coast. Lastly, offshore sites close to population centers can offset higher electricity production cost in levelized cents per kWh through lower transmission costs.

As of 2014, all commercially operating offshore wind sites were in Europe and Asia, with no offshore wind farms in North America, although plans were underway for sites in Massachusetts (Cape Wind project) and Delaware (Bluewater Wind project). As of the end of 2012, European countries have installed a total of 5000 MW of offshore wind in the North Sea, Irish Sea, and other bodies of water.

Offshore wind installations currently require shallow depths, on the order of 30 m (100 ft) or less, so that the tower can be sunk directly into the sea floor below the surface. Turbine manufacturers are developing "floating" turbine designs where the tower would rest on a submerged but buoyant platform that is in turn tethered to the sea floor using an anchoring system (albeit at a cost premium beyond the nonfloating offshore technology). These designs would greatly expand the accessible offshore wind resource, especially in North America where resources with water shallower than 30 m are limited. Norwegian company Statoil Hydro inaugurated a test of a 2.3-MW turbine on a floating platform off the Norwegian coast in 2009. Other floating turbines have since been deployed off the coast of Portugal in Europe and the states of Maine and Oregon in the United States.

13-2-3   Alternative Turbine Designs: Horizontal versus Vertical Axis

In the best-known large or small turbine designs currently in use, the blades rotate in a vertical plane around a horizontal axis, so the design is given the name HAWT. However, it is not the only design option available. Wind engineers continue to be interested in the possibility of a turbine operating on a vertical axis, also known as a VAWT (vertical axis wind turbine).

The most prominent airfoil design for the VAWT is the *Darrieus VAWT*, patented by Georges Darrieus in France in 1931. The VAWT has a number of potential advantages over the HAWT, for example, that it receives wind equally well from any direction, and therefore does not need any mechanism to change orientation with changing wind direction. Also, the VAWT avoids the need to put much of the energy conversion equipment in a nacelle atop a tower, as in the HAWT, so it can be more convenient for maintenance. At the present time, there is just a small number of VAWT makers selling equipment in the 100–250 kW range that might compete for utility-scale generation, and almost all of the utility-scale market is made up of large HAWTs, so we do not consider VAWTs further in this chapter. However, the VAWT technology remains an active area of research, and some conceptual and prototype designs are currently emerging that may lower the cost per kWh for this technology in the future, so it is entirely possible that new designs will successfully capture some significant market share from HAWTs as the wind energy industry evolves over the coming decades. In addition to 50–250 kW devices, small-scale VAWTs are commercially available at the present time in rated capacities from 500 W to 20 kW (see Fig. 13-8).

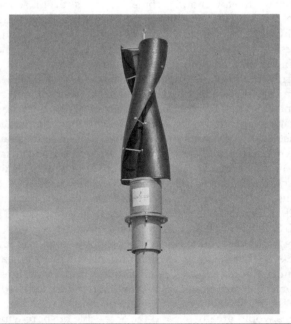

FIGURE 13-8   Small-scale VAWT: 1 kW GUS vertical axis device.
(*Source:* Tangarie Alternative Power. Reprinted with permission.)

13-3 Using Wind Data to Evaluate a Potential Location

The viability of wind energy at a given site depends on having sufficient wind speed available at the height at which you intend to install the turbine. Long-term data gathering of wind speed data at many sites over multiyear periods shows that once the average and variance of the wind speed are known, these values are fairly consistent from year to year, varying no more than 10% in either direction in most locations.

There are two levels of detail at which to measure wind speed. The first approach is to measure the average wind speed with a limited number of readings, or to obtain such a measure from a *statistical wind map* that uses wind data from nearby locations and information about terrain and prevailing winds to estimate the average wind speed. The other approach is to measure the wind continuously at the site throughout the year (or through multiple years), and assign each of the 8760 h in the year to a wind speed "bin" based on the average speed for that hour. Naturally, the latter approach costs more and takes longer to record, but allows the analyst to more accurately predict how well the turbine will perform. In this section, we will start with a data set based on detailed wind measurement and then compare the results with others based on use of the average wind speed for the same site.

The detailed wind data set for a hypothetical site (see Table 13-1) is based on continuous wind data gathering for an entire year. Note that the table is formatted as shown for reasons of brevity and simplicity; in the wind industry, bins are often broken down in 0.5 m/s increments and centered on half or whole m/s average speeds, for example,

Bin	Wind Speed		Hours/Year	Frequency (pct)	Bin Avg. Speed (m/s)
	Min. (m/s)	Max. (m/s)			
1	0	0	80	0.9%	0
2	0	1	204	2.3%	0.5
3	1	2	496	5.7%	1.5
4	2	3	806	9.2%	2.5
5	3	4	1211	13.8%	3.5
6	4	5	1254	14.3%	4.5
7	5	6	1246	14.2%	5.5
8	6	7	1027	11.7%	6.5
9	7	8	709	8.1%	7.5
10	8	9	549	6.3%	8.5
11	9	10	443	5.1%	9.5
12	10	11	328	3.7%	10.5
13	11	12	221	2.5%	11.5
14	12	13	124	1.4%	12.5
15	13	14	60	0.7%	13.5
16	14	No upper bound	2	0.02%	—

TABLE 13-1 Wind Data Distributed by Bins for Hypothetical Site (Average Wind Speed 5.57 m/s, or 12.2 mph)

Class	Average Annual Wind Speed	
	m/s	mph
Marginal	4–5	9–11.3
Fair	5–6	11.3–13.5
Good	6–7	13.5–15.8
Excellent	7–8	15.8–18
Outstanding	Over 8	Over 18

Note: Values shown in this table are representative; for example, for commercial applications related to the development of utility-scale wind facilities, an analyst may seek higher average speeds than those shown in order to classify a site as "excellent" or "outstanding."

TABLE 13-2 Classification of Wind Resource by Wind Speed Range in m/s and mph at Hub Height of Turbine

a bin might be created between 0.75 and 1.25 m/s, with a 1 m/s average speed. For this site, the winds range between 0 and 14 m/s, except for 2 h/year where they exceed this value. The wind speed distribution is typical of a flat area with no obstructions close to the turbine, and would be characterized as "fair" in terms of its potential for wind power development (see Table 13-2). In other words, a private entity such as a household or small business might be able to develop the resource, since they would then avoid paying grid charges to bring electricity in, but the average wind speed is not high enough to compete with the best sites for *wind farms* (i.e., facilities with multiple large turbines, sometimes also called *wind parks*). For comparison, the site used by the utilities at Fenner, New York, near Syracuse, New York, in the United States, averages 7.7 m/s (17 mph) year round at the turbine hub height of 65 m. The data are divided up into "bins," where the number of hours per year that the wind speed is in the given bin is shown in the table (e.g., for bin 1, 80 h/year of no wind, and the like). The hours per year can be translated into a percentage by dividing the number of hours in the bin by 8760 h total per year. The last column gives the average wind speed for the bin. Taking a weighted average of the bin average wind speed gives the following:

$$U_{\text{average}} = \sum_{i=1}^{n} p_i \cdot U_{i,\text{average}} \tag{13-1}$$

Here U_{average} = average speed for the site, p_i = percentage of year that wind speed is in bin i, and $U_{i,\text{average}}$ = average speed for the bin i. The calculation gives U_{average} = 5.57 m/s for the year, as shown. Note that the 2 h for bin 16 are not included in the average speed calculation since the bin average speed is unknown.

13-3-1 Using Statistical Distributions to Approximate Available Energy

The distribution of wind speeds in Table 13-1 provides a solid basis for calculating available energy at this site. However, suppose we only knew U_{average} = 5.57 m/s, and did not have the hourly bin data for the year. Observations have shown that in many locations, if the average wind speed is known, the probability of the wind speed being in a given range can be predicted using the *Weibull* or *Rayleigh* distributions from probability and statistics, the Rayleigh being a special case of the Weibull.[2]

The probability density function (PDF) for the Weibull distribution has the following form:

$$f(x) = \frac{k}{\sigma}\left(\frac{x}{\sigma}\right)^{k-1} \exp[-(x/\sigma)^k], \text{ for } x \geq 0$$

$$f(x) = 0, \text{ for } x < 0$$

(13-2)

where k and σ are shape and scale parameters, respectively, and x is the independent variable for which the PDF is to be evaluated. Integrating from 0 to x gives the cumulative distribution function (CDF) as follows:

$$F(x) = 1 - \exp[-(x/\sigma)^k], \text{ for } x \geq 0$$

(13-3)

If we set the shape factor to $k = 2$, we arrive at the particular instance of the Weibull called the Rayleigh function, which has the following PDF:

$$f(x) = 0, \text{ for } x < 0$$

$$f(x) = 2x \cdot \frac{\exp[-(x/\sigma)^2]}{\sigma^2}, \text{ for } x \geq 0$$

(13-4)

The CDF for the Rayleigh is then

$$F(x) = 0, \text{ for } x < 0$$

$$F(x) = 1 - \exp[-(x/\sigma)^2], \text{ for } x \geq 0$$

(13-5)

For analysis of the distribution of wind speed, it is convenient to rewrite the CDF in Eq. (13-5) in terms of a given wind speed U and the average wind speed $U_{average}$. From probability and statistics it is known that the expected value of a Rayleigh function is $\bar{x} = \sigma \cdot (\pi/4)^{1/2}$. Substituting $U_{average}$ for $\bar{x}$ and rearranging to find the scale factor in terms of known values gives:

$$\sigma = U_{average} \cdot (4/\pi)^{1/2}$$

(13-6)

---

[2] A complete review of statistical distributions is beyond the scope of this book. The reader is referred to the wide variety of available books on the subject, e.g., Devore (2009) or Mendenhall et al. (1990).

Substituting Eq. (13-6) into Eq. (13-5) and simplifying gives:

$$F(U) = 1 - \exp\left[-\left(\frac{U}{U_{average}(4/\pi)^{1/2}}\right)^2\right]$$

$$= 1 - \exp\left[-\left(\frac{1}{(4/\pi)}\right)\left(\frac{U}{U_{average}}\right)^2\right] \tag{13-7}$$

$$= 1 - \exp\left[(-\pi/4)\left(\frac{U}{U_{average}}\right)^2\right]$$

The CDF is used to calculate the probability that the wind speed will be at or below a given value U, given a known value of $U_{average}$:

$$p(wind\ speed \le U) = 1 - \exp[(-\pi/4)(U/U_{average})^2] \tag{13-8}$$

So, for example, substituting the value $U = 4$ m/s into Eq. (13-4), the probability with $U_{average} = 5.57$ m/s that the wind speed is less than or equal to this value is 32%. The application of the Rayleigh distribution to working with wind speed bins is illustrated in Example 13-1.

Example 13-1   Using the Rayleigh distribution and $U_{average} = 5.57$ m/s, calculate the probability that the wind is in bin 6.

Solution   The probability that the wind is in the bin is the difference between the probability of wind at the maximum value for the bin, and the probability of the minimum value. Bin 6 is between 4 and 5 m/s. From Eq. (13-8) above

$$P(wind\ speed \le 4) = 1 - \exp[(-\pi/4)(4/5.57)^2] = 0.333 = 33.3\%$$

$$P(wind\ speed \le 5) = 1 - \exp[(-\pi/4)(5/5.57)^2] = 0.469 = 46.9\%$$

Therefore, the bin probability is 46.9% − 33.3% = 13.6%. This is the value obtained using the Rayleigh distribution; the observed value is 1254 h/8760 h/year = 14.3%.

To show how well the estimate using the Rayleigh distribution fits the above data, we can plot them next to each other, as shown in Fig. 13-9. From visual inspection, the Rayleigh estimated curve fits the observed data fairly well. Note that the estimating technique must be used with caution: although it happened to fit this data set well, there is no guarantee that for another location, the fit might not be quite poor, for example, in a case where the observed distribution has more than one peak.

In order to understand the impact of using an estimated versus observed distribution for wind speed, we will calculate the estimate of energy available in the wind using

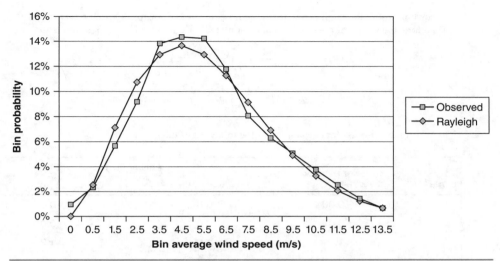

FIGURE 13-9 Comparison of observed and Rayleigh estimated probabilities of wind speeds in a given bin for wind speeds up to 14 m/s. *Conversion:* meters/second × 2.25 = miles/hour.

both methods. For a given wind speed U and swept area of a turbine A, the power P in watts available in the wind is calculated as follows:

$$P = 0.5\rho U^3 A \qquad (13\text{-}9)$$

Here ρ is the density of air in kg/m³, which often has values on the order of 1 kg/m³, depending on elevation above sea level and current weather conditions. From Eq. (13-9), the amount of energy available in the wind grows with the cube of the wind speed, so there is much more energy available in the wind at high wind speeds. For example, per unit of swept area, at $U_{average}$ and $\rho = 1.15$ kg/m³, the available power density is 99 W/m², but at the upper range of the bins at 14 m/s, the power density is 1578 W/m². Variation in air density plays a significant role as well: values as low as 0.9 and as high as 1.3 kg/m³ are observed, depending on the site. With air density at the low end of this range, P_i is reduced by 31% compared to the high value, all other things being equal.

Based on the actual bin data, we can calculate the annual energy available at the site by calculating the power available in kW in each bin, based on the bin average speed, and multiplying by the number of hours per year to obtain energy in kWh. Example 13-2 tests the accuracy of the Rayleigh function compared to the observed data.

Example 13-2 Suppose a wind analyst calculates the wind energy available at the site given in Table 13-1, knowing only $U_{average}$ for the site and using a Rayleigh function to calculate probabilities for the given wind bins. By what percent will the estimated energy differ from the value obtained if the bin data are known? Use an air density value of $\rho = 1.15$ kg/m³.

Solution Using Eq. (13-9) to calculate the power available in each bin and the percentage of the year in each bin to calculate the number of hours, it is possible to generate a table of observed and estimated energy values by bin. We will use bin 6, for which we calculated statistical probability in Example 13-1, as an example. The average speed in this bin is 4.5 m/s; therefore, the average power available is

$$P = (0.5)(1.15)(4.5)^3 = 52.4 \text{ W/m}^2$$

According to the Rayleigh estimate, the wind is in the bin for (0.136)(8760 h/year) = 1191 h/year. Therefore, the values of the observed and estimated output in the bin are, respectively

$$E_{observed} = \left(1254\frac{h}{year}\right)\left(52.4\frac{W}{m^2}\right) = 65.7 \text{ kWh/m}^2$$

$$E_{estimated} = \left(1191\frac{h}{year}\right)\left(52.4\frac{W}{m^2}\right) = 62.4 \text{ kWh/m}^2$$

Repeating this process for each bin gives the following results:

Bin	Power Density (W/m²)	Annual Output	
		Observed (kWh/m²)	Estimated (kWh/m²)
1	0.00	0.0	0.0
2	0.07	0.0	0.0
3	1.94	1.0	1.2
4	8.98	7.2	8.5
5	24.65	29.9	27.9
6	52.40	65.7	62.4
7	95.67	119.2	108.2
8	157.91	162.2	155.9
9	242.58	172.0	194.2
10	353.12	193.9	214.1
11	492.99	218.4	212.2
12	665.63	218.3	191.2
13	874.50	193.3	158.0
14	1123.05	139.3	120.5
15	1414.72	84.9	85.1
Total		1605.1	1539.3

From the observed bin data, this value is 1605 kWh/m², while for the Rayleigh estimate the value is 1539 kWh/m², excluding any wind above bin 15. So the Rayleigh underestimates the energy available at the site by a factor of 4.1%.

Example 13-2 shows that the Rayleigh approximation gives an energy estimate quite close to the observed value, in this case; for example, the difference is less than the year-to-year variability value of ±10% discussed earlier. Alternatively, it is possible to make a rough approximation of the available power by using an *adjusted* value of the average speed that takes into account the effect of cubing the wind speed in the power equation. For any group of N numbers $n_1, n_2, \ldots, n_N$, it can be shown that the relationship between the average of the cube and the cube of the averages is the following:

$$\lim_{N \to \infty} \left[\sum_{i=1}^{N}(x_i^3)/N\right]\Big/\left[\left(\sum_{i=1}^{N}x_i/N\right)^3\right] = 6/\pi \tag{13-10}$$

On the basis of Eq. (13-10), it is possible to adjust the average wind speed by multi-plying by $(6/\pi)^{1/3}$ and using the resulting value in the power equation to quickly estimate available power. For example, for the data from Table 13-1, the adjusted speed is 6.91 m/s, and the energy available at this wind speed if it is supplied continuously year round is 1662 kWh/m², or 3.6% more than the observed value. On the other hand, if we simply use the unadjusted wind speed of 5.57 m/s for all 8760 h of the year, the estimated energy available would be only 870 kWh/m². This comparison shows the important contribution of winds at higher than the average wind speed to overall wind energy output.

13-3-2 Effects of Height, Season, Time of Day, and Direction on Wind Speed

To complete the discussion of available energy in the wind, we must consider several factors, starting with height above the ground. The effect of height above the ground can be approximated by the following equation:

$$U(z) = U(z_r)\left(\frac{z}{z_r}\right)^{\alpha} \tag{13-11}$$

where z is the height above the ground, z_r is some reference height above the ground for which wind speed is known, $U(z)$ is the wind speed as a function of height, and α is the wind shear coefficient. A typical value of the wind shear in flat locations might be $\alpha = 0.2$. Thus as height above ground is increased, wind speed increases rapidly, and then less rapidly above a certain height. Example 13-3 illustrates the influence of height on wind speed.

Example 13-3 Suppose that the wind data shown in Table 13-1 was gathered at 30 m elevation. A wind analyst evaluates the data and surmises that if the turbine height is increased, the site may be viable for a commercial wind venture. The analyst believes that an average wind speed of 7.5 m/s will make the site commercially viable. At what height will this wind speed be obtained, if the wind shear value of $\alpha = 0.2$ is used?

Solution From above, the average wind speed at the measured height is 5.57 m/s. Rearranging Eq. (13-11) to solve for the necessary height to achieve a speed of 7.5 m/s gives

$$z = z_r\left[\frac{U(z)}{U(z_r)}\right]^{1/\alpha}$$

Substituting all appropriate values and solving for z gives $z = 133$ m, to the nearest meter. Thus a significant increase in hub height is needed to achieve the desired wind speed.

Figure 13-10 shows the relative values of growth in wind speed and power in the wind with increasing height, based on Eq. (13-11). Observe that the wind speed increases rapidly from a height of 0 to 10 m, and then more slowly with additional height, while the power increases more rapidly between 10 and 70 m since it is a function of the cube power of the wind speed.

The choice of tower height plays a role in the cost-effectiveness of a wind project. To use a wind industry saying, "installing a turbine on a tower that is too low is like install-ing solar panels in a shady location." In general, increasing tower height improves the economics of a project up to some point of diminishing returns, after which the cost of additional height outweighs the benefit in terms of added output. From Fig. 13-10, power output as a function of increasing height begins to taper off, whereas, beyond a

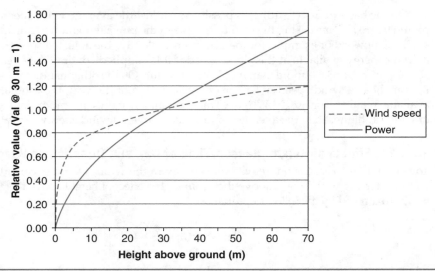

FIGURE 13-10   Relative value of wind speed and power as a function of height above the ground, indexed to value at 30 m = 1.00.

At 30 m, in this example, wind speed is 8.86 m/s, and power is 396 W/m².

point, tower construction cost begins to increase exponentially due to both structural engineering and installation complexities of erecting a very tall tower. It is therefore necessary to optimize the chosen height, weighing out both sides of this trade-off.

Season of Year and Time of Day

The effect of season of the year on average wind speed varies from location to location. Many, though not all, sites experience increased winds in the winter and decreased winds in the summer. Figure 13-11 shows average hourly wind speed by month for a proposed wind farm site in the Town of Enfield, New York. The average monthly values varied between 5 m/s in August and 7.5 m/s in December. The data are gathered using an anemometer on a meteorological tower at a height of 58 m above ground; although this is not as high as the typical turbine nacelle height of 80 m, average wind speeds at 80 m would be higher but the pattern would be the same. Similarly, a site near Toronto, Canada, averages 5 m/s in December but only 3.5 m/s in June.[3]

The effect of time of day is more ambiguous than that of season, with some sites observing higher average speeds at night but others having higher speeds during the day, possibly due to local heating of the air by the sun, leading to an upturn in daytime wind speeds. Again using the Enfield, New York, site as an example, the average value using the hourly average for all 365 days of the year results in the variation shown in Fig. 13-12 between 5.6 m/s at noon and 6.6 m/s at 1 a.m. Sites such as this can play a complementary role with investments in solar energy, which has peak output during the middle of the day and in the summer, just when the wind resource is at its lowest point.

---

[3]A complete review of statistical distributions is beyond the scope of this book. The reader is referred to the wide variety of available books on the subject, e.g., Devore (2009) or Mendenhall et al. (1990).

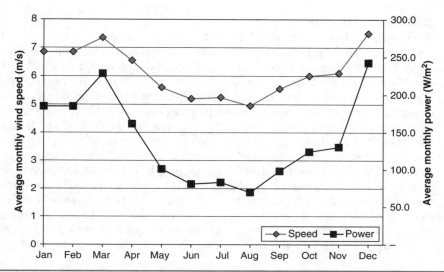

Figure 13-11 Seasonal distribution of wind speed in m/s and average monthly power in W/m^2 by month of year at proposed Enfield, New York, wind farm site (height = 58 m; U_{avge} = 6.15 m/s; P_{avge} = 140.8 W/m^2)[4].

Note: Power calculation assumes air density of 1.15 kg/m^3.

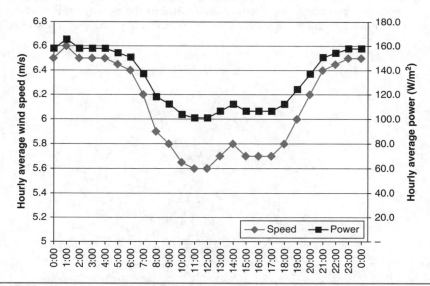

Figure 13-12 Hourly distribution of wind speed and power per unit of area for example 365 days of year at proposed Enfield, New York, wind farm site (height = 58 m; U_{avge} = 6.11 m/s; P_{avge} = 132.3 W/m^2).

Note: The *y*-axis scale does not start at 0 m/s, to make variation more discernable. Power estimation assumes air density 1.15 kg/m^3.

[4]Acknowledgment: Thanks to Jun Wan, Kim Campbell, Nicole Gumbs, Sandeep George, Tim Komsa, Tyler Coatney, Christina Hoerig, Karl Smolenski, and Happiness Munedzimwe for their input into the underlying project on which the Enfield wind distribution figures are based.

Effect of Wind Direction

Along with height, season, and time of day, compass direction has an impact on average wind speed. Wind direction affects the orientation to which the turbine will most likely be facing when it is operating, and also the layout of individual turbines in a wind farm, since the downwind wake of one turbine can affect the operation of another. Wind direction data are presented graphically using a *wind rose*, in this case illustrated using an example from a feasibility study for a proposed 1.5-MW wind turbine to provide electricity for Ithaca College in Ithaca, New York (see Fig. 13-13). The various wedges of the wind rose represent the compass directions (e.g., top represents north, bottom represents south, and so on), and the size of the wedge corresponds to the prevalence of the given direction. The wind rose may be divided into 12 wedges as in this case, or 16 wedges in other cases (N, NNE, NE, etc.). Thus for the Ithaca College example, the most commonly observed directions are S and approximately NNE, each occurring 15% of the hours of the year. It is also possible to divide the individual wedges into segments corresponding to the percent of hours at a given direction and average wind-speed range (e.g., 0–5 m/s, 5–10 m/s, etc.) although this was not done in Fig. 13-13.

Using Regional Long-Term Wind Data to Adjust 1-year Average Values

In cases where the wind analyst has gathered a 1-year data set as was done for the Enfield site in Figs. 13-11 and 13-12, the question arises as to whether the chosen year is representative of the long-term average for the site. The analyst can compare and adjust 1-year values by comparing them to other sites in the nearby region where wind data are continuously gathered, such as meteorological stations ("met stations") at one or

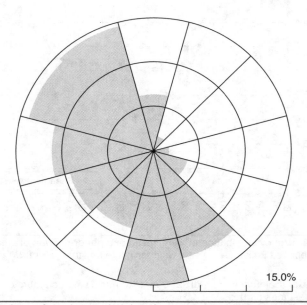

15.0%

Figure 13-13   Wind rose from 1-year wind data for proposed wind turbine site at Ithaca College, Ithaca, New York. (*Source:* Courtesy of Ithaca College, reprinted with permission. Thanks to Prof. Beth Clark, Dept. of Physics, Ithaca College for assistance with information about the Ithaca College wind project.)

more of the airports in the surrounding regions. For example, in the case of Enfield, nearby airports with met stations include those of Ithaca, Binghamton, and Elmira, New York. Variability in annual average wind speed is roughly consistent between these met stations, that is, if a given calendar year had an average wind speed at one met station X% above or below the long-term average, the percentage will be approximately the same at the other two met stations. By this logic, the same observation should hold for the site at which the wind data are gathered, and we can apply the same derived adjustment factor to the average wind speed data at the site as we would to the airport meteorological stations. Example 13-4 illustrates the long-term adjustment technique.

Example 13-4   Suppose the annual average value for the Enfield site is 6.1 m/s. For the same 12-month period, the average at the nearby Ithaca-Tompkins Regional Airport is 5.12 m/s. The 30-year average for the airport up to and including the chosen 12-month period is 5.26 m/s. What is the long-term adjusted average wind speed at the Enfield site?

Solution   The long-term adjustment factor is 5.26/5.12 = 1.027. Therefore, the long-term adjusted Enfield average is $6.1 \times 1.027 = 6.27$ m/s.

To summarize this section, we have thus far considered the effect of variability in wind speed on available energy in the wind, and shown that the calculation of available energy is not as simple as taking the power at the average wind speed and multiplying by the number of hours in a year. To convert this information into the energy output from the turbine, we turn to the performance of the turbine at different wind speeds in the next section.

13-4   Estimating Output from a Specific Turbine for a Proposed Site

Output from a wind turbine as a function of wind speed is represented by the *power curve* of the device, which gives output over the operating range of the turbine. Figure 13-14 gives the power curve for air density of 1.15 kg/m^3 of two hypothetical turbines, one with swept area radius of 34.5 m and a maximum output of 1.5 MW and another with radius 50 m and output 1.7 MW, along with percentage of energy extracted from the wind at the given wind speed, also referred to as the coefficient of performance C_p. For the turbines shown, efficiency at first increases with wind speed. However, once the turbine reaches the "knee" of the power curve where the output begins to taper toward the rated maximum, efficiency begins to fall as the turbine captures less of the power available in the wind. After the turbine reaches its maximum output, its value falls further since the device output is fixed but power available continues to increase. Above 21 m/s wind, the device shown stops generating output due to excessive force in the wind activating the cut-out mechanism.

It is the responsibility of the turbine manufacturer to measure the power curve for each device produced and to provide this information as a basis for estimating output from a turbine for a site with known wind speed distribution. Turbine performance is measured both empirically on the basis of numerous instantaneous readings of output as a function of estimated wind speed and theoretically by simulating turbine response to wind conditions in a computer model. However, since it is not possible to measure exactly the average wind speed passing through the entire swept area of the turbine at any one moment, manufacturers use the collection of output readings mixed with modeled turbine output based on design parameters to create a best-fit power curve that predicts output for each wind speed at a constant air density value. Therefore, some amount of

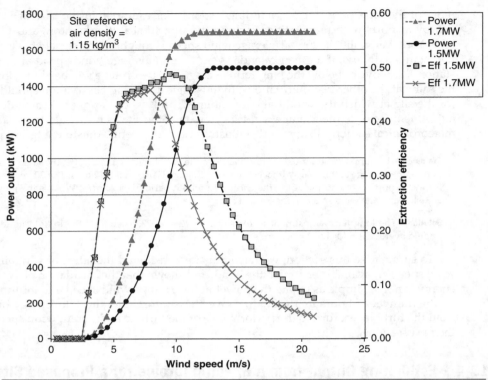

FIGURE 13-14   Power curve for 1.5-MW and 1.7-MW turbine for wind speeds from 0 to 21 m/s, with extraction efficiency (power output divided by power available in the wind).

imprecision is introduced during this process. When purchasing large turbines for a utility-scale wind facility, the turbine maker and facility owner agree on a range of acceptable annual output values for the site, and the owner accepts that actual performance may be somewhat lower than the value predicted using the power curve. In return, the maker guarantees that the output will not fall below some agreed minimum threshold.

Turbine manufacturers offer different sizes of turbines with different power curves, so that they can market devices that fit well with different shapes of wind distributions. For example, small-scale turbines are typically installed in locations with lower average wind speeds than utility-scale devices in part due to tower height, so it is beneficial for a small-scale turbine to have a steeper power curve in the lower wind speed range (e.g., from 4 to 10 m/s wind).

The difference between power curves for different turbine designs complicates the comparison of performance between competing turbines. In terms of the cost-effectiveness, the engineer cannot simply divide turbine cost by maximum output in kW or MW to calculate a measure of cost per installed kW or MW, and choose the device with the lowest unit cost. Instead, since the range of annual wind energy available as a function of wind speed varies from site to site, it is necessary to evaluate competing devices using power curves and available wind data specific to the site, as demonstrated next using the power curve from Fig. 13-14. Note that although curves for two turbines are given, we use the 1.5-MW power curve throughout.

Once the power output P_i in kW is known for each wind speed bin i, the total gross annual energy output from the device AEO in kWh can be estimated by multiplying by the predicted number of hours in the wind speed range of each bin and then summing across all bins:

$$AEO = \sum_i P_i \cdot t_i \qquad (13\text{-}12)$$

This method is approximate, since the average wind speed of the bin is used to calculate output per hour for that bin, whereas instantaneous wind speed within the bin may vary by ± 0.25 to 0.5 m/s, depending on the increment chosen. Suppose we apply the power curve in Fig. 13-14 to an empirically measured set of annual hours in each bin from a hypothetical site, as shown in Table 13-3. Here the wind speed increment is 1 m/s per bin; however, an increment of 0.5 m/s may be preferable so as to more accurately predict device output, as discussed earlier. The wind speed distribution given in the table has a higher annual average than that of Table 13-1, and is well suited for the 1.5 MW device evaluated, since output above 1 MW is predicted for over 2700 h/year. For 2 h/year, winds are strong enough that the device stops generating electricity so as to avoid damage to any part of the wind conversion system, while for 780 h/year the wind speeds are too low and the turbine does not rotate. Also note that the large number of hours for bin 12 is because this bin includes a wider range of wind speeds where output is constant at 1.5 MW, from 13 to 21 m/s in this case. The evaluation assumes an average air density of 1.15 kg/m³ for the site; for sites with higher or lower values of air density, the power curve should be adjusted to reflect higher or lower values of power in the wind at each

Bin	Minimum (m/s)	Maximum (m/s)	Bin Avg. (m/s)	Frequency (hours)	Power (kW)	Energy (1000 kWh)
1	0	3	n/a	780	0	0
2	3	4	3.5	537	14	8
3	4	5	4.5	672	60	40
4	5	6	5.5	807	155	125
5	6	7	6.5	836	269	225
6	7	8	7.5	833	420	350
7	8	9	8.5	831	625	519
8	9	10	9.5	685	900	616
9	10	11	10.5	579	1195	692
10	11	12	11.5	473	1395	659
11	12	13	12.5	366	1485	544
12	13	21	17	1359	1500	2039
13	21	And above	n/a	2	0	0
			Total	8760		5816

TABLE 13-3 Calculation of Annual Output for Turbine with Power Curve from Fig. 13-14 and Empirically Measured Distribution at a Hypothetical Site with Average Wind Speed $U_{average} = 8.4$ m/s

wind speed setting. Note that the value of AEO is output prior to downstream losses and before taking out parasitic loads required to operate the wind facility.

Based on the 1.5-MW capacity of the device and the AEO value of 5.8 GWh predicted for the site, the turbine would likely be considered very suitable for the location. This procedure can be repeated for other turbine/site combinations to evaluate feasibility of large or small wind projects, as long as the power curve and either empirical or estimated wind distribution are known and are reasonably accurate.

The power curve for the 1.7-MW turbine in Fig. 13-14 indicates the possible role for a turbine design with a relatively large swept area and relatively low nameplate capacity at sites with relatively low average wind speed. The 1.7-MW gives a 13% increase in nameplate capacity and a 111% increase in swept area; it has 216 kW of capacity for each 1000 m² of swept area, versus 402 kW per 1000 m² for the smaller turbine. As the two turbines begin to turn with increasing wind speed, the output from the 1.7-MW turbine rises more rapidly toward the rated output. On the other hand, this device incurs extra cost for longer turbine blades and a higher tower. Therefore, when choosing a turbine, the project manager must consider that the capacity factor will be higher with 1.7-MW turbine, but the average efficiency will be lower because over much of the range of wind speeds individual efficiency values are lower.

13-4-1   Rated Capacity and Capacity Factor

In the introduction it was mentioned that the worldwide installed capacity of wind energy reached approximately 370 GW at the end of 2014. This figure is calculated using the maximum output of the turbines installed, also known as the *rated capacity* or *nameplate capacity*. In no location does the wind blow at or above the rated wind speed at all times, however, so the actual output from wind installations is substantially lower than what would be generated at the rated wind speed continually. For example, the United States had 65.9 GW of capacity at the end of 2014. This amount of capacity could generate (65.9 GW) × (8760 h/year) = 57.8×10^5 GWh/year, or approximately 578 TWh per year of gross electrical output (for comparison, U.S. annual output from wind was on the order of 182 TWh in 2014).

The *capacity factor* for a wind turbine is the actual annual output divided by the annual output that would be obtained if the devices functioned at rated capacity for the entire year. For the device installed in the location represented in Table 13-3, the total energy produced at 1.5 MW of output for an entire year is 13.1 GWh, so the capacity factor is

$$\text{Capacity Factor} = \frac{5.8 \text{ GWh}}{13.1 \text{ GWh}} = 0.444 = 44.4\%$$

For the world wind energy example, if the capacity factor were on average 25% (which is closer to the world average than the 44% value above), all the wind turbines around the world would produce 809 TWh per year.

Capacity factors for wind turbines typically were in the range of 20–35% in earlier years; with new low-wind turbines being added to the mix, industry-wide averages could be expected to increase. However, comparisons between wind installations to other alternatives should consider productivity and not just nameplate capacity. For example, returning to the example of fossil-fuel-powered plants in the United States, the 870 GW of capacity produced 2744 TWh of electricity in 2013, for a capacity factor of 36%. The 2014 U.S. wind figures above result in a capacity factor of 33% (i.e., 182 TWh/578 TWh, where the latter value is based on the mid-2014 estimated capacity of 63.5 GW, rather than the year-end value of 65.9 GW). Thus fossil-fuel-powered plants on average produce a larger electric output per GW of installed capacity than do wind plants. Furthermore, fossil plants are not constrained by the availability of the wind and their capacity factor is largely determined by demand.

The calculation of capacity factors can be applied to three other purposes for analyzing wind energy. The first is to compare output both at different wind farm locations and across the months of the year. Figure 13-15 takes the available monthly power at the Enfield, NY site (see Fig. 13-11) and compares predicted monthly output at that site to output at two western U.S. wind farms in Colorado and Wyoming. Average annual capacity factor ranges between 30.3% (Enfield) and 48.2% (Wyoming). However, there is substantial variation depending on the month of the year, with all three sites peaking in the winter months and declining in the summer months.

A second purpose is to compare capacity factors from year to year, as shown for a wind farm at Lake Benton, Minnesota (Fig. 13-16). For this site, the National Renewable Energy Laboratory has published a 10-year series of normalized annual production, where the values range from 82% to 114% of the 10-year average. The long-run average capacity factor was not available from the data source, so an average capacity factor of 33% based on prevailing conditions in the upper Midwest has been applied to the relative production percentages, resulting in the time series shown. The site had somewhat higher variability than typical wind farms, as there were 3 years out of 11 where the variation away from the long-run average was greater than 10% (2001, 2002, and 2004).

A third purpose is to evaluate the overall productivity of the wind industry as a whole, in this case focusing on the situation in the United States. Since the American Wind Energy Association (AWEA) publishes time series data on installed capacity, and the U.S. Department of Energy provides annual output, it is possible to create a national

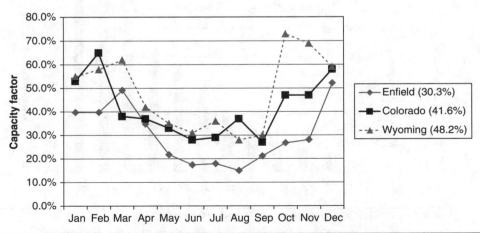

Figure 13-15 Variation in monthly capacity factor for three U.S. wind farms.

Source: National Renewable Energy Laboratory, for Colorado and Wyoming data.

Note: Percentage in parentheses in legend is the overall annual average capacity factor for the site.

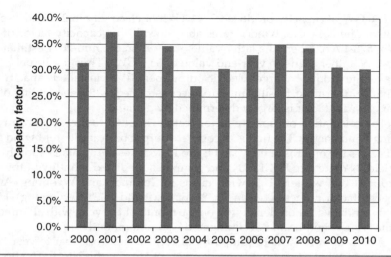

FIGURE 13-16   Example of variation in capacity factor by year for wind farm at Lake Benton, MN, 2000–2010.

Note: For 11-year data set, $\mu = 0.330$, $\sigma = 0.0326$; see text for explanation.

Source: Wan (2012).

average capacity factor (Fig. 13-17). For each year, the mid-year average capacity, or average of the year-end values of the two consecutive year-end values, is used to calculate the output that would have been observed at 100% capacity factor. Then the actual capacity factor is calculated from annual output for the given year published by USDOE.

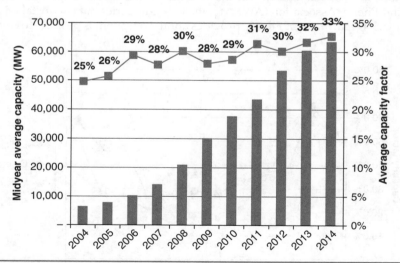

FIGURE 13-17   U.S. average wind energy capacity factor, 2004–2014.

Sources: American Wind Energy Association, for capacity; USDOE, for wind output.

Note: Midyear average capacity is the average of the two surrounding year-end values. Example: Capacity on 12/31/03 is 6327 MW and on 12/31/04 is 6719 MW. Therefore, 2004 average is 6523 MW, as shown.

As shown, average capacity factor rose from 25% in 2004 to 33% in 2014, influenced both by the introduction of "low-wind" turbines that have a higher capacity factor in low wind conditions, and improved turbine management leading to less downtime.

13-5 Turbine Design

The main functional goal of the turbine design is to convert as much wind energy as possible into mechanical energy of the rotating drive shaft, over the operating range of wind speeds of the turbine. Rotational speed of the shaft changes with changing wind speed, so the power electronics of the generator is used to match electrical current leaving the turbine with the correct frequency, phase, and voltage demanded by the electrical grid. The operating range consists of several regimes (which are illustrated graphically by the device in Fig. 13-14):

- Below the *cut-in speed* of the wind (3 m/s in the example turbine), there is insufficient power in the wind to turn the blades, and the device does not operate.
- In the range of wind speeds from the cut-in speed upward, turbine output increases rapidly toward its maximum value.
- Since power increases with cube of wind speed, at some wind speed above the cut-in speed (9.5 m/s in the example 1.5-MW turbine), it becomes difficult to extract the maximum available power from the wind, due to the large force exerted on the blade. From this point upward, as wind speed continues to increase, the fraction of the available energy extracted by the turbine begins to decline. Output continues to grow with increasing wind speed, but at a slower rate. This limiting of output is done for engineering reasons, as cost of fabricating a turbine that would be sufficiently strong to convert all of the wind force into rotational energy in the shaft would be prohibitive. Also, in many locations wind speed rarely reaches the very highest values, so it would be of little commercial value to be able to capture a large fraction of the available energy. Reducing load on the turbine blades at high wind speeds can be accomplished in one of two ways:
  - In *fixed-pitch* devices, where the position of the blade in the hub is fixed in one position, the wind passing over the turbine blade enters a *stall* regime, where the stalling of the blade slows its rotation and reduces the power extracted from the airstream. (The *pitch* of the blade is the angular orientation of the blade around its own axis, relative to the plane of rotation of the entire turbine. See Sec. 13-5-3 for a more detailed description.) These devices are also known as "stall-regulated."
  - In *variable-pitch* devices, the blades can rotate in the hub so that they extract less power from the wind as it passes. While smaller devices currently available in the market include a mix of fixed- and variable-pitch devices, utility-scale devices currently being installed are all variable pitch.
- At some wind speed that is usually above that of maximum extraction efficiency, the device reaches its *rated wind speed*, or wind speed where the device first puts out its maximum output (13 m/s in the 1.5-MW example). Above the rated

wind speed, output may either hold approximately constant or decline with increasing wind speed, with pitch-regulated turbines typically able to achieve the former, and stall-regulated turbines subject to the latter.

- In order to protect from damage at the highest wind speeds, many devices have a *cut-out speed* of the wind (21 m/s in the example), above which the turbine drive shaft rotation is stopped by some combination of changing the yaw or pitch so that the blades are no longer catching wind, or through dynamic braking of the generator. Thereafter, a mechanical brake is applied to prevent the device from rotating.

The device must also be capable of a controlled stop in case of failures of the electric grid, other emergencies, or other routine events such as scheduled maintenance. Rotation of the nacelle around the vertical axis to face the wind is accomplished either by the force of the wind itself in smaller devices, or by active mechanical rotation in utility-scale devices.

13-5-1   Theoretical Limits on Turbine Performance

One important observation about kinetic energy in wind is that it is not physically possible to design a device that extracts all energy from the wind, for if one did, the physical mass of air being moved by the wind would stop in the catchment area downstream from the device, preventing upstream air from reaching the device and acting on it. (This observation is analogous to the Carnot limit on heat engines, which states that it is not possible to achieve 100% efficiency in a practical heat engine because the exhaust gases are always expelled to the cold reservoir at a temperature above 0 K.) It is of interest to quantify this limit on energy that can be extracted from the air in a theoretical model of an optimal wind device, to be used as a benchmark against which to compare empirically measured turbine performance in actual devices.

Theoretical Evaluation of Translating Devices

This concept can be explored by first considering the case of a simple translating wind-catching device, which moves in the same direction as the wind. Examples of such devices include a sailing vessel "running" with the wind (i.e., being pushed from behind), an iceboat being pushed across an icy surface, or vertical-axis grain-grinding devices used in antiquity. In the case of the sailboat or iceboat, the force acting to overcome friction is a function of the drag coefficient of the device against the surface over which it is translating, and is a function of the difference between the wind speed and the speed at which the device is traveling, for example, per unit of area of the surface struck by the wind:

$$F = [0.5\rho(U-v)^2]C_D \qquad (13\text{-}13)$$

where F is the force in N/m², v is velocity at which the device is traveling in m/s, and C_D is the dimensionless drag coefficient of the device against the supporting surface.

The power P extracted from the wind per unit area, expressed in W/m², can then be written as

$$P = [0.5\rho(U-v)^2]C_D v \qquad (13\text{-}14)$$

Since the power available in the wind is equal to 0.5 ρU^3, we can now write the *power coefficient* C_p as the ratio between the power extracted by the device and the power available in the wind:

$$C_P = \frac{[0.5\rho(U-v)^2]C_D v}{0.5\rho U^3} = \frac{[(U-v)^2 C_D v]}{U^3} \qquad (13\text{-}15)$$

Using differential calculus, it is possible to show that the maximum value of C_p that can be achieved by such a device is $4/27C_D$ or $0.148C_D$ (the solution is left as an exercise at the end of the chapter).

This low value for the translating device is not a fair measure of the value of a sailing device used for transportation, since the passenger is concerned with travel between points and is not concerned with how much power the device is extracting from the wind. For devices intended to do stationary work, such as a mechanical windmill or wind turbine, however, it is clear that the translating device will not be effective for converting wind kinetic energy into mechanical or electrical energy. We therefore turn our attention to rotating devices.

Rotating Devices and the Betz Limit

In this section, we will again solve for the maximum power extraction possible, relative to the power available in free wind with speed U, this time for a device that rotates in a plane perpendicular to the direction of the wind. An "actuator disk" model of the turbine is used for the purpose of theoretical modeling. The actuator disk can be thought of as a rotating disk with an infinite number of blades that converts the translation of air in the wind into rotation of the disk. An actuator disk is therefore like a Carnot heat engine, in that it is a device that cannot be built in practice but is useful for setting the upper limits on efficiency.

As shown in Fig. 13-18, airflow through the actuator disk starts with speed upstream from the disk equal to the free wind speed, that is, $U_1 = U$. As the kinetic energy is removed from the wind at the disk, the speed is reduced to $U_2 = U_3$, that is, the airflow speed is the same just upstream and just downstream of the disk. Finally, the wind speed is further reduced to the "far wake" speed, U_4.

For the actuator disk, the following assumptions apply

1. Incompressible fluid flow

2. No frictional drag on the device

3. Uniform thrust over the entire disk area

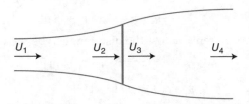

FIGURE 13-18 Diagram of airstream flow through the actuator disk.

4. Nonrotating wake
5. Steady-state operation, that is, constant wind speed velocity at points along the flow through the disk, constant disk rotational speed

The ratio of reduction in airflow speed at turbine to free wind speed $(U_1 - U_2)/U_1$ is called the axial induction factor a. The reduction in airflow speed from the free stream speed U_1 to the far wake speed U_4 is the result of the wind exerting thrust on the actuator disk, that is,

$$T = M_{air}(U_1 - U_4) = \rho A U_2 (U_1 - U_4) \tag{13-16}$$

Here M_{air} is the mass of air flowing through the disk of area A in units of kg/s.

Another fundamental starting point for analysis of the actuator disk is Bernoulli's law, which states that the sum of energy measured in terms of pressure, velocity, and elevation is conserved between any two points along a fluid stream:

$$p_1 + \frac{\rho V_1^2}{2} + \rho g z_1 = p_2 + \frac{\rho V_2^2}{2} + \rho g z_2 \tag{13-17}$$

where p is pressure, V is airflow velocity, g is the gravitational constant, and z is height.

It can be shown using the relationship between thrust and airstream velocity, and the relationship between velocity and pressure (Bernoulli's law), that the reduction in wind speed relative to U in the far wake is always twice the amount at the plane of the disk (this is left as an end-of-chapter exercise). Thus by rearranging terms the following relationships hold

$$\frac{U_1 - U_4}{U_1} = 2\left(\frac{U_1 - U_2}{U_1}\right)$$

$$a = 1 - U_2/U_1$$

$$U_2 = U_1(1 - a) \tag{13-18}$$

$$U_4 = U_1(1 - 2a)$$

In the remainder of this section, we will use Bernoulli's law and the function of the actuator disk to calculate the value of a at which the maximum power is extracted from the wind, and the resulting maximum value. Assuming that change in height is negligible, we can write equations for the airflow on the upside and downside of the disc, as follows:

$$p_1 + \tfrac{1}{2}\,\rho U_1^2 = p_2 + \tfrac{1}{2}\,\rho U_2^2$$

$$p_3 + \tfrac{1}{2}\,\rho U_3^2 = p_4 + \tfrac{1}{2}\,\rho U_4^2 \tag{13-19}$$

The pressure difference across the disk can then be rewritten as

$$p_2 - p_3 = p_1 + \tfrac{1}{2}\,\rho U_1^2 - \tfrac{1}{2}\,\rho U_2^2 + \tfrac{1}{2}\,\rho U_3^2 - p_4 - \tfrac{1}{2}\,\rho U_4^2$$

$$= p_1 - p_4 + \tfrac{1}{2}\,\rho(U_1^2 - U_4^2) \tag{13-20}$$

$$= \tfrac{1}{2}\,\rho(U_1^2 - U_4^2)$$

since $U_2 = U_3$ and the pressure in the far wake and the free wind are both equal to the ambient pressure. The result of Eq. (13-20) can be used to write an equation for the thrust T in units of N acting on the disk in terms of U_1 and U_4, as follows:

$$T = A_2(p_2 - p_3)$$
$$= 0.5\rho A_2(U_1^2 - U_4^2) \tag{13-21}$$

where A_2 is the area of the disk. Next, since power P is the product of thrust and wind velocity at the disk, we can rewrite the power in the device as follows:

$$P = (T)\, U_2 = \tfrac{1}{2}\, \rho A_2\, (U_1^2 - U_4^2)\, U_2 = \tfrac{1}{2}\, \rho A_2\, U_2\, (U_1 + U_4)\, (U_1 - U_4) \tag{13-22}$$

Substituting for U_2 and U_4 in terms of U_1 and a, and replacing U_1 with the free stream velocity U gives:

$$P = \tfrac{1}{2}\, \rho A_2\, U(1 - a)[U + U(1 - 2a)][U - U(1 - 2a)] = \tfrac{1}{2}\, \rho A_2\, U^3\, 4a\, (1 - a)^2 \tag{13-23}$$

The power coefficient C_p of the disk is then

$$C_P = \frac{0.5\rho A_2 U^3 4a(1-a)^2}{0.5\rho A_2 U^3} = 4a(1-a)^2 = 4a^3 - 8a^2 + 4a \tag{13-24}$$

To find the maximum value of Eq. (13-24), we differentiate with respect to a and set equal to zero:

$$\frac{dC_p}{da} = 12a^2 - 16a + 4$$

$$12a^2 - 16a + 4 = 0$$

This expression simplifies to

$$3a^2 - 4a + 1 = 0$$

The roots of the equation are then $a = 1/3$ and $a = 1$. The maximum value of $C_P = 0.593$ occurs when $a = \tfrac{1}{3}$. Thus the maximum theoretically possible power extraction of 59.3% occurs when the wind velocity has been reduced by $\tfrac{1}{3}$ at the disk and hence by $\tfrac{2}{3}$ at the far wake. This result is attributed to Betz and is given the name *Betz limit*.

The actuator disk model can be made more accurate by adding realism to the underlying assumptions, which has the effect of lowering the maximum energy theoretically available from the wind below the value of 59.3% derived above. For example, in a more realistic model of turbine rotation, the wake from the turbine rotates in the opposite direction from the rotation of the rotor. Since the energy transferred to the rotating wake is not available for transfer to the turbine, wake rotation will tend to reduce the maximum value of C_P.

Maximum Thrust and Maximum Efficiency

The maximum value of thrust as a function of axial induction is derived in a similar way. The value of the thrust can be rewritten in terms of a as

$$T = 2\rho A_2 a U^2 (1-a) \qquad (13\text{-}25)$$

The maximum value occurs when $a = 0.5$, so that wind speed in the far wake is reduced to zero. The *efficiency* η_{disk} of the actuator disk is defined as the power extracted divided by the power available in the airstream moving at the turbine. The power available at the turbine is thus a function of the mass flow rate $\rho A_2 U_2$, rather than the free wind mass flow rate $\rho A_2 U$. The value of efficiency is

$$\eta_{\text{disk}} = [\tfrac{1}{2}\,\rho A_2\,(U_1^2 - U_4^2)\,U_2]\, /\, \tfrac{1}{2}\,\rho A_2\,U_1^2\,U_2 = 4a\,(1-a) \qquad (13\text{-}26)$$

Thus the value of η_{disk}, like the thrust, reaches a maximum value at $a = 0.5$, when the far wake wind speed is 0. The maximum value of η_{disk} is 1. At $a = 0.5$, $C_P = 0.5$, while at $a = \tfrac{1}{3}$, the values of T/T_{max} and η_{disk} are both 0.888.

13-5-2   Tip Speed Ratio, Induced Radial Wind Speed, and Optimal Turbine Rotation Speed

In this section, we move from discussion of the actuator disk with infinite number of blades to a practical wind turbine with finite number of airfoil-shaped blades (typically two or three). The turbine responds to wind by rotating at an angular velocity Ω, measured in rad/s, that is determined by the airfoil design, load on the turbine's generator, and other factors.[5]

Tip Speed Ratio, Advance Ratio, and Effect of Changing Tip Speed Ratio on Power Extracted by Blade

The ratio of the translational velocity of the tip of the turbine blade to the wind speed is called the *tip speed ratio* (TSR), and is calculated as follows:

$$\lambda = \Omega R / U \qquad (13\text{-}27)$$

where λ is the TSR, R is the radius of the swept area of the turbine, and U is the free wind speed. A related measure to λ is the *advance ratio*, denoted J, which is the inverse of the tip speed ratio, or $J = 1/\lambda$. The effect of the force of the wind on a cross section of the airfoil of the turbine blade varies with λ. This effect is examined in Fig. 13-19.

The rotation of the rotor about its axis induces a wind speed v in the opposite direction of the direction of rotation. Therefore, the blade encounters a relative wind velocity V_r, which is the vector sum of U and v. The *angle of attack*, α, is the angle between V_r and the center line of the airfoil cross section.

A real-world airfoil has nonzero lift coefficient C_L and drag coefficient C_D; the ratio of C_L to C_D is called the *lift-to-drag* ratio. (In some calculations, the inverse *drag-to-lift*

---

[5]Conversion: 2π radians = 1 revolution = 360°.

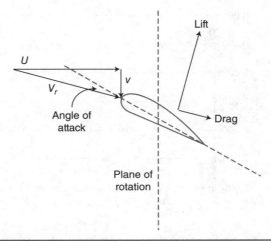

Figure 13-19   Winds and forces acting on a cross section of a turbine blade, showing angle of attack α between midline of cross section and relative wind velocity V_r

Note that in this example the blade is rotating up the page.

ratio C_D/C_L is used instead.) A high lift-to-drag ratio is desirable for efficiently converting energy in the free wind into motion of the blade. Using the geometric relationship between the directions of lift, drag, and the direction of motion of the blade, power output from the blade can be written in terms of C_L, C_D, and the ratio of induced to free wind speed v/U as follows:

$$P = 0.5\rho U^3 A \ (v/U)[C_L - C_D(v/U)][1 + (v/U)^2]^{1/2} \qquad (13\text{-}28)$$

where A is the cross-sectional area of the blade facing the direction of the wind. The power coefficient C_P of the blade is the ratio of the power delivered to the rotation of the blade cross section, divided by the power available in the free wind passing through the projected area of the cross section.[6] Solving for C_P, the ratio simplifies to

$$C_P = (v/U)[C_L - C_D(v/U)][1 + (v/U)^2]^{1/2} \qquad (13\text{-}29)$$

In Fig. 13-20, $C_P/C_{P\text{-}max}$ is plotted as a function of the ratio of v/U to $(v/U)_{max}$, where $(v/U)_{max}$ is the value where C_P is reduced to zero due to the effect of the drag term in Eq. (13-29). As v/U increases from 0, at first C_L dominates the value of Eq. (13-29), so that C_P increases. Beyond a maximum point for C_P at $v/U = 2/3(v/U)_{max}$, however, the effect of drag dominates and C_P decreases at higher values of v/U.

Note that the power coefficient calculated in Eq. (13-29) is a comparison of the force acting locally on a cross section of the blade, and not the force acting on the entire swept area of the turbine. Therefore, values greater than 1 are possible, because induced values of v much greater than U are possible. The entire device will not, however, violate the Betz limit, and the overall value of C_P for an entire device with nonzero C_D will always be less than 0.593.

---

[6]Note that C_P in Eq. (13-29) is different from the power coefficient introduced in Eq. (13-24).

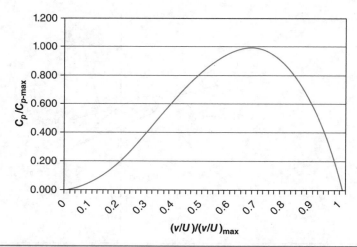

Figure 13-20 Relative value of power coefficient $C_p/C_{P\text{-max}}$ as a function of ratio of v/U to $(v/U)_{\text{max}}$, using Eq. (13-29).

Rotor Power Coefficient

The *rotor power coefficient* is the power coefficient of a physical rotor with finite number of blades, and is denoted $C_{P,r}$ as opposed to C_p defined for the actuator disk in Eq. (13-24). The rotor power coefficient is illustrated in this section using a hypothetical turbine with 10 m radius, $C_L = 1.0$, and $C_D = 0.1$. $C_{P,r}$ is the ratio of output P from the turbine to power in the wind, and is a function of λ:

$$C_{P,r}(\lambda) = \frac{P(\lambda)}{0.5\rho U^3 A_{\text{turbine}}} \tag{13-30}$$

where A_{turbine} is the swept area of the turbine. Let us assume a constant wind speed and vary the turbine rotation speed and hence λ. Here we will plot a curve for $C_{P,r}(\lambda)$ based on a sample of empirically observed data points for $P(\lambda)$. For wind speed $U = 7$ m/s, the following performance data points are given in Table 13-4. The four points can be used to characterize the curve of $C_{P,r}$ versus λ as shown in Fig. 13-21.

Since λ is proportional to v, as λ increases and U remains fixed, the ratio v/U will also increase. Therefore, in keeping with Eq. (13-29), as rotation speed increases from a minimum of 2.8 rad/s at point 1, the value of $C_{P,r}$ at first increases due to the increased lift exerted by the wind, then reaches a maximum at point 3 (7.9 rad/s), then declines as the force of drag comes to dominate.

Relationship between Ideal Turbine Operation and Tip Speed Ratio

As discussed above, for a given fixed wind speed, there is a TSR value that maximizes output, on either side of which output declines as TSR moves away from this ideal range. Thus for each combination of wind speed and output in Fig. 13-14, there is a corresponding single TSR value since there is a single turbine rotational speed associated with that point on the power curve. Looking at the entire range of wind speeds to which

Point	Angular velocity (rad/s)	TSR	Power (kW)	$C_{P,r}$
1	2.8	4.0	4.0	0.064
2	4.2	6.0	11.5	0.185
3	5.6	8.0	20.0	0.323
4	7.0	10.0	26.4	0.426
5*	7.9	11.3	27.9	0.450
6	8.4	12.0	27.4	0.442
7	9.8	14.0	19.6	0.317

Note: *Maximum rotor power coefficient of 0.450 achieved at $\Omega = 7.9$ rad/s, TSR = 11.3.

TABLE 13-4   Turbine Performance as a Function of λ for Representative Turbine with 10-m Blade Radius and Fixed Wind Speed $U = 7.0$ m/s

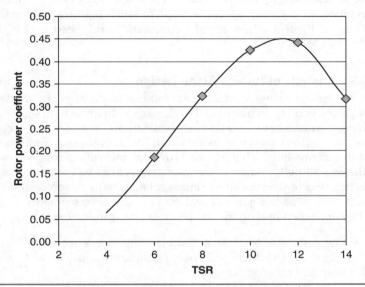

FIGURE 13-21   Rotor power coefficient $C_{P,r}$ as function of tip speed ratio λ.

the turbine is exposed, the range can be divided into two, based on the relationship between wind speed and TSR, as follows:

1. *Maximum efficiency range:* As wind speed increases from the cut-in speed, ideal turbine operation seeks to keep TSR in an ideal range to maximize output. For a utility-size turbine, TSR values between $\lambda = 6$ and $\lambda = 8$ are typical. Note that due to continuous slight variations in exact wind speed and direction, it is not possible to maintain a precise value for TSR; however, as long as the value is kept within a narrow range, the effect on conversion efficiency is acceptable.

2. *Maximum output range:* As wind speed continues to increase, at some point before the device reaches the rated wind speed, the goal for ideal turbine operation shifts from maximizing efficiency to maintaining the rated output of the device without excessive physical stress on the various components, at the expense of lowering TSR. Thus wind speed continues to increase but the tip speed (product of swept area radius and angular velocity) increases only modestly compared to the highest values achieved in the maximum efficiency range, and approaches an asymptotic value.

To illustrate the two regimes, we can revisit the representative 1.5-MW turbine in Fig. 13-14, with a radius of 34.5 m. The maximum efficiency range occurs between wind speeds of approximately 4 and 10 m/s, which for ideal λ values of 6 or 8 gives tip speeds of between 24/32 and 60/80 m/s. Thereafter, the regime shifts to the maximum output range, where output stabilizes at the rated 1.5 MW and efficiency declines eventually to values below 10%. Although tip speed at and above the rated wind speed cannot be deduced from Fig. 13-14, it remains in the range of 80–100 m/s, with TSR falling correspondingly. For example, at the cut-out speed of 21 m/s, if $\lambda = 8$ were maintained, tip speed would reach 176 m/s, or nearly 400 mph. Because it is not practical to engineer a turbine blade for such high translational speeds, a lower TSR is maintained: at a tip speed of 100 m/s and wind speed of 21 m/s, $\lambda = 4.8$ and $\Omega = 2.9 \text{ s}^{-1}$ (28 RPM).

13-5-3　Analysis of Turbine Blade Design

Up to this point we have considered the rudimentary design of a wind extraction system and set some bounds on its optimal performance. In order to design the actual blades of a turbine, it is necessary to model them explicitly and consider how their shape will affect the ability to extract energy from the wind. For this purpose, we introduce an approach called *strip theory*, in which the blade is divided into infinitesimal elements called strips, and the overall performance of the blade is taken to be the sum of strip performance. Thereafter, we consider an approximate solution to the blade design problem in which the blade is divided into a finite number of equally sized segments.

The application of strip theory depends on the following assumptions:

1. The performance of the individual strip in a "stream tube" of air (i.e., annular ring through which air is passing uniformly) can be analyzed independently from any other stream tube.

2. Spanwise flow along the length of the turbine blade is negligible, so that the two-dimensional cross section of the blade appearing in the strip can be used as a basis for analysis without loss of accuracy.

3. Uniformity of flow conditions around the circumference of the stream tube of air and uniformity of blade geometry within the strip or element.

The reader should understand the limitations of these assumptions. For example, in practice, flow conditions do vary for a turbine from top to bottom of its swept area due to changing flow conditions that vary with height from the ground, especially for large devices where the span of the swept area may be 50 m or more. Also, the use of strip theory introduces two new aspects of turbine behavior that were ignored previously.

First, the possibility of *wake rotation*, which was ignored in the derivation of the Betz limit, is considered, in the form of an *induced tangential velocity v*. Second, the blade can be thought of as a cantilever beam extending out from the turbine hub, so that under the force of the wind, the blade will deflect in a downwind direction from its plane of rotation.

In order to evaluate a turbine blade design, we need to know lift and drag coefficients for the blade. These coefficients are evaluated empirically and published for standard airfoil shapes. For example, starting in the 1930s, the National Advisory Council on Aeronautics (NACA) in the United States carried out testing to develop "families" of airfoil shapes denoted by the term NACA followed by a number (e.g., NACA 0012, NACA 63(2)-215, and so on). Other organizations besides NACA have also developed standard airfoil shapes. Regardless of the source, choice of a standard shape saves time for the engineer since he or she does not need to carry out empirical or theoretical work to derive drag and lift coefficients for an original airfoil shape.

Continuous Solution to Blade Element Design Problem Using Strip Theory

The purpose of this section is to develop a methodology for calculating available power based on airflow and turbine blade shape. The diagram in Fig. 13-22 introduces several new variables that are required for the methodology. The figure shows a strip of width dr at a location r along the length of the blade from 0 to R, deflecting by a *coning angle θ* in response to free wind U. In this example, the wind is traveling perpendicular to the plane of rotation of the blade, that is, along the x-axis of the system. In practice, there will often be a *yaw error $\Delta\Psi$*, not shown in Fig. 13-22, which is introduced due to the turbine not remaining perfectly aligned with the wind at all times. To take account of the coning angle, the distance along the axis of the blade from the strip to the axis of rotation is given the value s such that $r = s\cos\theta$.

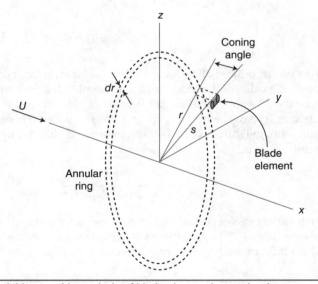

FIGURE 13-22 Variables used in analysis of blade element in annular ring.

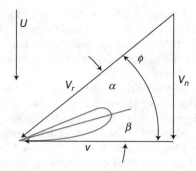

FIGURE 13-23   Relationship between angle of attack α, pitch β, and wind angle ϕ.

Figure 13-23 shows the relationship between the axial and radial winds, V_n and v. The relative wind V_r seen by the blade is the vector sum of V_n and v. The angle between V_r and the centerline of the cross section of the blade is called the *angle of attack* α. The angle between the plane of rotation and the centerline is called the *blade pitch* β. The sum of α and β is the *wind angle* ϕ, that is,

$$\phi = \alpha + \beta \qquad (13\text{-}31)$$

We now introduce the *radial induction factor* a', which is the amount by which v exceeds the angular velocity at the point of analysis r along the blade, that is, $r\Omega$. The axial and radial winds can be written in terms of free wind speed U, rotation speed Ω, r, and a', as follows:

$$V_n = U(1-a)\cos\theta \qquad (13\text{-}32)$$

$$v = r\Omega\,(1+a') = s\cos\theta\,\Omega\,(1+a') \qquad (13\text{-}33)$$

Having thus introduced all necessary variables for strip theory, we develop a method for estimating the value of the local thrust coefficient C_t for the strip that is being considered. This analysis assumes that the local value of axial induction a is known or can be fixed in advance. The case where a is not known is considered hereafter.

To begin with, the thrust coefficient is the thrust dT divided by the available force in the streamtube, that is,

$$C_t = \frac{dT}{0.5(\rho)(U^2)(2\pi r\,dr)} \qquad (13\text{-}34)$$

where $2\pi r\,dr$ is the cross-sectional area of the stream tube.

The magnitude of dT is then the force acting on an area the width of the chord of the airfoil c, acting in the lift and drag directions, multiplied by the number of blades B, that is,

$$dT = 0.5\,\rho V_r^2\,Bc\,(C_L\cos\phi + C_D\sin\phi)\,dr \qquad (13\text{-}35)$$

C_t then simplifies to

$$C_t = (B/2\pi)(c/r)(V_r/U)^2(C_L \cos\phi + C_D \sin\phi) \tag{13-36}$$

From the definition of relative wind, V_r can be rewritten as

$$V_r = V_n/\sin\phi = U(1-a)\cos\theta/\sin\phi \tag{13-37}$$

Substituting for V_r in Eq. (13-36) gives

$$C_t = (B2\pi)(c/r)(1-a)^2(\cos^2\theta/\sin^2\phi)(C_L\cos\phi + C_D\sin\phi) \tag{13-38}$$

In order to evaluate ϕ, it is first necessary to evaluate the induced radial velocity a'. From the angular relationship between axial and lateral wind components, two equations for ϕ can be written in terms of a and a', as follows:

$$\tan\phi = U(1-a)\cos\theta/[r\,\Omega\,(1+a')] \tag{13-39a}$$

$$\tan\phi = a'\,r\,\Omega/(a\cos\theta\,U) \tag{13-39b}$$

Define x to be the *local speed ratio*, or $x = r\,\Omega/U$. Note that the tip speed ratio introduced above is in fact a special case of the local speed ratio where $r = R$. Substituting x for $r\,\Omega/U$ and setting Eqs. (13-39a) and (13-39b) equal to each other gives

$$a(1-a)\cos^2\theta = a'(1+a')x^2 \tag{13-40}$$

Now we can evaluate a' in terms of a, and ϕ from a and a'. We therefore have sufficient information to evaluate local C_t. The overall thrust coefficient for the turbine can be evaluated by integrating over the span of the blade from the root (point where blade joins the turbine hub) to the tip, that is,

$$T = \int_0^R dT \tag{13-41}$$

In practice, C_t is adjusted by a *tip-loss factor* in regions close to the outer end of the blade. The tip loss factor takes into account the losses incurred by the interaction of the airstream near the tip of the blade with air outside the swept area of the device. This factor has the strongest effect at points along the radius at within 3% of the length of the blade from the tip. Quantitative evaluation of tip loss is beyond the scope of this chapter, and the reader is referred to other works on wind energy engineering in the references.

Example 13-5 applies Eqs. (13-39) to (13-41) to the calculation of C_t at a point along a blade.

Example 13-5 A turbine has three blades of 10 m length, and rotates at 6 rad/s in a 7 m/s wind. At a point 7.5 m from the axis of rotation $a = 0.333$, $C_L = 1$, $C_D = 0.1$, and chord ratio = 0.125. Assume that coning is negligible. Calculate C_t at that point.

Solution First calculate the local speed ratio, $x = r\,\Omega/U = 7.5(6)/7 = 6.43$. Since there is no coning, $\theta = 0$. Use Eq. (13-40) to solve for a':

$$(0.333)(1-0.333)\cos^2(0) = a'(1+a')(6.43)^2$$

Substituting known values reduces this equation to $a'^2 + a' - 0.00537$, for which the only feasible solution is $a' = 0.00535$. Substituting into Eq. (13-39a) gives

$$\tan\phi = a'r\,\Omega/a\cos\theta = (0.00535)(7.5)(6)/(0.333 \cdot 7) = 0.1032$$

Therefore, $\phi = 5.89°$. Using ϕ in Eq. (13-37) gives the following value for C_t:

$$C_t = (B2\,\pi)(c/r)(1-a)^2(\cos^2\theta/\sin^2\phi)(C_L\cos\phi + C_D\sin\phi)$$

$$= (3 \times 2\pi)(0.125/7.5)(1-0.333)^2[1/\sin^2(5.89)][(1)\cos(5.89) + 0.1\sin(5.89)] = 0.338$$

Ultimately, it is of interest to calculate the overall power coefficient C_P for the entire blade. By integrating across the range of possible values for x from 0 to λ, it is possible to calculate C_P using the following:

$$C_P = (8/\lambda^2)\int_0^\lambda x^3 a'(1-a)[1-(C_D/C_L)\cot\phi]\,dx \tag{13-42}$$

In some instances, C_L and C_D may be known as a function of α, based on the chosen airfoil, but both a and a' may be unknown. For such a situation, an iterative procedure based on the following equation can be used:

$$C_L = 4\sin\phi(\cos\phi - x\sin\phi)/[\sigma'(\sin\phi + x\cos\phi)] \tag{13-43}$$

Here σ' is the *local solidity* of the blade, defined as $\sigma' = Bc/2\pi r$. The procedure involves first guessing at values of a and a', then calculating ϕ from Eq. (13-39a) or (13-39b), then calculating angle of attack from $\alpha = \phi - \beta$, then calculating C_L, then calculating C_L and C_D using α, then using C_L and C_D to calculate new values of a and a'. This process is repeated until the incremental change in a and a' is within some predetermined tolerance.

Approximate Value of Maximum Power Coefficient

When carrying out an initial assessment of a turbine design, it is sometimes desirable to quickly calculate the value of $C_{P,\max}$, as shown in Example 13-6. The following empirical equation has been derived for this purpose:

$$C_{P,\max} = 0.593\left[\frac{\lambda B^{0.67}}{1.48 + (B^{0.67} - 0.04)\lambda + 0.0025\lambda^2} - \frac{1.92\lambda^2 B}{1+2\lambda B}C_D/C_L\right] \tag{13-44}$$

Here C_D/C_L is the drag to lift ratio at the design angle of attack α of the turbine.

Example 13-6   Refer to the hypothetical 1.5-MW turbine in Fig. 13-14. Assume $C_D/C_L = 0.01$ and that the turbine has 3 blades. (a) If the turbine rotates at 1 rad/s in a 6 m/s wind, what is the predicted value of $C_{P,\max}$ at that wind speed? (b) Produce a graph of $C_{P,\max}$ as a function of tip speed ratio λ up to $\lambda = 15$ for both $C_D/C_L = 0.01$ and $C_D/C_L = 0$, and state the maximum value of each curve. (c) Compare the maximum value in part (b) to the value observed in Fig. 13-14.

Solution
   (a) Based on the turbine radius of 34.5 m, the tip speed ratio is

$$\lambda = \frac{\Omega r}{U} = \frac{1(34.5)}{6} = 5.75$$

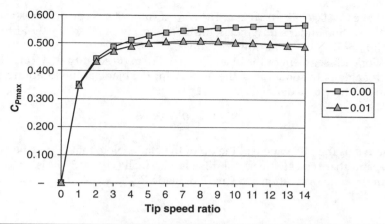

FIGURE 13-24 $C_{P,\max}$ as a function of tip speed ratio for $C_D/C_L = 0.00$ or $C_D/C_L = 0.01$.

Substituting other known values into Eq. (13-43) to solve for $C_{P,\max}$ gives

$$C_{P,\max} = 0.593 \left[\frac{\lambda B^{0.67}}{1.48 + (B^{0.67} - 0.04)\lambda + 0.0025\lambda^2} - \frac{1.92\lambda^2 B}{1 + 2\lambda B} C_D/C_L \right]$$

$$= 0.593 \left[\frac{5.75(3)^{0.67}}{1.48 + (3^{0.67} - 0.04)5.75 + 0.0025(5.75)^2} - \frac{1.92(5.75)^2 3}{1 + 2(5.75)3} (0.01) \right] = 0.5019$$

(b) Repeating the calculation in part (a) yields the graph shown in Fig. 13-24. For the case of $C_D/C_L = 0.01$, the curve reaches a maximum value of $C_{P,\max} = 0.5050$ at $\lambda = 8$ (to the nearest whole number value of λ), and declines thereafter for increasing λ. For the case of $C_D/C_L = 0$, since there is no drag coefficient term, as λ increases, the value of $C_{P,\max}$ approaches the Betz limit without bound and there is no fixed maximum value.

(c) The turbine represented in Fig. 13-14 has a maximum value of $C_P = 0.4893$ at $U = 9.5$ m/s, so the curve for $C_D/C_L = 0.01$ is a reasonable predictor of the upper bound on C_P, although the actual turbine is slightly less efficient since the approximation in Eq. (13-41) does not fully capture real-world turbine behavior.

Approximate Solution to Blade Design Problem by Blade Element Model Analysis

As an alternative to integrating thrust over the length of the turbine blade as proposed in Eq. (13-41), it is possible to perform a discrete approximation of the thrust or power extracted by the turbine under given wind and blade conditions by dividing the blade into a limited number of sections and analyzing the performance at the midpoint of each section. This approach is known as blade element model (BEM) analysis. Another approach known as computational fluid dynamics (CFD) models are used for detailed modeling of utility-scale turbines since the high cost per turbine justifies the additional time and expense of building the models. CFD models are beyond the scope of this chapter, but BEMs provide a stepping stone toward understanding the more complex development of CFD models.

The first step is to evaluate the lift coefficient as a function wind angle, local speed ratio, and solidity, using Eq. (13-45), and then the characteristic equation for lift as a function of α, based on the airfoil shape. The lift coefficient equation for the airfoil will typically be some function of α, for example, a linear function for a representative airfoil might be $C_L = 0.46352 + 0.11628\alpha$. The correct value of α is one that satisfies both equations for the section under study, that is, the lift coefficient predicted by the airfoil design equals the lift predicted from airflow conditions.

For each segment of the blade, we need data on solidity σ', local speed ratio x, and pitch angle β. In some cases, the *twist* θ_T for the blade at the location r is given, rather than β, where θ_T is defined as

$$\theta_T = \beta - \beta_0 \tag{13-45}$$

where β_0 is the pitch angle at the tip of the blade. Note that β_0 may be positive or negative, and that typically its absolute value is small ($|\beta_0| < 3°$). Also, x can be calculated from the ratio of location r to length of blade R

$$x = (r/R)\lambda \tag{13-46}$$

Calculation of β, σ', and x makes it possible to calculate α for the segment. Once α, ϕ, C_L, and C_D are known, it is possible to calculate the local contribution to power coefficient $C_{P,i}$ using the following:

$$C_{P,i} = \left(\frac{8\Delta\lambda}{\lambda^2}\right)\sin^2\phi_i(\cos\phi_i - x\sin\phi_i)(\sin\phi_i + x\cos\phi_i)[1 - (C_d/C_l)\cot\phi_i]x^2 \tag{13-47}$$

Here $\Delta\lambda$ is the change in x from one segment to the next, that is, $\Delta\lambda = x_i - x_{i-1}$, which is a constant value since the width of the segments is uniform. The overall power coefficient for the blade is the sum of all the local values of $C_{P,i}$ for each segment i for each segment $1, 2, \ldots, N$. The total power produced by the blades is then the power in the wind multiplied by the overall power coefficient, that is, $P_{prod} = C_P \times P_{available}$. Example 13-7 illustrates the calculation of the local contribution from a segment to C_P; the calculation for an entire blade is left as an exercise at the end of the chapter.

Example 13-7 A two-bladed wind turbine is designed with the following characteristics. The blades are 13 m long, and the pitch at the tip is –1.970°. The turbine has a design TSR of $\lambda = 8$. Lift coefficient is a function of α as follows (note, α is in degrees): For $\alpha < 21°$: $C_L = 0.42625 + 0.11628\alpha - 0.00063973\alpha^2 - 8.712 \times 10^{-5}\alpha^3 - 4.2576 \times 10^{-6}\alpha^4$. For $\alpha > 21$: $C_L = 0.95$. For the entire blade, the drag coefficient is $C_D = 0.011954 + 0.00019972\alpha + 0.00010332\alpha^2$. For the *midpoint* of section 6 ($r/R = 0.55$), find the following for operation at a tip speed ratio of 8: (a) angle of attack α; (b) angle of relative wind θ; (c) C_L and C_D; (d) the local contribution to C_P. Ignore the effects of tip losses. The chord at that point is 0.72 m and the twist is 3.400°.

Solution First calculate factors needed for calculating α: $r = 0.55(13) = 7.15$ m; $\sigma' = Bc/2\pi r = 2(0.72)/2\pi(7.15) = 0.0321$; $x = r\lambda/R = 0.55(8) = 4.4$; $\beta = \theta_T + \beta_0 = 3.400 - 1.970 = 1.430$. We can now solve for α using the lift coefficient equation for the airfoil design and Eq. (13-43). The exact solution using a computer solver is $\alpha = 7.844°$, which gives $C_L = 1.241$ from both equations. In other words, given the values of r, σ', x, and β, and the derived value of $\phi = \alpha + \beta = 7.844 + 1.430 = 9.274°$, the following hold:

$$C_L = 4\sin\phi(\cos\phi - x\sin\phi)/[\sigma'(\sin\phi + x\cos\phi)] = 1.241$$

$$C_L = 0.42625 + 0.11628\alpha - 0.00063973\alpha^2 - 8.712 \times 10^{-5}\alpha^3 - 4.2576 \times 10^{-6}\alpha^4 = 1.241$$

Table 13-5 shows values of C_L as a function of α around the solution.

Angle of Attack α (deg.)	Lift Coefficient C_L	
	From Eq. (13-43)	From Airfoil Design
7.70	1.2716	1.2289
7.75	1.2610	1.2331
7.80	1.2503	1.2372
7.85	1.2394	1.2413
7.90	1.2283	1.2454
7.95	1.2170	1.2495
8.00	1.2055	1.2535

TABLE 13-5 Comparison of Approaches to Calculating Lift Coefficient C_L as a Function of Angle of Attack α

Thus answer to part (a) is $\alpha = 7.844°$. Part (b), $\phi = \alpha + \beta = 7.844 + 1.43 = 9.274$. Part (c), $C_L = 1.241$, $C_D = 0.011954 + 0.00019972\alpha + 0.00010332\alpha^2 = 0.01988$. Part (d), for the contribution to $C_{P,i}$ we first calculate $\Delta\lambda = x_i - x_{i-1} = 4.4 - 3.6 = 0.8$. Then solve for $C_{P,i}$ using $\Delta\lambda$, λ, ϕ, x, C_L, and C_D in Eq. (13-47), which gives $C_{P,i} = 0.0567$.

13-5-4 Steps in Turbine Design Process

In the previous section, we investigated the analysis of the performance of a given blade design. It is also necessary to be able to start with a desired level of performance from a turbine and design a blade that meets the requirements.

The selection of a blade design can be seen in the larger context of designing the overall wind turbine system. This design process can be summarized in the following steps:

1. Determine the radius of the rotor based on the desired power at typical wind speeds.

2. Choose a desired tip speed ratio based on function, that is, for electric generation $\lambda > 4$ is typical, for mechanical applications such as water pumping a lower value of λ is chosen.

3. Choose a desired number of blades.

4. Select an airfoil, and obtain all necessary data (e.g., C_L and C_D data).

5. Create an ideal shape for the blade, based on the chosen airfoil and dividing into 10 to 20 elements. Note that optimal airfoil shapes are typically too difficult to fabricate exactly, so that some modification will be necessary.

6. Create an approximation of the ideal shape that provides ease of fabrication without greatly sacrificing performance.

7. Quantitatively evaluate blade performance using strip theory.

8. Modify shape of blade, based on results of step 7, so as to move blade design closer to desired performance.

9. Repeat steps 7 and 8, trading off blade performance and ease of production, until a satisfactory design is obtained.

r/R	Rotor 1 $\lambda = 6, B = 3$		Rotor 2 $\lambda = 10, B = 2$	
	ϕ	c/R	ϕ	c/R
0.15	32.0	0.191	22.5	0.143
0.25	22.5	0.159	14.5	0.100
0.35	17.0	0.128	10.6	0.075
0.45	13.5	0.105	8.4	0.060
0.55	11.2	0.088	6.9	0.050
0.65	9.6	0.076	5.8	0.042
0.75	8.4	0.067	5.1	0.037
0.85	7.4	0.059	4.5	0.033
0.95	6.6	0.053	4.0	0.029
Overall solidity		0.088		0.036

Source: After Manwell et al. (2002), p. 124. John Wiley & Sons Limited. Reproduced with permission.

TABLE 13-6   Representative Three-Blade and Two-Blade Designs, in Relative Dimensions

A complete example of this procedure from start to finish is beyond the scope of this chapter. A partial example is presented as a problem at the end of the chapter. As a starting point for the design procedure outlined in steps 1 to 9, Table 13-6 provides parameters for two representative ideal blade designs. The data given in the table provide the information necessary for steps 2, 3, 5, and 6 in the turbine design process. The two representative ideal blades shown in the table have constant $C_L = 1.00$ and $C_D = 0$.

Note that values are given here independent of the exact radius of the turbine, and the *overall solidity* is the percent of the swept area that is covered by the area of the blades. For example, if rotor 1 has a blade length of 10 m, then the total swept area is ~314 m², and of this area, a viewer facing the turbine at rest would see 8.8% or 27.6 m² of blade, and the rest open space. A typical application of rotor 1 might be in a medium to large turbine (10 kW to 2.5 MW), whereas rotor 2 would typically be applied to a small turbine, with a rated capacity of 2 kW or less. The turbine shapes above are ideal in the sense that they are designed without taking into account drag on the blade ($C_D = 0$); they assume a constant lift coefficient of $C_L = 1.0$.

13-6   Economic and Social Dimensions of Wind Energy Feasibility

The economic analysis of a wind power project, whether small- or large scale, focuses on the ability of the revenues from electricity generation to pay capital and operations/maintenance (O&M) costs. Wind developers must also consider the social acceptability of wind farms near both human habitation and natural areas.

On the economic side, O&M includes routine care for the turbines and ancillary facilities, insurance, and possibly royalties paid to the landowner on whose property the turbines are erected. For utility-scale wind energy, a common practice is for farms and ranches to enter into a long-term agreement for the use of their land for wind energy generation. In addition to royalties paid, the farmer or rancher continues to use the land for its original purpose, except in the direct vicinity of the turbine(s). O&M costs tend to be small relative to capital repayment costs in the case of wind power (on the order of 2–3% of capital cost). Although treated as a constant annual cost stream, maintenance costs may in fact be spent sporadically on worn components, such as transmission gearboxes that may last on the order of 5–7 years due to the rugged nature of the duty cycle that they endure.

Cost of onshore wind installation varies significantly between locations, with wind farm owners seeing costs in the range of $1500 to $2000 per rated kW for large installations with numerous turbines, and costs as high as $3000 per kW for large turbines (1.5 MW and higher) when installed individually, including turbines, towers and foundations, and all ancillary costs. Costs also fluctuated widely during the decade from 2000 to 2010. Prior to the recession of 2009, costs rise with high demand for turbine orders and rising costs of raw materials such as steel and concrete, but then fell again with the recession and drop-off in new orders to turbine makers. Another factor is the ratio of swept area to nameplate capacity, as illustrated in Fig. 13-14. A turbine with a relatively large swept area may incur a higher cost on the order of $3000/kW, but it will also have higher economic productivity per kW, so both sides must be factored into the economic analysis. Turbine purchasers pay an additional premium for offshore instal-lations. Here again total installed cost varies greatly by location, but as one indication of the difference between onshore and offshore, the USDOE's feasibility study of 20% of electricity from wind by 2030 gives values for levelized cost per kWh of production of $0.10–0.15 for offshore versus $0.06–0.09 for onshore wind (USDOE, 2009).

Within an individual installation site, the total cost (combining materials and labor) is typically split as follows: 40% for turbines, including nacelles, hubs, and blades; 40% for supporting structures, including towers and foundations; and the remaining 20% for ancillary costs, including access roads, underground cables, electrical substation, and high-voltage connection to existing grid. The single most expensive component of the turbine system is the steel tower, which costs more than the nacelle, hub, or collec-tion of blades.

13-6-1   Comparison of Large- and Small-Scale Wind

In this section, we compare large-scale wind farm installations with individual installa-tion of small turbines. Starting with utility-scale wind, suppose a collection of 20 tur-bines with rated output of 1.5 MW is installed in a location with 7.7 m/s average wind speed (17.0 mph). Each turbine costs $1750 per kW, or $2.625 million per device, includ-ing turbines, towers, and balance of system costs, bringing the total project cost to $52.5 million. Upon completion, the farm is eligible for a U.S. federal government 30% rebate designed to encourage wind development, which is worth $15.7 million and reduces the net cost to $36.7 million. The installer might also take a $0.015/kWh production tax credit over the first number of years of the project in lieu of the one-time rebate. Although we do not consider the calculations here, the incentive program is designed so that the one-time rebate or production tax credit gives approximately the same level of incentive. The farm produces an average of 80 million kWh per year, for a 30% capac-ity factor. The investment lifetime is considered to be 20 years.

Net value of the project over its lifetime for a range of MARR values from 3% to 7% is shown in Fig. 13-25. Here we are assuming that the turbine operators are able to sell

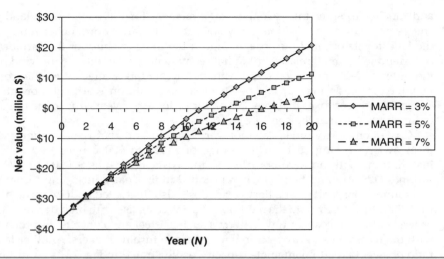

FIGURE 13-25   Net value of wind park investment in constant dollars for MARR in range 3% to 7%. Analysis based on $1750/kW cost including balance of system, 30% one-time initial rebate, and $0.05/kWh paid wholesale for electricity; see comments in text.

electricity to the grid wholesale at an average rate of $0.05/kWh in constant dollars, and that revenues are reduced by a 3% per annum O&M cost. The sales revenue assumption is simplistic, since, in a market-driven grid environment, the actual price paid for wind-generated electricity will fluctuate up and down with peaks and troughs in demand, and the owner must accept the price paid at the time, as detailed in the Power Purchase Agreement into which owner and utility enter, given that wind is an intermittent resource. However, the rate of $0.05/kWh is realistic in many locations as an average. The project is profitable for the range of MARR values used in the figure, with a payback (i.e., net value reaching the level of ±$0) ranging from 11 years for MARR = 3% to 16 years for MARR = 7%. This calculation also does not consider a premium that many customers are willing to pay in order to purchase "green" electricity, which would increase the value of each kWh of production above $0.05 and hasten the breakeven point.

The economics of the investment in Fig. 13-25 represents the situation in many electricity markets around the world today. With growing demand for electricity, and relatively small penetration of wind in terms of overall market share of kWh, there is plenty of room for the grid operator to buy wind energy from the turbine operator whenever it is generated, and there are plenty of windy sites available in many countries where wind turbines can be added. Some exceptions also arise: where surplus amounts of competing electricity such as hydroelectricity are generated, or the grid is congested, wind farms at times may be unable to deliver electricity to consumers even when wind is available.

We now contrast the investment in commercial-scale wind power with an investment in a residential or small-commercial size 10-kW turbine. Suppose the residence or small business is grid connected and has net metering, meaning that real-time electricity production from the wind turbine beyond what is being consumed on the premises is transmitted to the grid, and the system owner receives credit for this electricity against bills for power bought from the grid. The turbine output then displaces purchases of electricity from the grid whenever it can be generated from the wind, at an assumed rate of $0.13/kWh. The complete system costs on the order of $50,000 to $60,000 before and $25,000 after applicable incentives and tax relief. Note the higher cost per kW compared to commercial wind, for example, $5000 to $6000 per kW for the small system before incentives. The wind resource is whatever wind is available immediately adjacent to the building, which, since people generally are reluctant to live or work in extremely windy locations, is lower than that of the commercial turbines, namely 4.3 m/s (9.4 mph) at the hub height of the device. (The hub height is also lower than that of a commercial turbine, given both cost and zoning restrictions on the height of the turbine tower.)

Output from this small-scale system at a favorable site might be on the order of 12,000 kWh/year, for a capacity factor of approximately 14%. This amount of output is worth $1560/year, so that the simple payback is approximately 16 years. At an MARR of 5%, the payback increases to 35 years, at or near the expected lifetime of the device.

Thus the cost disadvantage for small wind comes both from the relatively smaller manufacturing market for small turbines, when measured in total rated kW manufactured per year, and from diseconomies of small scale when converting wind energy with a smaller swept area per turbine. The example also shows the importance of incentive programs for both utility-scale and small wind. For investors it is important that the incentives be available, and that the government makes a long-term commitment to them. At times through the 2000s, the short duration of incentive programs has discouraged investment in wind.

From this comparison, it is clear that on the basis of price alone, the consumer is better off buying wind-generated power from the grid than purchasing a small-scale wind system, unless they happen to live or work in an extremely windy location or are not already connected to the grid. Nevertheless, on-site wind energy from small-scale wind may have its own advantages that make it justifiable:

1. *Potential for greater reliability:* A grid-tied system can be installed in such a way that if grid power is lost, the system can switch into a stand-alone mode and continue operating, given additional investment in batteries and controls. In certain applications, for example, businesses where loss of merchandise or sales opportunities may be critical, not being vulnerable to extended blackouts has real economic value.

2. *Marketing, public relations, or morale-boosting value:* Having a small wind turbine on site makes an obvious statement about the building owner's commitment to renewable energy, and visitors to such a site may respond positively. For example, a Wal-Mart store in McKinney, Texas, United States (30 mi north of Dallas, Texas) recently installed a 50-kW wind turbine along with other amenities to promote an image of a green store. The state of Texas has some of the best wind resources in the entire United States, and it is unlikely that a small-scale turbine like this one can compete with grid-purchased wind power on the basis of price alone in the Texas electricity market. However, as part of a larger program of green technologies used to help attract customers and increase sales, the investment makes sense. A number of turbine manufacturers also see the potential to bring down the cost of small-scale wind turbines if enough units can be purchased.

3. *The right combination of green, local, and affordable:* When choosing a clean energy option, a building owner may look at the available options and choose to invest in an on-site renewable system, even though it may cost more than other options. In this case, the owner may be looking to make a more personal, local contribution to the sustainable generation of energy, and if the cost is within the allowable budget, the owner may choose to spend the extra cents per kWh to generate wind energy on site.

4. *Reduced nonrenewable fuel costs:* In some remote locations that rely on diesel generators to supply electricity to the local grid, it can be cost-effective to install a small wind system and use it whenever possible, with the diesel generator used as a backup source.

13-6-2 Integration of Wind with Other Intermittent and Dispatchable Resources

With the information presented in this chapter about wind energy it is now possible to combine background on wind with solar (Chaps. 10 and 11) and dispatchable resources such as fossil or biomass (Chap. 6) to consider an integrated system of energy sources for meeting demand. In the short run, new resources such as wind or solar are added incrementally by verifying that the grid and the market can support a new wind or solar farm and then permitting the farm. In this section, we instead take a long-term, strategic view and consider optimal choice of different resources to meet predicted total demand with minimal total cost at some point in the future. The demand in question could be for the entire market, or for some fraction of it that is to be allocated to renewable energy sources.

Integrating resources such as wind and solar is of interest because in many areas these two resources are complementary on a seasonal basis, as is the case in the area around Ithaca, NY (Fig. 13-26). In this figure, the capacity factor for the wind farm site in Enfield, NY has been used to represent the wind energy potential, and the solar productivity data for Binghamton, NY from NREL's PV Watts database has been used to predict solar energy output. The raw monthly capacity factor curves are then normalized so that the points in the figure represent the percent capacity factor relative to the annual average of 30.3% for wind and 14.0% for solar PV. The figure shows that in this location wind and solar are complementary since wind peaks in the colder months and declines in the summer, whereas solar has the opposite characteristic. The figure does not consider variability in energy demand by month of year, but to the extent that demand does not vary greatly with changing seasons, it is beneficial to install a combination of solar and wind.

Combined Capital and Operating Cost Optimization Model

An optimization model can be applied to solving the problem of finding the minimum cost combination of energy sources to meet year-round demand, taking into account variations in demand and capacity factor by month of year. The model allocates electricity production to different sources, and at the same time invests in sufficient amounts of each resource type so that capacity is available to generate the required energy.

In the model, let i be a subscript for resource type (e.g., wind, solar, dispatchable) t a subscript for time period (e.g., month of the year), and NH_t as the number of hours in time period t (e.g., 744 hours in the month of January, etc.). Further, define $CCAP_i$ as the annualized capital cost per kW, COP_i as the operating cost per kWh, η_{min} as the minimum fraction of demand allocated to each source, D_i as the demand in period i, and

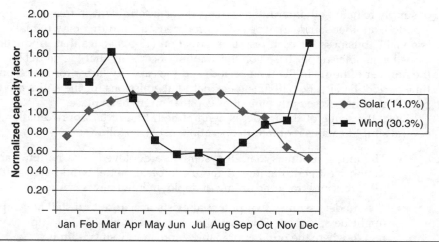

FIGURE 13-26   Comparison of normalized capacity factor for wind and solar in region around Ithaca, NY.

Source: National Renewable Energy Laboratories, for solar data.

Note: Figure in parentheses is average annual capacity factor. Normalized capacity factor is monthly value compared to annual average, e.g., January wind actual capacity factor is 39.8%, so its normalized value is 1.31.

$CF_{i,t}$ as the capacity factor for source i in period t. The decision variables are then $X_{i,t}$ for output allocated to source i in period t in kWh and Y_i for the amount of capacity installed of type i. The model is then formulated as:

$$Min\ Z = \sum_{i,t} COP_i \cdot X_{i,t} + \sum_i CCAP_i \cdot Y_i \qquad (13\text{-}48)$$

$$\overset{s.t.}{\sum_i} X_{i,t} = D_t, \forall t \qquad (13\text{-}49)$$

$$X_{i,t} \geq \eta_{min} \cdot D_t, \forall i,t \qquad (13\text{-}50)$$

$$X_{i,t} \leq CF_{i,t} \cdot Y_i \cdot NH_t, \forall i,t \qquad (13\text{-}51)$$

$$\sum_{i,t} \mu_t X_{i,t} \leq E_{max} \qquad (13\text{-}52)$$

$$X_{i,t} \geq 0, Y_i \geq 0, \forall i,t \qquad (13\text{-}53)$$

The objective function minimizes the combined operating and capital cost in dollars. The constraints are as follows. Equation (13-49) requires that the combination of production in period t is sufficient to meet demand, while Eq. (13-50) requires a minimum amount to go to each available source. This requirement provides both diversity of supply to help with intermittency (e.g., if one intermittent source under-produces in a period, another source may be able to compensate) and diversity to make sure there is a viable business for each resource. Equation (13-51) states that production cannot exceed available supply, based on the amount of capacity that the model has specified, the number of hours in the time period, and the capacity factor for the resource in the time period. Finally, Eq. (13-52) states that total emissions for the entire modeled time period must not exceed any emissions cap that might be in place.

Since the scope of the model is a simplified, first-order approximation, there are several simplifications required to operationalize it, as follows:

- The model assumes that all systems are deployed at same time, and for the same investment lifetime. A more complete model would start with existing resources and track the timing of deployment of new resources.

- The model assumes that potential exists in nature to install any capacity that might be specified. This assumption may not always hold, for example in a region where the number of viable wind energy sites is limited.

- The model ignores the impact of intermittency and assumes that a fixed available supply will be able to meet deterministic demand for a given time period. A more realistic model would require an additional investment in storage or smart grid technology, since both instantaneous output of intermittent resources may fall short, and the actual capacity factor in a given time period may be less than the fixed value given in the model. This limitation is, however, partially addressed by having diverse energy supplies, since another resource may have extra capacity available to make up for the shortfall.

- The model considers only the minimization of cost and does not consider maximizing profit, which is the actual financial objective of a commercial enterprise.

- The model does not consider standby charges or grid regulation services that would be paid, whereas in a real-world grid the electricity producers are paid for both of these services, adding to the total cost required to meet demand.

The model is demonstrated in Example 13-8 for a single time period, so as to make the technique more transparent. The model can be applied to a single period by suppressing the subscript t throughout. (A multiperiod 12-month model is left as an end-of-chapter exercise.)

Example 13-8   A region must meet 2500 GWh of electricity demand in a 1-year period (8760 h) from a mixture of wind, solar PV, and natural gas resources. Annual capital cost per GW, fuel and operating cost per GWh, and CO_2 emissions per kWh are given in the table below. The capacity factors for the three resources are 30%, 14%, and 90%, respectively. Each resource must contribute at least 20% of the total demand, and emissions are capped at 250,000 tonnes per year. What is the minimum-cost allocation of production to the three resources, how much capacity is installed, how much CO_2 is emitted, and what is the total cost?

Source	Ann Cap Cost (mil.$/GW)	Op Cost (mil.$/GWh)	Fuel Cost (mil.$/GWh)	Emissions (gCO$_2$/kWh)
Solar	$179	$0.004	$0.000	0
Wind	$245	$0.020	$0.000	0
NatGas	$113	$0.017	$0.030	325

Solution   The problem can be formulated as a cost-minimizing optimization with six decision variables, three capacity variables Y, and three production variables X. It can also be solved by inspection. We first note that natural gas is the least expensive option, so it will produce up to its limit of 250,000 tonnes per year. The emissions rate is equivalent to 325 tonnes per GWh, so the model chooses a maximum production of 769 GWh, requiring 0.098 GW of capacity (i.e., 98 MW). Next, solar is the most expensive, so the model allocates exactly 20% of the output or 500 GWh, requiring 0.408 GW (408 MW) of capacity. Lastly, the remaining unmet demand is 2500 − 769 − 500 = 1231 GWh that must be met by wind. The corresponding capacity requirement is 0.468 GW or 468 MW. The solution can be verified by multiplying capacity by hours and capacity factor to calculate output:

$$X_{gas} = (98\ \text{MW})(8,760\ \text{h})(0.9) = 769,000\ \text{MWh} = 769\ \text{GWh}$$

$$X_{solar} = (408\ \text{MW})(8,760\ \text{h})(0.14) = 500,000\ \text{MWh} = 500\ \text{GWh}$$

$$X_{wind} = (468\ \text{MW})(8,760\ \text{h})(0.3) = 1,231,000\ \text{MWh} = 1231\ \text{GWh}$$

The total cost calculation of $261.9 million per year is shown in the table below, in millions of dollars:

Source:	CapCost	OpCost	TotCost
Solar	$73.1	$2.0	$75.1
Wind	$114.9	$24.6	$139.6
NatGas	$11.1	$36.2	$47.2
Total	$199.1	$62.8	$261.9

Note that the total cost of $261.9 million represents an overall average cost of $0.1048/kWh, which could be compared to a cost of $0.0614/kWh if the electricity had been generated using 100% natural gas, although emissions would also rise to 812,500 tonnes per year.

13-6-3   Public Perception of Wind Energy and Social Feasibility

Utility-scale wind energy development inherently has a high public profile, so public perception and successful integration into the community and region are very important for its success. Large-scale wind achieves efficient energy conversion at a low cost precisely because the devices are large and high off the ground, typically located on ridge tops as opposed to deep valleys, and hence visible from a great distance. For instance, from the 30 MW Fenner wind farm in Fenner, New York, it is possible to see the 10 MW Madison wind farm in Madison, New York, a linear distance of some 16 miles. It is therefore almost certain that for a radius of many miles around either installation, the turbines will be visible on the horizon.

Furthermore, taking New York State (NYS) as an example, with a large population and high demand for electricity, onshore wind will have a major profile in the state if the desire is to generate a significant fraction of total electricity demand from this source. As wind installations have grown in size, so has their output: the 112 MW High Sheldon farm in Varysburg, New York, aims to produce 260 million kWh per year. Even with 75 turbines distributed over a wide area around Varysburg, the output is equivalent to just 0.2% of the state's total demand of roughly 150 billion kWh per year. Many more such farms would be required to achieve 5% or 10% penetration into the market.

Faced with the arrival of large-scale wind energy, communities across the state have responded at times with support, but at other times with concern and opposition. A survey of popular responses in the press and on the Internet in northern and western NYS in 2009 found numerous statements of concern about the nuisance noise from living close to active turbines, or low-level health effects, possibly from low-frequency vibrations emanating from rotating turbine blades.[7] Similarly, in 2007 a representative of a wind energy design and installation service company studied the more than 60 sites around NYS planned or under construction for large-scale wind, and found some level of community opposition at every single one.[8]

Concerns about local negative impacts are real and should be taken seriously. For instance, if local noise is a concern, wind farms can develop curtailed operating regimens that reduce noise at certain times of day or when the wind is from a certain direction. At the same time, there is a need to forge ahead with wind energy development because the overall wind resource is vast, the cost is low compared to other renewable energy options, and because wind can play a complementary role to other resources such as solar because it is often strongest at night or during the winter. When entering a new region to develop wind farms, the industry can maximize chances of success with the following steps:

- *Establish good connections with the community:* Approach both elected officials and the community at large. Make available channels of communications so that residents can ask questions and voice concerns. Respond candidly and in a timely manner.

- *Be prepared to educate:* Explain how wind works, how it benefits both the environment and the community. One misperception commonly heard is that the energy is generated locally in the community but then transported off site without any local benefit. The industry should explain that because electricity is generated in large quantities by wind farms (e.g., the generation at High Sheldon is equivalent to the needs of 50,000 homes) and must be consumed in real time, it is not practical to isolate the wind farm for strictly local energy use. The community does benefit, however, from the economic activity that the wind farm generates both in constructions and operations.

- *Provide economic benefits to both leaseholders and the community at large:* Direct beneficiaries of the wind farm include property owners such as farmers who receive royalty payments from the wind farm. This may leave other owners who are not in or near enough to the farm to be paid royalties feeling "left out," and has led to conflicts in a number of instances. Therefore, an annual payment to local government that helps to reduce the tax burden of all residents can restore trust and help the entire community to appreciate the benefits of the wind installation.

- *Treat offset requirements carefully:* "Offsets" are the minimum distance required between the turbine tower and nearby residential buildings. Even if a turbine meets the offset requirement as stated, the company developing the site might in some instances choose not to construct a turbine out of concern that the device may be too close to residents, leading to public outcry and backlash later once the facility is operating.

---

[7]Review of media and Internet conducted by Cornell University student Megan Giles during Spring semester, 2009. Megan's contribution is gratefully acknowledged.
[8]Loren Pruskowski, Sustainable Energy Development, speaking at Regional Renewable Energy Conference, Ithaca College, November 2007.

- *Consider community wind development:* "Community wind" development is a financing model in which local communities finance part or all of a wind project and reap the financial benefits. The state of Minnesota in the United States has become a leader in this area and passed the Community Based Development, or C-BED, statute in 2005 to facilitate community wind. By design, community wind gives communities a sense of connection to a wind project.

In the long term, the development of remote and very windy locations in the U.S. Great Plains (or similarly in the western steppes of China or in remote Siberia) coupled with highly efficient long-distance dc transmission may obviate these problems by placing wind generation far away from population centers. Similarly, development of floating offshore wind turbines might provide wind-generated electricity without the aesthetic concerns that arise with projects such as Cape Wind in Nantucket Sound in Massachusetts. Expansion of long-distance transmission would of course bring its own share of local resident concerns and the need to work with communities to reach an acceptable solution. Even as these technologies develop, however, there will for the foreseeable future be a demand for large-scale wind farms closer to population centers in North America and Europe, as well as in emerging economies like China and India, so the need to work with communities to satisfy the demand for social sustainability will remain.

13-7 Representative Levelized Cost Calculation for Electricity from Utility-Scale Wind

The following calculation of levelized cost per kWh is based on values and discussion throughout this chapter. It represents the construction of a newly built wind farm built in the United States at recent costs. We assume a capacity factor of 30%, which is close to the U.S. national average at this time (32.7% for the year 2014 according to Fig. 13-17).

We assume a capital cost of $2600/kW for a 100-MW plant, based on a recent survey of project capital cost values, resulting in a $260 million capital cost. The required discounting factor is (A/P, 20 years, 7%) = 0.0944 (see Chap. 6, Sec. 6-8). This factor gives a capital cost per year of $24.5 million. The chosen capacity factor results in 263 GWh/year of output. Dividing capital cost among output gives a value of $0.0934/kWh. Note that because of the conservative assumptions this value is higher than the range of $0.05–$0.08/kWh suggested by USDOE's "20% Wind by 2030" report (USDOE, 2009). Noncapital operating cost is set to $0.01/kWh, derived from a survey of recent typical wind farm operating cost values, reflecting the need to occasionally inspect and maintain plants.

The final value is $0.1034/kWh, as summarized in the calculations below:

$$Output = (100\ MW)(8{,}760\ h/y)(0.3) = 263\ GWh/y$$

$$CapCost = \frac{(\$260\ M)(0.0944)}{263\ GWh/y} = \frac{\$24.5\ M}{263} = \$0.0934/kWh$$

$$TotCost = CapCost + EnergyCost + OpCost = \$0.0934 + \$0 + \$0.01$$

$$= \$0.1034/kWh$$

13-8   Summary

The successful application of wind energy technology requires both knowledge of the wind resource in a location and a suitable turbine design. The wind resource can be evaluated accurately by continually gathering wind speed data at a site for at least 1 year and then assigning each hour of the year to a wind speed "bin" based on hourly average wind speed. In many locations, a statistical pattern of a Weibull or Rayleigh form can be fitted to the distribution of wind speeds, which provides a convenient way of representing the wind speed in a functional form or approximating the wind speed distribution if the year-round average wind speed for the site is known but bin data are not available. Turbine designs can be analyzed in a preliminary way using the "actuator disk" model, from which the Betz limit, which states that the theoretical upper bound on extraction of available wind energy by a wind device is 59.3%, is derived. For more detailed analysis and design of turbine blades to maximize power conversion within manufacturing limitations, techniques such as strip theory, blade element modeling, or computational fluid dynamics modeling are required. Among different sizes of turbines, it is generally only the commercial-scale turbines that can produce at prices competitive with other resources available on the grid; nevertheless, small-scale wind may also be attractive to some owners for reasons of energy self-reliance or for promoting a green image, or in off-the-grid locations.

References

Brown, C., and D. Warne (1975). *Wind Power Report.* Ontario Research Foundation, Report P.2062/G. ORF, Toronto.

Manwell, J. F., J. G. McGowan, and A. L. Rogers (2010). *Wind Energy Explained: Theory, Design, and Application,* 2nd ed Wiley, Chichester, West Sussex.

Mendenhall, W., D. Wackerly, and R. Scheaffer (1990). *Mathematical Statistics with Applications,* 4th ed. Duxbury, Belmont, California.

Devore, J. (2009). *Probability and Statistics for Engineering and the Sciences,* 7th ed. Brooks/Cole, Belmont, California.

Sorensen, Bent (2000). *Renewable Energy: Its Physics, Engineering, Environmental Impacts, Economics and Planning,* 2nd ed. Academic Press, London.

U.S. Department of Energy (2009). *Reaching 20% Wind Energy by 2030: Increasing Wind Energy's Contribution to U.S. Electricity Supply. Report,* USDOE, National Renewable Energy Laboratories, Golden, CO.

Wan, Y. (2012). *Long-term wind power variability.* Report, National Renewable Energy Laboratories, Golden, CO.

Further Reading

American Wind Energy Association (2011). *AWEA Turbine Installation Data Set.* Online resource, available at www.awea.org. Accessed Jun. 15, 2011.

Burton, T., D. Sharpe, and N. Jenkins (2001). *Wind Energy Handbook.* Wiley, Chichester, W Sussex.

Danish Wind Industry Association (2007). *Informational Website.* Web resource, available at http://www.windpower.org/en/core.htm. Accessed Nov. 9, 2007.

Gipe, P. (2004). *Wind Power: Renewable Energy for Home, Farm, and Business,* revised edition. Chelsea Green Publishing Company, White River Junction, Vermont.

Global Wind Energy Council (2011). *Annual Wind Energy Data.* Online resource, available at www.gwec.net. Accessed July 5, 2011.

Spera, D., Editor (1994). *Wind Turbine Technology: Fundamental Concepts of Wind Turbine Engineering.* ASME Press, New York.

Stecky, N. (2004). "Renewable Energy for High-Performance Buildings in New Jersey: Discussion of PV, Wind Power, and Biogas and New Jersey's Incentive Program." *ASHRAE Transactions,* pp. 602–610.

Vanek, F., and L. Vogel (2007). "Clean Energy for Green Buildings: An Overview of On- and Off-Site Alternatives." *Journal of Green Building,* Vol. 2, No. 1, pp. 22–36.

Exercises

13-1. The following wind measurements by bin are given for a proposed wind turbine location. Assume the air has a density of 1.15 kg/m³.

Bin	Wind Speed (m/s)		Hours/year
	Minimum	Maximum	
1	0.0	0.0	80
2	0.0	1.0	200
3	1.0	2.0	501
4	2.0	3.1	850
5	3.1	4.1	1300
6	4.1	5.2	1407
7	5.2	6.2	1351
8	6.2	7.2	990
9	7.2	8.2	641
10	8.2	9.3	480
11	9.3	10.3	375
12	10.3	11.3	291
13	11.3	12.3	180
14	12.3	13.4	78
15	13.4	14.4	32
16	14.4	And above	4

a. Calculate the total estimated energy available in the wind from the bin data, in kWh/m². For bin 16, assume an average wind speed of 16 m/s.

b. Calculate the estimated energy available in the wind using the average wind speed and the Rayleigh function. For bin 16, assume an average wind speed of 16 m/s.

c. Calculate the estimated energy using the method of Eq. (13-10) in the chapter.

d. Calculate the error between the actual value from (a) and the estimated values from (b) and from (c).

e. Plot both estimated and actual curves on a single graph, with hours per year as a function of bin number.

13-2. Estimating output from a turbine: the following data for a large turbine designed for high-wind areas are provided. The turbine swept area has a diameter of 80 m and you can assume an air density of 1.15 kg/m³. In the data below, bin 1 spans from 0 m/s up to the cut-in speed of 6.25 m/s, and thereafter each bin spans a 0.5 m/s increment, for example, bin 2 is 6.25 to 6.75 m/s, and so on. Bin 18 spans from 14.25 to 22.25 m/s, and above 22.4 m/s the machine stops (i.e., this is the cut-out speed). The device is being considered for use in a site with average wind speed of 9.6 m/s. Note that bin 18 has an approximate average wind speed of 17.25 m/s, reflecting the fact that lower speeds are more probable than higher ones within this bin. (Source: ASME Table 5-1 from ASME Press, "Wind Turbine Technology: Fundamental Concepts of Wind Turbine Engineering," David Spera, p. 22.)

Bin Num.	V-avg. (m/s)	P-out (kW)
1	—	0
2	6.5	175
3	7	318
4	7.5	460
5	8	603
6	8.5	745
7	9	949
8	9.5	1153
9	10	1316
10	10.5	1479
11	11	1642
12	11.5	1805
13	12	1968
14	12.5	2120
15	13	2263
16	13.5	2385
17	14	2500
18	17.25	2500

a. Plot C_p for the turbine as a function of bin number, using the $V_{average}$ value for each bin as a basis for calculating power in the wind.

b. Calculate the predicted total annual output and capacity factor for this device in the chosen location, using the site average wind speed and assuming the wind follows a Rayleigh distribution.

c. Based on (b), is the device suitable for the site?

13-3. Apply the power curve of Fig. 13-14 to a site with average wind speed of $U = 7.7$ m/s, with bin speed estimated using a Rayleigh distribution. Using 0.5 m/s bin size increments, estimate total output from the device in this location in kWh and the capacity factor.

13-4. Suppose a small turbine rated at 10 kW has approximately the following power curve: $U \leq 4$ m/s, $P_i = 0$ kW; 4 m/s $\leq U \leq 10$ m/s, $P_i = (10/6)(U-4)$ kW; 10 m/s $\leq U \leq 20$ m/s, $P_i = 10$ kW; above 20 m/s, no output. Apply this power curve to the wind distribution of Table 13-1. What is the predicted annual output in kWh?

Section	r/R	Radius (m)	Chord (m)	Twist (°)
1	0.05	0.248	Hub	Hub
2	0.15	0.743	0.428	35.3
3	0.25	1.238	0.415	20.65
4	0.35	1.734	0.347	13.05
5	0.45	2.229	0.285	8.9
6	0.55	2.724	0.241	5.95
7	0.65	3.219	0.207	3.6
8	0.75	3.715	0.18	2.05
9	0.85	4.210	0.152	0.9
10	0.95	4.705	0.122	0.2
Tip	1.00	4.953	0.107	0

13-5. You are given the following geometry for a turbine blade, divided into 10 sections with the dimensions for the midpoint of each section given:

The pitch at the tip of the blade is $-2°$. The turbine has 3 blades, an ideal tip speed ratio of 7, and a rated wind speed of 11.62 m/s. For $\alpha < 12$, $C_L = 0.368 + 0.0942\alpha$, $C_D = 0.00994 + 0.000259\alpha + 0.0001055\alpha^2$, where α is given in degrees. The hub occupies the innermost section, so it does not contribute to the performance of the turbine.

   a. For each section, calculate (i) angle of attack, (ii) angle of relative wind, (iii) lift and drag coefficient, and (iv) the local contribution to C_P. Ignore tip loss effects.
   b. Find the overall power coefficient.
   c. How much power do the blades produce at 11.62 m/s?

13-6. You are asked to carry out a preliminary design of a turbine that will deliver 250 kW at a design speed of 10 m/s. Use the data for ideal rotor 1 in Table 13-6 to carry out the design. Assume constant $C_L = 1.00$ and negligible coefficient of drag, and design each blade based on the division of the blade into 10 even increments, each with uniform aerodynamic characteristics.

13-7. Using Eq. (13-15) as a starting point, show that the maximum power that can be extracted by a translating device is 0.148 C_D.

13-8. Using Eq. (13-16) and Bernoulli's law as a starting point, show that the axial induction between the free wind and far wake speed is always twice the reduction from the free wind to the actuator disk.

13-9. A wind turbine rated at 250 kW costs $450,000 and has a swept area of 1200 m². It is installed in a location with year-round average power density available in the wind of 320 W/m², and on average converts 25% of the energy into electricity. The electricity is sold to the grid at $0.039/kWh, paid at the location of the turbine (i.e., no transmission cost). The project has a 20-year time horizon with salvage value of $20 thousand, and the MARR is 8%. Calculate the net present value (NPV) of this investment.

13-10. Consider a site where either of the turbines in Fig. 13-14 could be installed. The average wind speed at a hub height of 80 m is 6.74 m/s. If the 1.5-MW turbine is chosen, its output could be modeled with a Rayleigh distribution. Alternatively, the 1.7-MW turbine could be chosen, in which case a 95-m tower would be required with corresponding increase in average wind speed, using Eq. (13-11) and a wind shear coefficient of $\alpha = 0.2$, again using a Rayleigh distribution. (a) Use the power curve in Fig. 13-14 and the Rayleigh distribution to calculate the capacity factor for each option, and calculate the percentage point increase by choosing the larger turbine. (b) Since the large turbine would be more productive but also increase project cost, discuss qualitatively some of the elements that would contribute to the higher cost.

13-11. You are to choose an optimal combination of wind, solar PV, and biomass to meet 12-month demand in a region. Annual capital cost and operating cost are the same as in Example 13-8, except that biomass replaces natural gas, and has a capital cost value of $250/kW/year, and a combined fuel and nonfuel operating cost of $0.123/kWh. Monthly demand and wind/solar capacity factor values are given in the table below. Biomass maximum capacity factor is 80% in each month. Each source is required to produce a minimum of 20% in each month. Since there is no end-use CO_2 emissions, there is no emissions cap in this case. Create an optimization model and solve for the amount of capacity installed, the generation by source and month, and the total annual cost.

Month	Demand (mil.kWh)	Wind CF (pct)	Solar CF (pct)
Jan	698	40%	7%
Feb	653	40%	8%
Mar	608	49%	13%
Apr	563	35%	15%
May	550	22%	17%
Jun	589	17%	16%
Jul	602	18%	18%
Aug	640	15%	18%
Sep	563	21%	16%
Oct	570	27%	13%
Nov	576	28%	7%
Dec	589	52%	6%
Total	7201	n/a	n/a
Average	600	30.3%	13.5%

13-12. Suppose the turbine and site from Problem 13-3 are used as the basis for an investment in a wind park. Each turbine has a total cost of $1000/kW, plus 20% for all site and grid connection costs. Ignore operating and maintenance costs. If the project life is 25 years and the MARR is 5%, what is the cost per kWh to produce electricity at this facility, before considering any transmission costs?

13-13. Suppose the 10-kW turbine from Problem 13-4 is installed in a remote location to provide off-the-grid power. Including battery and thermal storage system, the cost is $70,000. If the system lasts for 25 years and the MARR is 6%, what is the levelized cost per kWh? You can assume all electricity produced is equally valuable, whether it is consumed in electrical applications or goes into battery or thermal storage.

CHAPTER 14

Bioenergy Resources and Systems

14-1 Overview

This chapter focuses on bioenergy generation from biomass (mostly plant material), or bioenergy generation with a bioprocessing step that uses microbes, including yeast, algae, or bacteria, to convert inorganic or organic materials into an energy carrier. We predominantly discuss technologies that are already used at an industrial scale or that are very close to be implemented. It is important to note that we cannot be comprehensive—there are simply too many technologies that generate bioenergy. For instance, we do not discuss combustion of wood chips. On the other hand, we do discuss anaerobic gasification (anaerobic means without oxygen), because of the possibility to generate liquid biofuels as an energy carrier from one of the gasification products. We explain several bioprocessing systems with yeast or bacteria that generate biofuels. Since the bulk of biofuels is generated for transportation, a link between this chapter and Sec. 15-4-3 exists with a focus on transportation fuels. We also describe the production of the gaseous energy carriers—methane and hydrogen. These gases can be used as transportation fuels, but may be more practical as energy carriers to power stationary energy plants.

14-2 Introduction

Biofuels for transportation include any products derived from living organisms, ranging from food crops to plant and tree material to microbes, such as algae, which can be processed into substitutes for liquid transportation fuels. The use of biofuels for transportation dates back to the early years of the modern transportation vehicles, when Rudolf Diesel, who is the inventor of the diesel engine [i.e., compression-ignition (CI) engine], used peanut oil as a fuel in his early engine prototypes. One of Diesel's motivations was to create a source of mechanical power that could utilize a wide variety of fuels, so that small businesses of the day would not be captives of the coal industry for their energy supplies. In recent years, interest in biofuels has surged as nations, such as Brazil, Germany, and the United States, have sought to reduce petroleum imports. Also, many countries see biofuels as a way to prepare for anticipated dwindling petroleum reserves by developing a renewable alternative. For more information on biofuels, see, for example, Drapcho et al. (2008); Nag (2008); Olsson (2007); Sorensen (2002) Chap. 4; or Tester et al. (2012) Chap. 10.

To provide real benefit to society, a biofuel must achieve at least the following two objectives:

1. It must measurably reduce emissions of CO_2 and other pollutants compared to petroleum when taking into account life-cycle energy inputs and emissions. These inputs include energy and emissions resulting from growing and harvesting crops, processing crops into finished fuels, and transportation of raw materials and finished fuels.

2. It must be available in sufficient quantities to displace a measurable fraction of the fuel currently derived from petroleum without curtailing sufficient grain for nutritional purposes. It is especially important to safeguard the survival of the poorest people in the world, who depend on food imports at times when locally produced food supplies fall short. Thus, even if it were economically attractive to shift a large part of the grain harvest of countries, such as the United States from exports to domestic biofuel production, it would be morally untenable to pursue such a policy.

The presence of a positive net energy balance (NEB) ratio is often used as a criterion for evaluating a biofuel, meaning that a biofuel that delivers more energy to the vehicle than it requires in the production life cycle. A biofuel with a considerable positive NEB ratio is considered sustainable. While useful as a surrogate for a more complete analysis of environmental benefit and impact, the NEB ratio is imperfect because it does not consider the relative CO_2 emissions of the energy source (e.g., coal versus natural gas for process heat). It also does not consider land use changes that might be necessary for producing the biofuel; for example, clearing forests to grow crops for biofuels releases a large quantity of CO_2. Thus, if possible, measures of the target environmental concern (CO_2, air pollutants, water usage, and pollution) should be sought to evaluate the net environmental benefits of a biofuel. NEB ratios are also subject to change over time due to improvements in technology and agricultural/industrial practices, so it is important to work with the most up-to-date values available when assessing biofuels. For example, the NEB ratios for the corn-to-ethanol industry in the United States have been a moving target and are now more favorable than 10 years ago due to improved processing technologies at the ethanol plant, including lower energy needs for distillation of ethanol—the most energy-consuming process for ethanol production (Shapouri et al., 2010).

14-2-1 Policies
Developing a bioenergy industry cannot be uncoupled from local, national, and/or international policies. Communities, societies, and intergovernmental organizations may have very different reasons why they choose to develop bioenergy system rather than to rely on fossil sources for energy carriers. These reasons may include: (1) environmental: to tackle problems, such as air pollution of fossil fuel use (including greenhouse gas release), or impact from oil exploration; (2) socio-economical: to equalize the difference between the rich and the poor by focusing on decentralized renewable energy systems rather than centralized fossil fuel exploration by multinational organizations; and (3) geostrategical: to prevent national dependency on fossil fuels from other nations or intergovernmental organizations, such as the organization of petroleum exporting countries (OPEC). A society could also choose a laissez-faire policy to have the economic market dictate whether energy should come from fossil or renewable sources.

It is important to note that bioenergy research, development, and production occur within a complex and dynamic policy environment. Policies shape the bioenergy industry mainly through allocation of government funds and renewable energy mandates. Government programs, such as the U.S. Department of Energy Biomass Program, provide direct funding to biofuel companies, amounting to federal investment in bioenergy. Other bioenergy investments are made in the form of feed-in tariffs, which are essentially long-term government contracts for renewable energy producers that include cost-based compensation (e.g., in the form of a set price for a biofuel for a certain length of time). This directly decreases the pressure of competition and allows renewable energy companies to grow and develop. Feed-in tariffs also encourage technological advancement and cost reduction through a planned and gradual decrease in the price paid for the energy. The feed-in tariff system has been particularly successful in Germany and has resulted in dramatic increases in renewable energy (Mitchell et al., 2006).

In contrast, mandates do not directly support bioenergy development, although they encourage investment from private interests and other government programs. For example, the European Union has passed a directive requiring each member state to obtain at least 10% of total energy used from renewable sources by 2020. Some European Union members have also made more ambitious renewable energy goals, such as Denmark, which has mandated a minimum of 30% renewable energy by 2020. These mandates will make renewable energy companies more attractive to investors and more competitive in markets. However, not all policies are favorable toward bioenergy development. On a broad scale, subsidies and tax benefits for fossil fuel companies make it more difficult for biofuels to compete. Indeed, the Organization for Economic Cooperation and Development has estimated that removal of fossil fuel subsidies in some countries could lower global greenhouse gas emissions by up to 10% through a combination of energy conservation and increased use of renewable (Burniaux and Chateau, 2011). In addition, the bioenergy industry is subject to all zoning and local regulations just like any other industry. This means that before embarking on any bioenergy enterprise it is always important to consider the current policy situation to determine the technologies and locations that make sense not only scientifically and industrially, but also politically.

14-2-2　Net Energy Balance Ratio and Life-Cycle Analysis

The *NEB ratio* is a metric used to evaluate bioenergy systems by comparing the energy available for consumption in the produced energy carrier (e.g., ethanol) to the energy (e.g., petrodiesel and natural gas) required for growing, harvesting, and processing. For renewable energy systems, the available energy is seen as renewable, while the consumption of energy is often nonrenewable. A NEB ratio greater than one indicates that more energy is available for consumption than is required to produce the biofuel, while a NEB ratio less than one indicates that more energy is required to produce the fuel than is available for consumption, which is unattractive. This metric has been widely used in the biofuel debate, but, as we mentioned earlier, cannot stand alone. Life-cycle assessment (LCA) is a method of product assessment that considers all aspects of the product's life cycle. It is also referred to as cradle-to-grave analysis, and evaluates a product from raw material extraction through disposal and/or recycling. LCA was developed to account for additional factors missing from conventional economic analysis, such as environmental impacts of the production, use, and disposal of a product. For example, think about a grocery store in Virginia that needs to decide whether to purchase oranges from a farm in Florida or a farm in California. Using traditional economic analysis, the grocery store would evaluate the cost of the oranges (including transportation costs), while an LCA would include additional factors, such as increased greenhouse gas emissions associated with shipping the oranges across a larger distance (California versus Florida). As our society becomes increasingly aware of environmental issues, LCA provides a tool for a holistic comparison of alternative processes. Example 14-1 explores LCA considerations.

Example 14-1   A corn-to-ethanol plant is located 40.2 km (25 miles) from farm A and 160.9 km (100 miles) from farm B. Farm A will sell corn to the plant for \$289.36/metric ton (\$6.30/bushel), while farm B will sell corn at a price of \$248.02/metric ton (\$5.40/bushel). Each truckload can carry 10.9 metric tons (500 bushels), and the truck emits 212.3 g CO_2eq/metric ton-km (310 g CO_2eq/ton-mile). The plant needs 130.6 metric tons/year (6000 bushels/year), and the truck weighs 9.1 metric tons empty. Examine the two farms, considering both economics and greenhouse gas emissions. From which plant is it better to buy corn grain?

Solution   The corn-to-ethanol plant needs

$$130.6 \text{ metric tons/year @ } 10.9 \text{ metric tons} = 12 \text{ truckloads/year}$$

The economic return for farm A is

$$130.6 \text{ metric tons/year} \times \$289.36/\text{metric ton} = \$37{,}800/\text{year}$$

The total length for transportation between farm A and the plant for one way is

$$40.2 \text{ km/trip (one way)} \times 12 \text{ trips/year} = 482.4 \text{ km/year full (same amount empty to return)}$$

Greenhouse gas emissions for farm A are then a combination of transportation with an empty truck:

$$482.4 \text{ km/year} \times 9.1 \text{ metric tons} \times 212.3 \text{ } CO_2\text{eq/metric ton-km} = 0.93 \text{ Mg } CO_2\text{eq}$$

and a full truck:

$$482.4 \text{ km/year} \times 20 \text{ metric tons} \times 212.3 \text{ } CO_2\text{eq/metric ton-km} = 2.05 \text{ Mg } CO_2\text{eq}$$

The economic return for farm B is

$$130.6 \text{ metric tons/year} \times \$248.02/\text{metric ton} = \$32{,}400/\text{year}$$

The total length for transportation between farm B and the plant for one way is

$$160.9 \text{ km/trip (one way)} \times 12 \text{ trips/year} = 1930.8 \text{ km/year full (same amount empty to return)}$$

Greenhouse gas emissions for farm B are again a combination of transportation with an empty truck:

$$1930.8 \text{ km/year} \times 9.1 \text{ metric tons} \times 212.3 \text{ } CO_2\text{eq/metric ton-km} = 3.72 \text{ Mg } CO_2\text{eq}$$

and a full truck:

$$1930.8 \text{ km/year} \times 20 \text{ metric tons} \times 212.3 \text{ } CO_2\text{eq/metric ton-km} = 8.18 \text{ Mg } CO_2\text{eq}$$

Discussion   Based on purely the corn price difference between farms A and B, choosing farm B over farm A will result in a 14% economic savings per year (\$5400), but will result in a 300% increase (9 Mg) in annual greenhouse gas emissions. Obviously it will also be more expensive to drive a longer distance, but the cost for that may not be as considerable as the increase in greenhouse gas emissions. Regardless, from this simple assessment it seems reasonable that most corn-to-ethanol plants are located in the growing areas of corn.

14-2-3 Productivity of Fuels per Unit of Cropland per Year

Before choosing a regional crop for biofuel production, it is important to understand the productivity of fuel per unit of cropland per year. This analysis does not only take into account the yields of the crop at the regional soil and climate conditions, it also includes the efficiencies of harvest and the crop-to-fuel conversion technology used. This is especially important because of a relatively low retention rate of solar energy in plants and trees for later conversion into biofuels. Of the incoming energy, which may average 100 to 250 W/m^2 year round, taking into account diurnal cycles and variations in regional climate and latitude, less than 1% is available in the starches or oils as a raw material for conversion to fuel. This limit affects the ability of agriculture or forestry to provide sufficient raw materials to meet the greater part of the world transportation energy demand, especially as our societies must also generate enough nutrition to feed the world. In addition, because the crops for biofuels accumulate energy at such low density, a large energy expenditure is required to gather the material together for processing, which hurts the balance of energy produced relative to energy input requirements (i.e., NEB ratio). Much research work is, therefore, under way to find ways to capture more of the solar energy in the biofuel crop and to use more of the biomass (i.e., lignocellulose) of the entire plant as a raw material. Breakthroughs in research along these lines are, in fact, critical if biofuels are to become a major source of energy for transportation; otherwise, using current crops and technology, they can play at best a moderate role in displacing petroleum consumption and reducing CO_2 emissions.

It is also important to note that simply integrating crop yields per acre with harvest and conversion efficiencies is not enough. The decision maker should include data on the changes to the land that must be made to sustain a large-scale biofuels program. Noteworthy to mention is a 2008 study,[1] which showed that the conversion of nonagricultural land, such as rainforests, peatlands, savannas, or grasslands to produce crop-based biofuels in Brazil, Southeast Asia, and the United States creates much more CO_2 (i.e., carbon debt) than the annual greenhouse gas reductions that these biofuels would, for a very long time, provide by displacing fossil fuels (i.e., carbon credit). This vast release of CO_2 during land changes is from burning biomass or slow microbial decomposition of organic carbon stored in biomass (including roots) and soils. Such a CO_2 release does not occur when biomass is grown on degraded and abandoned agricultural lands. Similarly, the use of organic wastes to generate biofuels does also not have a carbon debt and offers immediate greenhouse gas credits. Besides the carbon debt from land conversion, the decision maker should also take into consideration data on the quantity of inputs of water, nutrients, and nonrenewable energy that is necessary to grow the bioenergy crops to account for the CO_2 release during the agricultural activity. Taking both the carbon credit for land changes and for agricultural activity into account, analyses may show that it is, from a greenhouse gas reduction standpoint, better to maintain low-intensity prairie grasses for biomass production than high-intensity monoculture crops, such as corn, on, for example, abandoned cropland.[2] It is, therefore, important to remain both critical and have an open mind when it comes to the development of bioenergy systems. Constant increases in yields and efficiencies and

---

[1]Fargione et al. (2008) estimated the carbon debt for land clearing activities.
[2]Fargione et al. (2008) also compared the greenhouse gas reduction from prairie biomass versus monoculture crops.

technological breakthroughs may change the outlook of a technology over time, and therefore periodic re-evaluation of bioenergy system alternatives is also important.

14-3 Biomass

Biomass comes in many different forms; plant materials, including their extracted sugars, starches, and oils and lignocellulosic stalks or wood, algae, and organic waste materials, including agricultural wastes, food wastes, and urban wood wastes. Currently, most ethanol is produced either from sugar from sugarcane or from pretreated cornstarch. The sugarcane and corn crops are specifically grown to produce ethanol. In tropical climates, growing sugarcane is attractive because of its high growth rates and for the sugars that are easily extractable without any other conversion needs (Fig. 14-1). Sugarcane can consist of up to 20% of sugar in the form of sucrose (i.e., a dimer of a glucose and a fructose molecule each with six carbon atoms). Yeast cells, which convert sugars with six carbon atoms into ethanol, can use sucrose in their metabolic pathways, and thus no conversion is needed before feeding it to yeast cells. However, the leftover lignocellulosic stalks from sugarcane (bagasse) are not used very efficiently, and this encompasses the largest part of the carbon in sugarcane. In most climates, the predominant quantities of biomass as a feedstock for bioenergy systems can come from lignocellulosic biomass, such as corn stover (i.e., leftover maize plant materials after removal of the corn grain), soft woods (e.g., poplar), and grasses, such as mixed prairie grasses or perennial grasses, such as switchgrass or miscanthus. Below, we discuss three feedstocks for bioenergy systems in more depth: lignocellulose materials, organic wastes, and algae. This section ends with some basic information on physical/chemical/thermal pretreatment steps that are available to make lignocellulose materials accessible to hydrolysis (i.e., breakdown of polymers into monomers and oligomers).

FIGURE 14-1 Sugarcane plantation, which surrounds an ethanol plant, in the Northeastern region of Brazil (state of Paraíba). (*Source:* Lars Angenent.)

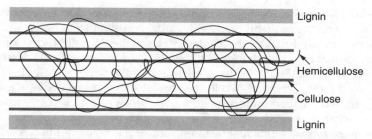

Figure 14-2 Simplified cartoon of the lignocellulosic matrix. Lignin forms a protective coat with hemicellulose acting as a connective network to keep the cellulosic fibers and lignin together.

14-3-1 Sources of Biomass

Lignocellulose Feedstocks

Lignocellulosic feedstocks are materials derived from plants, including trees, and they form the most abundant biomass source on land. For the United States, estimates were made of 1.3 billion tons of biomass per year that could be harvested (a large majority consists of lignocellulose materials, such as wood, corn stover, and grasses) to offset 30% of the fossil fuel for the transportation sector.[3] These numbers were further verified and updated in 2011.[4] In very general terms, lignocellulose consists of three components—cellulose, hemicellulose, and lignin (Fig. 14-2). Cellulose is a polymer of glucose molecules, which is the reason why so much attention has been placed on using lignocellulose for biofuel generation. Cellulose consists of crystalline and amorphous cellulose to form cellulose fibers. Hemicellulose is a complex material that consists of sugars with five carbon atoms (including xylan) and sugars with six carbon atoms. Due to its branched nature, hemicellulose is easier to break down than cellulose and its function is to form a connective network to keep the cellulose bundles and lignin together. Finally, lignin is an aromatic and nonsoluble compound that gives the plant its structure and protection against stresses. It is very hard to break down and is often toxic at high concentration to microbes. The relative composition of these three major components can vary considerably between different lignocellulosic feedstocks and this will have important consequences for the choice of effective conversion processes. Hydrolysis of polymeric lingocellulosic compounds is seen as the rate-limiting step in bioprocessing of feedstocks. This makes evolutionary sense, because lignocellulosic materials form the scaffolds for plant metabolism and must resist microbial attack.

Organic Wastes

Organic wastes are diverse in composition and are abundant; often they include relatively large quantities of lignocellulosic materials. Examples of organic wastes are animal wastes, such as dairy waste; agricultural leftover materials, such as corn stover; and urban food and wood wastes. Because of the relatively easy breakdown of *food waste*, this source of biomass has been getting more attention for bioenergy systems,

---

[3]Perlack et al. (2005) used conservative estimates to investigate if enough harvestable biomass was present in the United Stated to sustain a biofuel industry that could offset fossil fuel use for transportation.
[4]Perlack et al. (2011) updated and nuanced the Perlack et al. (2005) report.

especially for renewable methane production in Europe. Data is available from a recent EPA report on the availability of food, yard, and urban wood wastes.[5] In this report, the authors estimated that in the United States alone 26 million dry tons of waste from urban sources could produce 8.7 billion liters (2.3 billion gallons) of ethanol-equivalent fuel per year. This is a considerable fraction of the required 61 billion liters (16 billion gallons) of fuels derived from cellulosic feedstocks as mandated by the U.S. government (see more explanation on this fuel mandate below). One currently mostly unexploited and growing renewable energy source is urban wood biomass residues (i.e., *urban wood*) from trees, yards, municipalities, and construction/demolition sites. In the United States, this biomass source is the fifth largest projected source of biomass. In certain areas of the United States, however, including the Northeastern region (state of New York), this is the largest source of biomass (32%) with a total of 2 million tons produced per year.[6]

Algae

Microalgae are unicellular organisms capable of growing in inorganic salt media by fixing CO_2 and deriving energy from light harnessed by photosynthesis. Microalgae have high photosynthetic efficiencies (at least an order of magnitude higher than terrestrial crop plants), and thus a faster growth rate compared to traditional terrestrial plants. The significance of their rapid biomass assimilation is seen in global element cycling, where microalgae have been found to be responsible for upwards of 30% of atmospheric carbon fixation. Unlike terrestrial biomass, problems with freshwater consumption can be averted because many strains of algae are capable of growth in saline or wastewater. The combination of rapid growth and low nutritional demands could make algae an attractive biomass feedstock option for sustainable biofuels.

Currently, of the three primary components in algal biomass (cellular lipids, carbohydrates, and proteins), biodiesel derived from algal lipids is the most popular studied form of biofuel being produced from algae. By-products of this biodiesel process include glycerol (derived from triglycerides), and the remaining algal carbohydrates and proteins. Anaerobic digestion, however, is capable of using all components from algal biomass for the production of energy, and thus can generate more renewable methane (bioenergy) with the leftover components after lipid removal. During LCAs it has become clear that the NEB ratio for biodiesel production from microalgae is much lower than 1, which makes this process unattractive as a bioenergy system. However, one of the LCAs also shows that a best-case scenario exists with: (1) special photobioreactors rather than algae ponds (Fig. 14-3); (2) modern separation technology to extract lipids; and (3) anaerobic digestion to convert leftover biomass into methane gas, and that this would result in a NEB ratio that is close to 1, albeit still lower than 1 (and thus still unattractive).[7] This does show, however, that considerable improvements are possible during the maturation of the conversion technology.

---

[5]The EPA (2010a) report has a section on organic wastes as a considerable source of biomass.
[6]Milbrandt (2005) explored the regional availability of biomass sources in the United States.
[7]Brentner et al. (2011) compared a base-case with a best-case scenario for algae-to-biodiesel conversion and found it to be both unattractive as bioenergy systems at this point in time.

FIGURE 14-3   Photograph of a flat-plate photobioreactor, size approximately 50 cm × 50 cm.

14-3-2   Pretreatment Technologies

Yields and rates for converting biomass are greatly enhanced by pretreating the lignocellulosic material prior to enzyme hydrolysis and/or fermentation. Biomass is pretreated by subjecting it to physical, chemical, and/or thermal conditions that open up the cell wall structure, displace the xylan and lignin (when present) away from the cellulose polymers, and decrystallize (i.e., soften) the cellulose structure. Pretreatments, thereby, serve to give cellulase (enzymes to breakdown cellulose) access to the cellulose polymer, allowing for rapid and complete hydrolysis. A wide variety of pretreatments have been described which have been widely reviewed (Hendriks and Zeeman, 2009). Here, we mention four pretreatment schemes:

1. *Milling:*   This is done to increase the surface area of the biomass feedstock, to decrease the level of polymeric structure, and to shear the biomass. Milling increases the speed of hydrolysis and also the yields of production.

2. *Dilute-sulfuric acid:*   Pretreating with dilute-sulfuric acid is very effective at releasing starch and xylan sugars as monomers (i.e., degrading polymeric hemicellulose sugars) and leaving the more digestible cellulose in an accessible form. Major shortcomings of this pretreatment method are the generation of toxic by-products from sugar decomposition products (e.g., furfural) that inhibit, for example, yeast cells. Often this method is combined with hot-water

schemes. The produced toxic by-products have less of an effect on bioenergy systems with mixed consortia of microbes (e.g., anaerobic digesters) because by-products can be degraded.

3. *Hot water:*   Liquid hot-water pretreatment has been developed to solubilize the starch and xylan polymers and to produce digestible cellulose. This pretreatment only involves the use of process water and avoids the use of chemicals. However, this method does not completely degrade polymeric hemicellulose sugars.

4. *Alkaline:*   Alkaline pretreatment can be performed by adding different types of chemicals. Examples include ammonia fiber expansion (AFEX) and lime addition. The chemical basis of these pretreatments is the dissolution of xylan at a pH of 10 and above, which makes it highly effective at producing digestible cellulose. Similar to the hot-water scheme, it does not completely degrade polymeric hemicellulosic sugars.

14-4   Platforms

Within the biorefinery concept of integrating biomass conversion processes and equipment to generate bioenergy and biochemicals, several different platforms have been identified. We will discuss four different platforms—sugar platform, syngas platform, bio-oil platform, and carboxylate platform. The difference between these four platforms is the method of hydrolysis to untangle the lignocellulosic matrix of biomass, and thus the platforms are classified based on the most difficult step of biomass conversion. Each of the different methods of hydrolysis results in different platform chemicals (i.e., sugar, syngas, bio-oil, and carboxylate), which gives the platforms their names. Subsequent conversion steps into bioproducts are interchangeable between platforms. For example, anaerobic fermentation is able to convert both sugars from the sugar platforms and syngas from the syngas platform into ethanol. It is important to note that all platforms may be compatible and should be used in different and variable combinations with the goal to maximize product formation and economic viability at the biorefinery. In the end, a biorefinery is the same as a refinery; the bottom line is to maximize the revenue from the conversion of a raw product into refined products.

14-4-1   Sugar Platform

The best-known biorefinery platform is the sugar platform. The hydrolysis is performed with special enzymes (including cellulases) to convert the polymers of sugars (cellulose and hemicellulose) into their derived six carbon and five carbon-atom sugars after which bacterial or yeast fermentation can convert the sugars into alcohols. Before enzymes can be added, most often a pretreatment step is necessary to soften the feedstock and to make it accessible to the enzymes. Several companies in the United States are scaling this technology up for conversion of, for example, sustainable soft woods, such as poplar trees, into ethanol. The technology has been slow to develop due to technical and economical difficulties, including slow enzyme kinetics, costly bulk addition of extracted enzyme, and inhibiting effects of pretreatment by-products, among other factors. However, continued improvements may be making the sugar platform a promising technology.

14-4-2   Syngas Platform

When certain types of lignocellulosic biomass are hard or impossible to degrade by enzymatic processes due to their high lignin content and/or strong cellulosic matrix (Fig. 14-2), alternative strategies for biomass to liquid biofuel conversion must be used. A very promising strategy is to first convert these biomass feedstocks to synthesis gas (*syngas*, which is mostly H_2, CO, and CO_2) by gasification (Fig. 14-4) and then to use syngas as building blocks to synthesize a fuel. In many ways, gasification can be seen as a combustion process without oxygen. Besides syngas, the process also generates bio-char (a charlike solid material) and bio-oil (a viscous solution). The gasification process can be operated in many different configurations and operating conditions, which changes the ratios of different products and their consistencies. Syngas can be used for fuel production by biologic or abiotic processes. Biologically, bacteria of the class Clostridia can convert CO and H_2 from syngas into alcohols. Several companies in the United States are currently scaling up bioreactors with an enhanced transfer of these gases into solution so that the bacteria can convert them at sufficient rates. In the past, syngas from coal has mostly been converted by metal catalysts (an abiotic process), and this is currently also under investigation for the conversions of syngas from biomass. Besides producing fuels, syngas can also be used to power engines or fuel cells for combined heat and power generation. Finally, organic wastes, such as agricultural wastes or urban wastes, can be converted to syngas, adding to the potential of the syngas platform.

14-4-3   Bio-oil Platform

This platform also uses anaerobic gasification technology. To maximize *bio-oil* production from biomass feedstocks, rather than syngas and bio-char production, the gasification process is operated at an operating temperature of 500°C and a relatively

Figure 14-4   A gasification system at a chicken farm in the Eastern region of the United States (state of West Virginia). (*Source*: Johannes Lehmann, Cornell University.)

short biomass residence time of around 2 s. This process is often called fast pyrolysis due its speed of conversion. The main product is a viscous solution that is referred to as bio-oil. Some researchers have compared bio-oil to crude oil that can be further upgraded in biorefineries. However, problems exist—the bio-oil contains a lot of water (20%), may be corrosive, and is very complex in composition. Numerous fast pyrolysis configurations have been developed to convert biomass into bio-oil, and this is a thriving area of research.

14-4-4   Carboxylate Platform

For the carboxylate platform, mixed consortia of microbes hydrolyze the biomass. These microbes produce enzymes themselves, which are needed for hydrolysis, and therefore the enzymes do not have to be purchased. A proxy for the carboxylate platform is the animal gut in which complex biomass is converted to volatile fatty acids (short-chain carboxylic acids) that are taken up by the host. In bioenergy systems, the microbial process takes place in large-volume tanks. To speed up hydrolysis, pretreatment of the biomass may be necessary. The end products of this platform, which can be generated with mixed consortia, are either volatile fatty acids (acetic acid, propionic acid, lactic acid, or n-butyric acid) or the products from these acids. End products, then, can consist of methane, hydrogen, and medium-chain carboxylic acids. We will discuss the production of methane from volatile fatty acids later in this chapter (anaerobic digestion). The use of mixed consortia is advantageous because it has the genetic depth to handle a complex and variable feedstock and it is an open culture without the requirement to sterilize feedstock.

14-5   Alcohol

Ethanol, which is the best-known alcohol and often used synonymously with alcohol, is the name for a chemical compound, which resembles ethane (C_2H_6) with two carbon atoms, but has one hydrogen atom replaced by a hydroxide ion (C_2H_5OH). Other alcohols have a shorter (i.e., methanol with one carbon atom) or longer carbon chain (e.g., n-butanol with four carbon atoms and n-hexanol with six carbon atoms) compared to ethanol. Most ethanol is produced at present by fermenting crop sugars (such as sugarcane in Brazil or sugar-derived cornstarch in the United States) with pure-cultures of yeasts. The ethanol is produced in highly aqueous solutions of up to 15% ethanol (150 g/L). A higher concentration of ethanol would not be possible because the yeasts that carry out the fermentation would not survive. To produce a transportation fuel that is 99% or more ethanol, the producer must distill the solution to remove water, which is a highly energy-intensive process. In modern plants, distillation produces a solution with 95% ethanol and next a water-adsorbing system (i.e., molecular sieve) removes more water to produce a 99% ethanol solution. Ethanol is completely miscible, which means that at any concentration it remains in solution when water is present. The total energy consumption at any ethanol plant is high because of the need to distill—for a modern corn-to-ethanol plant in the United States, 20% of the energy content of ethanol itself may be required to separate it from the reaction solution (Shapouri et al., 2013). When natural gas is used as a source for steam to power the distillation systems, than the corn-to-ethanol technology, in fact, converts a gaseous fossil fuel in an easier-to-store liquid fuel. Fortunately, increasing amounts of biomass, including agricultural wastes, are used to power the steam boilers to considerably decrease the use of nonrenewable fuels at the plant, and thus the NEB ratio.

The energy content per unit volume for ethanol is lower than that of gasoline (21.3 MJ/L or 75,700 BTU/gal net energy content versus 32.5 MJ/L or 115,400 BTU/gal, respectively), so any comparisons of ethanol and gasoline must be conducted on an equivalent energy content basis. It is possible to combust pure ethanol in spark-ignition (SI) engines with appropriate modifications, or to blend it with gasoline and sell to consumers under the label "EXX," where XX stands for the percentage ethanol in the blend. SI engines can combust up to 10% ethanol without modification. Above this ratio, the engine must be modified to function well with the higher proportion of ethanol. In flex-fuel ethanol vehicles, the engine detects the proportion of ethanol and adjusts combustion accordingly. Different countries have taken different approaches. In Brazil, the passenger car fleet includes a mixture of vehicles that run on 100% ethanol, vehicles that use a fuel made up of a fixed ratio of gasoline to ethanol, and flex-fuel vehicles that can adapt to changing proportions (Fig. 14-5). In the United States, vehicles that can run on over E10 ethanol are flex-fuel vehicles, and can combust ethanol mixtures up to E85.

While world ethanol production is currently only a small fraction of the total output of gasoline, there has been robust growth in recent years. U.S. production of ethanol grew from 7.6 to 53 billion liters from 2001 to 2010 (2.0 to 14 billion gallons). Brazil produced 26 billion liters (6.9 billion gallons) in 2010, making it the second largest producer in the world after the United States. There is room for further growth in output before limitations on the total size of sugarcane or corn crops would curtail production. However, most of the future production growth in the United States will have to come from ethanol that is produced from other biomass sources than cornstarch because this

FIGURE 14-5 In Brazil, a car driver can choose between ethanol fuel or conventional gasoline based on their seasonal-depending costs, because their cars are flex-fuel vehicles. (*Source:* Lars Angenent.)

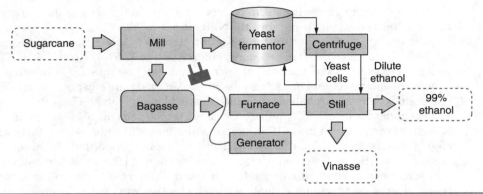

Figure 14-6   Simplified schematic of the sugarcane-to-ethanol process.

is mandated by the Energy Independence and Security Act of 2007 (EISA). This Act specifies that in 2020, about 136 billion liters (36 billion gallons) of ethanol-equivalent fuel should be produced from renewable sources of which 79 billion liters (21 billion gallons) from noncornstarch sources. A total of 61 billion liters (16 billion gallons) is expected to come from cellulosic feedstocks. Thus, the United States has almost maximized the ethanol that can be produced from cornstarch (56 billion liters or 15 billion gallons). In 2010, this close-to-maximized value resulted in the use of approximately 30% of U.S. corn produced allocated to ethanol.[8]

14-5-1   Sugarcane to Ethanol

Sugarcane is a tall perennial grass with fibrous stalks that are rich in sugar (sucrose) and may reach up to five meters (16 ft) tall (Fig. 14-1). In the Northeastern area of Brazil, which has a tropical climate, this grass is harvested once a year. Some irrigation and nutrient addition may be necessary to increase the crop yields depending on the location. After harvesting, stalks of the sugarcane are transported to the ethanol plant (often the fields with sugarcane are surrounding the plant) and crushed by milling to recover the sugar (Fig. 14-6). The solution with high sugar content (i.e., juice) is transported to fermentation tanks with yeast to generate ethanol at concentrations of approximately 8%. Off gases from the yeast fermentation tanks are scrubbed for ethanol recovery.

The fermentation process may take half a day and yeast cells are recycled for a next batch (Fig. 14-6). The miscible ethanol from the fermentation tank is then pumped to distillation towers (stills) for ethanol recovery. Distillation to recover ethanol is energy intensive, and the required energy is generated in a furnace by burning the fibrous material (i.e., bagasse) that is left after crushing sugarcane (Fig. 14-6). The leftover solution without ethanol is called vinasse and can be partly recycled, used for irrigation, sold, or used to generate methane in anaerobic digesters. In the past, the furnace delivered just enough energy to power the ethanol plant, including pumps to irrigate the surrounding fields (Fig. 14-7). More modern and efficient energy recovery systems

---

[8]The United States EPA (2010a) Renewable Fuel Standard Program (RFS2) Regulatory Impact Analysis, estimated the maximum of renewable fuel that may be generated from cornstarch.

Figure 14-7 Fermentation tanks (foreground) and bagasse furnace (left in background) at an ethanol plant in the Northeastern region of Brazil (state of Paraíba). (*Source:* Lars Angenent.)

have made the Brazilian ethanol industry a net electricity exporter, but more can be done to harvest energy from sugarcane, for example, by converting some of the bagasse into liquid fuels.

14-5-2   Corn Grain to Ethanol

In the United States, ethanol is produced from corn grain, which is also referred to as corn kernel. Two different processes are used for ethanol processing—dry milling and wet milling. The former process is less capital intensive and is simpler to operate, resulting in utilization of dry milling for many of the small- to mid-size ethanol plants that have been built over the last 5–10 years in the United States (Fig. 14-8). Wet milling, on the other hand, is more capital intensive, but also generates more diverse and higher value products. Additional fractionation and separation technologies are used for wet milling to first completely separate all different compounds of corn grain. Only the purified cornstarch is used for ethanol production. Here, we discuss dry milling in more detail. More of the corn grain compounds than just cornstarch are used in dry milling, which increases the ethanol yield for dry milling versus wet milling.

Dry Milling

The entire corn grain is milled and then cooked during which added enzymes (α-amylases) liquefy the cornstarch into dextrins, which are polymers of glucose (Fig. 14-9). The solution with dextrin is further heated with steam and then cooled again to body temperature. In the fermentor, yeast cells, antibiotics (to control bacterial infections), and enzymes (glycoamylases) are combined for simultaneous saccharification (formation of glucose) and fermentation. After 2–3 days, a 15% ethanol concentration is achieved in the fermentation broth and the process is terminated. The off gases from the fermentor are treated to

FIGURE 14-8 Four fermentation tanks at a dry-milling plant in the Northeastern region of the United States (state of New York). This plant is powered by natural gas and electricity (hydropower). (*Source:* Lars Angenent.)

recover ethanol, and possibly to harvest carbon dioxide for the bottling industry. The final solution in the fermentor is called beer and is pumped to the distillation towers (stills) to recover ethanol at a concentration of approximately 95%. Molecular sieves increase the concentration of ethanol to 99% after which the ethanol is stored on site.

The solution without ethanol from the liquid outflow of the stills is called whole stillage and this is pumped to a centrifuge for recovery of the leftover corn grain solids and yeast cells. A wet cake is generated that is dried in large rotating dryers (Fig. 14-9). The liquid outflow of the centrifuges has still large quantities of solids, albeit at much lower concentrations, and this is pumped to evaporators that are powered with waste heat from the ethanol plant. Evaporators have two products—evaporated water and

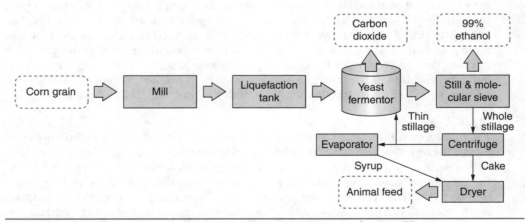

FIGURE 14-9 Simplified schematic of the corn grain-to-ethanol process for dry milling.

syrup. The evaporated water is recovered and recycled back to the start of the process to reduce the consumption of fresh water. Syrup is combined with the wet cake during drying. The combined dried material generated in the dryer is sold as animal feed and is called dried distillers grain with solubles (DDGS) in the ethanol industry. The quantities of DDGS are large and the economic viability of the ethanol industry is dependent on this co-product. This also shows that the discussion about fuel versus food is a gray area and not as black and white as it may seem upon first introduction, because ethanol plants generate a large value of pretreated animal feed. In fact, in LCAs, the production of animal feed results in energy credit (increases the NEB ratio).

Life-Cycle Assessments

Many LCAs have been performed for the United States corn-to-ethanol industry. Several years ago, the average NEB ratio that was published was in the range of 1.25 and was a reason for criticism of the entire ethanol industry. The LCAs were broad (as they should be) and included production of ammonia for plant uptake, production of farm equipment, and farm and household energy use, among many other factors. It also, on the other hand, included a credit for the production of animal feed. The most recent assessments show an improved NEB ratio between 1.9 and 2.3,[9] by modernizing the ethanol plants, including more efficient ethanol separation technology (stills and molecular sieves). Some plant locations are already incorporating agricultural-waste biomass for energy production rather than using natural gas or coal, and the estimated NEB ratio for such plants would be 2.8 with 50% biomass power.[10] More can be done, though, and new ideas are emerging. For example, thin stillage can be converted to renewable biogas by integration of anaerobic digesters in the plant,[11] or miscible ethanol can be converted to a different chemical that does not need as much energy to be separated from solution.

Even though the NEB ratio is not high in terms of sustainability, one attraction of ethanol production in the United States is that it serves as a petroleum multiplier because it takes a relative small input of liquid fuel (petroleum) as part of its life cycle (e.g., for farming and for truck or rail transportation to move the corn crop or ethanol product), and converts this input into a larger quantity of liquid fuel (ethanol) in terms of energy equivalent. This process requires large inputs of other fossil energy sources, but since natural gas and coal are used for these inputs, and the United States primarily relies on domestic reserves for these fuels, ethanol helps U.S. balance of trade. In addition, the 10% addition of ethanol to gasoline adds oxygen to the fuel mix and this results in a more efficient car engine. Already, this shows that the issues are more complex than just the discussion of what an appropriate NEB ratio should be. Additional factors that are important to take into consideration in the debate about U.S. ethanol industry are:

- The published NEB ratios are based on average numbers after evaluating different ethanol plants. There is a large amount of variability in ethanol production processes, and it would be difficult to create a study that truly represented the "average" process. Rather than attempting to generalize, we should attempt to identify best practices.

---

[9]Shapiro et al. (2010) evaluated several dry-milling plants based on 2008 data gathering.
[10]This is the same study from Shapiro et al. (2010) but now by adding data from biomass energy inputs.
[11]Agler et al. (2008) performed a lab-based study by integrating anaerobic digestion into the dry-milling process to estimate the NEB ratio to increase from 1.3 to 1.7 due to the formation of renewable methane—this included a correction for a lower production of animal feed, and lowered the animal feed credit.

- Even though the NEB ratio for ethanol is modest, producing ethanol should be seen as a stepping-stone toward producing more effective biofuels in the future, rather than an end in itself. Different fuels and processes could be adopted when these would emerge out of research. When that occurs, the United States has in the meantime gathered infrastructure and a trained work force that is able to design and build facilities to produce biofuels.

- The debate over energy inputs and outputs is a distraction from a much more important concern, namely, whether diverting too much corn to ethanol, and achieving modest environmental gains, will eventually have repercussions on basic food supplies that the United States as a country comes to regret.

Overall, it is clear that, for ethanol production in temperate regions to have a real positive effect on the environmental bottom line, the ethanol production process will need to be changed to reduce energy input required and use nonfossil resources.

14-5-3 Cellulosic Ethanol

In the future, a relatively large part of the renewable fuel mix will have to come from cellulosic feedstocks as mandated by the United States government. Currently, technology is being scaled up that can convert cellulosic biomass into ethanol by using either sugars with five and six carbon atoms from the sugar platform or syngas from the syngas platform. Bacteria and yeast cells can convert sugars into ethanol. When bacteria produce ethanol, the maximum ethanol concentration they can achieve is approximately 4–5%. When wild-type yeasts are used for sugar conversion, a higher concentration of ethanol may be achieved (15%), but the problem then is their inability to use the majority of sugars that stem from hemicellulose (five carbon-atom sugars), which results in a lower lignocellulose-to-ethanol conversion efficiency. For syngas fermentation, only bacteria are being used with an anticipated ethanol concentration that is approximately 4%. Thus, for cellulosic ethanol, a lower ethanol concentration of 4–5% is anticipated than the 15% that is achieved for the corn ethanol industry. With a lower concentration of ethanol, a larger amount of energy is needed to distill the product, which will result in a higher requirement for energy at the plant to distill ethanol. Estimations are that about 24% of the energy content in ethanol is used for distillation (Galbe and Zacchi, 2002) when cellulosic ethanol is distilled to 95%, while this is 20% for corn ethanol. Even though the energy requirement to distill ethanol for the corn-to-ethanol industry is lower, the NEB ratios for cellulosic ethanol are anticipated to be considerably higher than corn ethanol. We will have to wait until the cellulosic ethanol industry matures to find out what these NEB ratios will be.

14-5-4 *n*-Butanol

n-Butanol is an alcohol with a carbon length of four atoms in a straight carbon chain. Compared to ethanol, the chemical properties of *n*-butanol make it more similar to gasoline and a better fuel for SI engines. *n*-Butanol is produced from fermentation of sugars by bacteria from one class (Clostridia), which co-produce acetone and ethanol as by-products. The process to generate *n*-butanol is called acetone-butanol-ethanol (ABE) fermentation and was first publicized in 1861 by the famous microbiologist Louis Pasteur, who gave pasteurization its name. An industrial-scale fermentation process to supply acetone for gunpowder for the British army was developed during the World War I. The co-produced *n*-butanol was used as a solvent for lacquer in the car industry.

Later, the ABE fermentation process was also used to produce *n*-butanol as a fuel. With the rise of the petroleum industry after the World War II, the industrial production of *n*-butanol and acetone by fermentation gradually declined due to price competition until the last commercial ABE fermentation plant in South Africa closed in 1982. Today, the ABE-fermentation process has seen a small revival due to higher fuel prices and concerns about the future availability of fossil fuels. But, although *n*-butanol is a better engine fuel, producing it in large quantities as a biofuel for transportation comes with even greater challenges than ethanol. *n*-Butanol is considerably more toxic to bacteria than ethanol is to yeast, resulting in much lower maximum concentrations that are achievable during ABE fermentation, and a resulting higher costs for energy-intensive distillation. Another problem is that the sugars for fermentation should come from lignocellulosic feedstock rather than from starch (as mandated by the United States government). Even though *n*-butanol production from lignocellulosic feedstocks is studied in research labs, a widespread *n*-butanol industry is still missing.

14-6 Biodiesel

Both biodiesel and petrodiesel fuels consist of complex hydrocarbon strings with an average composition of $C_{12}H_{26}$, but with the number of carbon atoms varying around this average (10 to 15 carbons). The chemical composition of biodiesel is substantially different from that of petrodiesel, and it has lower energy content per unit volume than petrodiesel (33.0 MJ/L or 117,000 BTU/gal net energy content versus 36.2 MJ/L or 128,700 BTU/gal, respectively). Most biodiesel in use today fits this description, although work is ongoing to develop a bio-derived synthetic diesel that originates from plant or animal sources, but is chemically much closer to petrodiesel. Total sales of biodiesel in the United States began rising rapidly in the late 1990s, increasing from 1.9 million liters (500,000 gallons) in 1999 to an estimated 3 billion liters (0.8 billion gallons; this is equivalent to 1.2 billion gallons of ethanol) in 2011. Germany produced an estimated 5.7 billion liters (1.5 billion gallons) of biodiesel in 2010, mostly from rapeseed oil. Feedstocks for biodiesel production in Germany and the United States come primarily from agricultural crops. At the same time, biodiesel is produced from local waste oil streams by individuals and small-scale businesses, which also produce straight vegetable oil (SVO) or vegetable fat (i.e., not converted to diesel fuel using the process described above) that can be combusted in a CI engine when preheated so as to reduce viscosity.

The most common form of biodiesel is as a 20% biodiesel/80% petrodiesel blend known as "B20," since it requires very little adaptation of the vehicle or fuel supply system for use in existing vehicles. With adaptation, B100 (100% biodiesel) can also be used. Diesel (CI) engines are in general more sensitive to cold weather operation than gasoline (SI) engines, and with use of biodiesel, this sensitivity is increased. Therefore, biodiesel operators must take precautions to avoid gelling of fuel in supply lines, especially when using biodiesel at higher concentrations (greater than B20). Otherwise, no significant changes are needed to use biodiesel in CI engines. In some ways, biodiesel can be beneficial to the engine and fuel supply system; for example, adding 2% biodiesel to fuel increases the lubricating capability of the fuel inside the engine.

To eliminate the price gap between petrodiesel and biodiesel, a U.S. federal tax incentive went into effect in 2005, which provided a \$0.20/gal incentive for B20 made by blending #2 diesel with virgin biodiesel and a \$0.10/gal incentive for B20 made by blending #2 diesel with reused animal- or vegetable-based oil. Here, virgin biodiesel is made from either crops, such as soybeans, or animal fat from the meat-packing industry, while reused biodiesel comes from the purification and reacting of waste oil products, such as cooking oil. Improving efficiency in the biodiesel industry and rising prices of crude oil on the world market since 2005 have made biodiesel more competitive. Nevertheless, the industry continues to depend on price supports. In the United States, this became very apparent in 2010–2011, when their tax incentive was removed only to see a considerable decrease of biodiesel production. In 2011, production was back up due to reinstatement of the tax incentive.

14-6-1   Production Processes

Biodiesel is considered renewable because it is derived from agriculturally produced and industry-wasted plant oils, such as palm oil, soybean oil, and rapeseed oil; from algae-extracted triglycerides; or from animal fats, such as tallow or lard. The biological starting materials for biodiesel production are triglycerides: an organic molecule consisting of three fatty acid chains attached to a common glycerine backbone at its three hydroxyl groups with an ester bond (Fig. 14-10). In the best-known biodiesel manufacturing process, the triglyceride molecules are cleaved into their constituent fatty acid molecules and glycerol molecules through a thermo-chemical reaction known as transesterification. This reaction involves adding a catalyzing agent (typically a strong base, such as sodium hydroxide or potassium hydroxide), which mediates the replacement of the fatty acid–glycerine bond with that of an alcohol, such as methanol or ethanol. This results in the formation of biodiesel and glycerol.

The mixture of biodiesel and glycerol is then allowed to settle and the denser glycerol is removed by gravity separation. On a weight-to-weight basis, one part of glycerol is generated for every 10 parts of biodiesel produced. This biodiesel process is relatively easy to operate and the diesel engines are so forgiving that many small and decentralized plants have emerged with homeowners performing the production in their own garages. Glycerol, which is seen as a waste material from biodiesel production, is an ideal substrate to be mixed in with, for example, animal wastes for the production of renewable methane in anaerobic digesters (discussed below).

Although less known, full-scale manufacturing plants exist with a different manufacturing process with the same feedstock as the transesterification process, but based on a thermo-chemical reaction known as dehydrogenation. This process yields significant

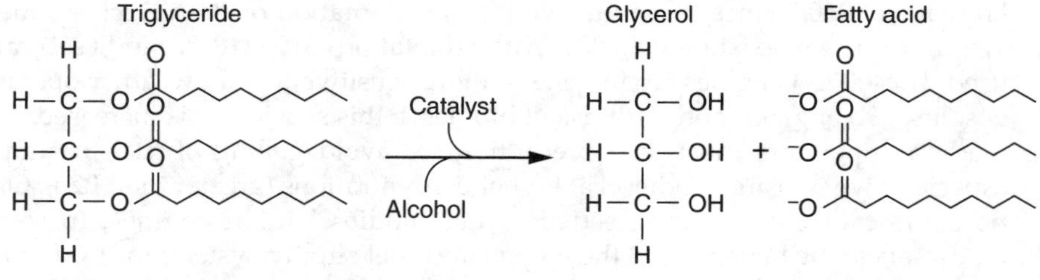

FIGURE 14-10   The reaction of the best-known biodiesel production process. This thermo-chemical reaction, which is known as transesterification, cleaves triglycerides in glycerol (by-product) and long-chain fatty acids (biodiesel). The carbon tails (right side of molecule) for the triglycerides and fatty acids can be of variable length.

volumes of biopropane as a by-product rather than glycerol (here the glycerine chain of the triglyceride is hydrogenated to produce propane). Biopropane is a small molecule with three carbons (C_3H_8) that can be used directly as liquid-petroleum gas for transportation. Liquid-petroleum gas is a generic name for mixtures of hydrocarbons [predominantly propane and butane (C_4H_{10})] that change from a gaseous to liquid state when compressed at moderate pressure or chilled. An alternative use for biopropane, which is especially important for rural areas that are not connected to the natural gas and/or electricity grid, is for cooking and heating of houses. This is the case for both developing and developed countries.

14-6-2 Life-Cycle Assessment

Most, though not all, recent studies of the life-cycle energy inputs and outputs from biodiesel production have estimated a substantially positive NEB ratio, which is greater than that of corn ethanol. One study performed the assessment for both corn grain ethanol and soybean biodiesel production as a comparative effort (Hill et al., 2006). Their study showed a NEB ratio of 1.9 for biodiesel (1.25 for ethanol) when co-products are included in the analysis. The apparent improvement in the NEB ratio compared to corn ethanol comes from the reduced energy requirements in the processing stage, since the used vegetable oils are closer to biodiesel in their chemical formulation, and require processing with a chemical catalyst, but do not require distillation. When greenhouse gases other than CO_2 are taken into account (e.g., N_2O from agriculture—note that N_2O has a much higher greenhouse gas equivalent than CO_2), biodiesel may release on the order of 60% of the greenhouse gas equivalent of petrodiesel over its life cycle per unit of energy delivered to the vehicle. Again, for LCAs, it is important to include all greenhouse gases during the entire life cycle of the biofuel even when this includes the impact from agriculture or waste management.

Although biodiesel from crops, such as soybeans, may have a considerably higher NEB ratio compared to corn ethanol, soybeans as a crop suffer from lower yields per land area compared to corn. Thus, the maximum soybean biodiesel production that could be achieved worldwide may be more limited in terms of total energy content. For example, in the United States, without substantial changes to the allocation of planting or noticeable reduction in soybean exports, the maximum biodiesel output is thought to be about 5 to 6% of current petrodiesel consumption on an energy content basis. Ethanol output from corn grain, on the other hand, is already generating approximately 10% of the gasoline consumption in the United States. Again, decision makers must take into account many factors besides just the NEB ratios.

14-7 Methane and Hydrogen (Biogas)

The gaseous fuels methane and hydrogen are often seen as fuels to power stationary power generation systems, such as generators (SI engines) and fuel cells. However, methane is used to power cars, buses, and trains, while hydrogen powers cars and buses. Renewable methane is produced from organic materials, often part of wastes, by anaerobic digestion. Hydrogen can be used as a fuel to generate electricity and some heat in a fuel cell, but because of the presence of expensive metal catalysts and their sensitivity to toxicity by many different molecules, the hydrogen gas has to be ultra clean to ensure a long lifetime of the fuel cell.

14-7-1 Anaerobic Digestion

Anaerobic digestion consists of a tank with an enclosed (air-tight) environment in which diverse consortia of microbes degrade organic material to generate biogas. The advantage of this technology is that besides mixing no other power sources are necessary because the microbes work under strict anaerobic conditions and aeration is not necessary. The process is very different from, for example, the ethanol fermentation systems that operate with a pure culture. Anaerobic digesters are open systems, which means that diverse types of microbes can come in with the waste streams to circumvent the need to sterilize the inflow streams, and this eventually results in thousands of microbial species being present in a relatively stable microbial consortium. Within anaerobic digesters, microbes comprise a food web, which means that a product from one microbe is the substrate (food) for another one. The biogas, which consists typically of 60–70% of methane, can be combusted to generate heat and/or electricity in a boiler or combined heat and electric power system.

Anaerobic digestion is a mature technology and, if operated under stable conditions, is a very efficient energy recovery system because the final products—methane and carbon dioxide—are automatically and constantly removed from solution by degassing (bubble formation) to increase the thermodynamic potential for the biological reactions. In contrast to ethanol production, no energy is needed to separate the fuel product. Another reason why this is an efficient process is because microbial cell yields for anaerobic microbial growth are low and all intermediate fermentation products are maintained at very low concentrations (i.e., no other product that lowers efficiency is formed). In the United States, anaerobic digestion technology is currently used on farms, at municipal wastewater treatment plants to treat biosolids, and in industrial wastewater treatment applications. In countries, such as Brazil and India, domestic wastewater is also treated with anaerobic digester technology.

In 2010, U.S. EPA estimated that there were only 157 digesters operational at commercial-scale livestock facilities in the United States, while 8000 farms across the country are good candidates for biogas installations with a combined capacity of 1670 MW.[12] In another estimate, the increased value of both methane reduction credits (currently, untreated manure in piles generates methane, which considerably adds to greenhouse gas release from the agricultural sector) and electricity prices will result in a competitive economic model for anaerobic digestion in 2025. Under these favorite conditions, anaerobic digesters on farms could generate 5.5% of U.S. electricity consumption.[13] This may even be a conservative estimate of the total capacity because over 5000 farm-based digesters have already been built in Germany alone over the last 10 years due to the favorable feed-in tariff system, albeit the German digesters are co-fed with animal manures and energy crops, such as ensilaged corn plants, to increase capacity in excess of 2000 MW. Figure 14-11 shows one of the 157 farm-based digesters that are in operation in the United States.

Anaerobic digestion offers many benefits to farmers in addition to its use as an on-site energy source. Some examples include odor reduction and the production of a stabilized fertilizer that can be applied to cropland. Why then are not more digesters

---

[12]AgStar through the EPA (2010b) has estimated the opportunity for anaerobic digesters on commercial live-stock facilities (farms).

[13]Zaks et al. (2011) have discussed under what circumstances farm-based digesters have an economic opportunity.

FIGURE 14-11 In-ground anaerobic digester on a 1000-head dairy farm in the Northeastern region of the United States (state of New York). The biogas from this co-digestion process, which treats animal manure and food wastes, generated over 330 KW of electricity by two SI engines at the moment this picture was taken. (*Source:* Lars Angenent.)

built in the United States? The combination of low natural gas/electricity prices and the high capital cost for digesters results in a too long return-of-investment period. Farmers can increase the revenue potential of their digesters, and therefore shorten the return-of-investment period, by charging tipping fees to treat other organic wastes from, for example, food industries in addition to their manure (co-digestion of combined wastes). There is a renewed interest in using anaerobic digestion at domestic wastewater treatment plants in the United States, because co-digestion of food waste from the community with the biosolids that these digesters were designed to treat originally may generate more methane (and thus electricity when a combined heat and power system is present). Such addition of organic materials can generate enough electricity to cover all energy needs at the wastewater treatment plant to make them carbon neutral.

Anaerobic digesters are typically run in either mesophilic (25–37°C) or thermophilic (55–65°C) temperature ranges, with each temperature range having its own advantages and disadvantages in terms of performance. The food web underlying the anaerobic digestion process can be broken down into four stages: hydrolysis, acidogenesis, acetogenesis, and methanogenesis. During the first stage, hydrolyzing bacteria mediate the degradation of large, complex organic molecules, such as proteins, carbohydrates, and lipids, into smaller, simpler molecules of amino acids, sugars, and fatty acids, respectively. Next, during acidogenesis, these compounds are further broken down by bacteria into volatile fatty acids and other intermediates. Acetogenesis constitutes the next degradation stage whereby the predominant intermediates (volatile fatty acids) are converted by bacteria to acetate, hydrogen, and carbon dioxide (CO_2). Finally, during methanogenesis, two possible pathways exist [splitting acetate to methane (CH_4) and carbon dioxide (CO_2) or combining hydrogen (H_2) and CO_2 to

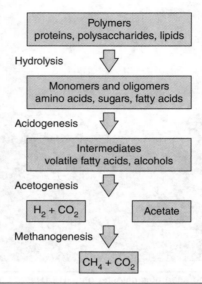

FIGURE 14-12   Simplified schematic of the anaerobic food web. The product from one group of microbes is the substrate for another group until methanogenens complete the conversion by producing methane and carbon dioxide.

produce CH_4] (Fig. 14-12). The group of microbes that perform these pathways is called methanogens, which are not bacteria, but rather archaea. Archaea are the third domain of life and are microscopic. Example 14-2 illustrates the function of anaerobic digestion.

Example 14-2   There are several digester configurations used for anaerobic digestion, including the continuously stirred anaerobic digester (CSAD), the plug-flow digester, the anaerobic sequencing batch reactor, and the upflow anaerobic sludge blanket reactor. One type of anaerobic digester system that is commonly used for animal waste treatment is CSAD. In an ideal CSAD system at steady state, the digester contents are assumed to be uniformly mixed, and the concentration of a compound coming out in the reactor effluent will equal the concentration of that compound within the reactor itself. Often, it is useful to estimate the final compound concentration leaving the reactor by performing a mass balance on the CSAD system. Sticks and Stones Farm has built a new anaerobic digester on their farm to process their dairy manure. They designed a digester with a volume (V) and flow rate (Q) to achieve a hydraulic residence time (θ) of 25 days. However, they just realized that a chemical (used in the United States to increase the milk yields) at a concentration (C_o) of 6 mg/L in the dairy manure will also be entering the digester. The farmers want to ensure that the effluent (C) leaving the reactor has 0.5 mg/L or less of the chemical. Assume that the antibiotic undergoes first-order decay in the reactor with a decay rate constant (k) of 3.5×10^{-4} min^{-1}. Determine whether the final antibiotic concentration leaving the steady-state reactor would be less than the desired 0.5 mg/L (assume an ideal CSAD system).

Solution   First, set up a mass balance on the system.

Rate (mg/d) of change of antibiotic conecentration within digester		Rate (mg/d) of the antibiotic entering digester		Rate (mg/d) of antibiotic leaving digester		Rate (mg/d) of decay of antibiotic within digester
	=		−		−	

Therefore,

$$V\frac{dC}{dt} = Qc_0 - QC - kCV \qquad (14\text{-}1)$$

Assume the digester is at steady state,

$$V\frac{dC}{dt} = 0 \qquad (14\text{-}2)$$

Also, we know that

$$\text{hydraulic residence time, } \theta = \frac{v}{Q} \qquad (14\text{-}3)$$

Substituting Eqs. (14-2) and (14-3) into Eq. (14-1) and rearranging yields

$$C = C_0\left(\frac{1}{1+k\theta}\right) \qquad (14\text{-}4)$$

Finally, putting in provided values into Eq. (14-4), we get

$$C = 6 \text{ mg/L}\left(\cfrac{1}{1+(3.5\times10^{-4}\,\text{min}^{-1})\left(\cfrac{1440\ \text{min}}{\text{days}}\right)(25\ \text{days})}\right) = 0.4 \text{ mg/L}$$

Note that this value is just below the maximum allowed concentration for Sticks and Stones Farm (!).

14-7-2 Anaerobic Hydrogen-Producing Systems

Researchers have explored the use of mixed consortia of microbes in engineered systems to produce hydrogen gas. When the methanogenic archaea are inhibited within the mixed consortia, hydrogen and volatile fatty acids, such as acetate, butyrate, and propionate, are the fermentation end products from organic material conversion. Most of the chemical energy is amassed in the volatile fatty acids, and unfortunately not in the produced hydrogen gas. Of the 12 hydrogen atoms that are present in, for example, glucose (i.e., $C_6H_{12}O_6$), only a maximum of four hydrogen atoms (one-third of the maximum yield) can be generated with bacteria, but in reality only approximately two hydrogen atoms (one-sixth) is generated. From a renewable energy point of view, focusing solely on hydrogen production with mixed consortia may, therefore, not have much future. However, two anaerobic digesters that are placed in series (outflow stream of the first digester is the feed stream of the second digester) to convert an energy-rich organic waste may generate some hydrogen in the off gas from the first digester. In the second digester, the soluble volatile fatty acids are then converted into methane. By mixing the off gases from both digesters, a biogas with some hydrogen gas is produced and this has a higher energetic value compared to biogas without hydrogen gas when burned in a generator (SI engine) or boiler. One problem with generating hydrogen that always needs to be considered, however, is the very diffusive nature of the small molecule hydrogen; it leaks through most plastics, and therefore out of holding tanks, tubing, and connectors (special metal components do not leak hydrogen, but these add costs). This diffusive character of hydrogen gas is also a problem for other renewable energy systems that have been proposed, such as using algae in special photobioreactors to generate hydrogen gas from sunlight.

14-8   Summary

In this chapter, we have discussed several different bioenergy systems to produce energy carriers. We have focused not only on liquid biofuels but also on stationary power, and have spent most of our text on systems that are already being used at an industrial scale, including ethanol production from sugarcane or corn grain, or that have potential in the short term. We have not discussed novel biofuels that are developed by using synthetic biology (reshaping metabolic pathways on a cellular level), longer-chain carboxylic acids with mixed consortia of microbes, and drop-in biofuels that are similar in behavior to an existing fossil fuel and that can be used without adaptation to fuel systems. Many new ideas are being investigated and hopefully the biofuel industry will look very different 20 years from now. This is especially pertinent because for most of the discussed bioenergy systems, the NEB ratio is only slightly above 1, while for a true sustainable bioenergy system, this ratio may need to be 10–20. Of course, we must look further than just the NEB ratio and evaluate the land and water uses and never forget that we must produce enough human food.

References

Brentner, L. B., M. J. Eckelman, and J. B. Zimmerman (2011). "Combinatorial Life Cycle Assessment to Inform Process Design of Industrial Production of Algal Biodiesel." *Environmental Science & Technology*, Vol. 45, pp. 7060–7067.

Burniaux, J., and J. Chateau (2011). "Mitigation Potential of Removing Fossil Fuel Subsidies: a General Equilibrium Assessment." OECD Economics Department Working Papers, No. 853. OECD Publishing. DOI:10.1787/5kgdx1jr2plp-en.

Drapcho, C., J. Nghiem, and T. Walker (2008). *Biofuels Engineering Process Technology*. McGraw-Hill, New York.

EPA (2010a). Renewable Fuel Standard Program (RFS2) Regulatory Impact Analysis, EPA-420-R-10-006. U.S. Environmental Protection Agency, Washington, DC.

EPA (2010b). US Anaerobic Digester Status Report, AgSTAR, October 2010. U.S. Environmental Protection Agency, Washington, DC.

Fargione, J., J. Hill, D. Tilman, et al. (2008). "Land Clearing and the Biofuel Carbon Debt." *Science*, Vol. 319, pp. 1235.

Galbe, M., and G. Zacchi (2002). "A Review of the Production of Ethanol from Softwood." *Applied Microbiology and Biotechnology*, Vol. 59, pp. 618.

Hendriks, A.T.W.M., and G. Zeeman (2009). "Pretreatments to Enhance the Digestibility of Lignocellulosic Biomass." *Bioresource Technology*. Vol. 100, pp. 10.

Hill, J., E. Nelson, D. Tilman, et al. (2006). "Environmental, Economic, and Energetic Costs and Benefits of Biodiesel and Ethanol Biofuels." *Proceedings of the National Academy of Sciences*. Vol. 103, pp. 11206.

Milbrandt, A. (2005). *A Geographic Perspective on the Current Biomass Resource Availability in the United States*. National Renewable Energy Laboratory, Golden, CO.

Mitchell, C., D. Bauknecht, and P. M. Connor (2006). "Effectiveness through Risk Reduction: a Comparison of the Renewable Obligation in England and Wales and the Feed-in Tariff System in Germany." *Energy Policy*. Vol. 34, pp. 297.

Nag, A. (2008). *Biofuels Refining and Performance*. McGraw-Hill, New York.

Olsson, L., ed. (2007). *Biofuels (Advances in Biochemical Engineering/Biotechnology)*. Springer-Verlag, Berlin.

Perlack, R., L. Wright, A. Turhollow, et al. (2005). *Biomass as Feedstock for a Bioenergy and Bioproducts Industry: the Technical Feasibility of a Billion-Ton Annual Supply*. Sponsored by the U.S. DOE and USDA.

Perlack R. D., and B. J. Stokes (2011). *U.S. Billion-Ton Update: Biomass Supply for a Bioenergy and Bioproducts Industry*. Rep. ORNL/TM-2011/224, Oak Ridge National Laboratory, Oak Ridge, TN.

Sorensen, B. (2002). *Renewable Energy: Its Physics, Engineering, Use, Environmental Impacts, Economy and Planning Aspects*, 2nd ed. Academic Press, London.

Shapouri, H., P. W. Gallagher, W. Nefstead, et al. (2010). *2008 Energy Balance for the Corn-Ethanol Industry*. USDA: 16, Washington, DC.

Tester, J., E. Drake, M. Driscoll, et al. (2012). *Sustainable Energy: Choosing among Options Second Edition*. MIT Press, Cambridge, Massachusetts.

Zaks, D. P. M., N. Winchester, C. J. Kucharik, et al. (2011). "Contribution of Anaerobic Digesters to Emissions Mitigation and Electricity Generation under U.S. Climate Policy." *Environmental Science & Technology*, Vol. 45, pp. 6735–6742.

Further Reading

Kwon, E., H. Yi, and H. Kwon. (2013) Urban energy mining from sewage sludge. *Chemosphere* 90, pp. 1508–1513.

Exercises

14-1. As an exercise in exploring the impact of excessive energy input in ethanol production on NEB ratio, as discussed in the body of this chapter, consider the following process, in which each liter of ethanol requires 2.69 kg of corn as a feedstock. The corn must first be grown on arable land; per hectare (100 m × 100 m), the corn field has the following energy input requirements:

Input	Energy (GJ)
Machinery (embodied)	4.97
Energy products (diesel, gasoline, electricity)	6.04
Nitrogen	10.25
Balance of inputs	12.72

The hectare yields approximately 9000 kg of corn. The corn is then processed in an ethanol plant to make almost pure (99.5%) ethanol. Per 1000 liters of ethanol, the ethanol production process requires the energy inputs shown in the table below, in addition to the corn. Assume that a liter of ethanol contains 21.3 MJ net. For simplicity, ignore the energy impact of by-products resulting from corn ethanol production. Calculate the ratio of energy available in the resulting ethanol to the total energy input. Does the ethanol provide more energy than it consumes?

Input	Energy (GJ)
Transportation	1.35
Distillation steam	10.66
Electricity	4.87
Balance of inputs	0.59

14-2. Repeat Problem 14-1, but this time for biodiesel production from soy, using U.S. customary units. The energy requirement per acre for soy is the following:

Input	Energy (1000 BTU)
Fuels	1025
Fertilizer	615
Embodied energy	205
All other	205

Each acre yields 452 pounds of soy, and each gallon of soy biodiesel requires 7.7 pounds of soy as input. A gallon of biodiesel contains 117,000 BTU of net energy content. Production energy requirements per 1000 pounds of soy beans processed are in the table below. Calculate ratio of output to input, again ignoring by-product energy impact.

Input	Energy (1000 BTU)
Process heat and electricity	1784
Embodied energy	595
Transportation	297
All other	297

14-3. Repeat exercise 14-1, but this time for biodiesel production from soy. The energy requirement per hectare for soy is the following:

Component	Energy (GJ)
Fuels	2.67
Fertilizer	1.60
Embodied energy	0.53
All Other	0.53

Each hectare yields 508 kg of soy, and each liter of soy biodiesel requires 0.925 kg of soy as input. A liter of biodiesel contains 32.6 MJ of net energy content. Production energy requirements per 1000 kg of soy beans processed are in the table below. Calculate ratio of output to input, ignoring by-product energy impact.

Component	Energy (GJ)
Process heat and electricity	4.14
Embodied energy	1.38
Transportation	0.69
All other	0.69

14-4. Repeat the calculation of exercise 14-1 for corn ethanol for the advanced process with higher efficiency discussed in the chapter. Suppose that the net energy input in agriculture per hectare is cut by 50% thanks to advances in practices for growing corn. Suppose also that the "transportation" and "balance of inputs" figures for energy consumption remain fixed. What must the combined energy value per 1000 L of production for distillation and electricity be to achieve the net energy balance ratio of 2:1, which is the average of the reported current values discussed in the chapter?

14-5. Repeat the calculation of net energy balance for corn ethanol using U.S. customary units. Each gallon of ethanol requires 22.4 lb of corn as a feedstock. The corn must first be grown on arable land; per acre, the corn field has the following energy input requirements measured in millions of BTU (MMBTU):

Input	Energy (MMBTU)
Machinery (embodied)	1.906
Energy products (diesel, gasoline, electricity)	2.317
Nitrogen	3.932
Balance of inputs	4.879

The acre yields approximately 8013 lb of corn. The corn is then processed in an ethanol plant to make almost pure (99.5%) ethanol. Per 1000 gal of ethanol, the ethanol production process requires the energy inputs shown in the table below, in addition to the corn. Assume that a gallon of ethanol contains 75.7 MBTU(1 MBTU = 1000 BTU). For simplicity, ignore the energy impact of by-products resulting from corn ethanol production.

Input	Energy (MMBTU)
Transportation	4.84
Distillation steam	38.24
Electricity	17.47
Balance of inputs	2.12

Calculate the ratio of energy available in the resulting ethanol to the total energy input. Does the ethanol provide more energy than it consumes?

14-6. Repeat the calculation of exercise 14-5 for corn ethanol for the advanced process with higher efficiency discussed in the chapter. Suppose that the net energy input in agriculture per acre is cut by 50% thanks to advances in practices for growing corn. Suppose also that the "transportation" and "balance of inputs" figures for energy consumption remain fixed. What must the combined energy value per 1000 gal of production for distillation and electricity be to achieve the net energy balance ratio of 2.1, which is the average of the reported current values discussed in the chapter?

14-7. Many municipal wastewater treatment systems include anaerobic digesters to treat wastewater and recover energy. Your company has been hired by Greene City Council to design a continuously stirred anaerobic digester (CSAD) system at their municipal wastewater treatment facility. The CSAD system must be designed to handle an influent wastewater flow rate, Q, of 1 MGD (1 million U.S. gallons/day). The city wants you to ensure that $E.\ coli$, an indicator organism for pathogens, (initially at a concentration, C_0, of 2×10^7 CFU/g TS (colony forming units/grams total solids) in the influent wastewater) will be reduced to below a certain level (concentration, C), so that the effluent sludge from the system meets the EPA's pathogen reduction standards for Class B biosolids. Assume that $E.\ coli$ will undergo first-order inactivation in the CSAD system with a decay rate constant (k) of 0.13 d^{-1}. The city council has drawn up a plan in which one digester (volume, V, of 70 million U.S. gallons) will achieve the desired $E.\ coli$ reduction. Prove to the Greene City Council that they can reduce the tank volume required for the CSAD system and achieve the same treatment level if they arrange *two* equally sized digesters in series. What total volume is required to reduce $E.\ coli$ to the desired level if two digesters are arranged in series? Assume an ideal CSAD system. (Acknowledgment: Thanks to Catherine Spirito for providing this exercise.)

14-8.   Corn Inc. intends to build a corn-to-ethanol plant in the town of Mazeville. Corn Inc. requires 100 tons/year of corn for their plant and has a budget of $32,000/year to spend on purchasing the corn. They have already signed a contract with Belmont Farm to buy 20 tons of corn from the farm per year for $350/ton. However, they just found out that Ashville Farm has cheaper corn for $300/ton. Ashville Farm and Belmont Farm are located exactly 300 km apart along a straight highway that runs through the town. Corn Inc. can locate their plant anywhere along this highway between these two farms. However, they also must consider the cost of their greenhouse gas emissions in their budget, as the local government has recently implemented carbon pricing (at $100/ton CO_2 eq emitted). Corn Inc.'s trucks can carry 10 tons of corn per trip and weigh 10 tons when they are empty. The trucks emit 2×10^{-4} tons CO_2 eq/ton-km. At what location along the straight highway (i.e., what distance between farm A and farm B) should Corn Inc. build their plant if they intend to spend their full budget of $32,000/year? (Acknowledgment: Thanks to Catherine Spirito for providing this exercise.)

14-9.   Land used to grow soy beans for biodiesel receives on average 175 W/m² of solar insolation arriving from the sun. It is harvested once per year and the resulting soy beans yield 548.8 liters per hectare. Assume an energy content value of 32.6 MJ/L for pure biodiesel. A) Of the solar energy arriving per year, what percent is made available as energy content in the biodiesel? B) In a sentence or two, discuss the implications of this calculation.

CHAPTER 15

Transportation Energy Technologies

15-1   Overview

This chapter focuses on technological options for sustainably delivering energy to motor vehicles and other mechanized transportation equipment. Each specific propulsion technology can be categorized among a limited number of generic "endpoint technologies," and the strengths and weaknesses of each are considered. While the goal of sustainability may drive society toward adoption of endpoint technologies that will meet future expectations for transportation energy, the basic design of motor vehicles for the consumer market of today is driven by the need to trade off energy consumption against performance, size, and other consumer amenities. Therefore, a number of design factors are introduced and analyzed as a foundation for discussion of all technology alternatives. Thereafter we take a closer look at four major alternative propulsion platforms and fuels, namely, battery-electric, hybrid-electric, biofuel, and hydrogen vehicles. The chapter concludes with a discussion of "well-to-wheel" analysis of energy efficiency in transportation vehicles. Material in this chapter provides a basis for consideration of systems issues related to transportation energy in Chap. 16.

15-2   Introduction

In the modern world of motorized cities, long-distance travel by jet or limited-access highway, and global trading of manufactured goods and commodities, it comes as no surprise that the transportation sector has become an enormous end user of energy. This sector is the single largest consumer of petroleum resources in the world today, and the second largest consumer of nonrenewable fossil fuels next to electric power conversion. Worldwide, the transportation sector accounted for approximately one-fourth of the total end-use energy consumption value of 553 EJ (524 quads) in 2012. In the United States alone, in that year the transportation sector accounted for 28.3 EJ (26.8 quads), which constitutes 28.2% of the total U.S. energy end-use budget of 100 EJ (95.0 quads), or 5.1% of the world total. To put these values in context, the U.S. transportation energy consumption rate is equivalent to 9 billion 100-W lightbulbs burning continuously 24 h/day, 7 days a week, all year long, or 29 lightbulbs for every one of the United States' approximate population of

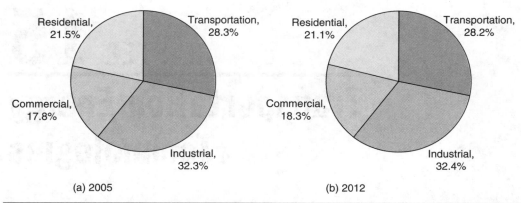

(a) 2005   (b) 2012

FIGURE 15-1   Transportation's share of U.S. primary energy end use by sector, 2005 and 2012.

Total value 106 EJ (100 quads) and 100 EJ (95 quads) for 2005 and 2012, respectively. (*Source:* U.S. Energy Information Administration.)

300 million people.[1] The world transportation energy consumption figure is equivalent to 44 billion lightbulbs, or 6 lightbulbs per person.

In terms of its role in the U.S. economy, the transportation sector was the second largest user of energy behind the industrial sector in both 2005 and 2012 (Fig. 15-1). Overall energy consumption declined during this 7-year period, and transportation's share stayed approximately constant, so that the absolute amount consumed for transportation fell as well. Due to heavy dependence on petroleum products as an energy source, the transportation sector emits a disproportionate fraction of U.S. CO_2 emissions, constituting 35% of total emissions for 28% of total energy use in 2012 (Fig. 15-2).

Information from the International Energy Agency (IEA) in Paris sheds further light on the relative value of transportation energy consumption when comparing different countries, as shown for the year 2009 in Table 15-1. World energy consumption in 2009 totaled 480 quads (506 EJ), according to the U.S. Energy Information Agency. Energy reports from the IEA break out end-use energy consumption between transportation and nontransportation sectors for some countries of the world of the world but not others, so it is not possible to present a comprehensive comparison of transportation versus nontransportation energy use. However, as an indication of transportation's role, Table 2-1 gives percent of total end use energy consumption for the transportation sector for select countries, whose combined total energy consumption of 184 quads (194 EJ) represents 38% of the 2009 world total. Percentage values range from 11% for China and India to 40% for the United States. The eight countries in the table together consume 44 quads for transportation, so that transportation consumes 24% of this total. The comparison of countries suggests that on average wealthy countries have a higher fraction of energy dedicated to transport, perhaps because residents of have more

---

[1]In both Chaps. 15 and 16, we will extensively use the United States as an example for transportation energy consumption. Although the per capita energy consumption is higher than most other countries, the relative allocation of energy to transportation (e.g., percent of energy for transportation, percent of transportation energy dedicated to different modes such as highway vehicles or railroads, and the like) is similar to that of many other industrialized countries. Some differences between the United States and peer countries in Europe and Asia do arise, and these are discussed in Chap. 16 as systems issues.

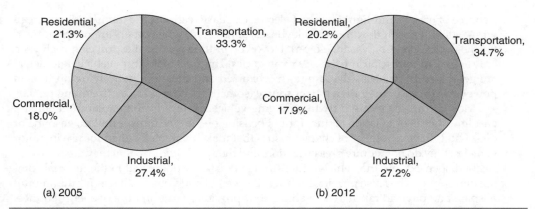

FIGURE 15-2 Transportation's share of U.S. CO_2 emissions from the combustion of fossil fuels for energy by sector, 2005 and 2012.

Total value of pie is 5697 MMT and 5061 MMT for 2005 and 2012, respectively. *Source:* U.S. Energy Information Administration.

disposable income for travel in energy intensive modes (car, aircraft), although Brazil appears to be an exception.

A unique challenge facing the ongoing use of various energy sources for transportation is the need to store energy in a concentrated form onboard the vehicle. One objective of vehicle design is to store a large quantity of energy per unit of weight and volume displaced on the vehicle, lest the weight or volume of the fuel should limit the capacity to carry passengers or cargo. Some vehicles bypass this problem by using the electric grid as

Country	Transportation	Total	Percent
Brazil	2.48	7.57	33%
China	6.38	56.87	11%
Germany	2.14	8.89	24%
India	2.04	17.83	11%
Japan	3.02	12.44	24%
Russia	3.56	16.78	21%
UK	1.66	5.24	32%
USA	22.93	58.04	40%
Sum of 8	44.21	183.66	24%
WORLD	n/a	480.0	n/a

Source: International Energy Agency (2012). Note that both transportation and total energy figures exclude energy losses in conversion and distribution. For example, for 2009, USEIA reports 94.5 quads for the United States in 2009 versus ~58 quads reported above from IEA, thus 36.5 quads or 39% of gross energy consumption are lost in conversion and distribution, and the remaining 61% delivered for end use.

TABLE 15-1 Percentage of 2009 End-Use Energy Consumed by Transportation Sector for Select Countries (Quads)

a source of energy, through the use of electric catenary (e.g., electric trains with overhead wires, subways with third rails, trolleybuses, and so on), while others are nonmotorized (e.g., bicycles, cycleshaws, and so on). However, of the vehicle-miles traveled each year, only a small minority fall into these two categories. The great majority of vehicle-kilometers are generated by vehicles that are (a) mechanized and therefore not relying on human power and (b) carry their own power source (sometimes referred to as "free-ranging").

Most of the energy used by free-ranging vehicles comes from petroleum in either a gasoline, diesel, or aviation jet fuel form. Focusing on surface transportation, gasoline or diesel fuels have a number of characteristics that make them well-suited for use in motor vehicles: they do not require pressurization, and they are liquids, so that they are relatively easily dispensed into the vehicle's fuel storage tank and combusted in the internal combustion engine. They also provide high specific weight and volume, that is, for the amount of space and payload taken in the vehicle, they provide a large amount of energy storage. Ideally, any alternative fuel must match these characteristics. Otherwise, society will need to radically rethink the way motorized transportation systems are used so that fuels that are less convenient or provide shorter range become the norm.

As discussed in Chaps. 4 and 5, the twin arguments of peaking of conventional petroleum production and CO_2-induced climate change make a compelling case for exploring alternative fuels and alternative technologies. Many such alternative technologies exist on a small scale or in concept, including natural gas vehicles, fuels derived from coal, nonconventional crude oil sources such as oil shale, conversion of renewable energy sources into a form transferable to vehicles, and the like. However, in order to succeed, such technologies must be technically robust and financially viable. Under pressure from declining petroleum supplies, society may be willing to pay more for an alternative fuel source than they currently pay for gasoline, but people will refuse to support such a fuel if its price is exorbitant, or they may simply be unable to afford it. Furthermore, the energy source must be developed in parallel with the vehicles that use it and the infrastructure to distribute it, and all these things must fall into place in a short amount of time, or at least in a way that keeps pace with the decline in availability of gasoline and diesel from the market. Mature technologies that use petroleum more efficiently, such as hybrid drivetrains, are already expanding in the marketplace, but they are not a permanent solution unless the energy source is changed from petroleum. Cost of any new distribution infrastructure system is also a concern, although given the high value of the transportation fuels market—at an average cost of $3.00/gal, including taxes, the approximately 200 billion gallons of gasoline, diesel, and jet fuel purchased in the United States in 2014 would have a retail value of some $600 billion—it is clear that companies that deliver transportation energy products should be able to recoup a significant investment in new infrastructure through continued sales.

Compared to the challenge of finding a clean and abundant source of energy from which to make a future energy source for transportation, the ability to transfer that energy source onto the vehicle poses the more daunting of the two challenges. For example, nuclear energy and large-scale wind turbines are two proven technologies for generating electricity without CO_2 emissions that have a similar cost per kWh to fossil fuel energy generation. However, we do not yet have an infrastructure to manufacture fuel cell or battery technologies on a scale to take nuclear or wind energy onboard the vehicle as a substitute for gasoline (though in the case of electricity, home recharging might provide a partial solution). We also do not have a network of refueling stations available to the public to distribute and dispense electricity or hydrogen. These obstacles may favor instead the development of a petroleum substitute, such as a biofuel, that behaves like petroleum so that we can use our existing distribution and dispensing infrastructure, but that does not depend on nonrenewable resources and that does not contribute to climate change.

To summarize, the development of a clean, abundant, and economical substitute for the petroleum-based transportation energy system is one of the major challenges facing the nations of the world today. Society would likely experience significant disruption from passing the peak oil point and not having a carefully prepared alternative waiting in the wings (see Chap. 5). Also, if the alternative were to be nonconventional petroleum sources, we might prolong for a time the worldwide transportation system based on liquefied fossil fuels, but greatly aggravate the climate change problem if no suitable system for mitigating CO_2 emissions is in place. This challenge is arguably one of the most difficult technological and systems problems that we face in the pursuit of sustainable energy.

15-2-1   Definition of Terms

The transportation literature refers to vehicles that run on a fuel other than gasoline or diesel as an *alternative fuel vehicle* (AFV); in some cases use of diesel in passenger cars is considered to be an alternative fuel, since, in markets such as that of the United States, it is somewhat unusual. Some alternative options for transportation energy, such as hybrid vehicles, do not use an alternative fuel but instead use an alternative propulsion "platform" to reduce the requirement for gasoline. Some sources refer to alternatives such as hybrids as "advanced" propulsion systems. It is also possible to have options that are a mixture of both, for example, a hybrid vehicle that runs on an alternative fuel. In this chapter, we use the umbrella term "alternative propulsion platforms and fuels" to describe the complete range of alternatives. We also interpret the term *fuel* broadly, so as to include electricity, even though electricity does not have the characteristics usually associated with the term *fuel*, such as combustibility.

Engines that use gasoline ignite the fuel using a spark, and are therefore called spark-ignition (SI) engines. Diesel-burning engines rely on the compression of the fuel alone to ignite the fuel, without the use of a spark, and are called compression-ignition (CI) engines. Although exceptions may occasionally arise, in general, when applied to conventionally fueled vehicles, the use of the term SI engine implies gasoline combustion, and the use of the term CI engine implies diesel combustion.[2]

15-2-2   Endpoint Technologies for a Petroleum- and Carbon-Free
Transportation System

Transportation energy technologies that replace the use of petroleum for transportation as currently practiced must (1) be based on a more abundant supply of energy and (2) avoid permanently increasing the concentration of carbon in the atmosphere to avert climate change. Note that the second requirement does not preclude a carbon-based energy source. It also does not preclude emitting CO_2 from the vehicle tailpipe to the atmosphere at some point, as long as carbon emitted to the atmosphere is later removed so that the overall atmospheric concentration does not permanently increase.

Although there might appear to be a wide range of technologies competing to play this role, each can in its essence be reduced to one of the five "endpoint

---

[2]The performance of SI and CI engines can be modeled theoretically using temperature-entropy diagrams of the type presented for stationary combustion systems in Chap. 6. These diagrams are not presented in this chapter, but are widely available in thermodynamics texts, for example, Wark (1983) or Cengel and Boles (2006).

	Name	Description
Endpoint 1	Battery—electric	Zero-carbon electricity* distributed to vehicles, which run on electricity between recharges
Endpoint 2	Hydrogen	Zero-carbon hydrogen distributed to vehicles, which run on hydrogen between refills[†]
Endpoint 3	Sustainable hydrocarbons	Hydrocarbon fuels resembling petroleum products made using biological processes[‡], or CO_2 emissions to atmosphere removed by separate process
Endpoint 4	Alternative onboard energy storage	Alternative systems such as compressed air or flywheels used to power vehicle between recharges
Endpoint 5	Eliminate free-ranging, mechanized vehicles	All mechanized vehicles are connected to grid via catenary. All free-ranging vehicles are human-powered[§]

*"Zero-carbon electricity" means that the electricity is generated and delivered to the vehicle without increasing CO_2 concentration in the atmosphere. Thus electricity generated from biofuels could be used in this endpoint technology.

[†]Although it is likely that this endpoint technology would combine use of hydrogen with fuel cell technology in order to achieve high efficiency, other conversion technologies (e.g., combustion in piston engine) are also possible.

[‡]Biological processes include crops, plant matter, or microbes. For sustainability, energy inputs should be renewable rather than from fossil fuel resources. Alternatively, fossil-fuel-based energy source could be used if the resulting CO_2 is captured and sequestered.

[§]Animal power might be substituted for human power.

TABLE 15-2   Alternative Endpoint Technologies for Petroleum-Free, Carbon-Free Transportation

technologies"[3] that meets the objectives of abundant supply and protecting the climate. The five endpoint technologies are shown in Table 15-2. Some appear more promising at this time than others. For example, much research effort is going into developing hydrogen and biofuels as energy sources, into developing improved battery technology for vehicular use, and into sequestering carbon from the use of fossil fuels. However, given that we are at a very early stage of this transition and that it will take many years to complete, other less well-emphasized options are included in the table for thoroughness.

All five endpoint technologies share a few common characteristics:

- They all present substantial technical, organizational, and financial challenges.

- Whether or not the endpoint technologies require the introduction of new vehicle technologies or use existing ones, they all require large new infrastructure systems to generate energy and distribute it to vehicles. Some also require infrastructure to transform existing energy sources into a form that can be stored on a vehicle, and others require infrastructure to process CO_2 already in the

[3]The use of the term *endpoint technology* in this chapter is similar to the term *backstop technology*. The latter was not adopted because the term in this chapter has a specific meaning in regard to the limited number of options, and because it was felt that the latter term is not well defined in the literature, and may mean different things to different readers.

atmosphere, or to sequester CO_2 that is a by-product of conversion to the energy currency used onboard the vehicle.

- The endpoint technologies are not mutually exclusive. It is possible that one will eventually become dominant, and it is also possible that multiple ones will each claim some niche in meeting transportation energy demand. An analogy could be made with today's situation, where most surface transportation (road, rail, ship) is propelled by petroleum-fueled internal combustion engines, but a minority of rail service uses electricity supplied from outside the vehicle.

We first consider endpoint technologies 4 and 5. Endpoint 4 bypasses the need for a fuel or currency altogether by storing and releasing energy in some way that does not involve combustion or the flow of electric current. Two forms currently available in an experimental form are compressed air, which is pumped into the vehicle and then released through an engine that turns a driveshaft, and an onboard flywheel that is set spinning by an external force prior to operation and then powers the vehicle's drivetrain when it is in motion. For example, vehicles using compressed air were launched experimentally into the European market between 2009 and 2013 by Motor Development International of France, and were intended to fill a small niche for an emissions-free micro-car. The MDI prototype that weighs 550 kg and can travel 80 km or more on a full tank of compressed air. Due to lack of market success, the MDI offering was eventually withdrawn. In any case, this is an "umbrella" category in that other options for unconventional onboard storage might be developed in the future, that we do not currently anticipate. While the reader should be aware of these developments, the great majority of R&D effort is currently aimed at endpoints 1 to 3, so we do not consider endpoint 4 further in this chapter.

Turning to endpoint technology 5, relying on externally powered vehicles, the goal is to enhance or expand existing electrical catenary-powered transportation systems. Transportation outside the catenary network would require human or animal power, and all free-ranging mechanized vehicles would be discontinued. Research aimed at developing *personal rapid transit* (PRT) in the 1970s showed that it is prohibitively expensive to provide an electrically powered guideway for vehicles on every street currently accessible to motor vehicles. Therefore, the catenary network would necessarily be sparser than the network currently accessible to motor vehicles. Mechanized vehicles would operate on main arterials in urban areas and on certain feeder streets, but side streets would have no powered vehicles. Similarly, in rural areas some, but not all, roads would have mechanized vehicles operating.

From a societal point of view, complete elimination of mechanized free-ranging vehicles is arguably the most draconian of all the endpoint technologies, because it would involve such a large change in social and economic patterns, including land use, business practices, and so on. The resulting urban landscape might resemble that of the late nineteenth century in certain cities in Europe and North America before the advent of the motor car, where electric tramways, horse-drawn vehicles, bicyclists, and pedestrians dominated the streets. Expanding the availability of catenary can, however, make an important contribution to sustainable energy for transportation. Some amount of extension of electric catenary is underway in the world at present, such as recent projects to electrify railway lines between New Haven and Boston in the United States, or the electrification of the East Coast Main Line in Great Britain, as well as additions and conversions of transit lines to electric trolleys or streetcars in various cities around the world. To the extent that these systems currently use zero-carbon electricity, or could do so in the future, they can contribute to reducing the CO_2 burden from transportation. However, we do not consider the underlying technology required for endpoint 5 further in this chapter (see Chap. 16 for a discussion of their role in an overall systems solution to transportation energy needs).

We now turn to the technologies that are the focus of the remainder of this chapter, namely, endpoints 1 to 3. These technologies have in common the use of some carrier that is stored onboard the vehicle, either electricity, hydrogen, or some type of hydrocarbon. It is thought that these three are, in fact, the only feasible terrestrial transportation energy sources, based on unsuccessful attempts to identify others. As will be seen in the following discussion, all three have substantial technical hurdles, but also show real potential over the long term, so we are justified in considering them in some depth here.

Endpoint technologies 1 and 2 have in common that the energy generation component is relatively well-developed, whereas the distribution, dispensing, and onboard technology are less developed. Large-scale wind and nuclear technologies are already able to produce electricity (and by extension, hydrogen through electrolysis) in large quantities without emitting CO_2. In some circumstances, the electricity from these sources costs more than the electricity from fossil fuels, but a concerted effort to build a transportation energy system around them would likely result in a reduction in cost so that any increase in cost per unit of energy equivalent delivered to the vehicle, compared to the current system, could be absorbed by consumers as part of the transition. Alternatively, with widespread and cost-effective carbon sequestration, the electricity or hydrogen for these technologies could come from fossil fuel combustion and conversion, which is another mature technology. In the case of endpoint technology 1, electrically powered vehicles might recharge at night when both power plant and grid usage are at a low point, so that the recharging function might be added to the tasks of the electric grid without requiring major expansion of generating and transmission infrastructure.

Other technologies in the chain from energy source to vehicle propulsion are less well developed. The list of these technologies includes the following:

1. *Adaptations to the electric grid:* Night recharging (as described in the previous paragraph) notwithstanding, the ability to recharge vehicles during the day would require a network of new charging stations, as well as possibly expansion of the transmission and distribution grid in certain locations, in order to deliver electricity to vehicles on a very large scale.

2. *Capacity for short-term storage:* In the case of endpoint 1, especially where intermittent energy sources such as solar or wind are used, if the electricity cannot be dispensed to the vehicle as it is generated, some means of short-term storage would be necessary to retain the energy content until it could be dispensed. For example, electricity could be converted to hydrogen for short-term storage.

3. *Long-distance hydrogen infrastructure:* In the case of endpoint 2, it is envisioned that hydrogen would be produced in large central facilities in order to benefit from economies of scale. These facilities would require a new distribution grid for the product to reach the end users.

4. *Infrastructure for long-lasting and cost-effective batteries or fuel cells:* These technologies must be perfected for their respective pathways to succeed, and once perfected, a new manufacturing infrastructure to produce, distribute, and recycle batteries or fuel cells would be necessary.

5. *Onboard hydrogen storage technology:* For endpoint 2, the hydrogen fuel cell must be coupled with a means of storing sufficient hydrogen onboard the vehicle in order for it to have sufficient range, while meeting cost and safety requirements.

Depending on which exact solution eventually takes shape, not all of the technologies on the list would be required, but some of them would be, and each requires substantial R&D to perfect.

By contrast, endpoint 3 takes advantage of the existing distribution and onboard conversion of the transportation energy resources more or less unchanged, and instead requires the development of a new energy generation technology that creates from renewable resources a fuel that closely resembles, or is chemically identical to, today's gasoline or diesel. In many of the available variants on this pathway, the original source of energy is the sun, with conversion of sunlight taking place in agricultural fields, in water-based resource "farms," or in controlled facilities that use microorganisms to generate the raw materials for liquid fuels. Coal might also be used as an energy source instead of the sun, by converting the carbon content of the coal into a synthetic liquid fuel, combusting this fuel in the vehicle, and then capturing the resulting CO_2 from the atmosphere.[4] The conversion of coal to synthetic fuel has already been used in the past in Germany and South Africa during times of war or political isolation. The attraction of endpoint 3 is that, once the hurdle of making the fuel is overcome, companies that currently refine, distribute, and dispense motor fuels, or manufacture vehicles, could continue to use familiar technologies with only minor adaptations.

As a variation on endpoint 3, synthetic fuels from coal might be combusted in vehicles with release of CO_2 to the atmosphere, and a separate system of devices, not powered by fossil fuels, would capture the CO_2 from the atmosphere in a location where it was convenient to sequester it (see Chap. 7). In such a system, much of the infrastructure cost would entail the construction of a worldwide network of "sequestration centers," in addition to facilities for synthesizing a vehicular fuel from coal. The latter technology already exists in the form of, for example, Fischer-Tropsch diesel, which can be produced from gasified coal.

Lastly, combinations of endpoints 1 and 3 are also possible. For example, a plug-in hybrid vehicle (see Sec. 15-4-2) might recharge using electricity from renewable resources from the grid (endpoint 1), and then refuel using a biofuel (endpoint 3).

15-2-3   Competition between Emerging and Incumbent Technologies

In considering the transition to alternative transportation technologies, we must recognize the influence of the starting point of today and the existing worldwide fleet of motor vehicles, aircraft, and other consumers of transportation energy from petroleum, and the infrastructure that supplies this energy. While alternative pathways are desirable in the long run, today they must be introduced in the context of a mature petroleum-based system that is the "incumbent" technology, and the expectations in terms of price, reliability, performance, and so on, which this system has created in consumers. The situation is therefore different from the early days of the automobile at the beginning of the twentieth century. At that time, use of horses had fallen out of favor for urban transportation, due to problems with fouling of streets. When "horseless carriage" designs emerged that used either the internal combustion engine, electricity, or steam for propulsion, each new technology had to compete with the other two for dominance of the new market, but horse propulsion was no longer a strong incumbent

---

[4]Although petroleum or natural gas might also be used as an energy source in this process, we only consider coal in this example because it is the only fossil fuel with sufficiently large remaining resources to justify the investment in a new infrastructure for the long term.

"technology" that could resist the rise of the automobile. Once the internal combustion engine had outvied electricity and steam, the market for both cars and gasoline could grow at a pace determined by the vehicles' success alone.

Is it a certainty that consumers will under no circumstances accept an alternative transportation energy option that has diminished performance but compensates with some other positive feature, for example, that it is better for the environment? Historically, there have been examples where citizens made radical changes in personal transportation patterns in order to contribute to some other objective. For example, at the time of World War II, many countries experienced great changes in this area, such as the discontinuation of passenger car production in the United States between 1942 and 1944, or the dramatic rise in use of public transportation in many countries around the world. Although governments required citizens to participate through laws, rationing, and so on, these programs were in the end successful in nations on both sides of the conflict because, by and large, individuals recognized the contribution that the savings of energy and raw materials could make to the war effort. In the same way, it is within the realm of possibility that citizens might shift en masse from the current technology to some other, the way that they have recently been shifting from landlines to cell phones, if the danger from climate change were clear enough, and the anticipated benefit of radically changing vehicle technology were convincing enough. Indeed, on an individual level, a small fraction of the population in various countries has already made such a switch for environmental reasons—for example, taking up regular commuting to work by bicycle rather than by car.

This scenario of rapid, radical shifting of technology has some possibility of occurring. However, because it presents a great challenge in terms of making social adjustments, and because there are strong candidates for vehicles that use energy sustainably without a great loss of performance, we focus in the remainder of this chapter exclusively on options that can "oust the incumbent" technology (petroleum-driven internal combustion engine vehicles, or ICEVs) by delivering equivalent performance while still achieving the desired reductions in CO_2 emissions and other environmental gains.

On the alternative fuel infrastructure side, there must also be support for the growth in the number of vehicles sold with an equal growth in energy outlets to supply them, which clearly requires a leading role for regional and national governments. As a cautionary tale about what can go wrong when support for AFVs is inadequate or ill conceived, consider the case of efforts in the United States to introduce flex-fuel vehicles (i.e., able to use mixtures of gasoline and biofuels, or diesel and compressed natural gas, in different percent combinations) in the 1980s. Through a system of mandates and financial incentives, several hundred thousand flex-fuel vehicles were sold, so that the potential existed for these vehicles to use biofuels and thereby reduce gasoline consumption.

Although the individual vehicles were technically proficient, the technology failed to take root because a business case was not established for building the infrastructure to deliver the alternative fuel. Many vehicles were driven in areas where they had no access to gasoline-biofuel mixtures, and never once during their driving life took advantage of the flex-fuel capability. The public judged the flex-fuel capability not to be useful, and it became unmarketable as soon as government incentives and mandates were removed. Fleet owners of flex-fuel vehicles found that, if they had paid a premium for this capability when the vehicle was new, they could not pass this premium on in the used vehicle market because the buyers saw no added value in it.

Changing performance of the gasoline technology in terms of air quality played a role as well. A major driver of introducing AFVs at the beginning of the program was the perception that they produced less air pollutants than gasoline vehicles, and that their introduction would help to improve urban air quality. However, over the lifetime of the AFV program, vehicle manufacturers were able to greatly reduce emissions per mile of many important pollutants, so that the air quality motivation no longer existed by the end of the program.

In the final analysis, the 1980s AFV program may have had some benefit, in terms of forcing the emissions control technology for gasoline engines to improve, or providing a technological platform on which biofuel programs that have been expanding since the year 2000 could build. Nevertheless, it failed as an attempt to create and sustain a permanent presence of flex-fuel vehicles using an alternative fuel dispensing network. This experience can be contrasted with that of Brazil, where an alternative fuel industry based on the production of ethanol from sugarcane has for many years provided a substantial fraction of the motor fuel consumed; see Chap. 14 or Sec. 15-4-3 on biofuels.

15-3 Vehicle Design Considerations and Alternative Propulsion Designs

The preceding example of AFVs shows the importance of considering the interaction between the alternative and incumbent technologies when introducing alternative fuels and propulsion systems. In Sec. 15-4, we will consider electric vehicles, hybrid vehicles, biofuels, and hydrogen fuel cell technology. Each technology must be seen in the context of the petroleum-based technology it seeks to replace. An understanding of vehicle design helps to provide some of that context, and is the focus of this section.

Regardless of the energy source or the propulsion technology used, design of any vehicle is based on meeting certain performance requirements, of which saving energy (by extension leading to reduced CO_2 emissions) is just one. Based on the power, weight, and aerodynamic characteristics of a vehicle, it is possible to predict a number of its performance measures. These performance measures provide an indication of the success of the engineer in meeting the customer's desire for a vehicle that performs well, for example, stopping, accelerating, climbing hills, and so on. Of course, most customers are aware that there is usually a trade-off between performance on the one hand and fuel economy/reduced cost of fuel on the other, so that vehicle performance may be only one of several factors in choosing a vehicle for purchase. Also, aspects of driving style, including typical cruising speed or rates of braking and accelerating, affect delivered fuel economy, so even if a customer anticipates a certain level of fuel economy from a given vehicle, the way they drive the vehicle may affect the actual fuel consumption. Lastly, customer preferences change in response to outside information, so it is to be expected that as they learn more about the role of vehicle choices in affecting climate change and other environmental issues, customers will give environmental concerns more weight in their decision making.

For their part, vehicle manufacturers have responded to government requirements and consumers' desire to reduce fuel expenses by using technological innovations to improve energy efficiency of vehicles. For example, Table 15-3 shows how drag coefficients have improved over generations of vehicle designs. The drag coefficient relates the velocity of the car to the amount of effort required to overcome aerodynamic drag, so as this coefficient is reduced, cars become more efficient, other things being equal.

Vehicle	Drag Coefficient (C_D)
1970s or 1980s standard passenger cars	0.5–0.6
1970s or 1980s sports cars	0.4–0.5
Post-2000 high fuel economy passenger cars	0.25–0.3
Highly aerodynamic concept cars	0.15–0.18
Theoretical minimum drag (teardrop shape)	0.03

Table 15-3   Evolution of Drag Coefficient Values, 1970s to Present

15-3-1   Criteria for Measuring Vehicle Performance

Typical performance measures used in vehicle design include the power requirement at cruising speed, maximum speed, maximum gradability, and maximum acceleration. Each is explained in turn below. The measures of performance can be applied to internal combustion engines (ICEs), electric motors, or vehicles which combine both (i.e., hybrids).

Power requirement at cruising speed: The power requirement for a vehicle to maintain cruising speed on a level road is the power provided from the transmission that just equals the rolling and aerodynamic resistance of the vehicle, so that it neither accelerates nor decelerates. Let P_{TR} be the required tractive power, ρ be the density of air, A_F the frontal cross-sectional area of a vehicle, C_D the aerodynamic drag coefficient (as represented in Table 15-3), V the vehicle speed, and C_o the coefficient of rolling resistance. The relationship between P_{TR} and V is then

$$P_{TR} = 0.5\rho A_F C_D V^3 + mgVC_o \tag{15-1}$$

where m is the mass of the vehicle and g is the gravitational constant. For a vehicle climbing a constant grade, Eq. (15-1) can be modified to include a term that incorporates the work done to move the mass of the vehicle against gravity. It can also include a term to capture the effect of acceleration and deceleration on tractive power requirement, although this term is not included here (see, e.g., Albertus et al., 2008).

Maximum speed: By extension from Eq. (15-1), the maximum speed for a vehicle is the speed V when the transmission in an ICE vehicle is in highest gear, or motor output in an electric vehicle (EV) is at its maximum value, P_{TR} is in equilibrium with aerodynamic and rolling resistance, and an increase in engine or motor rotational speed, in rpm, would lead to a drop in P_{TR}, so that the vehicle cannot accelerate to a higher speed. For a given desired maximum speed V_{max}, Eq. (15-1) tells us the required tractive power that must be provided by the drivetrain. Alternatively, for a given amount of available tractive power, we can predict V_{max} for the vehicle.

Maximum gradability: The maximum gradability is the grade of slope at which the gravitational force acting downward on the vehicle is just balanced by the maximum tractive force F_{TR} of the engine or motor acting upward, so that upward motion at an infinitesimal rate is just possible (see Fig. 15-3). Here slope is the ratio of distance of rise to distance of run, measured in percent, that is, 10 m vertical over 100 m horizontal is a 10% slope. Maximum gradability GR_{max} is a function of F_{TR} and m, as follows:

$$GR_{max} = 100 \times \left[\frac{F_{TR}}{\sqrt{(mg)^2 - F_{TR}^2}} \right] \tag{15-2}$$

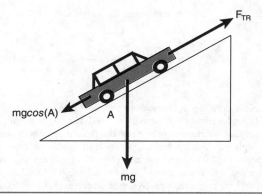

FIGURE 15-3 Force balance of gravitational and tractive forces acting on a vehicle on a grade.

Maximum acceleration: The maximum acceleration A_{max} achievable on a level surface by the vehicle is based on its maximum available tractive force F_{TR} and its total mass. In a simple form, ignoring the effect of drag and assuming constant force across the range of vehicle speeds, this relationship can be derived from Newtonian mechanics, as follows:

$$A_{max} = \frac{F_{TR}}{m} \tag{15-3}$$

Thus, in order to maximize acceleration, the engineer seeks to maximize force while at the same time reducing mass. In practice, increasing force tends to increase engine size, thereby increasing mass and lowering A_{max}, so a balance must be struck between the two factors. Examples 15-1 and 15-2 illustrate the use of these performance measures for evaluating vehicle designs at a basic level.

Example 15-1   A representative passenger car that is designed for fuel economy has a frontal area of 2.6 m² , drag coefficient of 0.3, rolling resistance coefficient of 0.01, curb weight of 1200 kg, maximum tractive force at low speeds of 3000 N, and tractive power at maximum speed of 30 kW. The vehicle has five-passenger capacity, plus rear end space behind the second seat for storage. Calculate the cruising power requirement at 96 km/h; the maximum speed; and the maximum gradability. (In standard units, the vehicle has a cross-sectional area of 28.1 ft² , weighs 2640 lb, and the desired cruising speed is 60 mph.)

Solution   Assume air density of 1 kg/m³ . The cruising speed is equivalent to 26.7 m/s. Therefore, using Eq. (15-1), the power requirement is

$$P_{TR} = 0.5(1)(2.6)(0.3)(26.7)^3 + (1200)(9.8)(26.7)(0.01) = 10.5 \text{ kW}$$

Next, the tractive power at maximum speed of 30 kW will determine the maximum speed V_{max}. Plugging in known values gives

$$30 \text{ kW} = 0.5(1)(2.6)(0.3)V_{max}^3 + (1200)(9.8)V_{max}(0.01)$$

Solving using a numerical solver gives $V_{max} = 40.2$ m/s, or 145 km/h (90.4 mph).

Maximum gradability is determined using the maximum tractive force at low speed and the mass of the vehicle:

$$GR_{max} = 100 \times \left[\frac{3000}{\sqrt{(1200 \cdot 9.8)^2 - (3000)^2}} \right] = 0.264 = 26.4\%$$

Example 15-2   Now consider a representative sport-utility vehicle (SUV) also with five-passenger capacity. The SUV rides higher off the road, and so has a larger frontal area of 3.1 m² and a higher drag coefficient of 0.4. Due to more rugged tires, the rolling resistance coefficient increases to 0.015. The curb weight is higher at 1500 kg, but the power train is also stronger, delivering a maximum tractive force at low speeds of 4500 N, and tractive power at maximum speed of 50 kW. (In standard units, the vehicle has a cross-sectional area of 33.4 ft² and weighs 3300 lb.) (a) Calculate the cruising power requirement at 96 km/h; the maximum speed; and the maximum gradability. (b) Suppose both the SUV and the economy car in Example 15-1 accelerate from a standstill at full force in a frictionless vacuum, and that they can maintain the maximum low-speed tractive force over the entire range of speeds. How fast will each of them reach the cruising speed of 96 km/h or 60 mph?

Solution
(a) Repeating the calculations from Example 15-1 but using new parameters for the SUV, the answers are $P_{TR} = 17.6$ kW, $V_{max} = 146$ km/h (91.0 mph), and $GR_{max} = 32.2\%$.

(b) Since there is no resistance and tractive force is constant, we can apply Eq. (15-3) to the case of both vehicles. For the car:

$$A_{max} = \frac{F_{TR}}{m} = \frac{3000}{1200} = 2.5 \text{ m/s}^2$$

In order to reach cruising speed of 26.7 m/s, the car must accelerate for $t = (26.7)/(2.5) = 10.7$ s. By similar calculation, for the SUV, $A_{max} = 3$ m/s², $t = (26.7)/(3) = 8.9$ s. Note that because the assumptions about vehicle specifications are simplistic, the comparison is not transferable to actual vehicles having approximately the same dimensions.

Discussion   A comparison of the results for the two vehicles is given in the table below. The percent change column gives the percent change up or down for the value for the SUV, relative to that of the car.

	Units	Compact	SUV	Change
Cruise	kW	10.5	17.6	67%
Maximum speed	km/h	144.6	145.7	1%
	mph	90.4	91.0	1%
Gradability	percent	26.4%	32.2%	22%
Acceleration	seconds	10.7	8.9	−17%

From the table, it is clear that the design features of the SUV give it the superior performance in the categories that consumers seek in such a vehicle, but also worsen its fuel economy. In practical terms, the maximum number of passengers is the same, but the SUV has presumably more cargo space, rides higher off the ground, and is heavier, giving it the impression of being a structurally stronger and hence safer car. However, cruising fuel consumption is 67% more; although it is not shown, the heavier mass will also lead to greater fuel consumption when accelerating. Increased drag and rolling resistance coefficients as well as mass lead to fuel consumption at constant speed being greater. On a per unit of mass basis, the SUV can provide more tractive force when accelerating and more tractive power at maximum speed, so acceleration, maximum gradability, and maximum speed are superior, which are desirable features for this type of vehicle.

The results for the two generic vehicles presented in the above table can be compared to real-world vehicles to examine the validity of using engineering formulas to predict performance. For instance, the 2005 Toyota Corolla and Toyota RAV-4 small SUV are comparable to the example vehicles in terms of curb weight and drag coefficient (1150 kg/0.3 and 1448 kg/0.4, respectively). Using current estimates of highway fuel economy from the USEPA, the increase in predicted fuel consumption for driving a RAV-4 in place of a Corolla is 37%. Though not as large as the predicted value in the table of 67%, the difference is nevertheless significant. Also, if we were to assume for the example SUV that the tires had the same rolling resistance as for the passenger car ($C_o = 0.1$), the fuel increase percentage in the table would be 49%, which is closer to the real-world vehicles. Given that Eq. (15-1) is calculating power requirement at one speed and does not take into account many variables that affect overall fuel economy (highway driving cycle, power train efficiency in converting gasoline into power, effects of accelerating and braking, and so on), we can see that the use of Eqs. (15-1) and (15-2) as shown in this example is a reasonable way to make first-order predictions about differences in performance and energy efficiency.

Complicating Factors in Vehicle Design

It is possible to use Eqs. (15-1) to (15-3) to make the calculations necessary at a basic level to make broad comparisons between major groups of light-duty vehicles, such as compact cars versus SUVs and minivans, or hybrids versus ICE vehicles. In practice, complicating factors come into play, which make accurate comparisons between competing vehicle alternatives, or interpretation of experimental results from a vehicle test track, a much more complex enterprise.

As an example, let us focus just on the question of acceleration. First, in real-world driving, the force acting to accelerate the vehicle is the net difference between tractive force from the power train and resistance acting on the vehicle. Since the resistance is a function of V, this component of the net force will change as the vehicle accelerates. Furthermore, F_{TR} changes with changing speed. Taking the example of the ICE, F_{TR} is related to the flywheel torque T_{FW}, measured in newton-meters or $N \cdot m$, of the engine, which is itself changing as a function of rpm, as follows:

$$T_{AX} = T_{FW} \times G_{TR} \times G_{FD} - L_{DR} \tag{15-4}$$

$$F_{TR} = T_{AX} \times r_{tire} \tag{15-5}$$

Here T_{AX} is the axle torque in $N \cdot m$, G_{TR} is the transmission gear ratio (which varies depending on which gear the transmission is in), G_{FD} is the final drive gear ratio (i.e., between the drive shaft and the transaxle), L_{DR} are the various drivetrain losses, and r_{tire} is the radius of the tire. This calculation assumes a manual transmission; additional losses are incurred due to slippage in an automatic transmission.

The practical result of Eqs. (15-4) and (15-5) is that, since the engine has an rpm range where T_{FW} reaches a peak value, and then above or below that range T_{FW} falls off, the ability to contribute to F_{TR} will diminish once the rpm is above the range. Presence of a multispeed gearbox allows the driver to compensate by shifting into a higher gear where once again the engine will operate in ideal rpm range. However, with each higher gear, G_{TR} is decreased, so that F_{TR} and hence maximum acceleration decreases. The effect of changing F_{TR} with changing speed is shown in Fig. 15-4 for the relationship between F_{TR} in the top gear in an ICE, resistance forces, and maximum velocity. At V_{max}, F_{TR} and resistance forces are in balance, so as V approaches V_{max}, resistance is increasing with the third power of V while F_{TR} is decreasing with increasing rpm, so that the vehicle will approach V_{max} asymptotically. This behavior is observed in vehicles traveling at very high speeds. A test driver can typically accelerate from a standstill to expressway speeds of 100 to 140 km/h relatively rapidly, depending on the vehicle in question, but thereafter

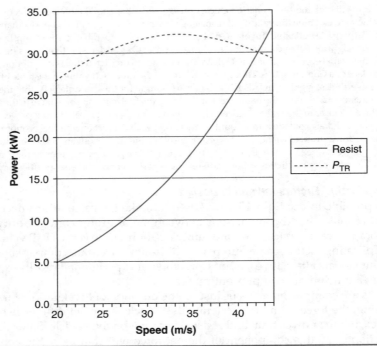

FIGURE 15-4   Tractive power P_{TR} versus resistance force as a function of velocity V. *Conversion*: 10 m/s = 22.5 mph.

finds that the increase of speed to the rated maximum speed of the vehicle on level ground happens much more slowly.

One practical outcome of the complex nature of relationship between design parameters and delivered performance is that manufacturers carry out performance testing using a mixture of theoretical modeling and empirical testing. As illustrated in Examples 15-1 and 15-2, theoretical evaluation can be used to make general predictions about the performance of classes of vehicles in terms of cruising fuel consumption, maximum gradability, and so on. On the other hand, where the goal is to create a transparent benchmark by which a discerning consumer will make choices between specific makes and models within a vehicle class, it is too complicated to create a defensible theoretical model that makes meaningful comparisons between vehicles, so the makers use empirical testing instead. For example, a vehicle's ability to accelerate is evaluated and published for promotional purposes in terms of "time from 0 to 30 mph" (0 to 48 km/h) or "time from 0 to 60 mph" (0 to 96 km/h), and so on, using professional drivers on a test course, rather than through theoretical analysis.

15-3-2 Options for Improving Conventional Vehicle Efficiency

From the discussion in the previous section, overall vehicle weight, maximum power, aerodynamic drag, and other parameters directly influence the energy consumption of gasoline and diesel ICEVs. It follows that making changes to the parameters, such as curb weight, aerodynamic drag, or rolling resistance, through improvements in vehicle design, provides a means to improve fuel economy. Some of the parameters are already evolving in this direction. As shown in Table 15-3, typical drag coefficient values have been decreasing steadily, with values as low as $C_D = 0.25$ reported for some 2015 passenger car models. Makers have also reduced curb weight per unit of passenger compartment volume through more efficient use of space, advances in materials, and the abolition of the underbody chassis that was common to vehicles in the 1960s and before. On the other hand, in markets such as that of the United States, larger vehicles such as SUVs have become popular, putting upward pressure on the average curb weight of the light-duty vehicles in the fleet. Nevertheless, further improvements in ICEV weight should be possible, especially if higher fuel prices discourage buyers from purchasing the largest vehicles. Through these changes, some improvement in fuel economy is attainable, without sacrificing performance or vehicle comfort. These incremental changes have a lower up-front cost to the maker than full-scale changes to alternative platforms, so they pose less of a financial risk.

Incremental improvements of this type have a limit, however. First, as time passes, it becomes harder to wring additional savings out of an ICEV platform that has already been substantially improved. Second, as long as the fuel for these vehicles remains gasoline or diesel, the makers and car buyers cannot fully achieve the resource and climate goals laid out at the beginning of the chapter if only this option is pursued. Therefore, in Sec. 15-3-3, we consider power requirements for nonhighway modes, some of which are more efficient or give the transportation system access to noncarbon fuels. Then in Sec. 15-4, we continue the discussion of alternative fuels and propulsion platforms as another alternative solution.

15-3-3   Power Requirements for Nonhighway Modes

Similar principles can be applied to the estimation of power requirements at cruising speed and maximum speed for other modes such as rail, marine, and air, as well as maximum capacity to climb grades or accelerate. For brevity, we focus only on cruising at constant speed on level ground with no positive or negative impact of winds or (in the case of marine) water currents. Also, the pipeline mode is not included in this discussion.

Passenger and Freight Rail

For rail, the power requirement for a given train is a function rolling and aerodynamic resistance, with rolling resistance dominant at low speeds and aerodynamic resistance dominant at higher ones. Parameter values for rail rolling resistance reflect the advantage of the steel rail wheel on a steel rail compared to rubber tires on an asphalt road, and result in lower rolling resistance for a given speed, all other factors equal. On the other hand, rail vehicles on steel rails suffer from significantly lower maximum gradability than road vehicles; for this reason, some urban underground mass transit systems have adopted rubber-tired vehicles on concrete tracks. Rail vehicles also require a wider turning radius compared to road vehicles. Lastly, a fully detailed equation governing rail power requirement considers not only cross-sectional area, speed, and coefficients, but also the length of the train, since as trains grow in length both aerodynamic and rolling resistance increase. Examples in this section are limited to intercity rail movements.

Formulaic treatment of power requirements here focuses on high-speed rail trainsets, since they are of particular interest at present as an alternative to either driving or flying in both industrial and emerging countries. Because trainset designs, internal components, lengths of trains, and operating conditions for HSR systems are different both within national networks and in comparing one country to another, there is no one theoretical equation that can be used to evaluate all HSR trains. Instead, empirical equations for individual classes of HSR trainsets (note that a "trainset" is a single unit consisting of one or more locomotives and multiple passenger cars) are used. In this case, we use a French TGV ("Train a Grand Vitesse") as an example (see Hopkins et al., 1999). Let D = drag on TGV train in kilonewtons (kN) and V = velocity in m/s. Then drag as a function of velocity is

$$D(V) = a + bV + cV^2 \qquad (15\text{-}6)$$

where a, b, c are parameters used to fit a curve to observed values of drag as a function of speed, in this case having the values $a = 3.82$, $b = 0.1404$, and $c = 0.006532$. Alternatively, if V is given in km/h, the parameters are $a = 3.82$, $b = 0.039$, and $c = 0.000504$, or if given in miles per hour and D in kilopounds force, the values are $a = 0.9587$, $b = 0.01403$, and $c = 0.00029$.

Regardless of units, the drag on the trainset consists of three components: (1) static, (2) linear, and (3) nonlinear. Hereafter we use V in units of m/s, since it is convenient for converting drag to power. The power requirement P in units of kW to maintain a given speed V is

$$P = D \cdot V$$
$$= aV + bV^2 + cV^3 \qquad (15\text{-}7)$$

Using drag values $a, b,$ and c given above, and a chosen velocity value, we can calculate total power requirement for different speeds, including the contribution of static, linear, and nonlinear components to the total value. For example, a typical steady-state or "cruising" speed might be 200 mph or 320 km/h. This value is equivalent to 88.9 m/s, so substituting into Eq. (15-7) and solving gives a power requirement of $P = 6037$ kW. For comparison, the maximum speed of current HSR trainsets is on the order of 100 m/s, equivalent to 360 km/h or 225 mph, which would give $P = 8318$ kW. By the time speed reaches 40 m/s (144 km/h or 90 mph), which is well below HSR cruising speeds, the nonlinear component of the power requirement dominates, reflecting the importance of aerodynamic drag in determining propulsion power required.

On the freight rail side, total resistance and hence power requirement is based on the number of cars and locomotives in motion, so the approach is to calculate resistance per car and per locomotive, and sum the total. In this case, resistance of both rail cars and locomotives are evaluated in U.S. customary units and are a function of car weight and rolling/aerodynamic resistance. As with HSR trainsets, there is no single theoretical formula for resistance, so we are limited to a specific empirical example in this instance. Let R = resistance in pounds, T = weight in tons, V = speed in mph, N = number of axles on locomotives (does not apply to railcars). The resistance of a railcar on level surface traveling at constant speed is then:

$$R_{CAR} = aT + b + cTV + dV^2 \qquad (15\text{-}8)$$

For a representative railcars, values $a = 1.5$, $b = 72.5$, $c = 0.015$, and $d = 0.055$ are appropriate. Next, for the locomotive, the following equation applies:

$$R_{LOCOMOTIVE} = aT + bN + cTV + dV^2 \qquad (15\text{-}9)$$

For a generic locomotive, representative values are $a = 1.3$, $b = 29$, $c = 0.03$, and $d = 0.312$. Note that in Eq. (15-9) the number of axles contributes to resistance, unlike the case of railcars. Lastly, the maximum tractive effort of the locomotive TE is measured in pounds and is a function of the maximum horsepower P of the locomotive and speed V. As V increases, the maximum tractive effort decreases according to

$$TE = aP/V \qquad (15\text{-}10)$$

Here the parameter value $a = 308$ is representative. Note that Eq. (15-10) is not intended for calculations at very low values of V, since as V approaches zero, TE approaches infinity. However, since in this example we are interested in steady-state energy consumption at line haul speeds, this limitation is not a concern.

From a railroad operations perspective, the key planning criterion is to provide sufficient locomotives capable of delivering sufficient value TE at desired speed V to overcome the total resistance calculated using Eqs. (15-8) and (15-9) (see Armstrong, 1993). The process entails adding locomotives until the total value TE exceeds the required amount of force in pounds. Excess TE beyond what is required is not a concern and is actually useful since climbing grades or acceleration increases power requirements. Indeed, Eqs. (15-8) and (15-9) could be extended by including the effect of instantaneous grade (ascending or descending) and acceleration (positive or negative).

Once the total resistance is known, the power required to maintain speed can be calculated by converting from standard to metric units and using the relationship of Eq. (15-7), now applied to freight rail instead of HSR. The needed conversion factor is 4.448 newtons per pound of force. Example 15-3 illustrates these calculations for a representative freight train.

Example 15-3 A freight trains consists of 50 freight cars, each weighing 30 tons empty and carrying a load of 90 tons (typical of, e.g., a fully load hopper car for carrying coal), at 50 mph on a level track. The train is to be pulled by a string of locomotives each with maximum power of 2500 horsepower, weight of 120 tons, and 4 axles. (a) What is the total resistance of just the pulled load of the train, not including the locomotives, in lb? (b) What is the minimum number of locomotives required to pull the load at the desired speed, and the resulting total resistance of the train in lb? Assume all parameter values for Eqs. (15-8) to (15-10) as given in the passage above. (c) What is the power required to maintain the desired speed?

Solution Below, equations are shown with parameter values substituted. (a) The resistance per car is calculated using Eq. (15-8) based on $V = 50$ mph and $T = 120$ tons total for the combined car and freight:

$$R_{CAR} = 1.5T + 72.5 + 0.015TV + 0.055V^2$$

$$= 1.5(120) + 72.5 + 0.015(120 \cdot 50) + 0.055(50)^2 = 480 \text{ lb}$$

Therefore, the resistance for all 50 cars is $(50)(480) = 24{,}000$ lb.

(b) The resistance per locomotive is calculated using Eq. (15-9):

$$R_{LOCOMOTIVE} = 1.3T + 29N + 0.03TV + 0.312V^2$$

$$= 1.3(120) + 29(4) + 0.03(120)(50) + 0.312(50)^2 = 1232 \text{ lb}$$

The tractive effort at $V = 50$ mph per locomotive is calculated as

$$TE = 308(2500)/50 = 15{,}400 \text{ lb}$$

From further calculations, a train with one locomotive has a resistance of $24{,}000$ lb $+ 1232$ lb $= 25{,}232$ lb and maximum tractive effort of $15{,}400$ lb, and will therefore not be able to travel at 50 mph since the resistance exceeds the available tractive effort. A train with two locomotives, however, has a resistance of $26{,}464$ lb and $TE = 30{,}800$ lb, so that it will be able to attain the required speed. Therefore, the number of locomotives required is 2, and $R_{TOT} = 26{,}464$ lb.

(c) The speed of 50 mph is equivalent to 22.2 m/s. Converting to metric and calculating P gives:

$$F = 26,464 \text{ lb}(4.448 \text{ N/lb}) = 117.7 \text{ kN}$$

$$P = FV = (117.7)(22.2) = 2,620 \text{ kW}$$

The ratio of maximum power to total weight pulled is called *horsepower per trailing ton* and is used as a measure to indicate the likely maximum speed or ability of a train to accelerate rapidly once it has been dispatched for the line haul. For instance, in Example 12-3, the train has 5000 hp total available and is pulling 6000 tons, and therefore has 0.833 hp per trailing ton.

Marine Vessels

For marine vessels, the drag created by the friction between the hull of the vessel and the surrounding water dominate the generation of physical drag; air resistance can also be included, but is relatively small due to the low travel speeds involved, and we will ignore it in the following first-order explanation. Speed is measured in "knots," with 1 knot = 1.15 mph = 1.86 km/h. A given hull design will have a specific "hull speed" to which the length of the hull contributes, with a longer hull being proportional to a higher hull speed. A significant discontinuity in the equation of drag as a function of speed occurs at the hull speed, with drag increasing greatly above this threshold. Speeds above hull speed are prohibitively expensive for commercial shipping operations such as transoceanic container shipping, and are instead the domain of other maritime applications, such as the military or competitive sailing.

In its simplest form, for speeds at or below hull speed, power requirement is a function of speed V in knots and an empirical resistance parameter K in units of BTU h^{-1} kn^{-3}, where kn is an abbreviation for knots:

$$P = KV^3 \tag{15-11}$$

The value of K depends on the length of the ship in question, hull design, efficiency of drivetrain from ship engine through drive shaft to propeller, and other factors. For example, a representative container ship capable of carrying 4000 40-foot containers might have a value of $K = 18,850$ BTU h^{-1} kn^{-3}. Hull speed for such a ship might be 25 knots. If the ship travels at a speed such as $V = 21$ knots that is below hull speed, application of Eq. (15-11) gives a power requirement of $P = 174.6$ MMBTU/h, equivalent to 184.2 GJ/h or 51.2 MW. Although this quantity is much larger than the power requirement of the freight train in Example 15-3, we observe that the load on the ship is on the order of two orders of magnitude larger.

Aircraft

Power requirement equations for aircraft differ from other transportation modes in two respects. First, once the aircraft is airborne there is no rolling resistance, so the frictional drag comes in the form of aerodynamic drag as the surface of the aircraft passes through the air. A new component of power requirement is introduced, however, namely, the force required to generate the aerodynamic lift that keeps the craft aloft. Since drag increases with velocity and the power requirement for lift decreases, there is an optimal cruising speed at which power requirement is minimized. Secondly, in Eq. (15-1) the power requirement to accelerate any vehicle from standstill to cruising speed has generally been ignored, but in the case of aircraft the power requirement and hence energy consumption to accelerate the craft from zero speed on the runway to being airborne and eventually climbing to cruising altitude

Make	Model	Length (m)	Width (m)	Seats*	Max Wt* (tonnes)	Range (km)
Airbus	A319	33.84	34.1	124	64.4	6850
	A330	63.69	60.3	295	230.9	11900
	A380	72.72	79.75	525	562	15700
Boeing	737	33.6	35.8	126	70.1	6370
	747	64.4	70.6	416	363	13450
	787	57	60	250	227.9	15200

*Notes: Values shown are for a representative variant of the given model. Each model is offered in the form of several different variants, each with slightly different specifications, e.g., seating, maximum weight, range, etc. Seating value shown is for a typical multi-class (either 2- or 3- passenger classes) configuration; a single economy-class configuration will have higher total number of seats. Maximum weight is the weight before takeoff with full fuel tanks, weight on landing or with no fuel on board will be less than the value shown due to fuel consumption in flight.

TABLE 15-4 Basic Specifications for Select Commercial Aircraft from Airbus and Boeing Companies in 2013. *Sources:* Airbus Corporation; Boeing Corporation.

is a substantial component of the energy requirement for the entire flight. Therefore, a linear model of energy consumed as a function of distance traveled is more problematic for aviation than for, e.g., rail or marine travel. Instead, comprehensive estimation of energy consumption for flying must take into account the relative proportion of the flight distance spent climbing to cruising altitude versus cruising, although this dimension is not considered here for reasons of brevity.

The performance of commercial jet aircraft can be evaluated based on physical dimensions, weight, maximum range, and other factors. Table 15-4 gives specifications for several models of aircraft from the two largest global suppliers of aircraft, the European-based Airbus and the U.S.-based Boeing. As shown in the table, each maker offers models to compete in the various air travel markets. For example, the Airbus A319 and Boeing 737 compete in the short-haul market, the A330 and 787 compete in the long-haul market with seating requirements in the range of 250–300 persons, and the A380 and 747 compete in the long-haul market with the largest seating requirements.

In equation form, total power required P_{total} is a combination of drag and lift, each of which can be calculated based on cruising speed and various other parameters:

$$P_{total} = P_{drag} + P_{lift}$$
$$= 0.5\rho C_d A_p V^3 + 0.5\frac{(mg)^2}{\rho V A_s} \tag{15-12}$$

Here ρ is the density of air, C_d is the aerodynamic drag coefficient specific to the particular aircraft, A_p is the frontal area of the aircraft, V is the speed, m is the mass of the aircraft, g is the gravitational constant, and A_s is the effective area of the aircraft supported by aerodynamic lift (which is the square of wingspan width). Since V increases P_{drag} but decreases P_{lift}, there is a tradeoff in choosing the design cruising speed of the aircraft. Note that aerodynamic drag decreases at cruising altitudes thanks to lower air density, on the order of 0.4 kg/m³ compared to ~1.0 kg/m³ encountered by road or rail vehicles at or near sea level. Example 15-4 illustrates the computation of power and energy requirements for a specific aircraft, namely, a Boeing 747, in comparison to HSR.

Example 15-4 A Boeing 747 has the dimensions given in Table 15-4. From Mackay (2009) this aircraft has an aerodynamic drag coefficient of 0.03, a frontal area of 180 m², and jet engines with an efficiency of 33%. At cruising altitude with air density of 0.4 kg/m³ it travels at a speed of 254 m/s (914 km/h). (a) What is the instantaneous power requirement for the 747 to maintain the given speed and altitude? (b) If the 747 travels a distance of 1000 km with a full load of passengers under these conditions, what is the energy content of the required amount of aviation fuel, ignoring take-off and landing energy requirements? (c) Suppose a TGV train travels 1000 km with a full capacity of 384 passengers at a speed of 320 km/h, with 75% electrical efficiency from overhead catenary to power delivered to the wheels, what is the energy consumption per passenger in this case?

Solution (a) Supported area A_s can be calculated from the width of the 747 from the table:

$$A_s = W^2 = (64.4)^2 = 4147.36 m^2$$

All other values are given so P_{total} can be calculated using Eq. (10-4):

$$P_{total} = 0.5\rho C_d A_p V^3 + 0.5\frac{(mg)^2}{\rho V A_s}$$

$$= 0.5(0.4)(0.3)(180)(254)^3 + 0.5\frac{(363,000 \text{ kg} \cdot 9.8 \text{ m/s}^2)^2}{(0.4)(254)(4147)}$$

$$= 1.76\times10^7 W + 1.37\times10^7 W = 3.26\times10^7 W = 32.6 \text{ MW}$$

(b) Since the distance traveled is 1000 km and the speed is equivalent to 914 km/h, the time required is:

$$\frac{1,000 \text{ km}}{914 \text{ km/h}} = 1.1 \text{ h}$$

The output energy E_{out} required to maintain the needed power for 1.1 h and the input energy E_{in} based on the efficiency of the jet engines are calculated as follows:

$$E_{out} = P_{total}t = (32.6 \text{ MW})(1.1 \text{ h}) = 35.9 \text{ MWh}$$

$$E_{in} = \frac{E_{out}}{\eta_{engines}} = \frac{35.9}{0.33} = 109 \text{ MWh}$$

$$E_{person} = \frac{E_{in}}{N_{occ}} = \frac{109 \text{ MWh}}{416} = 261 \text{ kWh}$$

(c) In the case of the HSR train, the speed is 320 km/h and the distance is 1000 km, so the time required is 3.1 h. Repeating a similar pattern from part (b) gives the following:

$$E_{out} = P\cdot t = (6,037 \text{ kW})(3.1 \text{ h}) =\sim 18,700 \text{ kWh} = 18.7 \text{ MWh}$$

$$E_{in} = \frac{E_{out}}{\eta_{trainset}} = \frac{18.7}{0.75} = 24.9 \text{ MWh}$$

$$E_{person} = \frac{E_{in}}{N_{occ}} = \frac{24.9}{384} = 64.8 \text{ kWh}$$

In conclusion, Example 15-4 raises several observations about intermodal comparisons between air and rail. On the one hand, the significantly lower energy per person figure for HSR illustrates the possibility for more energy-efficient middle-distance intercity passenger travel if HSR infrastructure can be developed. This possibility is further enhanced by access to CO_2-free electricity that could reduce CO_2 per passenger per trip. On the other hand, the significantly longer time (approximately triple that of air, not including stops at intermediate stations) and requirement that HSR travel occur over land and not over water shows the limitations. For very long distances (e.g., transoceanic or transcontinental in the case of countries such as Russia or the United States) it is usually not within travelers' available time in modern society to travel for such a long duration, even where HSR infrastructure exists.

15-4   Alternatives to ICEVs: Alternative Fuels and Propulsion Platforms

Vehicles that use an alternative fuel or are based on an alternative propulsion platform, which are the focus of this section, have the potential to radically reduce petroleum consumption and CO_2 emissions, and in some cases there are already vehicles on the road that are achieving this goal. Vehicles that use alternative fuels, including electric vehicles, biofuel-powered vehicles, and hydrogen fuel cell vehicles, can obtain their energy from sources other than fossil fuels. Hybrid vehicles take advantage of a fundamentally different propulsion platform to greatly reduce energy requirements, regardless of the energy source. All of the major makers in North America, Europe, and Asia, have active research in one or more of these technology fields.

Like an ICEV, an alternative vehicle must deliver acceptable performance to the customer to succeed in the market, while reducing energy consumption and greenhouse gas (GHG) emissions to succeed on environmental goals. Not only must it deliver sufficient power, but it must also do so without incurring a large weight or volume penalty, since otherwise the handling of the vehicle or its capacity to carry cargo may be compromised. This standard applies to the alternative vehicles considered throughout this section.

15-4-1   Battery-Electric Vehicles

The battery-electric vehicle (hereafter EV) has existed since the early days of the history of the automobile in a niche-market and prototype form. After automobile makers settled on the ICE as the propulsion system of choice in the 1900s, experimental work continued with the vehicles at a low level throughout the twentieth century, and by the 1970s prototypes existed with a range of 130 km (81 mi) per charge and acceleration of 0 to 96 km/h (0 to 60 mph) in 16 s. The World Solar Challenge between Adelaide and Darwin, Australia, in 1987 was a catalyst for developing a new generation of more advanced EVs, and positioned the winning entrant, General Motors, to develop a prototype EV called the "Impact" launch in January 1990 that achieved a 145 km range between charges, a 150-km/h top speed, and acceleration of 0 to 96 km/h (0 to 60 mph) in 8.6 s. Since it coincided with California's regulatory deliberations for tighter tailpipe standards, GM's introduction of the Impact concept EV had the unintended consequence of stimulating the zero emission vehicle (ZEV) mandate.

In the late 1990s, a number of American and Japanese makers entered EVs in the commercial passenger car and light-truck market in California in the United States, and leased or sold several thousand vehicles, ranging from the Saturn EV-1 two-seat coupe (the production version of the Impact), to the Toyota RAV-4 small SUV, to the Ford Ranger pickup truck. In 2001, the automakers succeeded in weakening a mandate that would have required the expansion of EV sales in the California market, and by 2002, new EVs were no longer for sale from the major manufacturers. This unfolding of events remains controversial even now. The major manufacturers claim that their efforts to sell the vehicle proved that a sufficient market did not exist for a vehicle that lacked adequate range for intercity driving, meaning that an owner would most likely need to own a second ICE-powered vehicle in order to meet all driving needs. Supporters of the EV countered that there was, in fact, a sufficiently large market of drivers who could afford to own an EV as a second car, and that automakers had not made a good faith effort to promote the technology.

Electric Vehicles Today

The experience with EVs in the late 1990s and early 2000s proved that technically, such a car could be built so as to operate well on urban road networks otherwise dominated by ICEVs, even if the economics were questionable. It also provided a stepping stone toward the HEV technology that has been expanding steadily in the North American and Japanese markets since 1997, and at a faster rate since 2003. Mass-produced EVs re-entered the U.S. vehicle market during the 2009–2011 period with entrants such as the Tesla Roadster and Nissan Leaf. These vehicles were later joined by the Tesla Model S sedan (2012 launch) and additional entries from automakers such as Ford, Volkswagen, Mitsubishi, and BMW, as well as others (see Figs. 15-5 and 15-6). As of 2015 automakers are targeting several segments of the light-duty vehicle market, from very compact EVs (Mitsubishi) to high-performance sport sedans (Tesla Model S). (Plug-in hybrid-electric vehicles and "range-extended" EVs such as the Chevy Volt or Plug-in Toyota Prius are discussed later in the chapter.)

Other EVs continue to enter the marketplace by various means. Limited-use vehicles (LUVs) powered by batteries that travel at a maximum speed of 50 to 60 km/h (30 to 35 mph) are available in the U.S. market and elsewhere; while not legal for travel on highways, they are useful for niche applications such as national and state parks, corporate office parks, or academic campuses (see Fig. 15-7). Also, there are a number of small companies and individuals who provide retrofitting services to a few thousand ICEVs each year, removing the ICE drivetrain and replacing it with a battery system and motor.

Electric Vehicle Drivetrain Design Considerations

The most recent generation of EVs has opted for ac motors due to superior performance. In such a vehicle, the drivetrain consists of a battery bank, a power controller

FIGURE 15-5   Nissan Leaf, September 2013, showing charging cable attached to front charging port.

FIGURE 15-6   Tesla Model S sedan at 2013 New York Auto Show. *Source:* Ricardo Daziano. Reprinted with permission.

FIGURE 15-7   Limited-use vehicle (LUV) from the GEM (Global Electric Motors) division of Chrysler.

The turn-signals, headlights, seatbelts, and license plate seen in the image allow for legal operation on public roadways. The letters "LU" in the New York State license plate signify that the vehicle is only legal to operate on roads with a speed limit of 35 mph (approx. 60 km/h) or less.

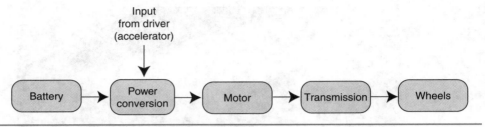

FIGURE 15-8   Schematic of components in EV drivetrain system.

that converts electrical energy from dc to ac, a motor, and a transmission to the driven wheels, as shown in Fig. 15-8. Control over the vehicle is transmitted from the accelerator pedal electronically to the power converter, which then controls the rate of energy flow to the motor. Note that, although not shown in the figure, the EV drivetrain may also incorporate regenerative braking. Instead of dissipating kinetic energy as heat in the brake pads, as occurs in an ICEV, the EV can run the transmission in reverse, using the electric motor as a generator that creates drag on the wheels to slow the vehicle, while at the same time recharging the batteries.

The *charge density* of the battery technology has a strong influence on the overall range of the vehicle. Charge density can be measured either in terms of energy per unit of weight or energy per unit of volume. A related measure is *power density*, which is the maximum power that the battery technology can achieve per unit of weight or volume. Charge density is important for range, while power density is important for maximum acceleration or sustained speed when climbing a grade.

For example, the differences in charge density and power density how two main types of batteries, namely lithium ion (Li-ion) and nickel metal hydride (NiMH), are used. Lithium ion has high charge density and attractive cost per unit of charge, and is therefore used in EVs, where it is important to store large amounts of charge as cheaply as possible. However, Li-ion has low power density, so for an application such as hybrid-electric vehicles (Sec. 15-4-2) where the battery system is small but needs on occasion to be able to discharge large amounts of power quickly, NiMH is chosen because it has high power density, and its relatively high cost per unit of charge stored is acceptable thanks to low overall size.

Suppose we take a simplified model of an EV in which charge availability is constant at all states of charge of the battery (i.e., whether full, half-full, and the like) down to the maximum depth of discharge DD_{max} (as a fraction of 100%). Assume further that the vehicle has constant average energy intensity in terms of the energy required per unit of mass of vehicle moved 1 km. The range R, in km, can be estimated as a function of battery weight as follows:

$$R = \frac{CD \cdot DD_{max} \cdot W_{battery}}{\mu(W_{vehicle} + W_{battery})} \qquad (15\text{-}13)$$

where CD is the charge per mass of battery in watthours/kg (or Wh/kg), $W_{battery}$ is the mass of batteries in kg, $W_{vehicle}$ is the mass of remainder of the vehicle in kg, not including the batteries, and μ is the energy intensity of the vehicle, in Wh/kg·km (i.e., average watthours of battery capacity required to move 1 kg of the vehicle for 1 km). This equation can be rewritten to solve for $W_{battery}$ as a function of range R:

$$R\mu(W_{vehicle} + W_{battery}) = CD \cdot DD_{max} \cdot W_{battery}$$

$$R\mu W_{vehicle} = CD \cdot DD_{max} \cdot W_{battery} - R\mu W_{battery} \tag{15-14}$$

$$W_{battery} = \frac{R \cdot \mu \cdot W_{vehicle}}{(CD \cdot DD_{max} - R \cdot \mu)}$$

If we take the fixed cost of the vehicle to be C_{fixed} before adding the cost of the batteries, the total cost of the vehicle TC can be written as a function of range as shown:

$$TC = C_{fixed} + W_{battery}C_{battery}$$

$$= C_{fixed} + \frac{R \cdot \mu \cdot W_{vehicle}}{(CD \cdot DD_{max} - R \cdot \mu)}C_{battery} \tag{15-15}$$

where $C_{battery}$ is the unit cost of the batteries in \$/kg. The use of these equations to estimate battery requirements and total cost as a function of range is illustrated in Example 15-5.

Example 15-5   A representative EV can be built with either lead-acid (Pb) or lithium-ion batteries. Regardless of battery type, the vehicle has the following values: $W_{vehicle} = 920$ kg, $\mu = 0.128$ Wh/kg-km, and $C_{fixed} = \$14,000$. Lead acid batteries have values of $CD = 55$ Wh/kg and $C_{battery} = \$125$/kWh. Li-ion batteries have values of $CD = 128$ Wh/kg and $C_{battery} = \$400$ per kWh. For both options assume a maximum depth of discharge of 80%. What mass of batteries is required, and what is the total cost, of a vehicle that has a range of (a) 200 km and (b) 300 km?

Solution   Mass of batteries for case (a), EV with 200 km range, from Eq. (15-1) the weight requirement for the lead-acid battery is:

$$W_{battery} = \frac{R \cdot \mu \cdot W_{vehicle}}{(CD \cdot DD_{max} - R \cdot \mu)} = \frac{200 \cdot 0.128 \cdot 920}{(55 \cdot 0.8 - 200 \cdot 0.128)} = 1280 \text{ kg}$$

Repeating for Li-ion batteries:

$$W_{battery} = \frac{R \cdot \mu \cdot W_{vehicle}}{(CD \cdot DD_{max} - R \cdot \mu)} = \frac{200 \cdot 0.128 \cdot 920}{(128 \times 0.8 - 200 \cdot 0.128)} = 307 \text{ kg}$$

Mass of batteries for case (b): For a distance of 300 km, the above calculations are repeated but with $R = 300$ km, giving values of 6309 kg for lead and 552 kg for Li-ion. Note that the value for lead-acid at 300 km range is unrealistic, as the battery system ways more than six times as much as the remainder of the vehicle. This result is discussed further below.

Turning to total cost of vehicle, it is necessary to compute $C_{battery}$, the cost per kg for batteries, from the given data. These values are computed for lead and Li-ion:

$$C_{battery.Pb} = (\$125/\text{kWh})(0.055 \text{ kWh/kg}) = \$6.88/\text{kg}$$

$$C_{battery.Li} = (\$400/\text{kWh})(0.128 \text{ kWh/kg}) = \$51.20/\text{kg}$$

We can solve for the total cost of the vehicle using Eq. (15-3) for the case of 200 km range:

$$TC_{Pb} = C_{fixed} + W_{batt}C_{battery} = \$14000 + (1280)(\$6.88) = \$22,800$$

$$TC_{Li} = C_{fixed} + W_{batt}C_{battery} = \$14000 + (307)(\$51.20) = \$29,718$$

For case (b) with 300 km range, repeating the above calculations gives $57,374 for the EV with lead-acid batteries, and $42,262 for the EV with Li-ion batteries.

Example 15-1 shows how the higher charge density of the lithium-ion battery translates into lower additional mass to the vehicle, especially at longer ranges. This comparison is made visually in Fig. 15-9, where the range of the EV is plotted as a function of battery mass using Eq. (15-1) for values from 0 to 2000 kg and other parameters taken from Example 15-5. At a range of 235 km, the batteries for the lead-acid system weigh 2000 kg, which comprises more than 2/3 of the total mass of the vehicle in the case of lead-acid. Note also that this estimate is an optimistic simplification, since increasing vehicle weight by such a large amount will have a multiplier effect in terms of requiring additional strengthening of the vehicle body, a stronger electric motor, heavier brakes, and so on, to be able to support the extra batteries, further increasing weight. These limitations are considered below in the discussion of a more complete mass-range model.

Using cost, weight, and charge density figures from Example 15-5, it is also possible to compare total vehicle cost figures for the two battery technologies as a function of range (Fig. 15-10). At low range values, there is a cost premium associated with the light weight of lithium-ion batteries, which is why applications such as golf carts for which cost rather than range per charge is the primary concern will continue to use lead-acid. However, at range values that begin to approach those of ICEVs (250 km and above), the cost premium narrows, and around 275 km the two technologies are equivalent in price. Above this range value, the cost for lead-acid increases rapidly, and the cost curve approaches a vertical asymptote around 344 km range, so that the vehicle cannot achieve this range for any cost. The problem of rising cost compounds the issue of excessive weight for lead-acid at significant range, as shown in Fig. 15-9. For instance, the Tesla Model S sedan with 85 kWh of storage capacity in theory adds 664 kg of battery mass to the vehicle, at a cost of $34,000, which compares favorably to lead-acid.

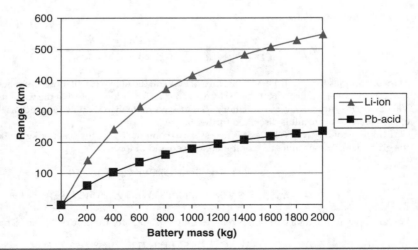

Figure 15-9 Maximum range for a representative lead-acid or Li-ion battery EV in km on a single charge, as a function of mass of batteries installed.

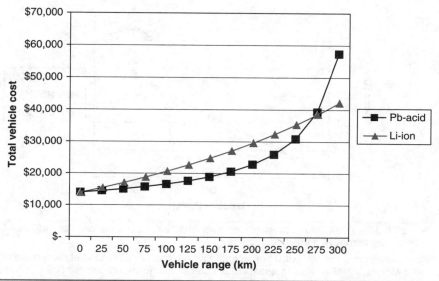

FIGURE 15-10 Vehicle cost as a function of range for lead-acid or Li-ion EV.

Application of Range Model to Case of Tesla Model S Sedan

Since the Tesla Model S Sedan comes in different configurations in terms of the size of the battery system (measured in kWh), it provides a useful application for estimation of range based on battery system size and weight and for comparison to other published sources. Range estimates are obtained from three sources: (1) the model of Eq. (15-13), (2) the USEPA at the website fueleconomy.gov, and (3) the manufacturers' claimed range per charge. The Model S sedan is examined in 40-kWh, 60-kWh, or 85-kWh versions. Tesla decided to discontinue development of the 40-kWh model before it could be put into production, but the company did publish specifications and range estimates were made available, so those are used. The range model estimate uses a charge density value of 105 Wh/kg including the entire weight of the battery system and maximum depth of discharge of 95% (both of which are estimated from the Tesla Motors website, and are different from those of Example 15-5). The intensity value μ is not known and must therefore be computed. Using above-mentioned data, curb weight of 2112 kg given by Tesla, and range given by USEPA of 261 km per charge gives a figure of $\mu = 0.104$ Wh/kg-km.

The results in Fig. 15-11 show that, having calibrated the range model using USEPA's estimated range for the 60-kWh version, the model's prediction for range are in good agreement with USEPA for the 40-kWh and 85-kWh versions. For the 40-kWh version, the model predicts a range of 190.9 km compared to 190.0 km for USEPA, while for the 85-kWh version the respective value are 331.7 and 340.0. Note that the claimed "Tesla Co." values in Fig. 15-11, which are higher than either the USEPA or model values, are not necessarily inaccurate; rather they reflect driving under conditions that may be more favorable to longer range per charge, such as a larger mix of highway driving.

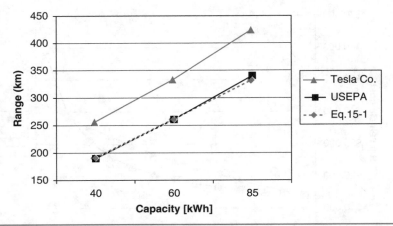

FIGURE 15-11 Prediction of average range per charge for three battery sizes of Tesla Model S from three alternative prediction sources.

Note: Y-axis does not start at 0 km range, to better show differences between curves. *Sources:* Tesla Motors, for "Tesla Co." values; USEPA fueleconomy.gov website, for "USEPA" values.

Desired Characteristics of More Complete EV Range Model in the Future

It is desirable to have an enhanced analytical model that goes beyond the basic model of Eqs. (15-13) to (15-15) to predict EV range based on battery system characteristics, balance of vehicle characteristics, and operational conditions (terrain, driver behavior, and so on). Specifically, a range model might incorporate the following parameters:

- *Effect of increasing battery size on balance of vehicle mass*: As battery mass increases, at some point the various components of the vehicle will need to measurably increase in strength and weight to function correctly. The enhanced model could therefore include a multiplier factor or other means to adjust the mass of balance of vehicle ($W_{vehicle}$) upward as battery mass increases.

- *Effect of aerodynamic drag*: In the basic model, the impact of aerodynamic drag is subsumed into the intensity factor μ, since a vehicle with higher aerodynamic drag will require more energy to move 1 kg of its mass 1 km, other things equal. However, in an enhanced model, the aerodynamic drag coefficient could be broken out so that for fixed combined weight $W_{vehicle} + W_{battery}$ changes in drag would affect overall range per charge.

- *Effect of rolling resistance*: Similar to aerodynamic drag, the enhanced model could capture impact of changing rolling resistance on range for fixed total weight (see Chap. 5 for mathematical relationship between aerodynamic drag, rolling resistance, and tractive power requirement).

- *Effect of operating conditions*: (e.g., driving speed, flat versus rolling terrain): Incorporating operating conditions would allow the model to predict range based on predicted conditions. For instance, in hilly terrain, electric vehicle range will tend to be reduced as large amounts of charge are expended to climb elevation. Some of this charge is returned to the battery system using regenerative braking, but there are limits due to both inherent losses in the system and the maximum rate at which the motor working backward can recharge the batteries. The driver's chosen highway cruising speed, aggressiveness of driving, use of air conditioning, and outside air temperature also have an effect.

At present, with the limited number of EVs on the road and lack of experience with the technology, it is premature to operationalize such a model for lack of real-world data. One solution in the interim is the use of a computerized driving cycle simulator, which takes a representative drive cycle or route between charges, measures power requirements at frequent intervals based on speed, acceleration, and grade, calculates instantaneous power requirement, and then adds up total energy consumption (see, e.g., Albertus et al., 2008). Although they ultimately represent the most accurate way of measuring range and other figures of merit for proposed EV designs, such models are time-consuming to implement and not accessible to a broad audience in the transportation community. There is therefore scope for a "middle ground" modeling option between the basic model already presented and the full simulator.

15-4-2 Hybrid Vehicles

Hybrid vehicles feature a hybrid drivetrain, which uses a combination of internal combustion engine and electric motor to optimize energy use for propulsion. These drivetrains have been introduced primarily into passenger cars to date, with additional use of hybrid drivetrains in midsize delivery vehicles that operate in stop-and-go conditions as well as full-length urban transit buses. The most efficient hybrid-electric vehicle (HEV) in the U.S. market, the Toyota Prius, has a rated fuel economy value of 21 km/L (49 mpg) for combined city/highway fuel economy, versus 10 to 14 km/L (23 to 33 mpg) for typical conventional drivetrain vehicles. The Prius is not as efficient as the best-performing car in the European market, the Audi A2 diesel (fuel economy of 36 km/L or 85 mpg in gasoline equivalent; taking into account the 11% greater energy content per gallon in diesel than gasoline, the fuel economy in gallons of diesel is 40 km/L or 95 mpg but the Prius also has a larger interior.

HEVs currently constitute a small part of the world passenger car market; however, concern about the environment and rising fuel costs have sparked interest, and sales have risen steadily, as shown in Fig. 15-12 for the sales of hybrids in the United States for the period 1999 to 2012. At the same time, with each passing year, automakers introduce new hybrid models in different market niches (sedans, minivans, SUVs, and so on), further increasing the opportunity to grow sales.

The HEV uses a combination of ICE and battery-electric technology to capture the advantages while avoiding some of the drawbacks of each. There are three basic energy flow concepts for the drivetrain hybridization (Fig. 15-13):

1. *Series hybrid:* All mechanical power from the ICE is first converted to electrical energy, then either to a motor for driving the wheels, or to a battery system for storage.

2. *Parallel hybrid:* Mechanical power from the ICE goes to a transmission, where it is divided between driving the wheels and driving a generator for storing energy, depending on operating conditions (used by Honda).

3. *Parallel-series hybrid:* An adaptation of the parallel hybrid, the drivetrain introduces a separate "power splitter" between the engine + motor on one side of the transmission and the driven wheels on the other, and has a separate motor for driving the wheels and generator for charging the battery (used by Toyota).

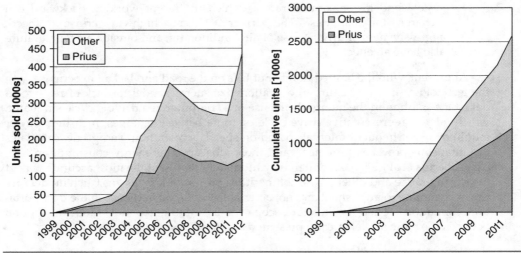

FIGURE 15-12 Annual (left) and cumulative (right) growth in U.S. annual hybrid vehicle sales, 1999 to 2012.

In other words, the parallel and parallel-series drivetrains make possible a direct mechanical connection between ICE and wheels, the same as an ICEV, and the series system does not. In practice, fuel economy evidence suggests that the Toyota parallel-series system as used in the Prius and now many other vehicles (including versions adopted by Ford and Nissan) adds complexity to the drivetrain but is able to achieve higher fuel economy. Automakers can develop hybrid vehicles from the ground up, or by adapting existing ICEV models to incorporate a hybrid drivetrain (see Figs. 15-14 and 15-15).

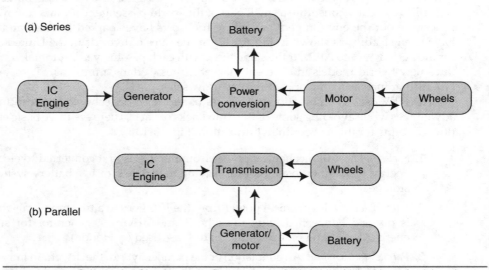

FIGURE 15-13 Schematic of basic energy flow concepts for hybrid vehicle drivetrain: (a) series and (b) parallel.

FIGURE 15-14 Two approaches to marketing a hybrid passenger car: (a) Toyota Prius, developed "from the ground up" as a hybrid, (b) Honda Civic Hybrid, adapted from the conventional Honda Civic ICEV, and (c) for comparison, Audi A2 diesel subcompact.

[*Source:* For image (c), Rosie Vanek. Reprinted with permission.]

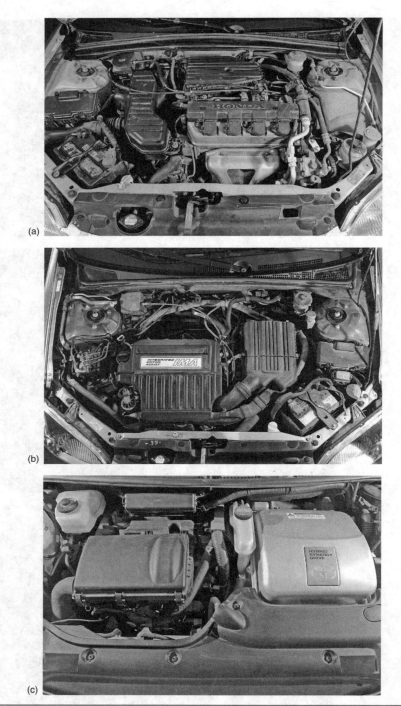

(a)

(b)

(c)

Figure 15-15   Birds-eye view of three engine compartments: (a) conventional 2004 Honda Civic, (b) 2004 Honda Civic Hybrid, and (c) 2005 Toyota Prius.

In Figs. 15-15(b) and (c), the additional components of the hybrid drivetrain do not require any significant increase in the volume of engine compartment. (*Source:* Philip Glaser, 2007, philglaserphotography.com.)

In most hybrid drivetrains, some fraction of the power from the engine is used to move the vehicle when the load is low, and the remainder is used to charge the battery or other storage device. When the load is too great for the power originating from the engine alone, the drivetrain withdraws energy from the battery system so that the combination of ICE and electrical power together can meet the load. Note that all hybrids use a *secondary battery system* to store and dispense electric charge to the drivetrain, which is separate from the primary battery system used in both HEVs and conventional ICEVs for auxiliary functions (lights, sound system, and the like).

A number of variations on the hybrid drivetrain are possible. A *through-the-road* hybrid (e.g., 2006 Toyota Highlander SUV, 2009 Dodge Durango SUV) uses the ICE to drive one set of wheels (front or rear) and the electric motor to drive the other. A *mild hybrid system* (e.g., 2007 Chevrolet Silverado truck) does not allow the vehicle to run entirely on electric power, but instead adds electric power as needed to smooth out load on the ICE. For heavy-duty vehicle applications where stop-and-go driving is common, such as in delivery vehicles or municipal waste haulage, companies such as Eaton and Parker-Hannifin provide hydraulic hybrid systems that store energy hydraulically on level ground or when braking, and release it in times of increased load. Diesel-electric hybrid drives are also in use in delivery vehicles belonging to companies such as FedEx and UPS.

Fuel Economy Advantage of the Hybrid System

The hybrid drivetrain has at least the three following advantages:

1. *Independence from external generating source:* All of the electric charge in the batteries comes originally from the energy in the fuel (gasoline or diesel), so it is not necessary to spend time recharging the vehicle from an off-board source of electricity while it is stationary, as in an EV. Also, since the electric charge is usually put to use in propulsion relatively soon after being generated (either from the ICE or from the drivetrain), the battery storage requirements are greatly reduced, in terms of space, weight, and cost.

2. *Optimal ICE efficiency at the expense of maximum power output:* Since the ICE in a hybrid can depend on the battery-motor system for additional power under peak load, it need not be optimized for maximum power output at the expense of efficiency, as occurs in many ICEVs. To this end, some hybrids (e.g., Toyota Prius) use a different thermodynamic cycle known as the Atkinson cycle, which emphasizes higher efficiency under average load at the expense of reduced peak power. Such a trade-off would be unacceptable to most consumers in an ICEV, because they would not have a satisfactory amount of power available for rapid acceleration, climbing steep grades at speed, and other high demand driving regimes. However, this arrangement works well in a hybrid, allowing it to improve fuel efficiency and satisfy performance requirements.

3. *Regenerative braking:* The HEV can use regenerative braking to recover kinetic energy from the wheels back into the secondary battery when decelerating. Measured in "round-trip" energy efficiency, up to 40% of the energy originally available as kinetic energy in the rotation of the wheels is returned to the wheels after conversion from mechanical to electrical and back to mechanical energy.

It is interesting to note that although point 3 is the feature that is more familiar to most drivers, it is point 2 that contributes the most to the improvement in fuel economy in many hybrids. In fact, the growth of the hybrid market has opened up new opportunities for the development of the Atkinson engine. The Atkinson cycle was originally invented by James Atkinson in 1882 as a way of circumventing the patent on the Otto cycle of Nicholas Otto. Although it did not come into widespread use at the time of its invention, the Atkinson cycle does have a combustion efficiency advantage, so that it was appropriate for use in the development of the hybrid drivetrain in the 1990s.[5]

In some models of HEVs, the hybrid drivetrain has been used to take larger models of cars that are fairly powerful and add additional power while at the same time modestly improving fuel economy. For example, the 2005 Honda Accord hybrid with a six-cylinder engine had a maximum output (combined ICE + motor) of 190 kW (255 hp) and a combined city/highway average fuel economy of 11.7 km/L (28 mpg), versus 179 kW (240 hp) and 9.8 km/L (23.5 mpg) for the ICEV version with a six-cylinder engine. Thus the hybrid platform gives the automakers flexibility to either pursue maximum fuel efficiency, or maximum performance with no loss in fuel economy. From an energy conservation or climate change point of view, the former option is desirable; however, the latter option may encourage some consumers, who might not consider the compact hybrids that achieve the best fuel economy, to purchase a hybrid and save some amount of fuel. (In the case of Honda, the Accord hybrid was withdrawn after several model years due to low sales; however, Honda continues to market both HEV and ICEV versions of the Civic.)

Hybrids in Comparison to Other Propulsion Alternatives

The potential of the hybrid drivetrain to reduce energy consumption across a wide range of vehicle applications can be exemplified by considering several current and recent passenger vehicle models (Table 15-5). Most vehicles are currently available in the U.S. market, although the Audi A2 is available only in Europe, and the EV-1 has been discontinued. The Ford Escape is a small SUV; otherwise, all vehicles are five-passenger coupes, sedans, or hatchbacks. The Toyota Echo is an ICEV that most closely resembles the 2001 Prius among Toyota's line of ICEVs, and is therefore included in the table for comparison; it has also been discontinued. The VW Passat and Audi A2 are diesel vehicles, the EV-1 and Nissan Leaf are EVs, the Chevrolet Volt is a PHEV, and the others are gasoline-powered.

A comparison of HEV and ICEV vehicles in the table shows that, in each case, the HEV vehicle delivers substantially improved fuel economy over the vehicle closest to it in size and performance. This is achieved despite a weight penalty for the additional drivetrain components, batteries, and so on. For example, the 2001 Prius weighs over 300 kg more than the Echo, but is approximately 4 km/L better in terms of fuel economy.

In the case of the HEV and PHEV models (Volt, Civic, and Prius), the ICE power per unit of engine volume is low compared to other models in the table, consistent with the discussion above of the use of the Atkinson cycle. For the four vehicles shown, the power per unit displacement range is 28.0 to 54.8 kW/L, versus a range of 46.2 to 64.8 for the ICEVs, not including the 1976 Accord. (The Accord is included for historical reasons, to show how weight, power, and fuel economy have evolved since the first generation of the sale of this vehicle in the United States.) Both Prius models compensate for this lack of specific power with an electric motor that provides additional torque and power when needed.

---

[5]A variation on the Atkinson cycle is the *Miller cycle*, which was developed in the 1940s by adding supercharging (forced intake of air into the cylinder) to the Atkinson cycle. Both are referred to sometimes as *Atkinson-Miller cycle engines*.

Make and Model	Eng Size (1) (cc)	Max Power (2) (kW)	(hp)	Power/L (3) (kW/L)	Curb wt (kg)	Overall Fuel Economy (4) (km/L)	(mpg)
2007 Audi A2	1190.0	55.0	73.8	46.2	930	35.7	85.5
2011 Chevrolet Volt	1400.0	111.9	150.0	42.6	1264	27.2	65.0
2005 Ford Escape (man)	2300.0	149.1	200.0	64.8	1575	9.9	23.7
2005 Ford Escape Hybrid	2260.0	99.2	133.0	n/a	1719	12.2	29.1
1976 Honda Accord (auto)	1600.0	50.7	68.0	31.7	909	16.0	38.3
2005 Honda Accord (man)	3000.0	179.0	240.0	59.7	1522	9.8	23.5
2005 Honda Accord Hybrid	3000.0	190.2	255.0	n/a	1591	11.7	28.0
2012 Honda Civic (man)	1798.0	104.4	140.0	58.1	1185	13.2	31.6
2012 Honda Civic Hybrid	1497.0	99.2	133.0	54.8	1297	18.4	44.0
2011 Nissan Leaf	n/a	80.0	107.3	n/a	1521	41.4	99.0
2001 Saturn EV-1	n/a	116.0	155.6	n/a	1350	25.1	60.0
2011 Tesla Roadster*	n/a	215.0	288.0	n/a	1238	56.4	135.0
2013 Tesla 60 kWh Model S (5)	n/a	310.0	415.3	n/a	2112	39.7	94.0
2001 Toyota Echo (man)	1500.0	78.3	105.0	52.2	927	13.6	32.6
2001 Toyota Prius	1500.0	72.0	96.6	28	1259	17.4	41.6
2011 Toyota Prius	1800.0	133.0	178.4	40.6	1383	20.8	49.7
2013 Toyota Prius PHEV (6)	1800.0	133.0	178.4	40.6	1439	40.2	95.0
2008 VW Passat TDI	2000.0	126.8	170.0	63.4	1527	13.6	32.6

Notes:
1) Engine size values are displacement of internal combustion engine, where applicable. Size of electric motor not considered.
2) Maximum power value for hybrids includes power from electric motor (specs not shown) in addition to power from ICE. Due to differences in the way that ICEs and electric motors function, each achieves maximum power output in different driving conditions, limiting the comparability of maximum power ratings between ICEVs, HEVs, PHEVs and EVs in this table. The calculation of maximum power for the hybrid models shown is the arithmetic sum of ICE and electric motor power.
3) Power per liter of ICE. Values for 2001 and 2011 Toyota Prius, Chevy Volt, and Honda Civic Hybrid are based on 42 kW, 73 kW, 60 kW, and 82 kW, respectively, for the ICE only. Values for Ford Escape Hybrid not given due to ICE/motor breakdown not being available.
4) Fuel economy value shown is the 55% city and 45% highway weighted average of economy values from USEPA, where available. USEPA values are post-2008 adjustment to reflect more realistic driving cycle. Exceptions: for EV-1, electricity consumption has been converted to gallons of gasoline equivalent and then fuel economy calculated. Fuel economy values for 1976 Honda and Tesla Roadster are from Sperling and Gordon (2009) and Randolph and Masters (2008), respectively. Chevrolet Volt fuel economy is average of USEPA electric and ICE fuel economy. Fuel economy values for Audi and VW diesel vehicles have been adjusted downward to reflect lower energy content per gallon of gasoline (assuming 89% of energy in 1 gallon diesel).
5) Tesla Model S sedan with 60 kWh of battery capacity.
6) Fuel economy for PHEV Toyota Prius is USEPA rated value while charge is available in battery. Once charge is depleted, fuel economy is same as non-PHEV Prius.
Conversions and abbreviations: 1000 cc = 1 L. "man" = manual transmission, "auto" = automatic transmission.

TABLE 15-5 Performance Characteristics of Representative EV, HEV, PHEV, Gasoline ICEV, and Diesel ICEV Models

Some of the advantages of HEVs over EVs are borne out in the table as well. For example, the 2011 Prius carries more passengers than the EV-1 (which is a two-passenger coupe), but weighs less and achieves similar fuel economy once upstream losses from delivering the respective energy source to the vehicle are taken into account. Battery requirements are a major contributor to the difference in curb weight; an EV-1 carries 395 kg of batteries, versus 27 kg for the secondary battery in the Prius. The Nissan Leaf has five-passenger capacity but outweighs the Prius by 267 kg. Also, while range is a limitation for the EV-1, the range of the Prius is actually better than many ICEVs in its class: a 2011 Prius has a nominal range of 946 km (591 mi), based on the tank size of 45 L (11.9 gal) and the overall fuel economy shown.

The table also shows the potential for diesel vehicles such as the Audi A2 to achieve very high fuel economy even without using a hybrid drivetrain. This observation suggests that the diesel ICE could be combined with a hybrid drivetrain to achieve the best possible fuel economy. The U.S. automakers adopted this approach in a government-sponsored R&D program known as the Partnership for a New Generation of Vehicles (PNGV), and developed concept cars that achieved 70 to 80 mpg using diesel hybrid drives on light-weight platforms. Peugeot-Citroen, GM, and Volkswagen, among others, have developed high-efficiency diesel hybrid prototypes, although they have not launched production because the sticker price premium for the technology is prohibitive at this point.

In addition, projections about the energy savings benefits of diesel hybrids must be made carefully, since the diesel engine gains much of its efficiency advantage over gasoline in steady state, highway driving, where HEVs generally have less of an efficiency advantage over ICEVs. From Table 15-5, if an efficient ICEV passenger car such as the Honda Civic averages 13 km/L (32 mpg) and the comparable gasoline Civic Hybrid achieves 18 km/L (44 mpg), one might assume that HEVs give a 40% increase in fuel economy across the board. However, it is unlikely that "hybridizing" of an efficient diesel ICEV such as the Audi A2 would result in a 40% increase in fuel economy for this vehicle on top of the already excellent fuel economy rating for this vehicle.

Further Reductions in Petroleum Consumption and GHG Emissions with Plug-in Hybrids

Although the preceding comparison showed how the EV has a relatively large battery weight requirement for relatively short range, the EV does have one key benefit, namely, that it can operate on electricity made from a wide range of energy sources, including renewable or noncarbon ones. As long as the HEV continues to run on gasoline or diesel, it does not permanently solve this problem (although rapid uptake of HEV technology in the vehicle market could measurably extend the life of the world's petroleum supply). A promising next step toward independence from fossil fuels is the "plug-in" hybrid, or PHEV, in which the secondary battery system has increased capacity that allows the vehicle to drive up to 80 km (50 mi) on stored charge, without using the ICE. This system allows the user to charge at home at night in addition to refueling at the filling station, greatly increasing the "effective" gasoline fuel economy (i.e., total distance divided by total gasoline consumption) of the vehicle. Example 15-6 illustrates the effect of turning an HEV into a PHEV on fuel economy and energy cost per km of driving.

Example 15-6   A plug-in hybrid has a fuel economy of 20 km/L (47.3 mpg) when running on gasoline, and is able to travel 8 km (5 mi) on 1 kWh of charge. Its range on a full charge is 80 km (50 mi), before it begins to use the ICE. Gasoline costs $0.79/L ($3.00/gal) and electricity costs $0.12/kWh.

Over a two-week period, the vehicle accrues the following distance of travel each day:

Day	km	Day	km
1	60.8	8	182.4
2	27.2	9	144
3	105.6	10	33.6
4	22.4	11	38.4
5	62.4	12	70.4
6	72	13	25.6
7	427.2	14	52.8

Thus days 7 to 9 involve an extended intercity trip, and the other days involve driving in and around the vehicle's home base. Assuming that the vehicle has access to recharging each night, and that it only recharges the amount necessary to offset the kilometers traveled that day, calculate (a) the effective fuel economy, (b) the energy cost per km (for combined electricity and gas), and (c) the energy cost per km if it were only a hybrid with no plug-in capability.

Solution   For each day, we subtract off 80 km of driving, and if there is a positive amount remaining, assign it to ICE propulsion. Thus on day 1, there is 0 km ICE use, and the vehicle uses $(60.8)/(8 \text{ km/kWh}) = 7.6$ kWh of electricity that must be recharged the following night; on day 3, 25.6 km are powered with the ICE, consuming 1.3 L, and so on.

(a) On the basis of this type of daily calculation, in total 1325 km (828 mi) are driven in two weeks, of which 539 km (337 mi) require the use of the ICE. At the given fuel economy of 20 km/L, the fuel consumed is 27 L. Thus the (effective) fuel economy is $(1325 \text{ km})/(27 \text{ L}) = 49.1$ km/L (116 mpg).

(b) At the given cost, 27 L of fuel cost $21.37. The remaining 786 km (491 mi) are driven using electric power, requiring 98.2 kWh of electricity, which costs $11.78. Thus dividing the total $33.15 of energy cost among the total distance driven gives $0.025/km ($0.04/mi).

(c) Assuming that the gasoline-only HEV achieves the same 20 km/L fuel economy over the entire distance traveled as does the PHEV when driving on gas, the fuel consumption is 66.2 L (17.5 gal), which costs $52.50. Dividing energy cost by distance driven gives ($52.50)/(1325 km) = $0.04/km ($0.063/mi).

Discussion   The treatment of fuel economy in this example is simplistic, because, depending on the design of the specific PHEV drivetrain, the drive controller might occasionally activate the ICE while there is still charge in the secondary battery, in order to deliver sufficient total power for hard acceleration or climbing a grade. However, inclusion of this level of detail would likely not lead to a dramatic change in the results, so that the overall fuel economy would still exceed 42 km/L (100 mpg).

The focus on gasoline consumption as the figure of merit in this example stems from the concern about volatility of gas prices at the pump, and about energy security in countries such as the United States that depend heavily on imports. A more complete comparison of the HEV versus PHEV would take into account the total energy consumption for both gasoline and electricity, and the total CO_2 emissions. In some circumstances, a PHEV might save gasoline but increase energy and CO_2. The comparison in this case is left as an exercise at the end of the chapter.

Major automakers have since 2011 launched PHEVs in the U.S. market and elsewhere, such as the Chevrolet Volt (2011), Ford C-max Energy plug-in (2012), or Toyota Prius Plug-in HEV (2012) (Fig. 15-16). In addition, a number of after-market companies such

(a)

(b)

FIGURE 15-16   Examples of plug-in hybrid electric vehicles (PHEVs): (a) Chevy Volt and (b) Ford
C-max Electric.

Note electric charging portal just forward of driver's door in both vehicles. These two vehicles
represent two different PHEV approaches, with the Volt having a larger battery system and hence
longer range, approximately 40 miles versus 11 miles for the C-max. Acknowledgment: Thanks to
Bill King and Kathryn Caldwell, respectively.

as A123 Systems provide plug-in retrofit kits that can convert an HEV to a PHEV and give the vehicle a range of 20 to 50 km without using the ICE. Given that many passenger vehicle owners drive 50 km or less on 70 to 80% of the days of the year, it is clear that a transition to PHEVs could significantly reduce petroleum production, as electricity from a wide range of sources (but very little from oil-fired electric power plants, as only about 1% of all U.S. electricity is generated by burning oil or other petroleum-derived products), as shown in Fig. 6-2 would replace gasoline and diesel consumption in ICEs. The PHEV could also have security benefits, since, in the case of a disruption in oil supplies, fleets of PHEVs in major urban centers could still drive a fixed distance each day on electric charge (which presumably is less vulnerable to crises in international relations), preserving some measure of mobility for the public.

For the PHEV technology to give a clear advantage over ICEVs in terms of CO_2 emissions and impact on air quality, it is important that any fossil-fuel-generated electricity come from a source that has high thermal efficiency and up-to-date emissions controls to prevent the escape of pollutants into the atmosphere. In a scenario unfavorable to PHEVs, low efficiency or poor emissions controls might wash out any environmental advantage compared to the ICEV technology, even if the PHEV reduced consumption of petroleum.

Optimal use of PHEV technology to reduce foreign petroleum dependency and protect the environment brings into play several transportation factors at the systems level. First, to optimally charge PHEVs so as to provide significant amounts of petroleum-displacing energy for everyday driving, vehicles should be connected to the grid through their charging system for long periods of time when not in use, and then current should be fed to the PHEVs when it is convenient from the grid's perspective. This function entails developing the *smart grid* which can sense overall grid load and generating availability, and then signal the PHEV's charger to start and stop. In this way, PHEVs might become a buffer for bursts of intermittent electricity from renewable solar and wind. Lastly, vehicles might not only be charged efficiently in this way, but also sell electricity back to the grid in a *vehicle-to-grid* system, a service which could give the vehicle owner a revenue stream to offset the higher purchase cost of the vehicle. Since these features focus more on systems integration of technological components rather than specific transportation energy technologies, they are developed further in Chap. 16.

15-4-3   Biofuels: Adapting Bio-energy for Transportation Applications

Bio-energy as a resource has already been introduced in Chap. 14; so in this section, we focus on practical aspects of delivering biofuels for use in transportation. From an agricultural perspective, biofuels are one of several possible applications of bio-energy, and bio-energy is in turn one of several applications of arable land, along side crop raising, animal husbandry, selviculture, and so on. From a transportation technology perspective, however, biofuels can be envisioned as a mechanism for storing and transporting renewable energy from the sun comparable to powering the EV technology of Sec. 15-4-1 with renewables, as shown in Fig. 15-17. The storage function that takes place in the battery system, or possibly in an intermediate pumped storage or hydrogen conversion system, in the EV is instead achieved by the crop concentrating energy in seeds and other plant matter for later transfer to a biofuel.

In general, biofuels provide a renewable, liquid substitute for petroleum-derived gasoline or diesel, thus avoiding the large technological transition required for widespread adoption of EV technology. Several adaptations are necessary, however.

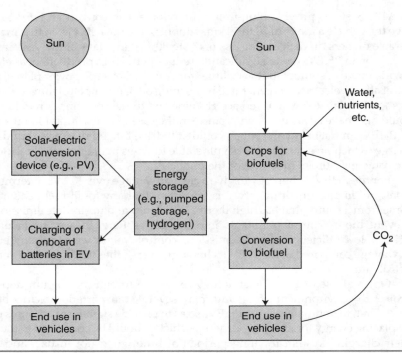

FIGURE 15-17 Comparison of pathways to solar-derived energy for vehicles from PV panels and EVs (left) and from biofuels and ICEVs (right).

One challenge is reduced energy content per liter or gallon: biodiesel and ethanol have 90% and 67%, respectively, of the energy content per unit of volume compared to petro-diesel and gasoline. Users must therefore stop more frequently to refuel unless they have increased tank capacity.

Biofuel-enabled vehicles must be specifically adapted to combust either pure biofuels or blends, above a certain threshold of biofuel content. An SI engine can routinely burn up to 10% ethanol (E10) and a CI engine 20% biodiesel (B20) without adaptation; although in the latter case, the frequency of certain routine maintenance may need to be adjusted in response to changed combustion behavior. In the case of ethanol, above E10 and up to E85, the engine must include an additional "flex-fuel" sensor that can measure the proportion of ethanol and adjust the combustion accordingly. Higher proportions than E85 combust poorly and are not used. Biodiesel blends in the range B20 to B100 can be used without a sensor but the fuel requires special care, for example, it gels at a relatively high temperature. Because the total supply of biodiesel in the United States is small relative to overall demand for diesel, there is little incentive to use blends higher than B20 and incur the added maintenance required.

Where ethanol blends can be provided in a fixed proportion at all times of the year, the vehicle can be simplified by omitting the flex-fuel system and pre-adjusting the vehicle to run at a set blend. This is done in Brazil, where sugar cane-derived ethanol is plentiful and many filling stations carry a uniform E70 blend in all seasons. In the U.S. market, however, because ethanol blends vary by both region and time of year, it is not practical to pre-adjust vehicles for a specific blend, and the fraction of the nation's light-duty vehicle fleet that can combust ethanol in blends higher than E10 has the E85 flex-fuel capability.

In recent years, improvements in biofuel processing technology have helped to decrease biofuel prices. Nevertheless, prices of both corn-based ethanol and soy-based biodiesel remain higher than their petroleum-based counterparts, leading to government subsidies per gallon of product to help make biofuels price competitive. These subsidies are paid in proportion to the biofuel blend: if a gallon of B100 brings a tax credit of $1.00, then a gallon of B20 will bring $0.20, and so on. Volatility and price spikes in the cost of crude oil that might have otherwise helped biofuels to increase market share have in general coincided with spikes in world demand for corn and soy, as well as increased production cost for biofuels because of increasing freight transportation costs, so crop-based biofuels have yet to see any large, sustained cost advantage over conventional gasoline and diesel, especially when measured in cost per unit of energy (as opposed to cost per gallon or liter). Nevertheless, the ability to blend in the future might protect transportation costs from volatility in the price of crude oil, and increasing biofuel content at a time of spikes in oil prices might protect motorists from exorbitant costs at the pump.

Subsidies for biofuels are a political issue as well, and as the serious debate in the U.S. Congress about these subsidies in 2011 indicates, these subsidies may someday be withdrawn, changing the landscape for these fuels. On the one hand, they have been criticized as interference in the normal function of the energy market that has failed to launch a biofuel industry that can stand on its own. On the other hand, they have been supported for diverting funds that might otherwise go to foreign oil exporters to domestic growers of grain and distillers of biofuels.

By far, the largest problem with biofuels, as already presented in Chap. 14, is their low productivity in terms of volume of biofuel produced per year per unit of crop area planted. Whereas the potential for other renewables such as wind or solar if one were to cover the open lands of the United States with devices is vast compared to demand, the maximum output for biofuels is only a fraction of fuel demand. Without a different biofuel feedstock that can substantially increase productivity, biofuels will remain in a supporting role as fuel for certain key long distance movements, as a minor component in a blend (e.g., E10 or B20), or as a regional fuel solution (such as for a sparsely populated region that also has abundant crops).

A comparison of U.S. energy demand in long-distance light-duty vehicle, heavy-duty vehicle, marine, rail, and aviation applications to energy provided in biofuels in 2010 shows the extent of the challenge but also the opportunity for market growth as biofuel technology advances (Fig. 15-18). The total biofuel energy content of 1.2 quads (the two columns to the right) is small compared to the total of 12.9 quads consumed by trucks, aviation, LDVs, marine, and rail (the five left columns).[6] The potential market for advanced biofuels in the future is therefore on the order of at least 10 quads, if the output can be adequately developed. The potential market for biofuels is large in part

---

[6]The figure of 6.1 quads shown for truck includes energy consumption for urban truck movements. Although the underlying data source from the U.S. Department of Energy does not distinguish between energy consumed in urban versus intercity truck movements, the fraction consumed in urban movements is not negligible and could be more readily powered by electricity in the form of battery-electric delivery vehicles.

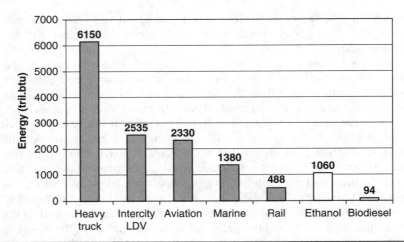

FIGURE 15-18   Comparison of biofuel output and transportation loads suited for bioenergy in the U.S., 2010. (Source: U.S. Department of Energy.)

because electricity and hydrogen are not suitable at present to meet this demand. Of the loads shown, only rail is suitable for electrification since it can be connected to overhead catenary to deliver power from the grid to locomotives. However, this load is only a small fraction of the total, at approximately 0.5 quad, and even then, the initial infrastructure costs would be high. Hydrogen is more flexible, as it can theoretically be applied to road, rail, marine or air applications. The only one where significant R&D progress has been made, however, is LDV; in particular, aviation and heavy truck applications are both very large energy loads where the technical challenges for hydrogen power are large. This situation leaves biofuels as the leading alternative energy source for each of the five loads in the figure. In each case, the loads in question have been powered by biofuels to some degree, either at a test case level (aviation) or in actual routine use in biofuel/petroleum blends (e.g., truck, rail). The remaining challenge is to create a global supply system that is sufficiently productive to meet the need.

15-4-4   Hydrogen Fuel Cell Systems and Vehicles

The hydrogen fuel cell (HFC) is one of a family of fuel cell designs that convert a range of fuels, such as methanol, solid oxides, or phosphoric acid, into electricity through a chemical reaction. Christian Schonbein developed the principle of the fuel cell in 1838, and in 1843 William Grove assembled the first working prototype. The first experimental application in a vehicle was the Allis-Chalmers Tractor Co.'s use of an alkaline fuel cell in a prototype farming tractor unveiled in 1959. The first commercial application was in the 1960s, when NASA used a fuel cell developed by General Electric in the Gemini 5 mission, demonstrating the availability of a working alternative to batteries for electricity supply onboard space flight. At about the same time, General Motors tested a fuel cell in a motor vehicle for the first time in its "Electrovan" prototype, retrofitting an ICE-powered van to run on a fuel cell powered by hydrogen and oxygen.

Since the time of the Gemini missions, fuel cells have been used primarily in space flight, where they are capable of providing power over long durations, in a weightless

Make	Model	Year	Number
Chevy	Equinox FCV	2014	119 (cumulative)
Honda	Clarity FCV	2016	250 (projected, Japan only)
Hyundai	Tucson FCV	2014	200 (cumulative)
Toyota	Mirai	2016	2000 (projected)

Note: Figures are for U.S. market, except for Honda Clarity, which is projected to launch in Japan in 2016 and in the U.S. in 2020. Data obtained from respective automakers.

TABLE 15-6   Selected Hydrogen Fuel Cell Vehicle Launch Examples: Historical and Projected Make and Model Data

environment, without the noise or emissions of other types of generators. In recent years, both vehicle manufacturers and energy companies have been taking an increasing interest in the fuel-cell-powered vehicle as an alternative to the ICEV powered by petroleum-derived liquid fuels. To this end, makers including Daimler, Ford, GM, Honda, and Toyota have been retrofitting production ICEVs to run on hydrogen fuel cells and also developing prototype hydrogen fuel cell vehicles (HFCVs) from the ground up, so as to take advantage of new design possibilities made available by the fuel cell platform. Energy companies have also begun installing demonstration hydrogen filling stations, dispensing liquid and/or compressed hydrogen, in cities such as Washington and Tokyo. Several automakers have been growing pilot programs to produce and launch HFCVs in the period 2005–2015. In 2005, Honda became the first automaker to lease an HFCV to a private individual for continuous use, providing a retrofitted version of a subcompact hatchback to a family in southern California. Other examples of initial HFCV launches include the Chevy Equinox FCV (2007) and the Hyundai Tucson FCV (2013) and Toyota Mirai (2014). Several of these models number in the hundreds of vehicles in use as of 2015 (Table 15-6). Toyota is projected to become the first maker to produce more than 1000 HFCVs in a model year with the goal of 2000 units in 2016. Among the examples in the table, the Mirai and Clarity are designed from the ground up to be HFCVs, while the Equinox and Tucson are retrofitted ICEVs.

Compared to passenger vehicles, operation of urban public buses may provide a more advantageous application for demonstrating the use of hydrogen (see Fig. 15-19). Public urban buses are operated, refueled, and maintained by professional staff; they can more easily accommodate the extra volume required for onboard hydrogen storage than passenger cars; they are usually refueled exclusively in a single location (typically the depot or maintenance facility where they are stored when not in service); and, on many urban bus routes, they are not required to travel great distances in a single day's service (typically not more than 160 to 240 km, or 100 to 150 mi). A number of bus vendors including Daimler, El Dorado National (United States), and Toyota-Hino have placed prototype HFCV buses in service in various cities, including Madrid, Stuttgart, Stockholm, Nagoya, Vancouver, and Palm Springs, California, among others.

In the remainder of this section, the basic thermodynamics governing the function of the fuel cell is first presented, since this technology is relatively new (e.g., compared

FIGURE 15-19   Prototype hydrogen fuel cell bus at the University of Delaware, United States, in January 2008.

The bus was in regular use in 2008 as a shuttle for carrying students between residences and campus.

to the internal combustion engine or electric motor) and may not be as familiar.[7] However, the reader should be aware that this technology is not as mature as that of the EV or HEV, and predictions of a future "hydrogen economy" are speculative at this point. The latter parts of this section discuss practical aspects of implementing fuel cells in vehicles, and prospects for successfully commercializing them.

Function of the Hydrogen Fuel Cell and Measurement of Fuel Cell Efficiency

The function of the individual fuel cell is shown in Fig. 15-20. In the diagram, the fuel cell is supplied with pure hydrogen on the right and ambient air containing approximately 21% oxygen on the left. At the anode, the hydrogen separates into protons and electrons (in the case of a deuterium atom, the neutron remains joined to the proton), and then the proton exchange membrane allows only the proton to pass. This leaves an excess of electrons on the anode side, which creates a voltage difference that can do work across an exterior load (e.g., electric motor in a vehicle). Since one fuel cell by itself does not create a sufficient voltage difference for most applications, multiple fuel cells are typically combined into a *fuel cell stack* to reach the desired voltage. On the cathode side of the membrane, electrons and protons recombine into hydrogen, and then the hydrogen combines with the oxygen in the air to form water, which is the main component of the fuel cell's exhaust.

---

[7]For full-length works on fuel cell technology, see, for example, Larminie and Dicks (2000) or Spiegel (2007).

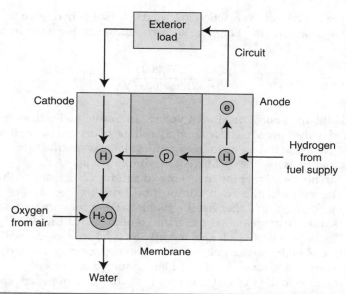

FIGURE 15-20   Schematic of hydrogen fuel cell function, showing anode, cathode, and proton exchange membrane.

Next, we focus on estimation of the maximum efficiency of the fuel cell, using a theoretical model that ignores the effect of imperfect function and losses. We begin by considering the *Gibbs free energy* g_f of a compound, which is a measure of its ability to do external work, given temperature and pressure conditions. Because the value of Gibbs free energy changes with the reference point chosen, we focus on the change in Gibbs free energy, or Δg_f, when a chemical reaction takes place, since the value of Δg_f is not affected by choice of reference point. For example, at ambient conditions of 298 K and 100 kPa pressure, the reaction of 1 mol of H_2 molecules with one-half mole of O_2 molecules to form 1 mol of water molecules has a change in Gibbs free energy value of −237.2 kJ/mol. (We use gram-moles and not kilogram-moles throughout this section.) In other words, the decrease in Gibbs free energy due to the transformation of hydrogen and oxygen into water is energy that is available to do other work, such as moving electrons in a circuit.

Next, we can compute the voltage difference created by the release of energy if all of the electrons available in the reaction flow from the anode through the load to the cathode in the fuel cell. Recall that *Faraday's constant F* is the amount of charge in 1 mol of electrons, or $(1.602 \times 10^{-19}$ coulomb/electron$) \times (6.02 \times 10^{23}$ electron/mole$) = 9.64 \times 10^4$ coulomb/mole. Furthermore, since 1 coulomb is equivalent to 1 J/V, dividing Δg_f among the total charge, as represented by the number of moles of electrons n in the reaction multiplied by F gives the potential E in units of volts, that is,

$$E = \frac{\Delta g_f}{nF} \tag{15-16}$$

Taking the case of the reaction in the HFC, 1 mol of hydrogen molecules is equal to 2 mol of hydrogen atoms. Plugging known values at ambient conditions into Eq. (15-16) gives

$$E = \frac{237,200 \text{ J}}{(2)(96,400 \text{ J/V})} = 1.23 \text{ V}$$

Thus the theoretical maximum voltage from this fuel cell, assuming all energy is converted to the flow of electrons and that all the hydrogen is consumed in the reaction, is 1.23 V. Note that this value for maximum voltage is specific to the HFC; fuel cells that run on other fuels may have different maximum voltages.

We can now evaluate the theoretical maximum efficiency of the fuel cell relative to the energy available from combustion with oxygen of the same quantity of hydrogen. Using the higher heating value (HHV) for combustion, which includes the condensation of steam to liquid, the combustion of 1 mol of H_2 molecules yields $\Delta h_f = -285.8$ kJ/mol.[8] In practice, not all of the hydrogen fuel will be reacted in the fuel cell, so we take account of this limitation by factoring in the consumption rate of fuel μ_f, which is the ratio of mass of fuel consumed to mass of fuel input. Comparing energy released in the fuel cell to energy available in combustion, and including losses due to unreacted fuel, the theoretical maximum efficiency of the cell η_{FC} is

$$\eta_{FC} = \mu_f \frac{\Delta g_f}{\Delta h_f} \tag{15-17}$$

For a well-designed fuel cell in normal operation, a value of $\mu_f = 95\%$ may be achievable. At an operating temperature of 80°C (353 K), which is current practice with typical prototype HFCVs, change in Gibbs free energy is decreased slightly, to $\Delta g_f = -228.2$ kJ/mol. Thus the maximum achievable efficiency is

$$\eta_{FC} = 0.95 \frac{-228.2}{-285.4} = 0.759 = 75.9\%$$

This simple calculation explains much of the interest in HFCVs. If efficiency values close to the value shown can be achieved in vehicular fuel cells in the future, then the HFC will have a clear efficiency advantage over the ICE, which is limited by the Carnot limit. Note that the operation at relatively low temperature (less than the boiling point of water) is advantageous for the fuel cell. Temperature and pressure have a strong effect on the value of Δg_f, and as the temperature increases, the theoretical maximum efficiency decreases. For example, at 1000°C, the maximum with $\mu_f = 95\%$ is $\eta_{FC} = 59\%$. Also, this comparison of HFC and ICE efficiency does not consider balance of system requirements for loads that are required to operate either technology, such as pumps for controlling oxygen and air flows in the fuel cell. A more accurate comparison would require an evaluation of the net energy available for propulsion after taking into account all required loads.

---

[8]Alternatively, the lower heating value (LHV) of −241.8 kJ/mol could be used, which is the value prior to condensation. We use HHV throughout this section.

Along with the proportion of fuel consumed, the partial pressures of both reactants and products in the fuel cell have an effect on the performance of the cell. Changes to partial pressures affect the value of E as calculated in Eq. (15-15). This effect is evaluated by first setting a reference value for E, called E^0, that is equal to the ideal change in potential at ambient conditions, that is, $E^0 = 1.23$ V at 25°C, 0.1 MPa. Next, the partial pressures of reactants are introduced, P_{H_2} for hydrogen, P_{O_2} for oxygen, and P_{H_2O} for water, each in units of atmospheres (1 atm = 100 kPa = 0.1 MPa); for example, at ambient conditions and assuming the operation of the fuel cell as in Fig. 15-20, $P_{H_2} = 1$, $P_{O_2} = 0.21$, and so on. The adjusted potential E taking into account partial pressures can now be written

$$E = E^0 + \frac{RT}{2F} \ln\left(\frac{P_{H_2}P_{O_2}^{0.5}}{P_{H_2O}}\right) \tag{15-17}$$

This equation is known as the *Nernst equation*. In the equation, R is the ideal gas constant and T is the temperature of the fuel cell reaction in kelvins. Also, the partial pressure of a compound is raised to the same power as the number of molecules in the reaction. Therefore, $P_{O_2}^{0.5}$ appears in Eq. (15-17) because there is ½ molecule of O_2 reacted per mole of H_2.

For some situations, it is useful to be able to rewrite the partial pressures of compounds as follows:

$$P_{H_2} = \alpha P$$

$$P_{O_2} = \beta P$$

$$P_{H_2O} = \delta P$$

Here P is the total pressure in atmospheres, and α, β, and δ are factors corresponding to the relative value of the partial pressures of hydrogen, oxygen, and water, respectively, compared to the total pressure. Equation (15-17) can now be rewritten to separate the natural log term into separate components:

$$E = E^0 + \frac{RT}{2F} \ln\left(\frac{\alpha P \beta^{0.5} P^{0.5}}{\delta P}\right)$$

$$= E^0 + \frac{RT}{2F} \ln\left(\frac{\alpha \beta^{0.5}}{\delta}\right) + \frac{RT}{2F} \ln(P^{0.5}) \tag{15-18}$$

$$= E^0 + \frac{RT}{2F} \ln\left(\frac{\alpha \beta^{0.5}}{\delta}\right) + \frac{RT}{4F} \ln(P)$$

Equation (15-18) facilitates the evaluation of the effect of changes to individual fuel cell operating parameters on the value of E. We can write the value of the change ΔE as the difference between E^{new} and E^{old}, or the value of E after and before the change in parameter, respectively, that is,

$$\Delta E = E^{new} - E^{old} = E(\alpha^{new}, \beta^{new}, \delta^{new}, P^{new}) - E(\alpha^{old}, \beta^{old}, \delta^{old}, P^{old}) \qquad (15\text{-}19)$$

since the terms that do not change will cancel out. The evaluation of E using the Nernst equation and the effect of changing operating parameters are demonstrated in Example 15-7.

Example 15-7   A hydrogen fuel cell operates at 25°C and pressure of 100 kPa, using a supply of pure hydrogen and oxygen from ambient air. The partial pressure of the water that forms from the hydrogen and oxygen is 30 kPa. (a) What is the voltage drop across the fuel cell? (b) If the supply of air is replaced with a supply of pure oxygen, and nothing else changes, what is the new voltage drop?

Solution

(a) Given the initial data, partial pressures for each of the components of the reaction are $P_{H_2} = 1$, $P_{O_2} = 0.21$, and $P_{H_2O} = 0.3$. Substituting into Eq. (15-17) gives

$$E = 1.23 + \frac{8.314(298)}{2(96400)} \ln\left(\frac{1(0.21)^{0.5}}{0.3}\right) = 1.23 + 0.0054 = 1.24 \text{ V}$$

(b) If the fuel cell is operated with pure oxygen, $P_{O_2} = 1$. Thus $\beta^{new} = 1$ while $\beta^{old} = 0.21$. Rewriting Eq. (15-19) to include only nonzero terms gives

$$\Delta E = E^{new} - E^{old} = \frac{RT}{2F}\left\{\ln[(\beta^{new})^{0.5}] - \ln[(\beta^{old})^{0.5}]\right\}$$

$$= (0.5)\frac{RT}{2F}[\ln(1) - \ln(0.21)] = 0.01$$

Thus $E^{new} = E^{old} + \Delta E = 1.24 + 0.01 = 1.25 \text{ V}$

Actual Losses and Efficiency in Real-Word Fuel Cells

Real-world fuel cells in current use do not achieve the 75% efficiency value predicted above based on Eq. (15-16); instead, they achieve on the order of 50% in the best circumstances. Some losses come from incomplete fuel consumption as mentioned above. In addition, in an ideal fuel cell, constant partial pressure of hydrogen is maintained despite continuing consumption of the fuel as it is reacted across the proton exchange membrane. In an actual fuel cell, the amount of fuel available and hence partial pressure declines as the fuel is consumed, especially in regions close to the outlet from the cell. In this area, current will decline, so that the overall power output of the cell declines as well. Also, some of the energy released in reacting the fuel is released to the surrounding as heat, rather than being converted to electrical energy. While the theoretical upper bound on fuel cell voltage may be on the order of 1.2 V, in practice a value of less than 1 V is more realistic.

Implementing Fuel Cells in Vehicles

The core component of an HFCV is the *fuel cell stack*, in which a large number of fuel cells (on the order of 120 to 200) are connected in series so as to achieve a sufficient voltage to serve the propulsion load (see Fig. 15-21). Once the voltage is fixed, current flow from each cell must be adequate to meet overall power requirements. Once the best

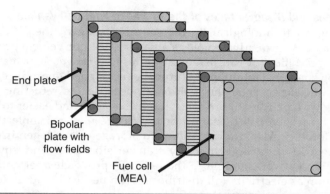

End plate

Bipolar
plate with
flow fields

Fuel cell
(MEA)

FIGURE 15-21   Exploded view of a fuel cell stack, made up of alternating units of fuel cells and bipolar plates. The connection is made to exterior loads from the end plates.

[*Source:* Spiegel (2007). Reprinted with permission.]

possible efficiency has been achieved, the stack designer can increase current flow from a single cell by increasing its cross-sectional area, so the stack is designed both with enough cells (for voltage) and with enough area per cell (for current) to provide sufficient power. From the fuel cell (FC) stack, electricity is delivered to the drivetrain to turn the wheels. One approach is to have the stack connected by wires to electric motors at each of the four wheels. This layout provides four-wheel drive, and also gives maximum design flexibility; since the motors are positioned close to the wheels, the stack takes up relatively little space, and the wire harnesses connecting stack to motors can be designed around other elements in the vehicle frame (e.g., dimensions of the passenger compartment). A regenerative braking system with secondary battery system for short-term storage of charge is also possible.

Systems for storage of hydrogen onboard the vehicle are quite different from those for gasoline or diesel in an ICEV. One kilogram of pure hydrogen has approximately the same energy equivalent as 2.8 kg of gasoline (1 gal or 3.8 L at ambient temperature and pressure). Assuming efficiency gains in mature HFCVs in the future, stored mass of 5 to 6 kg hydrogen might achieve sufficient range per refueling, on the order of 400 to 500 km. At low pressures, this amount of hydrogen would take up far too much space on the vehicle, so in order to achieve a realistic storage volume, the hydrogen must be either stored in a liquid form at cryogenic temperatures, or compressed to 34 to 68 MPa (5000 to 10,000 psi). These storage specifications require a mixture of efficient insulation and strong containers that can withstand the shock of accidents. They also require careful design of the interface between the vehicle and the refueling facility.

Ancillary systems on the HFCV also have different requirements. For example, with the fuel cell operating at low temperatures (on the order of 80 to 100°C), there is no longer a source of high-temperature exhaust gases for use in heating the passenger compartment as in the ICEV. Another significant issue is that the low temperature complicates heat rejection, requiring large heat exchangers. Even a 60% efficient FC delivering 20 kW has to throw away 8 kW of low-grade heat, and at peak vehicle loads the heat rejection task is much greater.

Advantages and Disadvantages of the Hydrogen Fuel Cell Vehicle

Assuming all technological hurdles can be bridged, the HFCV may be able to achieve a number of advantages as an alternative for vehicle propulsion. First, like the electric vehicle, it can make use of a wide range of primary energy sources, including some that might in the future be redesigned to produce hydrogen directly, such as nuclear or solar-thermal. Next, because the refueling process resembles refueling with petroleum products, it may be faster and easier to refuel than the EV, and the hydrogen filling station of the future may resemble the gasoline/diesel station of today. Also, transmission and storage of hydrogen in a mature distribution network, with large intercity transmission lines and smaller urban-region distribution networks, might in the long run prove cheaper and more robust than distribution as electricity (all distribution can be carried out at low pressures, with compression occurring at the refueling station in a way to minimize time kept in high-pressure storage before discharging to the vehicle). As discussed above, at low temperatures, FCs have a much higher theoretical efficiency limits than Carnot-limited ICEs. In addition, since the HFCV dispenses with the mechanical driveshaft, there is a new flexibility in locating the hydrogen storage, fuel cell stack, and network of electric motors that might enable automotive designers to radically reshape the passenger car so as to create new design possibilities. Lastly, the presence of a high-quality electricity source onboard the vehicle in the form of the FC stack may expand the role of the vehicle as a mobile power generator.

There are also substantial challenges with the HFCV, as follows:

1. *Cost-competitive and reliable fuel cell:* At present prices per kW of capacity, fuel cell stacks are too expensive to compete with ICEs, to the point that they make the entire HFCV too expensive to market competitively with current ICEVs or HEVs. Typical ICEs cost on the order of $25 to $35/kW of capacity, or $3000 to $4000 per engine. Fuel cell costs dropped from $500,000/kW for NASA in the 1960s to roughly $500/kW by the year 2000 and $60 to $150 per kW in large-scale production in 2015, depending on the estimate used, but parity with ICEs has not yet been reached. Current-generation fuel cell stacks are reliable enough in prototype form to function the course of a 160,000-km (100,000 mi) vehicle lifetime without needing a complete overhaul, but the reliability of fuel cell stacks in the field with thousands of HFCVs on the road remains to be proven. Platinum loading (total platinum or similar precious metal requirements) in current generation are also higher than desired. Although major gains have been made in reducing the amount of platinum required per kW of capacity from ~100 g/kW in the 1960s to 0.2 g/kW in 2015, at current rates, further reductions to values on the order of ~0.1 g/kW are needed to make the technology sustainable in terms of platinum availability.

2. *Safe and cost-effective storage of hydrogen on board:* Cryogenic cooling and storage of hydrogen incurs a large energy penalty (on the order of 30% of the initial energy in the hydrogen), so compressed hydrogen storage is the preferred option at present. However, storage at very high pressures requires containers that will not rupture in an accident and that are not vulnerable to fire. Dispensing the fuel, while faster than taking hours to charge an EV, is a riskier and more complex task than with an ICEV, perhaps requiring professional attendants at all hydrogen refueling facilities.

3. *New infrastructure needed to distribute/dispense hydrogen:* Although energy companies have experience with making and handling hydrogen, as a society we have very few of the infrastructure pieces in place to use hydrogen as a transportation fuel on a large scale. This includes an adequate number of manufacturing plants, a distribution grid, and dispensing stations. Although some demonstration filling stations have been opened, energy companies and vehicle makers have been criticized for launching public refueling stations too soon, when the basic viability of the vehicles themselves has not yet been proven (see Fig. 15-22).

Given the extent of these challenges, and the emergence of competing technologies such as the PHEV or cellulosic ethanol, it is possible that we will create an end-game technology that runs on stored electric charge from nonfossil sources whenever possible, and on advanced biofuels when not, so that there will be no place for the HFCV in the future. On the other hand, if other technologies such as PHEVs cannot completely solve the sustainability problem, the HFCV may yet emerge as the endpoint technology that is technically superior to all other options, and eventually becomes the permanent

FIGURE 15-22 Co-author Vanek at Hydrogen dispenser at Shell's Benning Road station in Washington, DC, November 2006. Although the experience of launching the Benning Road hydrogen station provided proof-of-concept that a prototype fleet of HFCVs could be refueled on a continuing basis at a commercial outlet, it did not by itself launch a local market for HFCV purchases among the general public.

solution for sustainable transportation energy. Such an evolution would likely take a long time, on the order of decades. In the meantime, we can expect that vehicle makers and others will continue R&D at a steady pace to improve fuel cells and other components of this system, and especially to lower production cost so they can sell to the consumer at a more affordable price.

One potential outcome in the short to medium term is that, rather than either the EV/PHEV or HFCV emerging as the single solution to replace the gasoline-fueled ICEV, the two incoming technologies might exist side by side, each creating a niche around its respective advantage. Vehicle owners who can rely on other options for traveling long distances might opt for an EV and rely entirely on recharging. Other owners who occasionally travel long distances might opt for PHEVs and use primarily the charging capability, while only occasionally using the liquid fuel option (possibly in the form of a biofuel) to travel long distances. HFCVs would then enter the market by appealing to drivers who frequently travel long distances: if a hydrogen refueling network can grow up alongside the market penetration of these vehicles, HFCV drivers would rely entirely on hydrogen for both short and long trips, simplifying their fuel purchasing requirements.

15-5   Well-to-Wheel Analysis as a Means of Comparing Alternatives

It is clear from the variety of alternative fuel and propulsion platform options reviewed in the previous section that it is essential to have some means of making an "apple-to-apple" comparison of quantitative effectiveness of each option in converting energy resources into vehicle propulsion while minimizing CO_2 emissions. The *well-to-wheel* (WTW) approach to energy efficiency is just such an objective measure of the overall energy efficiency of a transportation energy technology. The WTW analysis includes both performance onboard the vehicle and also energy consumed upstream in extracting, transforming, and delivering an energy product to the vehicle. The WTW η_{WTW} efficiency can be written as

$$\eta_{WTW} = \eta_{WTT} \times \eta_{TTW} \tag{15-20}$$

where η_{WTT} is the well-to-tank (WTT) process efficiency and η_{TTW} is the tank-to-wheel (TTW) process efficiency. The WTT process includes any energy expended to extract an energy product, transport it to a processing center, transform it into a fuel, and transporting the fuel onward to the dispensing point. The TTW process includes transforming the fuel into mechanical energy inside the vehicle and transmitting it to the wheels.

WTW values for four representative cases, namely, ICEV, HEV, EV, and HFCV are given in Table 15-7. The ICEV and HEV alternatives use compressed natural gas, under the assumption that the ICE would be adapted for gas combustion, as basis for calculating WTW efficiency, while the EV uses electricity generated from the combustion of natural gas in a combined-cycle power plant to charge its onboard batteries. Electrical generation includes not only losses in electrical plants (where overall system efficiencies are on the order of 40 to 45%), but also energy expenditure in extracting and transporting natural gas to the plant and line losses in distribution of electricity from the power

Option	Well-to-Tank	Tank-to-Wheel	Well-to-Wheel
Conventional ICEV	88%	16%	14%
Natural gas HEV (high efficiency)	88%	30%	26%
EV w/electricity from gas	48%	43%	21%
HFCV w/hydrogen from gas	55%	45%	25%

TABLE 15-7   Well-to-Wheel Comparison of Propulsion Technologies for Light-Duty Vehicles

plant to the point of recharging the vehicle. Hydrogen for the HFCV is assumed to be produced from natural gas through steam reformation, which is the most common method for producing hydrogen today.

The WTW comparison illustrates a number of useful points about comparing alternative transportation technologies for the future. First, since the ICEV and HEV have the same WTT efficiency, the effect of changing to the hybrid drivetrain, shown in the increase in TTW efficiency from 16 to 30%, translates directly into an increase in overall WTW from 14 to 26%. The conventional ICEV is limited by a low TTW value. This drivetrain has already been the subject of many decades of research, and it is unlikely that future improvements would improve the TTW value by more than one or two percentage points above the current 16%. In order to achieve a major improvement while still keeping onboard gasoline as a starting point, a more radical transformation is required, as exemplified by the conversion to HEV technology. For example, Toyota has set a goal of an HEV that can achieve a WTW efficiency of 42%, or triple that of the ICEV, all by improving the TTW efficiency.

Second, the result for the EV shows the importance of looking at the whole energy cycle from the source. Looking at TTW efficiency alone, it appears that either the EV or the HFCV is the most efficient approach, with 43% or 45% efficiency, respectively. Taking into account power plant and other distribution losses makes the comparison more realistic, showing that the HEV, EV, and HFCV are all fairly close together, although each is substantially better than the ICEV. Note that the numbers used in the table are intended to make general observations about the comparison between ICEVs, HEVs, EVs, and HFCVs; values for a specific vehicle and energy supply chain will of course vary somewhat from the values shown.

WTW analysis is the most useful when comparing mature technological alternatives over a short to medium term time horizon. All of the alternatives should derive their energy from a single source. For example, all of the options in Table 15-7 use natural gas as a primary energy source. Another possibility is a WTW comparison of solar energy converted to either electricity for EVs or biofuels for ICEVs/HEVs. However, if some of the sources do not emit GHGs and others do, then the relative WTW efficiency may be less important. For instance, if the EV in Table 15-7 were to derive its electricity from wind power instead of fuel oil, a low WTW efficiency in the table (due to the relatively low efficiency of wind turbines to convert the kinetic energy of the wind into electricity, compared to the efficiency of a combined-cycle power plant to convert the chemical energy in methane into electrical energy) might not matter, because the end result of greatly reducing CO_2 emissions would have been achieved.

15-6   Summary

In the future, sustainable transportation energy technologies will exploit energy sources that, unlike conventional petroleum, are available for the long term and will not contribute to the net increase of CO_2 in the atmosphere. Although many specific technologies are emerging today that pursue this goal, they can each in their essence be reduced to one of five endpoint technologies, of which the three most promising are (1) electricity, (2) hydrogen, and (3) sustainable carbon-based fuel. Energy technologies must be developed with both these long-term goals in mind, and also the short-term goal of providing an attractive alternative to the petroleum-driven conventional vehicle. Consumers evaluate vehicles on the basis of fuel economy and also performance, including such parameters as maximum speed, acceleration, and gradability, so vehicles that use alternative fuels or propulsion platforms should be designed with these parameters in mind.

Major emerging alternative fuels and platforms include electric vehicles (EVs), (plug-in) hybrid electric vehicles (HEVs and PHEVs), biofuel-powered internal combustion engine vehicles (ICEVs), and hydrogen fuel cell vehicles (HFCVs). Each faces substantial technological challenges. EVs require battery systems that are cheaper and lighter, and carry more charge, in order to have the range between recharging and total vehicle cost that customers seek. HEVs and PHEVs in their present form are capable of reducing but not eliminating reliance on petroleum. Biofuels suffer from high input requirements of fossil fuel energy per unit of energy output, and also from large land area requirements per unit of fuel produced. HFCVs are faced with both technical and cost challenges, especially with the fuel cell stack and onboard hydrogen storage. As these technologies continue to evolve, well-to-wheel analysis provides a convenient tool for making objective comparisons between alternatives on the basis of life-cycle energy efficiency from the energy source to the delivery of propulsion.

References

Albertus, P., J. Coutsa, and V. Srinivasan (2008). "A Combined Model for Determining Capacity Usage and Battery Size for Hybrid and Plug-in Hybrid Electric Vehicles." *Journal of Power Sources* 183:771–782.

Armstrong, J. (1993). *The Railroad: What it Is, What it Does.* Simmons-Boardman, Omaha, NE.

Hopkins, T., J. Silva, and B. Marder (1999). "Maglift Monorail: A High-Speed, Low-Cost, and Low Risk Solution for High-Speed Ground Transportation." *High Speed Ground Transportation Annual Conference*, Seattle, WA.

Cengel, Y., and M. Boles (2006). *Thermodynamics: An Engineering Approach*, 5th ed. McGraw-Hill, Boston.

Larminie, J., and J. Dicks (2000). *Fuel Cell Systems Explained.* John Wiley, Chichester, West Sussex.

MacKay, D. (2009). *Sustainable Energy: Without the Hot Air.* UIT Press, Cambridge, UK.

Randolph, J., and G. Masters (2008). *Energy for Sustainability.* Island Press, Washington, DC.

Sperling, D., and D. Gordon (2009). *Two Billion Cars: Driving Toward Sustainability.* Oxford University Press, Oxford, England.

Spiegel, C. (2007). *Designing and Building Fuel Cells.* McGraw-Hill, New York.

Wark, K. (1983). *Thermodynamics,* 4th ed. McGraw-Hill, New York.

Further Reading

An, F., and D. Santini (2003). *Assessing Tank-to-Wheel Efficiencies of Advanced Technology Vehicles*. SAE technical paper series report 2003-01-0412. Society of Automotive Engineers, Warrendale, Pennsylvania.

Andreoli, S., and P. De Souza (2006). "Sugarcane: The Best Alternative for Converting Solar and Fossil Energy into Ethanol." *Energy & Economy*, Year 9, No. 59. Electronic resource, available on-line at http://ecen.com/. Accessed Nov. 8, 2007.

Braess, H., and U. Seiffert, eds. (2004). *Handbook of Automotive Engineering*. Society of Automotive Engineers, Warrendale, Pennsylvania.

Commission on Engineering and Technical Systems (2000). *Review of the Research Program of the Partnership for a New Generation of Vehicles*, 6th report. National Academies Press, Washington, DC.

Davies, J., M. Grant, and J. Venezia (2007). Greenhouse gas emissions of the U.S. transportation sector: Trends, uncertainties, and methodological improvements.

Davis, S., and S. Diegel (2006). *Transportation Energy Data Book*, 26th ed. Oak Ridge National Labs, Oak Ridge, Tennessee.

Daziano, R. and F. Vanek (2013). On the gap between the willingness to pay for and the marginal cost of electric battery vehicles with improved range. Paper presented at the Kuhmo Nectar Conference on Transportation Economics, Chicago, July 2013.

DeCicco, J. (2004). "Fuel Cell Vehicles." *Encyclopedia of Energy*, Vol. 2. Elsevier, Amsterdam.

Drapcho, C., J. Nghiem, and T. Walker (2008). *Biofuels Engineering Process Technology*. McGraw-Hill, New York.

Fairley, P. (2004). "Hybrid's Rising Sun." *MIT Technology Review*, April issue, pp. 34–42.

Fay, J., and D. Golomb (2002). *Energy and the Environment*. Oxford University Press, New York.

General Motors Corp. (2001). *Well-to-Wheel Energy Use and Greenhouse Gas Emissions of Advanced Fuel/Vehicle Systems—North American Analysis*. Published by Argonne National Laboratories, Argonne, Illinois. Available on internet at http://www.ipd.anl.gov/anlpubs/2001/08/40409.pdf. Accessed Sep. 14, 2007.

Gillespie, T. D. (1999). *Fundamentals of Vehicle Dynamics*. Society of Automotive Engineers, Warrendale, Pennsylvania.

Hill, J., E. Nelson, D. Tilman, et al. (2006). "Environmental, Economic, and Energetic Costs and Benefits of Biodiesel and Ethanol Fuels." Proceedings of the National Academy of Sciences of the USA. Vol. 103, No. 30, pp. 11206–11210.

Husain, I. (2003). *Electric and Hybrid Vehicles: Design Fundamentals*. CRC Press, Boca Raton, Florida.

Kreith, F., R. E. West, and B. Isler (2002). "Legislative and Technical Perspectives for Advanced Ground Transportation Systems." *Transportation Quarterly*, Vol. 56, No. 1, pp. 51–73.

Nag, A. (2008). *Biofuels Refining and Performance*. McGraw-Hill, New York.

Olsson, L., ed. (2007). *Biofuels (Advances in Biochemical Engineering/Biotechnology)*. Springer-Verlag, Berlin.

Pimentel, D., and T. Patzek (2005). "Ethanol Production Using Corn, Switchgrass, and Wood; Biodiesel Production Using Soybean and sunflower." *Natural Resources Research*, Vol. 14, No. 1, pp. 65–76.

Romm, J. (2004). *The Hype about Hydrogen: Fact and Fiction in the Race to Save the Climate*. Island Press, Washington, DC.

Sorensen, B. (2002). *Renewable Energy: Its Physics, Engineering, Use, Environmental Impacts, Economy and Planning Aspects*, 2nd ed. Academic Press, London.

Sperling, D., and J. Cannon, eds. (2004). *The Hydrogen Energy Transition: Moving toward the Post-Petroleum Age*. Elsevier, Amsterdam.

Stone, R., and J. Ball (2004). *Automotive Engineering Fundamentals*. SAE International, Warrendale, Pennsylvania.

Tester, J., E. Drake, M. Driscoll, et al. (2006). *Sustainable Energy: Choosing among Options*. MIT Press, Cambridge, Massachusetts.

Union of Concerned Scientists (2007). "HybridCenter.org: a project of the UCS." Web resource, available at www.hybridcenter.org. Accessed Sep. 7, 2007.

U.S. Dept. of Energy, Energy Efficiency and Renewable Energy Office (2007). "Alternative Fuels Data Center." Informational website, available at http://www.eere.energy.gov/afdc/altfuel/biodiesel.html. Accessed Sep. 14, 2007.

Vanderburg, W. (2006). "The Hydrogen Economy as a Technological Bluff." *Bulletin of Science, Technology, and Society*, Vol. 26, No. 4, pp. 299–302.

Vanek, F., S. Galbraith, and I. Shapiro (2006). *Final Report: Alternative Fuel Vehicle Study for Suffolk County, New York*. Technical report, New York State Energy Research and Development Authority (NYSERDA), Albany, New York.

Wang, M., C. Saricks, and D. Santini (1999). *Effects of Fuel Ethanol Use on Fuel-Cycle Energy and Greenhouse Gas Emissions*. Report ANL/ESD-38, Argonne National Laboratories, Argonne, IL. Available on internet at http://www.ethanol-gec.org/information/briefing/10.pdf. Accessed Sep. 12, 2007.

Exercises

15-1. A high-efficiency passenger car has a frontal area of 2.4 m², a drag coefficient of 0.25, a rolling resistance value of 0.01, maximum tractive force at low speed of 2800 N, maximum power at high speed of 32 kW, and a curb weight of 1200 kg. Calculate: (a) power requirement at a cruising speed of 100 km/h, (b) maximum speed, and (c) maximum gradability.

15-2. Use the data in Table 15-5 to make a scatter chart of power versus fuel economy for the vehicles shown. How strong is the correlation between these two measures? For individual vehicles, are there any reasons you can give for deviation from the trend? Discuss.

15-3. A fleet manager is considering replacing a fleet of gasoline ICEVs with equivalent EVs. One factor in the decision is the CO_2 emissions from the vehicles. Each vehicle in the fleet drives an average of 10,000 mi/year, and the EVs have an average fuel economy of 20 mpg. The comparable EV consumes 274 Wh of electricity per mile. Emissions from electricity generation in the fleet manager's municipality are given in Example 7-2 in Chap. 7. You can assume that upstream CO_2 emissions (i.e., either from extracting petroleum and converting it to gasoline provided at the pump, or from extracting and transporting coal or gas to the electric plant) are even between the two options and that therefore they are outside the scope of your analysis. (a) How many pounds of CO_2 does each vehicle emit per year? (b) Does the EV reduce CO_2 emissions? (c) Repeat the calculation for your own location, based on local emissions per kWh of electricity. Does the result in (b) change?

15-4. Plot efficiency limit as a function of temperature Celsius for a fuel cell and a Carnot heat engine, as follows. For the fuel cell, use the data in the chapter for the efficiency limit at 80°C and 1000°C to plot a line as a function of temperature. For the Carnot engine, assume that the engine exhausts to a low-temperature reservoir at ambient temp (20°C). According to your plot, what is the predicted "cross-over" temperature above which the Carnot engine has a higher efficiency limit than the fuel cell?

15-5. A fuel cell stack consists of 75 fuel cells arranged in series, each one 12 cm^2 in size. The stack operates at 80°C and the amount of current generated is 4.257 A/cm^2. The fuel cell operates at a system pressure of 1 MPa using a supply of pure hydrogen and oxygen in ambient air. The partial pressure of the water in the outflow is 0.5 MPa. The reference voltage for a fuel cell at 80°C is $E^0 = 1.18$ V. Assume any losses to be negligible and that the voltage at which it operates is the open circuit voltage; in reality, the voltage would be lower due to the current flow. What is the output of the fuel cell stack when operating at full capacity, in kW?

15-6. A car buyer is considering whether or not to spend extra for a hybrid in order to save money in the long run on gasoline expenditures. The buyer drives 25,000 km/year. The cost difference between the two vehicles is $4000, and the fuel consumption is 7.9 L/100 km for the ICEV and 5.3 L/100 km for the HEV. The cost of fuel is $0.80/L (about $3/gal). (a) What is the simple payback for buying the HEV, in years? (b) If the buyer expects a 5% return on this investment, and either car will last for 10 years with negligible resale value at the end of that time, what is the NPV? (c) What is the NPV in (b) if the cost of fuel rises by 3% per year for the lifetime of the vehicle?

15-7. Consider total energy consumption and CO_2 emissions from the PHEV car in Example 15-6, compared to an HEV which is identical except that it uses only gasoline. The CO_2 emissions from electricity generation are the same as in Example 7-2 in Chap. 7. Ignore energy use and emissions upstream from the gasoline retailer or power plant. (a) Does the PHEV reduce energy consumption or CO_2 emissions compared to the HEV? (b) If the electricity is generated $1/3$ from gas and $2/3$ from other fossil fuels, does the answer in (a) change?

15-8. A future midsize EV has achieved higher battery performance and reduced vehicle weight, but at the expense of higher cost. The production cost and weight of the vehicle, before adding any batteries, are $18,000 and 750 kg, respectively. The batteries are available with a charge density of 85 Wh/kg and a cost of $300/kWh. The energy intensity of the vehicle is 0.11 Wh/kg · km. (a) Produce two figures for this vehicle, one plotting vehicle range as a function of total mass, and the other plotting total cost as a function of range. (b) Compare this vehicle to an ICEV in the same class that costs $30,000 and has a typical range per tank of 400 km. What would the EV cost if it were to have the same range? By what percent would the range be reduced compared to the ICEV if it had the same cost?

15-9. Figure 15-4 shows that in the vicinity of an automobile's maximum speed, power available P_a decreases with increasing speed even as total tractive power required increases. In this problem you will solve for maximum speed taking this factor into account. A large sport-utility vehicle (SUV) has a frontal area of 3.8 m^2, a drag coefficient of 0.45, a rolling resistance value of 0.011, and a curb weight of 3500 kg. Power available P_a in the highest gear measured in kW is a function of velocity v in m/s as shown in the function below. (A) Draw a figure of P_a and P_{TR} over the relevant range. (B) What is the predicted maximum speed in both m/s and mi/h? Use a numerical solver to compute, and then verify the value by substituting back into the appropriate equations.

$$P_a = (-6.53 \times 10^{-7})v^3 - (2.221)v^2 + (222.1)v - 5289.6$$

CHAPTER 16

Systems Perspective on Transportation Energy

16-1 Overview

This chapter uses the understanding of transportation energy technologies from the previous chapter as a basis for studying overall energy consumption and energy efficiency of transportation from a systems perspective. There are a number of possible factors that mitigate the ability of technological interventions to affect energy efficiency, and even ways in which technological changes can be undercut by the "rebound" effect. For many different categories of transportation system, the past several decades have seen a mixture of positive and negative effects at a systems level. Given the pressing energy and climate change issues of the twenty-first century, and the prominent role of transportation as a contributor to those problems, it is of growing importance to use systems tools to improve the energy efficiency and environmental performance of the transportation sector. Two possible tools for this purpose are (1) shifting transportation to more environmentally friendly "modes" (i.e., types) of transportation, and (2) rationalizing the system so that it uses fewer resources. The chapter concludes with a discussion of issues related to making a transition to a more sustainable system in the future.

16-2 Introduction

By its essential nature, transportation, and in particular the use and conservation of energy in service of transportation, lends itself to taking a systems approach. All energy applications interact with each other to some extent as they function in their surroundings; for example, they compete for finite resources, and they emit wastes that the natural environment has a finite capacity to absorb. In the case of transportation systems, however, these interactions take on a special importance, because the sharing of common infrastructures (e.g., roads, railroads, seaports, airports, and so on) leads to the various units in the system influencing each other's function and performance—sometimes significantly—whether it is motorists on an urban expressway, passengers in a train, or freight shipments moving through a distribution center. These interactions in turn affect how much energy is required to meet the needs of the system, as dictated by the level of congestion, the quality of maintenance of the system, or other factors. Over time, transportation system users make changes to the vehicles or the infrastructure in order to adapt to changing conditions in the network. Here again, systems effects will influence how well these adaptations work. Therefore, a systems perspective on transportation

577

energy use is a necessary and useful complement to the consideration of transportation technologies in Chap. 15.

Transportation energy consumption originates in the propulsion system of the vehicle, whether it is an internal combustion engine, a jet engine, or the electric motor of a railway locomotive that is supplied with electricity from the grid. It is further influenced by other design choices of the vehicle, such as materials' choices, which affect its overall weight, or styling, which affects its aerodynamic drag. Engineers in an R&D setting of a laboratory or design facility appreciate the benefit of designing a vehicle to be efficient, since reduced manufacturing cost or operating cost will make the product more appealing to the management of the company and to the prospective customer, respectively. However, the pursuit of efficiency must be weighed against other priorities, such as power or performance, and often in the pursuit of product sales it is the latter two criteria that are favored. The way in which the service of moving people and goods in the real world is delivered also affects total energy use, so that an identical vehicle may achieve different levels of energy efficiency in different situations, depending on the circumstances. Land use planning (i.e., the geographic location of amenities in a region), availability of transportation infrastructure, extent of congestion, and other factors all play a role.

As an example of how a systems perspective can help us understand more accurately the likely outcome of changes in technology intended to address transportation energy problems, we can consider the *rebound effect*, as shown in Fig. 16-1. A common response to rising energy use in the transportation sector is to introduce policies aimed at

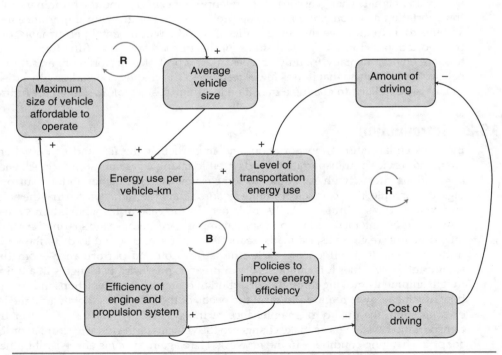

FIGURE 16-1   Causal loop diagram of the relationship between energy efficiency policies, energy use per vehicle·km or vehicle·mi traveled, average vehicle size, and demand for driving.

improving the energy efficiency of vehicles. These policies cause manufacturers to develop more efficient engines and other drivetrain components, so that the drivetrain is able to move a vehicle the same distance with less fuel consumption. One part of the effect of this change is to reduce energy consumption per unit of distance, thereby influencing total energy use in a downward direction. However, there are two other effects. One effect is that the more efficient drivetrain makes larger vehicles more affordable, increasing average vehicle size. The second effect is that, from the laws of economics, if we make an activity such as driving cheaper, demand for that activity will increase, driving up the total amount of driving. These latter two effects both influence the amount of energy consumption in an upward direction.

The causal linkages in Fig. 16-1 show us three possible pathways from the step of increasing the level of "policies to improve energy efficiency" back to "level of transportation energy use," but they do not tell us whether, in the end, a net improvement in the amount of energy use will result. The outcome depends in part on the circumstances in each situation where such a policy is tried. In many cases, a government that enacts such a policy may reduce overall energy consumption compared to a baseline "do-nothing" scenario, even after taking losses due to the rebound effect into account. However, some erosion of energy efficiency gains due to the rebound effect is almost inevitable, unless other policies specifically aimed at curbing growth in vehicle·km or vehicle size are instituted at the same time.

Up until recently, the overall long-term baseline in most industrialized and emerging countries has been a steady increase in transportation energy consumption year after year. In the 2005–2010 period in the United States, the higher cost of oil coupled with the recession of 2009 has led to an absolute decline in both transportation energy consumption and volume of passenger and good movements. Nevertheless, U.S. transportation energy consumption remains approximately 50% higher than it was in 1970, and although the growth trend may have been arrested, it is difficult to substantially reduce it while both the population and the economy continue to grow. Energy efficiency policies may make reductions against the business-as-usual (BAU) situation but not be able to reduce overall energy consumption toward something resembling 1970 levels. Individual technologies such as hybrid vehicles or policies such as the encouragement of use of alternative modes might suggest in theory that 33% reductions should be possible, but once they have run their course, these changes often have less impact than what was anticipated. This observation shows the value of looking at transportation energy use from a systems perspective, as will be further demonstrated in the remainder of this chapter.

16-2-1 Ways of Categorizing Transportation Systems

Transportation systems can be categorized in a number of ways, and the category to which each system belongs influences how it functions. In turn, the function of the system strongly influences energy requirements. The following is a typology of four transportation categories, where a classification on one level can be made independent of the other three (an illustrative example follows):

- *Function—passenger or freight:* One general way to classify transportation is between *passenger* and *freight* transportation. Passenger transportation constitutes any movement of people (e.g., for work, errands, and tourism), including all luggage or personal effects pertaining to their travel, and freight transportation constitutes the unaccompanied movement of goods other than luggage (e.g., bulk commodities, finished products, livestock, and mail and parcels). In general, vehicles are dedicated to one form or the other, although there are exceptions to this rule, for instance, in the case of aviation, large commercial airliners frequently carry passengers on the main deck and airfreight on the lower deck (also known as the "lower lobe") of the aircraft.

- *Modes—road, rail, water, air, and pipeline:* Passenger and freight transportation can be further divided into one of these five major modes. For example, the road mode consists of cars, buses, and motorcycles on the passenger side, and vans and trucks on the freight side, which is sometimes called the "truck" mode. Broadly defined, the road mode can also include nonmotorized modes such as bicyclists and pedestrians. The water mode (also called the "marine" mode) includes movements of boats and ships on both inland bodies of water such as rivers, lakes, and canals, and on the open seas. All five modes exist in both a passenger and freight form, except for pipeline, which is used exclusively for freight, and primarily for the transport of energy products such as oil and natural gas. In addition to these major modes, a number of niche modes exist as well, such as an aerial tramway in a mountainous region that carries passengers and/or goods from the outside world to a remote community.[1] The five major modes are, however, the only ones that consume a significant fraction of the world's transportation energy budget.

- *Geographic scope—urban or intercity:* Both passenger and freight transportation can be divided between urban and intercity transportation movements. On the passenger side, the word "urban" is used as an umbrella term that covers movements both in large urban areas and in small communities, but that in each case are characterized by movements for work, commerce, or recreation over a short distance and carried out on a daily or regular basis. Intercity passenger trips are typically of a distance of 50 to 100 km (approximately 30 to 60 mi) or more between two distinct towns or cities. On the freight side, intercity movements entail the long-distance "trunk" movement of goods between population centers, where the goods are often combined into a larger unit such as a tractor-trailer or a freight train with multiple cars. Urban movements are those (at the endpoints of a trunk movement) that are used to gather shipments together for long-distance movement, or distribute them once they have arrived at their destination. In some cases, smaller vehicles (such as delivery vans or light trucks[2]) are used to carry out urban distribution movements, while in other cases, the same truck may be used for both intercity movement and urban collection and delivery activities.

- *Ownership—private or commercial:* Transportation activity can be divided between *private transportation movements,* which consist of any activity in which the entity that generates the demand for transportation service also owns the vehicle, and *commercial movements,* which entails the selling of the transportation service to

---

[1]Means of conveyance, such as aerial tramways, and cruise ships in which the purpose is sightseeing or tourism, and the passenger travels in a circuit without the intention of reaching the endpoint of the circuit as a destination, are generally not counted in transportation statistics.
[2]In the remainder of this chapter, we use the term *light truck* to refer to pickup trucks, vans, minivans, and sport-utility vehicles (SUVs).

passengers or shippers of freight by professional transportation providers (e.g., taxi companies, airlines, and for-hire trucking companies). In the case of freight transportation by road, a fleet of trucks that is owned by a company for movement of products that it makes or sells itself is considered to be private transportation, even though the driver of the truck does not personally own the vehicle. For example, some large food retailers that operate chains of supermarkets may have their own private fleet, while others contract with for-hire firms to have this service carried out.

As examples of application of these categories, a driver in her/his own personal automobile to work represents passenger transportation using the road mode for urban transportation in a privately owned vehicle. A railroad moving a shipping container from a port to an inland city represents freight transportation using the rail mode for intercity transportation in a for-hire vehicle.

16-2-2   Influence of Transportation Type on Energy Requirements

The distinction between passenger and freight transportation is important because the movement of freight is strictly commercial in nature, while the movement of passengers entails more of a balance between competing factors including not only cost but also, in many cases, the comfort of the passenger, pleasure derived from the route of travel, or the image or status conveyed by the vehicle. When manufacturers, retailers, or other parties responsible for the overall cost of a product make decisions about freight, they determine the amount of protection needed for the product (e.g., protective packaging, and refrigeration) as well as the desired speed and reliability of delivery, and then seek out the least-cost solution that meets these requirements. There is therefore no incentive to spend extra on more expensive solutions, since the difference in cost will come directly out of the profitability of the product. When passengers make transportation choices, on the other hand, they may be more likely to spend extra on a larger, more comfortable vehicle, or more distant vacation destinations, so long as they have the economic means to do so. Especially in the case of middle- and upper-class populations in the industrial countries, some of whom have in recent decades greatly increased purchasing power, the attraction of energy-intensive transportation choices has made curbing the growth in overall transportation energy use more challenging.

The effect of increasing wealth and more demanding requirements for transportation service spills over into the choice of mode as well. For example, in the case of freight, modes such as water and rail are on average more energy efficient because they allow the vehicle operator to consolidate more goods on a vehicle, they move with less stopping and starting, and also water and rails create less rolling resistance than, for example, rubber tires on roads. However, these modes also require greater coordination at the terminal points to consolidate and break apart shipments. From a transportation management point of view, shipment by road and air is a more *agile* option because the shipment reaches the destination more quickly and reliably, although the energy requirement is on average greater. An analogous argument can be made in the case of passenger transportation. Overall, the marketplace for both passenger and freight transportation has in recent years favored the higher service of road and air modes over the energy efficiency of rail and water, increasing total energy consumption.

Lastly, one of the main effects of the geographic scope of transportation is to limit energy source options. In the case of passenger transportation, the majority of transportation activity is generated in urban movements. For many of these trips, it would be possible for travelers to use alternative energy options such as electric vehicles with batteries or alternative liquid fuels that do not have as high an energy density, because the distances between opportunities to recharge/refuel are not too great. It is also easier to connect vehicles such as buses or urban rail vehicles to a catenary grid (e.g., overhead wires), because the density of passenger demand is high enough to justify the cost. By contrast, the majority of freight transportation activity happens over long distances between cities, where the current expectation is that the vehicle or aircraft can travel for long periods between refueling stops. For the most densely traveled rail routes, electric catenary may be justified, but for other routes, rail locomotives must rely on liquid fuels stored onboard the vehicle between refueling stops in the same way that trucks, aircraft, or ships do.

16-2-3   Units for Measuring Transportation Energy Efficiency

Transportation energy efficiency can be measured at various stages, from the testing of the equipment in the laboratory to the delivery of transportation services in the real world. Each successive stage introduces the possibility for ever greater numbers of intervening factors that can disrupt the smooth operation of a component, vehicle, or entire system, so that the potential for losses is increased, as shown in the following four stages:

1. *Technical efficiency of components:* A component of the drivetrain can be tested for its ability to transmit input to output energy. For example, the engine might be tested under steady-state, optimal conditions to determine what percent of the energy present in the fuel combusted is transferred to the rotation of the driveshaft. Engineers might evaluate losses in other drivetrain components in a similar way, that is, calculating efficiency on the basis of power out divided by power in.

2. *Laboratory vehicle fuel economy:* At this level of measurement of energy efficiency, vehicles are tested on a dynamometer to estimate fuel economy, where the dynamometer drive cycle is used to represent driving conditions in the real world. The results are given in units of city or highway mi/gal or km/L. For metric measurement of fuel consumption, the measure of "L/100 km" is commonly used. Laboratory testing recognizes that measuring technical efficiency of the drivetrain (approach #1) does not capture the use of the component in the vehicle, or the effect of parts of the vehicle that do not directly consume energy (e.g., the vehicle body). Estimation of fuel economy also results in a measure that incorporates times when the vehicle is not operating at optimal energy efficiency, for example, stop-and-go driving conditions. Lastly, prospective buyers of the vehicle want to know the effect of fuel economy on operating cost of the vehicle. Whereas a measure of technical efficiency of the drivetrain gives the buyer little information on this point, a measure of fuel economy can readily be translated into an estimated cost per year for fuel, if the buyer knows how far she/he typically drives in a year.

3. *Real-world vehicle fuel efficiency or intensity:* Government agencies typically report overall fuel efficiency (also referred to as *energy intensity*) for different vehicle

classes (e.g., passenger cars, light trucks, and heavy-duty vehicles), in terms of energy consumption per unit of distance traveled. Thus fuel efficiency or intensity is the inverse of fuel economy, which is distance per energy. (L/100 km, introduced under approach #2, is a measure of fuel intensity.) These agencies use vehicle counts on selected roadways and modeling techniques to estimate total vehicle·km of travel, and the allocation of total annual transportation fuel sales to different transportation applications, as a basis for estimating actual kJ/vehicle·km (BTU/vehicle·mi in standard units).[3]

4. *Real-world transportation service efficiency or intensity:* Use of energy per vehicle·km as a measure of technological progress is imperfect because the purpose of the transportation system is not the movement of vehicles but rather the delivery of transportation service, that is, the movement of passengers or freight. Movement of vehicles is a means to this end, but it is not the end itself. The quantity of transportation service is measured in units of passenger·km (pkm) and tonne·km (tkm, i.e., the movement of 1 passenger or 1 tonne of freight for a distance of 1 km, respectively) in order to incorporate the effect of distance on transportation intensity. In other words, the movement of 100 passengers for 100 km will require more energy, incur more wear-and-tear on transportation infrastructure, and so on, than the movement of 100 passengers for 1 km. As with vehicle·km statistics, total passenger·km and tonne·km are published by governments through a mixture of sampling and modeling. Thus it is possible to publish measures of transportation energy intensity in terms of kJ/passenger·km and kJ/tonne·km, respectively. These measures capture the effect of changing transportation practices on energy consumption: for example, if vehicles are not loaded as fully and all other factors remain the same, energy intensity measured in energy per passenger·km or tonne·km will increase.

The role of different measures in assessing the energy efficiency of the transportation system is summarized in Table 16-1.

Comparison of laboratory vehicle fuel economy values (approach #2) for an entire fleet of cars in a country and real-world fuel efficiency (approach #3) typically reveals a fuel efficiency shortfall or gap between the predicted and actual fuel consumption. The laboratory fuel economy values can be used to predict the annual fuel consumption of each vehicle in the fleet, based on its rated fuel economy and estimated distance driven. The sum of all the vehicles in the national fleet gives one possible value for the amount of fuel consumed. When the real-world fuel efficiency is converted to units of fuel economy, it is not as high a value as the estimate based on laboratory fuel economy figures. The shortfall occurs because laboratory tests typically do not fully capture the negative effect of high-speed driving, traffic congestion, decline of fuel economy due to aging vehicles, inadequate owner maintenance practices (e.g., failure to keep tires sufficiently inflated), and other factors.

---

[3]For brevity, metric units are used in the remainder of this chapter. Conversion between standard and metric units are as follows: 1 passenger·mi = 1.6 passenger·km, 1 ton·mi = 1.45 tonne·km, 1 vehicle·mi = 1.6 vehicle·km.

Name	Units		Description
	Metric	**Standard**	
Technical efficiency of components	Percent efficiency; kJ out per kJ in	Percent efficiency; BTU out per BTU in	Laboratory testing of drivetrain components (engine, transmission, tires, etc.)
Laboratory vehicle fuel economy	L/100 km, km/L	mi/gal	Estimate of real-world energy consumption performance based on laboratory drive-cycle test
Real-world vehicle fuel efficiency or intensity	kJ/vehicle·km	BTU/vehicle·mi	Real-world energy efficiency based on estimates of actual energy consumption and vehicle distance traveled
Real-world transportation service efficiency or intensity	kJ/passenger·km, for passenger; kJ/tonne·km, for freight	BTU/passenger·mi, for passenger; BTU/ton·mi, for freight	Real-world energy efficiency based on actual energy consumption and passenger or freight distance traveled

TABLE 16-1   Levels of Measuring Transportation Energy Efficiency

16-3   Recent Trends and Current Assessment of Energy Use in Transportation Systems

In many industrialized countries, transportation is one of the fastest growing energy users, on a percentage basis. For example, Fig. 16-2 shows the change from 1970 to 2013 for the United States, indexed to a value of 1.00 in 1970. As shown in the figure, freight transportation had the highest growth at 107%, while among the four major end uses shown (transportation, commercial, residential, and industrial), transportation was second only to commercial with 59% growth. For comparison, overall energy use ("All energy" in Fig. 16-2) grew by 41%.

The U.S. energy consumption data show that the trend of relatively rapid increase in transportation energy use has been continuing for some time. (Values in the figure were dampened in 2009 by the severe economic recession, and then rose as the economy recovered.) There are many possible contributors to this phenomenon, including the rise of more automobile-dependent land-use patterns; a faster paced working life in some sectors of the economy, forcing workers to travel more and longer distances in each working week; or greater wealth, allowing citizens to spend more on purchase of travel services. This situation can be contrasted with that of other sectors. In the industrial sector, manufacturers either improved energy efficiency so as to cut costs, or moved manufacturing activities out of the United States and hence off of the U.S. energy consumption records. In the residential sector, higher energy costs encouraged a significant shift to more efficient buildings, better insulation, and more modern appliances.

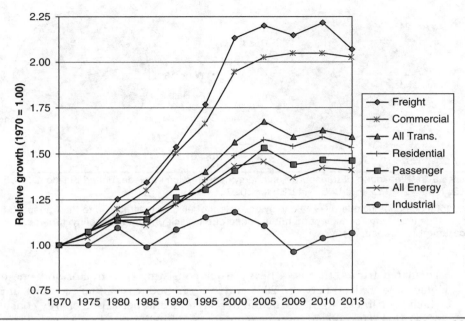

FIGURE 16-2   Relative growth of U.S. energy consumption 1970–2013, indexed to 1970 = 1.00.

Notes: 1970 values—Freight transportation = 3.7 EJ, Commercial = 9.3 EJ, All transportation = 17.8 EJ, Passenger transportation = 14.1 EJ, All energy = 70.2 EJ, Residential = 14.5 EJ, Industrial = 31.0 EJ. *Conversion:* 1 quad = 1.055 EJ. (*Source:* Own calculations based on data from U.S. Department of Energy.)

The pattern in many other industrial countries is similar to the one in Fig. 16-2, with growth in energy consumption of sectors of special importance to a service economy (e.g., freight and commercial activities) outpacing that of other sectors, such as the industrial sector. There are, however, wide variations between countries in the total amount of transportation energy consumption or energy efficiency of transportation. As an example, Figs. 16-3 and 16-4 show the breakdown of total energy consumption for the United Kingdom and the United States, respectively. The modal percentages in the figures are for the combination of passenger and freight energy consumption. In both cases, the road and air modes are the largest energy consumers, although for the United States, the percent share for the road mode is somewhat higher than in the United Kingdom. (Pipeline energy consumption data were not available for the United Kingdom, but we believe that their inclusion would not change this outcome, given the compact geography of the United Kingdom.) Also, the per capita energy consumption for transportation of the United States is over twice as much as that of the United Kingdom (approximately 88 GJ/person versus approximately 38 GJ/person). This difference can be explained partly in terms of factors beyond the control of the population, such as the large geographic expanse of the United States, and partly in terms of short- and long-term choices about land-use patterns, size of vehicles, and so on.

Focusing on changes in U.S. transportation energy consumption, Fig. 16-5 shows the change in fuel economy of passenger cars and combination trucks (also known as

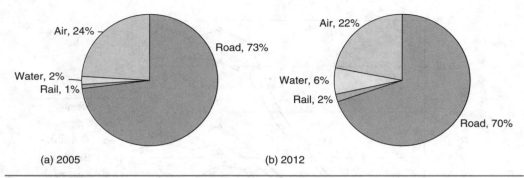

Figure 16-3 Breakdown of transportation energy consumption by mode, United Kingdom, 2005 and 2012. Total = 2.49 EJ, for 2005; 2.41 EJ, for 2012.

Note: Total does not include 0.03 EJ energy equivalent of electricity attributed to the transportation sector. Pipeline energy consumption not included due to lack of data availability. (*Source:* U.K. Department for Transport.)

articulated trucks, that is, a heavy truck consisting of a tractor and one or more detachable trailers) between 2005 and 2012. In both cases, fuel economy improved, which contributed to the reduction in both total and per-capita energy consumption in Fig. 16-4. The percent improvement for combination trucks was actual larger than that of passenger cars, at 12% versus 9% for the latter. These improvements came from improved technology and operating practices, as well as consumer choice of more efficient cars when purchasing new vehicles in the case of passenger cars.

16-3-1   Passenger Transportation Energy Trends and Current Status

The dominant user of passenger transportation energy in most industrial countries is the light-duty road vehicle (passenger cars and light trucks) (Fig. 16-6). In 2012, of the 17.7 EJ of energy attributed to passenger transportation in the United States, 86% was consumed by light-duty vehicles. Most of the remainder was consumed by passenger air travel.

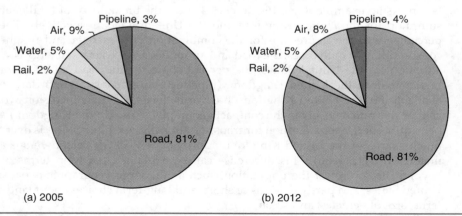

Figure 16-4 Breakdown of transportation energy consumption by mode, United States, 2005 and 2012. Total = 28.8 EJ for 2005; 27.6 EJ, for 2012. (*Source:* U.S. Department of Energy.)

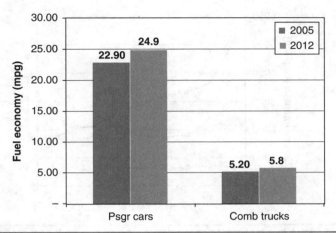

FIGURE 16-5   U.S. passenger car and combination truck fuel economy in miles per gallon, 2005 and 2012. (*Source:* U.S. Department of Energy.)

In other industrial countries, light-duty vehicles are almost always the largest consumer of energy, although in some cases railroads and not aircraft are the second largest user, depending on the intensity of use of the rail versus air system. Data on passenger transportation energy consumption for emerging countries are generally not available, but it is likely that in some countries with particularly low rates of private ownership of light-duty

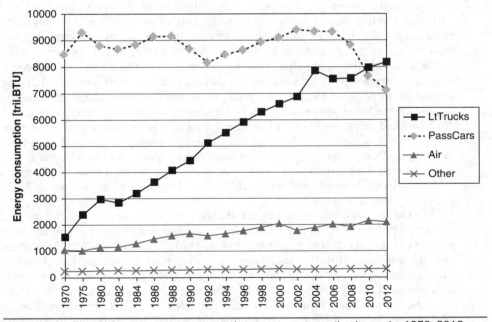

FIGURE 16-6   Total U.S. passenger transportation energy consumption by mode, 1970–2012. (*Source:* U.S. Department of Energy.)

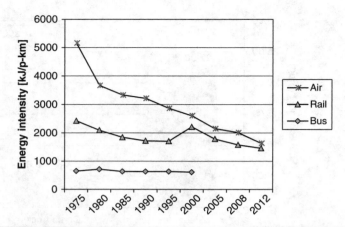

FIGURE 16-7   Intercity passenger transportation energy intensity in kJ/passenger · km, 1975 to 2012. *Note:* Data for bus energy intensity after 2000 were not available. (*Source:* U.S. Department of Energy.)

vehicles, the largest consumer of energy may be buses or passenger trains, since these modes become critical for urban or long-distance travel in the absence of personal cars.

Delivered energy intensity of passenger modes in kJ/passenger · km depends both on the inherent technical efficiency of the vehicle and the load factors of the service provided (i.e., percent of available seats filled). Figure 16-7 shows energy intensity for three intercity passenger modes in the United States, namely, bus, air, and rail. Although the rail mode is generally very efficient when load factors are high, intercity rail in the United States suffers from low occupancy, so that energy intensity per passenger · km is only slightly lower than that of airline travel. By contrast, the bus mode is consistently more efficient than the other two. The air mode has been able to reduce energy intensity by 68% during this time, due to both improved aircraft and engine technology, and the use of *yield management* to maximize seat occupancy on flights. Of the three modes, only air had a significant share of total passenger · km; in 2012, bus and rail accounted for less than 2% each, whereas air accounted for approximately 12%. Data on light-duty vehicle passenger energy intensity divided between intercity and urban geographic scope were not available, so intercity car travel does not appear in the figure. However, for comparison, combined urban and intercity car and light truck intensity values in 2012 were 2105 kJ/passenger · km and 2348 kJ/passenger · km, respectively.

For the two U.S. modes for which data were available, air and rail, the trend in load factors is shown in Fig. 16-8. For commercial aircraft, load factors are given in terms of percent of seats filled on average for the years 1970 to 2012. However, for rail, these data are not available directly, so instead a measure of passenger · km delivered per traincar · km moved is used. Data are also available on train · km of movement, or distance traveled by entire trains, but traincar · km are used as a basis for this figure since they more closely reflect the capacity provided. Because two different measures of load factors are used for the two modes, the values are presented in relative terms indexed to a value of 1970 = 1. The graph shows that load factors have steadily increased

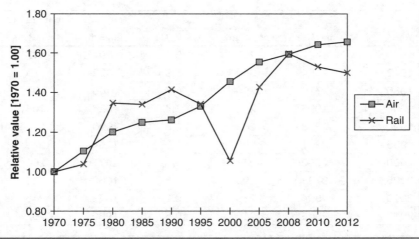

FIGURE 16-8   Relative usage of available capacity for U.S. commercial air and passenger rail modes 1970–2012, indexed to 1970 = 1.00.

Notes: For air, 49.7% of seats were occupied in 1970; for rail, each traincar · km of movement delivered 14.2 passenger · km of service. (*Source:* U.S. Department of Energy.)

for air from approximately 50% in 1970 to 82% in 2008, while for rail the number of passenger·km delivered per traincar·km of movement rose 50% over the same period (from 14.2 to 21.3), although with more variability. Note that for the rail mode the use of passenger·km/traincar·km is not entirely precise because the average size of traincars may change over time, but it is thought to be adequate for this comparison.

The United States is unusual among the industrial nations for having a particularly low share of passenger·km and passenger energy consumption for bus and rail. Most other industrial nations have a higher share for these two modes, and also lower energy intensity for rail, since load factors are higher.

Divisia Analysis of U.S. Car and Light Truck Energy Consumption: A Case Study

Among passenger modes in the United States, the light-duty vehicle is the dominant user of transportation energy at present, and it has also been driving the rapid growth in transportation energy consumption over the last few decades. In 1970, light-duty vehicles consumed a total of 9.7 EJ of energy; in 2004, this amount had grown to 16.6 EJ, an increase of 70%.

Divisia analysis (see Chap. 2) provides insight into factors that contributed to this trend, by considering the relative roles of the two main classes of light-duty vehicles, namely cars and light trucks. While both cars and light trucks in the United States may have reduced energy intensity in recent decades, in the case of cars, improving intensity has kept pace with growing vehicle·km, while in the case of light trucks it has not. As shown in Table 16-2, fuel consumption for cars rises only slightly from 257 to 293 billion L/annum in 2005 before falling to 219 billion L/annum in 2012, while fuel consumption for light trucks increased approximately fivefold from 47 to 251 billion L. A significant shift in purchasing habits by American drivers drove this trend: as Americans bought fewer passenger cars and more pickups, minivans, and SUVs, the

Year	Cars			Light Trucks		
	bn. Km	bn. Liters	km/L	bn. km	bn. Liters	km/L
1970	1467	257	5.7	197	47	4.2
1975	1654	280	5.9	322	75	4.3
1980	1779	265	6.7	466	90	5.2
1985	1995	271	7.4	626	104	6.0
1990	2253	263	8.6	920	135	6.8
1995	2301	258	8.9	1264	173	7.3
2000	2560	277	9.3	1477	200	7.4
2005	2733	293	9.3	1666	223	7.5
2010	2394	235	10.2	1843	245	7.5
2012	2302	219	10.5	1962	251	7.8

TABLE 16-2   Vehicle·km Traveled, Liters of Fuel Consumption, and Fuel Economy of U.S. Cars and Light Trucks 1970–2012

total number of vehicle·km traveled by light trucks increased almost tenfold, while the increase in vehicle·km for passenger cars was less than twofold.

To carry out the Divisia analysis, we need to know the kilometers, fuel consumption, and fuel economy of the combined fleet of cars and light trucks in the base year of 1970. These are obtained by adding together the respective values given in Table 16-3; values from 2012 are also included in the table, for comparison. As shown, the overall fuel economy increases from 5.5 km/L to 9.1 km/L. Based on the increase in activity (kilometers of vehicle travel), the trended fuel consumption with no change in fuel economy would have been 793 billion L:

$$\frac{4.264 \times 10^{12} \text{ km}}{5.5 \text{ km/L}} = 7.79 \times 10^{11} \text{ L}$$

The Divisia analysis is completed by calculating the effect of overall efficiency changes (the combined energy intensity of cars and light trucks) and structural changes (the relative share of the two vehicle types). The results are shown in Fig. 16-9. Of the two types of changes, efficiency changes have the stronger effect, amounting to a change of −363 billion L in 2012. The increasing share of vehicle·km for light trucks, from 12% to 46% of the total between 1970 and 2012, has an upward effect on fuel consumption

Year	bn. Km	bn. Liters	km/L
1970	1664	304	5.5
2012	4264	470	9.1

TABLE 16-3   Vehicle · km Traveled, Liters of Fuel Consumption, and Fuel Economy of All U.S. Light-Duty Vehicles, 1970 and 2012

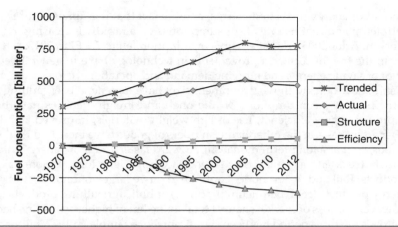

FIGURE 16-9 Divisia analysis of U.S. light-duty vehicle fuel consumption in billion liters, 1970 to 2008.

equivalent to 54 billion L in 2008. The two effects together explain the difference between actual and trended fuel consumption in the graph, that is,

$$E_{actual} = E_{trended} + \Delta E_{efficiency} + \Delta E_{structure}$$

$$470 = 779 - 363 + 54 \qquad \text{[billion L fuel]}$$

Note that the results of the Divisia analysis do not explain the substantial rise in both activity and energy consumption over the period 1970 to 2012, nor do they tell us to what extent the upward pressure on fuel consumption due to rising vehicle·km could have been offset through increases in efficiency. One can surmise that, since vehicle·km increased by 156% over this period, it would have been difficult to prevent some amount of rise in absolute fuel consumption in any case—the average fuel economy would have needed to improve to 14.0 km/L to keep pace. The analysis results do tell us that improvements in efficiency accounted for most of the improvement against the trended projection of where fuel consumption would have been in 2012. Although changes in the structure profoundly affected the nature of the light-duty vehicle market, with automakers selling many more light trucks, they only modestly increased fuel consumption relative to the trended value.

16-3-2   Freight Transportation Energy Trends and Current Status

As mentioned earlier, freight transportation energy consumption is growing rapidly around the world, due to both the move toward a more global economy, and changes to domestic distribution systems within many countries (For further discussion of freight energy trends during the globalization period 1970–2000, see Vanek and Morlok, 2000). Taking the United States as an example, the freight sector energy consumption has been increasing by more than 20% each decade since 1970, up to the 2009 recession. The absolute value for freight rose 120% from 3.8 EJ in 1970 to 8.4 EJ in 2005, as shown in Fig. 16-2; although the latter value is smaller than that of some other sectors, 8.4 EJ is a large amount of energy in absolute terms, and the fact that it is almost entirely derived from petroleum and that its rate of consumption is on a rapidly increasing trend is cause for concern. After 2005, freight energy consumption fell to 7.9 EJ in 2013.

In the freight sector, improvements in technology have in general been working in favor of saving energy, and growing demand and operational choices (e.g., by what mode to ship, how quickly, in what shipment size) have in general been putting upward pressure on energy consumption. On the one hand, freight vehicles have benefited from improved engine efficiency, use of lightweight materials, improved aerodynamics, and other advances. On the other hand, shippers of goods have increased their service expectations, so that a long-term modal shift toward faster and more energy-intensive modes, namely truck and airfreight, has taken place, especially for the more valuable finished products. Rail and water modes continue to move large volumes of bulk goods, such as energy products (coal, petroleum products) or bulk agricultural products (grains, feeds). However, the loss of market share of total tonne·km of higher-value finished products for railroads over this period has been significant. For example, in 1977, railroads carried 45% of all food product tonne·km, which include any value added foods that have been canned, packaged, or prepared in some other way. (Grains and other unprocessed agricultural output are classified as *agricultural products* and are therefore not included in this figure.) By 1997, rail's share of food product tonne·km had fallen to 23%. This type of modal shift has been seen in many European countries and Japan as well.

The increase in U.S. freight energy consumption is driven by the road mode, which grew from 1.6 EJ in 1970 to 5.8 EJ in 2012 (Fig.16-10). Other modes including water, rail, and pipeline held more or less constant due to modest increases in total freight tonne·km combined with gradually increasing energy efficiency. In the case of the air mode, rapid reductions in energy intensity offset the rapid growth in demand for this mode over

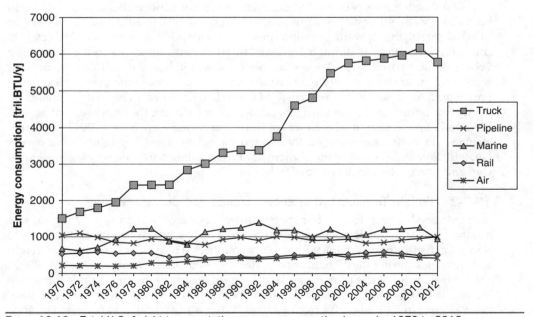

Figure 16-10 Total U.S. freight transportation energy consumption by mode, 1970 to 2012.

this time period, although there was an upturn in consumption from 1995 to 2008. Note that marine and airfreight time series data in Fig. 16-10 are subject to the following limitations. Overall marine energy consumption time series from the U.S. Department of Energy has been reduced by subtracting energy dedicated to recreational boating, which is assumed to be passenger transportation. The remaining marine energy consumption includes a small amount of energy dedicated to passenger transportation in passenger or car ferries, and so on, but it was not possible to remove these figures since no time series data was available. Marine energy consumption also includes some fraction of the energy used for international marine freight shipments to and from the United States, but it was not possible to distinguish between domestic and international energy use from the underlying source. The airfreight energy series was adapted from USDOE's combined passenger and freight aviation energy figure and using other sources to estimate a breakdown between passenger and freight of 80% to 20%, respectively, which is applied to each year in the series to arrive at an airfreight figure.

At present, the truck mode is on average much more energy intensive than the water or rail modes. In the year 2012, the average energy intensity values for these three modes were 3190, 152, and 213 kJ/tonne·km, respectively. Note that this measure masks the impact of bulk goods on energy efficiency of water and rail, since they tend to be densely packed and move slowly, allowing these modes to achieve efficiency values that are not possible for more high value, time-sensitive goods. It is still more efficient to move high-value goods by rail than by truck, but the efficiency gain is not as great as might be implied by the approximate 10:1 advantage base on the overall average modal energy intensities alone.

The influence of the type of product or commodity on the efficiency of freight movement leads to an alternative approach to understanding freight energy consumption, namely, to divide total energy among different types of products or commodities (coal, food products, paper products, etc.). The commodity-based approach sheds light on the role of different commodities in generating consumption of freight energy, including the intensity of freight energy requirements, the trend over time for the product in question, and the contribution of the different freight transportation modes to the total freight energy for the product. This information can then be used to carry out more targeted improvement of freight energy efficiency (see Secs. 16-4-1 and 16-4-2). For example, it may be of particular interest to work on improving energy consumption with a sector of the economy that either uses a large fraction of the total freight energy budget, or is energy intensive relative to the value or weight of goods moved, or is increasing its consumption rapidly.

The influence of value versus ton-mile on the size of the freight market can be seen in Fig. 16-11. When freight is disaggregated by ton-miles, high-value products such as electronics or pharmaceuticals are small in physical volume and therefore account for only a small fraction. When disaggregated by dollar value of product moved, however, these products are a much larger part of the freight market, while bulk commodities such as ores or grains that generate large volumes of ton-miles are much smaller.

A comparison of commodity-based analysis of intercity freight energy use in the United States and United Kingdom is given in Fig. 16-12(a) and (b). For the United Kingdom [Fig. 16-12(a)], 13 commodity groups are shown plus a miscellaneous shipments category ("Misc products"). The data did not support disaggregation of rail and water modes in the United Kingdom by commodity. Therefore, the contribution of these modes is not included in the commodity disaggregate values, and instead is shown separately on

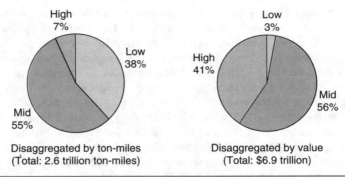

Disaggregated by ton-miles
(Total: 2.6 trillion ton-miles)

Disaggregated by value
(Total: $6.9 trillion)

FIGURE 16-11   Alternative approaches to disaggregating the U.S. freight transportation market: by ton-mile and by product value. (*Source:* U.S. Department of Commerce.)

the right side with hash-marked bars. The remaining values are then just for road energy consumption. For the United States [Fig. 16-12(b)], 14 commodity groups are disaggregated, including energy consumption for multiple modes (e.g., road, rail, and water) for each group. Both Figures 16-12(a) and (b) exclude pipeline energy consumption, energy use for urban movement of freight, and energy consumption in outbound international airfreight movements.[4]

A common point between the two figures is that in both countries, shipping of food products is a large consumer of energy, relative to other commodities. The agricultural products (United Kingdom) or farm products (United States) groups are also fairly large consumers of transportation energy, so the total energy balance for the entire delivery of food items from crops on the farm to the final consumer, as a fraction of total freight energy, is even larger: 24% and 22% of the total energy consumption covered in the two figures, respectively. This observation makes the case for working with the food industry to make sure that food distribution is as energy efficient as possible.

Applying Freight Energy Consumption by Product Type to Life-Cycle Energy Use

It is also possible to use commodity-disaggregate energy consumption to look at freight in context of the energy life cycle of products. Table 16-4 provides a partial life-cycle analysis for total energy consumption of selected commodity groups (the analysis is partial because energy data on extraction, end use, and disposal stages were not available). For 10 commodity groups where comparable numbers were available, the freight energy consumption values for different commodity values for the year 1993 were added to production energy consumption for the product sector in question, as estimated by the U.S. Department of Energy, to give the combined energy consumption shown. The percentage values are then the percent of the combined value attributed to each stage.

---

[4]Note that commodity disaggregate data for the United Kingdom are not as current as other data sets in this chapter, due to lack of time series data that would allow updating to more current values. The data in Fig.16-11 are nonetheless presented because they illustrate the concept of commodity disaggregate energy consumption and because absolute values by sector may have changed over time, but relative values have not changed since the 1990s, for example, food and agriculture remain very large consumers, and so on.

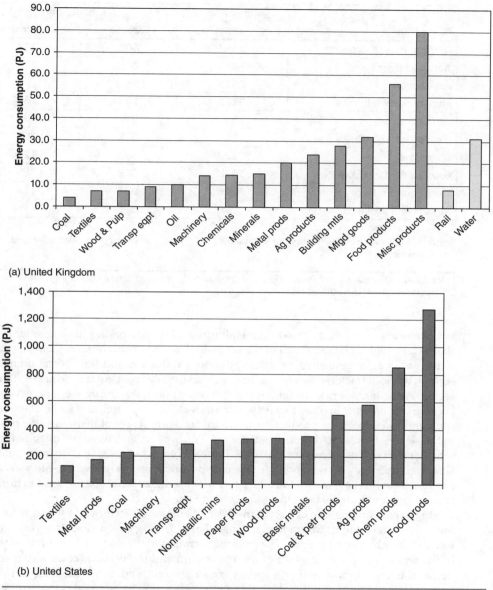

(a) United Kingdom

(b) United States

FIGURE 16-12   Freight energy consumption disaggregated by commodity for (a) United Kingdom, 1995, and (b) United States, 2012. Total energy: for U.K., 0.36 EJ; for U.S., 5.3 EJ.

Note: Not all freight energy consumption of respective countries is covered in these figures; see text.

As shown, the food, lumber, and apparel sectors have relatively high percent values for the transportation side, suggesting that attention to freight energy requirements is an important part of life-cycle energy efficiency in these sectors. By contrast, sectors requiring intensive manufacturing processes including paper products, chemical products, and

Commodity	Combined	Percent Share	
	PJ	Production	Transport
Fabricated metal products	405	80%	20%
Transport equipment	458	74%	26%
Pulp/paper	2831	93%	7%
Primary metal products	2775	94%	6%
Petroleum/coal products	2547	91%	9%
Lumber/wood products	784	61%	39%
Chemicals	2232	88%	12%
Food/kindred products	1618	62%	38%
Textile mill products	326	89%	11%
Apparel and textile products	78	59%	41%

Source: U.S. Department of Energy, for production energy consumption; own calculations, for freight energy consumption.

TABLE 16-4   Comparison of Production and Transportation Stage Energy Consumption Values for 10 Representative U.S. Commodity Sectors, 1993

various types of products based on metallurgy and metalworking have smaller percent shares for freight. The treatment of life-cycle analysis in this example is simplistic in that it deals with aggregate energy consumption for production and transportation in entire sectors, without tracking the specific life-cycle analyses of particular products within that sector. Also, boundaries around energy consumption are imprecise; some products become components in other products, such as textiles in apparel, and some of the energy consumption in both the production and transportation stages of many specific products in the sectors represented occurs outside the U.S. borders and therefore outside the data sets that are used as a basis for the table. A more complete life-cycle analysis would consider freight energy consumption for specific products as part of a whole-life view of the product from raw material to finished item, with a geographic scope not limited to energy consumption within the borders of the United States.

In conclusion, the role of freight modes and commodities moved by freight vehicles are important elements for understanding what is driving the current status and growth trend in freight transportation. For passenger transportation as well, an understanding of the amount of passenger·km of travel demand and the vehicles chosen to meet this demand is useful for explaining energy consumption trend. Since the long-term trend in energy consumption is upward, there is a clear scope for strategies that will reduce energy consumption, CO_2 emissions, and other types of pollution. In the next section, we explore some specific avenues for achieving this end for both freight and passenger transportation.

16-3-3   Estimated CO_2 Emissions Factors by Mode

Along with the preceding discussion of energy trends for both passenger and freight transportation, it is of interest to estimate CO_2 emitted per unit of transportation service provided, measured in either passenger-miles or ton-miles. Data on energy intensity

Mode	Units	BTU/unit	Lb.CO$_2$/1000 units
Car	psgr-mile	3192	542
Light truck	psgr-mile	3561	605
Motorcycle	psgr-mile	2475	420
Air	psgr-mile	2484	422
Transit bus	psgr-mile	4030	684
Intercity bus	psgr-mile	932	158
Rail transit	psgr-mile	2398	656
Commuter rail	psgr-mile	2838	776
Intercity rail	psgr-mile	2214	606
Truck	ton-mile	4398	765
Rail freight	ton-mile	294	51
Marine freight	ton-mile	210	37

Source: Own calculations based on energy intensity data from U.S. Department of Energy.
Note: Intercity bus energy intensity is from year 2000 due to discontinuation of time series data gathering.

TABLE 16-5   Emissions Factors for U.S. Passenger and Freight Modes in lb. CO$_2$ per Passenger-mile or Ton-mile, 2012

from the U.S. Department of Energy as published in the *Transportation Energy Data Book* (Davis et al., 2014) and known rates of emissions for gasoline, diesel, jet fuel, and electricity can be used to generate a first-pass estimate of tailpipe emissions (or power plant emissions in the case of electrified rail transportation.)

Table 16-5 gives the result of the estimation process, with the top rows dedicated to passenger modes and the bottom rows to freight. The third column gives the underlying intensity from USDOE in units of BTU either per passenger-mile or ton-mile, and the rightmost column the pounds of CO$_2$ emitted per 1000 passenger-miles or ton-miles. Emissions rates are based on typical values of pounds of CO$_2$ per gallon of fuel consumed (see Appendix E) or the national average rate of 0.933 pounds CO$_2$ per kWh for electrified modes. For intercity and commuter rail, the breakdown of passenger-miles pulled by either diesel or electric locomotives is not known, so a ratio of 50% for each is assumed. Upstream CO$_2$ emissions from either refining petroleum into fuels or line losses in transmission and distribution of electricity are ignored. Air freight and pipeline modes do not appear because the source did not estimate BTU/ton-mile for those modes, and the "Rail transit" mode is not broken out between heavy rail and light rail because a separate energy intensity for each mode is not provided.

Several limitations of the estimates in Table 16-5 should be recognized. First, the underlying energy consumption pieces in USDOE's energy intensity estimates are not provided in the source, so the use of energy content in either petroleum-derived fuels or electricity is an approximation. Verification against other metrics make the approach seem at least reasonable: a passenger car with intensity of 3192 BTU/ton-mile might average 25 miles per gallon fuel economy, leading to an emission rate of 0.784 pounds per vehicle mile and an average occupancy of 1.45, which is consistent with national data.[5] Next, the 50% ratio for electric and diesel power for commuter and intercity rail is uncertain, and since the underlying emissions factors are 0.776/0.494 and 0.606/0.385 for electricity/diesel, respectively, knowing the actual ratio would change the table value significantly. Furthermore, the use of the national average emissions rate could be made more accurate by using state-level emissions per kWh, in states such as California or New York, emissions per kWh are lower, and would correspondingly lower the emissions factor per passenger-mile (we have ignored the small amount of freight moved using electric traction). Perhaps most significant is the effect of using average rather than marginal emissions rates if one is considering modal choice for an additional trip. Choice of car or light truck for a trip often generates an additional vehicle movement and therefore actual additional emissions, but many of the public transportation modes in the table have spare capacity, so that when a traveler chooses these modes, the vehicle is already committed to the trip and added emissions from the weight of one more passenger are very small.

16-4   Applying a Systems Approach to Transportation Energy

In the previous section, we reviewed the ways in which total transportation energy consumption, for both passenger and freight transportation, is an interaction between the vehicle technology and the way in which it is used. In this section, we look at ways in which taking a systems approach to transportation energy can lead to possible solutions for curbing the growth in energy demand. The solutions covered include the following:

1. *Modal shifting:* Meeting a given demand for transportation services while increasing the modal share of the most efficient modes to reduce energy consumption.

2. *Rationalizing transportation services:* Meeting a given demand for transportation services while reducing total vehicle·km, passenger·km, or tonne·km, using optimization or other planning tools to deliver the service more efficiently.

3. *Integrating EVs and PHEVs with the electric grid:* Reducing petroleum consumption and CO_2 emissions by substituting electricity as an energy source for EVs and PHEVs, and at the same time improving technical and economic performance of the electric grid.

16-4-1   Modal Shifting to More Efficient Modes

The practice of shifting transportation demand to more efficient modes can lead to energy savings for both passenger and freight transportation. By taking a more holistic view of transportation systems, it is possible to use the most energy intensive modes such as cars and trucks less, and use more energy efficient modes such as buses or trains more. This practice assumes that the more efficient modes will have a high enough load factor (e.g., percent loading relative to maximum capacity) to achieve energy savings. As shown in Figs. 16-7 and 16-8 for the case of intercity rail in the United States, low

---

[5]Explanation: 25 mpg and 19.6 pounds CO_2 per gallon gives an emissions rate of 0.784 pounds per mile. From the table, the emissions value is 0.542 pounds per passenger-mile, so the implied occupancy is 0.784/0.542 = 1.45 passengers per vehicle.

load factors can reduce the energy efficiency of modes that are generally thought to be advantageous. In this section, the examples of (1) urban public transportation, (2) personal transportation choices, and (3) intermodal freight transportation are used to illustrate how modal shifting can be put into practice.

Modal Shifting to Public Transportation in Urban Regions

In recent times, many urban regions around the world have attempted to create a shift in modal choice by modernizing and expanding public transportation systems. Historically, as car ownership has grown over the past decades, public transportation systems have lost some share of passenger·km to the private automobile in many countries, so efforts are now being made to win back modal share for buses, subways, and other forms of public transportation.

The focus of many of these programs is not just on energy efficiency and protecting the environment, but also on improving livability of cities. Public transportation systems can move passengers rapidly from origin to destination, without the delays common to travel by car in congested urban arteries. In an ideal situation, the mix of public transportation and car travel is rebalanced to the point that congestion is eliminated. Residents then choose between public transportation and use of their own car, depending on the nature of their trip, but in either case, there is no congestion, and travel times are predictable. Since the public transportation system must be well utilized in order to achieve this rebalancing, overall energy consumption decreases, so that the system-wide energy per passenger·km is also reduced. As an additional benefit, some types of public transportation can use renewable energy, further reducing CO_2 emissions. Some types of public transportation can also reduce the amount of space required for transportation infrastructure, thereby preserving more green space. For example, at maximum capacity, a subway system can carry more passengers per track in each direction than an urban expressway can carry per lane in each direction.

Public transportation comes in several forms, each with specific characteristics in terms of cost and maximum capacity, as follows:

- *Heavy-rail transit:* These systems include subway and commuter trains that have few or no at-grade crossings with streets, and run underground or on elevated passageways. Subway systems in New York, London, Tokyo, and so on are examples.

- *Light-rail transit (LRT):* These systems comprise vehicles that run on rails but are typically smaller than those used in heavy-rail transit (see Fig. 16-13). They also use a mixture of guideways separated from street traffic and mixed in traffic, as well as occasional use of tunnels or overpasses to improve flow at key points. They are less expensive to build per kilometer of track than heavy rail. There are numerous examples around the world, including Geneva, Switzerland, as shown in the figure; Manila, Philippines; Manchester, England; and Portland, Oregon, United States.

- *Bus rapid transit (BRT):* These systems resemble light rail in their use of bus-only roadways separated from street traffic, but the use of buses instead of rail vehicles reduces cost and simplifies the extension of bus rapid transit routes from the busway to and from local streets (see Fig. 16-14). Examples include Curitiba, Brazil; Bogota, Colombia; and Cleveland, Ohio, United States, as shown in the figure.

FIGURE 16-13   Bombardier "Flexity" Light-rail transit (LRT) vehicle at a stop in Geneva, Switzerland.

In recent years, the LRT railcar industry has pushed the maximum length of LRT vehicles while still allowing them to navigate street traffic where necessary. This development increases both maximum passenger capacity during peak periods and financial productivity of the asset, since the driver's wages are distributed among a larger passenger base, all other things being equal.

FIGURE 16-14   Bus rapid transit, Cleveland, United States.

The busway for this BRT system includes typical BRT features such as separate right-of-way with one lane in each direction; distinctive stop layout with off-board fare collection; separate traffic signals at intersections; and doors on either side of bus for either right- or left-side boarding and alighting.

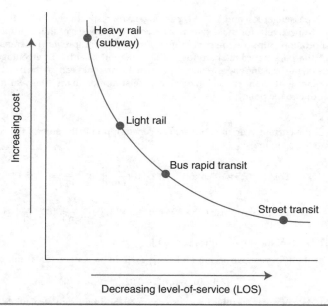

FIGURE 16-15   Cost versus level-of-service trade-off for public transportation systems.

- *Street transit:* These systems include buses, tramways, and trolleys (i.e., electrically powered rail vehicles that operate using overhead catenary), and trolleybuses (also known as *trackless trolleys*), that operate entirely in the presence of street traffic. These systems are the least expensive to build and maintain, but they also provide the slowest service and are the most susceptible to congestion.

Of the four types mentioned, heavy-rail and street transit are the oldest, while LRT and BRT are developed more recently. The reason for their emergence can be illustrated using a *cost versus level-of-service* (also known as LOS) diagram, as shown in Fig. 16-15. At the midpoint of the twentieth century, public transportation primarily offered only heavy-rail and street transit systems. Between these two lay a gap in terms of service offering, where, for many cities, heavy-rail systems were too expensive, and street transit was too slow, to compete with the private automobile for passengers. LRT and BRT reduce the total system cost compared to heavy-rail, but they also provide a measure of *rapid transit*, in which passengers travel faster than the stop-and-go speeds of street traffic. Since the 1960s, many cities of 1 million or less population that could not afford heavy-rail systems built LRT or BRT instead, and today, the number of cities in the world that operate LRT and/or BRT has surpassed the number with heavy-rail systems. In the very large cities as well, LRT and BRT can serve certain niches, for example on the periphery of a city where demand is not high enough to justify a heavy-rail system.

Example 16-1 illustrates the potential effect of modal shifting to public transportation on energy consumption and efficiency.

Example 16-1   A metropolitan region generates 7.5 billion passenger·km of transportation demand per year and is experiencing 2% per annum growth in total passenger·km, and a 0.5% decline in energy intensity of passenger·km. Current energy intensity of automobile travel and public transportation is 2200 kJ/passenger·km and 1000 kJ/passenger·km, respectively, and share of passenger·km is 90% and 10%, respectively, for those two modes. For the purposes of this example, ignore all other types of transportation. Suppose the government institutes a comprehensive transportation program that increases the passenger·km share for public transportation by 5 percentage points over 10 years. Calculate (a) the baseline energy consumption in the present year, (b) the baseline energy consumption after 10 years, and (c) the reduction in energy consumption in the 10th year relative to the baseline in that year due to the policy.

Solution

(a) For the current year, the total energy consumption is the sum of the consumption for the two modes, as follows:

$$\text{For car: } (7.5\times10^9 \text{ pkm})(0.9)\left(2200\frac{\text{kJ}}{\text{pkm}}\right)\left(10^{-12}\frac{\text{PJ}}{\text{kJ}}\right) = 14.85 \text{ PJ}$$

$$\text{For public trans.: } (7.5\times10^9 \text{ pkm})(0.1)\left(1000\frac{\text{kJ}}{\text{pkm}}\right)\left(10^{-12}\frac{\text{PJ}}{\text{kJ}}\right) = 0.75 \text{ PJ}$$

The sum of these two values is 15.6 PJ.

(b) For the 10th year, we account for the growth in passenger·km as follows:

$$(7.5\times10^9 \text{ pkm})(1+0.02)^{10} = 9.14\times10^9 \text{ pkm}$$

Modal energy intensities have also changed, and these are recalculated as

$$\text{For car: } \left(2200\frac{\text{kJ}}{\text{pkm}}\right)(1-0.005)^{10} = 2092\frac{\text{kJ}}{\text{pkm}}$$

$$\text{For public trans.: } \left(1000\frac{\text{kJ}}{\text{pkm}}\right)(1-0.005)^{10} = 951\frac{\text{kJ}}{\text{pkm}}$$

Recalculating the energy consumption by repeating the calculation in part (a) with the increased number of passenger·km and reduced energy intensity gives 17.22 PJ for car, 0.87 PJ for public transport, and a total value of $17.22 + 0.87 = 18.09$ PJ for the entire system.

(c) This alternative assumes that in the 10th year the modal split is 85% car/15% public transport. We can therefore recalculate as in part (a) using the new values:

$$\text{For car: } (9.14\times10^9 \text{ pkm})(0.85)\left(2092\frac{\text{kJ}}{\text{pkm}}\right)\left(10^{-12}\frac{\text{PJ}}{\text{kJ}}\right) = 16.26 \text{ PJ}$$

$$\text{For public trans.: } (9.14\times10^9 \text{ pkm})(0.15)\left(951\frac{\text{kJ}}{\text{pkm}}\right)\left(10^{-12}\frac{\text{PJ}}{\text{kJ}}\right) = 1.30 \text{ PJ}$$

The combined total is 17.56 PJ. Therefore, the energy reduction is $18.09 - 17.56 = 0.52$ PJ.

Discussion   The results of this example show the challenge of curbing growth in transportation energy consumption in the face of continually expanding passenger·km values. Despite a 50% increase in the amount of public transportation over 10 years, energy consumption is higher in year 10 than it is in the current year. Nevertheless, public transportation has made a measurable reduction in energy use of 0.52 PJ, which is equivalent to 21% of the 2.49 PJ growth in energy use that occurs with no increase in public transportation.

One limitation of this type of analysis is that it assumes each unit of passenger·km shifted to public transportation will achieve energy savings based on the difference between the average energy intensity of the two transportation options. For a more accurate estimate, the analyst might build a computer model of the transportation network that includes the modeling of residents' modal choices for different types of trips. The analyst would first verify that the model reproduces the baseline system in the real world within some degree of accuracy, and then impose the new expanded public transportation system on the model to evaluate the new modal split and total energy consumption. Such a modeling exercise gives more reliable results but requires a far greater amount of effort.

Modal Shifting of Personal Transportation Choices

In the previous section, the objective of modal shifting relied on a substantial commitment by local and regional governments to provide public transportation service as an alternative to travel by car. Even without using public transportation, however, the individual traveler can take steps to shift modes in her/his personal choices.

One of the simplest ways of achieving this end is for travelers who drive to own different vehicles for different purposes, assuming they have the necessary financial means. For these travelers, especially those with higher incomes, it may be practical to own a smaller vehicle for single-occupant work or nonwork trips, and a larger vehicle such as a van, SUV, or truck for occasions where the motorist is either carrying a large amount of goods or has several passengers on board. The situation of the driver traveling by himself or herself alone is termed a *single-occupant vehicle* (SOV), while a vehicle with a large number of passengers is a *high-occupancy vehicle* (HOV), which in many urban areas, has access to special HOV lanes that allow congestion-free travel on urban expressways. From an energy efficiency point of view, it is desirable to avoid larger vehicles, and especially light trucks, traveling as SOVs. For example, an SUVs that delivers 15 mpg (6.34 km/L or 15.8 L/100 km) during urban travel with a single occupant has an intensity of approximately 5000 kJ/passenger·km, which is much higher than the average U.S. value given above. There is anecdotal evidence that, with the volatility in gasoline costs in the U.S. market since 2005, drivers who own both a compact car and a light truck increasingly are choosing the compact car for single-occupant travel in order to reduce their fuel expenditures.

Outside of highway-capable light-duty vehicles (i.e., vehicles able to travel at the full range of speed limits, including expressway speeds), other options exist that can further allow the motorist to mode-shift toward the travel solution that meets the needs in the most efficient way possible:

- *Limited-use vehicles:* Small battery-electric or gasoline vehicles with limited top speed and range that are useful for short trips in an urban setting (see Chap. 15). Some of these vehicles have necessary signaling and safety equipment to be road legal. In other situations, a nonroad legal vehicle can be used on separate roadways, such as in golf communities where residents use golf carts within the community for golf, shopping, and other amenities.

- *Motorcycles, scooters, and other motorized two- and three-wheel vehicles:* Where the motorist is traveling alone and does not have many goods to carry, these vehicles provide a very energy efficient option for local travel.

- *Nonmotorized modes:* Travel by bicycle or on foot by definition uses no energy, and additionally provide health benefits to the traveler through cardiovascular exercise, whenever the weather is suitable. In many instances, it is not the lack of a vehicle but rather the lack of bicycle- and pedestrian-friendly infrastructure that prevents their greater use. In response, many urban regions have been expanding and improving sidewalks, bike lanes and bike paths, road shoulders, and multiuse trails. In some cases, national government funding is available, such as the Chester Creek Rail Trail project near Philadelphia, PA, which in 2007 received funding from the U.S. Environmental Protection Agency's *Congestion Mitigation and Air Quality* (CMAQ) program.

While the appeal of having different options for different types of trips is clear, many travelers find it impossible to create these opportunities for themselves, both for reasons of the additional capital cost of owning different vehicles and the practical requirement for space to store all the various vehicles in and around one's residence. The need to occasionally carry a large amount of goods or a large number of passengers often dictates that the individual who can only afford one vehicle purchase a large one, and then travel at most times with "excess capacity" in terms of passenger seats or volume of cargo space. The excess capacity in turn translates into additional weight that must be moved around when the vehicle is in use, which increases energy consumption.

To address this limitation, the practice of *vehicle sharing* is expanding in many urban areas of Europe and North America, including both *car-sharing* and *bike-sharing*, as shown in Fig.16-16(a)–(c). In the case of car-sharing, individuals join an organization that rents automobiles for short periods of time and for short distances, and makes them available in distributed locations around the urban area so that they are easy to access from one's home address, workplace, or any location where access may be useful. In this way, car-sharing is different from the rental car industry, which is geared more toward rental of cars for periods of 24 hours or longer, and for longer distances. Car-sharing gives the member access to specific types of vehicles at the times when they need them (e.g., compact cars for solo trips and minivans or light trucks for moving large amounts of goods) without the need to own and maintain the vehicle on one's own premises. Matching the size of the vehicle to the needs of the trip in this way helps to reduce energy consumption. Members of car-sharing organizations may join in addition to owning their own car, but for those who join as a substitute for car ownership, car-sharing encourages the individual to diversify choice of transportation modes, since they no longer have the "sunk cost" of owning a vehicle. Car-sharing also benefits from dedicated parking spaces for the vehicle, which allows the user to retrieve the vehicle from a guaranteed location and return the vehicle to the same location without the nuisance of needing to look for scarce parking [Fig. 16-16(a)].

Similar to the function of car-sharing, the goal of bike-sharing is to reduce auto dependence by conveniently providing bicycles for certain trips where the user needs flexible mobility and bicycling provides an attractive alternative to driving. The Paris bike-sharing station in Fig. 16-16(b) represents the commercial approach to bike-sharing: residents can pay a fee to join the system, and then are charged for each time they rent a bicycle from a station. Electronic tracking of each bicycle rental allows the user to rent at one location and return at another, facilitating one-way travel. The Bixi system first launched in Montreal and more recently in Ottawa and Toronto, Canada, allows tourists as well to swipe in to the system with a credit card and begin using bike-sharing immediately, all with the goal of making travel without a private car in the urban core more attractive. The Cornell University bike-sharing shown in Fig. 16-16(c) represents a university-based variant of bike-sharing. Bicycles are provided to students, faculty, and staff free of charge with the goal of encouraging them not to bring private cars to the central campus, and the bikes are administered by the library system, with the procedure for borrowing a bicycle resembling that of borrowing a book. Users may borrow at one location and return at another, and the university supports the program by periodically repositioning bicycles with a delivery vehicle if the supply becomes imbalanced. The next generation of bikesharing may bypass the fixed rental location altogether and build the infomatic system into the bicycle, so that trips can start and end anywhere, and users can use smart phones to locate and reserve bicycles from any location.

FIGURE 16-16   Variants on vehicle sharing systems: (a) car-sharing provided by Ithaca Carshare, including reserved parking space; (b) Paris "Velibe" bike-sharing system automated pubic rental facility; (c) Cornell "Big Red Bikes" university community bike-sharing system.

Modal Shifting of Freight Transportation

The most important concept in expanding the use of energy-efficient modes in freight transportation today is the concept of *intermodalism*, or the creation of a seamless freight transportation service that delivers the shipment from origin to destination with a high LOS while using different modes for different segments of the journey where each mode has a comparative advantage. The freight industry has come to recognize that, over long distances, modes such as rail (and in certain situations water) have attractive advantages, in terms not only of energy efficiency but also reduced cost and labor requirement. At the same time, most shipments today must at some point move by truck, since trucks provide the most convenient access to origins and destinations. Some locations, such as large manufacturing plants or mining facilities, have direct access to the rail and/or water network. However, in today's dynamic economy, locations change rapidly, and it is much more practical to link a new facility to the road network than to the rail or water network. Also, for small- and medium-sized facilities, it is often more financially attractive to both the business and the transportation operators to send shipments via truck to the nearest transit point to the rail or water modes, rather than incur the high cost of building a dedicated rail/water facility on site.

A number of systems exist to transfer shipments from the road to other modes at the intermodal transit point, as follows:

- *Roll-on roll-off:* The single-body truck or tractor-trailer enters and leaves the railcar or marine vessel as a complete unit.

- *Loading of truck trailers:* The trailer is separated from the truck tractor and loaded onto the railcar (also known as "piggyback" rail service). Separate loading of trailers onto marine vessels is less common, though possible.

- *Loading of shipping containers:* When arriving by road to a rail or marine intermodal facility, the chassis (wheeled underbody that allows a shipping container to move over the road) is removed and then the container is loaded onto the railcar or marine vessel. Direct loading/unloading between rail and marine is also possible. Modern intermodal railcars allow *double-stacking* of containers on railcars to maximize the productivity of each train.

On the basis of these different systems, a wide variety of applications are possible. For example, between Salerno, Italy and Valencia, Spain, the European Union has been supporting the development of a roll-on roll-off service aimed at trucks that would allow them to avoid driving along the perimeter of the Mediterranean Sea via France between the two countries. Although a motorway exists along the land route, it is circuitous and passes through a number of highly populated areas, so it is desirable to transfer some truck traffic to the seas. In a different European location, the Swiss government has developed a network of roll-on roll-off trains for trucks to allow them to transit the Alpine region without driving. The main driver of this program is the reduction of air pollution in an environmentally sensitive region, but there are also CO_2-reduction benefits, since the Swiss railways are almost entirely electrified and Swiss electricity comes from hydro and nuclear power.

One of the largest intermodal freight shipping operations in the world today is the shipping of containers and truck trailers in continental North America, along the rail networks of Canada, the United States, and Mexico (see Fig. 16-17). In this market, some freight shipments move very long distances (2000 to 3000 km or more) to and from population centers in the interior to shipping ports along the Atlantic and Pacific. Over these long distances, it makes financial sense to bundle a large number of shipments onto a single train so as to save energy and labor costs, as a single double-stack train with crew of three in the locomotive can replace 300 or more trucks each carrying one container. Through the 1990s, this industry saw double-digit percent annual growth in the number of containers moved as the railroads improved service quality and an increasing number of shippers took advantage of the cost savings available from this service.

FIGURE 16-17   Double-stack intermodal train carrying shipping containers. (*Source:* BNSF Railroad. Reprinted with permission.)

Energy savings from intermodalism come from moving a given shipment via rail or water instead of over the road, reducing energy consumption on each tonne·km of movement. Intermodal shipments sometimes require truck movements between the origin/destination and the intermodal terminal that are not necessary when the shipment moves by truck directly from origin to destination using the intercity highway network. Nevertheless, for most long distance movements where intermodal transportation is financially competitive with direct shipment by truck, the energy savings can be substantial. Taking the case of the U.S. intermodal network during the 1990s and 2000s, these movements may have had an energy intensity on the order of 880 to 1690 kJ/tonne · km (1200 to 2320 BTU/ton · mi), compared to an average value of 2100 kJ/tonne · km (2900 BTU/ton · mi) for trucking. Based on the growth in use of intermodal services over the period 1980 to 2012, energy savings attributed to moving shipments by intermodal transportation compared to moving the same shipments by truck under a midrange efficiency scenario (1285 kJ/tonne·km, or 1752 BTU/ton·mi) rose from 14 PJ to 117 PJ, as shown in Fig. 16-18. The latter quantity of energy is equivalent to approximately 19% of all rail energy use in the United States in 2012.

The expansion of intermodalism requires financial investment in the necessary equipment, including not only vehicles and rights-of-way (rail lines or waterways), but also intermodal transfer facilities that make possible the smooth transition from one mode to another. If either transfer points or long-distance corridors do not function correctly, then the entire system becomes unattractive to shippers, and the potential to save energy is lost. On the other hand, intermodal freight is a win-win situation for governments, shippers, rail and marine operators, and even for trucking firms, who can profit from transferring segments of truck freight movements to the rail or water modes: for all of these entities, it brings the benefits of taking pressure off the overburdened and overcongested road network. From an energy perspective, it also brings the benefit of

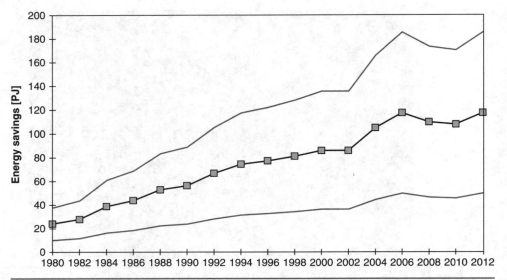

FIGURE 16-18   Low-, mid-, and high-range for energy savings achieved by shifting freight from road to intermodal in United States, 1980–2012.

moving freight at lower energy intensity. At the present time, there is every indication that governments and the private sector will continue to invest in these systems and at the same time advance the underlying technology to make it even more competitive.

16-4-2   Rationalizing Transportation Systems to Improve Energy Efficiency

The discussion of modal shifting in the previous section takes the demand for moving freight or passengers as measured in passenger·km or tonne·km as fixed, and seeks modal alternatives that will deliver the required transportation service with reduced energy consumption. In this section, we consider the "other half of the equation," namely, the reduction of vehicle·km, passenger·km, or tonne·km through strategies that streamline or "rationalize" transportation movements while still meeting the needs of the customer. As a simple illustration, bus transit system operators often rationalize their service on weekends by reducing the number of buses per hour on a bus route, since otherwise the number of bus vehicle·km required would be excessive compared to the number of customers using the bus.

Since the emergence of the railroads in the nineteenth century, it has been the goal of transportation planning at many levels of society to use assets (vehicles, railcars, and the like) in an optimal way so as to maximize the movement of passengers and freight, minimize travel time, minimize costs, and so on. In recent years, the development of information technology (IT) systems has greatly expanded this capability. IT systems can help transportation planners route freight trains over rail networks, assign empty taxi cabs to new customers or to key waiting points, match mobility impaired passengers to paratransit service, route delivery vehicles bringing parcels to addressees in a metropolitan area, or perform many other functions. While shipping clerks and other human operators developed pencil-and-paper solutions to these problems in the era prior to the digital age, IT systems can solve complex problems more quickly and with superior results, often by using some type of optimization (see Chap. 2). Rationalizing systems in this way translates into energy savings, since the required passenger·km or tonne·km of demand are met with fewer vehicle·km of movement. With very few exceptions, the solution to the optimal vehicle planning problem that minimizes energy consumption is also the one that minimizes financial cost by reducing capital, labor, and maintenance costs, and is therefore the most attractive from both a business and societal perspective.

In some cases, IT systems and optimization may also be used to reduce underlying demand for passenger·km and tonne·km. Although it was stated earlier that the latter measures are the underlying service that transportation service delivers, they are in turn derived from the public's need for access to goods and services. If this need can be met while reducing total passenger·km/tonne·km through locational choices, both cost and energy consumption can be further reduced. For example, a large retail firm that sells their products through a chain of outlet stores in a region with growing demand may be able to reduce total shipping cost by developing a source for certain products that is closer to the market. Assuming the new source does not cost more or consume more energy on the production side than the original one, the total cost of the product is reduced due to lower shipping costs, and energy is saved thanks to fewer tonne·km of movement required.

System-Wide Rationalization of Transportation Demand: A Case Study of U.S. Freight Transportation Patterns

Advances in vehicle technology and IT systems have acted as a two-edged sword in regard to their effect on energy consumption. The same technologies that allow transportation professionals to plan operations in an optimal way also enable passengers and goods to travel reliably over longer distances and at a lower cost. Even at the longest distances, journey times can be planned with greater precision than ever before. It is little wonder that not just the total number of trips being made or amount of goods being shipped has increased but the *average length of trip or shipment* has increased as well, driving up energy consumption.

Transportation statistics consistently point in this direction. To take the example of freight transportation in the United States, from 1993 to 2007, the average shipment distance grew from 230 to 387 km. Along with estimates of total tonne·km of activity, U.S. government agencies also track total tonnes of freight originated in the system, which grew from 10.9 to 12.5 billion tonnes over the same period, or 15%. At the same time, tonne·km grew from 3.5 to 4.9 trillion, or 40%. Thus the average number of tonne·km for every tonne originated is also growing, from 323 in 1993 to 427 in 2007. A similar phenomenon has been observed in the United Kingdom and other European countries, and the term *spatial spreading* has been coined to describe it.

A possible way to counteract this trend and curb the related growth in energy consumption is to take the scale of the optimization technique up to the next level, and consider not just optimizing the activities of individual firms internally, but optimizing the transportation choices of multiple firms that make up a product sector, so as to achieve larger gains. Analysis of government-gathered *commodity flow data*, or volume of freight of different types along major corridors between regions, can be used to develop a graphical representation of the spatial patterns of freight flows. The resulting database can in turn be used to identify opportunities to rationalize the system.

To illustrate the concept, we have created a demonstration network using 1993 data on paper product movements from the U.S. Bureau of Transportation Statistics between

Destinations	Origins			
	OR	**WI**	**GA**	**ME**
CA	1904	243	122	55
NY	67	257	424	435
FL	0	79	1258	55
IL	40	1230	335	404
NJ	0	85	293	114

Note: Abbreviations are as follows: Oregon = OR, Georgia = GA, Maine = ME, and Wisconsin = WI, along with five major destination states for paper products (California = CA, New York = NY, Florida = FL, Illinois = IL, and New Jersey = NJ).

TABLE 16-6   Flows of Paper Products between Selected U.S. Origins and Destinations in 1000 tonnes, 1993

four major producing regions and five major markets, as shown in Table 16-6. In large part, the data suggest a pattern where geographic distance already influences the amount of flow, and thus the volumes of flow are fairly rational: for example, the largest source of paper products for California is Oregon, and for Illinois is Wisconsin, which in both cases are adjacent states. However, there are also smaller volumes of flows moving very long distances, including from Maine to California or Oregon to New York.

The pattern in Table 16-6 suggests that, by producing the right mix of paper products closer to the market for which they are destined, total tonne·km of freight could be reduced while still delivering enough product to meet consumer demand. As a preliminary indication of the potential to reduce, Table 16-7 shows an alternative pattern where the flows have been rearranged using optimization such that the same tonnes of product are produced in each origin, and the same number delivered to the destinations, but the amount of generated tonne·km is minimized. Many of the very longest flows are eliminated in this alternative solution, leading to reduced tonne·km requirements. Based on typical modal share among road, rail, and water movements, and average energy intensities, the estimated energy requirements for moving the 7.4 million tonnes of paper products in the base case is 13.4 PJ. The resulting change in transportation pattern would,

To	From			
	OR	**WI**	**GA**	**ME**
CA	2011	0	313	0
NY	0	0	120	1064
FL	0	0	1393	0
IL	0	1894	115	0
NJ	0	0	491	0

TABLE 16-7   Flows of Paper Products after Optimizing to Reduce tonne·km Requirement

if it were instituted entirely, eliminate 2.3 billion tonne·km of paper products movement, which is equivalent to 3.1 PJ, or approximately 200,000 tonnes of CO_2 saved per annum.

As with any optimization modeling exercise, the above results must be interpreted carefully. In many instances, there may be commercial reasons why it is important to source a specific product in a given location, regardless of distance. Therefore, one would not expect the improved pattern to be reproduced exactly. Also, from an energy life-cycle perspective, there may be efficiency advantages to large-scale, centralized production that offset the extra energy requirements for shipping long distances. However, a process of studying spatial patterns, and then governments and private firms working together to transform the information contained therein into adjustments to rationalize the system can make a contribution to reducing energy consumption. Such changes could be beneficial, because the amount of energy consumption involved, even in a selection of product movements used in this example, is quite large. The figure of 13.4 PJ quoted is small relative to the total freight energy figure of 7.9 EJ for the year 2013 from Fig. 16-2, but it is equivalent to the entire energy budget of a small developing country. (Total energy consumption across all end uses for Sierra Leone in 2012 was 12 PJ.) Therefore, reducing energy consumption by 5 or 10% through a more geographically compact freight flow pattern could lead to worthwhile reductions in energy consumption.

Lastly, the examples presented in Tables 16-6 and 16-7 held the amount of paper products produced in each origin fixed, and reduced transportation energy consumption by rationalizing the destinations for those products. Additional gains could be reaped from expanding local and regional production, so that more demand can be met without going outside of a region for the supply. For example, in Table 16-6, 18% of the 2.3 million tonnes arriving in California in the base case come from other distant parts of the United States. With expansion of paper products manufacturing along the west coast, including states such as California, Oregon, and Washington, more of the demand for products could be met within the region, and the shipment of products from the East Coast to California might be reduced.

16-4-3   Integrating Light-Duty Vehicles and Electricity Supply to Optimize Vehicle Charging and Grid Performance

Integration of energy supply to light-duty vehicles with the function of a technologically updated smart grid through the growing use of PHEVs or EVs in the LDV fleet provides opportunities for improved energy efficiency and performance, both in the short- and long run. In the short run, PHEVs or EVs might be recharged at night to take advantage of unused generating capacity in large generating assets, providing energy to vehicles at low cost. In the long run, grid operators might match growing capacity to store and use intermittent renewable energy with the capacity to store and use energy provided by these vehicles in the fleet.

In its fully developed form, the grid-to-vehicle system allows the Independent System Operator (ISO) to coordinate the generation of electricity from both dispatchable and intermittent resources with the opportune charging of vehicles when they are connected at charge points, as shown in Fig. 16-19. For ideal operation in a system with a large percent of EV/PHEV penetration into the market, charging is available not only at night at home but also during the day at the workplace. This arrangement makes some portion of the fleet available to receive charge for most of the hours of the day, so that the ISO has maximum flexibility to allocate charge to vehicles.

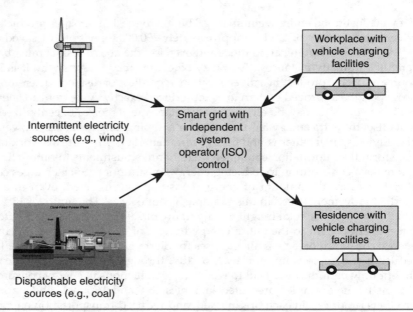

FIGURE 16-19 Components of smart grid-based electricity generation and vehicle charging system, including vehicle-to-grid (V2G) option and home- or workplace-based charging. (*Source:* Powerplant public domain image courtesy of Tennessee Valley Authority.)

Both communication technology and price policy have a role in the envisioned charging system. Land-based or wireless communications networks would allow the ISO to know, at any given time, the number of vehicles available and the remaining uncharged capacity spread across those vehicles. An ability to forecast output from intermittent solar and wind resources up to several hours ahead would also facilitate the ISO's coordinating role. Differential pricing could be used to encourage vehicle owner participation, in other words, reduced cost per kWh to charge vehicles in return for availability when the vehicle is charged for long periods of time at home or at work.

As a further option, the ISO might oversee a *vehicle-to-grid* (V2G) system that not only optimizes the charging of the EV/PHEV fleet but also allows the vehicles to sell electricity back to the grid to create a revenue stream for the owners, offsetting the incremental cost of including the battery system in an EV or PHEV. This incremental cost can be substantial: in 2011, the 24-kWh battery pack in the Nissan Leaf at $400/kWh costed $9600. The two-way arrows in Fig. 16-17 represent this potential flow both to and from the vehicles. For instance, the fleet of V2G-enabled vehicles might absorb a burst of electricity from renewable energy sources at low cost to the vehicle owners (since the incremental cost per kWh for installations such as solar or wind is low) and then sell it back to the grid later at a profit. V2G vehicles can also assist with *grid regulation*, that is, providing small increments of electricity to help supply more precisely match demand, a service currently provided by a certain subset of the large central power plants.

The smart grid and EV/PHEV fleet can function successfully without deployment of V2G, since the main requirement is that vehicles be connected through their chargers

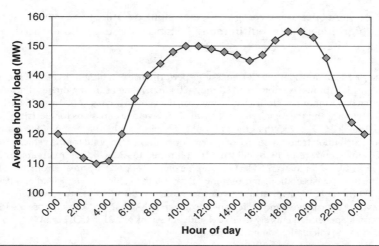

FIGURE 16-20 Average electricity demand over 24-hour period for Tompkins County region in western New York State.[6] Note that the *y*-axis scale does not start at zero so as to better show variation in hourly load values.

for extended periods of time so that the smart grid can provide charge at optimal times. In a non-V2G system, the flows from the grid in Fig. 16-19 become single-direction arrows going to the vehicles only. The addition of V2G may, however, make the system more effective and more affordable for both stakeholders in the grid and vehicle owners.

In the short run, EVs/PHEVs can take advantage of the uneven demand for electricity to charge at times of day when other sources of demand are low, even before deployment of large amounts of renewable generation or the V2G system. Figure 16-20 shows average demand by time of day for the year 2009 for the Tompkins County region surrounding Ithaca, New York, which has a population of approximately 100,000. In a pattern typical of most grid operations, a "trough" in demand exists between approximately 2200 and 0500 hours. Tompkins County has a cooler, more temperate climate than other parts of the United States, so there is less air-conditioning load and the trough is not as deep (29% decline from a peak of 155 MW at 1800 hours to low value of 110 MW at 0300 hours); in other regions, the trough may be more pronounced. In any case, low demand in the trough implies that plant capacity is available that could meet vehicle electrical charging requirements, so long as vehicles are controlled not to start charging before approximately 2200 hours. This control policy could be achieved through pricing or designing charging hardware that only allows the vehicle to charge at certain times. Figure 16-20 also shows the inherent risk in widespread electrification of the fleet without such controls: in the worst case scenario, vehicle owners might all return home around 1700 to 1900 hours and plug in their vehicles to begin charging just when household demand is at its peak, thus aggravating challenges with grid operations instead of smoothing out 24-hour demand.

---

[6]Attribution: Demand by hour of day originally compiled by Spring 2010 M.Eng project in Engineering Management, School of Civil & Environmental Engineering, Cornell University. Thanks to project team members Christina Hoerig, Dan Grew, Happiness Munedzimwe, Jun Wan, Karl Smolenski, Kim Campbell, Nicole Gumbs, Sandeep George, Tim Komsa, and Tyler Coatney.

Another concern is average CO_2 emissions per mile when vehicles switch from gasoline to grid electricity, but in many instances these can be reduced, as illustrated in Example 16-2.

Example 16-2   Consider the CO_2 emissions performance of a plug-in hybrid of the type introduced in Chap. 15, with fuel economy of 47.3 mpg (20 km/L) when running on gasoline and 5 mi (8 km) per kWh when running on electricity. In one year, the vehicle travels 12,000 mi, of which 60% are powered by electricity and the remaining 40% by the ICE. Average emissions for electricity from the local grid are 0.609 kg CO_2 per kWh at the point of generation, with an average of 8% losses in transmission and distribution from the source of generation to the charge point for the vehicle. For ICE propulsion, CO_2 emissions are 19.49 lb/gal from the tailpipe plus 2.66 lb/gal upstream, or 22.15 lb/gal total (2.66 kg CO_2/liter total = 2.34 kg CO_2 tailpipe plus 0.32 kg CO_2 upstream from the vehicle in petroleum extraction, transportation, and refining). Suppose the vehicle were instead an HEV that traveled all 12,000 mi on ICE power at the given fuel economy. (a) What are the respective CO_2 emissions for these two alternatives, and which one has lower emissions? (b) Now suppose the PHEV instead used electricity generated entirely from coal with emissions of 2.31 lb CO_2 per kWh (1.05 kg CO_2 per kWh). How do the calculations in part (a) change?

Solution

(a) We need to first adjust the emissions per kWh of electricity to reflect T & D losses, by factoring in the fraction of electricity delivered after losses are incurred:

$$\text{Emits}_{\text{Delivered}} = \frac{\text{Emits}_{\text{Plant}}}{1 - \text{Pct Loss}} = \frac{0.609}{1 - 0.08} = \frac{0.609}{0.92} = 0.662 \text{ lbCO}_2/\text{kWh}$$

Total emissions per year from EV operation of the PHEV are then

$$\text{Tot Emits} = \left(7200 \frac{\text{mi}}{\text{year}}\right)\left(\frac{1 \text{ kWh}}{5 \text{ mi}}\right)\left(0.662 \frac{\text{lbCO}_2}{\text{kWh}}\right) = 954 \frac{\text{lbCO}_2}{\text{year}}$$

Additional total emissions per year from ICE operation of the PHEV are the following:

$$\text{Tot Emits} = \left(7200 \frac{\text{mi}}{\text{year}}\right)\left(\frac{1 \text{ gal}}{47.3 \text{ mi}}\right)\left(22.15 \frac{\text{lbCO}_2}{\text{gal}}\right) = 2248 \frac{\text{lbCO}_2}{\text{year}}$$

Total emissions from the PHEV are therefore the sum of the two sources, or 3202 lb CO_2 per year, equivalent to 1.6 tons.
  For the equivalent HEV, emissions are

$$\text{Tot Emits} = \left(12000 \frac{\text{mi}}{\text{year}}\right)\left(\frac{1 \text{ gal}}{47.3 \text{ mi}}\right)\left(22.15 \frac{\text{lbCO}_2}{\text{gal}}\right) = 5619 \frac{\text{lbCO}_2}{\text{year}}$$

HEV emissions are therefore equivalent to 2.81 tons. Thus the reduction in emissions going from HEV to PHEV is 1.21 tons per year, a decrease of 43%.

(b) As in part (a), we first adjust the emissions per kWh of electricity:

$$\text{Emits}_{\text{Delivered}} = \frac{\text{Emits}_{\text{Plant}}}{1 - \text{Pct Loss}} = \frac{2.31}{1 - 0.08} = \frac{2.31}{0.92} = 2.51 \text{ lbCO}_2/\text{kWh}$$

Repeating the calculation for the 7200 mi driven in EV mode with an emissions factor of 2.51 lb CO_2 per kWh gives 3616 lbs CO_2 per year emitted. Emissions on the ICE side do not change, so total emissions are then:

$$3616 + 2248 = 5863 \frac{\text{lbCO}_2}{\text{year}}$$

This amounts to a 4% increase compared to the HEV alternative from part (a), or 244 lb CO_2 per year.

Discussion Whether the PHEV platform decreases CO_2 emissions compared to the HEV depends on many factors, including the exact fuel economy delivered by the vehicles compared, the extent of line losses in T & D, the emissions per kWh, and other factors. Some generalizations can be offered. First, for electrical grids with large amounts of gas or nonfossil energy generation, it is almost certain that PHEV will reduce emissions compared to HEV, irrespective of other factors. This is true because the primary energy source of the HEV is entirely petroleum, which is CO_2-emitting, giving PHEV an insurmountable advantage in reducing CO_2 as long as the share of carbon-free energy is large enough. Second, even in the case where coal dominates electricity production and PHEV has the potential to increase emissions, there may be motivation to introduce PHEVs. The grid provides an inroad to deliver carbon-free electricity in the future from renewables, nuclear, or fossil with CCS, so by the time PHEVs could significantly penetrate the light-duty vehicle market, emissions of CO_2 may have been partially or mostly eliminated from the grid, leading to net CO_2 reductions.

The impact of adding one PHEV at the margins resulting in a marginal change in power generation and emissions, as opposed to using the average emissions rate based on the generation mix, should be considered carefully as well. In general, renewables such as solar and wind are in scarce supply (see Fig. 6-2) and are already committed to meeting existing loads. Thus, if no new capacity is added, the additional load of recharging a PHEV will be met by conventional generation sources, affecting the outcome of the emissions analysis presented in Example 16-2. Therefore, to achieve true emissions reduction, renewable or CO_2-free generation must be developed commensurate with the influx of PHEVs. Even if the deployment is not made at first, the investment in PHEV infrastructure can be justified in the short run as preparation for the eventual deployment of increased noncarbon generation. Since one of the main motivations for electrification of the light-duty vehicle fleet is to phase out CO_2 emissions, it is reasonable to expect that growing presence of PHEVs in the fleet will serve as a catalyst for the deployment of new generation, although of course society must follow through on this expectation.

As an alternative to the smart grid system for selectively charging EVs/PHEVs, another approach is to charge batteries off-board of vehicles when electricity is available, and then swap charged batteries for discharged ones when vehicles stop at a "swapping station." This approach is being developed by the Better Place company, which opened the first demonstration swapping station in Japan in April 2010 and the first station in Denmark in June 2011. Swapping stations require compatibility between the station mechanism, the battery, and the vehicles, to which end Better Place collaborated with the Nissan/Renault conglomerate to develop Nissan Leaf and Renault Fluence Z.E. models that are compatible with the swapping system.

One potential advantage of swapping over recharging is that the swapping experience for the driver is similar to that of visiting a service station with an ICEV, where the entire operation is finished in a matter of minutes. On the other hand, since it is anticipated that EVs will at times require recharging as opposed to swapping, there is a growing market for vendors of public charge station equipment and efforts are underway to standardize the interface between charge points and vehicles. Challenges with either recharging or swapping suggest a potential advantage for PHEVs in the short to medium term, since the investment in the parallel ICE drivetrain gives the driver the security that if electrical charging or battery swapping is not available, the vehicle can still operate on liquid fuels until they are.

16-5 Understanding Transition Pathways for New Technology

Many analyses of potential energy savings from changes to the transportation system, including the preceding analysis of paper product movements, are carried out on a *comparative static* basis, meaning that a comparison is made between two static solutions, one before some change is made and one after. While this approach is useful for quickly obtaining a preliminary estimate of potential benefits of changes, and may be the only option where data of adequate quality do not exist to support a more sophisticated model, it also ignores the *transition effects* of going from one state to another.

Some of the limitations of "static analyses" that ignore transition effects include

- *Changes to baseline conditions during transition period are overlooked:* Transitions involving energy technology for transportation systems inevitably require long time horizons. During these time spans, all the underlying factors in a static analysis are subject to change, including total demand for energy, the efficiency of the incumbent technology, amount of greenhouse gas (GHG) emissions, and so on. Furthermore, in addressing situations such as climate change or potential petroleum shortages, the timing of when the energy savings and CO_2 reductions are achieved is important. A comparative static analysis does not shed light on these issues.

- *The nature of the transition itself can shape the eventual outcome:* Certain factors that act on the system during the transition may act as an obstacle to its completion in the way that is expected in the static analysis. Projections that do not consider the transition as a possible barrier may overstate the benefits of the change.

- *Where transitions depend on government policy for support, the transition may fail to take root permanently:* While superior technological performance drives some transition (e.g., from paper-and-pencil to IT systems in transportation operations planning and management), others require the intervention of government through tax policy, subsidies, or regulations. Some transitions may require the permanent intervention of government in order for the technology to attract customers, which may or may not be financially or politically sustainable. Others may require an intervention long enough for the transition to "take root," and will fail if it is too short. Where transitions fail for these reasons, the effect on energy consumption will not be the one that the static analysis predicts.

Fortunately, the research community has enough historical experience with technological transitions (e.g., the influx of a new, superior technology into a market, displacing an incumbent technology, or the government-led phasing out of an undesirable product) that we understand to some degree the shape that these transitions take, the forces at work during the transition, and even the future shape that a transition will take based on its early path. For example, transition pathways in the form of percent market penetration from first year of introduction are shown in Fig. 16-21 for major technologies such as radio, telephone, automobile, and grid electricity. Where analysts can obtain the necessary data, and have the time and resources to carry out a more careful, dynamic study of the transition, more accurate results are possible. Tools such as the logistics function or triangle curve are applied to mapping the transition pathway of new products and systems over time, leading to a more realistic understanding of when and to what extent energy savings will occur. Example 16-3 illustrates this process. (The impact of the slowdown in HEV sales after 2009 is discussed after the example.)

Example 16-3   Using the growth in hybrid sales in the United States shown in Fig. 15-12 as a starting point, consider the transition pathway to hybrid penetration into the fleet, in the following way. Suppose that the number of hybrid electric vehicles (HEVs) in the fleet, which starts in the year 2000 with 9350 vehicles added, eventually tapers off to 13,000,000 units, and that each hybrid averages 15 km/L fuel efficiency, versus 8.5 km/L for the internal combustion engine vehicle (ICEV) alternative. Consider the year 2000 to be the year $t = 0$. Sales from 2001 to 2009, in order, number 20,287; 35,000; 47,500; 88,000;

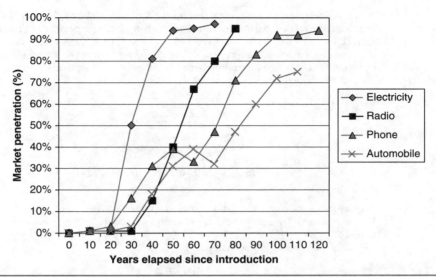

FIGURE 16-21 Market penetration in the U.S. market as a function of years elapsed since launch for electric grid connection, radio, landline telephone, and automobile.

Source: International Council on Systems Engineering (2006, p. 2–7).

207,000; 253,000; 355,000; 320,000; and 245,000. Note that the decline in the last few years is due to the severe recession of 2009. Assume that for each new sale another car is scrapped, so that the overall size of this segment of the fleet does not change, and that each car drives 16,000 km/year. (a) Use the logistics function to calculate in which year (call it year N) the number of hybrids in the fleet surpasses 99% of the 13 million target. (b) Calculate the cumulative fuel savings in liters for the shift to HEVs from years 0 to N, if the new cars are assumed to all be available on day 1 of each new year. For the years 2000 to 2009, use the sales estimated from the logistics curve model, rather than the actual sales, to calculate the number of HEVs in the fleet. (c) Compare the savings to the situation where a fleet of 13 million ICEVs is instantaneously transformed into HEVs at the beginning of year 0, and then driven to the end of year N.

Solution

(a) We begin by converting sales numbers into cumulative numbers in the fleet. The following table provides this information through year 2006:

Year	Number of Vehicles	
	Annual	**Cumulative**
2000	9350	9350
2001	20,287	29,637
2002	35,000	64,637
2003	47,500	112,137
2004	88,000	200,137
2005	207,000	407,137
2006	253,000	660,137
2007	355,000	1,015,137
2008	320,000	1,335,137
2009	245,000	1,580,137

We obtain all necessary parameters for the logistics function as follows. Based on full penetration of the 13 million market, we calculate fractional values of relative penetration, such as for 2000, $(9350)/(1.3 \times 10^7) = 0.000719$, for 2001, $(29{,}637)/(1.3 \times 10^7) = 0.00228$, and so on. The value of c_1 and c_2 is solved in a spreadsheet by calculating the error between observed and model values for years 2000 to 2009, and then minimizing the square of the error terms, resulting in $c_1 = -5.26$ and $c_2 = 0.374$. For example, for year $t = 2$,

$$f(t) = \frac{e^{(c_1 + c_2 \cdot t)}}{1 + e^{(c_1 + c_2 \cdot t)}}$$

$$f(2) = \frac{e^{(-5.26 + 0.374 \cdot 2)}}{1 + e^{(-5.26 + 0.374 \cdot 2)}} = 0.0109$$

$$HEVTotal_{yr=2002} = 0.0109 \cdot 1.3 \times 10^7 = 141{,}621$$

This estimate can be compared to the observed value of 64,637.

Plotting observed versus modeled penetration through 2009 gives the curve shown in Fig. 16-22.

In year $N = 17$, the penetration surpasses 99%:

$$p(16) = \frac{0.00072}{0.00072 + (1 - 0.00072)e^{(-16/1.38)}} = 0.988$$

$$p(17) = \frac{0.00072}{0.00072 + (1 - 0.00072)e^{(-17/1.38)}} = 0.994$$

(b) Since the total fleet size is constant at 13 million, savings accrue from replacing some fraction of the fleet with hybrids. In each year, the baseline fuel consumption with no HEV penetration is

$$\frac{(1.3 \times 10^7 \text{ vehicles})(16{,}000 \text{ km/year})}{8.5 \text{ km/L}} = 2.45 \times 10^{10} \text{ L/year}$$

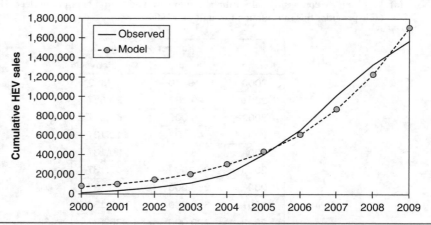

The savings in each year is then the baseline minus the actual consumption, based on the mix of HEVs and ICEVs in the fleet. Taking year 2 as an example again, the HEV and ICEV fuel consumption values are, respectively:

$$\frac{(3.992 \times 10^5 \text{ vehicles})(16,000 \text{ km/year})}{15 \text{ km/L}} = 4.26 \times 10^7 \text{ L/year}$$

$$\frac{(1.3 \times 10^7 - 3.992 \times 10^5 \text{ vehicles})(16,000 \text{ km/year})}{8.5 \text{ km/L}} = 2.44 \times 10^{10} \text{ L/year}$$

The total fuel consumption is therefore little changed from the baseline, at 2.444×10^{10} L.

Repeating the calculation of annual savings for each year and summing to quantify cumulative savings gives the following graph of fuel saved from 2000 through 2017, as shown in Fig. 16-23. Based on these calculations, at the end of year 17, the savings has reached 8.01×10^{10} L.

(c) An instantaneous transition to 13 million HEVs gives the following energy consumption per annum:

$$\frac{(1.3 \times 10^7 \text{ vehicles})(16,000 \text{ km/year})}{15 \text{ km/L}} = 1.39 \times 10^{10} \text{ L/year}$$

On this basis, the annual savings is $(2.45 \times 10^{10}) - (1.39 \times 10^{10}) = 1.06 \times 10^{10}$ L/year. Including the savings in year 0, the project has an 18-year time span. Therefore, the cumulative savings is $(1.06 \times 10^{10})(18 \text{ years}) = 1.91 \times 10^{11}$ L of fuel.

Discussion Comparing the results of parts (b) and (c) in the example shows that during the time of the project, considering the transition period results in a 58% reduction in the projected savings. Thus the benefits of reducing the consumption of petroleum or emissions of CO_2 would accrue much more slowly when one takes into account a realistic amount of time for the new vehicle technology to penetrate the market.

More realism could be added to this example by taking into account other real-world factors. Both HEV and ICEV fuel economy are likely to change over time. Also, stratifying the fleet by vehicle type,

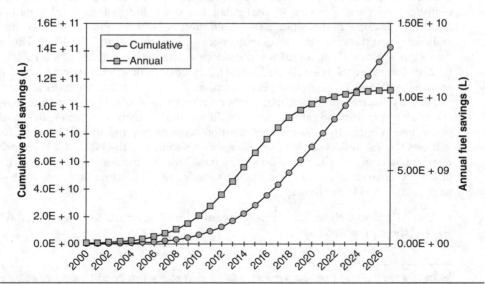

FIGURE 16-23 Annual and cumulative fuel savings, 2000–2027.

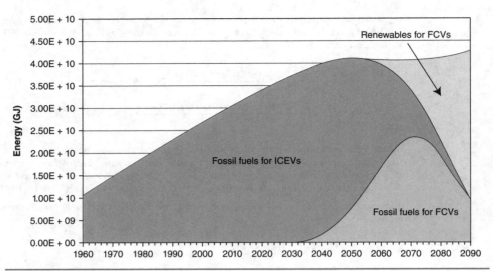

FIGURE 16-24   Scenario for the influx of hydrogen fuel cell vehicles and carbon-free transportation energy in the twenty-first century, with estimate of world energy requirement.

[*Source:* Ingvarsson et al. (2011).]

and then tracking sales and fuel economy of each vehicle type would make the projections of fuel savings more realistic. Furthermore, by the end of year 17, vehicles purchased in year 1 would likely be at the end of their useful lifetime, so an accounting of the turnover of both types of vehicles as part of the model would add accuracy.

Over very long time spans, it may not be possible to predict a single "correct" pathway for the unfolding of a technological transition. Indeed, the HEV pathway in Example 16-3 is illustrative of major technological transitions in general, but may not be realistic for hybrid vehicles in particular, based on the historical sales pathway in Fig. 15-12. It appears that after 2009 penetration of hybrids into the market may have stalled, so that the saturation level may be lower than the example predicts. Therefore, a "scenario approach" is useful for considering alternative combinations of factors to bracket the range of possible pathways for indicators of interest such as total energy required or total CO_2 emitted. For example, a complete transition to a propulsion technology for transportation that emits no net emission of CO_2 to the atmosphere but meets all of the demand on the planet could last until sometime between 2050 and 2100, or beyond. Figure 16-24 shows a transition scenario for the introduction of fuel cell vehicles (FCVs), followed by nonfossil resources to power the FCVs, for the worldwide fleet of passenger cars from 1960 to 2090, based on assumptions about the relative fuel efficiency of internal combustion engine vehicles (ICEV).[6] Thus there are three stages, each color coded in the figure:

- *FF (fossil fuel) for ICEV:* Energy required for fossil fuels to power ICEV, which is the current technology.

---

[7]*Acknowledgment:* This figure is a result of collaboration with Julien Pestiaux and Audun Ingvarsson, M.Eng students, 2003–2004, whose contribution is gratefully acknowledged.

- *FF for FCV:* During the first phase of ramping up the presence of FCVs in the world vehicle market, the hydrogen required is derived from fossil fuels.
- *RE (renewable energy) for FCV:* The hydrogen for the FCVs is derived from renewable energy, or some other source that does not emit net CO_2 to the atmosphere.

The pathway shows that if FCVs only enter the market in 2030 and it takes this technology 60 years until 2090 to completely replace ICEV, then energy required from fossil fuels will continue to grow until 2050, reaching a value of approximately 41 EJ, compared to 27 EJ in the year 2000. Furthermore, a transition to hydrogen from renewable energy starting in 2050 and lasting 60 years (off the right side of the figure) would still require 10 EJ from fossil fuels for making hydrogen in the year 2090. The overall lesson from this scenario is that if the transition to alternative-fuel vehicles (AFVs) without CO_2 emissions starts many decades from now, and present growth in worldwide vehicle·km continues, there is cause for concern that fossil fuel use and CO_2 emissions from transportation may grow substantially before they begin to decline.

The FCV is chosen for the scenario analysis in Fig. 16-24 for illustrative purposes only; other AFV technologies could be analyzed in the same way. Another good candidate is the PHEV, which might use a mixture of fossil and nonfossil electricity in a "bridge period" through the middle of the century until sufficient CO_2-free electricity generating capacity could be developed to power the world's fleet of PHEVs.

16-6   Toward a Policy for Future Transportation Energy from a Systems Perspective

In Chaps. 15 and 16, we have discussed a range of energy technologies for propulsion in transportation, and also various systems effects that influence the implementation of these technologies in a transportation system. In this concluding section, we combine technology and systems perspective into a discussion of policies for transportation energy in the future, focusing on two topics: (1) a portfolio of options for a metropolitan region seeking more energy-efficient transportation, and (2) a plan for allocating emerging energy resources and technologies to different transportation functions, modes, and geographic domains.

16-6-1   Metropolitan Region Energy Efficiency Plan

In this section, a preliminary concept is presented for incorporating all of the options that are discussed in the body of this chapter that are available in the near to medium term (on the order of 10 to 30 years) into a transportation energy plan for a metropolitan region. This plan applies to a typical metropolitan region (hereafter referred to as "the region") in an industrialized country in Europe, North America, or Asia, which seeks to reduce total transportation energy consumption across the board. The term metropolitan region implies the urban core of the city and surrounding suburban and exurban developments out to the perimeter of the developed area. For many cities, this definition includes both the area within the politically-defined city limits, and the built-up areas surrounding the city limits. The plan focuses on cities in industrialized countries because of their high per capita transportation energy use. Many cities in the emerging countries could also benefit from some parts of the plan. Notwithstanding the limitations on static analyses discussed in Sec. 16-5, the plan is static in nature, suggesting end targets for reducing energy consumption from each option in percentage terms, but not considering how long implementation might take or what might happen to baseline energy consumption in the mean time.

The plan incorporates measures for both passenger and freight transportation. On the passenger side, options including the transformation of the light-duty vehicle fleet to more efficient models of vehicles, expansion of the use of public transportation, development of the use of limited-use vehicles and nonmotorized modes, and changes to land use patterns that can be implemented in the short to medium term. On the freight side, the region works with firms that provide movement of goods to, from, and within the region to use more energy-efficient modes, and also works with businesses within and outside of the region to replace some of the distant sources of goods with others that are closer. The plan does not include technologies such as fuel cell vehicles that may not be available for some time.

The summary of the plan is shown in Table 16-8. In the table, the percentage values given are compared to the future baseline value of total transportation energy consumption. The range of values shown are the author's assessment of a plausible range of savings values, since no previous study was identified that considered all alternatives available to a metropolitan region; individual statistics are provided to support components of the plan, where possible. Several points can be made about the table:

- The plan in the table does not address the potential increase in baseline passenger·km and tonne·km that might occur during the 10 to 30 year implementation period. As in Example 16-1, it is possible for an efficiency option to reduce energy consumption relative to the future baseline, and still have the resulting future value be higher than the current value. In a region experiencing rapid population growth, increases in population and growing demand per capita might wash out all the improvements achieved by the plan. If this happened, the region would be better off than if no steps had been taken, but no closer to reducing the absolute value of its transportation energy consumption footprint.

- The implementation of not just one of the options, but most or all of them, however, makes it more likely that the absolute amount of energy consumption will decrease.

- The upper bounds of the percentage range for each option are intentionally made to be ambitious. It is likely that the twin challenges of reducing CO_2 emissions and preserving petroleum resources will motivate cities to move toward unprecedented levels of reductions.

- Although the projected percent savings are higher for some options than others, none of them are mutually exclusive of one another, so all are worthy of pursuit.

- No attempt is made to calculate a total energy savings value on the basis of the percent values provided. Such an analysis would require a careful treatment of the interaction between different options (e.g., once the fleet has been transformed into a more efficient one, the total energy savings available from modal shifting away from light-duty vehicles is less than the baseline), and is beyond the scope of this text.

In conclusion, the table does not result in a calculation of how much energy could be saved overall by the region. However, as an initial impression, the values in the table

Function	Efficiency Option	Potential Savings	Rationale
Passenger	Improved efficiency of light-duty fleet: more efficient ICEVs, and increase in share of EVs, HEVs, and PHEVs.	10–25%	Current and proposed improvements in fuel efficiency standards in all major auto markets of the world; efficiency improvement and rate of market penetration by HEVs during the period 2000–2007.
	Expanded (quasi-)public transportation: new rapid transit systems, expanded bus service, support for carpooling and vanpooling*	5–15%	Up to 50% reduction in energy use possible for U.S. public transportation compared to car/ passenger · km. 5–15% figure assumes limited ability to add transit and/or carpools to all areas of region. Ability to reduce energy consumptions assumes sufficient load factors in alternatives to cars.
	Motorcycles, LUVs and nonmotorized modes: motorcycles, mopeds, scooters, limited-use EVs, expanded use of bicycling and walking through improved infrastructure	2–10%	30–40% of all trips in Amsterdam, Netherlands, by bicycle. Rate of bicycle trips in Melbourne, Australia, increases fourfold in 20 years. Untapped market for electric LUVs in industrialized cities.
	Short and medium term land use changes: locally available shops and facilities, telecommuting centers, revitalizing traditional shopping districts	2–15%	California "smart growth" plan could achieve 3–10% reduction in vehicle · km and energy use compared to baseline by 2020.[8] Smart growth concept is widely applicable in the United States and other countries.
Freight	Encouraging modal shifting: greater use of intermodalism, supporting the development of intermodal terminals within the region	2–10%	Projection based on savings per tonne · km moved by intermodal vs. truck, and maximum applicability of intermodal (not all sources of goods are served by intermodal service).
	Rationalizing freight demand: substituting near for far sources, developing local resources, e.g., local food production	2–10%	Types of products most easily substituted (foods, farm products, certain building products, etc.) and their relative contribution to freight energy consumption—see Figs. 16-10(a) and (b).

Percentage reduction figures are for total urban transportation energy consumption relative to future value for baseline ("do-nothing") scenario; see explanation in text.

*Note that carpools and vanpools rely on collaboration between private individuals to share vehicles, but can be supported by local and regional governments, and are therefore labeled quasi-public in this option.

TABLE 16-8   Range of Possible Percent Reduction Values for Energy Consumption from Efficiency Options Available to a Metropolitan Region

[8]Estimated savings published by California Energy Commission (2001).

support a potential overall reduction ranging from 20 to 40%. These savings would lead to substantial reductions in CO_2 emissions, even if most of the remaining demand for motorized transportation was powered by fossil fuels. Over the longer term, the region might shift the remaining transportation energy consumption over to a source that did not increase CO_2 in the atmosphere (using the types of technologies discussed in Chap. 15).

16-6-2   Allocating Emerging Energy Sources and Technologies to Transportation Sectors

In the future, and especially over the longer term beyond 20 or 30 years, alternative fuels such as biofuels, electricity, and/or hydrogen are expected to expand into the marketplace to provide a carbon-free substitute for petroleum products. It makes sense to allocate each to its application of comparative advantage, especially taking into account the function, mode, and geographic scope of the transportation application in question.

Current technological characteristics suggest that electricity and/or hydrogen will hold an advantage over biofuels in urban transportation, where the distances traveled between opportunities to recharge are not as great. Whether an HFCV bus or a plug-in HEV, either one is well suited for urban movements during the day followed by refueling or recharging at night. These applications also favor the new infrastructure needed to support the energy source, since there is a high concentration of vehicles in each urban region that can more easily support the infrastructure cost of each new recharging station or hydrogen filling station. Assuming that a cost-effective means of generating carbon-free electricity or hydrogen can be developed, there will not be a space constraint on generating large quantities of these energy carriers for transportation. Therefore, it should be possible to meet the large energy demand that urban transportation requires with these sources.

Electricity and hydrogen in the urban setting may have additional benefits for the electric grid. As discussed in Sec. 16-4-3, plug-in hybrids might dock at workplace charging stations and discharge electricity stored the previous night into the grid during the middle of the day, helping to meet peak electric demand, especially on sunny summer days. Similarly, HFCVs might take a supply of hydrogen supplied to the workplace parking facility and convert it to electricity to be supplied to the grid rather than used for propulsion.

Biofuels and hydrogen, by contrast, have an advantage over electricity for long-distance use in that they can more easily match the long range per refueling of petroleum-derived gasoline or diesel, and are therefore well suited for intercity travel. In the United States, for example, there is a good match between the 5 to 6 EJ/year energy requirement for the surface intercity freight transportation system and the maximum output from robustly developed biofuel production facilities that may be possible in the future without harming the capability to grow enough food. A situation may emerge where the biofuel output is largely allocated to freight, and is able to meet most or all of the intercity demand. However, biofuels alone may not be able to meet both freight and intercity passenger energy demand. At the time of this writing, it is not clear how this gap might eventually be filled, but it is possible that the expanded use of electric catenary on the railroads might help to ease some of the pressure, since the electricity could come from diverse sources, and the railroads might increase their volume of both freight and passenger transportation. Hydrogen might fill some of this demand as well.

Turning to air transportation, the aviation industry is slowly making progress in developing an alternative fuel for jet-powered aviation, but substantial challenges remain. This is in part due to the difficult conditions under which jets operate at cruising altitude, where it is more difficult for alternative fuels to function well because of cold temperatures. Also, the higher cost of alternative fuel would have a large economic effect on the industry, since energy costs are especially a large part of the overall operating cost of commercial jetliners. Nevertheless, the industry is interested in developing alternatives over the long term to assure its survival. An early milestone was the use for the first time in 2006 of a mixture of traditional jet fuel and Fischer-Tropsch synthetic fuel derived from natural gas in a U.S. Air Force jet during a routine training flight. Other milestones include the first-time use of biofuels in a commercial airliner by Virgin Atlantic Airlines in February 2008, and the development and testing by Boeing in 2008 and 2009 of a biofuel-based aviation fuel that meets or exceeds jet fuel standards. By 2015, a number of airlines, including Virgin Atlantic and All-Nippon Airways, were using biofuels in small quantities to displace conventional jet fuel. These steps represent progress, but the challenge of developing an adequate supply of bio-feedstocks remains. Given the time required to develop a mature supply chain for aviation biofuel that can meet most or all of demand, one strategy is to give aviation priority access to petroleum resources and emphasize the development of alternative fuels for other transportation modes so as to extend the lifetime of the petroleum supply.

Lastly, barring unanticipated technological developments, or a change in world political and economic priorities, we can expect that progress will be slow in making large reduction in the absolute amount of CO_2 emitted by worldwide transportation up to the middle of the twenty-first century. More efficient technologies that use petroleum may curb or reverse the growth in emissions, and systems-wide changes such as modal shifting may contribute as well. However, carbon-free transportation energy systems that can meet the majority or most of the world's demand are still some time away. One possible interim strategy is to focus extra effort on nontransportation CO_2 emissions reduction, using options discussed in Chaps. 6 to 14, to offset emissions from the transportation sector. Ideally, the community of nations will accelerate reduction in nontransportation CO_2 while at the same time ramping up adoption of sustainable, carbon-neutral transportation energy systems.

16-7 Summary

Reducing energy consumption and GHG emissions from transportation has both a technological and a systems component, and systems effects strongly influence the total energy required. Transportation systems can be understood from many perspectives, including function, mode, geographic scope, and ownership. Energy efficiency can in turn be evaluated using measures such as kJ/vehicle · km, kJ/passenger·km, and kJ/tonne·km, depending on the circumstances. Although world energy use has been growing steadily in recent decades, both passenger and freight transportation energy consumption have been on an even more rapidly increasing trend, as modern land use patterns encourage the use of the automobile, and the global economy allows goods to move all around the world at increasing speeds. Much effort is currently being made to reverse this trend through government actions, for example, to encourage use of more environment-friendly modes such as rail and water. As we look to introduce new technologies in the future to reduce energy consumption, we should take into account transition effects that influence the rate at which technologies can enter the market and the rate at which energy can be conserved.

References

California Energy Commission (2001). *California Smart Growth Energy Savings: MPO Survey Findings.* Consultant Report Prepared by Parsons-Brinckerhoff. California Energy Commission, Sacramento, CA.

Davis, S., and S. Diegel (2014). *Transportation Energy Data Book,* 33rd ed. Oak Ridge National Laboratories, Oak Ridge, TN.

Ingvarsson, A., J. Pestiaux, and F. Vanek (2011). "A Global Assessment of Hydrogen for Future Automotive Transportation: Projected Energy Requirements and CO_2 Emissions." *International Journal of Sustainable Transportation*, Vol. 5, Num. 2, pp. 71–90.

International Council on Systems Engineering (2006). *Systems Engineering Handbook: A Guide for System Life Cycle Processes and Activities.* INCOSE, Seattle, WA.

U.K. Department for Transport (2015). *Transport energy and environment statistics.* Electronic resource, available at "http://www.gov.uk/government" www.gov.uk/government. Accessed June 12, 2015.

Vanek, F., and E. Morlok (2000). "Reducing US Freight Energy Use Through Commodity Based Analysis: Justification and Implementation." *Transportation Research Part D,* Vol. 5, No. 1, pp. 11–29.

Further Reading

Bhadra, D., and R. Schaufele (2007). Probabilistic forecasts for aviation traffic at FAA's commercial terminals. *Transportation Research Record* Number 2007, pp. 37–46.

Davis, S., and S. Diegel (2003). *Transportation Energy Data Book,* 22nd ed. Oak Ridge National Laboratories, Oak Ridge, TN.

Davis, S., and S. Diegel (2007). *Transportation Energy Data Book,* 26th ed. Oak Ridge National Laboratories, Oak Ridge, TN.

Department for Transport (2006). *Transport Statistics Great Britain.* DfT, London.

Earth Policy Institute (2002). *Eco-Economy Indicators: Bicycle Production breaks 100 million.* Earth Policy Institute, Washington, DC. Web resource, available at www.earth-policy.org/Indicators/indicator11.htm. Accessed Nov. 14, 2007.

Greene, D., and Y. Fan (1995). "Transportation Energy Intensity Trends: 1972–1992." *Transportation Research Record*, No. 1475, pp. 10–19.

Greene, D. (1996). *Transportation and Energy.* Eno Foundation, Washington, DC.

Kempton, W., and J. Tomic (2005). "Vehicle-to-Grid Power Fundamentals: Calculating Capacity and Net Revenue." *Journal of Power Sources*, Vol. 144, pp. 268–279.

Kempton, W., and J. Tomic (2005). "Vehicle-to-Grid Power Implementation: From Stabilizing the Grid to Supporting Large-Scale Renewable Energy." *Journal of Power Sources*, Vol.144, pp. 280–294.

Leiby, P., and J. Rubin (2004). "Understanding the Transition to New Fuels and Vehicles: Lessons Learned from Analysis and Experience of Alternative Fuel and Hybrid Vehicles." Chap. 14. In D. Sperling and J. Cannon, Eds., *The Hydrogen Energy Transition: Moving Toward the Post-Petroleum Age in Transportation*, pp. 191–212.

Lovins, A., and B. Williams (2001). "From Fuel Cells to a Hydrogen Based Economy: How Vehicle Design is Crucial to a New Energy Infrastructure. Fortnightly." *The North American Utilities Business Magazine*, Vol. 139, No. 4, pp. 12–22.

Schaefer, A. (1998). "The Global Demand for Motorized Mobility." *Transportation Research Part A*, Vol. 32, No. 6, pp. 455–477.

Schiller, P., E. Bruun, and J. Kenworthy (2010). *An Introduction to Sustainable Transportation: Policy, Planning, and Implementation*. Earthscan, London.

Schipper, L., L. Scholl, and L. Price (1997). "Energy Use and Carbon Emissions from Freight in 10 Industrialized Countries: An Analysis of Trends from 1973 to 1992." *Transportation Research Part D*, Vol. 2, No. 1, pp. 57–75.

Sperling, D., and D. Gordon (2008). *Two Billion Cars: Driving Toward Sustainability*. Oxford University Press, Oxford.

Theis, M., N. Cass, and J. Corbett (2004). Role of technology in achieving environmental policy goals in the maritime transportation sector. *Transportation Research Record* 1871, pp. 42–49.

U.S. Bureau of Transportation Statistics (1996). *Commodity Flow Survey 1993: United States*. U.S. Department of Transportation, Washington, DC.

U.S. Bureau of Transportation Statistics (2004). *Commodity Flow Survey 2002: United States*. U.S. Department of Transportation, Washington, DC.

U.S. Dept. of Energy (1994). *Manufacturing Consumption of Energy 1991*. Energy Information Agency, USDOE, Washington, DC.

Vanek, F., and J. Campbell (1999). "UK Road Freight Energy Use by Product: Trends and Analysis from 1985 to 1995." *Transport Policy*, Vol. 6, pp. 237–246.

Vanek, F. (2001). "Growth of Exports from Developing Countries: Implications for Freight Trends and Ecological Impact." *Futures*, Vol. 33, pp. 393–406.

Vuchic, V. (1999). *Transportation for Livable Cities*. Center for Urban Policy Research, New Brunswick, N.J.

Vuchic, V. (2005). *Urban Transit: Operations, Planning and Economics*. Wiley, New York.

Vuchic, V. (2007). *Urban Transit Systems and Technology*. Wiley, New York.

Exercises

16-1. A retail firm operates a *decentralized* distribution system (factory to warehouse to shop) in which the system uses six smaller warehouses distributed around a region to receive a product from manufacturer (called *primary distribution*) and then send product to retail outlets (called *secondary distribution*). The firm is offered the opportunity to shift to a *centralized* system in which each unit of product will still undergo primary and secondary distribution, but now there will only be one warehouse in the middle of the region. The transportation of the product incurs financial cost and energy consumption per vehicle·km (v·km) of movement. In the case of the warehouse costs, inventory costs are incurred by virtue of needing to keep stock on hand in the warehouse between the time that the stock is purchased and the time it is sold, and energy is consumed to operate the warehouse. Assume that all other costs and rates of energy consumption are the same for either option.

   a. Based on the data given, determine whether the decentralized or centralized system is preferred.

   b. What is the environmental dilemma underlying this decision? Discuss.

Warehouse Costs and Energy Use Per Warehouse		
	Inventory Cost ($1000/year)	Energy (GJ/year)
Decentralized	190	155
Centralized	710	730

Transportation Volume Generated Per Year Per Warehouse (1000 vehicle·km/year)		
	Primary	Secondary
Decentralized	35	82
Centralized	181	645

Transportation Cost and Energy Use		
	Cost ($/1000 v·km)	Energy (MJ/1000 v·km)
All shipments	1500	19,000

Data:

16-2.   The standard unit equivalent of the data given in Table 16-2 is given below. Carry out a Divisia analysis and produce a figure showing the influence of efficiency and structure, with years from 1970 to 2000 on the x-axis and billion gallons of fuel on the y-axis.

	Passenger Cars		Light Trucks	
	10^9 mi	10^9 gal fuel	10^9 mi	10^9 gal fuel
1970	917	68	123	12.3
1975	1034	74	201	19.8
1980	1112	70	291	23.8
1985	1247	71.5	391	27.4
1990	1408	69.6	575	35.6
1995	1438	68.1	790	45.6
2000	1600	73.1	923	52.9

16-3.   Show that the data in Table 16-2 result in the Divisia analysis graph shown in Fig. 16-7.

16-4.   Below are the data showing the trends for U.S. trucks and railroads for the period 1980 to 2008, showing volume in billion tonne·km and energy in EJ. Use Divisia decomposition to create a table and a graph for the period 1980 to 2008, showing four curves: (1) actual fuel consumption, (2) trended fuel consumption, and the contribution of (3) energy intensity, and (4) structural changes to the difference between actual and trended fuel consumption.

Year	Tonne·km		Energy	
	Truck (billion)	Rail (billion)	Truck (EJ)	Rail (EJ)
1980	836	1337	1.89	0.583
1985	925	1276	1.96	0.485
1990	1045	1504	2.17	0.478
1995	1194	1900	2.40	0.469
2000	1524	2132	3.24	0.555
2005	1873	2467	6.17	0.602
2008	1909	2585	6.33	0.572

16-5.   A food products company ships foods from three production plants to four markets as follows. The capacity of plants at Boise, Dubuque, and Charleston is 2200 tonnes, 3000 tonnes, and 2000 tonnes, respectively. The demand at San Francisco, New York, Miami, and St. Louis is 2002 tonnes, 1784 tonnes, 1355 tonnes, and 1972 tonnes, respectively. The mode of shipment is by truck, and the energy intensity is 2200 kJ/tonne·km. A table of distances between cities is given below.

   a. What is the allocation of shipments from plants to markets that meets all demands, does not exceed supplies available at any plant, and minimizes energy consumption? What is the value of energy consumption in this case?

   b. Suppose that for all routes with distance of 1500 km or more, an intermodal rail service is made available with energy intensity of 1400 kJ/tonne·km. Recalculate part (a). What is the new value of energy consumption? Does the shipment pattern from part (a) change?

Distances in km:

	Boise	Dubuque	Charleston
San Fran.	1427	3763	4299
New York	5038	1685	1782
Miami	5458	2262	548
St. Louis	3855	390	1431

16-6.   Park rangers in a region live in communities of the region and commute 5 days/week to four regional parks. In a day, each ranger must make one round trip from her/his home to one of the forests and back again. Due to the need for flexibility on the job, the rangers must drive in their own cars and cannot carpool. The number of rangers in Bayside, Mountainview, Springfield, and City Heights is 24, 28, 22, and 42, respectively. The number of rangers required in the parks of Sandy Beach, Lookout Mountain, Lone Tree, and Endless Forest is 23, 41, 19, and 33, respectively. The rangers' cars average 9.3 km/L fuel economy, and the one-way distances are given in the table below.

   a. In one allocation of rangers to parks, the rangers rotate through the four parks over a 4-week period such that the amount of time spent in each park is proportional to the percent of slots represented by that park. For example, if a park requires 20% of the rangers on any given day, then each ranger will spend 20% of their trips going to that park, to the nearest whole number. Calculate the liters of fuel consumption per month for this allocation.

b. The regional park supervisor decides to reduce energy consumption by reallocating rangers to minimize monthly driving. What is the new value of fuel consumption per month for this allocation? How much fuel is saved compared to the solution in part (a)?

c. From a job satisfaction point of view, what is the limitation on the solution in part (b)? How might it be addressed? Discuss.

d. The use of SOV driving in parts (a) and (b), while allowing flexibility, could be criticized for excessive fuel consumption, especially given the environmental focus of the regional parks organization. Describe a program that might facilitate carpooling while still meeting the demand for rangers and allowing some level of flexibility in vehicle use.

(One-way Kilometers)	Bayside	Mountainview	Springfield	City Heights
Sandy Beach	41	33	50	62
Lookout Mountain	50	23	22	51
Lone Tree	80	65	49	76
Endless Forest	33	48	45	19

16-7. Revisit Example 16-3, using the same fuel economy values for ICEVs and HEVs, but this time considering the entire U.S. passenger car fleet. Suppose that HEVs achieve 50% penetration of the national fleet by 2040. The fleet in 2000 consisted of 134 million vehicles; assume this number is fixed for the duration of the transition. Compare this transition to a scenario where there is no influx of HEVs.

a. Calculate the cumulative fuel savings for the period of 2000 to 2040 using a triangle function with peak rate of change in 2020.

b. Calculate the cumulative fuel savings for the period of 2000 to 2040 using a logistics function. Fit the logistics function to the data in years 2000, 2009, and 2040 only, with the assumed value in 2040 of 50% of the fleet, that is, 67 million vehicles.

c. Compare the results from calculations in parts (a) and (b). Discuss the differences.

16-8. Repeat Problem 16-7, but this time with changing overall fleet size and fuel efficiency. Assume that both the fleet size and average fuel economy grow linearly at the average annual rate from 1980 to 2000. Also, assume that HEV fuel economy improves by the same percent each year as the ICEV fleet for the period of 2000 to 2040. Obtain the necessary data to extrapolate rates of change from the Internet or other source.

16-9. Extend the logistics rationalization concept of Sec. 16-4-2 and Tables 16-6 and 16-7 by considering a situation where the assignment of freight from origin to destination is fixed but the routing can be varied to reduce energy consumption and CO_2 emissions. There are four cities in the network, with distances shown in the table below, as well as origin-destination demand levels. Trucks can be sent either directly from origin to destination, or shipments can be consolidated and sent to Syracuse, which is in the middle of the network and represents a "hub", where shipments are routed to their final destination on a different shipment leg. Trucks have a capacity of 125 arbitrary freight units, cost $1.50 per mile to operate, and average 5.8 mpg diesel. Each arbitrary unit of freight weighs 200 lb (in other words, the capacity of 125 units weighs 25,000 lb or 12.5 short tons). (a) Solve the model for three options, option 1 where there is no use of the hub and all shipments must go via direct truck, option 2 where all shipments must move via the hub, and option 3 where the lowest cost combination of running direct and via the hub can be chosen. Compute total cost, total energy consumption, total emissions, and average energy intensity per ton-mile for all three options. By how much does option 3 reduce these metrics compared to the other two? (b) Discuss ways in which the problem as posed is simplistic and overlooks real-world details that would make choice of the optimal solution more complex. There are several possible answers.

Demand:	Rochester	Syracuse	Watertown	Binghamton
Rochester	0	225	146	111
Syracuse	225	0	191	187
Watertown	146	191	0	97
Binghamton	111	187	97	0

Distances:	Rochester	Syracuse	Watertown	Binghamton
Rochester	0	87	134	140
Syracuse	87	0	70	73
Watertown	134	70	0	143
Binghamton	140	73	143	0

CHAPTER 17

Conclusion: Creating the Twenty-First-Century Energy System

17-1   Overview

This chapter first reviews and compares the motivations, tools, and technologies considered in this book. It also describes some emerging energy technologies that were not given extensive treatment in earlier chapters, but that show potential for the future. Next, scenarios for future growth of energy production are developed that illustrate how the world's energy systems might be transformed as a response to the challenges we face in the twenty-first century. Lastly, the chapter describes ways that the energy professional can support the development of a more sustainable energy future, both in the professional arena and through extracurricular activities.

17-2   Introduction: Energy in the Context of the Economic-Ecologic Conflict

The discussion of qualitative tools in Chap. 2 includes the role of stories, scenarios, and models in describing energy engineering challenges or evaluating possible solutions. We therefore initially consider a story with both economic development and environmental protection elements as a way to conclude the discussion of technologies and systems in this book.

Running deeply beneath all efforts to both deliver the required services of energy systems and protect the natural environment is a fundamental tension between the economic cost of foregoing business opportunities versus foregoing steps to protect the health and well-being of humans and other living beings, illustrated as follows:

A leather dealer builds a tannery just outside a small town that has no major business. Tanneries use chemicals and other materials that have unpleasant and penetrating odors, and the fumes can be irritating and dangerous to health. Following both law and common sense, the tannery was located downwind, to the east of town. The tannery became very successful and helped the town prosper. Eventually, as the town grew, houses were built to the east of the tannery, and the people who moved in found themselves breathing foul air. They brought a suit against the tannery, demanding that it either modify its operations or move further east. The tannery owner claimed that doing either would be too expensive and might force him to close his facility and move to another town.[1]

Several threads common to many challenges with sustainable energy are apparent from the parable. First we have the situation of the town prior to the arrival of the tannery, which could be likened to pre-industrial societies that existed in many parts of the world prior to the year 1800, including Europe, the Americas, Asia, and Africa. These societies developed varying degrees of sophistication but did not accumulate wealth to the extent that industrial societies of today do because they did not have access to modern energy sources. The growth of the tannery represents the initial success of industrialization in bringing prosperity.

The situation once the lawsuit is underway represents the conflicted situation of today. The plaintiffs in the lawsuit against the tannery owner might represent all people everywhere who are critical of the negative effects of energy systems, including air and water pollution, the negative effect of space occupied by energy infrastructure in the built environment, species loss due to habitat destruction, and other issues. Naturally they demand that decision-makers with authority over the energy system, represented in the parable by the tannery owner, take action to protect their health. These decision-makers include elected officials as the most visible element but also career civil servants involved with transportation and leaders in the private sector who provide energy resources, conversion devices, and infrastructure.

The paradox and dilemma is that the same tannery that is harming the plaintiffs is also providing their livelihood. Prior to the arrival of the tannery, many of the advantages of economic prosperity were not available to the town. The risk as laid out by the owner is that steps taken to modify or move the tannery will lead to its economic undoing. In the same way, there is a risk that changes to the energy system may be so expensive that the primary effect is not to improve health but to derail the economy.

An additional factor that would apply if the tannery were operated at any time in recent history is the question of fossil fuel consumption and greenhouse gas emissions. Assuming the tannery uses coal or natural gas for heat, or is connected to the electric grid, or receives supplies or ships finished product using motor vehicles, it is contributing to the release of CO_2 into the atmosphere. Conceivably, one could operate a tannery using only renewable energy, but if we consider tanneries that have operated around the world in the last 100 to 200 years, virtually all of them have used fossil fuels in one form or another. This situation becomes a further problem and challenge for both the tannery owner and the aggrieved townspersons: not only must they consider the direct pollution from the tannery in the form of noxious chemicals and fumes, but the tannery on which the town economy depends is contributing to climate change.

This brings us to the situation of today: an unprecedented size of human population, an unprecedented world per capita energy consumption value—and both growing steadily and inexorably from year to year. Before us lie three critical challenges: how to spread the benefits of access to energy more evenly, how to contend with the eventual exhaustion of nonrenewable energy resources, and how to prevent climate change, in large

---

[1]From Bentley, P. (1998),"Business and Environment: A Case Study," pp. 225–226, in *Ecology and the Jewish Spirit: Where Nature and the Sacred Meet*, copyright 1998 Ellen Bernstein (Woodstock, VT: Jewish Lights Publishing). Permission granted by Jewish Lights Publishing, P.O. Box 237, Woodstock, VT 05091, *www. jewishlights.com.*

part caused by increasing concentration of CO_2 in the atmosphere. It is the fundamental tenet of this book that a successful balance between the two sides of the dilemma can be found. In so doing, the goal of sustainable development with its three pillars of ecology, economy, and society can be achieved (Chap. 2).

17-2-1   Comparison of Three Energy System Endpoints: Toward a Portfolio Approach

For the purposes of this book, alternatives for energy systems are reduced to three fundamental energy system endpoints, namely, fossil, nuclear, and renewable, each of which is marked by fundamental differences from the other two. The three endpoints are compared in Table 17-1 in terms of longevity, economic cost, and environmental cost. Note that transportation energy is not considered to be a fourth endpoint, since ultimately all the energy that is used to propel vehicles must come from one of the three chosen endpoints. Instead, suitability for transportation energy is treated as an additional point of comparison in the table.

Several conclusions can be drawn from the table. First, all three pathways can provide energy over very long periods of time into the future, assuming the underlying technologies evolve to the full extent that is anticipated. In the case of renewable, or nuclear energy, this time period is almost indefinite. For fossil fuels, it is on the order of centuries, although it may be difficult to use them in an environmentally sustainable way over such a long time period.

Second, there are environmental impacts other than climate change associated with any of the three pathways that are not well quantified at present, but that are likely to be substantial. Often these impacts revolve around the effect of extracting raw materials or disposing of by-products, whether from the extraction of coal or uranium resources or of materials needed to manufacture renewable energy systems. For example, an extraction operation such as a mine or quarry disrupts the surrounding ecosystem and may poison water resources during its operational lifetime, even if it is restored to its natural state once the mining operation is complete. Related to these impacts are the risks associated with possible negative effects from radioactive by-products from the nuclear energy industry.

Lastly, there are advantages to a *portfolio approach*, in which energy is generated from a portfolio of different options. In the same way that a financial investment portfolio protects the investor from the failure of any one investment, diversity of energy sources protects us from the risk that a fundamental weakness in a single energy source might disrupt the meeting of energy requirements. This is especially relevant in the case of energy systems because of the uncertainties that surround each of the pathways. Many of the key technologies, such as carbon sequestration, or nuclear fusion, are still at a very early stage. Another type of portfolio approach is an exclusively-renewable portfolio. This approach would rely primarily on solar and wind as the most abundant and widely distributed, with a supporting role for other renewables, energy storage, and smart grid systems.

In a portfolio approach, each of the three pathways can benefit from the presence of the other two, as follows:

- *Fossil fuels:* Fossil fuels are not evenly distributed throughout the world, so it is beneficial for countries that do not possess them to develop other resources such as renewables so that they are not dependent on imports. Countries such as Spain or Denmark, for example, do not possess major fossil fuel reserves but do have a number of regions with good wind resources. Exploiting domestically

	Fossil Fuels with Sequestration	Renewables (including biofuels)	Nuclear
Description	Fossil fuels (principally coal) are extracted, converted to electricity and/or hydrogen, resulting CO_2 by-product is captured and sequestered.	Renewable energy sources are converted to electricity and/or hydrogen. Biofuels may also be converted directly to mechanical power in vehicles.	Energy released from fissile materials or from nuclear fusion.*
Long-term availability	200–300 years worldwide, assuming aggressive efforts to improve end-use efficiency.	Indefinite. Climate change may lead to changes in output (positive or negative) in specific locations.	In the case of deuterium-based fusion, almost indefinite, nearly as long as expected lifetime of planet earth.
Economic cost considerations	Combustion side is mature technology, inexpensive and reliable. Large-scale sequestration is not yet developed, cost unknown but likely to add significantly to life-cycle cost.	High cost per unit of energy, due to low capacity factors, low concentration of energy. Need for storage system due to intermittency increases cost (except for biofuels and geothermal). Costs have been steadily declining in recent years.	Capital costs higher relative to fossil fuel combustion; variable cost for fuel may be lower at present. Long-term disposal cost and cost of fusion technology, breeder reactor technology for U-238 are unknown.
Environmental cost considerations other than from GHGs†	Negative effect of extraction of coal on land and water resources, natural habitat, and so on.	Land requirements, solid waste issues with large-scale construction of infrastructure.	Effect of extraction of uranium; disposal of radioactive waste and decommissioned infrastructure.
Suitability as an energy source for transportation	Requires conversion to electricity or hydrogen, unless means can be found to extract large amount of CO_2 from atmosphere.	Requires conversion to electricity or hydrogen, except for biofuels.	Requires conversion to electricity or hydrogen.

*In order to replace current once-through U-235 cycle, must develop use of U-238, fusion from deuterium or tritium, or nonconventional U-235, for example, from oceans, must be developed.
†By implication, all pathways in this table have eliminated significant net emissions of CO_2 to the atmosphere.

TABLE 17-1   Comparison of Fossil, Renewable, and Nuclear Endpoints

available renewable energy sources for some fraction of energy demand allows these countries to avoid importing fossil fuels.

- *Nuclear energy:* Nuclear reactors change their rate of output of electricity very slowly, and are not well adapted to varying output in real time with changing demand for electricity. They are therefore well served by having a separate network of fossil-fuel-fired "load-following" and "peaking" plants that are more easily able to adjust output with changing demand. It would be awkward to meet all energy needs (electricity and otherwise) solely from nuclear power.

- *Renewable energy:* Most renewable energy sources are intermittent, and also "nondispatchable"—at times when the underlying resource (sun, wind, and the like) is not available, they cannot be turned on, or "dispatched," at will by the utility operator. Like nuclear energy, they are therefore well served by having as backup a network of dispatchable generating plants, which can easily be fired by a mixture of biofuel and fossil fuel, in the interim period while a full portfolio of renewable energy sources and storage systems can be deployed.

In the short run, it is practical to have a mixture of two or three of these major sources so that each can contribute to the grid in a way that exploits its advantage. Longer term, it appears from Table 17-1 that the problems of the fossil and nuclear pathways (e.g., sequestration, non-GHG impacts, managing nuclear waste, or developing fusion) are significant enough that by default the only pathway left will be renewables.

17-2-2   Summary of End-of-Chapter Levelized Cost Values

As a continuation of the comparisons of Table 17-1, a further comparison is made using the levelized cost calculations at the end of Chaps. 6, 8, 10, and 13, covering natural gas, nuclear, solar, and wind, respectively. Values are compared in Table 17-2; see individual chapters for detailed explanations of background data and assumptions. Presenting the four sources in a single table allows a comparison of cost per ton of CO_2 reduced, with natural gas used as a basis for calculating the amount of emissions saved for using a carbon-free source and its incremental cost.

The following observations can be made from the table:

1. At present, natural gas constitutes the benchmark against which the other sources are compared, since gas has significantly lower cost per kWh and also at 325 g/kWh reduces CO_2 per kWh compared to recent national average values (526 g/kWh in 2011, according to the U.S. Energy Information Administration). Natural gas is however not carbon-free, unlike the other sources, and the values

Source	Size (MW)	Cost (mil.$)	Cap. Fact.	Lev. Cost ($/kWh)	Cost/ton ($/tonCO$_2$)
Nat. Gas	1000	1050	65%	0.0644	$—
Nuclear	1000	8000	88%	0.1193	$169
Solar PV	100	200	17%	0.1308	$204
Wind	100	260	30%	0.1034	$120

Source	CapCost ($/kWh)	FuelCost ($/kWh)	OpCost ($/kWh)	Lev. Cost ($/kWh)
Nat. Gas	0.0174	0.039	0.008	0.0644
Nuclear	0.098	0.0043	0.017	0.1193
Solar PV	0.1268	0	0.004	0.1308
Wind	0.0934	0	0.01	0.1034

TABLE 17-2 Comparison of Representative Levelized Cost and Cost per Ton of CO_2 Reduction for Natural Gas, Nuclear, and Renewables

in the table do not take into account the carbon equivalent of methane emissions from the natural gas extraction and distribution process.

2. Although the other three sources in the table (nuclear, solar, and wind) are subject to wide variation from the representative values used in the table, the general finding that there is a step up in price to any of these three compared to gas holds even taking this variability into account. Therefore, the cost per ton values to go from gas to any of these three CO_2-free sources is positive and significant, although of the same order of magnitude as some proposals for a significant carbon tax.

3. Factors that might close the gap between natural gas and the other options include a rise in the price of natural gas, or the introduction of a nontrivial carbon tax. Reductions in the cost of nuclear or renewable technology might also close some of the gap, although physical limitations to the technology might limit possible cost reductions.

4. Solar PV might in general be expected to have a somewhat higher levelized cost than nuclear or wind due to its low capacity factor. On the other hand, it is not as visually intrusive as wind, can be installed in more locations, and does not produce radioactive waste, factors that are not captured in its price.

Several caveats should be noted as well:

1. Specific local circumstances around the United States would change the values from those shown in the table.

2. Values in the table show typical cost of production, and do not consider the eventual sale price. Thus a source that is better able to sell at times when prices paid are higher would have an advantage, although this is not reflected in the table.

3. Changing prices in the economy in the future can affect the relative cost of the options shown. A rise in the price of natural gas would increase the relative cost of electricity from this source. On the other hand, a rise in the price of commodities such as concrete, steel, or copper might particularly affect renewables such as solar and wind.

4. The nuclear cost value shown is conservative in the sense that the capital cost per kW is on the high end of estimates. If large-scale installation of nuclear plants were to take place, the cost per kW would likely fall.

5. Since the cost values shown are for production at a centralized facility before transmission and distribution, they do not capture the potential advantage of distributed solar PV generation, where the power is generated at the point of consumption and the transmission cost (often $0.05–0.06) is avoided. However, distributed generation with solar PV might incur higher costs per kW, so some of this cost advantage would be negated.

17-2-3   Other Emerging Technologies Not Previously Considered

In choosing topics for this book, it has not been possible to include all types of fossil, nuclear, and renewable energy technologies within the body of the book, so we have loosely applied a "systems approach" in choosing technologies on which to focus, as explained in Chap. 2. In terms of setting the scope or boundary around the project (step 3 of the systems approach), we have applied the following criteria to the choice of technologies:

1. Technologically proven
2. Widely distributed throughout the planet
3. Capable of significant expansion in the future

On this basis, we have included advanced fossil-fuel combustion systems, carbon sequestration (the Sleipner project in Chap. 7 provides a working example of the technology currently in use), several emerging nuclear energy technologies, solar energy, and wind energy. Solar and wind qualify on the basis that, even though PV panels or wind turbines are not major producers of electricity at present in terms of percent of world demand, their technical reliability is accepted, and the underlying physical resource is widely distributed around the world. For transportation, we have selected electric, biofuel, and hydrogen fuel cell vehicles, as well as hybrid platforms, as emerging alternative technologies that either already in commercial use or in a prototype form such that they might be commercialized in the future. All of these transportation technologies have the potential for widespread use around the world.

In the area of renewable energy in particular, a number of technologies have not yet been mentioned. Options such as large-scale hydropower systems that rely on dams are not mentioned because there appears to be little room for expansion beyond the network of large dams existing in the world today, although a limited number of large hydroelectric projects may be built in China, Latin America, and Central Asia in the next few decades. In many areas where potential exists to create a dam that could generate significant amounts of power, the amount of displacement of human population centers, loss of agricultural output, or loss of unique natural habitats would be unacceptable, so governments are unwilling to support future projects of this type.

Another energy form derived from the kinetic energy in water is tidal energy, which is extracted from changing tides along coastlines and can provide a widespread source of renewable energy. One possible device for capturing this energy is a "tidal turbine," that is, a device resembling a wind turbine but under water. Other designs that resemble water power turbines from hydroelectric stations are also under development for use with tidal energy. Locations for these turbines are chosen so that the underwater terrain combined with large tides provides a sufficient flow of water with each ebb and flow tide to produce significant amounts of energy.

As an example of tidal energy development, Marine Current Turbines Ltd. of the United Kingdom has operated an experimental 300 kW turbine (called "Seaflow") off the southwest coast of England since 2003, and a 1.2 MW SeaGen commercial tidal generator in Strangford Narrows, Northern Ireland since 2008 (see Fig. 17-1). The SeaGen unit is capable of delivering more than 6000 MWh per year, achieving a capacity factor of 60%. Because of the higher density of water compared to air, their tidal turbine can produce a larger amount of electricity per unit of swept area than a wind turbine. According to their estimates, the United Kingdom may have the potential for 10 GW of electricity generating capacity based on tidal resources all around the coast.[2] Other countries with long coastlines and strong tides can benefit from this resource as well.

---

[2]Mackay (2009) provides an extensive discussion of tidal power potential for the British Isles.

FIGURE 17-1   Illustration of tidal turbine of the type currently installed along the coast of England. Swept area of tidal turbines currently under development is 15 to 20 m in diameter. [*Source:* Copyright Marine Current Turbines TM Ltd. (2007). Reprinted with permission.]

Another form of energy available along coastlines is the energy in waves. One design for a wave energy station consists of a structure that channels the incoming wave into a chamber with an entry on one end for the waves, and an orifice for air to pass on the other end. As the air pressure in the chamber changes due to the entry and exit of the waves, air flows in and out of the chamber through the orifice, driving a turbine mounted on the other end of the orifice, thereby generating electricity. An experimental, grid-connected 500-kW station based on this principle has been operating on the island of Islay in Scotland since 2000.

Run-of-River and Small-Scale Hydropower

With available sites for large-scale hydroelectric dams largely exploited, future expansion of hydropower may come from smaller run-of-the-river hydroelectric plants, which divert a portion of a river's flow in order to turn a hydroelectric turbine, generating electricity. These projects are attractive because they do not require impounding (i.e., creating a reservoir to hold) large amounts of water behind a dam, inundating land, or relocating human settlements. On the other hand, they are susceptible to seasonal changes in water flow and cannot regulate output by storing up water to be released through the turbine

Figure 17-2   A 1.9-MW run-of-river hydroelectric plant adjacent to the campus of Cornell University. Water enters the plant after traveling 520 m (1700 ft) and dropping 42 m (140 ft) along the penstock pipe that connects to the upstream intake.

during periods of peak demand. Other sites may use water impoundment on a small scale to generate electricity with a rated capacity of as little as 1 kW, thus avoiding the negative consequences of large-scale impoundment.

Total small-scale hydroelectric capacity stood at approximately 70 GW worldwide in 2005, with some 38 GW installed in China, where it is allocated among thousands of village and rural systems. As an example of a run-of-the-river system in an industrialized country, Cornell University operates a 1.9-MW plant adjacent to its campus, which produces approximately 5 million kWh/year (see Fig. 17-2).

Example 17-1 illustrates the calculation of small-scale hydropower output based on volumetric flow rate of water.[3]

Example 17-1   Instantaneous output P of a hydropower system is a function of physical availability of energy in the stream of water passing through the system, called *head*, and the conversion efficiency

---

[3]Acknowledgment: Thanks to John Manning from Earth Sensitive Solutions, LLC, for providing the underlying figures and system background for this example.

of the hydropower conversion device, usually some type of turbine. In equation form, P is calculated as follows:

$$P = [\rho g Z + 0.5\rho\Delta(v^2)]Q\varepsilon$$

where ρ is the density of water (typically 1000 kg/m³ or 62.4 lb/ft³), g is the gravitational constant of 9.81 m/s², Z is the upstream vertical drop from the water intake to the turbine, v is the speed of the water stream either before or after the device, Q is the flow rate (in m³/s or ft³/s), and ε is the efficiency. Using the hydropower plant in Fig. 17-2 as an example, suppose for simplicity the contribution of $\Delta(v^2)$ is small compared to the contribution of the 42 m drop leading into the plant, so that it can be ignored and we can focus on just the first term in the equation for P, and that the efficiency of the plant is $\varepsilon = 70\%$. (a) What must the flow rate be in order to achieve the rated output of 1.9 MW? (b) Suppose the average flow rate over the course of a year is 2 m³/s. How much electricity will the plant produce during that year in kWh?

Solution

(a) Rearranging the equation for P, including only the effect of vertical drop, and substituting known values gives the following:

$$Q = \frac{P}{[\rho g Z + 0.5\rho\Delta(v^2)]\varepsilon}$$

$$= \frac{1.9\times10^6\,\text{W}}{(1000\,\text{kg/m}^3)(9.81\,\text{m/s}^2)(42\,\text{m})0.7 + 0.5(0)} = \frac{1.9\times10^6\,\text{W}}{2.884\times10^5\,\text{kg/m}\cdot\text{s}^2} = 6.59\ \text{m}^3/\text{s}$$

Thus the required flow is nearly 7 m³/s. For a run-of-the-river hydro system, the maximum and average flow requirement should be compared to average and standard deviation of flow from the originating river or stream. It should be ascertained that the amount of water diverted will not adversely affect the ecosystem of the river in a significant way, for example by threatening species therein by leaving an inadequate amount of water for healthy living.

(b) Substituting the given flow into the equation for power gives

$$P = (1000)(9.81)(42)(0.7)(2\ \text{m}^3/\text{s}) = 576{,}828\ \text{W} = 576.8\ \text{kW}$$

Annual output is then (576.8 kW)(8760 h/year) = 5.05 million kWh/year. This value is close to the typical output reported above for Cornell's plant.

Geothermal Energy Systems: Ground Source, Geologic, and Enhanced

Geothermal energy systems can be generally divided among three types: (1) systems which use shallow thermal resources and heat pumps for low-temperature heat applications, (2) systems that use sources of geologically produced steam near the earth's surface to produce electricity or for other utility-scale applications, and (3) systems that tap into thermal resources that are located deeper below the earth's surface in order to generate useful quantities of energy, known as enhanced geothermal systems (EGS).

In the first type of system, heating of a building is accomplished by pumping fluid through a horizontal trench or vertical well and then transferring it inside the building, where it heats either water or air. The heated water or air is then distributed to the rooms of the building in the same manner as a conventional heating system. These systems can also be designed to run in reverse during the hot season, in which case heat from inside the building is expelled into the trench or well. For many homeowners, relatively long payback periods similar to those of other small-scale renewable systems (solar or wind) limit access to these systems. However, this situation could change in the future if other types of space heating become more expensive. Shallow geothermal systems also require a space adjacent to the building for drilling or excavating the trench or well, which may not be possible in all locations. Example 17-2 shows how the productivity of a residential-size shallow geothermal system is calculated.

Example 17-2 In a simplified form, a ground source geothermal heating system takes heat from a low-temperature reservoir (the ground) and transfers it to a high-temperature reservoir (the insulated space, e.g., interior of a residence, that is to be heated). Within the system, fluid from the primary loop in the geothermal trenches or wells passes a heat exchanger where it transfers heat to a secondary loop. The fluid in the secondary loop is then compressed and passes through a second heat exchanger, where it transfers heat either to water or to air, depending on which is used to heat the space. The secondary loop then passes through an expansion valve and returns to the low-temperature heat exchanger. For a given rate of heat transfer, the energy flux on the high-temperature side $\dot{E}_H$ is greater than that of the low-temperature side $\dot{E}_L$, so the compressor in the secondary loop must supply the necessary work $W_{Compressor}$ to make up the difference (energy flux and compressor work are measured in BTU/h or kW):

$$\dot{E}_L + W_{Compressor} = \dot{E}_H$$

The *coefficient of performance* (COP) of the system then measures the ratio of energy flux delivered to pump work required:

$$COP = \frac{\dot{E}_H}{W_{Compressor}} = \frac{\dot{E}_H}{\dot{E}_H - \dot{E}_L}$$

Consider for example a system with forced air heating on the high-temperature side where the primary loop fluid passes through ground at 50°F at the rate of 15 gal/min, entering at 38°F and exiting at 44°F. Air passes through the high-temperature heat exchanger at the rate of 2000 ft³/min, entering at 68°F and leaving at 95°F. Calculate the work input required in the compressor in BTU/h, and the COP value. Ignore all losses.

Solution Because the primary loop fluid increases by 6°F as it passes through the geothermal bed, it must also decrease by the same amount in passing through the low-temperature heat exchanger. Taking into account the thermal capacity per gallon of water, we calculate $\dot{E}_L$ across this exchanger in units of BTU/h:

$$\dot{E}_L = \left(8.33\frac{BTU}{gal \cdot F}\right)\left(15\frac{gal}{min}\right)(6°F)\left(60\frac{min}{h}\right)$$

$$= 45,000\frac{BTU}{h}$$

On the high-temperature side, the temperature change is 27°F. Incorporating the rate of air flow and thermal capacity of air per cubic foot gives $\dot{E}_H$ for the high-temperature heat exchanger:

$$\dot{E}_H = \left(0.018\frac{BTU}{ft^3 \cdot F}\right)(2000 \text{ cfm})(27°F)\left(60\frac{min}{h}\right)$$

$$= 58,320\frac{BTU}{h}$$

Compressor work is therefore 58,320 − 45,000 = 13,320 BTU/h. The COP is then (58,320)/(13,320) = 4.38.

Discussion   The value of COP = 4.38 indicates the appeal of the geothermal system: in return for one unit of energy input into the compressor, the system delivers 4.38 units of heat. The performance of an actual system will be substantially lower, because the calculation has ignored energy input into pumps for circulating the two fluid loops and blowing the forced air in the heating system, as well as system-wide losses.

The system can nevertheless achieve a substantial energy efficiency advantage over alternative systems. For example, a homeowner might install electric resistance heating instead of geothermal. Both systems are then using electricity as the sole purchased energy input, the former for direct space heating and the latter to drive one or more pumps, fans, and compressors. The maximum possible COP that the electric heating can achieve is COP = 1, so to the extent that the geothermal system is able to exceed a COP value of unity, it can deliver financial savings on annual operating costs. On the other hand, the capital cost of the geothermal system is substantially higher. The homeowner must then evaluate the business case for upfront investment versus lower ongoing cost, and attempt to quantify payback period and life-cycle net present value—all in the face of highly uncertain future utility costs (!).

In the second type of geothermal system, steam that reaches the surface at a geologic fault is piped into a power plant and expanded in a turbine to generate electricity. This resource is attractive because of its low cost, ability to regulate output, and lack of emissions other than excess steam. The first such plant was opened at Lardarello, Italy, in 1904. Today the United States is the leading producer of geothermal electricity in the world, with 16 TWh produced in 2005. One of the largest geothermal systems in the world is the 750 MW facility at The Geysers in California, which was first developed in the 1920s.

The development of near-surface steam power generation requires access to sites that have the resource available, and unexploited sites are in limited supply in countries such as the United States. Therefore, interest is growing in an alternative that has more widespread potential, namely, EGS. In one approach to the design of an EGS site, a power plant is situated on the earth's surface above the resource, and the two are connected by an injection well and a production well, which go down in the range of 5000 to 20,000 ft below the surface in order to reach the underground high-temperature resource. Water is injected into the injection well and then permeates the region of hot rock at the bottom of the well. As the water boils, the pressure forces the steam up the production well to the top, where it is expanded in a steam turbine.

Unlike the surface steam resource, hot rock strata that are suitable for EGS are widely distributed beneath the U.S. land mass, and likely under other countries as well. EGS plants are also dispatchable, so that they are not subject to the intermittency of some other renewable energy sources. So long as the thermal resource is not extracted too quickly, heat from the earth's crust replenishes heat removed by the working fluid, and the process can continue. One large challenge with EGS is the cost of drilling, which must often take place in rock strata that are harder and deeper than those encountered by the oil and gas industry. Another challenge is that output from an EGS plant depends on the characteristics of the rock formation on which it is based. Since it is not possible to know with certainty the nature of the rock without actually drilling into the site and installing the system, there is a risk that a plant may underperform economically if the formation proves to be not as productive as desired. Therefore, in order to lower the cost and reduce the financial risk of EGS, efforts are under way to develop advanced techniques for deep drilling, such as replacing conventional mechanical drilling with "spallation drilling" that uses a high-temperature flame to chip or "spall" pieces of rock material from the borehole of the well.

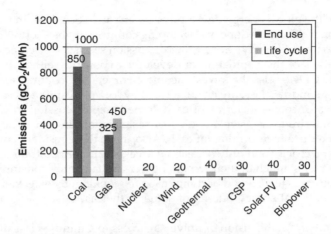

Figure 17-3   End-use and life cycle CO_2 emissions per kWh from different energy sources.
Source: U.S. Energy Information Administration, for end-use emissions; National Renewable Energy Laboratories, for life-cycle emissions. Note: CSP = Concentrating Solar Power.

17-2-4   Comparison of Life Cycle CO_2 Emissions per Unit of Energy

As stated in Table 17-1, one of the advantages of renewable and nuclear energy systems is that they do not emit CO_2 from fossil fuels at the end use stage of the life cycle. When other stages of the life cycle are taken into account (raw material extraction, freight transportation, construction of infrastructure, and end-of-life disposal) there are additional CO_2 emissions from both embodied energy and use of materials such as concrete. These emissions lead to nonzero CO_2 emissions even for nuclear and renewables. Figure 17-3 shows typical values per kWh for both the end-use stage and life cycle total emissions for coal, gas, nuclear, and renewable sources. When life cycle emissions are distributed across the entire amount of electricity produced by a facility in its lifetime, the emissions per kWh for nuclear and renewable are still small compared to the burning of fossil fuels to generate electricity, especially from coal. Note that the life cycle values in Fig. 17-3 are average numbers from the U.S. National Renewable Energy Laboratories, as the manufacture and installation of any type of energy system is very heterogeneous, so that actual numbers vary from installation to installation. However, even taking into account variability observed, the general observation that the end-use emissions of CO_2 from burning coal or gas in a plant to generate electricity is the dominant factor in determining whether overall emissions are high or low still holds.

17-3   Sustainable Energy for Developing Countries

Before looking at a scenario for global energy development in Sec. 17-4, we consider here the specific case of developing countries, referred to elsewhere in this book as "emerging" countries. Some areas of the developing world, such as parts of India and China, are rapidly adopting the technologies of the industrialized countries, such as modern building techniques, climate control systems, electric appliances and in some cases private cars. We focus on the implications of these amenities reaching all developing countries as part of the global energy development scenario in Sec. 17-4. In this section, we focus instead on what can be done in developing countries in the short to medium term using "appropriate technology" (for further discussion see, e.g., Haseltine and Bull, 2003).

Members of the population of developing countries, either those with the lowest per capita GDP or else the lowest-income members of a country such as India that also has a large middle class, are characterized by low per-capita access to energy and also low annual incomes in terms of cash earned (see discussion in Chap. 1). They may therefore desire to use sweat equity, subsistence farming, or noncash barter arrangements to reduce their outflow of cash. Also, since these residents may have little or no access to electricity and only occasional access to motorized transportation in the form of public buses, much of their remaining per-capita energy consumption may be in the form of biomass (e.g., wood or dung) used for cooking. By extension, a large fraction of the total national energy budget of some of theleast developed countries may be for cooking.

Separate from the vision of universal access to a middle-class lifestyle as discussed in Chap. 1 as a goal for the year 2100, the primary early objectives of this population is the basic requirements of sustainability, namely access to safe drinking water, sufficient nutrition, medical care, and cooking fuels and systems that do not cause indoor air quality problems. On the latter point, if an effective design can be found, it may be advantageous in this situation to build either an energy efficient wood stove or solar cooking device (see Chap. 11) rather than spending scarce cash to purchase a mass-produced version in the marketplace, even when devices may be available for a price.

On the other hand, devices specifically designed for use in developing countries that are carefully designed and mass produced may be advantageous in the long run, even if the initial cash purchase cost is higher. For instance, water pumping can be accomplished using hand or foot power, and if the use of electricity is avoided, both the cash cost and the additional burden on the national energy budget is avoided. One mass-produced option is the "Mark One" hand pump developed and manufactured in India. This pump represents an example of "south-to-south" trade, in which tropical countries trade with each other rather than the industrialized countries, with whom they often have trade deficits, leading to long-term structural economic problems. Another example is the "Money-Maker" foot-powered pump developed by ApproTec. This pump also requires no purchased electricity or diesel input, but has the advantage of using foot—instead of hand-power, increasing the pumping rate since legs are stronger than arms. The result is a pump that can pay back its initial $80 USD cost by increasing the irrigation capability and hence productivity of smallholder farmers in tropical countries.

Lastly, even the most impoverished developing countries have some measure of electricity and transportation fuel consumption, and here the development of renewable resources can have a major impact because total national energy budgets tend to be small by U.S. standards. For example, the country of Nicaragua in Central America consumed 90.7 PJ of energy in 2012. Nicaragua has an excellent wind resource along its Pacific coast, especially in a strip of land between the Pacific shore and Lake Nicaragua, where winds consistently blow since there are bodies of water on either side and the terrain is relatively flat. The wind resource is already under development with a number of Indian-made Suzlon turbines installed in two farms as of 2012. If this development could be expanded to several wind farms comprising a total of 400 2.5-MW turbines, for a combined capacity of 1000 MW, and a capacity factor of 30% could be maintained, the output would be 2628 GWh, equivalent to 9.5 PJ or more than 10% of the entire national energy budget. For the sake of comparison, the U.S. total primary energy consumption value in 2012 was 100 EJ. The same output from 1,000 MW of wind turbines would constitute just 0.009% of total consumption.

17-4   Pathways to a Sustainable Energy Future: A Case Study

Following from the advantages of the portfolio approach, as discussed above, most projections of primary energy production for the twenty-first century recognize that the three endpoint technologies of Table 17-1 will all play a role in meeting demand, although the relative contribution of each resource varies from projection to projection (see, e.g., Nakicenovic and Jefferson, 1995, or Beck, 1999). In this section, we return to the energy pathway presented in Fig. 1-11 in Chap. 1, in which both world population and total primary energy consumption stabilize by the year 2100. We further consider a possible scenario for transformation of energy production during the course of the century.

In analyzing the scenario, we focus on renewable energy resources as an option that is relatively small now (currently less than 15% of world total primary energy generation) but that might play a much larger role in the future. Of particular interest is the year-by-year addition of energy generating capacity required to reach total renewable output goals for the end of the century dictated by the scenario, and the maximum annual increment observed over the entire pathway. In principle, the same process could be applied to the nuclear or CCS (carbon capture and sequestration) components of the pathway in Fig. 1-11. For reasons of brevity, we do not include the details here, but return to the implications for these other two components at the end of this section.

We take as a starting point the growth of world renewable energy shown in Fig. 1-13 from 29 EJ (27 quads) in 2000 to 316 EJ (300 quads) in 2100. Note that although standard units are used in Chap. 1, metric units are used in the scenario as presented in Chap. 17. The growth of renewable energy follows approximately a triangle function path, with relatively slow growth in an early adoption phase currently under way (2000–2020), a maximum growth phase in the middle of the path (2020–2080), and a "tapering off" phase at the end of the process (2080–2100).

We make the following assumptions:

- For simplicity, we do not attempt to recalibrate the pathway from Fig. 1-13 based on the actual rollout of renewables in the periods from 2000 to 2013, which are known values and likely to be slightly different from the values in the figure.

- For renewable energy, the resource is added in units of either 100 MW wind farms or 100-MW solar farms, with output allocated 50%/50%. Capacity factors for the two resources are 30% and 15%, respectively, and installation costs are $3000 and $2000 per kW, respectively. In principle, some share could be allocated to other renewable types (e.g., biomass, geothermal) but for reasons of brevity this is not included.

- There are sufficient resources to build infrastructure and that the new capacity is located in such a way that the energy production can be distributed to end users, wherever they might be located, so as to meet world demand.

- Once made available from these new resources, primary energy can be converted to the appropriate form needed for the desired end-use application, in the correct proportions. For example, output from wind and solar could be distributed on the grid for stationary end uses, or converted to hydrogen and electricity to be used in vehicles.

- Practical solutions can be found to deal with the intermittency of renewables through energy storage or allocation of excess production to applications that are not time sensitive (e.g., recharging vehicles at night).

17-4-1 Renewable Scenario Results

In this section we convert the increase in solar and wind primary energy production into the increase in installed capacity required, and from there the number of units (100-MW solar or wind farms) that must be added each year. For simplicity, we assume that each decade has a linear rate of increasing installed capacity, which then adjusts up or down in the following decade as requirements change. For instance, the increase from 2000 to 2010 amounts to 9.4 EJ of new production, or 940 PJ per year added on average. Divided between the two sources gives 470 PJ, which corresponds to 99.5 GW of wind and 49.8 GW of solar added per year, or 995 and 498 farms added, respectively. The results of the pathway analysis are shown in Fig. 17-4. Each decade increment is labeled with the last year in the decade, thus the increment of added capacity for the period 2000–2010 is labeled "2010," and so on.

The required increases are very large by historical standards, peaking in the decade 2040–2050 with an average of 7475 solar farms and 3737 wind farms installed per year. After that point, global population has largely stabilized according to the assumptions, so the growth in need for renewable energy slows down and with it the rate of installation.

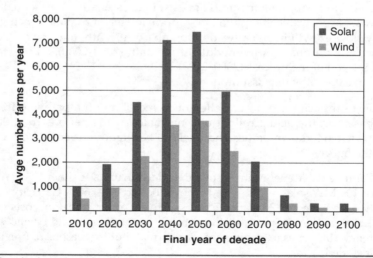

FIGURE 17-4 Required number of solar and wind farms installed per year to meet transition objectives 2000–2100.

Explanation: Year "2010" in figure represents decade 2000–2010, year "2020" represents 2010–2020, etc.

The number of solar farms installed could be reduced somewhat because some of the new capacity could be installed in a distributed form at residences and businesses, however for wind the requirement would need to be met with utility-scale turbines since that is the cost-effective and productive way to generate electricity from wind.

To put the peak rate of installation in context, the largest rate of installation on record for solar came in the year 2013 when approximately 38,000 MW of solar PV capacity was installed worldwide. The peak rate of 747,000 MW per year therefore represents approximately 20 times as fast a rate as the 2013 record. For wind, the average rate of installation from 2010 to 2014 was 44,000 MW per year, and the target peak rate is equivalent to 374,000 MW per year, so the ratio is on the order of 8:1.

Fortunately, there are sufficient locations around the world where the desired wind and solar capacity could be installed, if the project can be financed, and if the accompanying investment in grid infrastructure can be made. Also, the annual price tag for the required investment can be seen in the context of current expenditures on energy as a fraction of total world GDP. According to the International Monetary Fund (IMF), the nominal world GDP in 2014 was $77.6 trillion. Depending on how one draws boundaries between economic sectors, approximately one-ninth of world GDP is spent on energy, amounting to $8.6 trillion in that year. At a nominal installed price today of $2/watt and $3/watt for solar and wind, respectively, the total annual cost in the peak decade is $2.6 trillion per year, or within the size of the current world energy budget. Furthermore, by that time much renewable energy infrastructure would be in place, so that a substantial payment stream that would otherwise be spent on fuel (and, in particular, fossil fuel) could instead be invested in renewable energy infrastructure.

17-4-2   Comparison to Nuclear and CCS Pathways

The underlying scenario from Fig. 1-13 requires equal contributions from renewables, nuclear, and CCS, so the required rate of infrastructure would be on a similar very large scale. For nuclear energy, the number of new plants added per year would be smaller because each plant has a typical capacity of 1 GW and the capacity factors are usually much higher, on the order of 90% instead of 15–30% as in the case of renewables. However, the number of plants installed per year is still formidable, given that the total world fleet today is only on the order of 500 plants. Also, the rate of nuclear waste generation would accelerate. For CCS, the conversion plants, grid connections, and upstream supply chain to deliver coal and gas already exist, but the retrofitting of plants to capture CO_2 plus the capacity to transport and sequester such large volumes of the gas would pose another very large infrastructure investment challenge.

17-4-3   Comparison of Industrialized versus Emerging Contribution

Building on the preceding discussion of changes in global energy production, it is of interest to contrast the very different roles of industrialized and emerging countries in the transformation of world primary energy production. In this case, we consider all three final energy sources and not just renewables so that we can see the whole picture of required energy supply, and not just renewables. Also, some part of the transition is shifting of generation from fossil fuel without sequestration to CCS, rather than actual net new generation. The industrialized countries are those that historically have possessed the wealth needed to support an energy-intensive lifestyle, so their average energy intensity is on the order of three times as high as that of the emerging countries (Table 17-3). The emerging countries have smaller average per capita energy use but are

Item	Industrial	Emerging
Population, 2000 (billions)	1.3	4.7
Energy use, 2000 (EJ)	200	211
Energy/cap, 2000 (GJ/person)	154	50
Scenario assumption	Energy per capita decreases due to more efficient technology and per capita CO_2 limits	Energy per capita increases as citizens adopt efficient versions of modern technologies
Population, 2100 (billions)	1.2	7.8
Energy/cap, 2100 (GJ/person)	100	100
Energy use, 2100 (EJ)	120	780
Change in energy/cap	−35%	100%
Change in energy use	−40%	270%

Source for underlying data: U.S. Energy Information Administration.

TABLE 17-3   Allocation of Total and Per-capita Energy Production in Global Transition Scenario for Years 2000 and 2100: Emerging and Industrial Countries

somewhat more than three times larger in population, so that taken as a whole they produce slightly more energy.

The scenario sets a target of 100 GJ/person as the energy intensity for all of the world population, as a way to fairly distribute access to energy to all. Populations in many industrialized countries are expected to fall slightly, so that taken as a whole the population of the industrialized group declines, and since per capita energy consumption declines, total energy consumption does as well, as seen in Table 17-2. On the other hand, population in the emerging countries grows by 66%, and since per capita energy also doubles, the total growth in energy production is substantial. The net result is that most of the growth in energy demand occurs in the emerging countries, which by the year 2100 consume 87% of world energy demand.

Based on the growth in renewable, nuclear, and CCS in Fig. 1-13 and the relative growth in industrialized versus emerging countries, decade-by-decade energy growth can be allocated between these two groups of countries (Fig. 17-5). Note that the figures are given in terms of annual average energy growth, under the assumption that annual energy growth will be constant for each decade. Total energy growth over the entire century is much larger for the emerging countries, so that annual average energy growth is also much larger for this group, peaking at 20.2 EJ/year in the decade 2040–2050, versus 2.34 EJ/year for the industrialized countries in that same decade. Note that although there is a 40% decline in total energy consumption in Table 17-2 for the industrialized countries, the growth figures in Fig. 17-5 are still positive because the industrialized countries are transitioning from a mixture of mostly non-CCS fossil fuel use (85% of total demand) in the year 2000 to a mixture of sources in 2100 that is 100% CO_2 emissions free.

17-4-4   Discussion

The previous examples of pathways are preliminary projections of what the future might look like, and therefore the projections of new infrastructure required should be seen in terms of their orders of magnitude, and not their exact values. Even so, if the

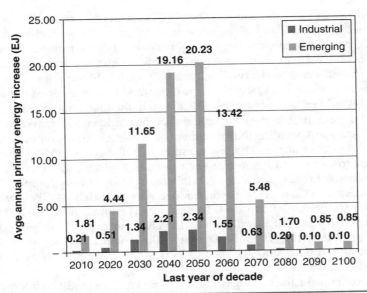

FIGURE 17-5   Comparison of annual increase in primary energy from carbon-free sources to meet transition objectives 2000–2100: Emerging and industrial countries.

Explanation: Year "2010" in figure represents decade 2000–2010, year "2020" represents 2010–2020, etc.

whole world population is to have access to modern energy and CO_2 emissions are to be eliminated, the path of energy system transformation will at least roughly resemble the paths shown, which raises challenging issues for discussion.

First, in the case of renewable energy, there is sufficient physical resource available in nature for the global wind and solar, possibly supported by other sources such as hydro and geothermal, to supply 317 EJ per year in the year 2100, assuming the necessary infrastructure can be built. However, the location of generation relative to the location of demand may pose a serious challenge. New distribution grids would be required between centers of renewable energy production (e.g., deserts for solar energy, or areas with large wind resources such as the Great Plains in United States and Canada) and population centers in order for the transition to succeed.

Second, a comparison between the pathways shown in Fig. 17-4 and the actual pathways of renewables between 2000 and 2010 shows that we are not adding capacity as fast as the rates shown. To have stayed on the pathway, the world would have needed to add an annual output of 9.4 EJ from renewable sources, respectively, over this 20-year time span; this rate of growth has not been achieved. It is possible that we might make up for this shortfall by adding capacity more rapidly than shown in the pathway during the rest of the period up to 2100. Alternatively, the pathway could be modified to reflect actual levels in 2010 and pushing back the endpoint to 2110 or 2120 instead of 2100, but this would delay the benefits of shifting away from CO_2-emitting sources accordingly.

Lastly, the most important point of discussion is the rate of increase of energy infrastructure required to meet the targets proposed. This is a concern for renewable energy, as it is for nuclear and CCS.

In the case of renewable energy, the necessary rate of increase in infrastructure is very challenging in terms of both cost and organizational logistics. As noted above, we would need to add the equivalent of 7500 solar farms and 3700 wind farms per year during the peak period. Taking wind as an example, if each turbine is rated at 2.5 MW on average, the number of farms is equivalent to 148,000 turbines per year.

Looking at finances, the cost per kW of wind turbines and solar farms is difficult to predict. With continued advances, costs might fall to $1/watt installed for each technology, but the result would still be on the order of $1 trillion invested per year. On the other hand, even if installed cost rises 50% for each, the total annual cost is still on the same order of magnitude as the economic size of the world energy market (~$9 trillion).

Cost aside, spreading the investment out among different types of renewable energy systems would simplify the organizational challenges of adding so much renewable energy to our overall resource each year, but adding so much renewable energy generation into the world's capacity mix remains logistically complex. For comparison, as discussed in Chap. 13, the world added approximately 44 GW of wind capacity per year for the period 2010 to 2014. This amount is equivalent 18,000 turbines per year on average. This rate of installation is therefore an order of magnitude less than the proposed peak rate. Thus the capacity to add wind turbines around the world has grown substantially since the 2000-to-2005 period, but is still not at the level envisioned in the scenarios.

In conclusion, the findings of the scenario analysis might be interpreted in different ways. The following three statements represent some of the possibilities:

1. *"The renewable pathway 2000 to 2100 is itself a compelling reason for nuclear and coal with sequestration during the twenty-first century."* Based on the rate of expansion required, targets of 33% or more of all primary energy from renewables by the end of the twenty-first century are not feasible. Realistically, renewable energy systems will develop at a slower rate, perhaps no more than 40,000 2.5-MW wind turbine equivalents per year (two times faster than presently) at the peak in even the most robust scenario. This assessment argues in favor of nuclear and fossil fuel with sequestration to carry us through the next 100 to 200 years while we more gradually add renewable capacity. It also assumes that we would be successful in managing a larger stream of high-level nuclear waste, in containing the threat of nuclear proliferation, and in exploiting fossil fuels with sequestration in an environmentally sustainable way, none of which are trivial tasks.

2. *"33% of 950 EJ from renewables in 2100 is the wrong problem to solve."* The flaw lies not with renewable energy expansion but with the magnitude of the targets for 2100 for total energy production. Rather than seeking to add so much renewable energy capacity, we need to completely overhaul the way we use energy so that we deliver a good quality of life to those who already have one, and achieve a much higher quality of life in the emerging countries, while also lowering the total amount of energy required, perhaps to a level of 300 EJ in 2100. This overhaul of energy use would reduce energy demand by (1) directing R&D toward developing much more efficient end-use technologies (appliances, electronics, vehicles, and so on) and (2) changing the way people use the technologies. If the 300 EJ target in 2110 were achieved, a 40 to 80% share for renewable energy systems would be more realistic. This assessment implies that we would turn over old inefficient energy end-use technologies into new efficient ones at a sufficiently high rate, since at present we already have a

population of several billion consumers who purchase and use energy-consuming devices at a certain rate, and who can only improve their personal efficiency as fast as technological improvement and education about efficient use of energy allows.

3. *"Let no person say it cannot be done."* We should not be deterred by the size of the target relative to our experience to date, but rather forge ahead and find a way to add these large quantities of renewable energy. Our energy technology is evolving very rapidly and 100 years (or 85 years from the year 2015) is a long time; our generation should not make presumptions about what can or cannot be achieved in the future. An analogy can be made between the scope of investment in renewable energy systems envisioned, and the output of the U.S. war effort during World War II.[4] During a period from 1942 to 1945, the U.S. manufactured 230,000 aircraft of all descriptions for use in fighting the war, a level of industrial output that was also unprecedented up to that time. By comparison, the target for renewables in the scenarios presented is larger than the output of aircraft during the war in terms of raw materials required or economic value, but our technology is also more advanced, and it will be a global effort, rather than that of just one country.

At this time, it appears that any of the three outcomes could emerge during the course of the twenty-first century. It is possible that, if sequestration or nuclear technologies develop favorably, much of the energy production during this time period could come from these sources, with a lesser role for renewable energy, as in point 1. Alternatively, renewable energy technologies may develop favorably, and gain a much larger share of the total energy production value than they currently possess, as in point 3. One expectation regarding the future of energy is that regardless of the eventual level of energy consumed or the mix of endpoint technologies used to deliver it, the improvement in efficiency of delivering desired services per unit of energy consumed envisioned in point 2 would help with any of the outcomes, since it takes financial and environmental pressure off the need to build new infrastructure. In the remainder of this section, we take a closer look at the role of renewable energy options, and the allocation of energy production between industrial and emerging countries.

The Case for Robust Development of New Renewable Energy Options

Regardless of the target for renewables, there are several reasons to develop them in a robust way. The first reason for emphasizing renewable energy is that the current rate of growth is insufficient. At present, the renewable energy industry is growing quickly relative to previous levels (see data in Chap. 10 on PV and in Chap. 13 on wind), but is nevertheless small relative to other sources, if we exclude large-scale hydropower as having limited scope for growth in the future. Furthermore, at current rates of growth, the renewable sector would not take up a large percent of total energy demand at any time before 2100. Therefore, support is needed to accelerate the research essential to improving the performance of renewable energy systems. Support is also needed to expand the infrastructure needed to build and deliver the equipment, since it is

---

[4]As presented in Brown (2006, p. 254). The quotation is based on "let no man say it cannot be done," a reference to targets for wartime production from President Roosevelt's State of the Union address, January, 1942.

capital-intensive and requires that the necessary devices be installed at the beginning of a project in order to convert a given amount of renewable energy to human uses.

A second reason to develop renewable energy is the risk associated with either the carbon with sequestration or nuclear options. As mentioned above, carbon with sequestration is already in use in the world today with enough success that it can be expanded and adapted in the future so that more carbon is sequestered. However, we do not yet have experience with sequestration on the massive scale that would be required to sequester most of the current 9 Gt of carbon in CO_2 per year from fossil fuels to stabilize concentration in the atmosphere, and there is a risk that the technique may not work at or near 100% effectiveness on such a large scale. In turn, expansion of nuclear energy carries with it the risk that eventually the world community will want to limit its growth to limit the risk of proliferation or the total size of the nuclear waste stream. Also, the amount of uranium that will eventually be recovered, or whether breeder reactors or fusion will ever work successfully on a large scale, is an unknown, and there is a risk that one or both of these requirements for long-term use of nuclear energy on a large scale will not materialize.

At the present time, renewable energy options are often relatively expensive in terms of levelized cost per kWh (with some exceptions, e.g., large-scale wind in locations with an excellent wind resource), but the underlying technologies themselves have been proven to work reliably. Furthermore, there are technologies such as large-scale solar thermal (see Chap. 11) or pumped storage to accommodate the intermittency of renewable energy, which are not in general use at present but could be put into service in the future if needed. Therefore, because the underlying technological components for renewable energy are relatively more proven, there is less risk associated with this option than with the other two. On this basis, it makes sense to develop renewable in a robust way so that it will be ready for the future. Some of the underlying technologies for coordinating large-scale use of renewable energy are not yet in place, but these are smaller technological challenges than the fundamental challenges that accompany the other two options, for example, successful development of nuclear fusion.

Already, a number of state governments in the United States as well as national governments of other countries are accelerating the growth of renewable energy through the use of *renewable portfolio standards* (RPSs). In these programs, the government body sets a target for percent of energy from renewable sources, and then provides financial incentives to help achieve the goals of the RPS. For example, California has a target of 33% of all electricity generation from renewables by 2020. Funds for the RPS come from a surcharge paid by customers on energy expenditures, with the revenues distributed among eligible projects.

Achieving Sustainable Energy by Closing the Loop on Energy and Material Flows

The dilemma of modern technology represented by the tannery parable at the beginning of the chapter is not solved easily or quickly. In pursuit of sustainable energy we can, however, look to the natural world for an example of a model that can be replicated in our manufacturing and recycling of products (Fig. 17-6).

Natural systems use energy and physical materials to create and recreate organisms. At the heart of this system are the various transformative processes that build up matter into living organisms and then convert organisms back into raw matter. Most of the energy comes originally from the sun in a high-quality form, meaning that it has low entropy and is capable of a wide array of applications. The energy may arrive into

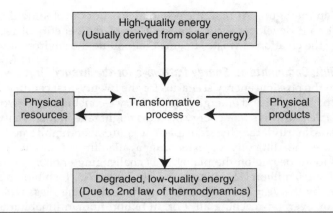

FIGURE 17-6 Transformative process that use high-quality energy to convert physical resources into physical products and vice versa.

processes on the earth's surface directly in the form of solar rays, but it may also arrive as wind energy, or the wind energy may in turn create waves, or solar energy can be stored in plants in the form of biomass energy. Energy is conserved as it is used to perform transformative processes, but even though the amount of energy in joules or BTUs remains constant, it exits in a lower quality form that is no longer as capable of acting as a catalyst. For instance, energy used once as a catalyst for photosynthesis in a leaf may eventually be stored in a plant and consumed by an animal when some part of the plant is eaten, but the same energy is no longer available for a second round of photosynthesis. However, because the sun continues to provide solar rays, the process can continue using newly provided high-quality energy.

At the level of life cycle of animals and plants, organisms may consume foods that provide both nutrients (proteins, fats, etc.) and energy (caloric content) for growth and maintenance of health. Organisms that are capable of locomotion also use the energy content in food to meet travel requirements of their daily life. At the end of their lives, energy is also used by the natural system for the reverse process of breaking down organisms into component materials, perhaps using the physical force of solar, wind, or water energy, or perhaps relying on microorganisms that have their own life cycle from matter to organism to matter. Over geologic time periods, some of the material is trapped under the earth's surface (hence the formation of fossil fuels), but most becomes available again for formation into new organisms.

Sustainable energy systems can emulate the natural world on two levels, as shown in Fig. 17-6. First, they can use high-quality incoming energy as a source for creating energy carriers of all types. The energy is dissipated as heat and friction as the system consumes it (thus becoming "degraded low-quality energy") but because a mechanism exists to continuously convert incoming energy in nature into energy carriers (i.e., the sun continues to supply primary energy every day), the system continues to operate. Second, energy is used to assemble physical resources into the physical products needed for meeting daily living requirements: mass produced products, infrastructure, and other supporting systems. Products are designed with recyclability in mind, so that at the end of their life, ideally 100% of the product can be broken down and returned to be used as a physical resource. Thus, there is no arrow from physical product to "waste stream" in the figure. At first, the economic cost of developing this system is prohibitive, but as it develops, society begins to reap the benefits of reusing materials and avoiding the creation of waste streams that pollute the environment and harm health.

Concluding Comments on Energy Pathways for the Twenty-First Century

Based on the possible energy scenarios for the twenty-first century and the allocation of energy to industrial and emerging countries, we conclude with two observations about the possible pathways that energy systems might take in the twenty-first century. First, the extraordinarily large physical capacity requirements highlighted in Figs. 17-4 and 17-5 underscore the difficulty of extending availability of modern, sustainable energy systems to all people on the planet. It is challenging enough to transform the energy system for 450 million U.S. citizens or, as in Table 17-3, 1.2 billion residents of industrial countries by the year 2100. The calculations underlying Figs. 17-4 and 17-5 are predicated, however, on reaching all 9 or 10 billion human inhabitants by the end of the century, in keeping with the goals of sustainable development laid out in Chap. 2. Thus in effect we are increasing the size of the challenge by an order of magnitude simply by expanding the geographic scope from just the industrialized countries to the whole world. Also, we have discussed only the physical pathway for installing sufficient capacity. The underlying policies that might achieve such a pathway (as discussed in, e.g., Leighty, 2012, or National Academy of Sciences, 2013) entails yet another level of complexity.

This leads to the second observation: whether renewable energy as favored by the discussion in this section, or more of a balanced mixture including nuclear and CCS as well, the options for primary energy remain just these three. The energy systems of the future must fulfill many criteria, including the need to eliminate net GHG emissions, to provide energy at steady and affordable prices, to be scalable in a reasonable time frame, and to meet total demand for the long term. All three sources face major challenges ahead if they wish to fulfill these criteria.

17-5   The Role of the Energy Professional in Creating the Energy Systems of the Future

We now transit from the question of "What needs to be done in the twenty-first century?" to the question of "Who is going to do this work?" It is clear that the readers of this book, namely, the energy professionals who design, operate, or manage energy systems, will serve at the center of shaping our energy future. They can serve in this role in both a professional and extracurricular context.

Turning first to the professional role, evaluation of the various technologies in the book shows us where technological developments will take place over the next several decades as we adapt our energy systems to make them more sustainable. Broadly speaking, these challenges can be categorized in the following areas:

- *Research and development:* Challenges include, among others, (1) the redesign of combustion systems to separate CO_2 and sequester it in reservoirs; (2) the development of nuclear fusion into a mature technology; (3) the development of lower cost solar photovoltaics or other systems for converting sunlight to electricity; (4) the use of structural engineering to reduce the cost of large-scale renewable energy systems, such as large offshore wind turbines or centralized solar power plants; (5) the development of more cost-effective and less land-intensive biofuels; and (6) the development of long-lived and cost-effective fuel cells.

- *Prospecting for resources:* Finding attractive locations for wind turbines, for tapping geothermal resources, for sequestration reservoirs for CO_2, or for other resources.

- *Streamlining the installation process:* Especially for renewable energy systems, developing and standardizing this process so that it is cheaper, faster, and more predictable. At the same time, making renewable energy systems more adaptable so that they are applicable in a greater range of locations.

It is clear from this list that almost any discipline of engineering as well as related disciplines can play a role in this process. For these efforts to succeed, decision makers in business and government must be well-informed. Therefore, energy professionals with the relevant experience will be needed in executive positions in energy businesses large and small, in policy decision-making positions in government, and as leaders in energy- and environment-related nongovernmental organizations.

17-5-1   Roles for Energy Professionals Outside of Formal Work

There are also many extracurricular roles for energy professionals to play outside of their formal work in engineering, management, policy, or advocacy. For example, in many spheres of daily life the energy professional can help to bridge the gap between efforts by government and professionals to provide sustainable energy and the public's participation in these efforts. While there are numerous examples of new, energy-efficient technologies being used in homes and workplaces, there are also many instances of energy used poorly: obsolete appliances, buildings that leak energy, poor indoor lighting that uses too much energy but still provides inadequate illumination, badly maintained motor vehicles, and so on. The level of "energy literacy" can continually be improved among energy consumers—whose numbers, in the industrial countries, include almost the entire population—and this will be an important part of the project to achieve global energy sustainability.

There are many ways in which an energy professional can help those outside the profession to understand energy better and make better decisions about it. One way is to lead by example, by upgrading home, workplace, and transportation choices to make them more efficient. Another way is to use alternative energy sources, such as solar, wind, or geothermal energy. Explaining to friends and acquaintances what the changes are, and the benefits that are expected, expands these efforts into an opportunity to encourage others to do the same.

In the case of one's workplace, there may be scope for serving in an extracurricular capacity as an *energy efficiency champion* that encourages efforts to use energy more efficiently. For example, a medium or large employer may be required to reduce travel to work in single-occupant vehicles by carpooling or public transportation. Taking leadership in these efforts can help to ensure their success. Related to outreach in the workplace is outreach to tradespeople who fabricate and maintain energy systems. Although the thrust of this book has been toward the design and financial management of energy systems, none can be built or maintained without the skills of electricians, plumbers, pipefitters, welders, carpenters, telecommunications workers, and a host of other trades that contribute to energy projects. Energy professionals should nurture strong relations with the trades, generate buy-in and enthusiasm for projects, especially innovative or experimental ones, and at the same time be open to practical input about how systems should be physically fabricated and assembled so that they work effectively and reliably.

Another way is to become involved in local government in an advisory capacity, to help public servants who do not have an energy background to better understand technology, policy, and how local laws, codes, and zoning ordinances affect energy efficiency. Many communities are interested in becoming more energy efficient, both to reduce costs and to help protect the environment. In some cases, local governments may be able to bid out contracts to professional energy service companies, but in other cases they may need to make decisions about energy and sustainability without paid professional help. As an example, new technologies such as small-scale wind or solar PV panels may require that new building or zoning ordinances be drawn up to give guidance to both property owners and building inspectors about site design issues. Often, when an energy professional gives a clear explanation of technical aspects of a project or decision, it takes relatively little of her/his time, but goes a long way toward assisting public servants to make better decisions. As another example, a community may be considering some new technology that on the surface appears to be able to deliver savings in energy cost, reduced pollution, or other benefits. However, the technical and/or economic strengths and weaknesses of the technology may be difficult to discern by those without a technical background. The energy professional can provide a valuable service by gathering and passing along this information.

Another critical area affecting communities is the media coverage of energy issues. The energy professional can contribute by monitoring this coverage for accuracy and clarity. Members of the media are no different from other parts of society: some have a solid grasp of energy issues and are able to present ideas accurately, but others may, even with the best of intentions, give incorrect information about energy, leaving the readership or listeners with misconceptions. Energy professionals can, when appropriate, respond to media reports and commentaries with letters to the editor or op-ed pieces. Not surprisingly, other community members often express their gratitude for these efforts to shed light on the vital yet confusing world of energy and energy systems.

Along with working individually in this way in one's community, it is also possible to join with others to support organizations that promote new energy solutions, energy efficiency, and the like. Along with financial contributions, one can also contribute to the outreach activities of these organizations, for example, by assisting with programming or joining a speakers' bureau. Also, community organizations such as Rotary, Soroptimist, or League of Women Voters, as well as religious organizations or retirement communities, may have regular meetings where the schedule occasionally includes a guest speaker on topics of interest. Given the awareness of energy issues at this time, such organizations may be very interested in a talk about energy.

Lastly, and perhaps most importantly, the energy professional can lend assistance by supporting young people's interest in the energy field. They can find opportunities to talk to children and young adults about personal experience such as the types of projects on which energy professionals work, or one's career path into the field. For example, it may be possible to spend a part of a day once a year in a local elementary school as a guest speaker on energy, and include a hands-on project related to energy that the children can experience. Many high schools and colleges have student teams that work on design projects that are related to energy, and the services of an energy professional who can act as an advisor are highly valued. Not only does their input help the students with the project at hand, but the entire interaction provides an opportunity for young people to understand the connection between the individual project and the larger world context in which it is set.

At the beginning of this book, it was stated that modern energy systems reach almost every human being on the planet: residents of the industrial countries all the time, poor residents of less developed countries on a less frequent basis, but even the most isolated residents from time to time. Energy professionals share the benefits of this system with a huge range of people all around the world. They can be seen as both leaders of society and partners with other users in creating the energy system of the future. They are leaders in the sense that they have the technical skills to create and deploy the systems, as well as educate about and advocate for them. They are partners in the sense that they interact with the rest of the public to learn how changes in the systems are received and what adaptations are needed to make the systems function better. At this time we cannot say exactly what mixture of systems will eventually emerge. But whichever specific system the energy professional works upon, this type of engagement with society is essential for the sustainability of the energy systems of the future. The sustainability of these systems is, in turn, essential for the well-being of humanity and of all life on our planet.

17-6   Summary

As discussed in the body of this book, the world faces a number of pressing energy problems revolving around (1) more equitable distribution of energy resources, (2) finding alternatives to nonrenewable resources that are dwindling in quantity, and (3) preventing climate change stemming from CO_2 emissions to the atmosphere. A portfolio approach combining renewable, fossil, and nuclear resources is attractive because it allows society to exploit the advantages of each of the three sources. In order for the portfolio approach to succeed, special emphasis should be given to developing renewable resources, so that they can play a larger role in the future. Engineers and other energy professionals can play a role in bringing about the transformation of the world's energy systems both in their professional work and through extracurricular activities.

References

Beck, P. (1999). "Nuclear Energy in the Twenty-First Century: Examination of a Contentious Subject." *Annual Review of Energy & Environment*, Vol. 24, No. 1, pp. 113–137.

Brown, L. (2006). *Plan B 2.0: Rescuing a Planet under Stress and a Civilization in Trouble.* Norton, New York.

Haseltine, B., and C. Bull, Eds. (2003) *Field Guide to Appropriate Technology*, Academic Press, New York.

Leighty, W. (2012) *Deep Reductions in Greenhouse Gas Emissions from the California Transportation Sector: Dynamics in Vehicle Fleet and Energy Supply Transitions to Achieve 80% Reduction in Emissions from 1990 Levels by 2050.* Technical report, Institute for Transportation Studies, U.C. Davis.

Mackay, D. (2009). *Sustainable Energy without the Hot Air.* Cambridge University Press, Cambridge, UK.

Nakicenovic, N., and M. Jefferson (1995). *Global Energy Perspective to 2050 and Beyond*. World Energy Council, London.

National Academy of Sciences (2013). *Transitions to Alternative Fuels and Vehicles*. National Academies Press, Washington, D.C.

Further Reading

Bardaglio, P., and A. Putman. (2009) *Boldly Sustainable: Hope and Opportunity for Higher Education in the Age of Climate Change*. Nacubo, Washington, DC.

National Renewable Energy Laboratories (2014). *Life Cycle Assessment Harmonization Results and Findings*. NREL, Golden, Colorado. Electronic Resource, available at http://www.nrel.gov/analysis/sustain_lca_results.html. Accessed Aug. 18, 2015.

Randolph, J., and G. Masters (2008). *Energy for Sustainability: Technology, Planning, Policy*. Island Press, Washington, DC.

REN21 (2006). *Renewables Global Status Report: 2006 Update*. Renewable Energy Policy Network for the 21st Century. Web resource, available at http://www.ren21.net/globalstatusreport/download/RE_GSR_2006_Update.pdf. Accessed Oct. 16, 2007.

Rockstrom, J, W. Steffen, and K. Noone. (2009). "A Safe Operating Space for Humans." *Nature*, 461, pp. 472–475.

Tester, J., E. Drake, M. Driscoll, et al. (2006). "Sustainable Energy: Choosing Among Options. Geothermal Energy," Chap. 11. *Geothermal Energy*. MIT Press, Cambridge, Massachusetts.

Toossi, R. (2009). *Energy and the Environment: Sources, technologies, and impacts*. Verve Publishers, Los Angeles.

von Weiszacker, U., A. Lovins, and H. Lovins (1997). *Factor Four: Doubling Wealth, Halving Resource Use*. Earthscan, London.

Exercise

17-1. Using the future scenario presented in this chapter as a starting point, create your own scenario for transition to more sustainable energy during the course of the twenty-first century. Include the following: (a) total energy production for the years 2050 and 2100, (b) source of production divided between fossil, nuclear, and renewable, (c) order of magnitude cost of new infrastructure, in constant 2000 dollars, (d) order of magnitude total CO_2 emissions, and (e) allocation of energy production between industrial and emerging countries. Discuss your choice of numbers for each calculation (a) to (e).

APPENDIX A

Perpetual Julian Date Calendar

Day	Jan	Feb	Mar	Apr	May	Jun	Jul	Aug	Sep	Oct	Nov	Dec
1	1	32	60	91	121	152	182	213	244	274	305	335
2	2	33	61	92	122	153	183	214	245	275	306	336
3	3	34	62	93	123	154	184	215	246	276	307	337
4	4	35	63	94	124	155	185	216	247	277	308	338
5	5	36	64	95	125	156	186	217	248	278	309	339
6	6	37	65	96	126	157	187	218	249	279	310	340
7	7	38	66	97	127	158	188	219	250	280	311	341
8	8	39	67	98	128	159	189	220	251	281	312	342
9	9	40	68	99	129	160	190	221	252	282	313	343
10	10	41	69	100	130	161	191	222	253	283	314	344
11	11	42	70	101	131	162	192	223	254	284	315	345
12	12	43	71	102	132	163	193	224	255	285	316	346
13	13	44	72	103	133	164	194	225	256	286	317	347
14	14	45	73	104	134	165	195	226	257	287	318	348
15	15	46	74	105	135	166	196	227	258	288	319	349
16	16	47	75	106	136	167	197	228	259	289	320	350
17	17	48	76	107	137	168	198	229	260	290	321	351
18	18	49	77	108	138	169	199	230	261	291	322	352
19	19	50	78	109	139	170	200	231	262	292	323	353
20	20	51	79	110	140	171	201	232	263	293	324	354
21	21	52	80	111	141	172	202	233	264	294	325	355
22	22	53	81	112	142	173	203	234	265	295	326	356
23	23	54	82	113	143	174	204	235	266	296	327	357
24	24	55	83	114	144	175	205	236	267	297	328	358
25	25	56	84	115	145	176	206	237	268	298	329	359
26	26	57	85	116	146	177	207	238	269	299	330	360
27	27	58	86	117	147	178	208	239	270	300	331	361
28	28	59	87	118	148	179	209	240	271	301	332	362
29	29		88	119	149	180	210	241	272	302	333	363
30	30		89	120	150	181	211	242	273	303	334	364
31	31		90		151		212	243		304		365

Note: Add 1 to Julian date after February 28 during leap years.

LCR Table

Portland, Maine Wall Designation	SSF = 0.1	SSF = 0.2	SSF = 0.3	SSF = 0.4	SSF = 0.5	SSF = 0.6	SSF = 0.7	SSF = 0.8	SSF = 0.9
TW A1	212	9	–	–	–	–	–	–	–
TW A2	86	22	8	–	–	–	–	–	–
TW A3	71	24	11	5	–	–	–	–	–
TW A4	62	25	13	7	4	2	–	–	–
TW B1	122	17	–	–	–	–	–	–	–
TW B2	73	23	10	5	–	–	–	–	–
TW B3	65	24	12	6	3	–	–	–	–
TW B4	63	23	12	7	4	2	1	–	–
TW C1	86	20	8	3	–	–	–	–	–
TW C2	69	22	10	6	3	1	–	–	–
TW C3	68	22	10	6	3	2	–	–	–
TW C4	76	21	10	5	3	2	1	–	–
TW D1	–	–	–	–	–	–	–	–	–
TW D2	84	36	20	12	8	5	3	2	1
TW D3	86	39	22	14	9	6	4	3	1
TW D4	97	51	32	21	15	11	8	5	3
TW D5	97	53	34	24	17	12	9	6	4
TW E1	107	50	30	19	13	9	6	4	2
TW E2	103	51	31	21	15	10	7	5	3
TW E3	120	69	45	21	23	17	12	9	6
TW E4	110	64	43	30	22	16	12	9	6

Baltimore, Maryland Wall Designation	SSF = 0.1	SSF = 0.2	SSF = 0.3	SSF = 0.4	SSF = 0.5	SSF = 0.6	SSF = 0.7	SSF = 0.8	SSF = 0.9
TW A1	405	49	21	12	7	4	3	2	–
TW A2	167	56	29	18	12	8	6	4	2
TW A3	136	57	32	20	14	10	7	5	3
TW A4	117	55	32	21	15	10	7	5	3
TW B1	231	48	23	12	8	6	4	2	1
TW B2	137	52	29	18	12	8	6	4	2
TW B3	121	51	29	19	13	9	6	4	3
TW B4	112	48	27	18	12	9	6	4	3
TW C1	157	46	24	14	9	6	4	3	2
TW C2	123	46	25	16	11	7	5	4	2
TW C3	119	43	24	15	10	7	5	3	2
TW C4	127	38	21	13	9	6	4	3	2
TW D1	79	27	13	7	4	2	–	–	–
TW D2	143	66	40	26	18	13	9	7	4
TW D3	146	73	45	30	21	15	11	8	5
TW D4	149	82	53	37	27	20	15	10	7
TW D5	143	81	54	38	28	21	15	11	7
TW E1	174	87	54	37	26	19	14	10	6
TW E2	161	85	54	37	27	20	14	10	7
TW E3	177	105	70	50	37	28	21	15	10
TW E4	161	96	65	47	35	26	19	14	10

Denver, Colorado Wall Designation	SSF = 0.1	SSF = 0.2	SSF = 0.3	SSF = 0.4	SSF = 0.5	SSF = 0.6	SSF = 0.7	SSF = 0.8	SSF = 0.9
TW A1	631	90	43	26	17	12	8	6	4
TW A2	267	96	54	35	24	17	13	9	6
TW A3	219	96	57	38	27	20	14	10	7
TW A4	188	92	56	38	28	20	15	11	7
TW B1	364	86	44	27	18	13	9	7	4
TW B2	217	88	50	33	23	17	12	9	6
TW B3	191	85	51	34	24	18	13	9	6
TW B4	176	78	47	31	22	16	12	9	6
TW C1	246	78	42	27	19	13	10	7	5
TW C2	192	76	43	28	20	14	11	8	5
TW C3	183	71	40	26	19	13	10	7	5
TW C4	191	63	35	22	16	11	8	6	4
TW D1	151	60	34	22	15	11	8	6	4
TW D2	220	106	65	44	32	24	17	13	9
TW D3	225	116	74	51	37	27	20	15	10
TW D4	219	123	81	57	42	32	24	18	12
TW D5	206	119	80	57	42	32	24	18	12
TW E1	262	136	86	60	43	32	24	17	12
TW E2	239	130	84	59	43	32	24	18	12
TW E3	255	154	104	75	56	43	32	24	17
TW E4	228	139	95	69	52	39	30	22	15

Buffalo, New York Wall Designation	SSF = 0.1	SSF = 0.2	SSF = 0.3	SSF = 0.4	SSF = 0.5	SSF = 0.6	SSF = 0.7	SSF = 0.8	SSF = 0.9
TW A1	142	–	–	–	–	–	–	–	–
TW A2	48	–	–	–	–	–	–	–	–
TW A3	40	–	–	–	–	–	–	–	–
TW A4	36	–	–	–	–	–	–	–	–
TW B1	76	–	–	–	–	–	–	–	–
TW B2	44	–	–	–	–	–	–	–	–
TW B3	40	–	–	–	–	–	–	–	–
TW B4	41	10	–	–	–	–	–	–	–
TW C1	55	–	–	–	–	–	–	–	–
TW C2	46	8	–	–	–	–	–	–	–
TW C3	47	10	–	–	–	–	–	–	–
TW C4	56	12	–	–	–	–	–	–	–
TW D1	–	–	–	–	–	–	–	–	–
TW D2	59	21	9	–	–	–	–	–	–
TW D3	60	24	11	–	–	–	–	–	–
TW D4	75	38	22	14	10	6	4	3	2
TW D5	78	41	26	17	12	8	6	4	3
TW E1	79	34	18	10	6	3	1	–	–
TW E2	78	37	21	13	13	8	5	3	2
TW E3	97	54	34	23	16	12	8	6	4
TW E4	90	51	33	23	16	12	8	6	4

Albuquerque, New Mexico Wall Designation	SSF = 0.1	SSF = 0.2	SSF = 0.3	SSF = 0.4	SSF = 0.5	SSF = 0.6	SSF = 0.7	SSF = 0.8	SSF = 0.9
TW A1	900	124	60	37	25	17	12	9	6
TW A2	361	130	73	48	33	24	18	13	8
TW A3	293	129	77	52	37	27	20	15	10
TW A4	249	123	76	52	38	28	21	15	10
TW B1	502	117	60	38	26	18	13	9	6
TW B2	291	118	68	45	32	23	17	12	8
TW B3	254	114	68	46	33	24	18	13	9
TW B4	233	104	63	42	30	22	16	12	8
TW C1	332	106	58	37	26	19	14	10	6
TW C2	255	101	58	39	27	20	15	11	7
TW C3	243	94	54	36	25	18	13	10	7
TW C4	254	84	46	30	21	15	11	8	5
TW D1	213	86	50	33	23	17	12	9	6
TW D2	287	139	86	59	43	32	24	17	12
TW D3	294	153	97	68	49	37	27	20	14
TW D4	281	158	104	74	55	41	31	23	16
TW D5	260	151	101	73	54	41	31	23	16
TW E1	339	177	113	78	57	43	32	23	16
TW E2	308	168	109	77	56	42	32	23	16
TW E3	323	195	133	96	72	55	42	31	21
TW E4	287	175	120	88	66	50	38	28	20

Edmonton, Alberta Wall Designation	SSF = 0.1	SSF = 0.2	SSF = 0.3	SSF = 0.4	SSF = 0.5	SSF = 0.6	SSF = 0.7	SSF = 0.8	SSF = 0.9
TW A1	251	21	–	–	–	–	–	–	–
TW A2	109	30	–	–	–	–	–	–	–
TW A3	89	31	12	–	–	–	–	–	–
TW A4	77	31	14	–	–	–	–	–	–
TW B1	149	24	–	–	–	–	–	–	–
TW B2	91	29	11	–	–	–	–	–	–
TW B3	81	29	13	–	–	–	–	–	–
TW B4	77	29	13	5	–	–	–	–	–
TW C1	105	26	9	–	–	–	–	–	–
TW C2	84	28	12	–	–	–	–	–	–
TW C3	82	27	12	5	–	–	–	–	–
TW C4	89	25	11	5	–	–	–	–	–
TW D1	–	–	–	–	–	–	–	–	–
TW D2	101	43	23	13	6	–	–	–	–
TW D3	101	46	25	14	7	–	–	–	–
TW D4	111	58	35	23	15	10	7	4	2
TW D5	110	60	38	25	18	12	8	5	3
TW E1	124	58	33	20	12	7	3	1	
TW E2	119	59	35	22	14	9	6	3	2
TW E3	134	76	49	33	23	16	11	7	4
TW E4	124	72	47	32	23	16	11	8	5

LCR Table for Six Selected North American Cities, Vented Trombe Walls.

Source: Data from Jones, R.W., ed. *Passive Solar Design Handbook Volume Three: Passive Solar Design Analysis.* National Technical Information Service, Springfield, Virginia.

APPENDIX C

CF Table

Portland, Maine Wall Designation	SSF = 0.1	SSF = 0.5	SSF = 0.8
TW A1	2.5	–	–
TW A2	2.3	–	–
TW A3	2.2	–	–
TW A4	2.1	4.9	–
TW B1	2.5	–	–
TW B2	2.3	–	–
TW B3	2.2	5.5	–
TW B4	2.2	4.7	–
TW C1	2.5	–	–
TW C2	2.3	5.6	–
TW C3	2.4	5.1	–
TW C4	2.5	5.1	12.7
TW D1	–	–	–
TW D2	1.8	3.1	4.9
TW D3	1.7	2.8	4.4
TW D4	1.4	2.1	2.8
TW D5	1.4	1.9	2.5
TW E1	1.5	2.3	3.3
TW E2	1.4	2.1	2.9
TW E3	1.2	1.6	2.2
TW E4	1.2	1.7	2.2

Baltimore, Maryland Wall Designation	SSF = 0.1	SSF = 0.5	SSF = 0.8
TW A1	1.7	3.3	5.4
TW A2	1.5	2.4	3.5
TW A3	1.5	2.2	3.1
TW A4	1.4	2.1	2.9
TW B1	1.7	3.0	4.4
TW B2	1.6	2.4	3.3
TW B3	1.5	2.3	3.2
TW B4	1.6	2.3	3.2
TW C1	1.7	2.7	3.9
TW C2	1.7	2.5	3.5
TW C3	1.7	2.6	3.6
TW C4	1.9	2.8	3.9
TW D1	2.1	4.9	–
TW D2	1.3	1.9	2.5
TW D3	1.2	1.7	2.3
TW D4	1.2	2.6	2.1
TW D5	1.1	1.5	2.0
TW E1	1.2	1.7	2.2
TW E2	1.2	1.6	2.1
TW E3	1.0	1.3	1.7
TW E4	1.0	1.4	1.8

Denver, Colorado Wall Designation	SSF = 0.1	SSF = 0.5	SSF = 0.8
TW A1	1.4	2.0	2.6
TW A2	1.2	1.6	2.1
TW A3	1.1	1.5	2.0
TW A4	1.1	1.5	1.9
TW B1	1.3	1.9	2.5
TW B2	1.2	1.7	2.1
TW B3	1.2	1.6	2.1
TW B4	1.2	1.7	2.2
TW C1	1.3	1.9	2.4
TW C2	1.3	1.8	2.3
TW C3	1.3	1.9	2.4
TW C4	1.5	2.0	2.6
TW D1	1.4	2.1	2.7
TW D2	1.0	1.4	1.8
TW D3	1.0	1.3	1.6
TW D4	0.9	1.2	1.5
TW D5	0.9	1.2	1.5
TW E1	0.9	1.2	1.5
TW E2	0.9	1.2	1.5
TW E3	0.8	1.0	1.3
TW E4	0.8	1.0	1.3

Buffalo, New York Wall Designation	SSF = 0.1	SSF = 0.5	SSF = 0.8
TW A1	3.5	–	–
TW A2	3.6	–	–
TW A3	3.3	–	–
TW A4	3.1	–	–
TW B1	3.7	–	–
TW B2	3.3	–	–
TW B3	3.1	–	–
TW B4	3.0	–	–
TW C1	3.4	–	–
TW C2	3.1	–	–
TW C3	3.1	–	–
TW C4	3.1	–	–
TW D1	–	–	–
TW D2	2.3	–	–
TW D3	2.1	–	–
TW D4	1.7	2.7	4.1
TW D5	1.6	2.3	3.3
TW E1	1.8	4.3	–
TW E2	1.7	3.0	4.9
TW E3	1.3	2.0	2.7
TW E4	1.4	2.0	2.7

Albuquerque, New Mexico Wall Designation	SSF = 0.1	SSF = 0.5	SSF = 0.8
TW A1	1.2	1.7	2.2
TW A2	1.0	1.4	1.8
TW A3	1.0	1.3	1.7
TW A4	1.0	1.3	1.6
TW B1	1.1	1.6	2.1
TW B2	1.0	1.4	1.8
TW B3	1.0	1.4	1.8
TW B4	1.1	1.4	1.8
TW C1	1.2	1.6	2.0
TW C2	1.1	1.5	2.0
TW C3	1.2	1.6	2.0
TW C4	1.3	1.8	2.3
TW D1	1.2	1.7	2.2
TW D2	0.9	1.2	1.5
TW D3	0.8	1.1	1.4
TW D4	1.4a	2.2a	2.8a
TW D5	1.2a	1.8a	2.5a
TW E1	1.5a	2.2a	3.6a
TW E2	b	b	b
TW E3	b	b	b
TW E4	1.1a	1.6a	2.2a

Note: a: Estimated; b: Not available.

Edmonton, Alberta Wall Designation	SSF = 0.1	SSF = 0.5	SSF = 0.8
TW A1	2.1	–	–
TW A2	1.9	–	–
TW A3	1.8	–	–
TW A4	1.8	–	–
TW B1	2.1	–	–
TW B2	1.9	–	–
TW B3	1.9	–	–
TW B4	1.9	–	–
TW C1	2.1	–	–
TW C2	2.0	–	–
TW C3	2.0	–	–
TW C4	2.2	14.2	–
TW D1	4.2	–	–
TW D2	1.6	3.7	–
TW D3	1.5	3.5	–
TW D4	1.3	2.1	3.4
TW D5	1.3	1.9	2.8
TW E1	1.3	2.5	6.7
TW E2	1.3	2.2	3.8
TW E3	1.1	1.7	2.4
TW E4	1.1	1.7	2.4

CF Table for Six Selected North American Cities, Vented Trombe Walls.

Note: A dash signifies not a recommended design.

Source: Data from Jones, R.W., ed. *Passive Solar Design Handbook Volume Three: Passive Solar Design Analysis.* National Technical Information Service, Springfield, Virginia.

APPENDIX D

Numerical Answers to Select Problems

Where appropriate, solutions to odd-number problems appear here. Only numerical solutions are included; open-ended essay-type answers are not for reasons of brevity.

Chap. 1
3. a. 0.736
7. India, China, United States, Japan

Chap. 2
3. 304.2 EJ
5. a. 392.6 GJ/pers, **b.** 3926 EJ/year

Chap. 3
1. NPV = $157,500; the investment is viable
3. 11%
5. Before = $97.06; After = $81.00

Chap. 4
1. 287.9 K
3. Mars = 3 K, Venus = 468 K
7. T = 202.7 K

Chap. 5
3. (a) 0.079 kg/mile, 3.95 MJ/mile; (b) 0.0991 kg input, 0.363 kg CO_2 released
5. 97%

Chap. 6
1. 42.1%
3. 30.2%
5. Thermal efficiency 46.5%; Carnot efficiency 67.2%
9. 10.41 tonnes CO_2 per hour

Chap. 7
3. a. 1.81 billion kWh/year, **b.** 25.3 million tonnes
5. a. $0.0238/kWh, **b.** $0.133/kWh

Chap. 8
1. 24.5%
3. $0.0417/kWh
5. $0.026/kWh
7. $0.0158/kWh, $0.0342/kWh

Chap. 9
1. (a) 6.18 kWh/m^2, (b) 16.9% loss
3. Angle formed by building: 51°; Solar azimuth: 68°; Solar azimuth is greater, so the building blocks the sun
5. Equality holds when $L = \pm 90$ and $\gamma = \omega$

Chap. 10
1. 0.0281 A/cm^2
3. 754.4 kWh/m^2; 800.8 kWh/m^2 with optimal tilt
7. $23,747

Chap. 11: Not included

Chap. 12
1. Base 18: 40 degree-days; Base 12: 14 degree-days

Chap. 13
1. a. 1529 kWh/m^2, **b.** 1609 kWh/m^2, **c.** 1597 kWh/m^2, **d.** 5.2%, 4.5%
3. 5.07 million kWh per year, 39%
9. NPV = ~ –$124,000
11. $0.195 per kWh

Chap. 14: Not included

Chap. 15
1. (a) 9.7 kW, (b) 160.8 km/h, (c) 24.5%
5. 4.57 kW

Chap. 16
1. Change: cost – $244,000/year, energy + 2156 GJ/year
5. a. 17.2 TJ, **b.** 14.4 TJ, no change in pattern

Chap. 17: Not included

APPENDIX E

Common Conversions

1 standard ton	= 2000 pounds (lb)
1 metric ton (aka "tonne")	= 1000 kilograms
1 kilogram (kg)	= 2.2 pounds
1 tonne	= 1.1 standard tons
1 gallon	= 3.785 liters (L)
1 mile	= 1.61 kilometers
1 meter	= 3.28 feet
1 British thermal unit (BTU)	= 1.055 kilojoules
1 kilowatt (kW)	= 1.341 horsepower (hp)
1 hp	= 2544 BTU
1 watt	= 1 joule/second
1 kilowatt-hour (kWh)	= 3.6 megajoules (MJ)
1 kWh	= 3412 BTU
1 gallon gasoline	= 115,400 BTU energy content
1 gallon diesel	= 128,700 BTU
1 gallon biodiesel (B100)	= 117,100 BTU
1 gallon ethanol	= 75,700 BTU
1 gallon gasoline combusted	= 19.6 lb CO_2 resulting emissions
1 gallon (petro-)diesel combusted	= 22.4 lb CO_2 resulting emissions
1 L gasoline	= 32.2 MJ energy content
1 L diesel	= 35.9 MJ
1 L biodiesel (B100)	= 32.6 MJ
1 L ethanol	= 21.3 MJ
1 L gasoline combusted	= 2.35 kg CO_2 resulting emissions
1 L (petro-)diesel combusted	= 2.69 kg CO_2 resulting emissions

Note: Where relevant, heat content values given above use the lower heating value.

Information about Thermodynamic Constants

Thermodynamic tables used in the book for in-chapter examples and end-of-chapter exercises for air, water, and steam are available at www.mhprofessional.com/energysystemsengineering, or are generally published in hard copy form in stand-alone reference books, or else in thermodynamic textbooks. They can also be obtained from online calculators.

One potential advantage of an online calculator is the ability to calculate or extract from database exact enthalpy or entropy values based on input conditions to the nearest degree of temperature or exact pressure value. In the case of thermodynamic tables in hard copy form, it is sometimes necessary to interpolate between nearby values to obtain the desired value to an acceptable degree of precision. For instance, in Example 6-4 in the body of the book, it was necessary to interpolate to find an enthalpy value of $h_4 = 706.5$ kJ/kg corresponding to a relative pressure value of $p_r = 27.85$. An online calculator can typically calculate the desired enthalpy value directly from the relative pressure, without needing any additional hand calculation steps.

Index